Core Curriculum

Introductory Craft Skills

Annotated Instructor's Guide
Third Edition

Upper Saddle River, New Jersey
Columbus, Ohio

National Center for Construction Education and Research
President: Don Whyte
Director of Product Development: Daniele Dixon
Core Curriculum Project Manager: Daniele Dixon
Production Manager: Debie Ness
Quality Assurance Coordinator: Jessica Martin
Editor: Cristina Escobar

Writing and development services provided by EEI Communications, Alexandria, Virginia.

Pearson Education, Inc.
Product Manager: Lori Cowen
Production Editor: Stephen C. Robb
Design Coordinator: Karrie M. Converse-Jones
Text Designer: Kristina D. Holmes
Cover Designer: Kristina D. Holmes
Copy Editor: Sheryl Rose
Scanning Coordinator: Karen L. Bretz
Scanning Technician: Janet Portisch
Production Manager: Pat Tonneman

This book was set in Palatino and Helvetica by Carlisle Communications, Ltd. It was printed and bound by Document Technology Resources. The cover was printed by Phoenix Color Corp.

Pearson Education Ltd.
Pearson Education Singapore Pte. Ltd.
Pearson Education Canada, Ltd.
Pearson Education—Japan
Pearson Education Australia Pty. Limited
Pearson Education North Asia Ltd.
Pearson Educación de Mexico, S.A. de C.V.
Pearson Education Malaysia Pte. Ltd.

10 9 8 7 6
ISBN 0-13-109192-1

Preface

TO THE INSTRUCTOR

We are proud to present the third edition of the highly successful *Core Curriculum: Introductory Craft Skills.* This Annotated Instructor's Guide offers an array of teaching tips and ideas that can be adapted to suit your instructional objectives and your students' individual needs. The Annotated Instructor's Guide is actually the Trainee Guide enhanced with specific directions to the instructor, space for the instructor's notes, suggestions for session break-outs, a comprehensive materials and equipment list, and teaching tips to correspond to the performance examinations, laboratories, and demonstrations. Additionally, each Annotated Instructor's Guide is packaged with a test booklet that includes module exams, performance tests, and answer keys. We hope you will find this book a valuable resource as you prepare your students for a rewarding career in the construction and maintenance industries.

NEW WITH CORE CURRICULUM: INTRODUCTORY CRAFT SKILLS

The full-color Trainee Guide not only includes the six integral modules for building foundation skills in construction, it also provides the tools necessary for achieving workplace success. *Core Curriculum* now contains two new modules, adapted from Prentice Hall's highly successful *Tools for Success* workbook, focusing on communications and employability skills—soft skills critical to advancing in the construction and maintenance industries.

To see where a job in construction could take your students, check out the photos of the construction project award winners from the two largest construction trade associations in the nation: Associated Builders and Contractors, Inc. and Associated General Contractors of America. These photos are found on the first page of each Trainee Guide module. From the historical renovation of the Oklahoma State Capitol Dome to the building of the General Motors Vehicle Engineering Center, these award winners show that there is much in the construction industry of which to be proud.

We invite you to visit the NCCER website at www.nccer.org for the latest releases, training information, newsletter, and much more. You can also reference the Contren® product catalog online at www.crafttraining.com. Your feedback is welcome. You may email your comments to curriculum@nccer.org or send general comments and inquiries to info@nccer.org.

CONTREN® LEARNING SERIES

The National Center for Construction Education and Research (NCCER) is a not-for-profit 501(c)(3) education foundation established in 1995 by the world's largest and most progressive construction companies and national construction associations. It was founded to address the severe workforce shortage facing the industry and to develop a standardized training process and curricula. Today, NCCER is supported by hundreds of leading construction and maintenance companies, manufacturers, and national associations. The Contren® Learning Series was developed by NCCER in partnership with Prentice Hall, the world's largest educational publisher.

Some features of NCCER's Contren® Learning Series are as follows:

- An industry-proven record of success
- Curricula developed by the industry for the industry
- National standardization, providing portability of learned job skills and educational credits
- Compliance with Apprenticeship, Training, Employer, and Labor Services (ATELS) requirements for related classroom training (CFR 29:29)
- Well-illustrated, up-to-date, and practical information

NCCER also maintains a National Registry that provides transcripts, certificates, and wallet cards to individuals who have successfully completed modules of NCCER's Contren® Learning Series. *Training programs must be delivered by an NCCER Accredited Training Sponsor in order to receive these credentials.*

Special Features of This Book

In an effort to provide a comprehensive user-friendly training resource, we have incorporated many different features for your use. Whether you are a visual or hands-on learner, this book will provide you with the proper tools to get started in the construction industry.

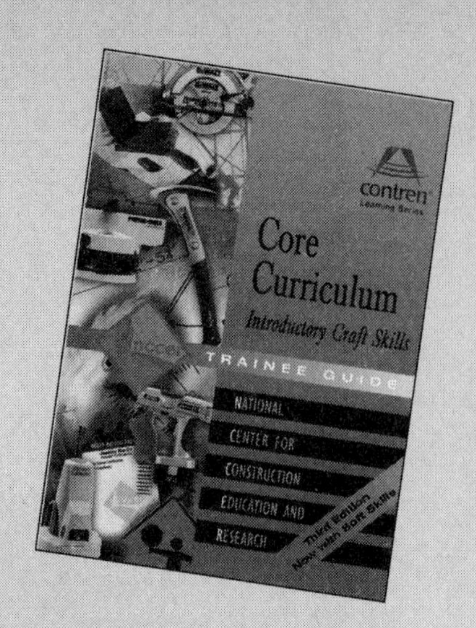

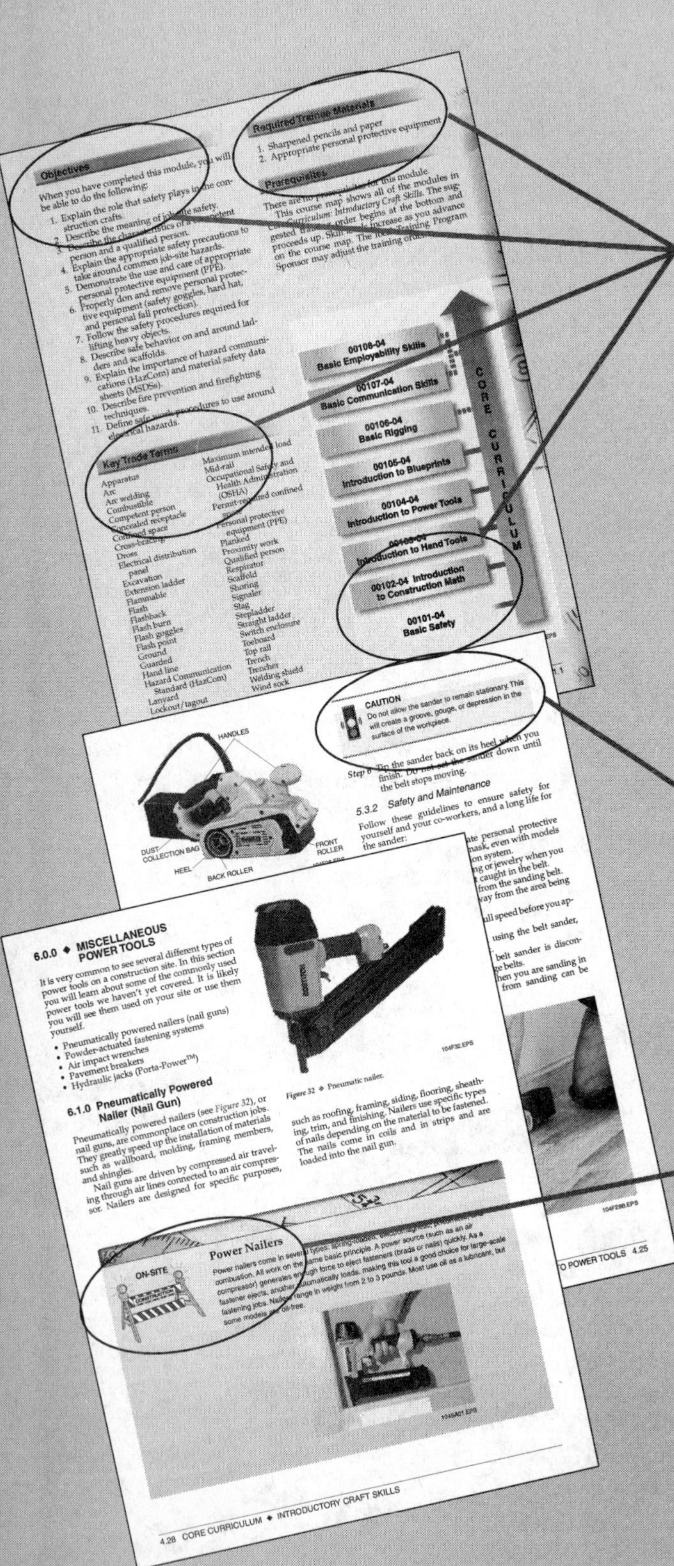

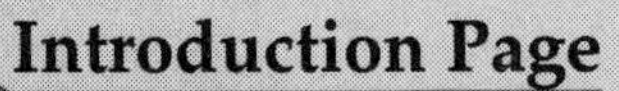

Introduction Page

This page is found at the beginning of each module and lists the Objectives, Key Trade Terms, Required Trainee Materials, Prerequisites, and Course Map for that module. The Objectives list the skills and knowledge you will need in order to complete the module successfully. The list of Key Trade Terms identifies important terms you will need to know by the end of the module. Required Trainee Materials list the materials and supplies needed for the module. The Prerequisites for the module are listed and illustrated in the Course Map. The Course Map also gives a visual overview of the entire course and a suggested learning sequence for you to follow.

Notes, Cautions, and Warnings

Safety features are set off from the main text in highlighted boxes and organized into three categories based on the potential danger of the issue being addressed. Notes simply provide additional information on the topic area. Cautions alert you of a danger that does not present potential injury but may cause damage to equipment. Warnings stress a potentially dangerous situation that may cause injury to you or a co-worker.

On-Site

The On-Site features offer technical hints and tips from the construction industry. These often include nice-to-know information that you will find helpful. On-Sites also present real-life scenarios similar to those you might encounter on the job site.

Did You Know?

The Did You Know? features introduce historical tidbits or modern information about the construction industry. Interesting and sometimes surprising facts about construction are also presented.

Step-by-Step Instructions

Step-by-step instructions are used throughout to guide you through technical procedures and tasks from start to finish. These steps show you not only how to perform a task but how to do it safely and efficiently.

Color Illustrations and Photographs

Full-color illustrations and photographs are used throughout each module to provide vivid detail. These figures highlight important concepts from the text and provide clarity for complex instructions. Each figure is denoted in the text in *italic type* for easy reference.

Review Questions

Review Questions are provided to reinforce the knowledge you have gained. This makes them a useful tool for measuring what you have learned.

Key Trade Terms

Each module presents a list of Key Trade Terms that are discussed within the text, defined in the Glossary at the end of the module, and reinforced with a Key Trade Terms Quiz. These terms are denoted in the text with **blue bold type** upon their first occurrence. To make searches for key information easier, a comprehensive Glossary of Key Trade Terms from all modules is found at the back of this book.

Profile in Success

Profiles in Success share the experiences of and advice from successful craftspersons. Also included are *ENR Next* highlights from the *Engineering News-Record*, the weekly news magazine of the construction industry. These articles focus on emerging engineers and contractors with fewer than 10 years of experience in the construction industry.

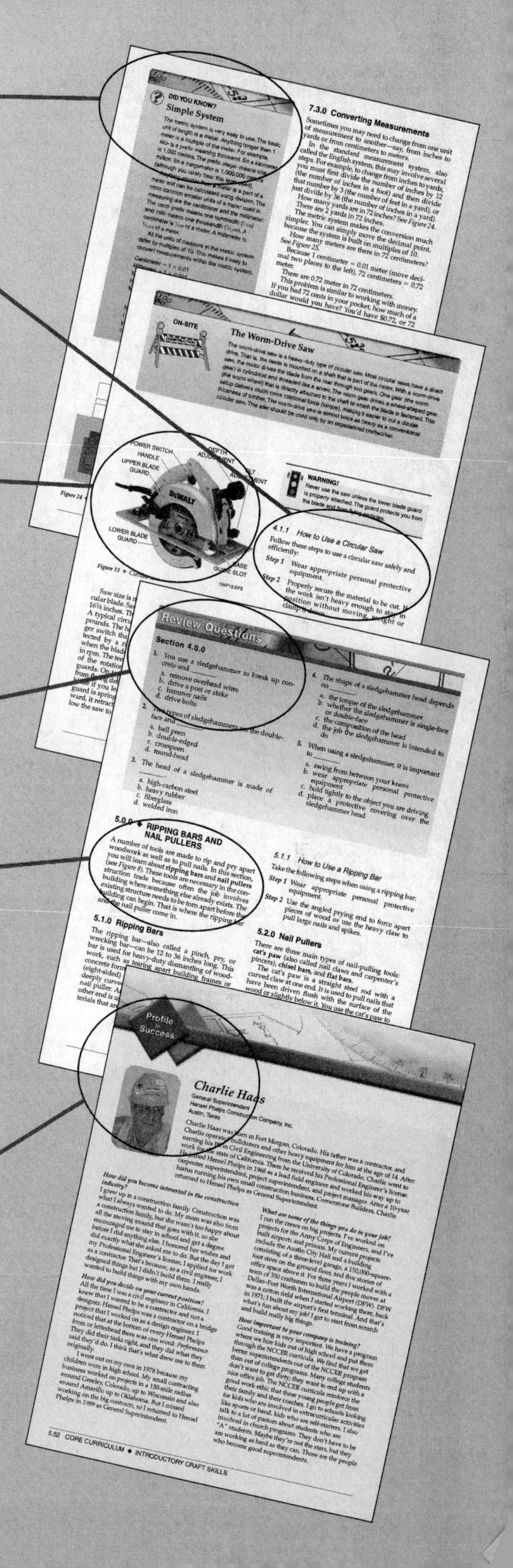

Supplemental Resources

Transparency Masters

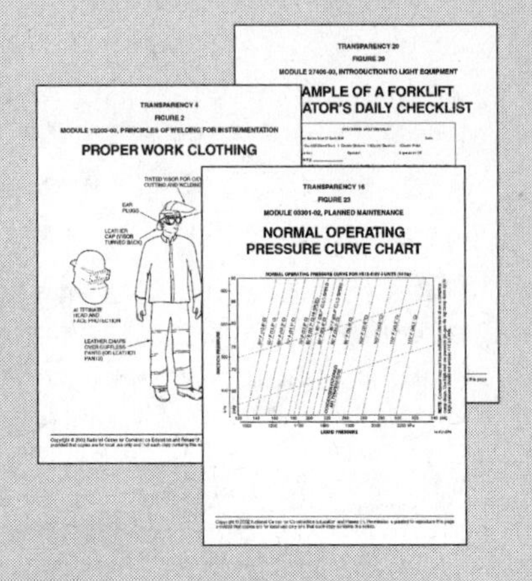

Spend more time training and less time at the copier by using the loose, reproducible copies of the overhead transparencies that are referenced in the Annotated Instructor's Guide. The transparency masters package includes most of the Trainee Guide illustrations, enlarged for projection and printed on loose sheets of paper for easy copying onto transparency film using your photocopier. To order, contact Prentice Hall at 1-800-922-0579 and ask for ISBN 0-13-109904-3.

PowerPoint® Presentation Slides

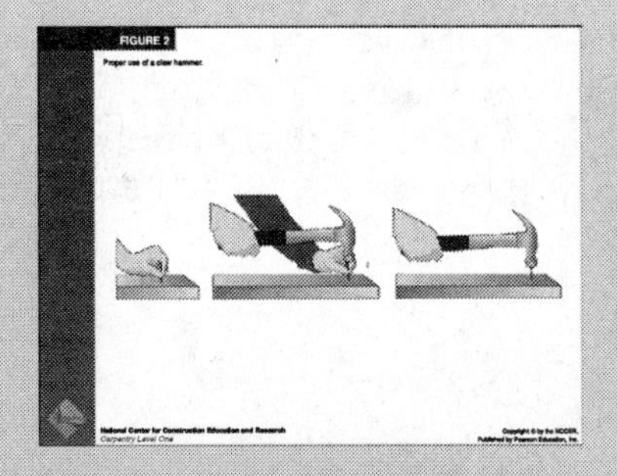

A powerful addition to your classroom discussion, NCCER's Microsoft® PowerPoint® presentation CD features all the drawings and photos from the Trainee Guide. To order, contact Prentice Hall at 1-800-922-0579 and ask for ISBN 0-13-160001-X.

Computerized Testing Software

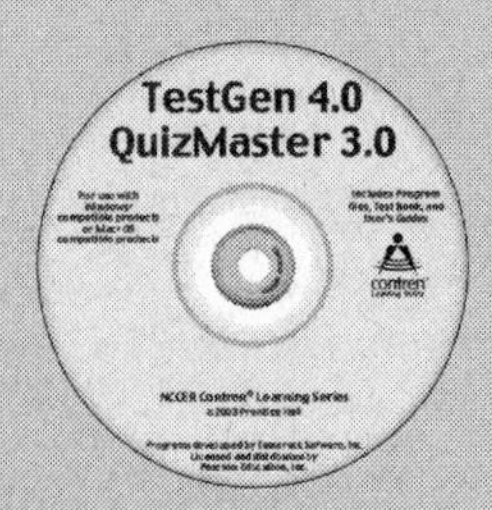

Ensure test security with NCCER's computerized testing software. This software allows instructors to scramble the module exam questions and answer keys in order to print multiple versions of the same test, customize tests to suit the needs of training units, add questions, or easily create a final exam. To order, contact Prentice Hall at 1-800-922-0579 and ask for ISBN 0-13-109917-5.

New! *Core National Construction Career Test (NCCT)*

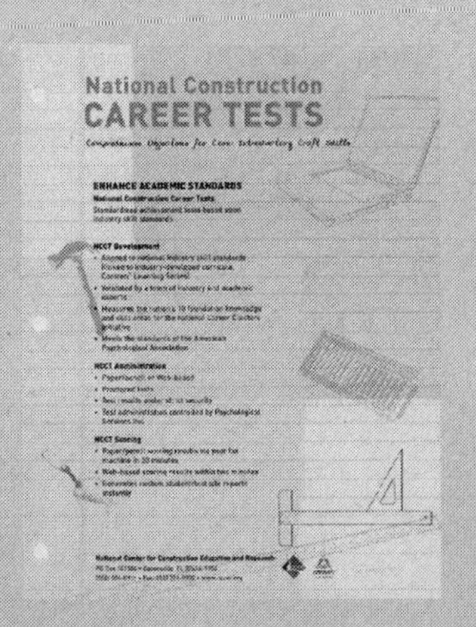

- Test student achievement on the topics covered in the Third Edition of *Core Curriculum: Introductory Craft Skills.*
- Give students the key to transition smoothly into post-secondary and apprenticeship programs.
- Support federal objectives for states to implement statewide accountability systems.
- Provide industry-recognized credentials.
- Support the national Career Clusters 10 foundation and knowledge skill areas in the architecture and construction pathway (see *www.careerclusters.org*).

These are just a few of the potential benefits for implementing NCCER's *Core NCCT*. For more information, and to learn more about other career and technical education achievement exams offered in the NCCT series, contact NCCER Customer Service at 1-888-NCCER-20 or visit *www.nccer.org*.

Revised! *Tools for Success Workbook*

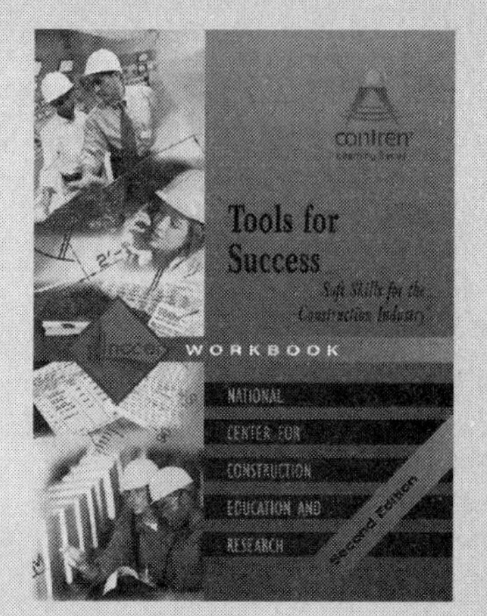

Provide your students with even more classroom activities to help them navigate their way through intangible workplace issues with this edition of the highly popular *Tools for Success* workbook! A complement to the two new soft skills modules contained in the Third Edition of *Core Curriculum: Introductory Craft Skills,* this revised workbook builds upon the 2000 edition and comes with an instructor's handbook. The handbook includes tried-and-true best practices from instructors teaching the material, as well as answer keys to the practice exercises provided in the student workbook. To order, call Prentice Hall at 1-800-922-0579.

Trainee Workbook	ISBN 0-13-109194-8
Workbook and Instructor's Handbook	ISBN 0-13-160000-1
Core/Tools Valuepack	ISBN 0-13-109933-7

Updated! *Safety Orientation Pocket Guide*

Provide your students with this handy pocket safety guide, now better aligned to topics covered in OSHA's 10-hour course. It is great as a reference book while working in the lab or in the field. The Guide is now available in an Annotated Instructor's Guide format, complete with exams and overhead transparency masters. To order, call Prentice Hall at 1-800-922-0579. Also inquire about the Spanish version of *Safety Orientation (Orientación Sobre Seguridad)* and a complete Safety PowerPoint® Presentation package coming soon!

Pocket Guide	ISBN 0-13-163612-X
Annotated Instructor's Guide	ISBN 0-13-163613-8
PowerPoint® Presentation	ISBN 0-13-163616-2
Spanish Pocket Guide	ISBN 0-13-163614-6
Spanish Instructor's Guide	ISBN 0-13-163615-4

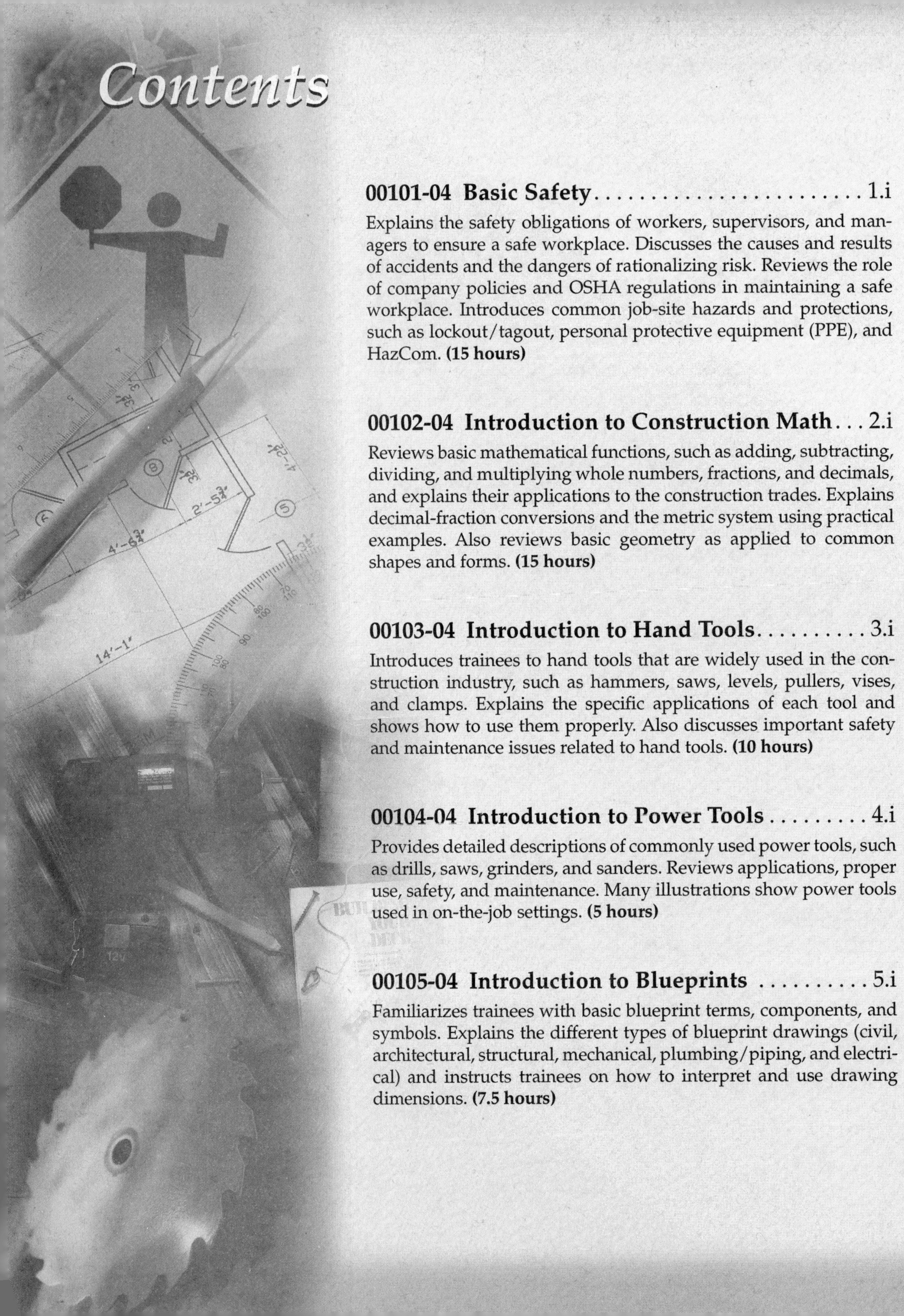

Contents

00101-04 Basic Safety 1.i

Explains the safety obligations of workers, supervisors, and managers to ensure a safe workplace. Discusses the causes and results of accidents and the dangers of rationalizing risk. Reviews the role of company policies and OSHA regulations in maintaining a safe workplace. Introduces common job-site hazards and protections, such as lockout/tagout, personal protective equipment (PPE), and HazCom. **(15 hours)**

00102-04 Introduction to Construction Math . . . 2.i

Reviews basic mathematical functions, such as adding, subtracting, dividing, and multiplying whole numbers, fractions, and decimals, and explains their applications to the construction trades. Explains decimal-fraction conversions and the metric system using practical examples. Also reviews basic geometry as applied to common shapes and forms. **(15 hours)**

00103-04 Introduction to Hand Tools 3.i

Introduces trainees to hand tools that are widely used in the construction industry, such as hammers, saws, levels, pullers, vises, and clamps. Explains the specific applications of each tool and shows how to use them properly. Also discusses important safety and maintenance issues related to hand tools. **(10 hours)**

00104-04 Introduction to Power Tools 4.i

Provides detailed descriptions of commonly used power tools, such as drills, saws, grinders, and sanders. Reviews applications, proper use, safety, and maintenance. Many illustrations show power tools used in on-the-job settings. **(5 hours)**

00105-04 Introduction to Blueprints 5.i

Familiarizes trainees with basic blueprint terms, components, and symbols. Explains the different types of blueprint drawings (civil, architectural, structural, mechanical, plumbing/piping, and electrical) and instructs trainees on how to interpret and use drawing dimensions. **(7.5 hours)**

00106-04 Basic Rigging 6.i

Explains how ropes, chains, hoists, loaders, and cranes are used to move material and equipment from one location to another on a job site. Describes inspection techniques and load-handling safety practices. Also reviews American National Standards Institute (ANSI) hand signals. **(20 elective hours)**

00107-04 Basic Communication Skills. 7.i

Provides trainees with techniques for communicating effectively with co-workers and supervisors. Includes practical examples that emphasize the importance of verbal and written information and instructions on the job. Also discusses effective telephone and email communication skills. (**5 elective hours.** ***Must be combined with 00108–04***)

00108-04 Basic Employability Skills 8.i

Identifies the roles of individuals and companies in the construction industry. Introduces trainees to critical thinking, problem-solving skills, and computer systems and their industry applications. Also reviews effective relationship skills, effective self-presentation, and key workplace issues, such as sexual harassment, stress, and substance abuse. **(15 elective hours.** ***Must be combined with 00107–04***)

Figure Credits. FC.1

Glossary of Key Trade Terms. G.1

Index I.1

Contren® Curricula

NCCER's training programs comprise more than 40 construction, maintenance, and pipeline areas and include skills assessments, safety training, and management education.

Boilermaking
Carpentry
Carpentry, Residential
Cabinetmaking
Concrete Finishing
Construction Craft Laborer
Construction Technology
Core Curriculum: Introductory Craft Skills
Currículum Básico
Electrical
Electrical, Residential
Electrical Topics, Advanced
Electronic Systems Technician
Exploring Careers in Construction
Fundamentals of Mechanical and Electrical Mathematics
Heating, Ventilating, and Air Conditioning
Heavy Equipment Operations
Highway/Heavy Construction
Instrumentation
Insulating
Ironworking
Maintenance, Industrial
Masonry
Millwright
Mobile Crane Operations
Painting
Painting, Industrial
Pipefitting
Pipelayer
Plumbing
Scaffolding
Sheet Metal
Site Layout
Sprinkler Fitting
Welding

Pipeline

Control Center Operations, Liquid
Corrosion Control
Electrical and Instrumentation
Field Operations, Liquid
Field Operations, Gas
Maintenance
Mechanical

Safety

Field Safety
Safety Orientation
Safety Technology

Management

Introductory Skills for the Crew Leader
Project Management
Project Supervision

Acknowledgments

This curriculum was revised as a result of the farsightedness and leadership of the following sponsors:

Associated Builders and Contractors, Inc.
Associated General Contractors
Calvert Cliffs Nuclear Power Plant
Constellation Power Source Generation
Guilford Technical Community College
Hensel Phelps Construction Company, Inc.
Holder Engineering Services
Manatee Technical Institute
Monroe #1 Board of Cooperative Education
North Carolina Department of Public Instruction
NE Florida Builders Association
Tabacon Systems, Inc./Beam Up, Inc.
Winter Construction

This curriculum would not exist were it not for the dedication and unselfish energy of those volunteers who served on the Authoring Team. A sincere thanks is extended to the following:

John Ambrosia
Joe Beyer
Richard W. Carr
Fred Day
Jim Evans
Doug Garcia
Charlie Haas
Chuck Hogg
R. P. Hughes
Connell Linson
Bruce Miller
Steven Miller
Denise Peek

A final note: This book is the result of a collaborative effort involving the production, editorial, and development staff at Prentice Hall and the National Center for Construction Education and Research. Thanks to all of the dedicated people involved in the many stages of this project.

NCCER PARTNERING ASSOCIATIONS

American Fire Sprinkler Association
American Petroleum Institute
American Society for Training & Development
American Welding Society
Associated Builders & Contractors, Inc.
Association for Career and Technical Education
Associated General Contractors of America
Carolinas AGC, Inc.
Carolinas Electrical Contractors Association
Citizens Democracy Corps
Construction Industry Institute
Construction Users Roundtable
Design-Build Institute of America
Electronic Systems Industry Consortium
Merit Contractors Association of Canada
Metal Building Manufacturers Association
National Association of Minority Contractors
National Association of State Supervisors for Trade and Industrial Education
National Association of Women in Construction
National Insulation Association
National Ready Mixed Concrete Association
National Systems Contractors Association
National Utility Contractors Association
National Vocational Technical Honor Society
North American Crane Bureau
North American Technician Excellence
Painting & Decorating Contractors of America
Plumbing-Heating-Cooling Contractors National Association
Portland Cement Association
SkillsUSA
Steel Erectors Association of America
Texas Gulf Coast Chapter ABC
U.S. Army Corps of Engineers
University of Florida
Women Construction Owners & Executives, USA
Youth Training and Development Consortium

Basic Safety

00101-04

NCCER STANDARDIZED CRAFT TRAINING PROGRAM

The National Center for Construction Education and Research (NCCER) provides a standardized national program of accredited craft training. Key features of the program include instructor certification, competency-based training, and performance testing. The program provides trainees, instructors, and companies with a standard form of recognition through a National Craft Training Registry. The program is described in full in the *Guidelines for Accreditation*, published by the NCCER. For more information on standardized craft training, contact the NCCER by writing us at P.O. Box 141104, Gainesville, FL 32614-1104; calling 352-334-0911; or e-mailing info@nccer.org. More information may be found at our Web site, www.nccer.org.

HOW TO USE THIS ANNOTATED INSTRUCTOR'S GUIDE

Each page presents two sections of information. The larger section displays each page exactly as it appears in the Trainee Module. The narrow column ties suggested trainee and instructor actions to each page and provides icons (detailed below) to call your attention to material, safety, audiovisual, or testing requirements. The bottom of each page includes space for your notes.

The **Audiovisual** icon indicates an appropriate time to show a transparency or other audiovisual aid.

The **Classroom** icon prompts you to define a term, stress a point, ask trainees to explain a concept, or give examples.

The **Demonstration** icon directs you to show trainees how to perform tasks.

The **Examination** icon tells you to administer the written module examination.

The **Homework** icon is placed where you may wish to assign reading for the next class, to assign a project, or to advise trainees to prepare for an examination.

The **Laboratory** icon is used when trainees are to practice performing tasks.

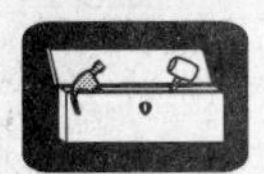

The **Materials** icon is a reminder for you to gather materials needed for classes, labs, and testing.

The **Performance Testing** icon tells you to administer a performance test or a portion thereof.

The **Safety** icon is used to emphasize safety issues. It is often keyed to *Caution* and *Warning* statements in the Trainee Module.

The **Teaching Tip** icon indicates additional guidance is available, such as how to conduct an exercise, get the most educational value from a field trip, or encourage class participation. Teaching Tips may expand on a feature (*Think About It, Did You Know?*) or provide Quick Quizzes or similar exercises. You will be referred to the Teaching Tips section at the back of the module if there is additional material.

The **Combination** icon indicates that the laboratory listed corresponds with a performance task. If desired, you can note the proficiency of the trainees during the laboratory and use it to satisfy performance testing requirements.

PREPARATION

Before teaching this module, you should review the Objectives, Performance Tasks, Materials and Equipment List, and the Module Outline. Be sure to allow ample time to prepare your own training or lesson plan and gather all required materials and equipment.

MODULE OVERVIEW

This module explains the role of safety in the construction crafts. Trainees will learn how to identify and follow safe work practices and procedures as well as how to properly inspect and use safety equipment. Trainees will be able to describe safe work procedures for lifting heavy objects, fighting fires, and working around electrical hazards.

PREREQUISITES

There are no prerequisites for this module.

OBJECTIVES

Upon completion of this module, the trainee will be able to:

1. Explain the role that safety plays in the construction crafts.
2. Describe the meaning of job-site safety.
3. Describe the characteristics of a competent person and a qualified person.
4. Explain the appropriate safety precautions to take around common job-site hazards.
5. Demonstrate the use and care of appropriate personal protective equipment (PPE).
6. Properly don and remove personal protective equipment (safety goggles, hard hat, and personal fall protection).
7. Follow the safety procedures required for lifting heavy objects.
8. Describe safe behavior on and around ladders and scaffolds.
9. Explain the importance of Hazard Communications (HazCom) and material safety data sheets (MSDSs).
10. Describe fire prevention and firefighting techniques.
11. Define safe work procedures to use around electrical hazards.

PERFORMANCE TASKS

Under the supervision of the instructor, the trainee should be able to:

1. Inspect PPE to determine if it is safe to use (PPE should include safety goggles, hard hat, gloves, safety harness, and safety shoes).
2. Properly don and remove PPE (safety goggles, hard hat, and personal fall protection).
3. Demonstrate safe lifting procedures.

MATERIALS AND EQUIPMENT LIST

Transparencies
Markers/chalk
Blank acetate sheets
Transparency pens
Pencils and scratch paper
Overhead projector and screen
Whiteboard/chalkboard
Copies of your local code
Variety of communication tags and signs
Dull and sharp cutting tools
Code of Federal Regulations Part 1910
OSHA 29 CFR 1926
OSHA Form 300
New and damaged hoses and regulators
Materials to create hypothetical fire hazards
Variety of safety tags
Variety of types of personal protective equipment, including:
- Hard hats
- Safety glasses, goggles, and face shields
- Safety harnesses
- Gloves
- Safety shoes
- Hearing protection
- Respiratory protection

Copies of your company's fall protection plan

Variety of ladders and scaffolds, including:
- Straight ladders
- Extension ladders
- Stepladders
- Manufactured scaffolds
- Rolling scaffolds

Sample material safety data sheet (MSDS)
Variety of fire extinguishers
Module Examinations*
Performance Profile Sheets*

*Located in the Test Booklet.

SAFETY CONSIDERATIONS

Ensure that the trainees are equipped with appropriate personal protective equipment. Always work in a clean, well-lit, appropriate work area.

ADDITIONAL RESOURCES

This module is intended to present thorough resources for task training. The following reference works are suggested for both instructors and motivated trainees interested in further study. These are optional materials for continued education rather than for task training.

Construction Back Safety. Videocassette. 10 minutes. Coastal Training Technologies Corp. Virginia Beach, VA.

Construction Confined Space Entry. Videocassette. 10 minutes. Coastal Training Technologies Corp. Virginia Beach, VA.

Construction Electrical Safety. Videocassette. 10 minutes. Coastal Training Technologies Corp. Virginia Beach, VA.

Construction Fall Protection: Get Arrested! Videocassette. 11 minutes. Coastal Training Technologies Corp. Virginia Beach, VA.

Construction Lockout/Tagout. Videocassette. 10 minutes. Coastal Training Technologies Corp. Virginia Beach, VA.

Construction Safety, 1996. Jimmie Hinze. Englewood Cliffs, NJ: Prentice Hall.

Construction Safety Council Home Page, http://buildsafe.org/home.htm.

Construction Safety Manual, 1998. Dave Heberle. New York: McGraw-Hill.

Construction Stairways and Ladders. Videocassette. 10 minutes. Coastal Training Technologies Corp. Virginia Beach, VA.

Construction Welding Safety. Videocassette. 10 minutes. Coastal Training Technologies Corp. Virginia Beach, VA.

Field Safety, 2003. NCCER. Upper Saddle River, NJ: Prentice Hall.

Handbook of OSHA Construction Safety and Health, 1999. James V. Eidson et al. Boca Raton, FL: Lewis Publishers, Inc.

HazCom for Construction. Videocassette. 11 minutes. Coastal Training Technologies Corp. Virginia Beach, VA.

NAHB-OSHA Jobsite Safety Handbook, 1999. Washington, DC: Home Builder Press. Available online at www.osha.gov.

Occupational Safety and Health Standards for the Construction Industry, latest edition. Washington, DC: Occupational Safety and Health Administration, U.S. Department of Labor, U.S. Government Printing Office.

Safety Orientation, 2003. NCCER. Upper Saddle River, NJ: Prentice Hall.

Safety Technology, 2003. NCCER. Upper Saddle River, NJ: Prentice Hall.

United States Department of Labor, Occupational Safety and Health Administration Home Page, http://www.osha.gov.

TEACHING TIME FOR THIS MODULE

An outline for use in developing your lesson plan is presented below. Note that each Roman numeral in the outline equates to one session of instruction. Each session has a suggested time period of 2½ hours. This includes 10 minutes at the beginning of each session for administrative tasks and one 10-minute break during the session. Approximately 15 hours are suggested to cover *Basic Safety*. You will need to adjust the time required for hands-on activity and testing based on your class size and resources. Because laboratories often correspond to Performance Tasks, the proficiency of the trainees may be noted during these exercises for Performance Testing purposes.

Topic	Planned Time
Session I. Introduction to Accidents	
A. Accident Causes and Results	________
B. Safety Policies and Regulations	________
C. Reporting Accidents	________
Session II. Job-Site Hazards	
A. Construction Job-Site Hazards	________
B. Working Safely with Job Hazards	________
Session III. Personal Protective Equipment and Lifting	
A. Personal Protective Equipment	________
B. Lifting	________
C. Performance Testing	________
1. Trainees must perform each task to the satisfaction of the instructor to receive recognition from NCCER. If applicable, proficiency noted during laboratory exercises can be used to satisfy the Performance Testing requirements.	
2. Record the testing results on Craft Training Report Form 200, and submit the results to the Training Program Sponsor.	
Session IV. Aerial Work	
A. Ladders and Scaffolds	________
B. Scaffolds	________
Session V. Hazard Communication and Fire Safety	
A. Hazard Communication Standard	________
B. Fire Safety	________
Session VI. Electrical Safety, Review, and Module Examination	
A. Basic Electrical Safety Guidelines	________
B. Working Near Energized Electrical Equipment	________
C. If Someone Is Shocked	________
D. Review	________
E. Module Examination	________
1. Trainees must score 70% or higher to receive recognition from NCCER.	
2. Record the testing results on Craft Training Report Form 200, and submit the results to the Training Program Sponsor.	

Build America Merit Winner— Building New

The McCormick Tribune Campus Center and the renovation of the historic Illinois Institute of Technology Commons Building in Chicago, Illinois, was completed by the Gilbane Building Company. This bold and innovative project included the replacement of the Chicago Transit Authority subway track supports, the construction of an elliptical acoustic tube around the elevated subway tracks, and the construction and renovation of the campus center and commons building.

00101-04
Basic Safety

Topics to be presented in this unit include:

Overview

Construction safety has advanced a great deal over the years. There was a time in the industry when very few safety precautions were taken. Workers often worked without hard hats, gloves, eye protection, fall protection, or safety shoes. This resulted in illnesses, injuries, and deaths. Today, with the help of the Occupational Safety and Health Administration (OSHA) and state and local rules and regulations, the construction industry is becoming safer.

Rules and regulations are only a part of what contributes to a safe work environment. Construction workers have the ultimate responsibility for their safety and the safety of others. One way to remain safe is always to use appropriate personal protective equipment (PPE). The type of personal protective equipment will vary depending on your job. For example, a heavy machine operator is not required to wear the same face shield a welder is required to wear.

Proper training is also an essential part of creating a safe work environment. Training must be provided to all workers to ensure that they perform work and use and maintain all tools and equipment safely. When a worker does not know how to use a tool or equipment or has not been properly trained to use personal protective equipment, he or she should immediately tell a supervisor and ask for the necessary training. Safe behavior and safe equipment help to prevent accidents and injury.

Instructor's Notes:

Objectives

When you have completed this module, you will be able to do the following:

1. Explain the role that safety plays in the construction crafts.
2. Describe the meaning of job-site safety.
3. Describe the characteristics of a competent person and a qualified person.
4. Explain the appropriate safety precautions to take around common job-site hazards.
5. Demonstrate the use and care of appropriate personal protective equipment (PPE).
6. Properly don and remove personal protective equipment (safety goggles, hard hat, and personal fall protection).
7. Follow the safety procedures required for lifting heavy objects.
8. Describe safe behavior on and around ladders and scaffolds.
9. Explain the importance of hazard communications (HazCom) and material safety data sheets (MSDSs).
10. Describe fire prevention and firefighting techniques.
11. Define safe work procedures to use around electrical hazards.

Key Trade Terms

Apparatus
Arc
Arc welding
Combustible
Competent person
Concealed receptacle
Confined space
Cross-bracing
Dross
Electrical distribution panel
Excavation
Extension ladder
Flammable
Flash
Flashback
Flash burn
Flash goggles
Flash point
Ground
Guarded
Hand line
Hazard Communication Standard (HazCom)
Lanyard
Lockout/tagout
Management system
Material safety data sheet (MSDS)
Maximum intended load
Mid-rail
Occupational Safety and Health Administration (OSHA)
Permit-required confined space
Personal protective equipment (PPE)
Planked
Proximity work
Qualified person
Respirator
Scaffold
Shoring
Signaler
Slag
Stepladder
Straight ladder
Switch enclosure
Toeboard
Top rail
Trench
Trencher
Welding shield
Wind sock

Required Trainee Materials

1. Sharpened pencils and paper
2. Appropriate personal protective equipment

Prerequisites

There are no prerequisites for this module.

This course map shows all of the modules in *Core Curriculum: Introductory Craft Skills.* The suggested training order begins at the bottom and proceeds up. Skill levels increase as you advance on the course map. The local Training Program Sponsor may adjust the training order.

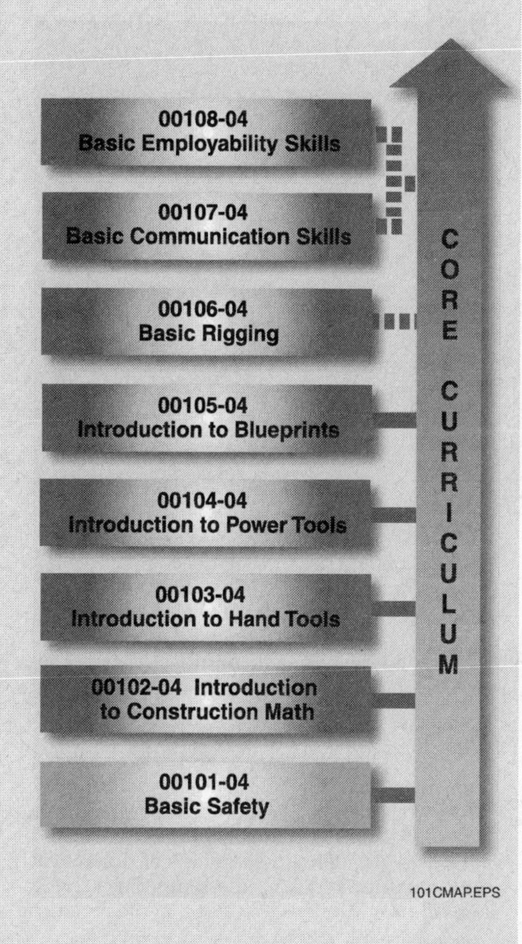

101CMAP.EPS

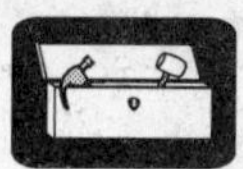

Ensure that you have everything required to teach the course. Check the Materials and Equipment List at the front of this module.

See the general Teaching Tip at the end of this module.

Show Transparency 1, Course Objectives.

Show Transparency 2, Performance Tasks.

Explain that terms shown in bold (blue) are defined in the Glossary at the back of this module.

Discuss the importance of working safely. Ask trainees to explain problems that can result if proper procedures are not followed.

Explain that the Occupational Safety and Health Administration (OSHA) enforces safety at residential and commercial work sites.

Ask trainees to explain how they can help prevent accidents. Provide examples of safe work habits from your own experience.

Discuss the most common causes of accidents. Ask trainees to discuss accidents that they may have witnessed as a result of these main causes.

Ensure that trainees understand when they are making assumptions. Discuss problems that may occur as a result of making assumptions.

Review the types of communication tags and signs that are used at work sites. Answer any questions trainees may have.

1.0.0 ◆ INTRODUCTION

When you take a job, you have a safety obligation to your employer, co-workers, family, and yourself. In exchange for your wages and benefits, you agree to work safely. You are also obligated to make sure anyone you work with is working safely. Your employer is likewise obligated to maintain a safe workplace for all employees. The ultimate responsibility for on-the-job safety, however, rests with you. Safety is part of your job. In this module, you will learn to ensure your safety and that of the people you work with by adhering to the following rules:

- Follow safe work practices and procedures.
- Inspect safety equipment before use.
- Use safety equipment properly.

To take full advantage of the wide variety of training, job, and career opportunities the construction industry offers, you must first understand the importance of safety. Successful completion of this module will be your first step toward achieving this goal. Other modules offer more detailed explanations of safety procedures and opportunities to practice them.

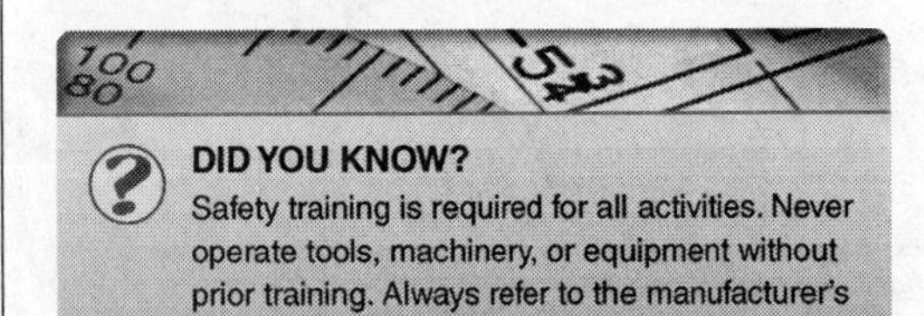

DID YOU KNOW?
Safety training is required for all activities. Never operate tools, machinery, or equipment without prior training. Always refer to the manufacturer's instructions.

2.0.0 ◆ ACCIDENTS: CAUSES AND RESULTS

Your boss might say, "I want my company to have a perfect safety record." What does that mean? A safety record is more than the number of days a company has worked without an accident. Safety is a learned behavior and attitude. Safety is a way of working. The time you spend learning and practicing safety procedures can save your life and the lives of others.

What causes accidents? Both poor behavior and poor working conditions can cause accidents. You can help prevent accidents by learning safe work habits and understanding what causes accidents. Accidents cost billions of dollars each year and cause much needless suffering. The National Safety Council estimates that the organized safety movement has saved more than 4.2 million lives since it began in 1913. This section examines why accidents happen and how you can help prevent them.

The lessons you will learn in this module will help you work safely. You will be able to spot and avoid hazardous conditions on the job site. By following safety procedures and being aware of the need for safety, you will help keep your workplace free from accidents and protect yourself and your co-workers from injury or even death.

2.1.0 What Causes Accidents?

You may already know some of the main causes of accidents. They include the following:

- Failure to communicate
- Poor work habits
- Alcohol or drug abuse
- Lack of skill
- Intentional acts
- Unsafe acts
- Rationalizing risks
- Unsafe conditions
- **Management system** failure

We will discuss each of these causes in the following sections.

2.1.1 Failure to Communicate

Many accidents happen because of a lack of communication. For example, you may learn how to do things one way on one job, but what happens when you go to a new job site? You need to communicate with the people at the new job site to find out whether they do things the way you have learned to do them. If you do not communicate clearly, accidents can happen. Remember that different people, companies, and job sites do things in different ways.

If you think that people know something without talking with them about it, then you are assuming that they know. Assuming that other people know and will do what you think they will do can cause accidents.

All work sites have specific markings and signs to identify hazards and provide emergency information (see *Figure 1*). Learn to recognize these types of signs:

- Informational
- Safety
- Caution
- Danger
- Temporary warnings

Instructor's Notes:

1.2 CORE CURRICULUM ◆ INTRODUCTORY CRAFT SKILLS

Bring in a variety of communication tags and signs. Ask trainees to identify the types of signs and explain what each sign is communicating. Review the colors used for each type of tag or sign.

Informational markings or signs provide general information. These signs are blue. The following are considered informational signs:

- No Admittance
- No Trespassing
- For Employees Only

Safety signs give general instructions and suggestions about safety measures. The background on these signs is white; most have a green panel with white letters. These signs tell you where to find such important areas as the following:

- First-aid stations
- Emergency eye-wash stations
- Evacuation routes
- **Material safety data sheet (MSDS)** stations
- Exits (usually have white letters on a red field)

Caution markings or signs tell you about potential hazards or warn against unsafe acts. When you see a caution sign, protect yourself against a possible hazard. Caution signs are yellow and have a black panel with yellow letters. They may give you the following information:

- Hearing and eye protection are required.
- **Respirators** are required.
- Smoking is not allowed.

Danger markings or signs tell you that an immediate hazard exists and that you must take certain precautions to avoid an accident. Danger signs are red, black, and white. They may indicate the presence of the following:

- Defective equipment
- **Flammable** liquids and compressed gases
- Safety barriers and barricades
- Emergency stop button
- High voltage

Safety tags are temporary warnings of immediate and potential hazards. They are not designed to replace signs or to serve as permanent means of protection. Learn to recognize the standard accident prevention signs and tags (see *Table 1*). Other important guidelines regarding communication are covered in the module, *Basic Communication Skills.*

101F01A.EPS

INFORMATION SIGN

101F01B.EPS

SAFETY SIGN

101F01C.EPS

CAUTION SIGN

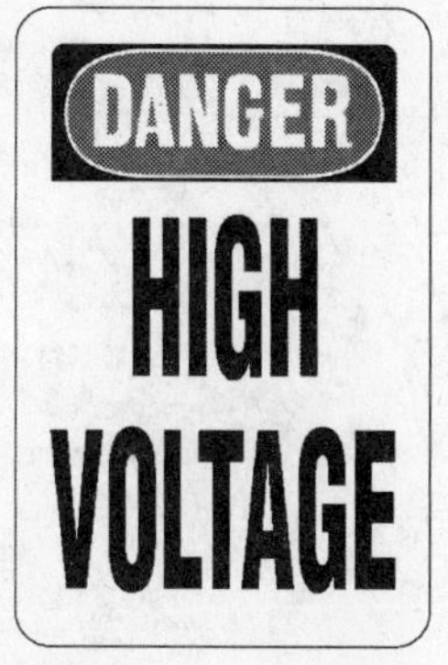

101F01D.EPS

DANGER SIGN

Figure 1 ◆ Communication tags/signs.

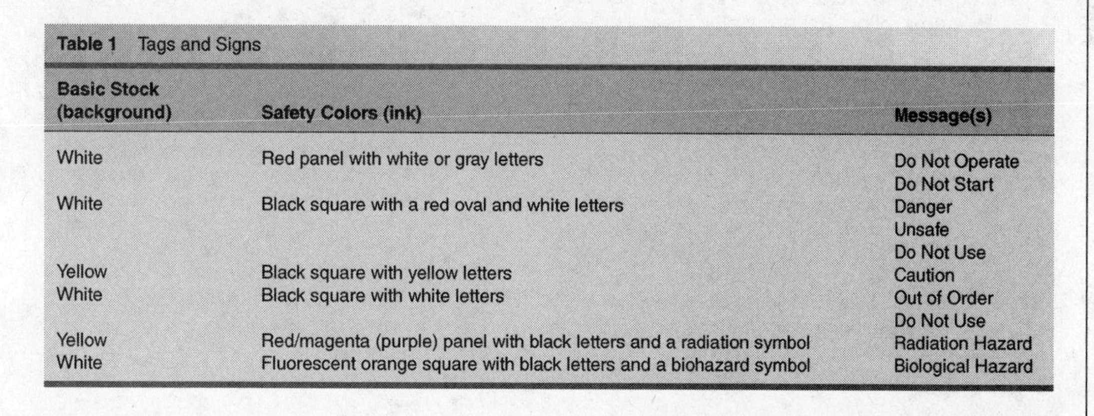

Table 1 Tags and Signs

Basic Stock (background)	Safety Colors (ink)	Message(s)
White	Red panel with white or gray letters	Do Not Operate Do Not Start
White	Black square with a red oval and white letters	Danger Unsafe Do Not Use
Yellow	Black square with yellow letters	Caution
White	Black square with white letters	Out of Order Do Not Use
Yellow	Red/magenta (purple) panel with black letters and a radiation symbol	Radiation Hazard
White	Fluorescent orange square with black letters and a biohazard symbol	Biological Hazard

MODULE 00101-04 ◆ BASIC SAFETY 1.3

Stress the importance of asking questions. Discuss the consequences of making assumptions.

WARNING!

Never assume anything! It never hurts to ask questions, but disaster can result if you don't ask. For example, do not assume that an electrical current is turned off. First ask whether the current is turned off, then check it yourself to be completely safe.

Explain how poor work habits can cause serious accidents. Ask trainees to provide examples of problems caused by procrastination.

2.1.2 Poor Work Habits

Poor work habits can cause serious accidents. Examples of poor work habits are procrastination, carelessness, and horseplay. Procrastination, or putting things off, is a common cause of accidents. For example, delaying the repair, inspection, or cleaning of equipment and tools can cause accidents. If you try to push machines and equipment beyond their operating capacities, you risk injuring yourself and your co-workers.

Machines, power tools, and even a pair of pliers can hurt you if you don't use them safely. It is your responsibility to be careful. Tools and machines don't know the difference between wood or steel and flesh and bone.

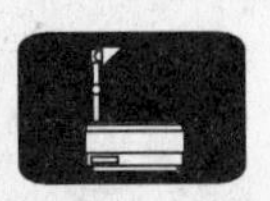

Show Transparency 3 (Figure 2). Explain mistakes the worker is making that could cause an accident. Ask trainees to identify problems illustrated in the picture.

Work habits and work attitudes are closely related. If you resist taking orders, you may also resist listening to warnings. If you let yourself be easily distracted, you won't be able to concentrate. If you aren't concentrating, you could cause an accident.

Your safety is affected not only by how you do your work, but also by how you act on the job site. This is why most companies have strict policies for employee behavior. Horseplay and other inappropriate behavior are forbidden. Workers who engage in horseplay and other inappropriate behavior on the job site may be fired.

Bring in a dull and a sharp cutting tool. Demonstrate how the two operate differently when cutting.

These strict policies are for your protection. There are many hazards on construction sites. Each person's behavior—at work, on a break, or at lunch—must follow the principles of safety.

DID YOU KNOW?

Dull blades cause more accidents than sharp ones. If you do not keep your cutting tools sharpened, they won't cut very easily. When you have a hard time cutting, you exert more force on the tool. When that happens, something is bound to slip. And when something slips, you can get cut.

The man pouring the water on his co-worker in *Figure 2* may look like he's just having fun by playing a prank on his co-worker. In fact, what he's doing could cause his co-worker serious, even fatal, injury. If you horse around on the job, play pranks, or don't concentrate on what you are doing, you are showing a poor work attitude. That can lead to a serious accident.

DID YOU KNOW?

Quick Quiz

Look at the man who's trying to play a prank on his co-worker in *Figure 2*. How many things are wrong in this picture? What other safety rules is he breaking?

- No hard hat
- No safety goggles
- No gloves
- Breaking shop rules

Discuss problems caused by stress. Ask trainees to discuss times they have experienced stress and how they dealt with it. Go over any questions trainees may have.

DID YOU KNOW?

Stress creates a chemical change in your body. Although stress may heighten your hearing, vision, energy, and strength, long-term stress can harm your health.

Not all stress is job-related; some stress develops from the pressures of dealing with family and friends and daily living. In the end, your ability to handle and manage your stress determines whether stress hurts or helps you. Use common sense when you are dealing with stressful situations. For example, consider the following:

- Keep daily occurrences in perspective. Not everything is worth getting upset, angry, or anxious about.
- When you have a particularly difficult workday scheduled, get plenty of rest the night before.
- Manage your time. The feeling of always being behind creates a lot of stress. Waiting until the last minute to finish an important task adds unnecessary stress.
- Talk to your supervisor. Your supervisor may understand what is causing your stress and may be able to suggest ways to manage it better.

Instructor's Notes:

1.4 CORE CURRICULUM ◆ INTRODUCTORY CRAFT SKILLS

101F02.EPS

Figure 2 ◆ Horseplay can be dangerous.

Show Transparency 4 (Figure 3). Ask trainees to identify the violations illustrated in the picture.

Emphasize the dangers associated with operating machinery while under the influence of drugs or alcohol. Discuss the risk that workers pose not only to themselves but also to their fellow workers.

Discuss the importance of understanding the side effects of medications. Ensure that trainees understand that even over-the-counter medications can cause drowsiness.

2.1.3 Alcohol and Drug Abuse

Alcohol and drug abuse costs the construction industry millions of dollars a year in accidents, lost time, and lost productivity. The true cost of alcohol and drug abuse is much more than just money, of course. Abuse can cost lives. Just as drunk driving kills thousands of people on our highways every year, alcohol and drug abuse kills on the construction site. Examine the person in *Figure 3.* Would you want to be like him or be working near him?

Using alcohol or drugs creates a risk of injury for everyone on a job site. Many states have laws that prevent workers from collecting insurance benefits if they are injured while under the influence of alcohol or illegal drugs.

Would you trust your life to a crane operator who was high on drugs? Would you bet your life on the responses of a co-worker using alcohol or drugs? Alcohol and drug abuse have no place in the construction industry. A person on a construction site who is under the influence of alcohol or drugs is an accident waiting to happen—possibly a fatal accident.

People who work while using alcohol or drugs are at risk of accident or injury, and their co-workers are at risk as well. That's why your employer probably has a formal substance abuse policy. You should know that policy and follow it for your own safety.

You don't have to be abusing illegal drugs such as marijuana, cocaine, or heroin to create a job hazard. Many prescription and over-the-counter drugs, taken for legitimate reasons, can affect your ability to work safely. Amphetamines, barbiturates, and antihistamines are only a few of the legal drugs that can affect your ability to work safely or to operate machinery.

CAUTION

If your doctor prescribes any medication that you think might affect your job performance, ask about its effects. Your safety and the safety of your co-workers depend on everyone being alert on the job.

MODULE 00101-04 ◆ BASIC SAFETY 1.5

Emphasize that trainees should never operate a power tool that they have not been trained to use. Have trainees discuss problems that can result if a power tool is used improperly.

Figure 3 ◆ How many violations can you identify?

Do yourself and the people you work with a big favor. Be aware of and follow your employer's substance abuse policy. Avoid any substances that can affect your job performance. The life you save could be your own.

DID YOU KNOW?

Quick Quiz

If the man in *Figure 3* doesn't kill himself first, he will almost certainly kill someone else eventually. How many violations can you identify? In this scene alone, the man is

- Consuming alcohol on the job site
- Not following proper safety procedures for operating a motorized vehicle near hazardous materials
- Not using both hands to drive the vehicle
- Not wearing a seat/safety belt
- Not wearing a hard hat
- Not wearing safety glasses or goggles

2.1.4 Lack of Skill

You should learn and practice new skills under careful supervision. Never perform new tasks alone until you've been checked out by a supervisor.

Lack of skill can cause accidents quickly. For example, suppose you are told to cut some 2 × 8s with a circular saw, but you aren't skilled with that tool. A basic rule of circular saw operation is never to cut without a properly functioning guard. Because you haven't been trained, you don't know this. You find that the guard on the saw is slowing you down. So you jam the guard open with a small block of wood. The result could be a serious accident. Proper training can prevent this type of accident.

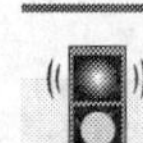

WARNING!

Never operate a power tool until you have been trained to use it. You can greatly reduce the chances of accidents by learning safety rules for each task you perform.

Instructor's Notes:

1.6 CORE CURRICULUM ◆ INTRODUCTORY CRAFT SKILLS

2.1.5 *Intentional Acts*

When someone purposely causes an accident, it is called an intentional act. Sometimes an angry or dissatisfied employee may purposely create a situation that leads to property damage or personal injury. If someone you are working with threatens to get even or pay back someone, let your supervisor know at once.

2.1.6 *Unsafe Acts*

An unsafe act is a change from an accepted, normal, or correct procedure that usually causes an accident. It can be any conduct that causes unnecessary exposure to a job-site hazard or that makes an activity less safe than usual. Here are examples of unsafe acts:

- Failing to use **personal protective equipment (PPE)**
- Failing to warn co-workers of hazards
- Lifting improperly
- Loading or placing equipment or supplies improperly
- Making safety devices (such as saw guards) inoperable
- Operating equipment at improper speeds
- Operating equipment without authority
- Servicing equipment in motion
- Taking an improper working position
- Using defective equipment
- Using equipment improperly

2.1.7 *Rationalizing Risk*

Everybody takes risks every day. When you get in your car to drive to work, you know there is a risk of being involved in an accident. Yet when you drive using all the safety practices you have learned, you know that there is a good chance that you will arrive at your destination safely. Driving is an appropriate risk because you have some control over your own safety and that of others.

Some risks are not appropriate. On the job, you must never take risks that endanger yourself or others just because you can make an excuse for doing so. This is called rationalizing risk. Rationalizing risk means ignoring safety warnings and practices. For example, because you are late for work, you might decide to run a red light. By trying to save time, you could cause a serious accident.

The following are common examples of rationalized risks on the job:

- Crossing boundaries because no activity is in sight
- Not wearing gloves because it will take only a minute to make a cut
- Removing your hard hat because you are hot and you cannot see anyone working overhead
- Not tying off your fall protection because you only have to lean over by about a foot

Think about the job before you do it. If you think that it is unsafe, then it is unsafe. Stop working until the job can be done safely. Bring your concerns to the attention of your supervisor. Your health and safety, and that of your co-workers, make it worth taking extra care.

DID YOU KNOW?
Most workers who die from falls were wearing harnesses but had failed to tie off properly. Always follow the manufacturer's instructions when wearing a harness. Know and follow your company's safety procedures when working on roofs, ladders, and other elevated locations.

2.1.8 *Unsafe Conditions*

An unsafe condition is a physical state that is different from the acceptable, normal, or correct condition found on the job site. It usually causes an accident. It can be anything that reduces the degree of safety normally present. The following are some examples of unsafe conditions:

- Congested workplace
- Defective tools, equipment, or supplies
- Excessive noise
- Fire and explosive hazards
- Hazardous atmospheric conditions (such as gases, dusts, fumes, and vapors)
- Inadequate supports or guards
- Inadequate warning systems
- Poor housekeeping
- Poor lighting
- Poor ventilation
- Radiation exposure
- Unguarded moving parts such as pulleys, drive chains, and belts

2.1.9 *Management System Failure*

Sometimes the cause of an accident is failure of the management system. The management system should be designed to prevent or correct the acts and conditions that can cause accidents. If the management system did not do these things, that

Discuss the differences between intentional and unsafe acts. Review examples of unsafe acts, and ask trainees to provide examples of accidents they have witnessed as a result of unsafe acts.

Ensure that trainees understand when it is appropriate to take risks on the work site. Provide examples of rationalized risks, and ask trainees whether they would consider taking the risk. Highlight problems that could result if the risk were taken.

Review the types of job-site conditions that are considered unsafe.

Explain differences between a successful and failing management system. Ask whether trainees have any questions.

Ensure that trainees understand the concept of chain of command within a company. Using an accident as an example, have trainees explain whom they would first contact and the procedures they would follow to report the accident.

See the Teaching Tip for Sections 2.1.0–2.1.9 at the end of this module.

Review the rules of good housekeeping. Have trainees explain how each rule helps to prevent accidents.

Distribute copies of the *Code of Federal Regulations Part 1910.* Have trainees review the code and determine which standards could apply to their jobs.

system failure may have caused the accident. What traits could mean the difference between a management system that fails and one that succeeds? A few important traits of a good management system follow:

- The company puts safety policies and procedures in writing.
- The company distributes written safety policies and procedures to each employee.
- The company reviews safety policies and procedures periodically.
- The company enforces all safety policies and procedures fairly and consistently.
- The company evaluates supplies, equipment, and services to see whether they are safe.
- The company provides regular, periodic safety training for employees.

2.2.0 Housekeeping

In construction, housekeeping means keeping your work area clean and free of scraps or spills. It also means being orderly and organized. You must store your materials and supplies safely and label them properly. Arranging your tools and equipment to permit safe, efficient work practices and easy cleaning is also important.

If the work site is indoors, make sure it is well lighted and ventilated. Don't allow aisles and exits to be blocked by materials and equipment. Make sure that flammable liquids are stored in safety cans. Oily rags must be placed only in approved, self-closing metal containers. Remember that the major goal of housekeeping is to prevent accidents. Good housekeeping reduces the chances for slips, fires, explosions, and falling objects.

Here are some good housekeeping rules:

- Remove all scrap material and lumber with nails sticking out from work areas.
- Clean up spills to prevent falls.
- Remove all **combustible** scrap materials regularly.
- Make sure you have containers for the collection and separation of refuse. Containers for flammable or harmful refuse must be covered.
- Dispose of wastes often.
- Store all tools and equipment when you're finished using them.

Another term for good housekeeping is pride of workmanship. If you take pride in what you are doing, you won't let trash build up around you. The old saying "A place for everything and everything in its place" is the right idea on the job site.

2.3.0 Company Safety Policies and OSHA Regulations

The mission of the **Occupational Safety and Health Administration (OSHA)** is to save lives, prevent injuries, and protect the health of America's workers. To accomplish this, federal and state governments work in partnership with the 111 million working men and women and their 7 million employers who are covered by the Occupational Safety and Health Act (OSH Act) of 1970.

Nearly every worker in the nation comes under OSHA's jurisdiction. There are some exceptions, such as miners, transportation workers, many public employees, and the self-employed.

2.3.1 The Code of Federal Regulations

The *Code of Federal Regulations* (CFR) *Part 1910* covers the OSHA standards for general industry. *CFR Part 1926* covers the OSHA standards for the construction industry. Either or both may apply to you, depending on where you are working and what you are doing. If a job-site condition is covered in the CFR book, then that standard must be used. However, if a more stringent requirement is listed in *CFR 1910,* it should also be met. Check with your supervisor to find out which standards apply to your job.

If a standard does not specifically address a hazard, the general duty clause must be invoked. This clause, found in *Section 5(a)1* of the OSH Act, summarizes the intent of the law:

> *Each Employer—shall furnish to each of his employees employment and a place of employment which are free from recognized hazards that are causing or are likely to cause death or serious physical harm to his employees.*

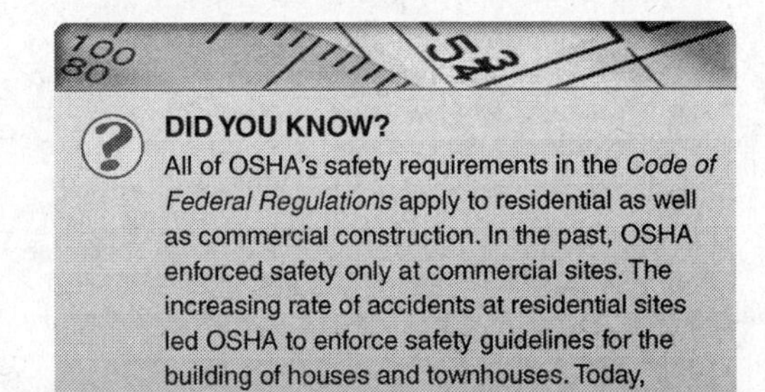

DID YOU KNOW?

All of OSHA's safety requirements in the *Code of Federal Regulations* apply to residential as well as commercial construction. In the past, OSHA enforced safety only at commercial sites. The increasing rate of accidents at residential sites led OSHA to enforce safety guidelines for the building of houses and townhouses. Today, however, OSHA still focuses its enforcement efforts on commercial construction.

Instructor's Notes:

1.8 CORE CURRICULUM ◆ INTRODUCTORY CRAFT SKILLS

2.3.2 Violations

Employers who violate OSHA regulations can be fined. The fines are not always high, but they can harm a company's reputation for safety. Fines for serious safety violations can be as high as $70,000 for each violation that was done willfully. In 2002, more than 78,000 fines were levied at a cost of $70,000 per violation.

2.3.3 Compliance

Just as employers are responsible to OSHA for compliance, employees must comply with their company's safety policies and rules. Employers are required to identify hazards and potential hazards within the workplace and eliminate them, control them, or provide protection from them. This can only be done through the combined efforts of the employer and employees. Employers must provide written programs and training on hazards, and employees must follow the procedures. You, as the employee, must read and understand the OSHA poster at your job site explaining your rights and responsibilities. If you are unsure where the OSHA poster is, ask your supervisor.

To help employers provide a safe workplace, OSHA requires companies to provide a **competent person** to ensure the safety of the employees. In *OSHA 29 CFR 1926*, OSHA defines a competent person as follows:

> *A person who can identify working conditions or surroundings that are unsanitary, hazardous, or dangerous to employees and who has authorization to correct or eliminate these conditions promptly.*

In comparison, *OSHA 29 CFR 1926* defines a **qualified person** as follows:

> *Someone who, by possession of a recognized degree, certificate, or professional standing, or who by extensive knowledge, training, and experience, has successfully demonstrated his ability to solve or resolve problems relating to the subject matter, work, or the project.*

In other words, a competent person is experienced and knowledgeable about the specific operation and has the authority from the employer to correct the problem or shut down the operation until it is safe. A qualified person has the knowledge and experience to handle problems. A competent person is not necessarily a qualified person.

These terms will be an important part of your career. It is important for you to know who the competent person is on your job site. OSHA requires a competent person for many of the tasks you may be assigned to perform, such as **confined space** entry, ladder use, and trenching. Different individuals may be assigned as a competent person for different tasks, according to their expertise. To ensure safety for you and your co-workers, work closely with your competent person and supervisor.

2.4.0 Reporting Injuries, Accidents, and Incidents

There are three categories of on-the-job events: injuries, accidents, and incidents. An injury is anything that requires treatment, even minor first aid. An accident is anything that causes an injury or property damage. An incident is anything that could have caused an injury or damage but, because it was caught in time, did not.

You must report all on-the-job injuries, accidents, or incidents, no matter how minor, to your supervisor (see *Figure 4*). Some workers think they will get in trouble if they report minor injuries. That's not true. Small injuries, like cuts and scrapes, can later become big problems because of infection and other complications.

101F04.EPS

Figure 4 ◆ All accidents, injuries, or incidents must be reported to your supervisor.

Emphasize that employers must be fined for violations.

Explain that it is the employee's responsibility to comply with regulations but that it is the employer's responsibility to identify and control hazards. Distinguish between a competent person and a qualified person.

See the Teaching Tip for Sections 2.3.0–2.3.3 at the end of this module.

Refer to Figure 4. Ensure that trainees understand the importance of reporting all injuries, accidents, and incidents. Distribute copies of OSHA Form 300. Have trainees complete the form using hypothetical information that you provide.

Review the four leading hazard groups. Provide an example of each type of hazard.

MODULE 00101-04 ◆ BASIC SAFETY 1.9

Review the procedures for an evacuation. Ask trainees to identify the signal that tells workers to evacuate their job site.

Go over the Review Questions for Section 2.0.0. Answer any questions trainees may have.

Have trainees review Sections 3.0.0–4.2.0 for the next session.

Many employers are required to maintain a log of significant work-related injuries and illnesses using OSHA Form 300. Employee names can be kept confidential in certain circumstances. A summary of these injuries must be posted at certain intervals, although employers do not need to submit it to OSHA unless requested. Employers can calculate the total number of injuries and illnesses and compare it with the average national rates for similar companies.

By analyzing accidents, companies and OSHA can improve safety policies and procedures. By reporting an accident, you can help keep similar accidents from happening in the future.

Table 2 shows an analysis of the causes of fatal accidents in the construction industry for 1997 through 1998.

Table 2 Causes of Fatal Accidents

Cause	Percentage of Fatalities
Falls from elevation	33%
Struck by . . .	27%
Caught in or caught between . . .	16%
Electrical shock	14%

All other accidents combined accounted for only 10 percent of the total. Because of these findings, OSHA developed its Focused Inspection Program to target these four high-hazard areas. OSHA hopes to reduce accidents, injuries, and fatalities in the construction industry.

Here are explanations of the four leading hazard groups:

- Falls from elevation are accidents involving failure of, failure to provide, or failure to use appropriate fall protection.
- Struck-by accidents involve unsafe operation of equipment, machinery, and vehicles, as well as improper handling of materials, such as through unsafe rigging operations.
- Caught-in or caught-between accidents involve unsafe operation of equipment, machinery, and vehicles, as well as improper safety procedures at **trench** sites and in other confined spaces.
- Electrical shock accidents involve contact with overhead wires, use of defective tools, failure to disconnect power source before repairs, or improper **ground** fault protection.

2.5.0 Evacuation Procedures

In many work environments, specific evacuation procedures are needed. These procedures go into effect when dangerous situations arise, such as fires, chemical spills, and gas leaks. In an emergency, you must know the evacuation procedures. You must also know the signal (usually a horn or siren) that tells workers to evacuate.

When you hear the evacuation signal, follow the evacuation procedures exactly. That usually means taking a certain route to a designated assembly area and telling the person in charge that you are there. If hazardous materials are released into the air, you may have to look at the **wind sock** to see which way the wind is blowing. Different evacuation routes are planned for different wind directions. Taking the right route will keep you from being exposed to the hazardous material.

Instructor's Notes:

____

____

____

1.10 CORE CURRICULUM ◆ INTRODUCTORY CRAFT SKILLS

Review Questions

Section 2.0.0

1. _________ is (are) *not* a main cause of accidents.
   a. Unsafe acts
   b. Alcohol or drug abuse
   c. Weather
   d. Poor work habits

2. Blue signs or markings that provide general information such as No Trespassing are _________ signs.
   a. caution
   b. informational
   c. warning
   d. safety

3. White and green signs or markings that give general instructions or suggestions about first-aid stations, exits, and evacuation routes are _________ signs.
   a. safety
   b. danger
   c. information
   d. MSDS

4. Many states have laws that prevent workers from collecting insurance benefits if they are injured while under the influence of alcohol or illegal drugs.
   a. True
   b. False

5. _________ is an example of an unsafe act.
   a. Taking an improper working position
   b. Using defective tools, equipment, or supplies
   c. Excessive noise
   d. Using a respirator

6. All of the following are good housekeeping guidelines *except* _________.
   a. keep aisles and exits clear
   b. clean up spills
   c. place oily rags in an uncovered container
   d. dispose of wastes often

7. The _________ summarizes the intent of the *OSHA Act of 1970.*
   a. competent person clause
   b. general duty clause
   c. hazardous duty clause
   d. qualified person clause

8. If a sign has a white background and a red panel with white or gray letters, you might see _________ on it.
   a. Out of Order
   b. Danger
   c. Biological Hazard
   d. Do Not Start

9. _________ must be reported to the employer.
   a. Only major injuries
   b. Only incidents and major injuries
   c. All injuries and incidents
   d. Only incidents in which a death occurred

10. An _________ is anything that could have caused an injury or damage if it hadn't been caught in time.
    a. incident
    b. accident
    c. injury
    d. intentional act

MODULE 00101-04 ◆ BASIC SAFETY 1.11

Ensure that you have everything required for laboratories during this session.

Review common job-site hazards. Ask trainees to make a list of hazards they think might be present at their job site. Discuss the potential hazards and possible solutions.

Ask a trainee to explain problems that could result if oxygen mixes with oil or grease.

Discuss the consequences of removing a protective cap if the cylinder is not secured.

Explain why it is essential to wear proper eye protection during an arc welding operation.

Show Transparency 5 (Figure 5). Review the personal protective equipment necessary when arc welding.

Discuss fire prevention programs, and explain how to implement a fire watch when welding or cutting.

Review the symptoms of flash burn to the eye. Emphasize that workers must never wear contact lenses when welding.

3.0.0 ◆ HAZARDS ON THE CONSTRUCTION JOB SITE

It's impossible to list all the hazards that can exist on a construction job site. This section describes some of the more common hazards and explains how to deal with them. You may want to make a list of other hazards you think could be present on your job site and discuss them with your instructor or supervisor.

For your safety, you must know the specific hazards where you are working and how to prevent accidents and injuries. If you have questions specific to your job site, consult your supervisor.

3.1.0 Welding

Even if you're not welding, you can be injured when you are around a welding operation. The oxygen and acetylene used in gas welding are very dangerous. The cylinders containing oxygen and acetylene must be transported, stored, and handled very carefully. Always follow these safety guidelines:

- Keep the work area clean and free from potentially hazardous items such as combustible materials and grease or petroleum products.

WARNING!

Keep oxygen away from sources of flame and combustible materials, especially substances containing oil, grease, or other petroleum products. Compressed oxygen mixed with oil or grease will explode. Never use petroleum-based products around fittings that serve compressed oxygen lines.

- Use great caution when you handle compressed gas cylinders.
- Store cylinders in an upright position and separate them by metal.

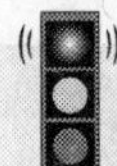

WARNING!

Do not remove the protective cap unless a cylinder is secured. If the cylinder falls over and the nozzle breaks off, the cylinder will shoot off like a rocket, injuring or killing anyone in its path.

- Never look at an **arc welding** operation without wearing the proper eye protection. The arc will burn your eyes.

WARNING!

In an arc welding operation, even a reflected arc can harm your eyes. It is extremely important to follow proper safety procedures at and around all welding operations. Serious eye injury or even blindness can result from unsafe conditions.

- If you are welding, use the proper PPE (see *Figure 5*), including the following:
  - Full face shield with proper lens
  - Earplugs to prevent flying sparks from entering your ears
  - All-leather, gauntlet-type welder's gloves
  - High-top leather boots to prevent **slag** from dropping inside your boots
  - Cuffless trousers that cover your ankles and boot tops
  - A respirator, if necessary

CAUTION

When welding on construction job sites, a hard hat with a full face shield may be required.

- If you are welding and other workers are in the area around your work, set up **welding shields.** Make sure everyone wears **flash goggles.** These goggles protect the eyes from the **flash,** which is the sudden bright light associated with starting a welding operation.
- A welder must be protected when the welding shield on the welder's headgear is down, because the shield restricts the welder's field of vision. A helper or monitor must watch the welder and the surrounding area in case of a fire or similar emergency, or rope off the area to prevent collisions and keep other workers away from the area.
- Welded material is hot! Mark it with a sign and stay clear for a while after the welding has been completed.

WARNING!

Post a fire watch when you are welding or cutting. One person other than the welding or cutting operator must constantly scan the work area for fires. Fire watch personnel should have ready access to fire extinguishers and alarms and know how to use them. Welding and cutting operations should never be performed without a fire watch. The area where welding is done must be monitored afterwards until there is no longer a risk of fire.

Instructor's Notes:

1.12 CORE CURRICULUM ◆ INTRODUCTORY CRAFT SKILLS

Figure 5 ◆ Personal protective equipment for welding.

Pay special attention to the safety guidelines about never looking at the arc without proper eye protection. Even a brief exposure to the ultraviolet light from arc welding can cause a **flash burn** and damage your eyes badly. You may not notice the symptoms until some time after the exposure. Here are some symptoms of flash burns to the eye:

- Headache
- Feeling of sand in your eyes
- Red or weeping eyes
- Trouble opening your eyes
- Impaired vision
- Swollen eyes

If you think you may have a flash burn to your eyes, seek medical help at once.

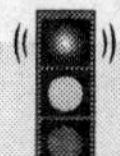

WARNING!

Never wear contact lenses while you are welding. The ultraviolet rays may dry out the moisture beneath the contact lens, causing it to stick to your eye.

3.2.0 Flame Cutting

Many of the safety guidelines for welding apply to flame cutting as well. Cutting is not dangerous as long as you follow safety precautions. Here are some of the precautions:

- Wear appropriate PPE, including a welding hood with a filter.

MODULE 00101-04 ◆ BASIC SAFETY 1.13

Discuss problems that could result from cutting galvanized metal without proper ventilation.

Ask a trainee to explain how hot dross can cause severe injuries.

Emphasize that if a torch goes out, the gas supply must be shut off immediately.

Show Transparency 6 (Figure 6). Ensure that trainees can distinguish between a fuel gas hose and an oxygen hose. Explain how to inspect a hose.

- Never open the valve of an acetylene cylinder near an open flame.
- Store oxygen cylinders separately from fuel gas cylinders.
- Store acetylene cylinders in an upright position.
- Always use a friction striker to light a cutting torch.

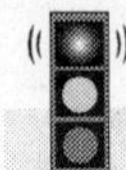

WARNING!

Never cut galvanized metal without proper ventilation. The zinc oxide fumes given off as the galvanized material is cut are hazardous. Also, use a respirator when you are cutting galvanized material.

The cutting process results in oxides that mix with molten iron and produce **dross.** The dross is blown from the cut by the jet of cutting oxygen. Hot dross can cause severe injury or can start fires on contact with flammable materials.

Before and during welding and cutting operations, you must follow certain safety procedures. *Figure 6* shows a typical oxyacetylene welding/cutting outfit. As the operator, you must check three things:

- Hoses
- Regulators
- Work area

3.2.1 Hoses and Regulators

Use the proper hose. The fuel gas hose is usually red (sometimes black) and has a left-hand threaded nut for connecting to the torch. The oxygen hose is green and has a right-hand threaded nut for connecting to the torch.

Hoses with leaks, burns, worn places, or other defects that make them unfit for service must be repaired or replaced. When inspecting hoses, look for charred sections close to the torch. These may have been caused by **flashback,** which is the result of a welding flame flaring up and charring the hose near the torch connection. Flashback is caused by improperly mixed fuel. Also check that hoses are not taped up to cover leaks (see *Figure 7*).

WARNING!

If the torch goes out and begins to hiss, shut off the gas supply to the torch immediately. Otherwise, a flashback could occur. Never relight a torch from hot metal. Doing so could cause an explosion.

New hoses contain talc and loose bits of rubber. These materials must be removed from the hoses before the torch is connected. If they are not removed, they will clog the torch needle valves. Common industry practice is to use compressed air to blow these materials out of the hose. Always

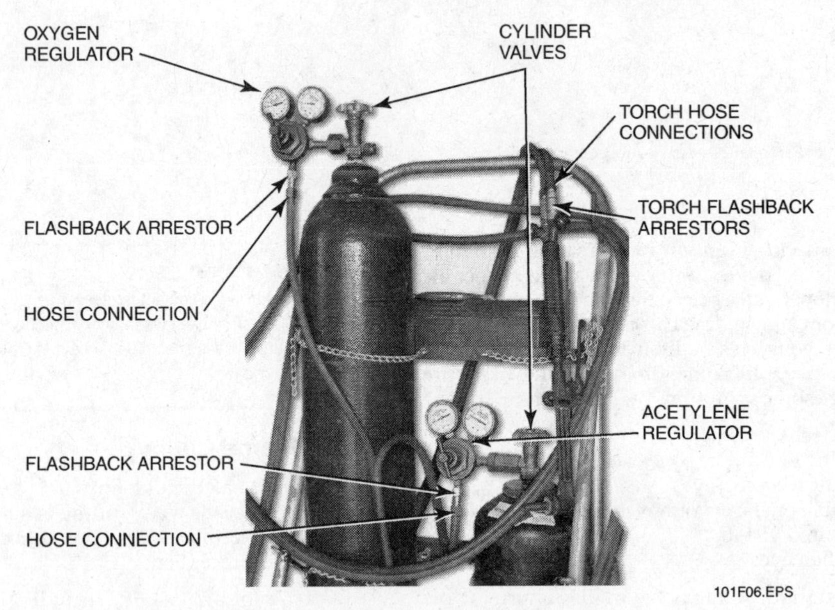

Figure 6 ◆ Typical oxyacetylene welding/cutting outfit.

Instructor's Notes:

1.14 CORE CURRICULUM ◆ INTRODUCTORY CRAFT SKILLS

make sure that the regulator valve is turned down to minimal pressures before using compressed air to clean a hose.

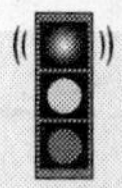

WARNING!

Never point a compressed air hose toward anyone. Flying debris and particles of dirt may cause serious injury.

Regulators are attached to the cylinder valve. They lower the high cylinder pressures to the required working pressures and maintain a steady flow of gas from the cylinder.

To prevent damage to regulators, always follow these guidelines:

- Never jar or shake regulators, because that can damage the equipment beyond repair.
- Always check that the adjusting screw is released before the cylinder valve is turned on.
- Always open cylinder valves slowly.

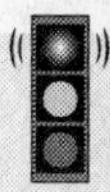

WARNING!

When opening valves, always stand to the side. Dirt that is stuck in the valve may fly out and cause serious injury.

- Once cutting or welding has been completed, fully release the adjusting screw to relieve line pressure.
- Never use oil to lubricate a regulator, because that can cause an explosion.
- Never operate fuel regulators on oxygen cylinders or oxygen regulators on fuel gas cylinders.
- Never use a defective regulator. If a regulator is not working properly, shut off the gas supply and have a qualified person repair the regulator.
- Never operate the fuel regulator above the recommended safe operating pressure.
- Never use pliers or channel locks to install or remove regulators.

3.2.2 Work Area

Before beginning a cutting or welding operation, check the area for fire hazards. Cutting sparks can fly 30 feet or more and can fall several floors. Remove any flammable material in the area or cover it with an approved fire blanket. Have an approved fire extinguisher available before starting your work.

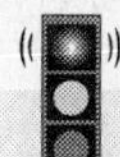

WARNING!

The slag and products that result from cutting and welding operations can start fires and cause severe injuries. Always wear appropriate personal protective equipment, including gloves and eye protection, when cutting or welding. Do not wear clothes made of polyester when welding or cutting. Observe the safety instructions of both the manufacturer and your shop.

Always perform cutting operations in a well-ventilated area. Heating and cutting metals with an oxyfuel torch can create toxic fumes.

Stress that a compressed hose must never be pointed toward another person.

Explain how to prevent damage to regulators. Ask a trainee to explain why it is important to stand to the side when opening valves.

Bring in a variety of new and damaged hoses and regulators. Ask trainees to identify which items are damaged and to specify the problems with each damaged part. Answer any questions trainees may have.

Explain how to inspect a work area for fire hazards. Emphasize the importance of wearing the appropriate personal protective equipment when cutting or welding.

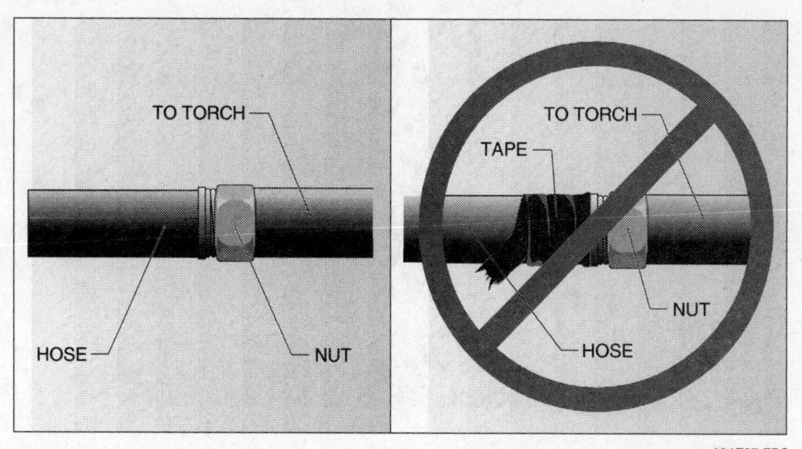

Figure 7 ◆ Proper hose connection.

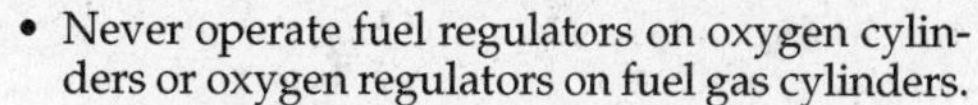

MODULE 00101-04 ◆ BASIC SAFETY 1.15

Create a hypothetical work area for a cutting or welding operation. Simulate a variety of fire hazards. Have trainees inspect the area and identify the hazards. Ask trainees to complete the tasks to maintain a clean and neat work area.

Review the safety rules for working around trenches and excavations.

Discuss the consequences of working on the face of an excavation in which concrete slabs have been used to flatten the surrounding area.

Discuss the role of a competent person in inspecting excavations. Ask a trainee to explain to whom they should speak and what they should do if an excavation has not been inspected before work is scheduled to begin.

Maintaining a clean and neat work area promotes safety and efficiency. When you are finished welding, be sure to do the following:

- Pick up cutting scraps.
- Sweep up any scraps or debris around the work area.
- Return cylinders and equipment to the proper places.
- Prevent fires by making sure that cut metals and dross are cooled before disposing of them.

3.3.0 Trenches and Excavations

In many construction jobs you will need to work in trenches or **excavations.** Cave-ins and falling objects are hazards in these areas. Obey the following safety rules when working around trenches and excavations:

- Never put tools, materials, or loose dirt or rocks within 2 feet of the edge of a trench. They can easily fall in and injure the people in the trench. Also, too much weight near the edge of a trench can cause a cave-in.
- Always walk around a trench; never jump over it or straddle it. You could lose your footing and fall in, or your weight could cause a cave-in.
- Never jump into a trench. Always use a ladder to get in and out.
- A stairway, ladder, or ramp must be provided for exit from any trench that is 4 or more feet deep.
- Put barricades around all trenches, as shown in *Figure 8*.
- Always follow OSHA regulations and your employer's procedures for **shoring** up a trench to prevent a cave-in. Never work beyond the shoring.
- Always follow OSHA regulations for determining the maximum allowable slope.

WARNING!

Never work on the face of either sloped excavations or excavations in which concrete slabs were used to flatten the surrounding area at levels above other workers. The workers below you may not be adequately protected from the hazard of falling, rolling, or sliding materials or equipment.

A competent person will inspect excavations daily and decide whether cave-ins or failures of protective systems could occur, and whether there are any other hazardous conditions. The competent person will conduct the inspection before any work begins and as needed throughout the shift. He or she will also inspect the excavations after every rainstorm or other hazard-increasing incident.

You cannot work in excavations that have standing water or in excavations where water is coming in unless you take precautions to protect yourself. A competent person will know what these precautions are. Always ask the competent person on site or your immediate supervisor if you have any questions about proper safety practices.

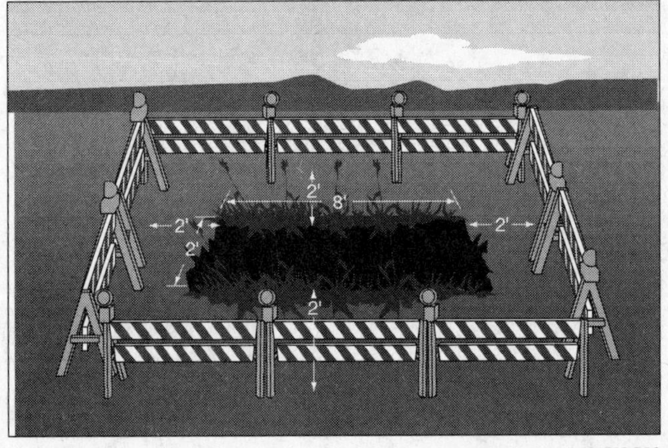

Figure 8 ◆ Barricade around a trench.

Instructor's Notes:

1.16 CORE CURRICULUM ◆ INTRODUCTORY CRAFT SKILLS

3.4.0 Proximity Work

Work that is done near a hazard but not in direct contact with it is called **proximity work.** Proximity work requires extra caution and awareness. The hazard may be hot piping, energized electrical equipment, or running motors or machinery (see *Figure 9*). You must do your work so that you do not come into contact with the nearby hazard.

You may need to put up barricades to prevent accidental contact. Lifting and rigging operations may have to be done in a way that minimizes the risk of dropping things on the hazard. A monitor may watch you and alert you if you are in danger of touching the hazard while you work.

Energized electrical equipment is very hazardous. Regulations and policies will tell you the minimum safe working distance from energized electrical conductors. You'll learn more about working with and around energized electrical equipment later in this module.

3.4.1 *Pressurized or High-Temperature Systems*

In many construction jobs, you must work close to tanks, piping systems, and pumps that contain pressurized or high-temperature fluids. Be aware of these two possible hazards:

- Touching a container of high-temperature fluid can cause burns (see *Figure 10*). Many industrial processes involve fluids that are as hot as several thousand degrees.
- If a container holding pressurized fluids is damaged, it may leak and spray dangerous fluids.

Any work around pressurized or high-temperature systems is proximity work. Barricades, a monitor, or both may be needed for safety (see *Figure 11*).

Show Transparency 7 (Figure 9). Explain the hazards associated with proximity work. Ask a trainee to explain when a barricade should be installed.

Refer to Figures 10 and 11. Review the hazards and accident prevention methods associated with pressurized or high-temperature systems.

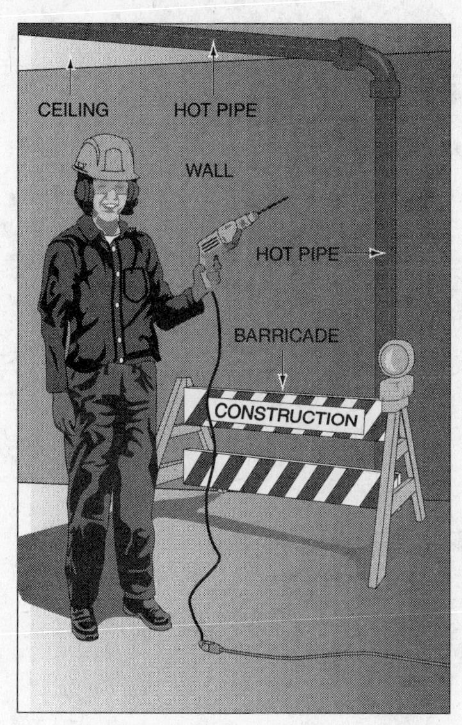

Figure 9 ◆ Proximity work.

Figure 10 ◆ Avoid touching high-temperature components.

MODULE 00101-04 ◆ BASIC SAFETY 1.17

Show Transparency 8 (Figure 12). Compare a confined space and a permit-required confined space. Have trainees explain the precautions for entering a permit-required confined space.

Discuss the consequences of entering a confined space without the proper training.

Emphasize that OSHA requires an attendant to remain outside a permit-required confined space. Ensure that trainees understand that they should never work alone in a confined space.

Figure 11 ◆ Work safely near pressurized or high-temperature systems.

3.5.0 Confined Spaces

Construction and maintenance work isn't always done outdoors. A lot of it is done in confined spaces. A confined space is a space that is large enough to work in but that has limited means of entry or exit. A confined space is not designed for human occupancy, and it has limited ventilation. Examples of confined spaces are tanks, vessels, silos, storage bins, hoppers, vaults, and pits (see *Figure 12*).

A **permit-required confined space** is a type of confined space that has been evaluated by a qualified person and found to have actual or potential hazards (see *Figure 13*). You must have written authorization to enter a permit-required confined space.

When equipment is in operation, many confined spaces contain hazardous gases or fluids. In addition, the work you are doing may introduce hazardous fumes into the space. Welding is an example of such work. For safety, you must take special precautions both before you enter and leave a confined space and while you work there.

Until you have been trained to work in permit-required confined spaces and have taken the needed precautions, you must stay out of them. If you aren't sure whether a confined space requires a permit, ask your supervisor. You must always follow your employer's procedures and your supervisor's instructions. Confined space procedures may include getting clearance from a safety representative before starting the work. You will be told what kinds of hazards are involved and what precautions you need to take. You will also be shown how to use the required PPE. Remember, it is better to be safe than sorry, so ask!

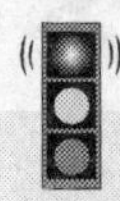

WARNING!

Without proper training, no employee is allowed to enter a permit-required or non-permit-required confined space. Employers are required to have programs to control entry to and hazards in both types of confined spaces.

Never work alone in a confined space. OSHA requires an attendant to remain outside a permit-required confined space. The attendant monitors entry, work, and exit.

Instructor's Notes:

1.18 CORE CURRICULUM ◆ INTRODUCTORY CRAFT SKILLS

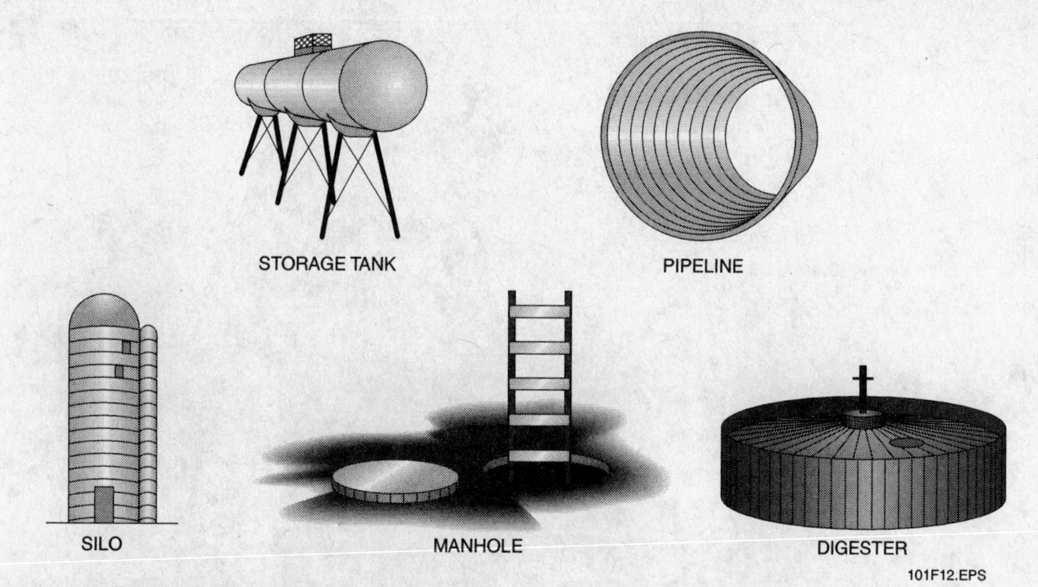

Figure 12 ◆ Examples of confined spaces.

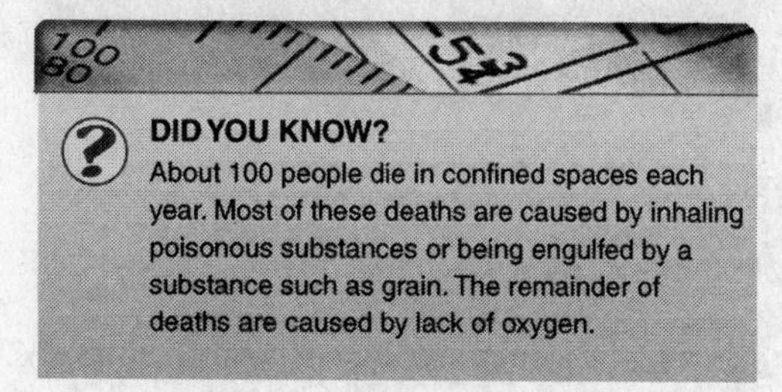

DID YOU KNOW?

About 100 people die in confined spaces each year. Most of these deaths are caused by inhaling poisonous substances or being engulfed by a substance such as grain. The remainder of deaths are caused by lack of oxygen.

3.6.0 Motorized Vehicles

Motorized vehicles used on job sites include trucks, forklifts, backhoes, cranes, and **trenchers.** Operators must take care when driving vehicles. Helpers, riggers, and anyone else working nearby must also be careful.

If a vehicle is used indoors, ventilation of the work area is especially important. All internal combustion engines give off carbon monoxide as part of their exhaust. You cannot see, smell, or taste carbon monoxide, but it can kill you. Make sure there is good ventilation before you operate any motorized vehicles indoors.

The operator of any vehicle is responsible for the safety of passengers and the protection of the load. Follow these safety guidelines when you operate vehicles on a job site:

- Always wear a seat belt.
- Be sure that each person in the vehicle has a firmly secured seat and seat belt.
- Obey all speed limits. Reduce speed in crowded areas.
- Look to the rear and sound the horn before backing up. If your rear vision is blocked, get a **signaler** to direct you. (Hand signals are covered in the *Rigging* module.)
- Every vehicle must have a backup alarm. Make sure the backup alarm works.
- Always turn off the engine when you are fueling.
- Turn off the engine and set the brakes before you leave the vehicle.
- Never stay on or in a truck that is being loaded by excavating equipment.
- Keep windshields, rearview mirrors, and lights clean and functional.
- Carry road flares, fire extinguishers, and other standard safety equipment at all times.

WARNING!

Driving a vehicle indoors without good ventilation can make you sick or kill you because of the carbon monoxide given off by the exhaust. Carbon monoxide is especially dangerous because you cannot see, smell, or taste it.

Review the types of vehicles used on a job site. Have trainees discuss the guidelines for operating a vehicle on the job site. Ask trainees which of these safety guidelines they follow when operating vehicles off site.

Ask a trainee to explain why it is important to have proper ventilation when driving a vehicle indoors.

See the Teaching Tip for Sections 3.0.0–3.6.0 at the end of this module.

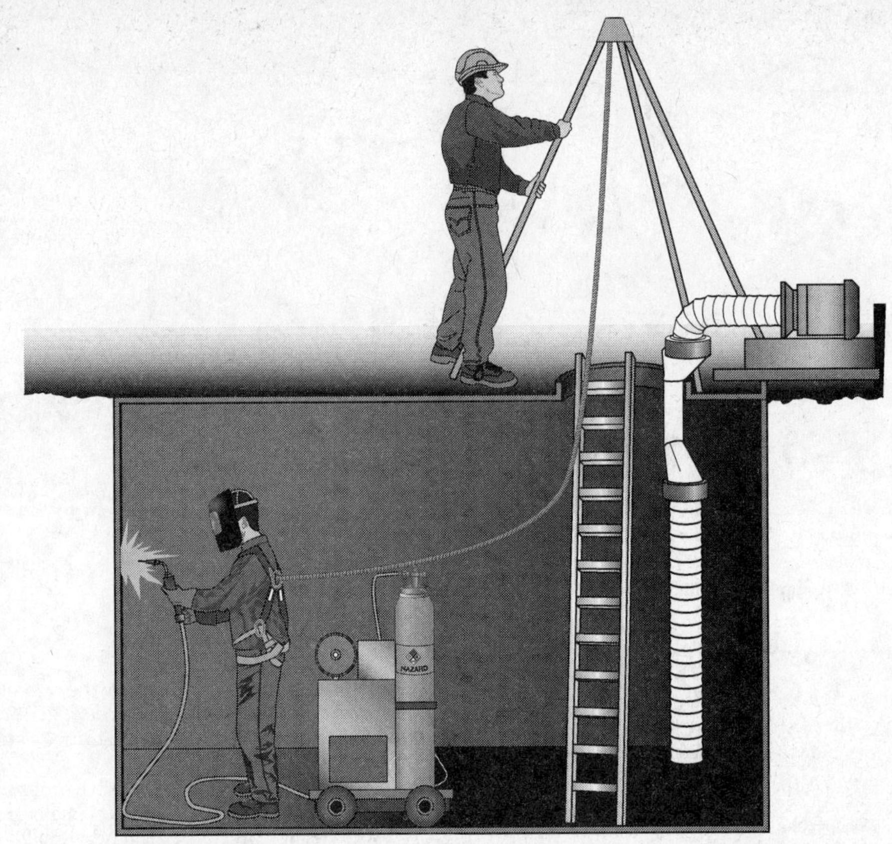

101F13.EPS

Figure 13 ◆ Permit-required confined space.

Instructor's Notes:

Review Questions

Section 3.0.0

1. Oil or grease in contact with _________ will cause an explosion.
   a. a gas cylinder
   b. acetylene
   c. compressed oxygen
   d. combustible materials

2. Flash burns are caused by exposing your eyes to _________.
   a. ultraviolet light
   b. oxygen and acetylene
   c. petroleum products
   d. zinc oxide fumes

3. A _________ is attached to a cylinder valve to reduce the high cylinder pressure to the required lower working pressure.
   a. safety valve
   b. regulator
   c. torch assembly
   d. compression hose

4. A confined space _________.
   a. has a limited amount of ventilation
   b. has no means of entry
   c. is too small to work in
   d. may be entered by untrained employees

5. All internal combustion engines give off a deadly odorless, tasteless, invisible gas called _________ as part of their exhaust.
   a. carbon trioxide
   b. carbon monoxide
   c. carbon dioxide
   d. carbon cyanide

Go over the Review Questions for Section 3.0.0. Ask whether trainees have any questions.

Explain how a lockout/tagout system protects workers from hazards.

Show Transparency 9 (Figure 14). Explain how tags and locks are used on switches, breakers, valves, and other components to prevent injury.

Bring in a variety of safety tags. Have trainees review the tags and identify the rules for the lockout/tagout system. Ensure that trainees are familiar with the lockout/tagout procedures at their job sites.

4.0.0 ◆ WORKING SAFELY WITH JOB HAZARDS

You can safely handle all the job hazards that you have learned about if you follow the rules. As long as everyone follows safety procedures, there is little risk of being hurt on the job site. In this section, you will learn about procedures and equipment used on construction sites to ensure worker safety.

4.1.0 Lockout/Tagout

A **lockout/tagout** system safeguards workers from hazardous energy while they work with machines and equipment. A lockout/tagout system protects workers from hazards such as the following:

- Acids
- Air pressure
- Chemicals
- Electricity
- Flammable liquids
- High temperatures
- Hydraulics
- Machinery
- Steam
- Other forms of energy

When people are working on or around any of these hazards, mechanical and other systems are shut down, drained, or de-energized. Tags and locks are placed on each switch, circuit breaker, valve, or other component to make sure that motors aren't started, valves aren't opened or closed, and no other changes are made that would endanger workers. Lockouts and tagouts protect workers from all possible sources of energy, including electrical, mechanical, hydraulic, thermal, pneumatic (air), and high temperature.

Generally, each lock has its own key, and the person who puts the lock on keeps the key. That person is the only one who can remove the lock. Tags have the words DANGER or CLEARANCE (see *Figure 14*).

Follow these rules for a safe lockout/tagout system:

- Never operate any device, valve, switch, or piece of equipment that has a lock or a tag attached to it.
- Use only tags that have been approved for your job site.
- If a device, valve, switch, or piece of equipment is locked out, make sure the proper tag is attached.

Show Transparency 10 (Figure 15). Review the types of barriers and barricades, and explain when this type of protection should be used.

Explain when a barricade can be removed.

Have trainees review Sections 5.0.0–6.0.0 for the next session.

Figure 14 ◆ Typical safety tags.

- Lock out and tag all electrical systems when they are not in use.
- Lock out and tag pipelines containing acids, explosive fluids, or high-pressure steam during maintenance or repair.
- Tag motorized vehicles and equipment when they are being repaired and before anyone starts work. Also, disconnect or disable the starting devices.

The exact procedures for lockout/tagout may vary at different companies and job sites. Ask your supervisor to explain the lockout/tagout procedure on your job site. You must know and follow this procedure. This is for your safety and the safety of your co-workers. If you have any questions about lockout/tagout procedures, ask your supervisor.

4.2.0 Barriers and Barricades

Any opening in a wall or floor is a safety hazard. There are two types of protection for these openings: (1) they can be **guarded** or (2) they can be covered. Cover any hole in the floor when possible. When it is not practical to cover the hole, use barricades. If the bottom edge of a wall opening is fewer than 3 feet above the floor and would allow someone to fall 4 feet or more, then place guards around the opening.

The types of barriers and barricades used vary from job site to job site (see *Figure 15*). There may also be different procedures for when and how barricades are put up. Learn and follow the policies at your job site.

Several different types of guards are commonly used:

- Railings are used across wall openings or as a barrier around floor openings to prevent falls (see *Figure 15a*).
- Warning barricades alert workers to hazards but provide no real protection (see *Figure 15b*). Typical warning barricades are made of plastic tape or rope strung from wire or between posts. The tape or rope is color-coded:
  - Red means danger. No one may enter an area with a red warning barricade. A red barricade is used when there is danger from falling objects or when a load is suspended over an area.
  - Yellow means caution. You may enter an area with a yellow barricade, but be sure you know what the hazard is, and be careful. Yellow barricades are used around wet areas or areas containing loose dust. Yellow with black lettering warns of physical hazards such as bumping into something, stumbling, or falling.
  - Yellow and purple means radiation warning. No one may pass a yellow and purple barricade without authorization, training, and the appropriate personal protective equipment. These barricades are often used where piping welds are being X-rayed.
- Protective barricades give both a visual warning and protection from injury (see *Figure 15c*). They can be wooden posts and rails, posts and chain, or steel cable. People cannot get past protective barricades.
- Blinking lights are placed on barricades so they can be seen at night (see *Figure 15d*).
- Hole covers are used to cover open holes in a floor or in the ground (see *Figure 15e*). They must be secured and labeled. They must be strong enough to support twice the weight of anything that may be placed on top of them.

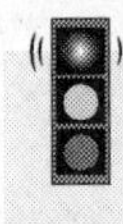

WARNING!

Never remove a barricade unless you have been authorized to do so. Follow your employer's procedures for putting up and removing barricades.

Instructor's Notes:

1.22 CORE CURRICULUM ◆ INTRODUCTORY CRAFT SKILLS

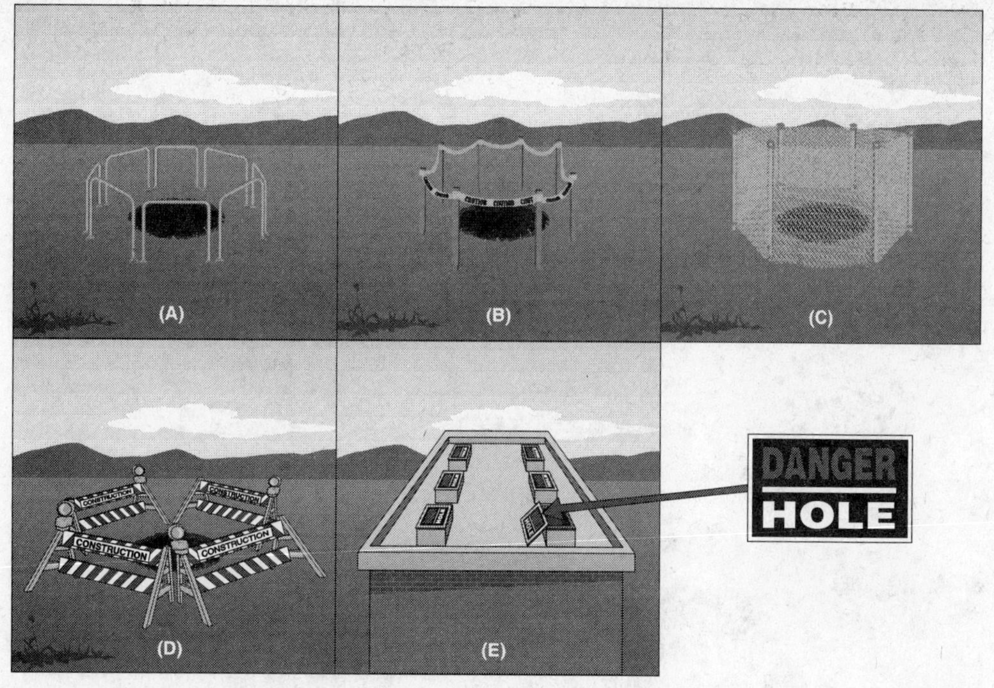

101F15.EPS

Figure 15 ◆ Common types of barriers and barricades.

Ensure that you have everything required for the demonstration, laboratories, and testing during this session.

Review the four basic elements of using and caring for personal protective equipment.

Discuss the importance of wearing a hard hat. Explain problems that could result from altering a hard hat.

5.0.0 ◆ PERSONAL PROTECTIVE EQUIPMENT

PPE is designed to protect you from injury. You must keep it in good condition and use it when you need to. Many workers are injured on the job because they are not using PPE.

5.1.0 Personal Protective Equipment Needs

You will not see all the potentially dangerous conditions just by looking around a job site. It's important to stop and consider what type of accidents could happen on any job that you are about to do. Using common sense and knowing how to use PPE will greatly reduce your chance of getting hurt.

5.2.0 Personal Protective Equipment Use and Care

The best protective equipment is of no use to you unless you do the following four things:

- Regularly inspect it.
- Properly care for it.
- Use it properly when it is needed.
- Never alter or modify it in any way.

The sections that follow describe protective equipment commonly used on construction sites and tell how to use and care for each piece of equipment. Be sure to wear the equipment according to the manufacturer's specifications.

5.2.1 Hard Hat

Figure 16 shows a typical hard hat. The outer shell of the hat can protect your head from a hard blow. The webbing inside the hat keeps space between the shell and your head. Adjust the headband so that the webbing fits your head and there is at least 1 inch of space between your head and the shell. Do not alter your hard hat in any way.

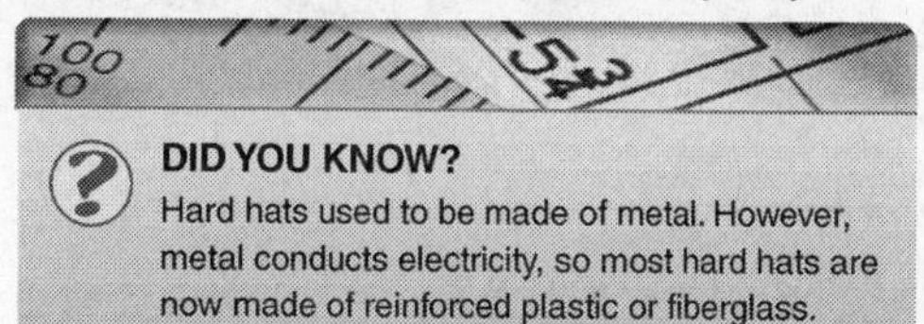

DID YOU KNOW?

Hard hats used to be made of metal. However, metal conducts electricity, so most hard hats are now made of reinforced plastic or fiberglass.

MODULE 00101-04 ◆ BASIC SAFETY 1.23

Bring in a hard hat. Demonstrate how to inspect a hard hat before use.

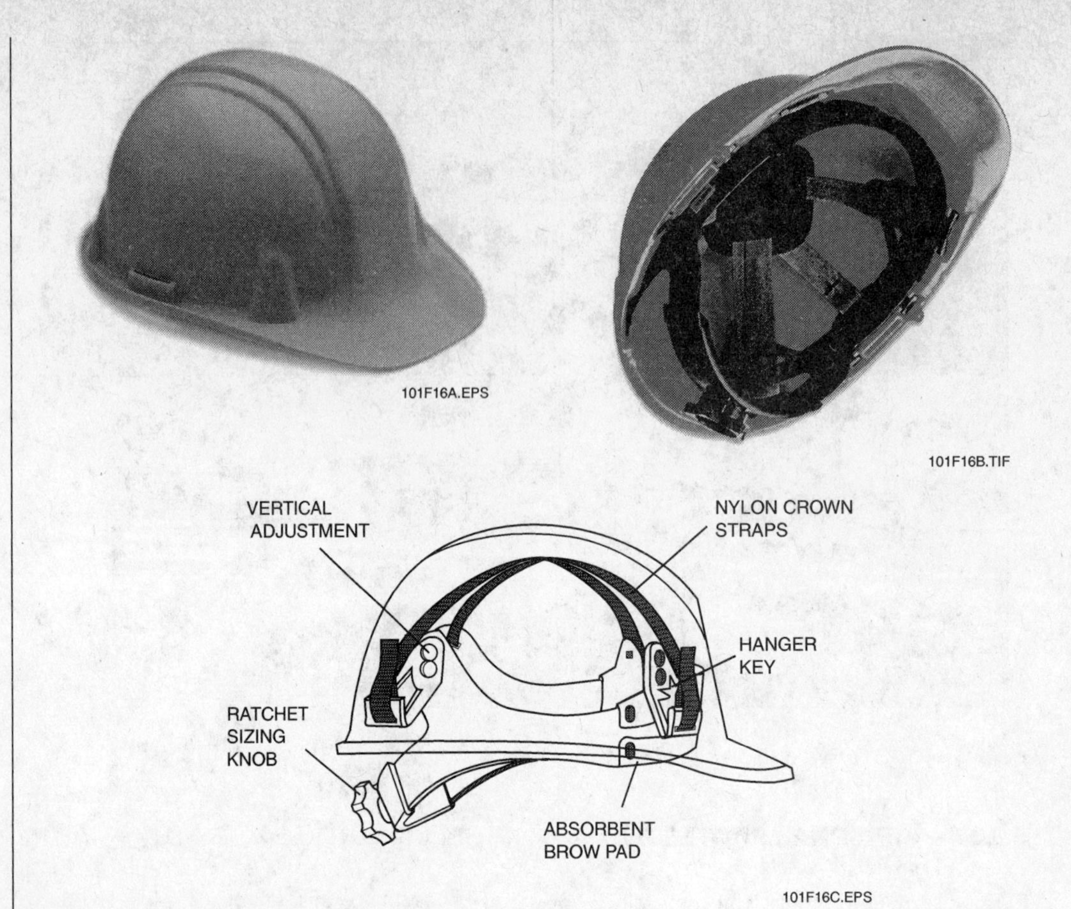

Figure 16 ◆ Typical hard hat.

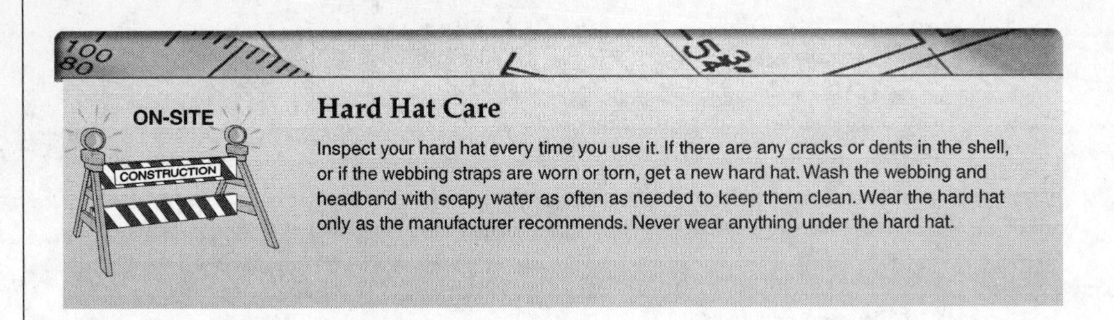

Hard Hat Care

Inspect your hard hat every time you use it. If there are any cracks or dents in the shell, or if the webbing straps are worn or torn, get a new hard hat. Wash the webbing and headband with soapy water as often as needed to keep them clean. Wear the hard hat only as the manufacturer recommends. Never wear anything under the hard hat.

Instructor's Notes:

1.24 CORE CURRICULUM ◆ INTRODUCTORY CRAFT SKILLS

5.2.2 *Safety Glasses, Goggles, and Face Shields*

Wear eye protection (see *Figure 17*) wherever there is even the slightest chance of an eye injury. Eye and face protection must meet the requirements specified in American National Standards Institute (ANSI) *Standard Z87.1-1968*. Areas where there are potential eye hazards from falling or flying objects are usually identified, but you should always be on the lookout for possible hazards.

Regular safety glasses will protect you from falling objects or from objects flying toward your face. You can add side shields for protection from the sides. In some cases, you may need a face shield. Safety goggles give your eyes the best protection from all directions.

Welders must use tinted goggles or welding hoods. The tinted lenses protect the eyes from the bright welding arc or flame.

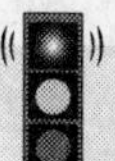

WARNING!

Handle safety glasses and goggles with care. If they get scratched, replace them. The scratches will interfere with your vision. Clean the lenses regularly with lens tissues or a soft cloth.

See Figure 17. Review the typical types of eye protection. Ask a trainee to explain when it is essential to wear eye protection.

Discuss the consequences of wearing safety glasses or goggles with scratches.

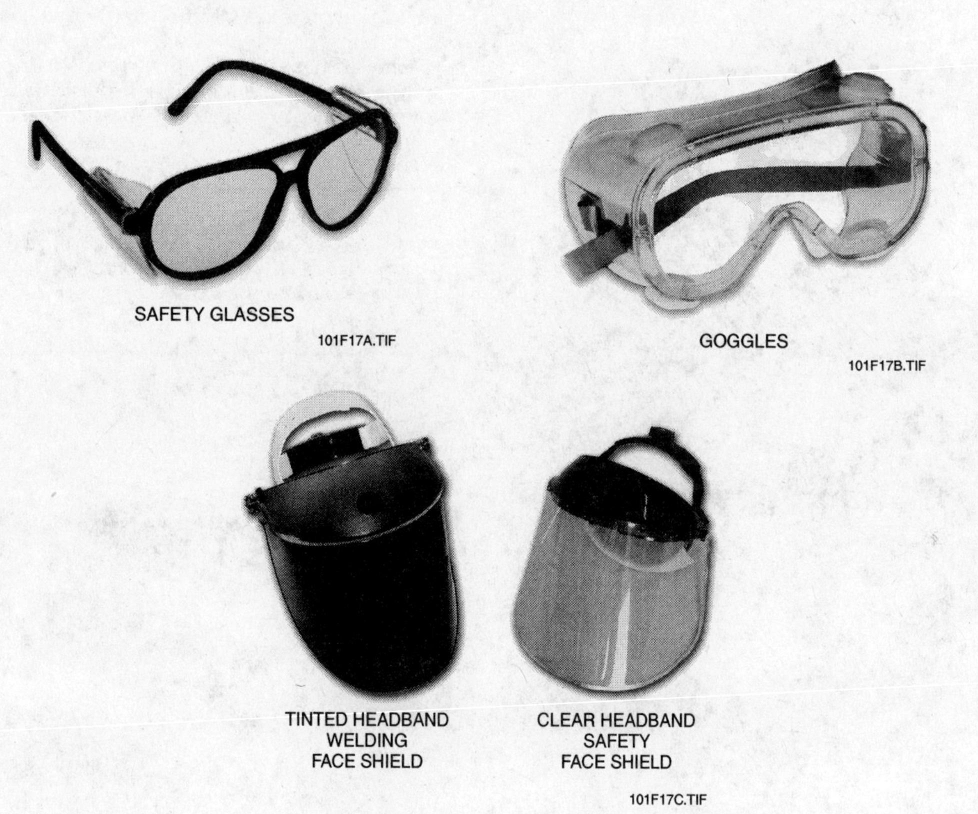

Figure 17 ◆ Typical safety glasses, goggles, and full face shields.

MODULE 00101-04 ◆ BASIC SAFETY 1.25

Bring in a variety of safety harnesses. Ask a trainee to demonstrate how to wear a safety harness and to identify the lanyard. Have trainees explain what was done correctly or incorrectly.

Explain when to use a safety harness and lanyard. Emphasize that safety harnesses and lanyards should always be used following the manufacturer's instructions.

Stress that nearly 70 percent of all job-site accidents occur because of improper use.

5.2.3 Safety Harness

Safety harnesses, like the one in *Figure 18,* are extra-heavy-duty harnesses that buckle around your body. They have leg, shoulder, chest, and pelvic straps.

Safety harnesses have a D-ring attached to one end of a short section of rope called a **lanyard** (see *Figure 19*). The other end of the lanyard should be attached to a strong anchor point located above the work area. (A qualified person will tell you what a strong anchor point is.) The lanyard should be long enough to let you work but short enough to keep you from falling more than 6 feet.

Use a safety harness and lanyard when you are working in the following situations:

- More than 6 feet above ground or according to company policy
- Near a large opening in a floor
- Near a deep hole
- Near protruding rebar

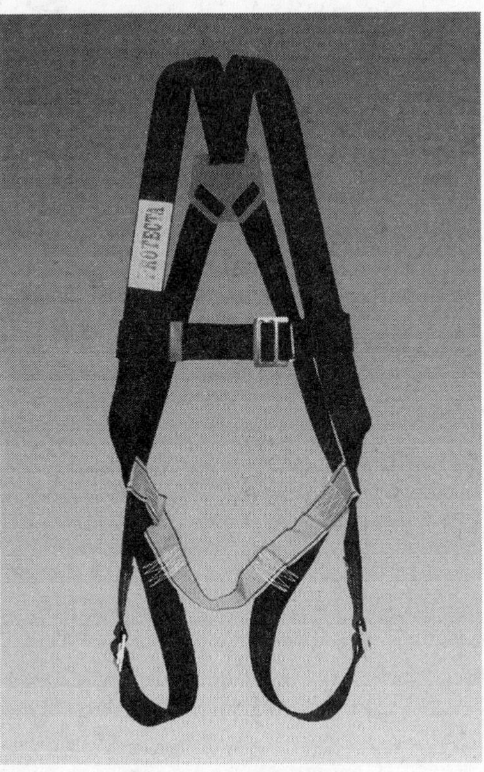

101F18.EPS

Figure 18 ◆ Typical full-body safety harness.

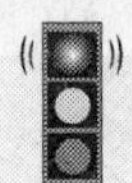

WARNING!

Never use a safety harness and lanyard for anything except their intended purpose. Always follow the manufacturer's instructions for hooking up a safety harness or lanyard.

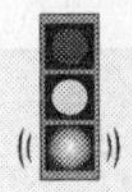

NOTE

The safety harness and lanyard are parts of a system that is known as the personal fall protection system. Workers must know how to properly inspect, don, and maintain their system.

Treat a safety harness as if your life depends on it, because it does! Carefully inspect the harness each time you use it. Check that the buckles and D-ring are not bent or deeply scratched. Check the harness for any cuts or rough spots. If you find any damage, turn in the harness for testing or replacement.

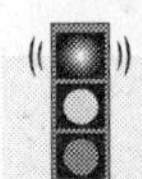

WARNING!

Always use a safety harness and lanyard properly. More than 70 percent of reported job-site accidents are caused by improper use of the lanyard and harness.

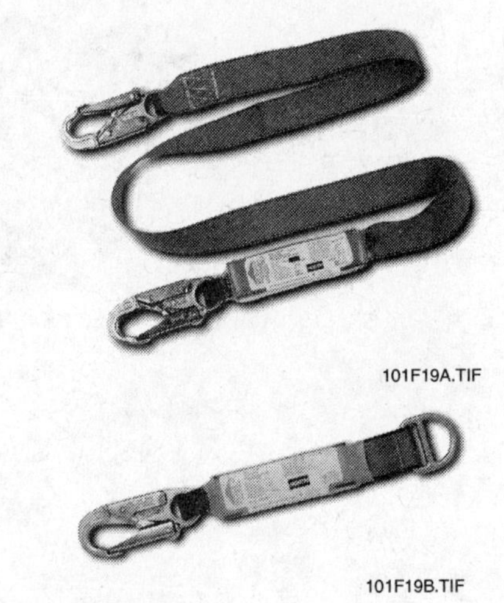

101F19A.TIF

101F19B.TIF

Figure 19 ◆ Lanyards.

Instructor's Notes:

1.26 CORE CURRICULUM ◆ INTRODUCTORY CRAFT SKILLS

5.2.4 Gloves

On many construction jobs, you must wear heavy-duty gloves to protect your hands (see *Figure 20*). Construction work gloves are usually made of cloth, canvas, or leather. Never wear cloth gloves around rotating or moving equipment. They can easily get caught in the equipment.

Gloves help prevent cuts and scrapes when you handle sharp or rough materials. Heat-resistant gloves are sometimes used for hot materials. Electricians use special rubber-insulated gloves when they work on or around live circuits.

Replace gloves when they become worn, torn, or soaked with oil or chemicals. Electrician's rubber-insulated gloves should be tested regularly to make sure they will protect the wearer.

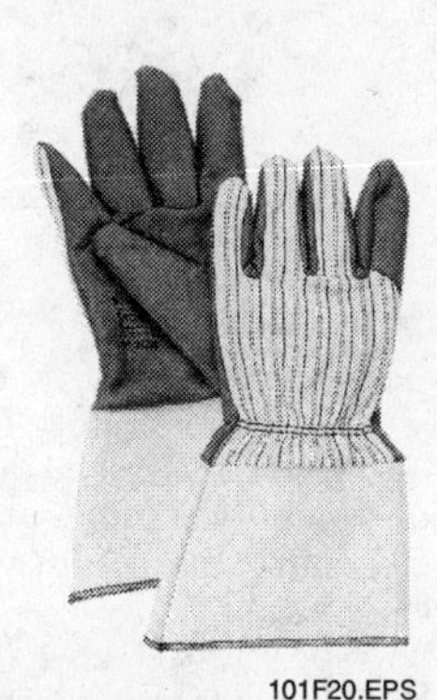

101F20.EPS

Figure 20 ◆ Work gloves.

5.2.5 Safety Shoes

The best shoes to wear on a construction site are ANSI-approved shoes (see *Figure 21*). The steel toe protects your toes from falling objects. The steel sole keeps nails and other sharp objects from puncturing your feet. The next best footwear material is heavy leather. Never wear canvas shoes or sandals on a construction site. They do not provide adequate protection.

Always replace boots or shoes when the sole tread becomes worn or the shoes have holes, even if the holes are on top. Don't wear oil-soaked shoes when you are welding, because of the risk of fire.

101F21.EPS

Figure 21 ◆ Safety shoe.

5.2.6 Hearing Protection

Damage to most parts of the body causes pain. But ear damage does not always cause pain. Exposure to loud noise over a long time can cause hearing loss, even if the noise is not loud enough to cause pain.

Most construction companies follow OSHA rules in deciding when hearing protection must be used. One type of hearing protection is specially designed earplugs that fit into your ears and filter out noise (see *Figure 22*). Clean earplugs regularly with soap and water to prevent ear infection.

Another type of hearing protection is earmuffs, which are large padded covers for the entire ear (see *Figure 23*). You must adjust the headband on earmuffs for a snug fit. If the noise level is very high, you may need to wear both earplugs and earmuffs.

Noise-induced hearing loss can be prevented by using noise control measures and personal protective devices. *Table 3* shows the recommended maximum length of exposure to sound levels rated 90 decibels and higher.

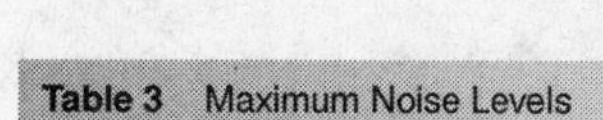

Table 3 Maximum Noise Levels

Sound Level (decibels)	Maximum Hours of Continuous Exposure per Day	Examples
90	8	Power lawn mower
92	6	Belt sander
95	4	Tractor
97	3	Hand drill
100	2	Chain saw
102	1.5	Impact wrench
105	1	Spray painter
110	0.5	Power shovel
115	0.25 or less	Hammer drill

Explain when and why heavy-duty gloves and ANSI-approved shoes must be worn on construction jobs. Answer any questions trainees may have.

Bring in earplugs and earmuffs. Ask trainees to wear the appropriate hearing protection while you simulate a variety of sound levels using mowers, sanders, saws, and drills. Ask trainees to describe the level of noise as the decibels increased.

MODULE 00101-04 ◆ BASIC SAFETY 1.27

See Figure 24. Emphasize when a respirator must be worn. Review the types of respirators, and discuss the situations in which each should be used.

Have trainees identify conditions that could interfere with a respirator's seal.

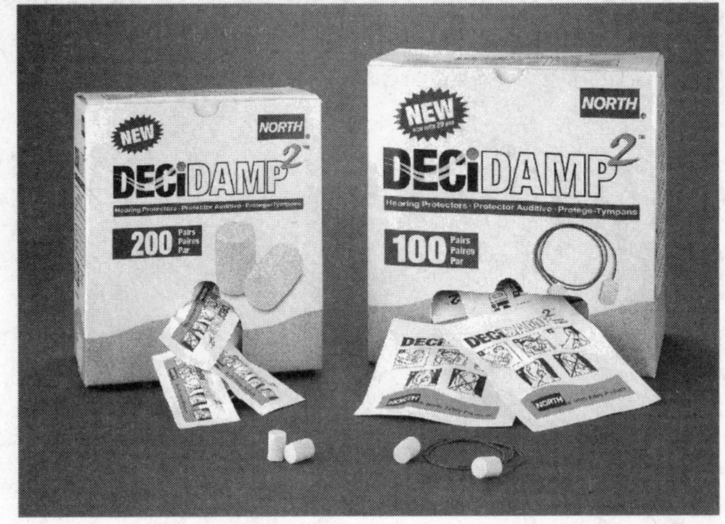

101F22.EPS

Figure 22 ◆ Earplugs for hearing protection.

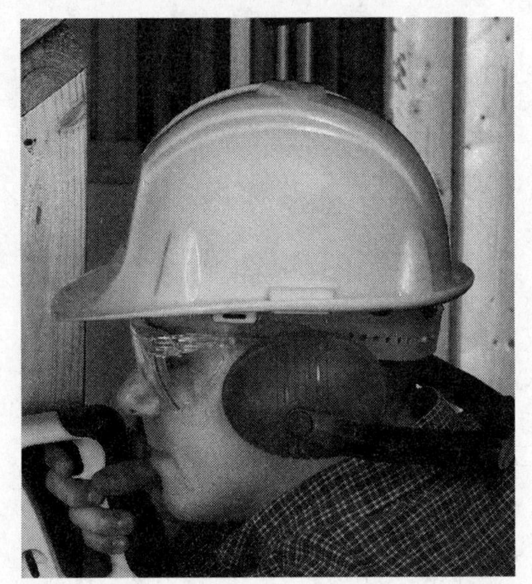

101F23.EPS

Figure 23 ◆ Earmuffs for hearing protection.

5.2.7 Respiratory Protection

Wherever there is danger of an inhalation hazard, you must use a respirator.

Follow company and OSHA procedures when choosing the type of respirator for a particular job.

Also be sure that it is safe for you to wear a respirator. Under OSHA's Respiratory Protection Standard, workers must fill out a questionnaire to determine any potential problems in wearing a respirator. Depending on the answers, a medical exam may be required.

Federal law specifies which type of respirator to use for different types of hazards. There are four general types of respirators (see *Figure 24*):

- Self-contained breathing **apparatus** (SCBA)
- Supplied air mask
- Full facepiece mask with chemical canister (gas mask)
- Half mask or mouthpiece with mechanical filter

An SCBA carries its own air supply in a compressed air tank. It is used where there is not enough oxygen or where there are dangerous gases or fumes in the air.

A supplied air mask uses a remote compressor or air tank. A hose supplies air to the mask. Supplied air masks can be used under the same conditions as SCBAs.

A full facepiece mask with chemical canisters is used to protect against brief exposure to dangerous gases or fumes.

A half mask or mouthpiece with a mechanical filter is used in areas where you might inhale dust or other solid particles. No medical exam is necessary to use this type of general respiratory protection.

Instructor's Notes:

1.28 CORE CURRICULUM ◆ INTRODUCTORY CRAFT SKILLS

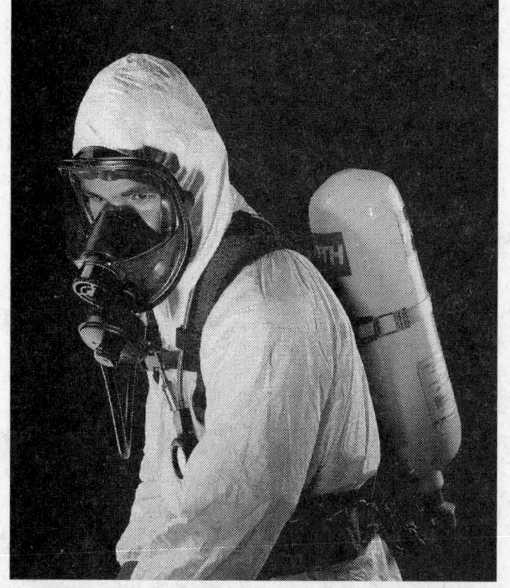

101F24A.EPS

SELF-CONTAINED BREATHING APPARATUS

101F24B.EPS

SUPPLIED AIR MASK

101F24C.EPS

FULL FACEPIECE MASK

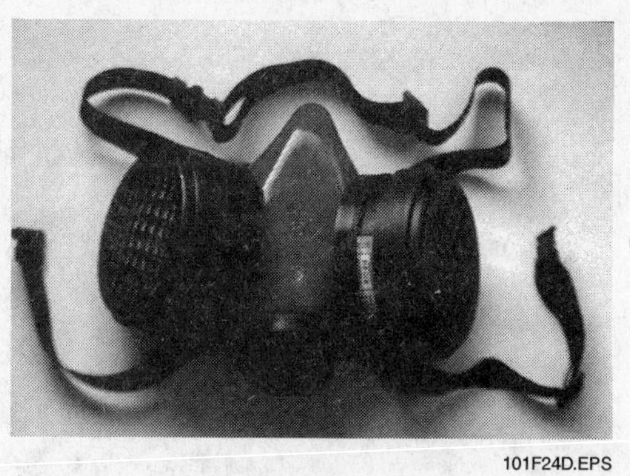

101F24D.EPS

HALF MASK

Figure 24 ◆ Examples of respirators.

MODULE 00101-04 ◆ BASIC SAFETY 1.29

Explain that, while contact lenses may now be worn with respirators, trainees must practice wearing them together to determine whether they experience any problems.

Ask a trainee to explain why personal monitoring devices are required when using a respirator.

See the Teaching Tip for Sections 5.0.0–5.2.7 at the end of this module.

Review the procedures to lift safely. Ask a trainee to carefully demonstrate how to lift safely. Have trainees observe the demonstration and identify what was done correctly or incorrectly.

Ensure that trainees review their company rules and regulations to determine the amount of weight they are allowed to carry.

It is very important to check a respirator carefully for damage and for proper fit. A leaking facepiece can be as dangerous as not wearing a respirator at all. All respirators must be fitted properly, and their facepiece-to-face seal must be checked each time the respirator is used. When conditions prevent a good seal, the respirator cannot be worn. The following conditions will interfere with the respirator's seal:

- Facial hair (such as sideburns or beards)
- Skullcaps that project under the facepiece
- Temple bars on glasses (especially when wearing full-face respirators)
- Absence of upper, lower, or all teeth
- Absence of dentures
- Gum and tobacco chewing

CAUTION

OSHA no longer bans the wearing of contact lenses with respirators (*29 CFR 1910.134 [g][1][ii]*). After sponsoring research and studies on the issue, OSHA recently concluded that wearing contact lenses with respirators does not pose an increased risk to the wearer's safety.

If you wear contact lenses, practice wearing a respirator with your contact lenses to see whether you have any problems. That way, you will identify any problems before you use the respirator under hazardous conditions.

Respirators used by only one person should be cleaned after each day of use, or more often if necessary. Those used by more than one person should be cleaned and disinfected (made germ-free) after each use.

When a respirator is not required, workers may voluntarily use a dust or particle mask for general protection. These masks do not require fit testing or a medical examination.

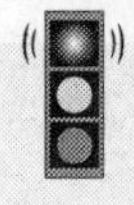

WARNING!

When a respirator is required, a personal monitoring device is usually also required. This device samples the air to measure the concentration of hazardous chemicals.

6.0.0 ◆ LIFTING

You may be surprised to learn that one-fourth of all occupational injuries happen when workers are handling or moving construction materials, especially lifting heavy objects. There is a right way and a wrong way to lift a heavy object. Lifting the wrong way can land you in the hospital. *Figure 25* shows the right way.

Step 1 Move close to the object you are going to lift. Position your feet in a forward/backward stride, with one foot at the side of the object.

Step 2 Bend your knees and lower your body, keeping your back straight and as upright as possible.

Step 3 Place your hands under the object, wrap your arms around it, or grasp the handles. To get your hands under an object that is flat on the floor, use both hands to lift one corner. Slip one hand under that corner. With one hand under, tilt the object to get the other hand under the opposite side.

Step 4 Draw the object close to your body.

Step 5 Lift by slowly straightening your legs and keeping the object's weight over your legs as much as possible.

Step 6 Pick the object up facing the direction you are going to go, to avoid twisting your knees or back.

These steps let you use your strongest muscles (those in your legs) instead of your weakest ones (those in your back) to lift. Practice with light objects. Then, when you've got it down, move on to heavier ones.

Many employers supply back belts to help reduce back injuries. You should be trained in the right way to use a back belt. Remember, a back belt is no substitute for using proper lifting techniques.

WARNING!

Always check your company's rules and regulations regarding the amount of weight one person is allowed to carry at a time.

Instructor's Notes:

1.30 CORE CURRICULUM ◆ INTRODUCTORY CRAFT SKILLS

Figure 25 ◆ How to lift safely.

MODULE 00101-04 ◆ BASIC SAFETY 1.31

Go over the Review Questions for Sections 4.0.0–6.0.0. Go over any questions trainees may have.

Have each trainee perform Performance Tasks 1–3 to your satisfaction. Fill out a Performance Profile Sheet for each trainee.

Have the trainees review Sections 7.0.0–7.2.4 for the next session.

Review Questions

Sections 4.0.0–6.0.0

1. The _________ keeps the key to a lock used for lockout/tagout.
   a. site supervisor
   b. person who puts on the lock
   c. site safety manager
   d. OSHA inspector

2. You may operate a device that is tagged out if there is no imminent danger.
   a. True
   b. False

3. A yellow and purple warning barricade means _________.
   a. caution
   b. danger
   c. physical danger
   d. radiation hazard

4. Hole covers must be strong enough to support _________ the weight of anything that may be placed on top of them.
   a. exactly
   b. twice
   c. four times
   d. ten times

5. Adjust the webbing of a hard hat so that there is _________ between your head and the shell.
   a. no space
   b. as much space as possible
   c. at least 1 inch of space
   d. less than 1 inch of space

6. _________ provide the best all-around protection for your eyes.
   a. Welding hoods
   b. Face shields
   c. Safety goggles
   d. Strap-on glasses

7. A _________ has its own clean air supply.
   a. half mask
   b. mouthpiece with mechanical filter
   c. self-contained breathing apparatus
   d. full facepiece mask

8. Whenever there is danger of an inhalation hazard, you must use a respirator.
   a. True
   b. False

9. _________ do(es) *not* interfere with the respirator's seal.
   a. Facial hair
   b. Gum and tobacco chewing
   c. Dentures
   d. Temple bars on glasses

10. When lifting heavy objects, keep as much weight as possible over your _________.
    a. hips
    b. shoulders
    c. legs
    d. back

Instructor's Notes:

7.0.0 ◆ AERIAL WORK

Working in elevated locations is common in the construction industry. If it is done properly and the proper equipment is used, it is safe. But falls from heights can cause serious injuries or even death. You must always have your supervisor's permission before working in an elevated location. In this section, you will learn about the equipment used for aerial work. You'll learn how to use it, inspect it, and maintain it.

7.1.0 Ladders and Scaffolds

Ladders and **scaffolds** are used to perform work in elevated locations. Any time work is performed above ground level, there is a risk of accidents. You can reduce this risk by carefully inspecting ladders and scaffolds before you use them and by using them properly.

Overloading means exceeding the **maximum intended load** of a ladder. Overloading can cause ladder failure, which means that the ladder could buckle, break, or topple, among other possibilities. The maximum intended load is the total weight of all people, equipment, tools, materials, loads that are being carried, and other loads that the ladder can hold at any one time. Check the manufacturer's specifications to determine the maximum intended load. Ladders are usually given a duty rating that indicates their load capacity, as shown in *Table 4*.

WARNING!

When you use a ladder, be sure to maintain three-point contact with the ladder at all times. Three-point contact means that either two feet and one hand or one foot and two hands are always touching the ladder.

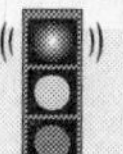

WARNING!

Use ladders and scaffolds for their intended uses only. Ladders are not interchangeable, and incorrect use can result in injury or damage.

Table 4 Ladder Duty Ratings and Load Capacities

Duty Ratings	Load Capacities
Type IA	300 lbs., extra-heavy duty/professional use
Type I	250 lbs., heavy duty/industrial use
Type II	225 lbs., medium duty/commercial use
Type III	200 lbs., light duty/household use

7.1.1 Portable Straight Ladders

Straight ladders consist of two rails, rungs between the rails, and safety feet on the bottom of the rails (see *Figure 26*). The straight ladders used in construction are made of wood or fiberglass.

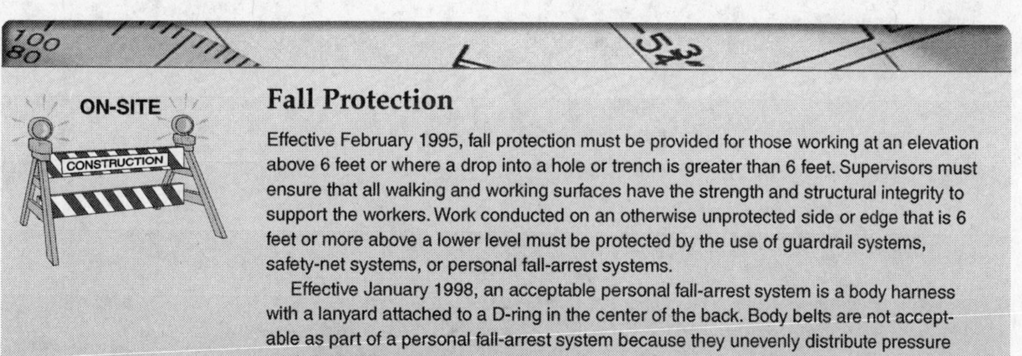

Fall Protection

Effective February 1995, fall protection must be provided for those working at an elevation above 6 feet or where a drop into a hole or trench is greater than 6 feet. Supervisors must ensure that all walking and working surfaces have the strength and structural integrity to support the workers. Work conducted on an otherwise unprotected side or edge that is 6 feet or more above a lower level must be protected by the use of guardrail systems, safety-net systems, or personal fall-arrest systems.

Effective January 1998, an acceptable personal fall-arrest system is a body harness with a lanyard attached to a D-ring in the center of the back. Body belts are not acceptable as part of a personal fall-arrest system because they unevenly distribute pressure on the wearer's midsection.

All companies are required by law to have a written Fall Protection Plan that addresses site-specific fall hazards and the steps taken to prevent each hazard. This information will normally be included in your training or regularly scheduled safety meetings. If you are unsure of what your company's plan is, or have questions about it, ask your supervisor before performing any aerial work.

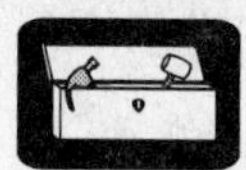

Ensure that you have everything required for the laboratory and demonstrations during this session.

Distribute copies of your company's fall protection plan. Have trainees review the plan and locate information that you specify. Ensure that trainees are familiar with their company's fall protection plan and understand the importance of following this plan.

Explain how to maintain three-point contact when using a ladder.

Emphasize the risks associated with exceeding maximum intended load of a ladder. Ensure that trainees understand that ladders and scaffolds should be used only for their intended purpose.

MODULE 00101-04 ◆ BASIC SAFETY 1.33

Three-Point Contact

When climbing or working from a ladder, you run the risk of falling. An important measure in safeguarding yourself against a fall is to maintain three-point contact with the ladder at all times. This means that you either have two hands and one foot or two feet and one hand touching the ladder constantly. Maintaining three-point contact with the ladder will help prevent you from falling and injuring yourself or a co-worker.

(A) 101SA01A.EPS

(B) 101SA01B.EPS

Instructor's Notes:

1.34 CORE CURRICULUM ◆ INTRODUCTORY CRAFT SKILLS

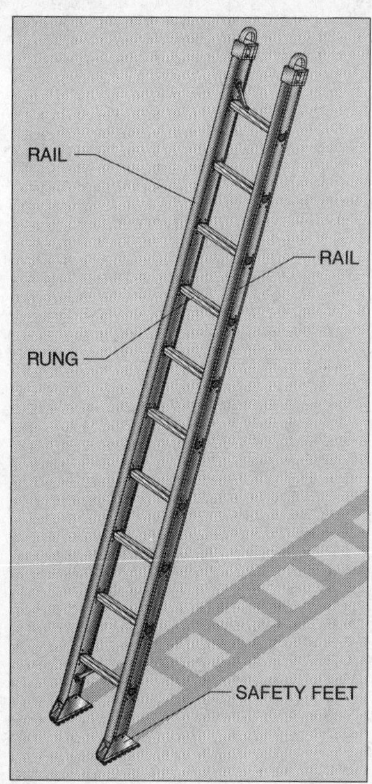

Figure 26 ◆ Portable straight ladder.

Metal ladders conduct electricity and should never be used around electrical equipment. Any portable metal ladder must have "Danger. Do Not Use Around Electrical Installations" stenciled on the rails in 2-inch, red letters. Ladders made of dry wood or fiberglass, neither of which conducts electricity, should be used around electrical equipment. Check that any wooden ladder is, in fact, completely dry before using it around electricity. Even a small amount of water will conduct electricity.

Different types of ladders should be used in different situations (see *Figure 27*). Aluminum ladders are corrosion-resistant and can be used where they might be exposed to the elements. They are also lightweight and can be used where they must frequently be lifted and moved. Wooden ladders, which are heavier and sturdier than fiberglass or aluminum ladders, can be used where heavy loads must be moved up and down. Fiberglass ladders are very durable, so they are useful where some amount of rough treatment is unavoidable. Both fiberglass and aluminum are easier to clean than wood.

7.1.2 Inspecting Straight Ladders

Always inspect a ladder before you use it. Check the rails and rungs for cracks or other damage. Also, check for loose rungs. If you find any damage, do not use the ladder. Check the entire ladder for loose nails, screws, brackets, or other hardware. If you find any hardware problems, tighten the loose parts or have the ladder repaired before you use it. OSHA requires regular inspections of all ladders and an inspection just before each use.

WARNING!

Wooden ladders should never be painted. The paint could hide cracks in the rungs or rails. Clear varnish, shellac, or a preservative oil finish will protect the wood without hiding defects.

Figure 28 shows the safety feet attached to a straight ladder. Make sure the feet are securely attached and that they are not damaged or worn down. Do not use a ladder if its safety feet are not in good working order.

7.1.3 Using Straight Ladders

It is very important to place a straight ladder at the proper angle before using it. A ladder placed at an improper angle will be unstable and could cause you to fall. *Figure 29* shows a properly positioned straight ladder.

The distance between the foot of a ladder and the base of the structure it is leaning against must be one-fourth of the distance between the ground and the point where the ladder touches the structure. For example, if the height of the wall shown in *Figure 29* is 16 feet, the base of the ladder should be 4 feet from the base of the wall. If you are going to step off a ladder onto a platform or roof, the top of the ladder should extend at least 3 feet above the point where the ladder touches the platform or roof.

Ladders should be used only on stable and level surfaces unless they are secured at both the bottom and the top to prevent any accidental movement (see *Figure 30*). Never try to move a ladder while you are on it. If a ladder must be placed in front of a door that opens toward the ladder, the door should be locked or blocked open. Otherwise, the door could be opened into the ladder.

Ladders are made for vertical use only. Never use a ladder as a work platform by placing it horizontally. Make sure the ladder you are about to climb or descend is properly secure before you do so. Check to make sure the ladder's feet are solidly

Show Transparency 11 (Figure 27). Explain when different types of ladders should be used, and describe applications for which each type is most suited.

Explain problems that could result if a wooden ladder is painted.

Show Transparency 12 (Figure 29). Review the proper positioning of a straight ladder. Ensure that trainees understand how to determine the appropriate distance between the ladder and the structure it is leaning against.

Show Transparency 13 (Figure 30). Ask a trainee to explain how to secure a ladder.

MODULE 00101-04 ◆ BASIC SAFETY 1.35

Using a properly secured straight ladder, demonstrate how to climb a ladder and how to use a hand line or tagline. Ask whether trainees have any questions.

Figure 27 ◆ Different types of ladders and applications.

Figure 28 ◆ Ladder safety feet.

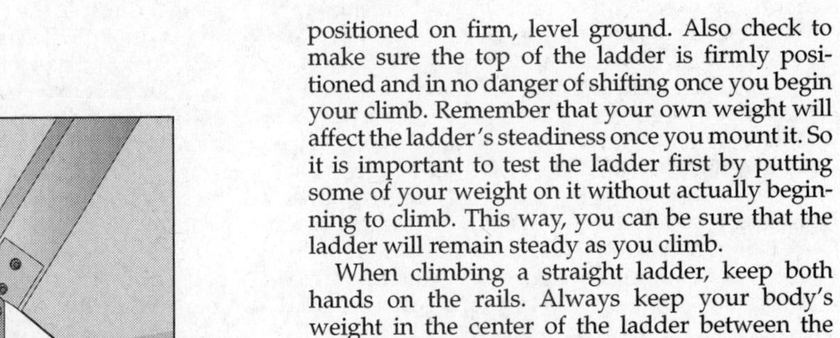

positioned on firm, level ground. Also check to make sure the top of the ladder is firmly positioned and in no danger of shifting once you begin your climb. Remember that your own weight will affect the ladder's steadiness once you mount it. So it is important to test the ladder first by putting some of your weight on it without actually beginning to climb. This way, you can be sure that the ladder will remain steady as you climb.

When climbing a straight ladder, keep both hands on the rails. Always keep your body's weight in the center of the ladder between the rails. Face the ladder at all times. Never go up or down a ladder while facing away from it (see *Figure 31*).

To carry a tool while you are on the ladder, use a **hand line** or tagline attached to the tool. Climb the ladder and then pull up the tool. Don't carry tools in your hands while you are climbing a ladder.

Instructor's Notes:

1.36 CORE CURRICULUM ◆ INTRODUCTORY CRAFT SKILLS

Figure 29 ◆ Proper positioning of a straight ladder.

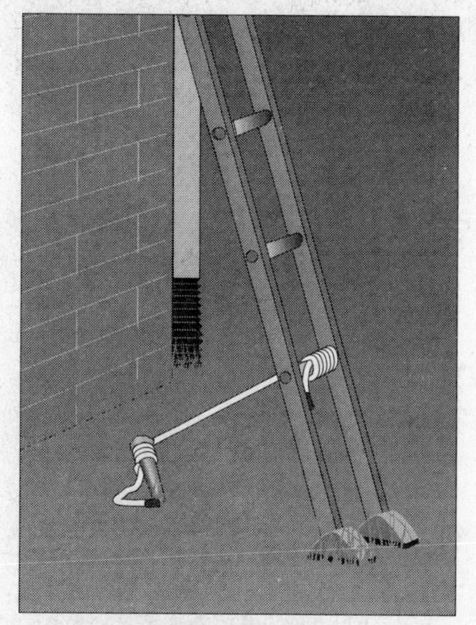

Figure 30 ◆ Securing a ladder.

Figure 31 ◆ Moving up or down a ladder.

MODULE 00101-04 ◆ BASIC SAFETY 1.37

Show Transparency 14 (Figure 32). Review the different types of extension ladders. Explain how they can be adjusted to change the length of the ladder.

Explain how to inspect extension ladders. Have a trainee explain how to safely use an extension ladder.

Discuss the consequences of removing the built-in extension stop mechanism.

Stress the importance of using extra caution when carrying anything on a ladder.

7.1.4 Extension Ladders

An **extension ladder** is actually two straight ladders. They are connected so you can adjust the overlap between them and change the length of the ladder as needed (see *Figure 32*).

7.1.5 Inspecting Extension Ladders

The same rules for inspecting straight ladders apply to extension ladders. In addition, you should inspect the rope that is used to raise and lower the movable section of the ladder. If the rope is frayed or has worn spots, it should be replaced before the ladder is used.

The rung locks (see *Figure 33*) support the entire weight of the movable section and the person climbing the ladder. Inspect them for damage before each use. If they are damaged, they should be repaired or replaced before the ladder is used.

7.1.6 Using Extension Ladders

Extension ladders are positioned and secured following the same rules as for straight ladders. When you adjust the length of an extension ladder, always reposition the movable section from the bottom, not the top, so you can make sure the rung locks are properly engaged after you make the adjustment. Check to make sure the section locking mechanism is fully hooked over the desired rung. Also check to make sure that all ropes used for raising and lowering the extension are clear and untangled.

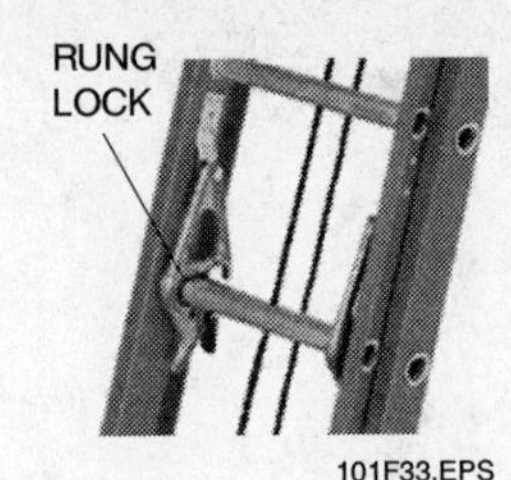

Figure 33 ◆ Rung locks.

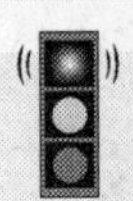

WARNING!
Extension ladders have a built-in extension stop mechanism. Do not remove it. This could cause the ladder to collapse under a load.

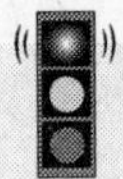

WARNING!
Haul materials up on a line rather than hand-carrying them up an extension ladder. Use extra caution when carrying anything on a ladder, because it affects your balance.

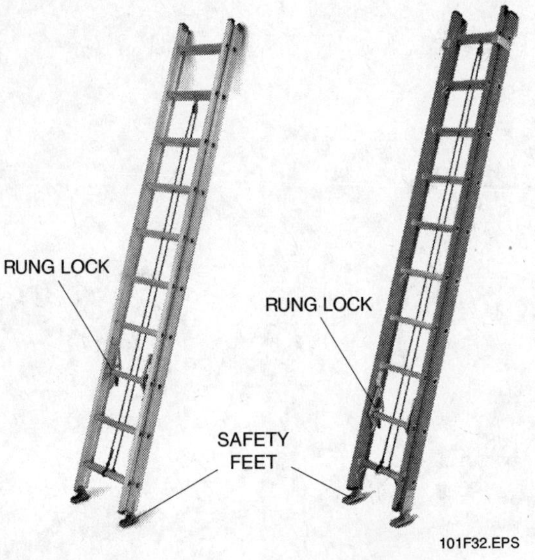

Figure 32 ◆ Typical extension ladders.

Instructor's Notes:

1.38 CORE CURRICULUM ◆ INTRODUCTORY CRAFT SKILLS

Never stand above the highest safe standing level on a ladder. The highest safe standing level on an extension ladder is the fourth rung from the top. If you stand higher, you may lose your balance and fall. Some ladders have colored rungs to show where you should not stand.

7.1.7 Stepladders

Stepladders are self-supporting ladders made of two sections hinged at the top (see *Figure 34*).

The section of a stepladder used for climbing consists of rails and rungs like those on straight ladders. The other section consists of rails and braces. Spreaders are hinged arms between the sections that keep the ladder stable and keep it from folding while in use.

7.1.8 Inspecting Stepladders

Inspect stepladders the way you inspect straight and extension ladders. Pay special attention to the hinges and spreaders to be sure they are in good repair. Also, be sure the rungs are clean. A stepladder's rungs are usually flat, so oil, grease, or dirt can build up on them and make them slippery.

7.1.9 Using Stepladders

When you position a stepladder, be sure that all four feet are on a hard, even surface. Otherwise, the ladder can rock from side to side or corner to corner when you climb it. With the ladder in position, be sure the spreaders are locked in the fully open position.

Figure 34 ◆ Typical fiberglass stepladder.

Never stand on the top step or the top of a stepladder. Putting your weight this high will make the ladder unstable. The top of the ladder is made to support the hinges, not to be used as a step. And, although the rear braces may look like rungs, they are not designed to support your weight. Never use the braces for climbing. And never climb the back of a stepladder. (For certain jobs, however, there are specially designed two-person ladders with steps on both sides.) *Figure 35* shows common dos and don'ts for using ladders.

7.2.0 Scaffolds

Scaffolds provide safe elevated work platforms for people and materials. They are designed and built to comply with high safety standards, but normal wear and tear or accidentally putting too much weight on them can weaken them and make them unsafe. That's why it is important to inspect every part of a scaffold before each use.

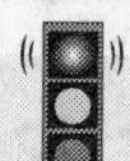

WARNING!

Never stand on a step with your knees higher than the top of a stepladder. You need to be able to hold on to the ladder with your hand. Also, keep your body centered between the side rails.

Two basic types of scaffolds—manufactured scaffolds and rolling scaffolds—are used in the construction industry. The rules for safe use apply to both of them.

7.2.1 Manufactured Scaffolds

Manufactured scaffolds are made of painted steel, stainless steel, or aluminum (see *Figure 36*). They are stronger and more fire-resistant than wooden scaffolds. They are supplied in ready-made, individual units, which are assembled on site.

7.2.2 Rolling Scaffolds

A rolling scaffold has wheels on its legs so that it can be easily moved (see *Figure 37*). The scaffold wheels have brakes so the scaffold will not move while workers are standing on it.

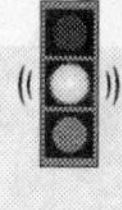

CAUTION

Only a competent person has the authority to supervise setting up, moving, and taking down scaffolding. Only a competent person can approve the use of scaffolding on the job site after inspecting the scaffolding.

Demonstrate how to position a stepladder. Demonstrate how to clean the rungs of a stepladder.

Show Transparency 15 (Figure 35). Review the safety dos and don'ts of using a stepladder.

Have a trainee explain why workers should never stand on a step with their knees higher than the top of the stepladder.

Compare the differences between manufactured and rolling scaffolds. Answer any questions trainees may have.

Ensure that trainees understand that only a competent person can supervise scaffold movement.

MODULE 00101-04 ◆ BASIC SAFETY 1.39

- Be sure your ladder has been properly set up and is used in accordance with safety instructions and warnings.
- Wear shoes with non-slip soles.

- Keep your body centered on the ladder. Hold the ladder with one hand while working with the other. Never let your belt buckle pass beyond either ladder rail.

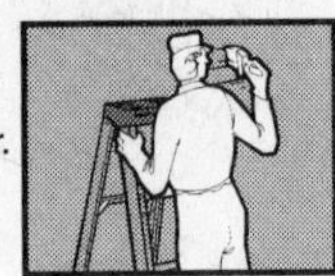

- Move materials with extreme caution. Be careful pushing or pulling anything while on a ladder. You may lose your balance or tip the ladder.

- Get help with a ladder that is too heavy to handle alone. If possible, have another person hold the ladder when you are working on it.

- Climb facing the ladder. Center your body between the rails. Maintain a firm grip.
- Always move one step at a time, firmly setting one foot before moving the other.

- Haul materials up on a line rather than carry them up an extension ladder.
- Use extra caution when carrying anything on a ladder.

Read ladder labels for additional information.

- DON'T stand above the highest **safe standing level**.
- DON'T stand above the second step from the top of a stepladder and the 4th rung from the top of an extension ladder. A person standing higher may lose their balance and fall.

- DON'T climb a closed stepladder. It may slip out from under you.
- DON'T climb on the back of a stepladder. It is not designed to hold a person.

- DON'T stand or sit on a step-ladder top or pail shelf. They are not designed to carry your weight.
- DON'T climb a ladder if you are not physically and mentally up to the task.

- DON'T exceed the Duty Rating, which is the maximum load capacity of the ladder. Do not permit more than one person on a single-sided stepladder or on any extension ladder.

- DON'T place the base of an extension ladder <u>too close</u> to the building as it may tip over backward.
- DON'T place the base of an extension ladder <u>too far away</u> from the building, as it may slip out at the bottom. **Please refer to the 4 to 1 Ratio Box.**

- DON'T over-reach, lean to one side, or try to move a ladder while on it. You could lose your balance or tip the ladder. **Climb down and then reposition the ladder closer to your work!**

4 TO 1 Ratio

Place an extension ladder at a 75-1/2° angle. The set-back ("S") needs to be 1 ft. for each 4 ft. of length ("L") to the upper support point.

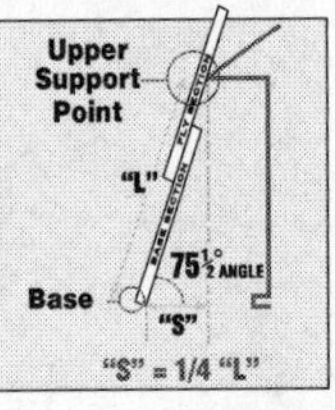

101F35.EPS

Figure 35 ◆ Ladder safety dos and don'ts.

Instructor's Notes:

1.40 CORE CURRICULUM ◆ INTRODUCTORY CRAFT SKILLS

Show Transparency 16 (Figure 38). Review the various tags used on scaffolds to indicate OSHA standards and safe use.

Discuss the roles of top rails, mid-rails, and toeboards on scaffolding.

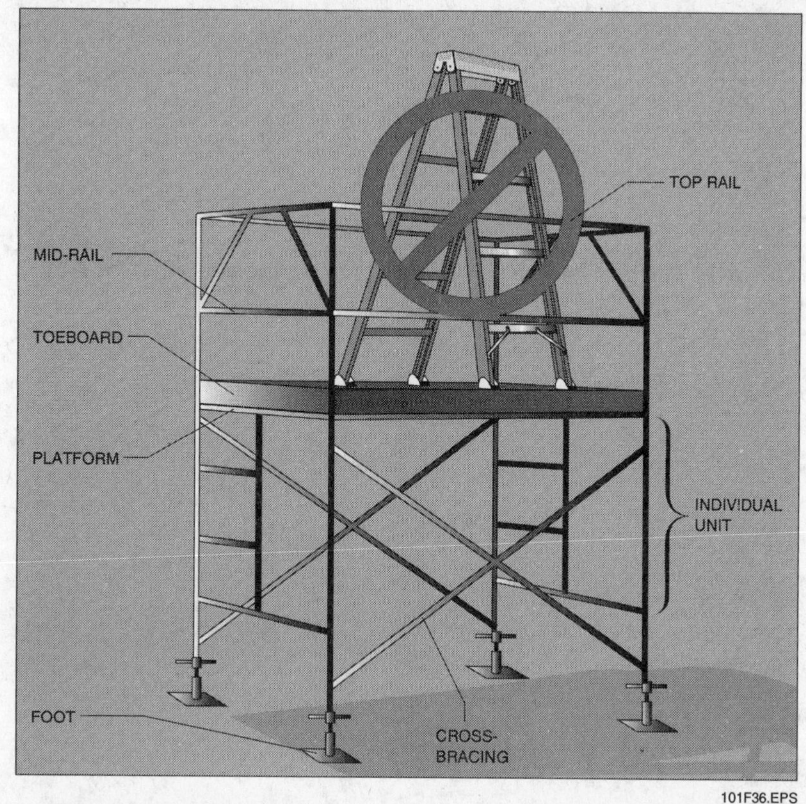

Figure 36 ◆ Typical manufactured scaffold.

7.2.3 Inspecting Scaffolds

Any scaffold that is assembled on the job site should be tagged. These tags indicate whether the scaffold meets OSHA standards and is safe to use. Three colors of tags are used: green, yellow, and red (see *Figure 38*).

A green tag means the scaffold meets all OSHA standards and is safe to use.

A yellow tag means the scaffold does not meet all OSHA standards. An example is a scaffold on which a railing cannot be installed because of equipment interference. To use a yellow-tagged scaffold, you must wear a safety harness attached to a lanyard. You may have to take other safety measures as well.

A red tag means a scaffold is being put up or taken down. Never use a red-tagged scaffold.

Don't rely on the tags alone. Inspect all scaffolds before you use them. Check for bent, broken, or badly rusted tubes. Check for loose joints where the tubes are connected. Any of these problems must be corrected before the scaffold is used.

Make sure you know the weight limit of any scaffold you will be using. Compare this weight limit to the total weight of the people, tools, equipment, and material you expect to put on the scaffold. Scaffold weight limits must never be exceeded.

If a scaffold is more than 10 feet high, check to see that it is equipped with **top rails, mid-rails,** and **toeboards.** All connections must be pinned. That means they must have a piece of metal inserted through a hole to prevent connections from slipping. **Cross-bracing** must be used. A handrail is not the same as cross-bracing. The working area must be completely **planked.**

If it is possible for people to walk under a scaffold, the space between the toeboard and the top rail must be screened. This prevents objects from falling off the work platform and injuring those below.

When you examine a rolling scaffold, check the condition of the wheels and brakes. Be sure the brakes are working properly and can stop the scaffold from moving while work is in progress. Be sure all brakes are locked before you use the scaffold.

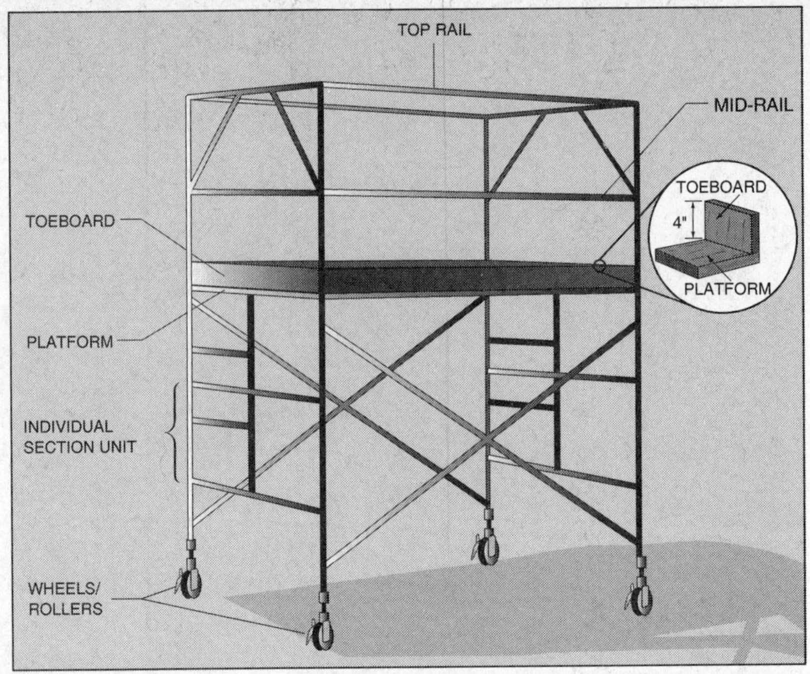

Figure 37 ◆ Typical rolling scaffold.

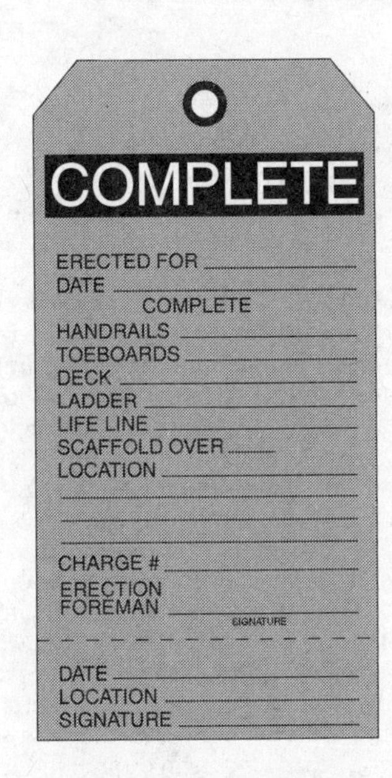

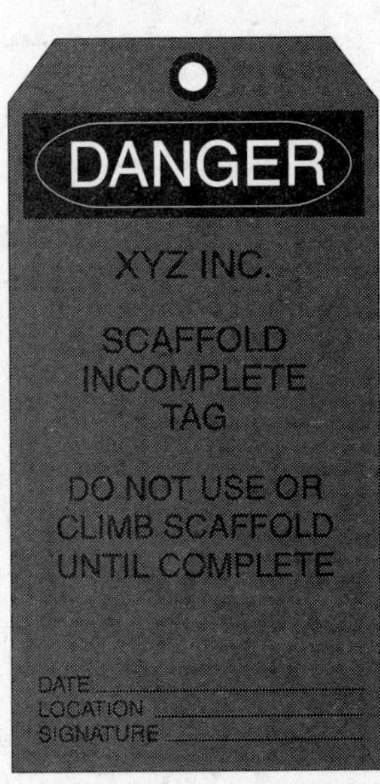

Figure 38 ◆ Typical scaffold tags.

Instructor's Notes:

1.42 CORE CURRICULUM ◆ INTRODUCTORY CRAFT SKILLS

7.2.4 Using Scaffolds

Be sure that a competent person inspects the scaffold before you use it.

There should be firm footing under each leg of a scaffold before you put any weight on it. If you are working on loose or soft soil, you can put matting under the scaffold's legs or wheels.

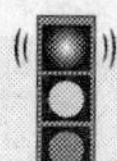

WARNING!
Keep scaffolds a safe distance from power lines in accordance with OSHA guidelines. Refer to *OSHA 1926.*

WARNING!
Falls from scaffolding and falls from ladders each account for 17 percent of deaths among construction workers. Prevent accidents by following OSHA and company guidelines.

When you move a rolling scaffold, always follow these steps:

Step 1 Get off the scaffold.

Step 2 Unlock the wheel brakes.

Step 3 Move the scaffold.

Step 4 Re-lock the wheel brakes.

Step 5 Get back on the scaffold.

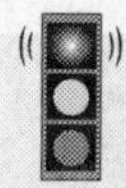

WARNING!
Never unlock the wheel brakes of a rolling scaffold while anyone is on it. People on a moving scaffold can lose their balance and fall.

Review Questions

Section 7.0.0

1. Never use a(n) _________ ladder anywhere near electrical current.
   a. fiberglass
   b. aluminum
   c. wooden
   d. straight

2. If you lean a straight ladder against the top of a 16-foot wall, the base of the ladder should be _________ feet from the base of the wall.
   a. 3
   b. 4
   c. 5
   d. 6

3. With a one-person stepladder, it is safe to _________.
   a. stand on the top step
   b. climb the back of it
   c. stand on the rear braces
   d. lock the spreaders in the fully open position

4. Never use a scaffold with a(n) _________ tag.
   a. blue
   b. red
   c. orange
   d. yellow

5. A scaffold must be equipped with top rails, mid-rails, and toeboards if it is more than _________ ft. high.
   a. 5
   b. 7
   c. 10
   d. 12

Ensure that trainees understand the distance scaffolds must be placed from power lines.

Emphasize the risks of working on scaffolds and ladders.

Discuss the consequence of unlocking the wheel brakes while someone is still working on a scaffold.

See the Teaching Tip for Sections 7.0.0–7.2.4 at the end of this module.

Go over the Review Questions for Section 7.0.0. Go over any questions trainees may have.

Have trainees review Sections 8.0.0–9.3.1 for the next session.

MODULE 00101-04 ◆ BASIC SAFETY 1.43

Ensure that you have everything required for the laboratory during this session.

8.0.0 ◆ HAZARD COMMUNICATION STANDARD

Ensure that trainees are familiar with and understand the Hazard Communication Standard (HazCom).

OSHA has a rule that affects every worker in the construction industry. It is called the **Hazard Communication Standard (HazCom).** You may have heard it called the "Right to Know" requirement. It requires all contractors to educate their employees about the hazardous chemicals they may be exposed to on the job site. Employees must be taught how to work safely around these materials.

Many people think that there are very few hazardous chemicals on construction job sites. That isn't true. In the OSHA standard, the term hazardous chemical applies to paint, concrete, and even wood dust, as well as other substances.

8.1.0 Material Safety Data Sheets

Distribute a typical material safety data sheet (MSDS). Ask trainees to identify information on the MSDS that you specify.

A material safety data sheet (MSDS) must accompany every shipment of a hazardous substance and must be available to you on the job site. Use the MSDS to manage, use, and dispose of hazardous materials safely. *Figure 39* shows part of a typical MSDS.

The information on an MSDS includes the following:

- The identity of the substance
- Exposure limits
- Physical and chemical characteristics of the substance
- The kind of hazard the substance presents
- Precautions for safe handling and use
- The reactivity of the substance
- Specific control measures
- Emergency first-aid procedures
- Manufacturer contact for more information

8.2.0 Your Responsibilities Under HazCom

Ensure that trainees understand their responsibilities under HazCom.

You have the following responsibilities under HazCom:

- Know where MSDSs are on your job site.
- Report any hazards you spot on the job site to your supervisor.
- Know the physical and health hazards of any hazardous materials on your job site, and know and practice the precautions needed to protect yourself from these hazards.
- Know what to do in an emergency.
- Know the location and content of your employer's written hazard communication program.

The final responsibility for your safety rests with you. Your employer must provide you with information about hazards, but you must know this information and follow safety rules.

Instructor's Notes:

1.44 CORE CURRICULUM ◆ INTRODUCTORY CRAFT SKILLS

Material Safety Data Sheet
May be used to comply with OSHA's Hazard Communication Standard, 29 CFR 1910 1200. Standard must be consulted for specific requirements.

U.S. Department of Labor
Occupational Safety and Health Administration
(Non-Mandatory Form)
Form Approved
OMB No. 1218-0072

IDENTITY (as Used on Label and List)	Note: Blank spaces are not permitted. If any item is not applicable or no information is available, the space must be marked to indicate that.

Section I

Manufacturer's name	Emergency Telephone Number
Address (Number, Street, City, State and ZIP Code)	Telephone Number for Information
	Date Prepared
	Signature of Preparer (optional)

Section II—Hazardous Ingredients/Identity Information

Hazardous Components (Specific Chemical Identity, Common Name(s))	OSHA PEL	ACGIH TLV	Other Limits Recommended	% (optional)

Section III—Physical/Chemical Characteristics

Boiling Point		Specific Gravity (H_20 = 1)	
Vapor Pressure (mm Hg)		Melting Point	
Vapor Density (AIR = 1)		Evaporation Rate (Butyl Acetate = 1)	
Solubility in Water			
Appearance and Odor			

Section IV—Fire and Explosion Hazard Data

Flash Point (Method Used)	Flammable Limits	LEL	UEL
Extinguishing Media			
Special Fire Fighting Procedures			
Unusual Fire and Explosion Hazards			

(Reproduce locally) OSHA 174 Sept. 1985

101F39A.EPS

Figure 39 ◆ Typical MSDS. (1 of 2)

MODULE 00101-04 ◆ BASIC SAFETY 1.45

Section V—Reactivity Data			
Stability	Unstable		Conditions to Avoid
	Stable		
Incompatibility (Materials to Avoid)			
Hazardous Decomposition or Byproducts			
Hazardous Polymerization	May Occur		Conditions to Avoid
	Will Not Occur		

Section VI—Health Hazard Data			
Route(s) of Entry	Inhalation?	Skin?	Ingestion?
Health Hazards (Acute and Chronic)			
Carcinogenicity	NTP?	IARC Monographs?	OSHA Regulated?
Signs and Symptoms of Exposure			
Medical Conditions Generally Aggravated by Exposure			
Emergency and First Aid Procedures			

Section VII—Precautions for Safe Handling and Use
Steps to Be Taken in Case Material Is Released or Spilled
Waste Disposal Method
Precautions to Be Taken in Handling and Storing
Other Precautions

Section VII—Control Measures		
Respiratory Protection (Specify Type)		
Ventilation	Local Exhaust	Special
	Mechanical (General)	Other
Protective Gloves		Eye Protection
Other Protective Clothing or Equipment		
Work/Hygienic Practices		

101F39B.EPS

Figure 39 ◆ Typical MSDS. (2 of 2)

Instructor's Notes:

1.46 CORE CURRICULUM ◆ INTRODUCTORY CRAFT SKILLS

Review Questions

Section 8.0.0

1. OSHA's Hazard Communication Standard (HazCom) rule requires all contractors to _________ on-site hazardous chemicals.
   a. store
   b. clean up
   c. remove all
   d. educate employees about

2. HazCom classifies all paint, concrete, and wood dust as _________ materials.
   a. hazardous
   b. common
   c. inexpensive
   d. nonhazardous

3. The information on an MSDS includes _________.
   a. cost and availability
   b. emergency first-aid procedures
   c. point of origin
   d. warranty limitations

4. Under HazCom, if you spot a hazard on your job site you must _________.
   a. report it to your supervisor
   b. leave immediately
   c. notify your co-workers
   d. clean it up

5. Although your employer must provide you with information about hazardous chemicals, the final responsibility for your safety rests with _________.
   a. your immediate supervisor
   b. your site foreman
   c. you
   d. your co-workers

Go over the Review Questions for Section 8.0.0. Ask whether trainees have any questions.

Explain how fires start. Ensure that trainees understand the concept of a flash point.

Show Transparency 17 (Figure 40). Review the fire triangle, and explain how each element needs to be present for a fire to start.

Review the basic safety guidelines for fire prevention.

9.0.0 ◆ FIRE SAFETY

Fire is always a hazard on construction job sites. Many of the materials used in construction are flammable. In addition, welding, grinding, and many other construction activities create heat or sparks that can cause a fire. Fire safety involves two elements: fire prevention and fire fighting.

9.1.0 How Fires Start

For a fire to start, three things are needed in the same place at the same time: fuel, heat, and oxygen. If one of these three is missing, a fire will not start.

Fuel is anything that will combine with oxygen and heat to burn. Oxygen is always present in the air. When pure oxygen is present, such as near a leaking oxygen hose or fitting, material that would not normally be considered fuel (including some metals) will burn.

Heat is anything that will raise a fuel's temperature to the **flash point.** The flash point is the temperature at which a fuel gives off enough gases (vapors) to burn. The flash points of many fuels are quite low—room temperature or less. When the burning gases raise the temperature of a fuel to the point at which it ignites, the fuel itself will burn—and keep burning—even if the original source of heat is removed.

What is needed for a fire to start can be shown as a fire triangle (see *Figure 40*). If one element of the triangle is missing, a fire cannot start. If a fire has started, removing any one element from the triangle will put it out.

Research has added a fourth side to the fire triangle concept, resulting in the development of a new model called the Fire Tetrahedron. The fourth element involved in the combustion process is referred to as the "chemical chain reaction." Specific chemical chain reactions between fuel and oxygen molecules are essential to sustaining a fire once it has begun.

9.2.0 Fire Prevention

The best way to ensure fire safety is to prevent a fire from starting. The best way to prevent a fire is to make sure that fuel, oxygen, and heat are never present in the same place at the same time.

Discuss the general fire precautions to ensure that a workplace is safe from fire.

Review fire prevention techniques for working with flammable or combustible liquids, flammable gases, and ordinary combustibles.

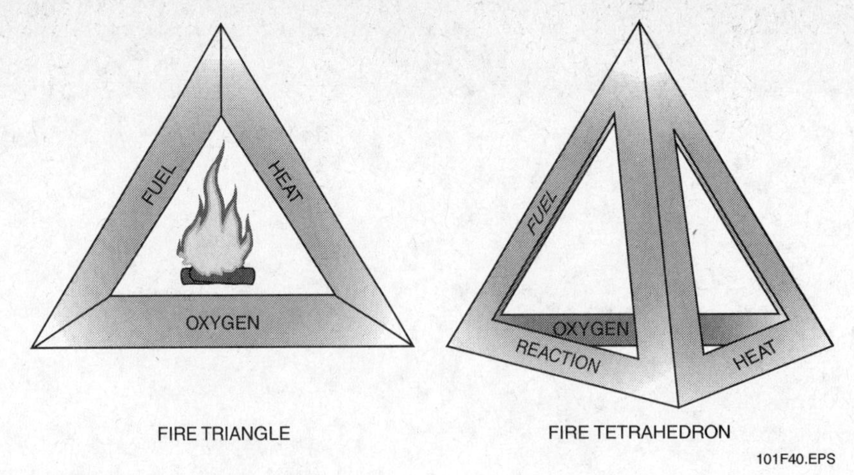

Figure 40 ◆ Basic fire requirements.

Here are some basic safety guidelines for fire prevention:

- Always work in a well-ventilated area, especially when you are using flammable materials such as shellac, lacquer, paint stripper, or construction adhesives.
- Never smoke or light matches when you are working with or near flammable materials.
- Keep oily rags in approved, self-closing metal containers.
- Store combustible materials only in approved containers.

9.2.1 Flammable and Combustible Liquids

Liquids can be flammable or combustible. Flammable liquids have a flash point below 100°F. Combustible liquids have a flash point at or above 100°F. Fire can be prevented by doing the following things:

- *Removing the fuel*–Liquid does not burn. What burns are the gases (vapors) given off as the liquid evaporates. Keeping liquids in an approved, sealed container prevents evaporation. If there is no evaporation, there is no fuel to burn.
- *Removing the heat*–If the liquid is stored or used away from a heat source, it will not be able to ignite.
- *Removing the oxygen*–The vapor from a liquid will not burn if oxygen is not present. Keeping safety containers tightly sealed prevents oxygen from coming into contact with the fuel.

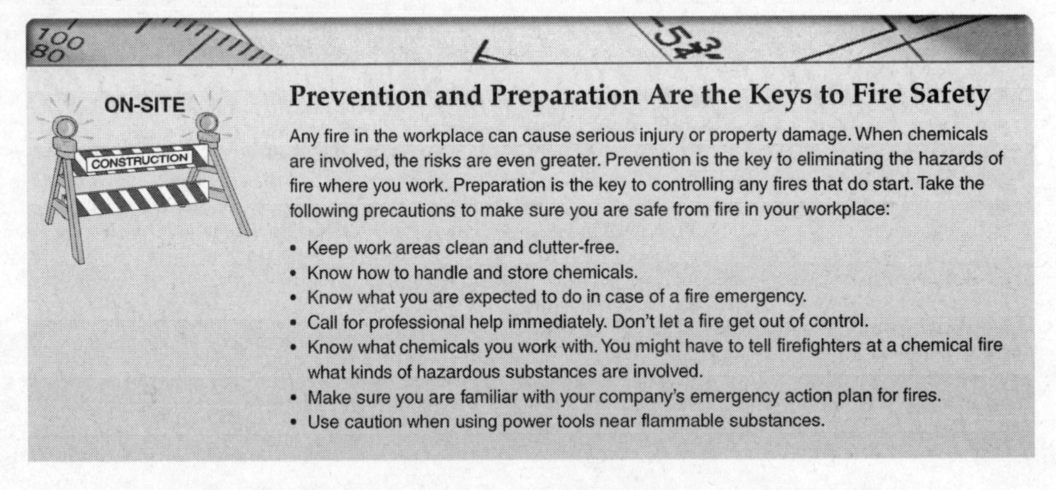

Prevention and Preparation Are the Keys to Fire Safety

Any fire in the workplace can cause serious injury or property damage. When chemicals are involved, the risks are even greater. Prevention is the key to eliminating the hazards of fire where you work. Preparation is the key to controlling any fires that do start. Take the following precautions to make sure you are safe from fire in your workplace:

- Keep work areas clean and clutter-free.
- Know how to handle and store chemicals.
- Know what you are expected to do in case of a fire emergency.
- Call for professional help immediately. Don't let a fire get out of control.
- Know what chemicals you work with. You might have to tell firefighters at a chemical fire what kinds of hazardous substances are involved.
- Make sure you are familiar with your company's emergency action plan for fires.
- Use caution when using power tools near flammable substances.

Instructor's Notes:

1.48 CORE CURRICULUM ◆ INTRODUCTORY CRAFT SKILLS

9.2.2 *Flammable Gases*

Flammable gases used on construction sites include acetylene, hydrogen, ethane, and propane (liquid propane gas, or LPG). To save space, these gases are compressed so that a large amount is stored in a small cylinder or bottle. As long as the gas is kept in the cylinder, oxygen cannot get to it and start a fire. The cylinders should be stored away from sources of heat.

If oxygen is allowed to escape and mix with a flammable gas, the resulting mixture will explode under certain conditions.

> **WARNING!**
> Never use grease or oil on the fittings of oxygen bottles and hoses. Never allow greasy or oily rags to come near any part of an oxygen system. Oil and pressurized oxygen form a very dangerous mixture that can ignite at low temperatures.

9.2.3 *Ordinary Combustibles*

The term ordinary combustibles means paper, wood, cloth, and similar fuels. The easiest way to prevent fire in ordinary combustibles is to keep a neat, clean work area. If there are no scraps of paper, cloth, or wood lying around, there will be no fuel for starting a fire. So establish and maintain good housekeeping habits. Use approved storage cabinets and containers for all waste and other ordinary combustibles.

9.3.0 Firefighting

You are not expected to be an expert firefighter. But you may have to deal with a fire to protect your safety and the safety of others. You need to know the locations of firefighting equipment on your job site. You also need to know which equipment to use on different types of fires. However, only qualified personnel are authorized to fight fires.

Most companies tell new employees where fire extinguishers are kept. If you have not been told, be sure to ask. Also ask how to report fires. The telephone number of the nearest fire department should be clearly posted in your work area. If your company has a company fire brigade, learn how to contact them. Learn your company's fire safety procedures. Know what kind of extinguisher to use for different kinds of fires and how to use them. Make sure all extinguishers are fully charged. Never remove the tag from an extinguisher—it shows the date the extinguisher was last serviced and inspected (see *Figure 41*).

Discuss the consequences of using grease or oil on the fittings of oxygen bottles and hoses.

Show Transparency 18 (Figure 41). Ensure that trainees know where the fire extinguishers are kept at their work site and understand their company's fire safety procedures.

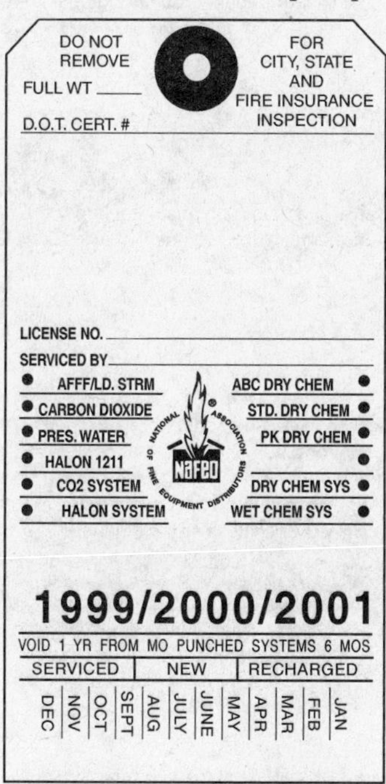

Figure 41 ◆ Fire extinguisher tag.

MODULE 00101-04 ◆ BASIC SAFETY 1.49

Refer to Table 5. Describe the materials involved in each class of fire and the way each should be extinguished. Have trainees identify each class of fire. Answer any questions trainees may have.

Show Transparencies 19 and 20 (Figure 42). Explain the information included on a typical fire extinguisher label.

See the Teaching Tip for Section 9.3.0 at the end of this module.

9.3.1 Classes of Fires

Four classes of fuels can be involved in fires (see *Table 5*). You've already learned about liquids, gases, and ordinary combustibles. Another fuel is metal. (You will learn about preventing electrical fires in another module.) Each class of fuel requires a different method of firefighting and a different type of extinguisher.

The label on a fire extinguisher clearly shows the class of fire on which it can be used (see *Figure 42*).

When you check the extinguishers in your work area, you will see that some are rated for more than one class of fire. You can use an extinguisher that has the three codes A, B, and C on it to fight a class A, B, or C fire. But remember, if the extinguisher has only one code letter, do not use it on any other class of fire, even in an emergency. You could make the fire worse and put yourself in great danger.

Table 5 Classes of Fires

Class	Materials and Proper Fire Extinguisher
Class A fires A ORDINARY COMBUSTIBLES 101SA02A.EPS A ORDINARY COMBUSTIBLES 101SA02B.EPS	These fires involve ordinary combustibles such as wood or paper. Class A fires are fought by cooling the fuel. Class A fire extinguishers contain water. Using a Class A extinguisher on any other type of fire can be very dangerous.
Class B fires B FLAMMABLE LIQUIDS 101SA02C.EPS B FLAMMABLE LIQUIDS 101SA02D.EPS	These fires involve grease, liquids, or gases. Class B extinguishers contain carbon dioxide (CO_2) or another material that smothers fires by removing oxygen from the fire.
Class C fires C ELECTRICAL EQUIPMENT 101SA02E.EPS C ELECTRICAL EQUIPMENT 101SA02F.EPS	These fires are near or involve energized electrical equipment. Class C extinguishers are designed to protect the firefighter from electrical shock. Class C extinguishers smother fires.
Class D fires D COMBUSTIBLE METALS 101SA02G.EPS	These fires involve metals. Class D extinguishers contain a powder that either forms a crust around the burning metal or gives off gases that prevent oxygen from reaching the fire. Some metals will keep burning even though they have been coated with powder from a Class D extinguisher. The best way to fight these fires is to keep using the extinguisher so the fire will not spread to other fuels.

Instructor's Notes:

1.50 CORE CURRICULUM ◆ INTRODUCTORY CRAFT SKILLS

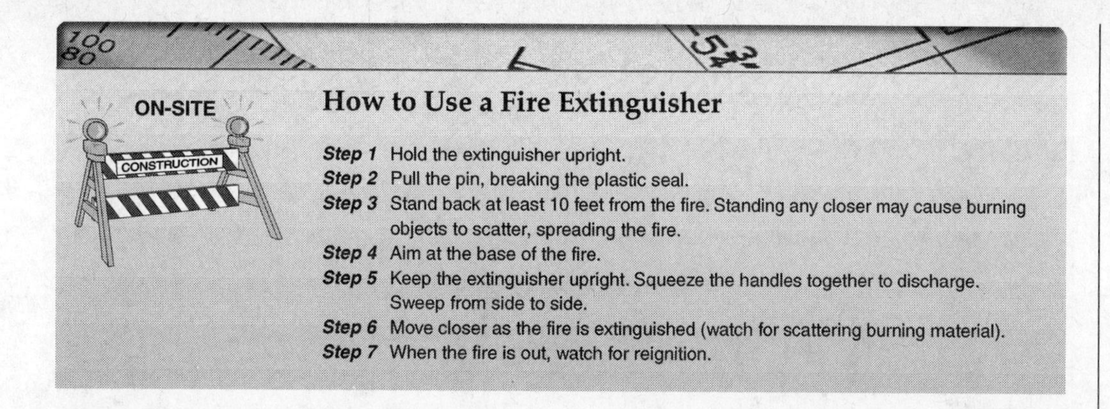

How to Use a Fire Extinguisher

Step 1 Hold the extinguisher upright.
Step 2 Pull the pin, breaking the plastic seal.
Step 3 Stand back at least 10 feet from the fire. Standing any closer may cause burning objects to scatter, spreading the fire.
Step 4 Aim at the base of the fire.
Step 5 Keep the extinguisher upright. Squeeze the handles together to discharge. Sweep from side to side.
Step 6 Move closer as the fire is extinguished (watch for scattering burning material).
Step 7 When the fire is out, watch for reignition.

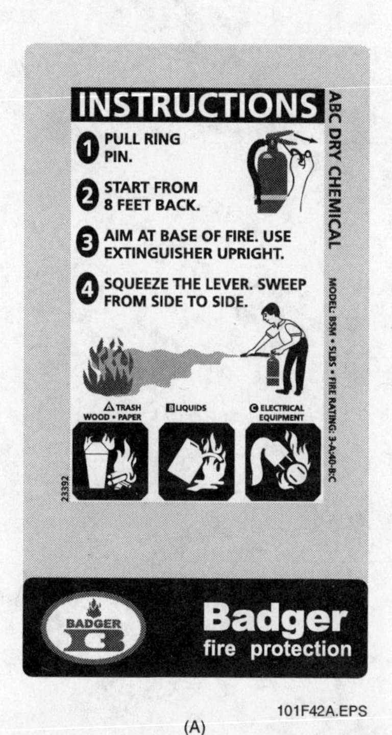

101F42A.EPS

(A)

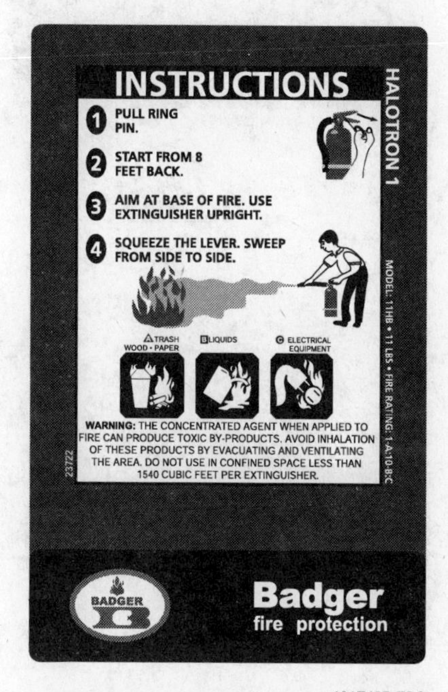

101F42B.EPS

(B)

Figure 42 ◆ Typical fire extinguisher labels. (1 of 2)

MODULE 00101-04 ◆ BASIC SAFETY 1.51

101F42C.EPS

(C)

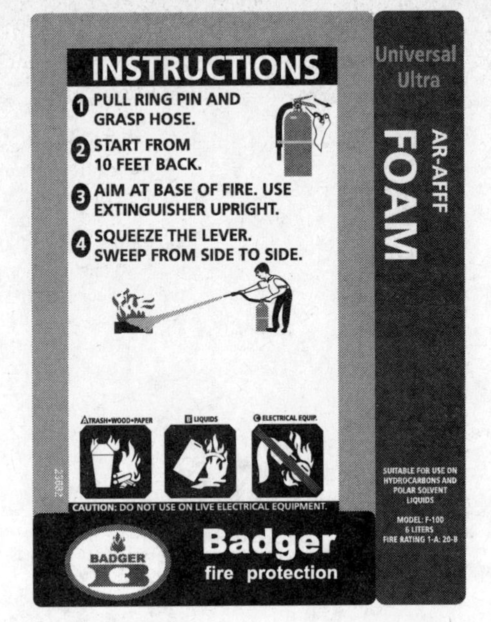

101F42D.EPS

(D)

COMBUSTIBLE METALS

FIRE EXTINGUISHER

1. HOLD UPRIGHT - PULL RING PIN.
2. HOLD NOZZLE OVER FIRE.
3. SQUEEZE LEVER AND COVER ALL BURNING METAL
4. REAPPLY AGENT TO HOT SPOTS.

CAUTION: FIRE MAY RE-IGNITE, ALLOW METAL TO COOL BEFORE CLEANUP.

101F42E.EPS

(E)

Figure 42 ◆ Typical fire extinguisher labels. (2 of 2)

Instructor's Notes:

1.52 CORE CURRICULUM ◆ INTRODUCTORY CRAFT SKILLS

Review Questions

Section 9.0.0

1. _________ must be present in the same place at the same time for a fire to occur.
   a. Oxygen, carbon dioxide, and heat
   b. Oxygen, heat, and fuel
   c. Hydrogen, oxygen, and wood
   d. Grease, liquid, and heat

2. _________ gas is flammable.
   a. Acetylene
   b. Carbon dioxide
   c. Peroxide
   d. Neon

3. A Class D fire involves _________.
   a. grease
   b. wood
   c. electrical equipment
   d. metal

4. Fire extinguishers that contain water for fighting fires involving ordinary combustibles are _________ extinguishers.
   a. Class A
   b. Class B
   c. Class C
   d. Class D

5. For a grease fire, use a _________ extinguisher.
   a. Class A
   b. Class B
   c. Class C
   d. Class D

10.0.0 ◆ ELECTRICAL SAFETY

Some construction workers think that electrical safety matters only to electricians. But on many jobs, no matter what your trade, you will use or work around electrical equipment. Extension cords, power tools, portable lights, and many other pieces of equipment use electricity. If you don't use this equipment safely and properly, the result could be death for you or a co-worker.

Electricity can be described as the flow of electrons through a conductor. This flow of electrons is called electrical current. Some materials—such as silver, copper, steel, and aluminum—are excellent conductors. This means that electrical current flows easily through them. The human body, especially when it is wet, is also a good conductor.

To create an electrical current, a path must be provided in a circular route, or a circuit. If the circuit is interrupted, the electrical current will complete its circular route by flowing along the path of least resistance. This means that it will flow into and through any conductor that is touching it. If it cannot complete its circuit, the electrical current will go to ground. This means that it will find the path of least resistance that allows it to flow as directly as possible into the earth. All of this takes place almost instantly.

If the human body comes in contact with an electrically energized conductor and is in contact with the ground at the same time, the human body becomes the path of least resistance for the electricity. This means the electricity flows through the body in less than the blink of an eye. You can't see that it's about to happen; it just happens. That's why safety precautions are so important when working with and around electrical currents. When a person's body conducts electrical current and the amount of that current is high enough, the person can be electrocuted (killed by electric shock). *Table 6* shows the effects of different amounts of electrical current on the human body and lists some common tools that operate using those currents.

NOTE

Electric shocks or burns are a major cause of accidents in the construction industry. According to the Bureau of Labor Statistics, electrocution is the fourth leading cause of death among construction workers.

WARNING!

All work on electrical equipment should be done with circuits de-energized and cleared or grounded. All conductors, buses, and connections should be considered energized unless proven otherwise.

Go over the Review Questions for Section 9.0.0. Go over any questions trainees may have.

Have trainees review Sections 10.0.0–10.3.0 for the next session.

Ensure that you have everything required for teaching and testing during this session.

Explain how to create an electrical current and how electrical currents flow through conductors.

Stress the importance of clearing and grounding electrical equipment.

Explain how even less than 1 amp of electrical current can kill.

Refer to Table 6, and review the effects of electrical current on the human body. Have trainees identify the body's reaction to currents produced by tools that you specify.

Discuss various methods of protection from electrocution. Review the different types of electrical accidents.

Discuss basic job-site electrical safety guidelines.

See Figure 43. Explain the role of the third wire in a three-wire cord. Ask whether trainees have any questions.

See Figure 45. Explain how protective guards prevent accidental contact.

Ensure that trainees understand how to use an assured grounding program with all electrical tools.

Table 6 Effects of Electrical Current on the Human Body

Current	Common Item/Tool	Reaction to Current
0.001 amps	Watch battery	Faint tingle
0.005 amps	9-volt battery	Slight shock
0.006–0.025 amps (women) 0.009–0.030 amps (men)	Christmas tree light bulb	Painful shock. Muscular control is lost.
0.050–0.9 amps	Small electric radio	Extreme pain. Breathing stops; severe muscular contractions occur. Death may result.
1.0–9.9 amps	Jigsaw (4 amps); Sawsall® or Port-a-Band® saw (6 amps); portable drill (3–8 amps)	Ventricular fibrillation and nerve damage occur. Death may result.
10 amps and above	ShopVac® (15-gallon); circular saw	Heart stops beating; severe burns occur. Death may result.

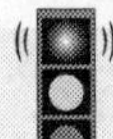

WARNING!

Less than 1 amp of electrical current can kill. Always take precautions when working around electricity.

Here's an example. A craftsperson is operating a portable power drill while standing on damp ground. The power cord inside the drill has become frayed, and the electric wire inside the cord touches the metal drill frame. Three amps of current pass from the wire through the frame, then through the craftsperson's body and into the ground. *Table 6* shows that this craftsperson will probably die.

A good method of protection from accidental electrocution is the use of a ground fault circuit interrupter (GFCI). The GFCI is a fast-acting circuit breaker that senses small (as little as approximately 5 milliamps) imbalances in the circuit caused by current leakage to ground. In as little as 1/40 of a second, the GFCI will interrupt the power.

Not all electrical accidents result in death. There are different types of electrical accidents. Any of the following can happen:

- Burns
- Electric shock
- Explosions
- Falls caused by electric shock
- Fires

10.1.0 Basic Electrical Safety Guidelines

OSHA and your company have specific policies and procedures to keep the workplace safe from electrical hazards. You can do many things to reduce the chance of an electrical accident. If you ever have any questions about electrical safety on the job site, ask your supervisor. Here are the basic job-site electrical safety guidelines:

- Use three-wire extension cords and protect them from damage. Never fasten them with staples, hang them from nails, or suspend them from wires. Never use damaged cords.
- Make sure that panels, switches, outlets, and plugs are grounded.
- Never use bare electrical wire.
- Never use metal ladders near any source of electricity.
- Never wear a metal hard hat.
- Always inspect electrical power tools before you use them.
- Never operate any piece of electrical equipment that has a danger tag or lockout device attached to it.
- Use three-wire cords for portable power tools and make sure they are properly connected (see *Figure 43*). The three-wire system is one of the most common safety grounding systems used to protect you from accidental electrical shock. The third wire is connected to a ground. If the insulation in a tool fails, the current will pass to ground through the third wire—not through your body.

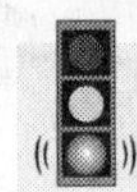

NOTE

It is becoming more common to use double-insulated tools because they are safer than relying on a three-wire cord alone.

- Never use worn or frayed cables (see *Figure 44*).
- Make sure all light bulbs have protective guards to prevent accidental contact (see *Figure 45*).

Instructor's Notes:

1.54 CORE CURRICULUM ◆ INTRODUCTORY CRAFT SKILLS

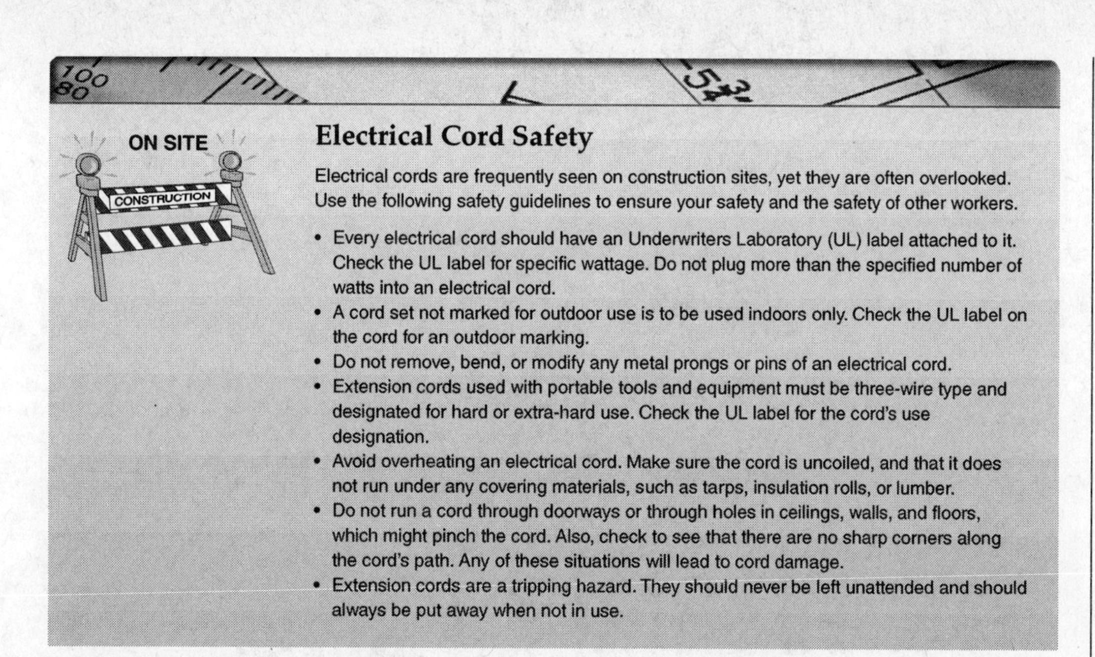

Electrical Cord Safety

Electrical cords are frequently seen on construction sites, yet they are often overlooked. Use the following safety guidelines to ensure your safety and the safety of other workers.

- Every electrical cord should have an Underwriters Laboratory (UL) label attached to it. Check the UL label for specific wattage. Do not plug more than the specified number of watts into an electrical cord.
- A cord set not marked for outdoor use is to be used indoors only. Check the UL label on the cord for an outdoor marking.
- Do not remove, bend, or modify any metal prongs or pins of an electrical cord.
- Extension cords used with portable tools and equipment must be three-wire type and designated for hard or extra-hard use. Check the UL label for the cord's use designation.
- Avoid overheating an electrical cord. Make sure the cord is uncoiled, and that it does not run under any covering materials, such as tarps, insulation rolls, or lumber.
- Do not run a cord through doorways or through holes in ceilings, walls, and floors, which might pinch the cord. Also, check to see that there are no sharp corners along the cord's path. Any of these situations will lead to cord damage.
- Extension cords are a tripping hazard. They should never be left unattended and should always be put away when not in use.

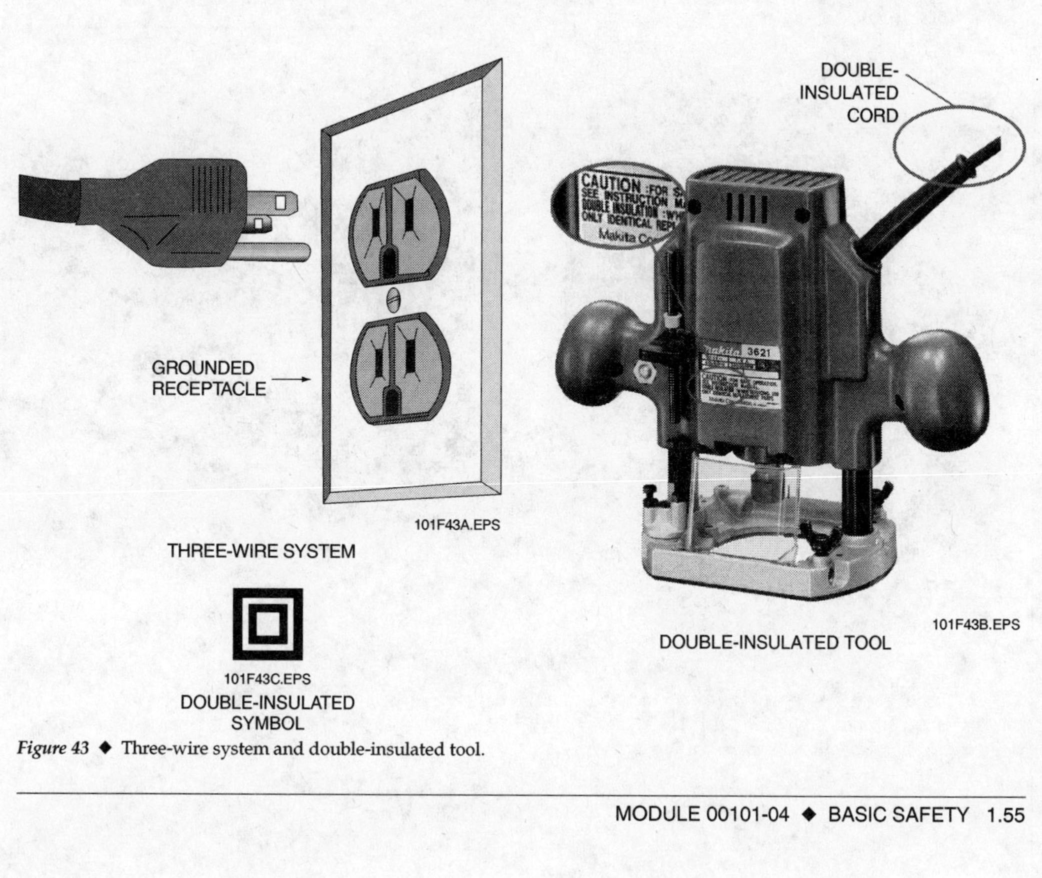

Figure 43 ◆ Three-wire system and double-insulated tool.

MODULE 00101-04 ◆ BASIC SAFETY 1.55

Explain that trainees must ensure that all tools are ground-fault protected.

See the Teaching Tip for Section 10.1.0 at the end of this module.

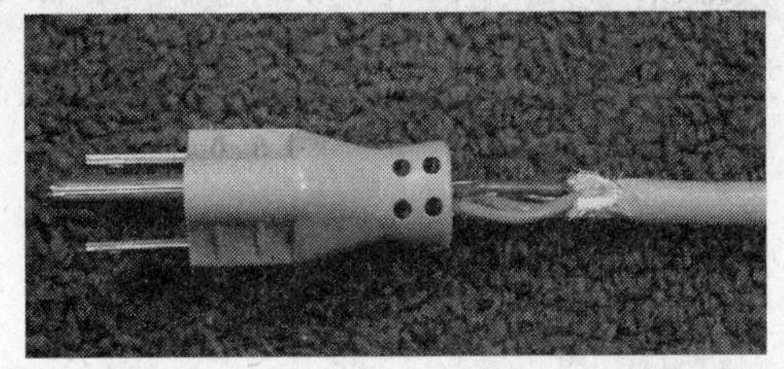

101F44A.EPS

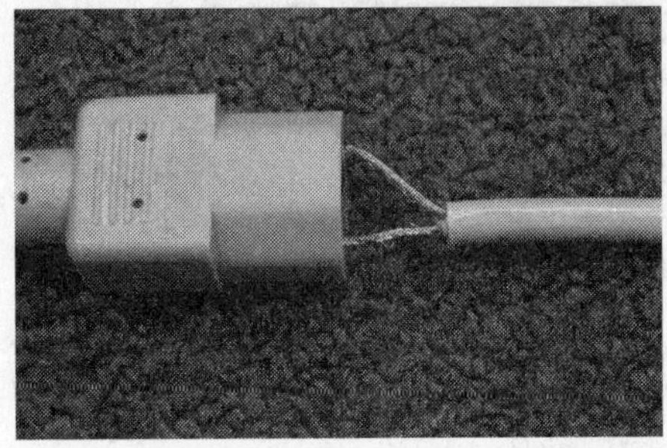

101F44B.EPS

Figure 44 ◆ Never use damaged cords.

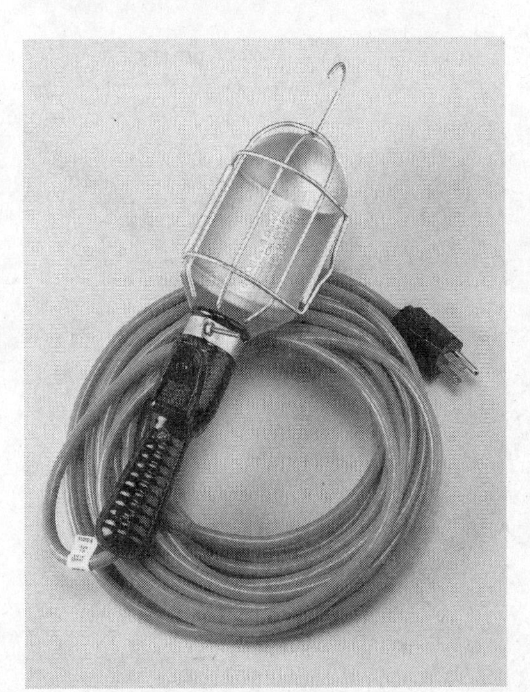

101F45.EPS

Figure 45 ◆ Work light with protective guard.

- Do not hang temporary lights by their power cords unless they are specifically designed for this use.
- Check the cable and ground prong. Check for cuts in the cords and make sure the cords are clean of grease.
- Use only approved **concealed receptacles** for plugs. If different voltages or types of current are used in the same area, the receptacles should be designed so that the plugs are not interchangeable.
- Any repairs to cords must be performed by a qualified person.
- Use a GFCI (see *Figure 46*) or an assured grounding program (see section 10.2.0) with every tool.
- Run extension cords overhead to avoid tripping, which can cause shock hazards and damage the cords.
- Always make sure all tools are grounded before use.

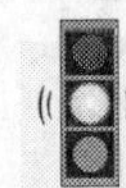

CAUTION

All tools used in construction must be ground-fault protected. This helps ensure the safety of workers.

101F46A.EPS

Figure 46 ◆ GFCI.

Instructor's Notes:

1.56 CORE CURRICULUM ◆ INTRODUCTORY CRAFT SKILLS

Review the minimum safe working distances from exposed conductors. Ensure that trainees understand their company policies and regulations.

Review the procedures to follow in the event someone is electrocuted.

Explain that today's kiln-dried lumber retains a high moisture content and can conduct electricity. It is unsafe to use this lumber to push someone away from the source of shock. It is safer to use a fiberglass ladder.

Stress that using a part of your body or another object to touch a victim of electrical shock could transfer the current to your body.

10.2.0 Working Near Energized Electrical Equipment

No matter what your trade, your job may include working near exposed electrical equipment or conductors. This is one example of proximity work. Often, **electrical distribution panels, switch enclosures,** and other equipment must be left open during construction. This leaves the wires and components in them exposed. Some or all of the wires and components may be energized. Working near exposed electrical equipment can be safe, but only if you keep a safe working distance.

Regulations and company policies tell you the minimum safe working distances from exposed conductors. The safe working distance ranges from a few inches to several feet, depending on the voltage. The higher the voltage, the greater the safe working distance.

You must learn the safe working distance for each situation. Make sure you never get any part of your body or any tool you are using closer to exposed conductors than that distance. You can get information on safe working distances from your instructor, your supervisor, company safety policies, and regulatory documents.

10.3.0 If Someone Is Shocked

If you are there when someone gets an electrical shock, you can save a life by taking immediate action. Here's what to do:

Step 1 Immediately disconnect the circuit.

Step 2 If you can't disconnect the circuit, do not try to separate the victim from the circuit. If you touch a person who is being electrocuted, the current will flow through you, too.

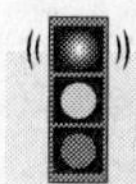

WARNING!

Do not touch the victim or the electrical source with your hand, foot, or any other part of your body or with any object or material. You could become another victim.

Step 3 Once the circuit is disconnected, give first aid and call an ambulance. If you cannot disconnect the circuit, call an ambulance.

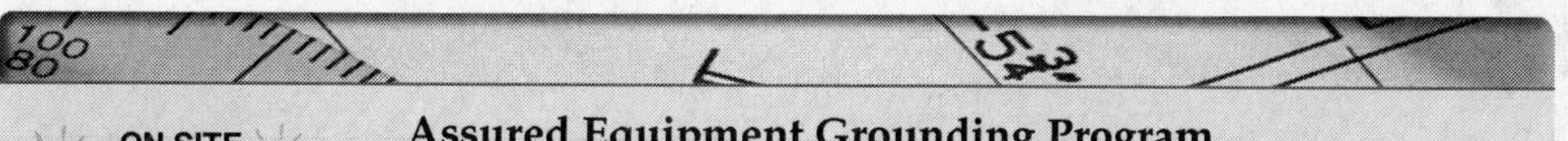

ON SITE

Assured Equipment Grounding Program

An assured equipment grounding program is an alternative to GCFIs. It covers all cord sets, receptacles that are not a part of the permanent wiring of the building or structure, and equipment connected by cord and plug that are available for use or used by employees. The requirements that the program must meet are stated in OSHA's *29 CFR 1926.404*, but employers may provide additional tests or procedures.

OSHA requires two tests. One is a continuity test to ensure that the equipment grounding conductor is electrically continuous. It must be performed on all cord sets, receptacles that are not part of the permanent wiring of the building or structure, and cord- and plug-connected equipment that is required to be grounded. This test may be performed using a simple continuity tester, such as a lamp and battery, a bell and battery, an ohmmeter, or a receptacle tester.

The other test must be performed on receptacles and plugs to ensure that the equipment grounding conductor is connected to its proper terminal. This test can be performed with the same equipment used in the first test.

These tests are required before first use, after any repairs, after damage is suspected to have occurred, and at three-month intervals. Cord sets and receptacles that are essentially fixed and not exposed to damage must be tested at regular intervals of no longer than six months. Any equipment that fails to pass the required tests must not be made available to or used by employees.

Go over the Review Questions for Section 10.0.0. Answer any questions trainees may have.

Review Questions

Section 10.0.0

1. Observing proper safety precautions when working with and around electrical current is important because the human body _________.
   a. resists the electricity's path
   b. can conduct electrical current
   c. won't conduct electricity
   d. doesn't offer electricity a circular route

2. The _________ system is one of the most common safety grounding systems used with portable power tools.
   a. distribution wire
   b. rubber cord
   c. three-wire cord
   d. insulation plug

3. The minimum safe working distance from exposed electrical conductors _________.
   a. depends on the voltage
   b. is at least 1 foot
   c. is a few inches
   d. is unlimited

4. If someone is being electrically shocked, the first thing you should try to do is _________.
   a. use a metal pole to separate the victim from the circuit
   b. disconnect the circuit
   c. give first aid
   d. call your supervisor

5. If someone is being electrically shocked and you cannot disconnect the circuit, the first thing you should do is _________.
   a. call an ambulance
   b. give first aid
   c. pull the victim away with your hands
   d. use a pole to separate the victim from the circuit

Instructor's Notes:

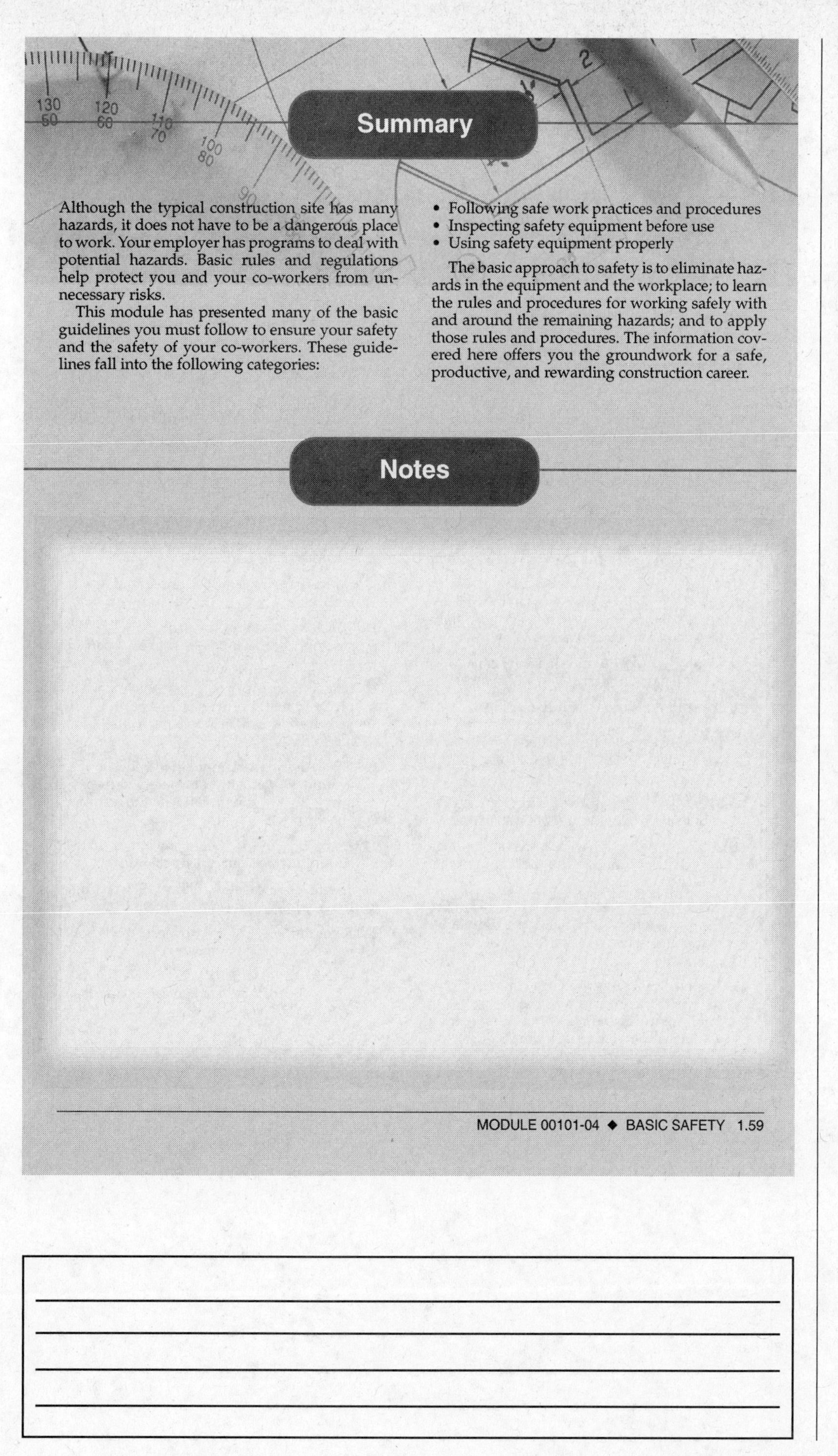

Summary

Although the typical construction site has many hazards, it does not have to be a dangerous place to work. Your employer has programs to deal with potential hazards. Basic rules and regulations help protect you and your co-workers from unnecessary risks.

This module has presented many of the basic guidelines you must follow to ensure your safety and the safety of your co-workers. These guidelines fall into the following categories:

- Following safe work practices and procedures
- Inspecting safety equipment before use
- Using safety equipment properly

The basic approach to safety is to eliminate hazards in the equipment and the workplace; to learn the rules and procedures for working safely with and around the remaining hazards; and to apply those rules and procedures. The information covered here offers you the groundwork for a safe, productive, and rewarding construction career.

Notes

Ask trainees to complete the Key Terms Quiz. Review trainees' answers, and go over any questions trainees may have.

Administer the Module Examination. Be sure to record the results of the Exam on Craft Training Report Form 200, and submit the results to the Training Program Sponsor.

Ensure that all Performance Tests have been completed and Performance Profile Sheets for each trainee are filled out. Be sure to record the results of the testing on Craft Training Report Form 200, and submit the results to the Training Program Sponsor.

Key Terms Quiz

Fill in the blank with the correct key term that you learned from your study of this module.

1. _________ is a formal procedure for taking equipment out of service and ensuring it cannot be operated until a qualified person has returned it to service.
2. _________ is the process of joining metal parts by fusion.
3. Because _________ scrap materials catch fire and burn easily, remove them regularly from the work area.
4. The cutting process results in oxides that mix with molten iron and produce _________.
5. A(n) _________ is any man-made cut, cavity, trench, or depression in an earth surface, formed by removing earth.
6. A(n) _________ identifies unsanitary, hazardous, or dangerous working conditions and has the authority to correct or eliminate them.
7. A(n) _________ is large enough to work in but has limited means of entry or exit; sometimes a permit is required to work in it.
8. Store _________ liquids in safety cans to avoid the risk of fire.
9. If a scaffold is more than 10 feet high, _________—or pieces of wood or metal placed diagonally from the bottom of one rail to the top of another rail that add support to a structure—must be used.
10. The _________ houses the circuits that distribute electricity throughout a structure.
11. A(n) _________ is basically two straight ladders that are connected so the length of the ladder can be changed.
12. Wear _________ to protect your eyes from the _____, which is the sudden, bright light that occurs when you start up a welding operation.
13. To save lives, prevent injuries, and protect the health of America's workers, _________ publishes rules and regulations that employees and employers must follow.
14. If fuel is improperly mixed, it can cause a(n) _________.
15. An opening in a wall or floor is a safety hazard and must be either covered or _________.
16. Even a brief exposure to the ultraviolet light from arc welding can damage your eyes, causing a(n) _________.
17. _________ is an OSHA rule requiring all contractors to educate their employees about the hazardous chemicals they may be exposed to on the job site.
18. The temperature at which a fuel gives off enough gases to burn is called the _________.
19. A(n) _________ is the conducting connection between electrical equipment or an electrical circuit and the earth.
20. When climbing a ladder or scaffold, use a tagline or _________ to pull up your tools.
21. SCBA is an example of a(n) _________, or an assembly of machines used to do a particular job.
22. If a work area is _________, that means it has pieces of material at least 2 inches thick and 6 inches wide used as flooring, decking, or scaffolding.
23. When doing work more than 6 feet above the ground, you must wear a safety harness with a(n) _________ that is attached to a strong anchor point.
24. A good _________ helps prevent or correct conditions that can cause accidents.
25. To prevent a cave-in, follow OSHA regulations for _________ up a trench.
26. Refer to the _________ to learn about how to handle hazardous substances.
27. Overloading, which means exceeding the _________ of a ladder, can cause ladder failure.

Instructor's Notes:

____

____

____

28. If you are operating a vehicle on a job site and cannot see to your rear, get a(n) _________ to direct you.

29. Before working in a(n) _________, you must be trained, obtain written authorization, and take the necessary precautions.

30. A(n) _________ is a narrow excavation made below the surface of the ground that is generally deeper than it is wide.

31. The _________ on a scaffold is placed halfway between the toeboard and the top rail.

32. _________ for welding includes a faceshield, ear plugs, and gloves.

33. A(n) _________ has proven his or her extensive knowledge, training, and experience and has successfully demonstrated the ability to solve problems relating to the work.

34. A(n) _________ provides clean air for breathing.

35. Manufactured and rolling are the two basic types of _________.

36. When doing _________, you must be careful not to come into contact with the nearby hazard.

37. A(n) _________ is a self-supporting ladder made of two sections hinged at the top.

38. Use only approved _________ for plugs.

39. A(n) _________ is nonadjustable and consists of two rails, rungs between the rails, and safety feet on the bottom of the rails.

40. A vertical barrier called a(n) _________ is used at floor level on scaffolds to prevent materials from falling.

41. A(n) _________ tells you which way the wind is blowing.

42. A horizontal board called a(n) _________ is used at top-level on all open sides of scaffolding and platforms.

43. Never look at the _________ caused by welding without proper eye protection.

44. An excavating machine called a(n) _________ is used to dig trenches, especially for pipeline and cables.

45. When you are welding and other workers are in the area, set up a protective screen called a(n) _________.

46. A(n) _________ houses electrical switches used to regulate and distribute electricity in a building.

47. _________ is the waste material from welding operations.

Key Terms

Apparatus
Arc
Arc welding
Combustible
Competent person
Concealed receptacle
Confined space
Cross-bracing
Dross
Electrical distribution panel
Excavation
Extension ladder
Flammable
Flash
Flashback
Flash burn
Flash goggles
Flash point
Ground
Guarded
Hand line
Hazard Communication Standard (HazCom)
Lanyard
Lockout/tagout
Management system
Material safety data sheet (MSDS)
Maximum intended load
Mid-rail
Occupational Safety and Health Administration (OSHA)
Permit-required confined space
Personal protective equipment (PPE)
Planked
Proximity work
Qualified person
Respirator
Scaffold
Shoring
Signaler
Slag
Stepladder
Straight ladder
Switch enclosure
Toeboard
Top rail
Trench
Trencher
Welding shield
Wind sock

MODULE 00101-04 ◆ BASIC SAFETY 1.61

Profile in Success

Doug Garcia

Chairman, Industrial Training Department
Manatee Technical Institute
West Bradenton, Florida

Doug grew up in Fall River, Massachusetts. After completing high school, he began his career as a union construction laborer. He decided to become a welder and worked in the power plant maintenance industry before turning to shipbuilding. Doug built ships at General Dynamics and submarines at Electric Boat. He worked his way up from shipfitter to lead man, foreman, and ultimately ship superintendent, responsible for the delivery of a ship from the keel on up. Doug became a welding instructor at Diman Regional Vocational Technical High School and then Manatee Technical Institute.

Doug has a BA in vocational education from Fitchburg State College and is a U.S. Department of Labor–approved general industry and construction outreach trainer. He is also an American Welding Society (AWS)–certified welding educator and a certified welding inspector, and holds a Level 2 certification from the American Society for Nondestructive Testing.

What aspect of the construction industry appeals to you most?
What I like the most is the personal satisfaction of seeing the finished product. When you look at a completed project, you are really seeing the result of a collaborative effort between a broad spectrum of trades. In the finished product, you can see how they all come together to make a whole.

How did you decide to become an educator?
Over the course of my career, I have really come to appreciate and understand the need for safety training. It bothers me most to see a great career cut short by a needless accident. Safety is one of the most important concerns in the construction industry. As a whole, the industry gives safety the attention it needs, but ultimately the responsibility still falls on the individual. No matter how good a safety program is, it is only as good as the person who's applying it. The company can only do so much; the employee has to finish the job. So I wanted to help people understand and apply safety on the job.

What do you think it takes to be a success in your trade?
I think that it takes a solid work ethic and pride in one's craftsmanship. You must also feel, or want to feel, a sense of satisfaction from seeing a completed project.

What are some of the things you do in your job?
I have a staff of 17 teachers, 2 of whom are approved construction safety instructors and 2 of whom are general industrial safety instructors. I review the existing curriculum and undertake curriculum development tasks. I try to ensure that the curriculum meets the needs of industry and that we remain current with the changes in construction technology and materials. I assist with job placement and maintain contact with area employers. We provide on-site training for local area employers and help them to develop and implement procedures that are required by the U.S. Occupational Safety and Health Administration (OSHA).

I teach safety instruction to all students in both the construction and general industry tracks at Manatee. I'm also a mentor for new vocational teachers. And I also coordinate the purchase of equipment and supplies. It's a full-time job, and then some!

What do you like most about your job?
I really enjoy seeing former students become successful. Many of them return to the program to serve on advisory committees. They come back to advise the staff and faculty on current trends in their craft area, and they get to make recommendations for the curricula and the purchase of equipment. That

Instructor's Notes:

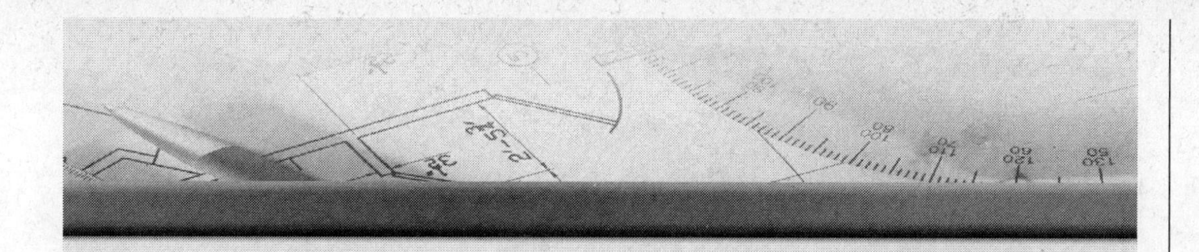

way, I get to work with them as professional colleagues and not just as students.

What would you say to someone entering the trades today?

In my classes I stress that no matter what your skill level is, if you are not doing your job safely, then you are jeopardizing all your hard work. Even the most talented individuals can't perform their jobs to the best of their ability if they have been injured or hurt.

I have personally witnessed two construction fatalities. The people most affected are usually the families. Your value as a craft worker is the ability to perform for your employer and to provide for your family. If you are injured or incapable of performing your craft, you are not only losing your earning potential and your potential for advancement, but you are also depriving the industry of a trained craft worker and reducing your family's quality of life.

So you have to ask yourself every day, "How would an accident affect my family?" When I tell my students that, it really gets them thinking. The industry has inherent dangers; there's no need for a careless worker to add to them.

MODULE 00101-04 ◆ BASIC SAFETY 1.63

Trade Terms Introduced in This Module

Apparatus: An assembly of machines used together to do a particular job.

Arc: The flow of electrical current through a gas such as air from one pole to another pole.

Arc welding: The joining of metal parts by fusion, in which the necessary heat is produced by means of an electric arc.

Combustible: Capable of easily igniting and rapidly burning; used to describe a fuel with a flash point at or above 100°F.

Competent person: A person who can identify working conditions or surroundings that are unsanitary, hazardous, or dangerous to employees and who has authorization to correct or eliminate these conditions promptly.

Concealed receptacle: The electrical outlet that is placed inside the structural elements of a building, such as inside the walls. The face of the receptacle is flush with the finished wall surface and covered with a plate.

Confined space: A work area large enough for a person to work, but arranged in such a way that an employee must physically enter the space to perform work. A confined space has a limited or restricted means of entry and exit. It is not designed for continuous work. Tanks, vessels, silos, pits, vaults, and hoppers are examples of confined spaces. See also *permit-required confined space.*

Cross-bracing: Braces (metal or wood) placed diagonally from the bottom of one rail to the top of another rail that add support to a structure.

Dross: Waste material resulting from cutting using a thermal process.

Electrical distribution panel: Part of the electrical distribution system that brings electricity from the street source (power poles and transformers) through the service lines to the electrical meter mounted on the outside of the building and to the panel inside the building. The panel houses the circuits that distribute electricity throughout the structure.

Excavation: Any man-made cut, cavity, trench, or depression in an earth surface, formed by removing earth. It can be made for anything from basements to highways. See also *trench.*

Extension ladder: A ladder made of two straight ladders that are connected so that the overall length can be adjusted.

Flammable: Capable of easily igniting and rapidly burning; used to describe a fuel with a flash point below 100°F.

Flash: A sudden bright light associated with starting up a welding torch.

Flashback: A welding flame that flares up and chars the hose at or near the torch connection. It is caused by improperly mixed fuel.

Flash burn: The damage that can be done to eyes after even brief exposure to ultraviolet light from arc welding. A flash burn requires medical attention.

Flash goggles: Eye protective equipment worn during welding operations.

Flash point: The temperature at which fuel gives off enough gases (vapors) to burn.

Ground: The conducting connection between electrical equipment or an electrical circuit and the earth.

Guarded: Enclosed, fenced, covered, or otherwise protected by barriers, rails, covers, or platforms to prevent dangerous contact.

Hand line: A line attached to a tool or object so a worker can pull it up after climbing a ladder or scaffold.

Hazard Communication Standard (HazCom): The Occupational Safety and Health Administration standard that requires contractors to educate employees about hazardous chemicals on the job site and how to work with them safely.

Lanyard: A short section of rope or strap, one end of which is attached to a worker's safety harness and the other to a strong anchor point above the work area.

Lockout/tagout: A formal procedure for taking equipment out of service and ensuring that it cannot be operated until a qualified person has removed the lockout or tagout device (such as a lock or warning tag).

Instructor's Notes:

1.64 CORE CURRICULUM ◆ INTRODUCTORY CRAFT SKILLS

Management system: The organization of a company's management, including reporting procedures, supervisory responsibility, and administration.

Material safety data sheet (MSDS): A document that must accompany any hazardous substance. The MSDS identifies the substance and gives the exposure limits, the physical and chemical characteristics, the kind of hazard it presents, precautions for safe handling and use, and specific control measures.

Maximum intended load: The total weight of all people, equipment, tools, materials, and loads that a ladder can hold at one time.

Mid-rail: Mid-level, horizontal board required on all open sides of scaffolding and platforms that are more than 14 inches from the face of the structure and more than 10 feet above the ground. It is placed halfway between the toeboard and the top rail.

Occupational Safety and Health Administration (OSHA): An agency of the U.S. Department of Labor. Also refers to the Occupational Safety and Health Act of 1970, a law that applies to more than more than 111 million workers and 7 million job sites in the country.

Permit-required confined space: A confined space that has been evaluated and found to have actual or potential hazards, such as a toxic atmosphere or other serious safety or health hazard. Workers need written authorization to enter a permit-required confined space. See also *confined space.*

Personal protective equipment (PPE): Equipment or clothing designed to prevent or reduce injuries.

Planked: Having pieces of material 2 or more inches thick and 6 or more inches wide used as flooring, decking, or scaffolding.

Proximity work: Work done near a hazard but not actually in contact with it.

Qualified person: A person who, by possession of a recognized degree, certificate, or professional standing, or by extensive knowledge, training, and experience, has demonstrated the ability to solve or prevent problems relating to a certain subject, work, or project.

Respirator: A device that provides clean, filtered air for breathing, no matter what is in the surrounding air.

Scaffold: An elevated platform for workers and materials.

Shoring: Using pieces of timber, usually in a diagonal position, to hold a wall in place temporarily.

Signaler: A person who is responsible for directing a vehicle when the driver's vision is blocked in any way.

Slag: Waste material from welding operations.

Stepladder: A self-supporting ladder consisting of two elements hinged at the top.

Straight ladder: A nonadjustable ladder.

Switch enclosure: A box that houses electrical switches used to regulate and distribute electricity in a building.

Toeboard: A vertical barrier at floor level attached along exposed edges of a platform, runway, or ramp to prevent materials and people from falling.

Top rail: A top-level, horizontal board required on all open sides of scaffolding and platforms that are more than 14 inches from the face of the structure and more than 10 feet above the ground.

Trench: A narrow excavation made below the surface of the ground that is generally deeper than it is wide, with a maximum width of 15 feet. See also *excavation.*

Trencher: An excavating machine used to dig trenches, especially for pipeline and cables.

Welding shield: (1) A protective screen set up around a welding operation designed to safeguard workers not directly involved in that operation. (2) A shield that provides eye and face protection for welders by either connecting to helmet-like headgear or attaching directly to a hard hat; also called a welding helmet.

Wind sock: A cloth cone open at both ends mounted in a high place to show which direction the wind is blowing.

MODULE 00101-04 ◆ BASIC SAFETY 1.65

This module is intended to present thorough resources for task training. The following reference works are suggested for further study. These are optional materials for continued education rather than for task training.

Construction Back Safety. Videocassette. 10 minutes. Coastal Training Technologies Corp. Virginia Beach, VA.

Construction Confined Space Entry. Videocassette. 10 minutes. Coastal Training Technologies Corp. Virginia Beach, VA.

Construction Electrical Safety. Videocassette. 10 minutes. Coastal Training Technologies Corp. Virginia Beach, VA.

Construction Fall Protection: Get Arrested! Videocassette. 11 minutes. Coastal Training Technologies Corp. Virginia Beach, VA.

Construction Lockout/Tagout. Videocassette. 10 minutes. Coastal Training Technologies Corp. Virginia Beach, VA.

Construction Safety, 1996. Jimmie Hinze. Englewood Cliffs, NJ: Prentice Hall.

Construction Safety Council Home Page, http://buildsafe.org/home.htm.

Construction Safety Manual, 1998. Dave Heberle. New York: McGraw-Hill.

Construction Stairways & Ladders. Videocassette. 10 minutes. Coastal Training Technologies Corp. Virginia Beach, VA.

Construction Welding Safety. Videocassette. 10 minutes. Coastal Training Technologies Corp. Virginia Beach, VA.

Field Safety, 2003. NCCER. Upper Saddle River, NJ: Prentice Hall.

Handbook of OSHA Construction Safety and Health, 1999. James V. Eidson et al. Boca Raton, FL: Lewis Publishers, Inc.

HazCom For Construction. Videocassette. 11 minutes. Coastal Training Technologies Corp. Virginia Beach, VA.

NAHB-OSHA Jobsite Safety Handbook, 1999. Washington, DC: Home Builder Press. Available online at www.osha.gov.

Occupational Safety and Health Standards for the Construction Industry. Washington, DC: Occupational Safety and Health Administration, U.S. Department of Labor, U.S. Government Printing Office.

Safety Orientation, 2003. NCCER. Upper Saddle River, NJ: Prentice Hall.

Safety Technology, 2003. NCCER. Upper Saddle River, NJ: Prentice Hall.

United States Department of Labor, Occupational Safety and Health Administration Home Page, http://www.osha.gov.

Instructor's Notes:

1.66 CORE CURRICULUM ◆ INTRODUCTORY CRAFT SKILLS

MODULE 00101-04 — TEACHING TIPS

The following are suggested activities or instructional methods to help you teach the material in this AIG.

General

When you call on someone to answer a question, the rest of the class relaxes or even tunes out because they expect that the question and answer will take place only between you and the trainee you called on. Instead, use this technique to involve more trainees in answering questions and to keep them on their toes.

1. Ask trainees to define a term or explain a concept.
2. After one trainee has answered, ask a trainee seated nearby if the answer is right. Then ask whether a trainee in the back of the room agrees.
3. Ask trainees to explain why they think an answer is right or wrong.
4. Use the session to clear up incorrect ideas, and encourage trainees to learn from their mistakes.

Sections 2.1.0–2.1.9

What Causes Accidents?

This exercise will familiarize trainees with the major causes of accidents. Trainees will need their Trainee Guides and pencil and paper to take notes for discussion. Allow between 15 and 20 minutes for this exercise.

1. Have trainees write a list of the nine causes of accidents discussed to this point.
2. For each cause, ask trainees to list one example.
3. Review and discuss trainees' examples.

Sections 2.3.0–2.3.3

Company Safety Policies and OSHA Regulations

This exercise will give trainees the opportunity to discuss safety policies and OSHA regulations. For this presentation, invite a competent or qualified person as defined by OSHA to speak with the class about safety policies and regulations. For this exercise, you will need copies of *OSHA 29 CFR 1926.* Trainees will need their Trainee Guides and pencil and paper to take notes for discussion. Allow between 45 and 60 minutes for this exercise.

1. Tell the trainees that a guest speaker will be presenting information on safety policies and OSHA regulations.
2. Have trainees brainstorm questions for the guest speaker before the speaker arrives. Suggested questions could address the following:
   - How does a competent person identify hazardous working conditions?
   - How does a competent person reduce dangers to employees?
   - How does a competent person eliminate hazards?
   - What qualifications does a qualified person possess?
   - How does a qualified person solve or resolve problems?
3. Introduce the guest speaker. Remember to include the speaker's name, title, company, and relevant credentials.
4. Allow between 30 to 45 minutes for the presentation and 15 minutes for questions.
5. Following the presentation, ask trainees to share their questions with the guest speaker.

Sections 3.0.0–3.6.0

Construction Job-Site Hazards

This exercise will give trainees the opportunity to observe job-site hazards and job-site safety. You will need to arrange a visit to a nearby construction site. If possible, select a job site that includes trench, excavation, proximity, or confined space work. Trainees will need their Trainee Guides and pencil and paper to take notes for discussion. Allow between 30 and 45 minutes for this exercise.

1. Ask trainees to tour the site and list potential dangers and hazards that they witness.
2. Ask to speak with the competent or qualified person at the site. Have trainees discuss their findings with this person.
3. Following the site visit, ask trainees to discuss their observations. Answer any questions trainees may have about job-site hazards.

Sections 5.0.0–5.2.7

Personal Protective Equipment

This exercise will familiarize trainees with a variety of personal protective equipment, which you will provide, that they will commonly use on the job site. Trainees will need their Trainee Guides and pencil and paper to take notes for discussion. Allow 30 minutes for this exercise.

1. At several stations, display a variety of types of personal protective equipment that trainees will commonly use on the job site.
2. Have trainees visit each station, identify the type of equipment, and explain when and why it should be used.
3. Review trainees' answers, and go over any questions they may have.
4. Ask trainees to demonstrate how to use each item.
5. Observe trainees' ability to use each item. Correct any mistakes that you witness.

Sections 7.0.0–7.2.4

Aerial Work

This exercise will give trainees the opportunity to practice aerial work. You will need a variety of ladders and scaffolds and the appropriate personal protective equipment. Trainees will need their Trainee Guides and pencil and paper to take notes for discussion. Allow 30 minutes for this exercise.

1. Set up a variety of ladders, stepladders, and scaffolds.
2. Have trainees demonstrate how to inspect and use each type of ladder or scaffold.
3. Ask trainees to note what was done correctly or incorrectly in each demonstration.
4. Answer any questions trainees may have about aerial work.

Section 9.3.0

Firefighting

This exercise will familiarize trainees with firefighting techniques using fire extinguishers. For this exercise, you will need a variety of extinguishers. Trainees will need their Trainee Guides and pencil and paper to take notes for discussion. Allow 30 minutes for this exercise.

1. Set up stations with a variety of fire extinguishers.
2. Have trainees explain when each class of fire extinguisher should be used.
3. Ask trainees to practice following the steps to use each fire extinguisher.
4. Go over any questions trainees may have.

Section 10.1.0 *Basic Electrical Safety Guidelines*

This informal quiz will allow trainees to test their electrical safety knowledge. For this exercise, trainees will need their Trainee Guides and pencil and paper to take notes for discussion. Allow between 15 and 20 minutes for this exercise.

1. On the whiteboard, write a list of basic job-site electrical safety guidelines. Include a variety of inaccuracies in the list.
2. Have trainees identify which items in the list are correct and which are incorrect.
3. Review trainees' answers, and ask trainees to correct the inaccuracies.
4. Answer any questions trainees may have about basic electrical safety guidelines.

MODULE 00101-04 — ANSWERS TO REVIEW QUESTIONS

Section 2.0.0

1. c
2. b
3. a
4. a
5. a
6. c
7. b
8. d
9. c
10. a

Section 3.0.0

1. c
2. a
3. b
4. a
5. b

Sections 4.0.0–6.0.0

1. b
2. b
3. d
4. b
5. c
6. c
7. c
8. a
9. c
10. c

Section 7.0.0

1. b
2. b
3. d
4. b
5. c

Section 8.0.0

1. d
2. a
3. b
4. a
5. c

Section 9.0.0

1. b
2. a
3. d
4. a
5. b

Section 10.0.0

1. b
2. c
3. a
4. b
5. a

MODULE 00101-04—ANSWERS TO KEY TERMS QUIZ

1. Lockout/tagout
2. Arc welding
3. combustible
4. dross
5. excavation
6. competent person
7. confined space
8. flammable
9. cross-bracing
10. electrical distribution panel
11. extension ladder
12. flash goggles; flash
13. the Occupational Safety and Health Administration
14. flashback
15. guarded
16. flash burn
17. Hazard Communication Standard (HazCom)
18. flash point
19. ground
20. hand line
21. apparatus
22. planked
23. lanyard
24. management system
25. shoring
26. material safety data sheet
27. maximum intended load
28. signaler
29. permit-required confined space
30. trench
31. mid-rail
32. Personal protective equipment (PPE)
33. qualified person
34. respirator
35. scaffold
36. proximity work
37. stepladder
38. concealed receptacles
39. straight ladder
40. toeboard
41. wind sock
42. top rail
43. arc
44. trencher
45. welding shield
46. switch enclosure
47. Slag

CONTREN® LEARNING SERIES — USER FEEDBACK

The NCCER makes every effort to keep these textbooks up-to-date and free of technical errors. We appreciate your help in this process. If you have an idea for improving this textbook, or if you find an error, a typographical mistake, or an inaccuracy in NCCER's *Contren®* textbooks, please write us, using this form or a photocopy. Be sure to include the exact module number, page number, a detailed description, and the correction, if applicable. Your input will be brought to the attention of the Technical Review Committee. Thank you for your assistance.

Instructors – If you found that additional materials were necessary in order to teach this module effectively, please let us know so that we may include them in the Equipment/Materials list in the Annotated Instructor's Guide.

Write: Product Development
National Center for Construction Education and Research
P.O. Box 141104, Gainesville, FL 32614-1104

Fax: 352-334-0932

E-mail: curriculum@nccer.org

Craft ________________ Module Name ________________________________

Copyright Date ____________ Module Number ____________ Page Number(s) ____________

Description

____

____

____

____

(Optional) Correction

____

____

____

(Optional) Your Name and Address

____

____

____

Introduction to Construction Math

00102-04

NCCER STANDARDIZED CRAFT TRAINING PROGRAM

The National Center for Construction Education and Research (NCCER) provides a standardized national program of accredited craft training. Key features of the program include instructor certification, competency-based training, and performance testing. The program provides trainees, instructors, and companies with a standard form of recognition through a National Craft Training Registry. The program is described in full in the *Guidelines for Accreditation*, published by the NCCER. For more information on standardized craft training, contact the NCCER by writing us at P.O. Box 141104, Gainesville, FL 32614-1104; calling 352-334-0911; or e-mailing info@nccer.org. More information may be found at our Web site, www.nccer.org.

HOW TO USE THIS ANNOTATED INSTRUCTOR'S GUIDE

Each page presents two sections of information. The larger section displays each page exactly as it appears in the Trainee Module. The narrow column ties suggested trainee and instructor actions to each page and provides icons (detailed below) to call your attention to material, safety, audiovisual, or testing requirements. The bottom of each page includes space for your notes.

The **Audiovisual** icon indicates an appropriate time to show a transparency or other audiovisual aid.

The **Classroom** icon prompts you to define a term, stress a point, ask trainees to explain a concept, or give examples.

The **Demonstration** icon directs you to show trainees how to perform tasks.

The **Examination** icon tells you to administer the written module examination.

The **Homework** icon is placed where you may wish to assign reading for the next class, to assign a project, or to advise trainees to prepare for an examination.

The **Laboratory** icon is used when trainees are to practice performing tasks.

The **Materials** icon is a reminder for you to gather materials needed for classes, labs, and testing.

The **Performance Testing** icon tells you to administer a performance test or a portion thereof.

The **Safety** icon is used to emphasize safety issues. It is often keyed to *Caution* and *Warning* statements in the Trainee Module.

The **Teaching Tip** icon indicates additional guidance is available, such as how to conduct an exercise, get the most educational value from a field trip, or encourage class participation. Teaching Tips may expand on a feature (*Think About It*, *Did You Know?*) or provide Quick Quizzes or similar exercises. You will be referred to the Teaching Tips section at the back of the module if there is additional material.

The **Combination** icon indicates that the laboratory listed corresponds with a performance task. If desired, you can note the proficiency of the trainees during the laboratory and use it to satisfy performance testing requirements.

PREPARATION

Before teaching this module, you should review the Objectives, Performance Tasks, Materials and Equipment List, and the Module Outline. Be sure to allow ample time to prepare your own training or lesson plan and gather all required materials and equipment.

MODULE OVERVIEW

This module introduces mathematical operations commonly used in construction and explains how the metric system and geometry are used in the trade. Trainees will learn how to add, subtract, multiply, and divide whole numbers, fractions, and decimals, as well as how to convert decimals, fractions, and percentages.

PREREQUISITES

Prior to training with this module, it is recommended that the trainee shall have successfully completed the following: *Core Curriculum: Introductory Craft Skills,* Module 00101-04

OBJECTIVES

Upon completion of this module, the trainee will be able to:

1. Add, subtract, multiply, and divide whole numbers, with and without a calculator.
2. Use a standard ruler and a metric ruler to measure.
3. Add, subtract, multiply, and divide fractions.
4. Add, subtract, multiply, and divide decimals, with and without a calculator.
5. Convert decimals to percentages and percentages to decimals.
6. Convert fractions to decimals and decimals to fractions.
7. Explain what the metric system is and how it is important in the construction trade.
8. Recognize and use metric units of length, weight, volume, and temperature.
9. Recognize some of the basic shapes used in the construction industry, and apply basic geometry to measure them.

PERFORMANCE TASKS

There is no performance testing for this module.

MATERIALS AND EQUIPMENT LIST

Transparencies
Markers/chalk
Blank acetate sheets
Transparency pens
Pencils and scratch paper
Overhead projector and screen
Whiteboard/chalkboard
Appropriate personal protective equipment
Copies of your local code
Sample work orders that require mathematical functions
60 straws
Calculator
Standard ruler (with $\frac{1}{16}$-inch markings)
Metric ruler (with centimeters [cm] and millimeters [mm])
Architect's scale
Blueprint
Machinist's rule
Protractors
Module Examinations*

*Located in the Test Booklet.

ADDITIONAL RESOURCES

This module is intended to present thorough resources for task training. The following reference works are suggested for both instructors and motivated trainees interested in further study. These are optional materials for continued education rather than for task training.

All the Math You'll Ever Need, 1999. Stephen Slavin. New York: John Wiley & Sons.

Basic Construction Math Review: A Manual of Basic Construction Mathematics for Contractor and Tradesman License Exams. Printcorp Business Printing. Construction Book Express.

Mastering Math for the Building Trades, 2000. James Gerhart. McGraw-Hill/TAB Electronics.

Math to Build On: A Book for Those Who Build, 1997. Johnny and Margaret Hamilton. Clinton, NC: Construction Trades Press.

Mathematics for the Million, 1993. Lancelot Thomas Hogben. New York: W.W. Norton & Company.

TEACHING TIME FOR THIS MODULE

An outline for use in developing your lesson plan is presented below. Note that each Roman numeral in the outline equates to one session of instruction. Each session has a suggested time period of 2½ hours. This includes 10 minutes at the beginning of each session for administrative tasks and one 10-minute break during the session. Approximately 15 hours are suggested to cover *Introduction to Construction Math.* You will need to adjust the time required for hands-on activity and testing based on your class size and resources.

Topic	Planned Time
Session I. Whole Numbers	
A. Parts of a Whole Number	______
B. Adding and Subtracting Whole Numbers	______
C. Multiplying and Dividing Whole Numbers	______
D. Using the Calculator	______
Session II. Measurements and Fractions	
A. Working with Measurements	______
B. Finding Equivalent Fractions and Reducing Fractions	______
C. Comparing Fractions and Finding the Lowest Common Denominator	______
D. Adding and Subtracting Fractions	______
E. Multiplying and Dividing Fractions	______
Session III. Decimals	
A. Reading a Machinist's Rule	______
B. Comparing Whole Numbers and Decimals	______
C. Comparing Decimals with Decimals	______
D. Adding and Subtracting Decimals	______
E. Multiplying and Dividing Decimals	______
F. Rounding Decimals	______
G. Using a Calculator	______
Session IV. Conversion Processes	
A. Converting Decimals to Percentages and Percentages to Decimals	______
B. Converting Fractions to Decimals	______
C. Converting Decimals to Fractions	______
D. Converting Inches to Decimal Equivalents in Feet	______

Session V. Introduction to the Metric System

A. Units of Weight, Length, Volume, and Temperature ____________
B. Using a Metric Ruler ____________
C. Converting Measurements ____________

Session VI. Introduction to Construction Geometry

A. Angles ____________
B. Shapes ____________
C. Area of Shapes ____________
D. Volume of Shapes ____________
E. Review ____________
F. Module Examination ____________
   1. Trainees must score 70% or higher to receive recognition from NCCER.
   2. Record the testing results on Craft Training Report Form 200, and submit the results to the Training Program Sponsor.

Introduction to
Construction Math
00102-04
Excellence in Construction Winner—
Commercial, $10–25 Million
Pinkard Construction Company built the 84,250 square foot Lakeshore Athletic Club in Broomfield, Colorado in two-thirds the time as other similar facilities. They accommodated more than $800,000 in owner-generated change orders and still completed the job by the owner's target opening date.

00102-04
Introduction to Construction Math

Topics to be presented in this module include:

Overview

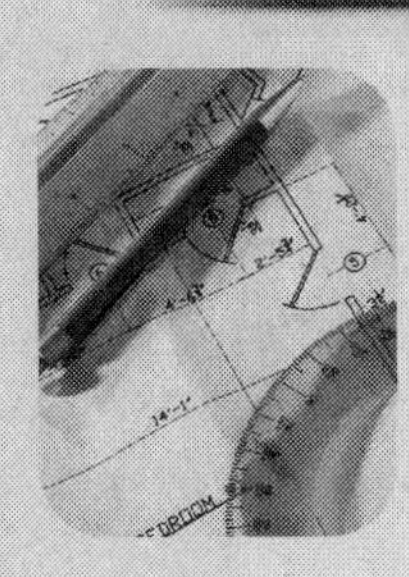

Basic math skills such as addition, subtraction, multiplication, and division are used every day in the construction industry. These skills are the foundation for more complex math skills that are used in construction such as working with fractions, percentages, decimals, and the metric system.

Math is used for tasks as simple as counting nails and as complex as calculating the dimensions of a building. All trades in the construction industry use math. For example, electricians use math to determine how much wire they need, pipefitters use math to accurately measure piping and fittings, and HVAC technicians use math to install and operate testing equipment.

Make math a priority. A good understanding of math also helps ensure safety on the site. Measurements and calculations must be accurate to avoid material and equipment failure, both of which can result in accidents and injuries. Take the time to do the math required for your job carefully and correctly.

Instructor's Notes:

Objectives

When you have completed this module, you will be able to do the following:

1. Add, subtract, multiply, and divide whole numbers, with and without a calculator.
2. Use a standard ruler and a metric ruler to measure.
3. Add, subtract, multiply, and divide fractions.
4. Add, subtract, multiply, and divide decimals, with and without a calculator.
5. Convert decimals to percentages and percentages to decimals.
6. Convert fractions to decimals and decimals to fractions.
7. Explain what the metric system is and how it is important in the construction trade.
8. Recognize and use metric units of length, weight, volume, and temperature.
9. Recognize some of the basic shapes used in the construction industry and apply basic geometry to measure them.

Key Trade Terms

Acute angle	Meter
Adjacent angles	Metric ruler
Angle	Mixed number
Area	Negative numbers
Bisect	Numerator
Borrow	Obtuse angle
Carry	Opposite angles
Circle	Percent
Circumference	Perimeter
Convert	Pi
Cubic	Place value
Decimal	Positive numbers
Degree	Radius
Denominator	Rectangle
Diagonal	Remainder
Diameter	Right angle
Difference	Right triangle
Digit	Scalene triangle
English ruler	Square
Equilateral triangle	Standard ruler
Equivalent fractions	Straight angle
Formula	Sum
Fraction	Triangle
Improper fraction	Vertex
Invert	Volume
Isosceles triangle	Whole numbers
Long division	
Machinist's rule	

Required Trainee Materials

1. Sharpened pencils and paper
2. Standard ruler (with 1/16-inch markings)
3. Metric ruler (with centimeters [cm] and millimeters [mm])
4. Machinist's rule
5. Calculator

Prerequisites

Before you begin this module, it is recommended that you successfully complete the following: *Core Curriculum: Introductory Craft Skills*, Module 00101-04.

This course map shows all of the modules in *Core Curriculum: Introductory Craft Skills*. The suggested training order begins at the bottom and proceeds up. Skill levels increase as you advance on the course map. The local Training Program Sponsor may adjust the training order.

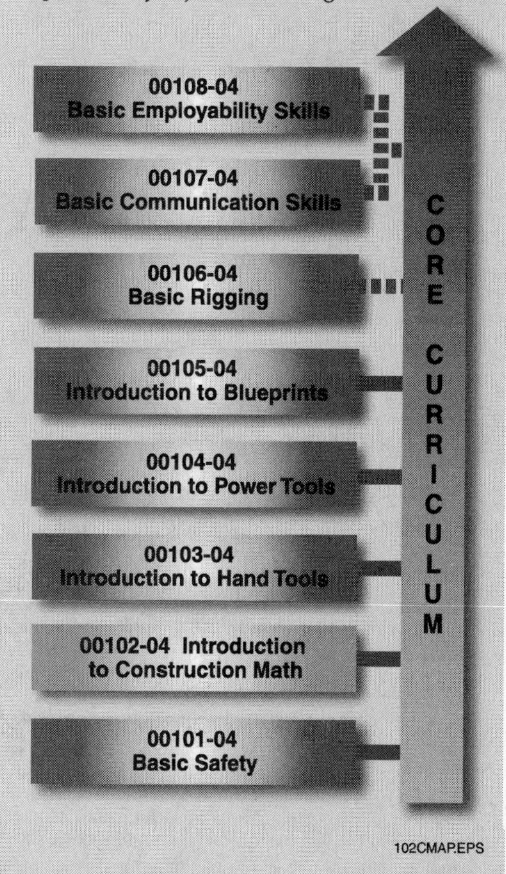

102CMAP.EPS

Ensure that you have everything required to teach the course. Check the Materials and Equipment List at the front of this module.

See the general Teaching Tip at the end of this module.

Show Transparency 1, Course Objectives.

Explain that terms shown in bold (blue) are defined in the Glossary at the back of this module.

MODULE 00102-04 ◆ INTRODUCTION TO CONSTRUCTION MATH 2.1

Review the types of mathematical procedures that trainees will use in the construction industry. Explain that all math for this section must be completed without a calculator.

Explain that whole numbers are numbers without fractions or decimals. On the whiteboard/chalkboard, write a variety of whole numbers, fractions, and decimals. Have trainees identify the whole numbers.

Refer to Figure 1. Explain each whole number's place value. On the whiteboard/chalkboard, write six-, seven-, and eight-digit whole numbers. Ask trainees to identify each number's place value.

On the whiteboard/chalkboard, draw a crossbar, and mark the center point with a zero. Write the positive numbers 1, 2, and 3 above the horizontal line in ascending order. Write the negative numbers −1, −2, and −3 below the horizontal line in descending order. Explain that negative numbers are used to represent values less than zero. Provide examples from your experience in which negative numbers are used in construction.

1.0.0 ◆ INTRODUCTION

Remember back in math class when the teacher said, "You will use this some day"? Today is the day! In the construction trades, workers use math all the time. Plumbers use math to calculate pipe length, read plans, and lay out fixtures. Carpenters use math to lay out floor systems and frame walls and ceilings. When you measure a length of material, fill a container with a specified amount of liquid, or calculate the dimensions of a room, you are using mathematical operations. This module provides practice in some basic mathematical procedures and gives you an idea of how they might apply in the construction industry. It also introduces you to some construction careers and their use of mathematics.

NOTE

You must do the math problems without using a calculator, except when the text specifically calls for calculator use or when you need to check the answers to your problems.

2.0.0 ◆ WHOLE NUMBERS

In this section, you will learn how to work with **whole numbers.** Whole numbers are complete units without **fractions** or **decimals.**

The following are whole numbers:

1 5 67 335 2,654

The following are *not* whole numbers:

½ ¾ 7⅛ 0.45 4.25

In this section, you will work only with whole numbers. Later, you will work with fractions and decimals.

2.1.0 Parts of a Whole Number

Let's look at the parts of a whole number. The whole number shown in *Figure 1* has seven **digits.** A digit is any of the numerical symbols from 0 to 9. The seven digits in *Figure 1* are the numbers shown across the top of that figure. If you read this seven-digit whole number out loud, you would say "five million, three hundred sixteen thousand, two hundred forty-seven."

Each of this whole number's seven digits represents a **place value.** Each digit has a value that depends on its place, or location, in the whole number. In this whole number, for example, the place value of the 5 is five million, and the place value of the 2 is two hundred.

Numbers larger than zero are called **positive (+) numbers** (such as 1, 2, 3). Numbers less than zero are **negative (−) numbers** (such as −1, −2, −3). Zero is neither positive nor negative. Except for zero, all numbers without a minus sign in front of them are positive.

Don't forget that some whole numbers may contain the digit zero—for instance, the whole number 7,093 has a zero in the hundreds place. When you read that number out loud, you would say "seven thousand ninety-three."

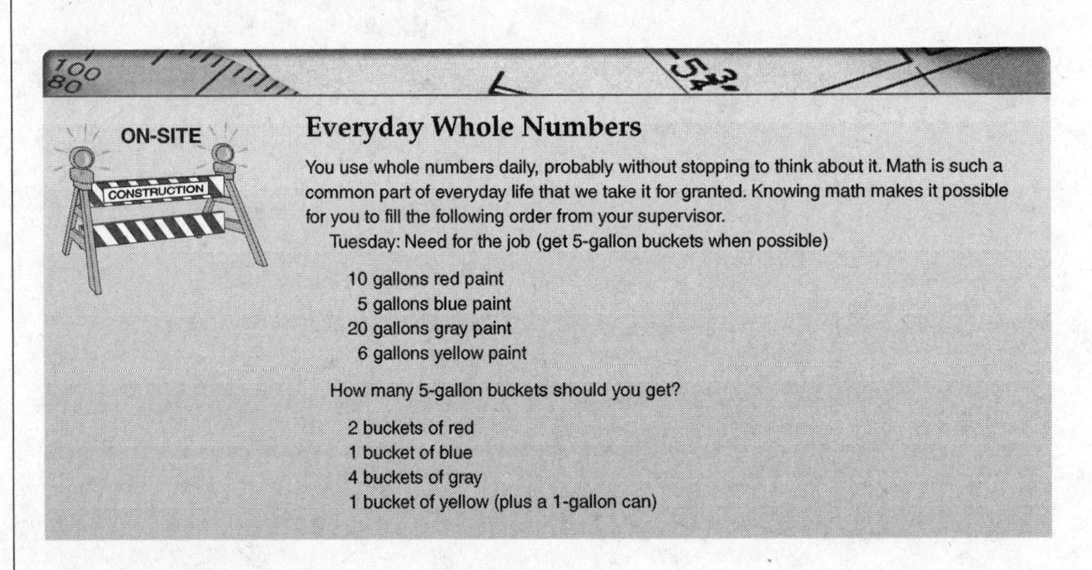

Everyday Whole Numbers

You use whole numbers daily, probably without stopping to think about it. Math is such a common part of everyday life that we take it for granted. Knowing math makes it possible for you to fill the following order from your supervisor.

Tuesday: Need for the job (get 5-gallon buckets when possible)

10 gallons red paint
5 gallons blue paint
20 gallons gray paint
6 gallons yellow paint

How many 5-gallon buckets should you get?

2 buckets of red
1 bucket of blue
4 buckets of gray
1 bucket of yellow (plus a 1-gallon can)

2.2 CORE CURRICULUM ◆ INTRODUCTORY CRAFT SKILLS

Instructor's Notes:

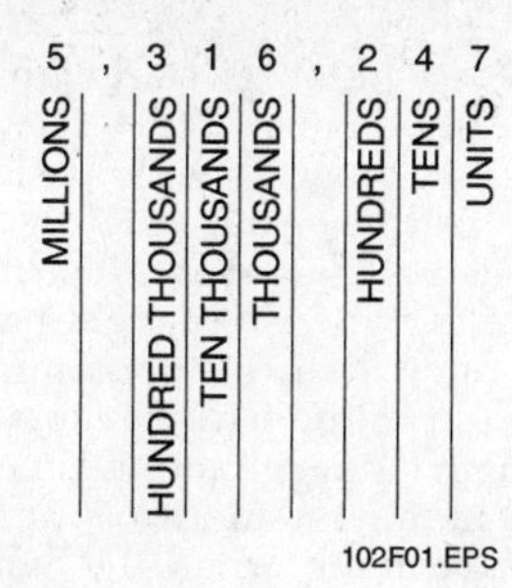

Figure 1 ◆ Place values.

2.1.1 Study Problems: Parts of a Whole Number

1. Look at this description of a number:

Digit in the units place:	4
Digit in the tens place:	6
Digit in the hundreds place:	9
Digit in the thousands place:	3

   This number would be written as _____.

   a. 3,964
   b. 4,693
   c. 30,964
   d. 39,064

2. In the number 25,718, the numeral 5 is in the _____ place.

   a. tens
   b. thousands
   c. units
   d. hundreds

3. The number for the word *eighty-five* is _____.

   a. 58
   b. 85
   c. 508
   d. 805

4. The number for the words *one hundred twenty-two* is _____.

   a. 122
   b. 212
   c. 221
   d. 1,022

5. The number for the words *two thousand four hundred ninety-seven* is _____.

   a. 2,079
   b. 2,479
   c. 2,497
   d. 4,297

2.2.0 Adding Whole Numbers

To add means to combine the values of two or more numbers together. When you add two or more numbers together, the total is called the **sum.** The sign for addition is the plus sign (+). Addition problems can be written vertically (up to down) or horizontally (left to right). Here are two examples:

$$\begin{array}{r} 6 \\ +\,3 \\ \hline 9 \end{array} \quad 6 + 3 = 9 \qquad \begin{array}{r} 5 \\ +\,2 \\ \hline 7 \end{array} \quad 5 + 2 = 7$$

In these examples, the sum of 6 + 3 is 9. The sum of 5 + 2 is 7.

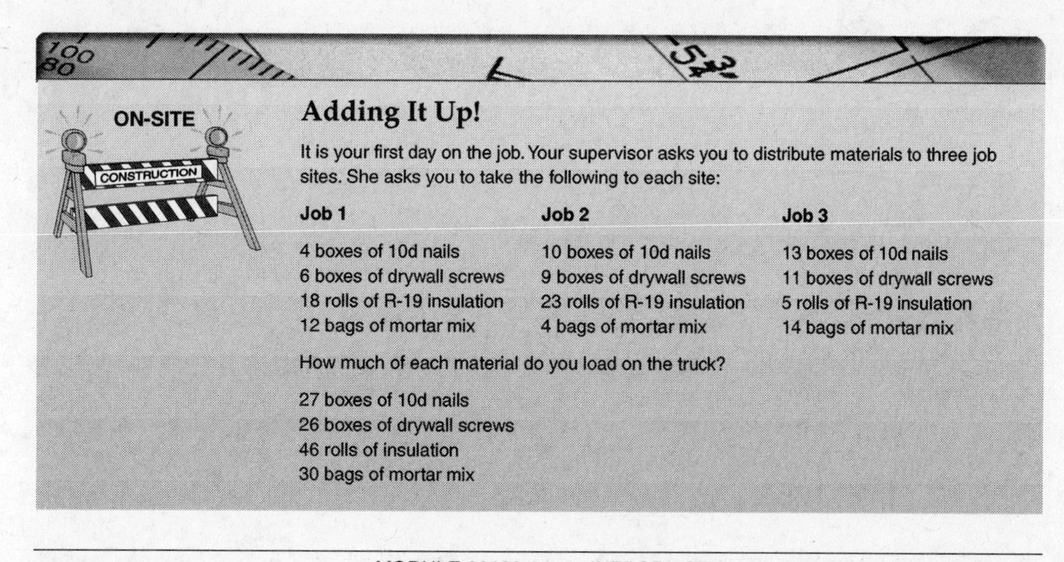

Adding It Up!

It is your first day on the job. Your supervisor asks you to distribute materials to three job sites. She asks you to take the following to each site:

Job 1	Job 2	Job 3
4 boxes of 10d nails	10 boxes of 10d nails	13 boxes of 10d nails
6 boxes of drywall screws	9 boxes of drywall screws	11 boxes of drywall screws
18 rolls of R-19 insulation	23 rolls of R-19 insulation	5 rolls of R-19 insulation
12 bags of mortar mix	4 bags of mortar mix	14 bags of mortar mix

How much of each material do you load on the truck?

27 boxes of 10d nails
26 boxes of drywall screws
46 rolls of insulation
30 bags of mortar mix

Ask trainees to complete the study problems for parts of a whole number. Review trainees' answers, and answer any questions trainees may have.

Using the example in the "Adding It Up!" On-Site, explain how to add the values of two or more numbers to find the sum.

MODULE 00102-04 ◆ INTRODUCTION TO CONSTRUCTION MATH 2.3

On the whiteboard/chalkboard, write the example 58 + 34. Demonstrate how to carry the digit in the tens place to the top of the tens column.

Ask trainees to complete the study problems for adding whole numbers. Go over any questions trainees may have.

2.2.1 Carrying in Addition

Adding single numbers probably seems pretty easy at this point. But what happens when you want to add larger numbers, such as 58 + 34? You still add up single numbers, but you do it in a certain order, and you may need to **carry** (see *Figure 2*) the tens part of a sum from the units column to the tens column and add it there.

Look at the following example:

58 pounds of nails
+ 34 pounds of nails

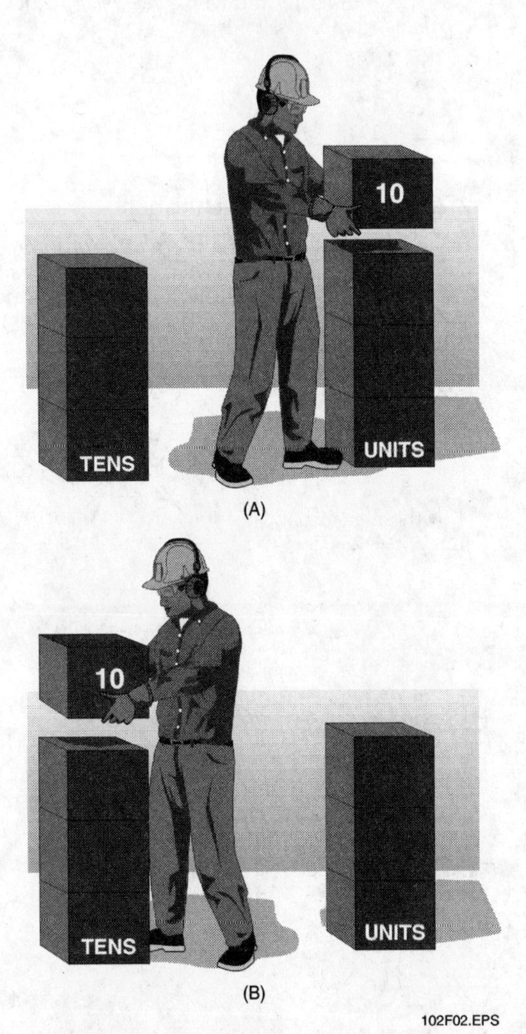

Figure 2 ◆ Carrying in addition.

To add these numbers, start with the two digits on the right-hand side (8 + 4), which add up to 12. But you don't write down 12 in the sum line at the bottom—you write down just the 2, which is the units amount. (Remember that the units place is the farthest right-hand column.) Then you carry the 1 (the tens amount from the 12) to the top of the tens column, which is the 5 + 3 column. That column then becomes 1 + 5 + 3. Those numbers add up to 9, and you put the 9 below that column, where the sum goes. Using 58 + 34:

Think	Write
5 tens + 8 units	58 *(1 ten carried to*
+ 3 tens + 4 units	+ 34 *tens column,*
(8 tens) + (1 ten + 2 units)	92 *2 units remaining in units column)*

So 58 + 34 = 92. You can use the same techniques for carrying when you are adding larger numbers. If a column adds to 10 or more, put only the right-hand digit below that column and carry the other digit to the next column to the left. If you have a total of 13 in the hundreds column, for example, put the 3 at the bottom of that column and carry the 1 to the thousands column.

2.2.2 Study Problems: Adding Whole Numbers

Solve the following addition problems on a separate piece of paper without using a calculator. Remember to show all your work.

1. 32 pounds of nails
+75 pounds of nails

2. 73 in. of molding
+45 in. of molding

3. 83 ft of cable
+53 ft of cable

4. 452 ft of baseboard
+74 ft of baseboard

5. 323 yds of binding
+758 yds of binding

Instructor's Notes:

2.4 CORE CURRICULUM ◆ INTRODUCTORY CRAFT SKILLS

2.3.0 Subtracting Whole Numbers

Subtraction means finding the **difference** between two numbers, or taking away one number from another. The subtraction sign (−) is also called the minus sign. The result (answer) of a subtraction problem is called the difference.

For example, you have a total of nine sockets to install today. You have installed five so far. How many more do you have to install today?

9	sockets to install today
−5	you've installed so far
4	sockets left to install

In some subtraction problems, you may have to subtract a larger digit from a smaller digit, such as in the following problem:

$$\begin{array}{r} 76 \\ -48 \\ \hline \end{array}$$

As with adding, when you are subtracting, you start with the units column, which is the right-hand side. In the units column in this problem, you have to subtract a larger number, 8, from a smaller number, 6. How can you do this? **Borrow** from the tens place.

Problem	Think	Write
76	7 tens + 6 units	76
−48	−4 tens + 8 units	−48

Step 1 Notice that there are not enough units to subtract from (you cannot subtract 8 from 6).

Step 2 Borrow 1 ten from the tens column.

(6 tens + 1 ten) +	(6 units)	76
−4 tens	+ 8 units	−48

Step 3 Add the borrowed 10 to the 6 in the units column (10 + 6 = 16), so that you have the larger number (16) you need there. Visualize a two-digit number in the units column. In the tens column, you are now left with 6 tens.

6 tens +	16 units	76
−4 tens	8 units	−48

Step 4 You now have enough in both columns to subtract from.

6 tens +	16 units	76
−4 tens	8 units	−48
2 tens	8 units	28

2.3.1 Study Problems: Subtracting Whole Numbers

Solve the following subtraction problems on a separate piece of paper without using a calculator. Remember to show all work.

1. 87 sockets
   −38 sockets

2. 26 connectors
   −17 connectors

3. 92 hours
   −34 hours

4. 246 ft of cable
   −18 ft of cable

5. 826 bricks
   −717 bricks

2.4.0 Multiplying Simple Whole Numbers

You have eight construction sites. You need to send four wrenches to each site. How many wrenches will you need in all? You could count out four wrenches eight times, put them in eight piles, and then count them all. *Figure 3* shows how you might do this.

DID YOU KNOW?

Numeral Systems

People in ancient Egypt, Babylon, Greece, and Rome developed different numeral systems or ways of writing numbers. Some of these early systems were very complex and difficult to use. A new numeral system came into use around 750 C.E. Originally developed by the Hindus in India, the system was spread by Arab traders. The Hindu-Arabic system uses only ten symbols—1, 2, 3, 4, 5, 6, 7, 8, 9, 0—and is still in use today. These ten symbols (also called numerals or digits) can be combined to write any number.

Explain how to borrow from the tens place when solving a subtraction problem.

On the whiteboard/chalkboard, write the steps to solve a subtraction problem. Demonstrate how to borrow from the tens column.

Ask trainees to complete the study problems for subtracting whole numbers. Ask whether trainees have any questions.

Show Transparency 2 (Figure 3). Using the example in Figure 3, ask a trainee to explain how to count out the wrenches needed at the construction site.

MODULE 00102-04 ◆ INTRODUCTION TO CONSTRUCTION MATH 2.5

See the Teaching Tip for Section 2.4.0 at the end of this module.

Ask trainees to complete the study problems for multiplying simple whole numbers. Answer any questions trainees may have.

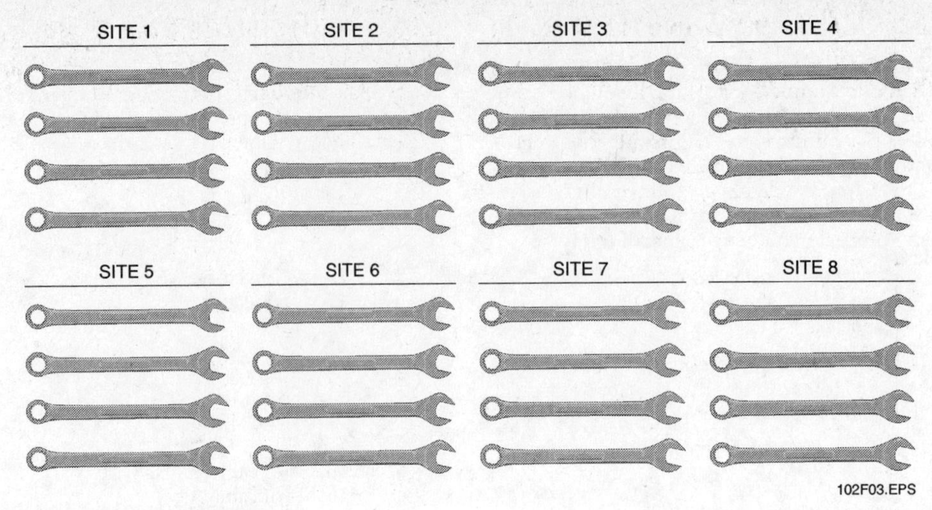

Figure 3 ◆ Counting out the wrenches.

Obviously, this is the hard way to add. It is much easier to figure out problems like this using multiplication. To do that, you will want to learn the times tables.

Multiplication is the quick way to add the same number together many times. It would be much easier, and more efficient, to multiply 4 times 8 than to add 4 eight times. In this example, you use multiplication to figure out 8 (construction sites) multiplied by 4 (wrenches), or 8 times 4.

The symbol for multiplication is the × sign. In this example, the operation would be written in either of the following two ways:

8 construction sites
×4 wrenches
32 wrenches needed

$4 \times 8 = 32$

The answer is 32. To work successfully in the construction industry, it is very helpful to know the answer to the most basic multiplication equations without having to write them out or think too hard about them. It is a good idea to spend some time memorizing the most basic times table or multiplication table, which means knowing automatically that, for example, $2 \times 4 = 8$, $5 \times 9 = 45$, and $12 \times 12 = 144$. *Appendix A* is a basic multiplication table that will help you.

2.4.1 Study Problems: Multiplying Simple Whole Number

1. Adding 4 + 4 + 4 + 4 + 4 can be simplified by multiplying _____.
   a. 4 × 5
   b. 4 × 4
   c. 12 × 8
   d. 20 × 4

Solve the following multiplication problems on a separate piece of paper without using a calculator. Remember to show your work.

2. 9 doors installed by
   ×8 workers

3. 9 bolts in
   ×6 frames

4. 7 fixtures
   ×2 plumbers

5. 8 beams hung by
   ×4 carpenters

Instructor's Notes:

2.6 CORE CURRICULUM ◆ INTRODUCTORY CRAFT SKILLS

2.4.2 *Multiplying Larger Whole Numbers*

These multiplication problems may have been easy, but what happens when you have a problem involving larger numbers, such as the following example? On a job, a pipefitter installs 75 feet of pipe in one hour. How many feet of pipe can this pipefitter install in 16 hours?

To solve that problem, you need to multiply 75 by 16. You do this by going through these steps:

Step 1 Write the numbers on two lines. As you did with addition and subtraction, place units under units, and tens under tens.

$$\begin{array}{r} 75 \\ \underline{\times\ 16} \end{array}$$

Step 2 Start with the digit in the units place of the bottom number (the 6 in the 16).

$$\begin{array}{r} 75 \\ \underline{\times\ 16} \end{array}$$

Step 3 Multiply every number in the top number (the 75) by the number in the units place of the bottom number (the 6), starting on the right with the units place of the top number (the 5).

carry the

$$\begin{array}{r} 3 \\ 75 \\ \underline{\times\ 16} \\ 450 \end{array}$$

6 times 5 is 30, or 3 tens and 0 units. Place the 0 in the units place. Carry the 3 tens to the tens place.

6 times 7 is 42, or 4 hundreds and 2 tens. Add the 3 tens that you carried, which gives you 45.

Step 4 Now multiply every individual number in the top number (the 75) by the number in the tens place of the bottom number (the 1). Start with the number in the units place (on the right) of the top number (the 5).

$$\begin{array}{r} 75 \\ \underline{\times\ 16} \\ 450 \\ 75 \end{array}$$

1 ten times 5 units = 5 tens. Place a 5 in the tens place.

1 ten times 7 tens = 7 hundreds. Place a 7 in the hundreds place.

Step 5 Now add the numbers in each column, beginning with the units place.

$$\begin{array}{r} 75 \\ \underline{\times\ 16} \\ 450 \\ \underline{+75} \\ 1{,}200 \end{array}$$

The answer to this multiplication problem is 1,200. In solving this problem, you have found that the pipefitter can install 1,200 feet of pipe in 16 hours.

2.4.3 *Study Problems: Multiplying Larger Whole Numbers*

Solve the following multiplication problems on a separate piece of paper. Remember to show all your work:

1. 12 wheelbarrows at
   ×21 sites

2. 11 ladders at
   ×15 sites

3. 30 barrels at
   ×25 sites

4. 452 square feet used at
   ×4 sites

5. 162 bricks at
   ×52 sites

2.5.0 Dividing Whole Numbers

Division is the opposite of multiplication. Instead of adding a number several times (5 + 5 + 5 = 15; 5 × 3 = 15), when you are dividing you subtract a number several times. But you can solve a problem much faster by using division instead of subtracting the same number over and over.

For example, you have 60 wrenches to distribute equally among 10 construction sites, and you need to find out how many wrenches should go to each site. You could make 10 piles (one for each site) and then count the number of wrenches in each pile. But it would be much easier and more efficient to divide 10 into 60. (Another way to state this is that you divide 60 by 10. Or you can say you are going to see how many times 10 goes into 60.) The answer, 6, would be the number of wrenches to send to each site.

Review the procedures to solve multiplication problems using larger whole numbers.

Have trainees complete the study problems for multiplying larger whole numbers. Ask whether trainees have any questions.

Review the steps to solve a division problem.

Bring in 60 straws. Using the straws, demonstrate how to divide by making piles and counting the number of straws in each pile.

MODULE 00102-04 ◆ INTRODUCTION TO CONSTRUCTION MATH 2.7

Ask trainees to complete the study problems for dividing whole numbers. Go over any questions trainees may have.

Review the procedure to solve complicated division problems using long division.

In this easy example, the numbers came out as whole numbers—60 divided by 10 equals 6. Some problems don't come out evenly, such as this one: You have 10 feet of cable and are asked to cut it into 4-foot pieces. How many 4-foot pieces will you have? How much of the 10-foot cable will be left over? (Note that the symbol ' is often used instead of the word *foot*, or *feet*, when you are writing out mathematical operations. This means that 4 feet will appear as 4' and 10 feet will appear as 10'.)

Step 1 A division problem is written like this:

$$4\overline{)10}$$

Four goes into 10 two times. Place a 2 on the line above the 10.

$$\begin{array}{r} 2 \\ 4\overline{)10} \end{array}$$

Step 2 Multiply 2 times 4. Place the answer, 8, under the 10.

$$\begin{array}{r} 2 \\ 4\overline{)10} \\ 8 \end{array}$$

Step 3 Subtract 8 from 10. The answer is 2. This is called the **remainder,** or the amount left over. The answer to this problem would be written 2 *r*2.

By dividing, you have found that you will have two 4-foot pieces of cable, with 2 feet of cable left over.

2.5.1 *Study Problems: Dividing Whole Numbers*

Solve the following division problems on a separate piece of paper without using a calculator. Include the remainder if there is one. Remember to show all your work.

1. $15 \div 3$

2. $36 \div 4$

3. $54 \div 5$

4. $6\overline{)17}$

5. $7\overline{)39}$

2.5.2 *Dividing More Complex Whole Numbers*

Now let's take a look at solving a more complicated division problem: Twenty-four people were a part of a lottery pool that just won $2,638 (after taxes). They decided to split it up evenly among themselves and then give whatever was left over (the remainder) to charity. How much did each person get? How much went to charity?

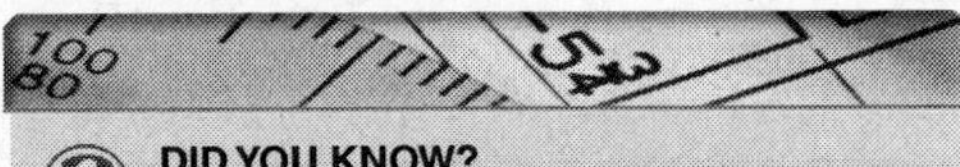

DID YOU KNOW?

In a division problem, the number you are dividing by is called the divisor, and the number being divided is the dividend. In the first example in this section, the number 10 is the divisor and the number 60 is the dividend. In the second example, 4 is the divisor and 10 is the dividend.

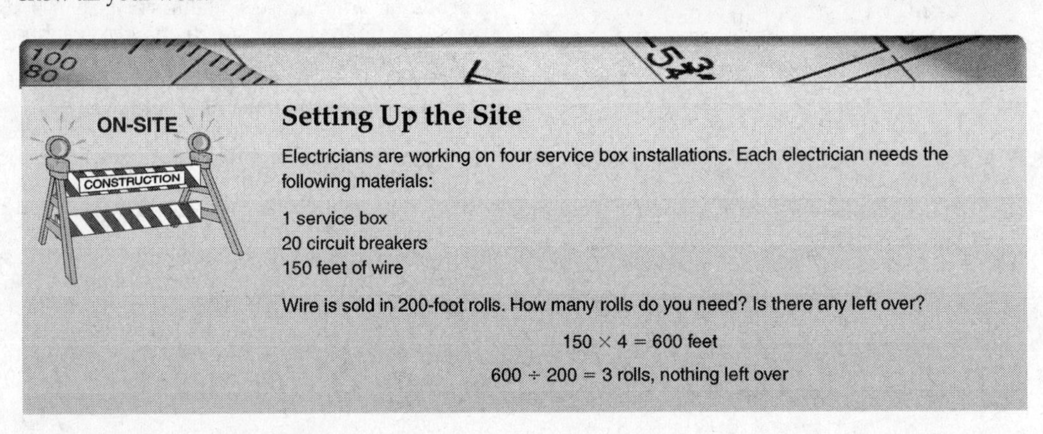

ON-SITE

Setting Up the Site

Electricians are working on four service box installations. Each electrician needs the following materials:

1 service box
20 circuit breakers
150 feet of wire

Wire is sold in 200-foot rolls. How many rolls do you need? Is there any left over?

$150 \times 4 = 600$ feet

$600 \div 200 = 3$ rolls, nothing left over

Instructor's Notes:

2.8 CORE CURRICULUM ◆ INTRODUCTORY CRAFT SKILLS

To solve this problem, you would use what is called **long division.**

The problem is $2,638 divided by 24 people.

Step 1 Estimate the answer. 24 is close to 25. 2,638 is close to 2,500. 25 goes into 2,500 one hundred times. Estimate: approximately $100 for each person.

Step 2 Now do the actual division, using the following steps:

24 goes into 26 one time, with a remainder of 2.

```
      1??
24 ) 2,638
    -24
     02
```

Bring down the next number, 3. Can 24 go into 23? It cannot, so put a 0 in the answer line at the top.

```
      10?
24 ) 2,638
    -24
     023
     000
```

Subtract 0 from 23. The answer is 23. Bring down the next number, the 8.

```
      10?
24 ) 2,638
    -24
     023
     000
     0238
```

How many times can 24 go into 238? To figure this, think: 24 is close to 25. 238 is close to 250. How many times can 25 go into 250? 10. 10 times 24 is 240, so that's too high. Let's try 9. 9 times 24 is 216. Subtract 216 from 238. The remainder is 22.

```
      109
24 ) 2,638
    -24
     023
     000
     0238
    -0216
     0022
```

So how much money did each of the 24 people get? $109 each. How much went to charity? $22. (Note that the answer, $109, is close to your estimate of $100.)

2.5.3 *Study Problems: Dividing More Complex Whole Numbers*

Solve the following division problems on a separate piece of paper without using a calculator. Include the remainder if there is one. Remember to show all your work.

1. 12) 263

2. 16) 4218

3. 15) 4532

4. Your supervisor sends you to the truck for 150 feet of electrical wire. When you get there, you find that all the coils of wire come in 15-foot lengths. You will need to bring back _____ coils of wire.
 a. 3
 b. 5
 c. 10
 d. 100

5. A plumbing job requires 100 feet of plastic pipe available in 20-foot sections. You will need _____ sections.
 a. 5
 b. 6
 c. 10
 d. 12

2.6.0 Using the Calculator to Add, Subtract, Multiply, and Divide Whole Numbers

It is important to be able to perform calculations (such as addition and subtraction) in your head even if you have a calculator. It allows you to estimate the answer before you use the calculator. Why is this important? So you can double-check the answer against your estimate. If you press a wrong key on the calculator, you might be out hundreds of dollars or off by several inches.

The calculator is a marvelous tool for saving time. Let's look at the most frequently used operations of the calculator: adding, subtracting, multiplying, and dividing whole numbers. *Figure 4* shows the parts of a common calculator.

Ask trainees to complete the study problems for dividing more complex whole numbers. Assign each of the problems to a different trainee, and have them show their work on the whiteboard/chalkboard. Review trainees' answers, and have trainees make corrections as needed.

Refer to Figure 4. Ask trainees to take out their calculators. Demonstrate how to identify and use the functions of a calculator.

MODULE 00102-04 ◆ INTRODUCTION TO CONSTRUCTION MATH 2.9

Ask trainees to complete the study problems for using the calculator to add, subtract, and multiply. Go over any questions trainees may have.

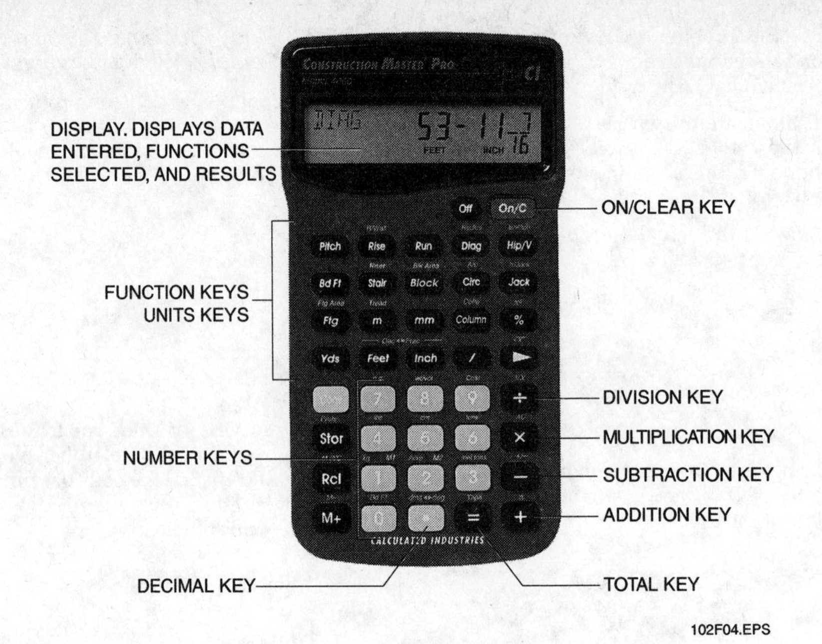

Figure 4 ◆ Parts of a calculator.

2.6.1 Using the Calculator to Add Whole Numbers

Adding numbers is easy with a calculator. Just follow these steps, using 5 + 4 to practice.

Step 1 Turn the calculator on. A zero (0) appears in the display.

Step 2 Press 5. A 5 appears in the display.

Step 3 Press the + key. The 5 is still displayed.

Step 4 Press the 4 key. A 4 is displayed.

Step 5 Press the = key. The sum, 9, appears in the display.

Step 6 Press the ON/C key to clear the calculator.

2.6.2 Study Problems: Using the Calculator to Add

Practice using your calculator by finding the answers to these addition problems:

1. 12 (press +)
24 (press +)
+33 (press =)

2. 67
46
+96

3. 83
35
+50

4. 34
938
24
+63

5. 67
774
983
+532

2.6.3 Using the Calculator to Subtract Whole Numbers

Subtracting with a calculator is as easy as adding. Here are the steps, using the problem 25 − 5 to practice.

Step 1 Turn the calculator on. A zero (0) appears in the display.

Step 2 Press the 2 key and then the 5 key. A 25 appears in the display.

Step 3 Press the − key. The 25 is still displayed.

Step 4 Press the 5 key. A 5 is displayed.

Step 5 Press the = key. The difference, 20, appears in the display.

Step 6 Press the ON/C key to clear the calculator.

Instructor's Notes:

2.10 CORE CURRICULUM ◆ INTRODUCTORY CRAFT SKILLS

2.6.4 Study Problems: Using the Calculator to Subtract

Practice using your calculator by finding the answers to these subtraction problems.

1. 97 − 5

2. 452 − 414

3. 1,254 − 557

4. 4,593 − 4,247

5. At 2:00 P.M., the thermometer showed 99°F. At 7:00 P.M., it showed 75°F. The temperature fell by _____ degrees between 2:00 P.M. and 7:00 P.M.

2.6.5 Using the Calculator to Multiply Whole Numbers

Multiplying with a calculator is as easy as adding and subtracting. Here are the steps, using the problem 6 × 5 to practice.

Step 1 Turn the calculator on. A zero (0) appears in the display.

Step 2 Press 6. A 6 appears in the display.

Step 3 Press the × key. The 6 is still displayed.

Step 4 Press the 5 key. A 5 is displayed.

Step 5 Press the = key. The answer, 30, appears in the display.

Step 6 Press the ON/C key to clear the calculator.

2.6.6 Study Problems: Using the Calculator to Multiply

Practice using your calculator by finding the answers to these multiplication problems:

1. 1,254 × 57

2. 943 × 93

3. 4,593 × 47

4. A machine produces 465 screws in one hour. It will produce _____ screws in 16 hours.
   a. 5,500
   b. 7,440
   c. 46,500
   d. 465,000

5. There are 12 inches to a foot. There are _____ inches in 18 feet.
   a. 122
   b. 216
   c. 220
   d. 236

2.6.7 Using the Calculator to Divide Whole Numbers

Dividing with a calculator is as easy as the other operations. Here are the steps, using 12 ÷ 4 to practice:

Step 1 Turn the calculator on. A zero (0) appears in the display.

Step 2 Press the 1 key and then the 2 key. A 12 appears in the display.

Step 3 Press the ÷ key. The 12 is still displayed.

Step 4 Press the 4 key. A 4 is displayed.

Step 5 Press the = key. The answer, 3, appears in the display.

Step 6 Press the ON/C key to clear the calculator.

2.6.8 Expressing a Remainder as a Whole Number

When one number does not go into another number evenly, you are left with a remainder. For example, use your calculator to figure this problem: There is a piece of wood 6 feet long. How many 4-foot pieces can a worker cut from it? How many feet will be left over?

Step 1 Turn the calculator on. A zero (0) appears in the display.

Step 2 Press 6. A 6 appears in the display.

Step 3 Press the ÷ key. The 6 is still displayed.

Step 4 Press the 4 key. A 4 is displayed.

Step 5 Press the = key.

Step 6 The total, 1.5, appears in the display (.5 is a decimal, a part of a number, represented by digits to the right of a point called the decimal point).

Step 7 Press the ON/C key to clear the calculator.

Step 8 To express the .5 as a whole number rather than a decimal, multiply it by the number you divided by (4). The remainder expressed as a whole number is 2 feet (see *Figure 5*).

The answer to the problem, "How many 4-foot pieces can a worker cut from a 6-foot piece of wood and how many feet will be left over?" is one piece with 2 feet left over.

Explain how to divide using a calculator, and review the procedure to express a remainder as a whole number.

Using the example in Figure 5, demonstrate how .5 is converted to a whole number.

Ask trainees to complete the study problems for using the calculator to divide. Ask whether trainees have any questions.

See the Teaching Tip for Sections 2.0.0–2.6.9 at the end of this module.

Ask trainees to complete the Review Questions for Section 2.0.0 and to show their work. Have trainees review Sections 3.0.0–4.7.1 for the next session.

Ensure that you have everything required for demonstrations and laboratories during this session.

Go over the Review Questions for Section 2.0.0. Answer any questions trainees may have.

DECIMAL REMAINDER — NUMBER DIVIDED BY — WHOLE NUMBER REMAINDER

$0.5 \times 4 = 2$

102F05.EPS

Figure 5 ◆ From decimal remainder to whole number remainder.

2.6.9 Study Problems: Using the Calculator to Divide

Practice using your calculator by finding the answers to these division problems. Include the remainder if there is one.

1. $20 \div 5 =$ ____
2. $54 \div 9 =$ ____
3. $16 \div 3 =$ ____
4. You have 100 gallons of liquid. You need to put the liquid into 20-gallon containers. You can fill ____ 20-gallon containers.
   a. 5
   b. 6
   c. 10
   d. 20
5. You need to cut two 3-foot lengths out of a board 8 feet long. You will have a piece ____ feet long left over when you are done.
   a. 2
   b. 2.5
   c. 2.9
   d. 3

Review Questions

Section 2.0.0

1. In the number 37,945, the numeral 7 is in the ____ place.
   a. units
   b. tens
   c. hundreds
   d. thousands
2. The number for the words *two thousand six hundred eighty-nine* is ____.
   a. 2,286
   b. 2,689
   c. 6,289
   d. 20,689

Solve problems 3–15 on a separate piece of paper without using a calculator.

3. A project has 8 workers on one job, 35 on another, and 18 on a third. The total number of people working on all three jobs is ____.
   a. 53
   b. 61
   c. 71
   d. 133
4. A bricklayer lays 649 bricks the first day, 632 the second day, and 478 the third day. The bricklayer laid ____ bricks in three days.
   a. 1,759
   b. 1,760
   c. 1,769
   d. 1,770
5. Four walls of a bathroom require 31, 46, 49, and 16 tiles. You will need ____ tiles to tile all four walls.
   a. 132
   b. 142
   c. 144
   d. 152
6. For eight different jobs, you need to fill a total of eight boxes with different numbers of a specified size of brightwood screw. The numbers of screws you need to put in the boxes are 142, 57, 35, 79, 32, 79, 53, and 95. You will need a total of ____ screws to fill all eight boxes.
   a. 562
   b. 572
   c. 582
   d. 592
7. You have already used 36,000 bricks, and your supervisor orders 1,500 more to complete the job. The job requires a total of ____ bricks.
   a. 35,500
   b. 36,500
   c. 37,000
   d. 37,500

Instructor's Notes:

2.12 CORE CURRICULUM ◆ INTRODUCTORY CRAFT SKILLS

Review Questions

8. You have a piece of wood that measures 86 inches. You cut off a 12-inch piece. The remaining piece of wood measures _____ inches.
   a. 70
   b. 74
   c. 76
   d. 78

9. Yesterday, a supply yard contained 93 tons of sand. Since then, 27 tons were removed. The supply yard now contains _____ tons of sand.
   a. 66
   b. 76
   c. 79
   d. 86

10. A total of 1,478 feet of cable was supplied for a job. Only 489 feet were installed. There are _____ feet of cable remaining.
    a. 978
    b. 980
    c. 989
    d. 1,099

11. A roof requires 11 trusses. A company is building 8 roofs. The company will need to order _____ trusses.
    a. 78
    b. 80
    c. 88
    d. 98

12. A worker has been asked to deliver 15 scaffolds to each of 26 sites. The worker will deliver a total of _____ scaffolds.
    a. 120
    b. 240
    c. 390
    d. 420

13. If cast-iron pipe weighs 8 pounds per foot, 65 feet weighs _____ pounds.
    a. 280
    b. 320
    c. 430
    d. 520

14. You have 5,814 feet of rope. You want to cut it into 27-foot sections. You will be able to cut _____ sections and will have _____ feet left over.
    a. 215; 9
    b. 218; 3
    c. 220; 6
    d. 222; 2

15. If tubing costs $4.00 a foot and you pay a total of $448.00, you have purchased _____ whole feet of tubing.
    a. 40
    b. 80
    c. 112
    d. 122

Use your calculator to find the answers to problems 16–20.

16. You need to lay a line of bricks that is 96 inches long. Each brick is 4 inches long. You will need _____ bricks to build the wall.
    a. 12
    b. 18
    c. 24
    d. 30

17. Your company is working on three projects. Project A requires 72 rolls of insulation, Project B requires 456 rolls of insulation, and Project C requires 125 rolls of insulation. You need to order a total of _____ rolls of insulation for all three projects.
    a. 653
    b. 749
    c. 820
    d. 943

18. You need to install 17 faucets in each of 5 buildings. You will install a total of _____ faucets.
    a. 75
    b. 85
    c. 95
    d. 105

MODULE 00102-04 ◆ INTRODUCTION TO CONSTRUCTION MATH 2.13

Show Transparency 3 (Figure 6). Review the types of measurement tools used in the construction trade. Compare a standard ruler with a metric ruler.

Review Questions

19. Your company ordered 8,000 carriage bolts. You used 560 bolts on Job A, 978 bolts on Job B, and 234 bolts on Job C. Your company has _____ bolts left.

    a. 1,772
    b. 3,789
    c. 5,463
    d. 6,228

20. You have a pipe that measures 628 inches. You can cut _____ 8-inch pieces of pipe. You will have _____ inches of pipe remaining.

    a. 75; 8
    b. 76; 5
    c. 77; 6
    d. 78; 4

3.0.0 ◆ WORKING WITH MEASUREMENTS

In the construction trade, you will need to use a ruler to measure various objects (see *Figure 6*). There are two types of rulers you may see on the job: the **standard ruler,** also called an **English ruler,** and the **metric ruler.** The markings on the standard ruler measure in inches and feet, while the markings on the metric ruler measure in millimeters and centimeters. A yardstick is a standard ruler that is one yard, or three feet, long.

In this section, you are working with the standard ruler and the architect's scale. Later you will work with the metric ruler. The standard ruler is divided into whole inches and then halves, fourths, eighths, and sixteenths. Some rulers are divided into thirty-seconds, and some into sixty-fourths. These represent fractions of an inch. In this unit, you will be working with a standard ruler and standard fractions to solve problems.

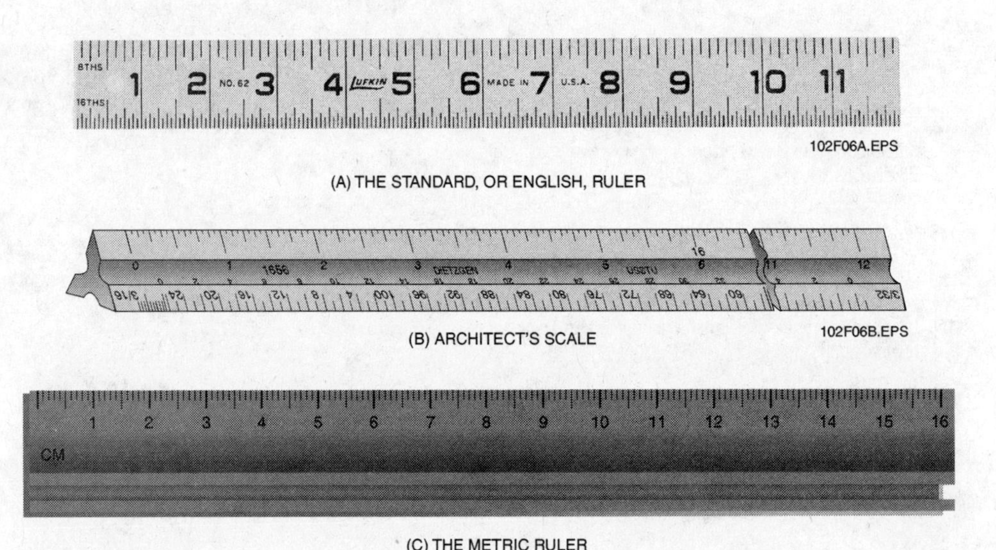

Figure 6 ◆ Types of measurement tools.

Instructor's Notes:

___

___

___

2.14 CORE CURRICULUM ◆ INTRODUCTORY CRAFT SKILLS

3.1.0 Using the Standard Ruler

You must learn to recognize and identify the distances shown (see *Figure 7*) on the standard ruler.

Note the varying lengths of the indicating marks. This feature is helpful when making precise measurements.

3.2.0 Architect's Scale

The architect's scale is used on all plans other than site plans, including blueprints. It is divided into proportional feet and inches. The triangular form is commonly used because it contains a variety of scales on a single tool. It can be read either from left to right or from right to left (see *Figure 8*).

Scales developed using an architect's scale include the following:

- ½" × 1'-0"
- ¾" × 1'-0"
- 1½" × 1'-0"
- 1" × 1'-0'

The fully divided scale at the top of the architect's scale shown in *Figure 8* represents inches, just like the English ruler.

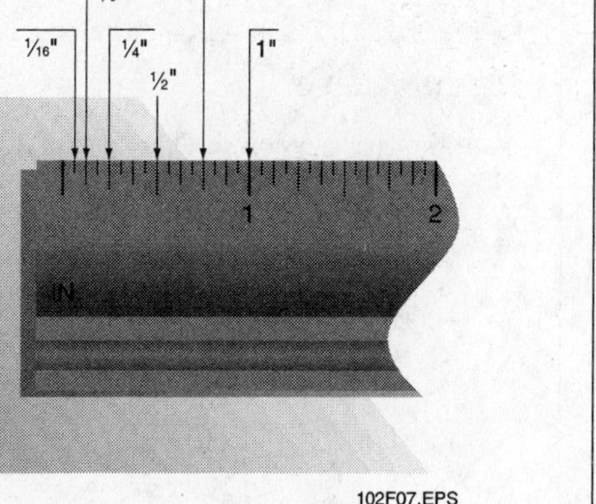
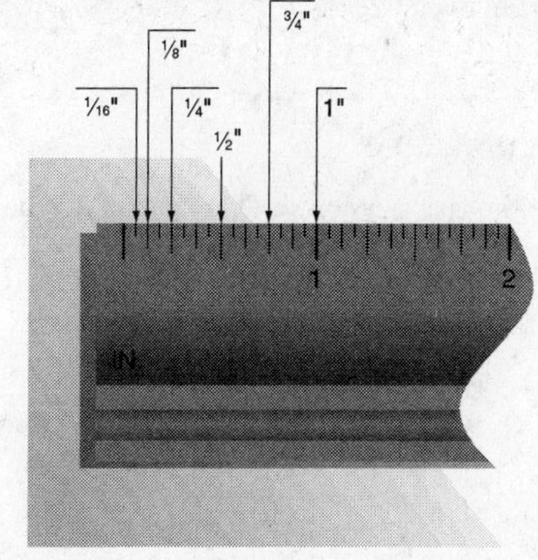

Figure 7 ◆ Distances on the standard ruler.

Demonstrate how to identify the distances shown on a standard ruler, including halves, fourths, eighths, and sixteenths.

Show Transparency 4 (Figure 8). Explain how the architect's scale can be read from left to right or from right to left.

Bring in an architect's scale and a blueprint. Demonstrate how to use the architect's scale on a blueprint.

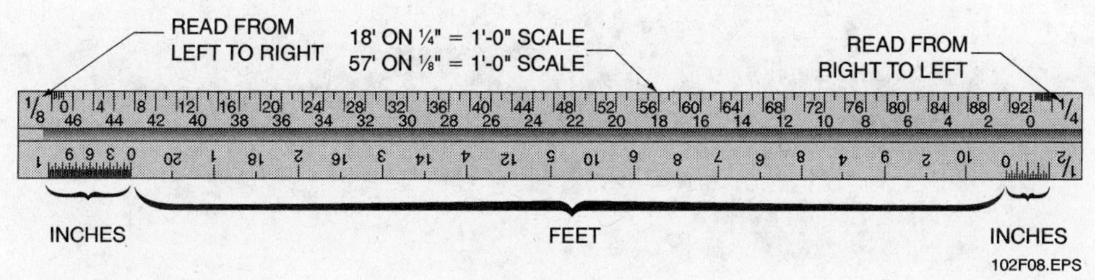

Figure 8 ◆ The architect's scale.

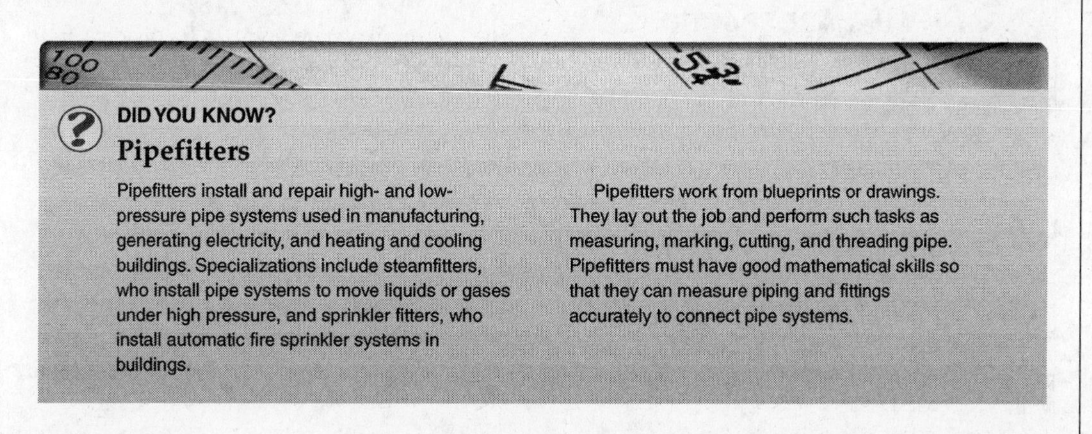

DID YOU KNOW?

Pipefitters

Pipefitters install and repair high- and low-pressure pipe systems used in manufacturing, generating electricity, and heating and cooling buildings. Specializations include steamfitters, who install pipe systems to move liquids or gases under high pressure, and sprinkler fitters, who install automatic fire sprinkler systems in buildings.

Pipefitters work from blueprints or drawings. They lay out the job and perform such tasks as measuring, marking, cutting, and threading pipe. Pipefitters must have good mathematical skills so that they can measure piping and fittings accurately to connect pipe systems.

MODULE 00102-04 ◆ INTRODUCTION TO CONSTRUCTION MATH 2.15

Go over the Review Questions for Section 3.0.0. Answer any questions trainees may have.

Explain how fractions divide whole units into parts.

Show Transparency 5 (Figure 10). Have trainees identify the measurement illustrated in Figure 10.

Review Questions

Section 3.0.0

Use the ruler shown in *Figure 9* to find the answers to the following questions.

1. A is at the _____-inch mark.
   a. ½
   b. ¾
   c. 1
   d. 1¼

2. B is at the _____-inch mark.
   a. 2½
   b. 2¾
   c. 2⅝
   d. 3

3. C is at the _____-inch mark.
   a. 1½
   b. 1¾
   c. 1⅝
   d. 1⅞

4. D is at the _____-inch mark.
   a. ¾
   b. ½
   c. ¼
   d. ⅞

5. E is at the _____-inch mark.
   a. ⅛
   b. ¼
   c. 15/16
   d. 1 5/16

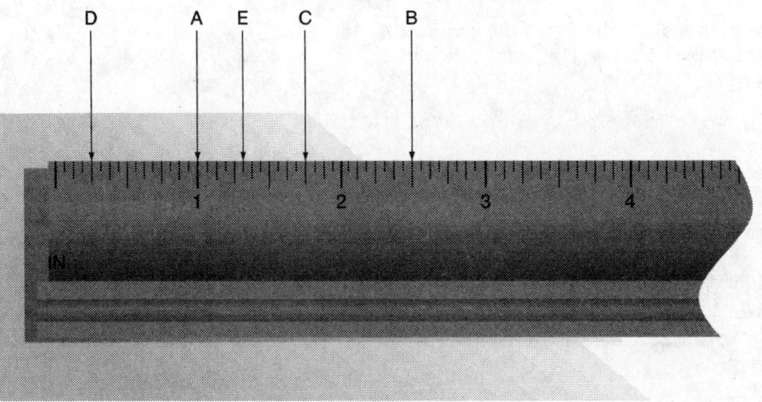

Figure 9 ◆ Review questions ruler.

4.0.0 ◆ WHAT ARE FRACTIONS?

A fraction divides whole units into parts. Common fractions are written as two numbers, separated by a slash or by a horizontal line, like this:

1/2 or ½

The slash or horizontal line means the same thing as the ÷ sign. So think of a fraction as a division problem. The fraction ½ means 1 divided by 2, or one divided into two equal parts. Read this fraction as "one-half."

The lower number (**denominator**) of the fraction tells you the number of parts by which the upper number (**numerator**) is being divided. The upper number is a whole number that tells you how many parts are going to be divided. In the fraction ½, the 1 is the upper number, or numerator, and the 2 is the lower number, or denominator. These numbers are also referred to as the terms of the fraction.

What measurement is the arrow in *Figure 10* pointing to?

Which is the correct answer?

A. 8/16
B. 2/4
C. 4/8
D. ½
E. All are correct!

The correct answer is E … all of the answers are correct! Let's find out why.

Instructor's Notes:

2.16 CORE CURRICULUM ◆ INTRODUCTORY CRAFT SKILLS

Figure 10 ◆ Identify the measurement.

On the whiteboard/chalkboard, write a variety of fractions. Demonstrate how to find equivalent fractions.

Ask trainees to complete the study problems for finding equivalent fractions. Review trainees' answers, and go over any questions trainees may have.

ON-SITE

Fractions on the Job

The first day you report to work on a construction site, you will realize how important it is to understand fractions. In the real world, most measurements are not whole numbers. Typically you will be measuring and cutting pipe or lumber to fractional lengths such as 3/8, 5/16, or 3/4. Being comfortable working with fractions is an important job skill.

4.1.0 Finding Equivalent Fractions

Note that 1/2 inch = 2/4 inch = 4/8 inch = 8/16 inch. These fractions are called **equivalent fractions.** Equivalent means that they have the same value or are equal. If you cut off a piece of wood 8/16-inch long, and then cut off a piece 1/2-inch long, both pieces of wood would be the same length.

When you measure objects, you often need to record all measurements as the same (common) fractions, such as sixteenths of an inch. Doing this allows you to easily compare, add, and subtract fractional measurements. This is why you need to know how to find equivalent fractions.

To find out how many sixteenths of an inch are equal to 1/2 inch, for example, you need to multiply both the numerator and the denominator by the same number. (Remember that a fraction in which both the numerator and the denominator are the same number is equal to 1.) Ask yourself what number you would multiply by 2 to get 16. The answer is 8, so you multiply both numbers by 8. (Remember that the fraction 8/8 is equal to 1.)

$$\frac{1 \times 8}{2 \times 8} = \frac{8}{16}$$

The answer is that 1/2 inch is equivalent to 8/16 inch.

4.1.1 Study Problems: Finding Equivalent Fractions

Find the equivalents of the following measurements:

1. 1/4 inch = ____/16 inch
a. 2 c. 6
b. 4 d. 8

2. 2/16 inch = ____/32 inch
a. 1 c. 4
b. 2 d. 8

3. 3/4 inch = ____/8 inch
a. 2 c. 5
b. 4 d. 6

4. 3/4 inch = ____/64 inch
a. 48 c. 52
b. 50 d. 54

5. 3/16 inch = ____/32 inch
a. 2 c. 6
b. 4 d. 8

Explain how to use division to reduce a fraction to its lowest term.

4.2.0 Reducing Fractions to Their Lowest Terms

If you find that the measurement of something is $\frac{4}{16}$, you may want to reduce the measurement to its lowest terms so the number is easier to work with. To find the lowest terms of $\frac{4}{16}$, you will use division:

Ask trainees to complete the study problems for reducing fractions to their lowest terms. Answer any questions trainees may have.

Step 1 To reduce a fraction, ask yourself, what is the largest number that I can divide evenly into both the numerator and the denominator? If there is no number (other than 1) that will divide evenly into both numbers, the fraction is already in its lowest form.

Step 2 Divide the numerator and the denominator by the same number.

In this example, you could divide both the numerator and the denominator by 4.

$$\frac{4 \div 4 = 1}{16 \div 4 = 4}$$

Therefore, $\frac{4}{16}$ may be reduced to $\frac{1}{4}$.

Refer to Figures 11 and 12. Ask trainees whether they can determine which portion is greater. Review the steps to find a common denominator.

Explain how to find the lowest common denominator.

DID YOU KNOW?

When you cannot think of a larger number to divide by, as long as the numerator and the denominator are both even, you can always divide each by 2.

4.2.1 *Study Problems: Reducing Fractions to Their Lowest Terms*

Find the lowest form of each of the following fractions without using a calculator.

1. $\frac{2}{16}$ = _____
2. $\frac{2}{8}$ = _____
3. $\frac{12}{32}$ = _____
4. $\frac{4}{8}$ = _____
5. $\frac{4}{64}$ = _____

4.3.0 Comparing Fractions and Finding the Lowest Common Denominator

Which measurement is larger, $\frac{3}{4}$ or $\frac{5}{8}$?

To find the answer, think about this question: Would you have more pizza if you had three pieces from a pie that was cut up in four equal slices (see *Figure 11*) or if you had five pieces of the same size pie that was cut up in eight equal slices (see *Figure 12*)?

As you can see, it's hard to compare fractions that do not have common denominators, just as it's hard to compare pizzas that are cut up differ-

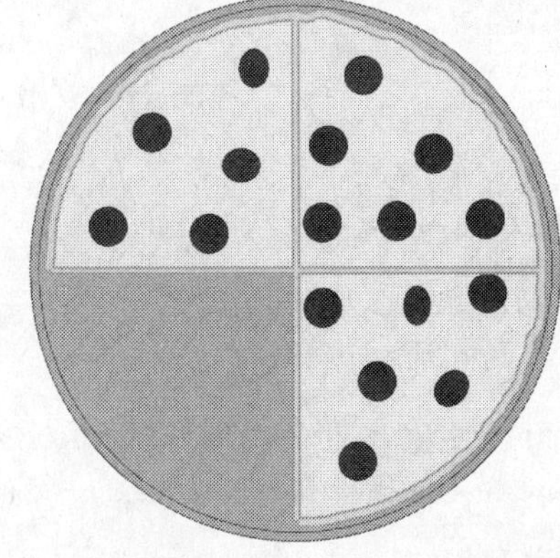

Figure 11 ◆ $\frac{3}{4}$ of a pizza.

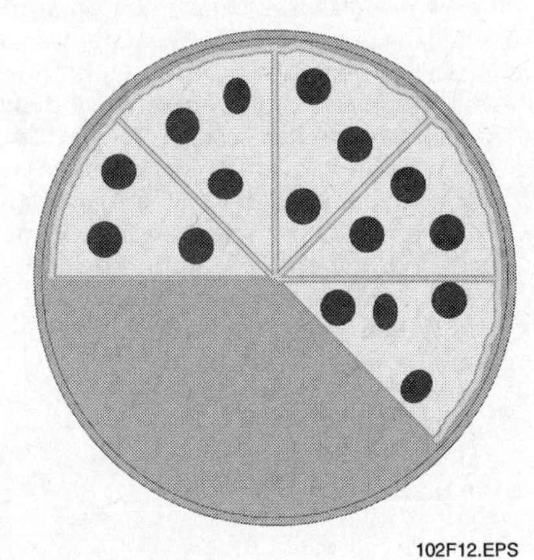

Figure 12 ◆ $\frac{5}{8}$ of a pizza.

Instructor's Notes:

____

____

____

2.18 CORE CURRICULUM ◆ INTRODUCTORY CRAFT SKILLS

ently. Using our pizza pie, remember that we are trying to determine which amount of pie is larger:

$$\frac{3}{4} \text{ or } \frac{5}{8}$$

To compare, you need to find a common denominator for the pizza slices. The common denominator is a number that both denominators can go into evenly.

Step 1 Multiply the two denominators together ($4 \times 8 = 32$). This is a common denominator between the two fractions.

You found a common denominator so that you can compare the pieces more easily. Now **convert** the two fractions so that they will have the same denominator.

Step 2 Convert each of the two fractions to fractions having the common denominator of 32.

$$\frac{3 \times 8}{4 \times 8} = \frac{24}{32}$$

$$\frac{5 \times 4}{4 \times 4} = \frac{20}{32}$$

Now it's easy to compare the two fractions to see which is larger. You'd have more pizza if you chose $3/4$ (you'd have $24/32$ instead of $20/32$ of the pizza).

You have found the common denominator for the pizza problem. However, working with fractions like $24/32$ or $20/32$ is difficult. To make this problem easier, you can find the lowest common denominator, which means reducing the fractions to their lowest terms.

To find the lowest common denominator, follow these steps:

Step 1 Reduce each fraction to its lowest terms.

Step 2 Find the lowest common multiple of the denominators. Sometimes this is as simple as one denominator already being a multiple of the other, meaning you can multiply by a whole number to get the larger number. If this is the case, all you have to do is find the equivalent fraction for the term with the smaller denominator.

Step 3 If neither of the denominators is a multiple of the other, you can multiply the denominators together to get a common denominator.

Let's look at the pizzas again. In this example, $3/4$ and $5/8$ are already in their lowest terms. When you look at the denominators, you see that 8 is a multiple of 4. So you should find the equivalent fraction for $3/4$ that has a denominator of 8.

$$\frac{3 \times 2}{4 \times 2} = \frac{6}{8}$$

You can now compare $6/8$ to $5/8$ and see that $6/8$ (which is the same as $3/4$) is the larger fraction.

Whether you find the lowest common denominator or just multiply the denominators to find a common denominator will depend on the situation. In some applications, you may want all fractions involved to have a particular denominator. But by finding the lowest common denominator this way, you can decrease the amount of multiplying you need to do and reduce the chances of making a mathematical error.

4.3.1 Study Problems: Finding the Lowest Common Denominator

Find the lowest common denominator for the following pairs of fractions.

1. $2/6$ and $3/4$ = _____.
a. 6
b. 10
c. 12
d. 16

2. $1/4$ and $3/8$ = _____.
a. 4
b. 8
c. 12
d. 18

3. $1/8$ and $1/2$ = _____.
a. 3
b. 5
c. 7
d. 8

4. $1/4$ and $3/16$ = _____.
a. 14
b. 16
c. 18
d. 20

5. $4/32$ and $5/8$ = _____.
a. 8
b. 14
c. 21
d. 32

Ask trainees to complete the study problems for finding the lowest common denominator. Ask whether trainees have any questions.

MODULE 00102-04 ◆ INTRODUCTION TO CONSTRUCTION MATH 2.19

Explain how to add fractions by finding the common denominator.

Have trainees complete the study problems for adding fractions. Go over any questions trainees may have.

Compare the procedure for adding fractions to the procedure for subtracting fractions.

Ask trainees to complete the study problems for subtracting fractions. Ask whether trainees have any questions.

On the whiteboard/chalkboard, demonstrate how to subtract a fraction from a whole number.

4.4.0 Adding Fractions

How many total inches will you have if you add ¾ of an inch plus ⅝ of an inch? To answer this question, you will have to add two fractions using the following steps.

Step 1 Find the common denominator of the fractions you wish to add. A common denominator for ¾ and ⅝ is 32.

Step 2 Convert the fractions to equivalent fractions with the same denominator. This is how to convert the fractions to equivalent fractions:

$$\frac{3 \times 8 = 24}{4 \times 8 = 32}$$

$$\frac{5 \times 4 = 20}{8 \times 4 = 32}$$

Step 3 Add the numerators of the fractions. Place this sum over the denominator.

$$\frac{24 \div 20 = 44}{32 \div 32 = 32}$$

Step 4 Reduce the fraction to its lowest terms. When you have done this, there will be no number other than 1 that can go evenly into both the numerator and the denominator.

When you reduce 44/32 to 11/8, it becomes a fraction in which no number other than 1 will go evenly into both the numerator and the denominator. But it is an **improper fraction** (meaning the numerator is larger than the denominator).

In this case, you need to reduce the improper fraction 11/8 to its lowest terms. We will soon learn how to convert improper fractions to **mixed numbers.** For now, try solving the problems with the fractions in the study problems for this section.

Don't forget! The fractions will need common denominators before they can be added.

4.4.1 *Study Problems: Adding Fractions*

Find the answers to the following addition problems. Remember to reduce the sum to the lowest terms.

1. ⅛ + 4/16 = _____

2. 4/8 + 6/16 = _____

3. 2/4 + ¼ = _____

4. ¾ + 2/8 = _____

5. 14/16 + ⅜ = _____

4.5.0 Subtracting Fractions

Subtracting fractions is very much like adding fractions. You must find a common denominator before you subtract. Say you have a piece of wood ⅞ of a foot long. You use ¼ of a foot. How much do you have left?

Step 1 Find the common denominator. In this case it is 8.

$$\frac{7}{8} \quad \frac{1}{4}$$

Step 2 Multiply each term of the fraction (¼) by 2 to get a fraction with the common denominator of 8.

$$\frac{1 \times 2 = 2}{4 \times 2 = 8}$$

Step 3 Subtract the numerators. ⅝ of a foot is left.

$$\frac{7}{8} - \frac{2}{8} = \frac{5}{8}$$

4.5.1 *Study Problems: Subtracting Fractions*

Find the answers to the following subtraction problems and reduce the differences to lowest terms.

1. ⅜ − 5/16 = _____

2. 11/16 − ⅝ = _____

3. ¾ − 2/6 = _____

4. 11/12 − 4/8 = _____

5. 11/16 − ½ = _____

4.5.2 *Subtracting a Fraction from a Whole Number*

Sometimes you must subtract a fraction from a whole number. For example, you need to take ¼ of a day off from a five-day workweek. How many days will you be working that week?

Here is how to set up this type of problem:

Step 1 To subtract a fraction from a whole number, borrow 1 from the whole number to make it into a fraction.

$$\begin{array}{cc} 5 = 4 + 1 \\ \underline{-1/4} \quad \underline{-1/4} \end{array}$$

Step 2 Convert the 1 to a fraction having the same denominator as the number you are subtracting.

$$\begin{array}{cc} 5 = 4 + 4/4 \\ \underline{-1/4} \quad \underline{-1/4} \end{array}$$

Instructor's Notes:

2.20 CORE CURRICULUM ◆ INTRODUCTORY CRAFT SKILLS

Step 3 Subtract and reduce to the lowest terms.

$$\begin{array}{rr} 5 = & 4 + \frac{4}{4} \\ -\frac{1}{4} & -\frac{1}{4} \\ \hline & 4 + \frac{3}{4} \text{ working days} \end{array}$$

4.5.3 *Study Problems: Subtracting a Fraction from a Whole Number*

Find the answers to the following subtraction problems and reduce the fractions to the lowest terms.

1. $8 - \frac{3}{4}$

2. $12 - \frac{5}{8}$

3. If two punches, one $4\frac{1}{64}$ inches long and the other $4\frac{3}{32}$ inches long, are made from a bar of stock $9\frac{7}{16}$ inches long, _____ inches of stock are not used.

a. $1\frac{1}{16}$
b. $1\frac{10}{32}$
c. $1\frac{15}{64}$
d. $1\frac{21}{64}$

4. If you saw $12\frac{1}{16}$ inches off a board $20\frac{3}{4}$ inches long, you'll have _____ inches left over.

a. $8\frac{1}{4}$
b. $8\frac{11}{16}$
c. $11\frac{1}{4}$
d. $16\frac{11}{8}$

5. A rough opening for a window measures $36\frac{3}{8}$ inches. The window to be placed in the rough opening measures $35\frac{15}{16}$ inches. The total clearing that will exist between the window and the rough opening will be _____ inch(es).

a. 1
b. $1\frac{7}{16}$
c. $\frac{7}{16}$
d. $1\frac{12}{16}$

4.6.0 Multiplying Fractions

Multiplying and dividing fractions is very different from adding and subtracting fractions. You do not have to find a common denominator when you multiply or divide fractions.

In a word problem, the word *of* usually lets you know you need to multiply. If a problem asks "What is $\frac{2}{8}$ of 9?" then think of the problem this way: $\frac{2}{8} \times \frac{9}{1}$. Note that any number (except 0) over 1 equals itself.

Using $\frac{4}{8} \times \frac{5}{6}$ as an example, follow these steps:

Step 1 Multiply the numerators together to get a new numerator. Multiply the denominators together to get a new denominator.

$$\frac{4 \times 5}{8 \times 6} = \frac{20}{48}$$

Step 2 Reduce if possible ($\frac{20}{48}$ reduces to $\frac{5}{12}$).

Although you can multiply fractions without first reducing them to their lowest terms, keep in mind that you can reduce them before you multiply. This will sometimes make the multiplication easier, since you will be working with smaller numbers. It will also make it easier to reduce the product to the lowest terms. What may seem like an extra step can save you time in the long run.

4.6.1 *Study Problems: Multiplying Fractions*

Find the answers to the following multiplication problems without using a calculator and reduce them to their lowest terms.

1. $\frac{4}{16} \times \frac{5}{8} =$ _____

2. $\frac{3}{4} \times \frac{7}{8} =$ _____

3. $\frac{2}{8}$ of 15 is _____

4. $\frac{3}{7}$ of 49 is _____

5. $\frac{8}{16}$ of $\frac{32}{64}$ is _____

4.7.0 Dividing Fractions

Dividing fractions is very much like multiplying fractions, with one difference. You must **invert,** or flip, the fraction you are dividing by. Using $\frac{1}{2} \div \frac{3}{4}$ as an example, follow these steps:

Step 1 Invert the fraction you are dividing by ($\frac{3}{4}$).

$$\frac{3}{4} \text{ becomes } \frac{4}{3}$$

Step 2 Change the division sign (÷) to a multiplication sign (×).

$$\frac{1}{2} \div \frac{3}{4} = \frac{1}{2} \times \frac{4}{3}$$

Step 3 Multiply the fraction as instructed earlier.

$$\frac{1 \times 4}{2 \times 3} = \frac{4}{6}$$

Step 4 Reduce if possible.

$$\frac{4}{6} \text{ reduces to } \frac{2}{3}$$

Have trainees complete the study problems for subtracting a fraction from a whole number. Answer any questions trainees may have.

Explain that you do not have to find a common denominator when multiplying and dividing fractions. Review the steps to multiply fractions.

Ask trainees to complete the study problems for multiplying fractions. Go over any questions trainees may have.

Explain the difference between dividing and multiplying fractions.

MODULE 00102-04 ◆ INTRODUCTION TO CONSTRUCTION MATH 2.21

Explain how to convert mixed numbers to improper fractions.

Have trainees complete the study problems for dividing fractions. Ask whether trainees have any questions.

Have trainees complete the Review Questions for Section 4.0.0 and review Sections 5.0.0–5.8.1 for the next session.

Ensure that you have everything required for demonstrations and laboratories during this session.

Go over the Review Questions for Section 4.0.0. Answer any questions trainees may have.

If you are working with a mixed number (for example, $2\frac{1}{3}$, you must convert it to a fraction before you invert it. Do this by multiplying the denominator by the whole number (3×2), adding the numerator [$(3 \times 2) + 1$], and placing the result over the denominator. It looks like the following:

$$2\tfrac{1}{3} = \frac{(3 \times 2) + 1}{3} = \frac{7}{3}$$

When dividing by a whole number, place the whole number over 1 and then invert it. Remember that $\frac{4}{1}$ is the same as 4.

For example:

$$\tfrac{1}{2} \div 4 =$$
$$\tfrac{1}{2} \div \tfrac{4}{1} =$$
$$\tfrac{1}{2} \times \tfrac{1}{4} = \tfrac{1}{8}$$

4.7.1 Study Problems: Dividing Fractions

Find the answers to the following division problems without using a calculator and reduce them to lowest terms (no improper fractions allowed).

1. $\frac{3}{8} \div 3 =$ ______

2. $\frac{5}{8} \div \frac{1}{2} =$ ______

3. $\frac{3}{4} \div \frac{3}{8} =$ ______

4. On a scale drawing, if $\frac{1}{4}$ of an inch represents a distance of 1 foot, then a line on the drawing measuring $8\frac{1}{2}$ inches long represents ____ feet.
   a. 34
   b. 36
   c. 38
   d. 40

5. You can cut ____ $\frac{7}{8}$-inch lengths from a 7-inch strip.
   a. 5
   b. 6
   c. 7
   d. 8

Review Questions

Section 4.0.0

Find the equivalents of the following measurements.

1. $\frac{3}{8}$ inch = _____/64 inch
   a. 3
   b. 6
   c. 24
   d. 36

2. $\frac{5}{16}$ inch = _____/32 inch
   a. 7
   b. 10
   c. 12
   d. 16

3. $\frac{1}{2}$ inch = _____/16 inch
   a. 1
   b. 2
   c. 4
   d. 8

Reduce these fractions to their lowest terms.

4. $\frac{16}{64}$ = _____
   a. $\frac{8}{32}$
   b. $\frac{4}{16}$
   c. $\frac{1}{4}$
   d. $\frac{1}{2}$

5. $\frac{8}{16}$ = _____
   a. $\frac{1}{8}$
   b. $\frac{1}{4}$
   c. $\frac{1}{2}$
   d. $\frac{2}{8}$

6. $\frac{2}{4}$ = _____
   a. $\frac{1}{8}$
   b. $\frac{1}{4}$
   c. $\frac{1}{2}$
   d. $\frac{2}{4}$

Instructor's Notes:

2.22 CORE CURRICULUM ◆ INTRODUCTORY CRAFT SKILLS

Review Questions

Find the lowest common denominator for the following pairs of fractions.

7. $\frac{8}{64}$, $\frac{1}{32}$

The answer is _____.

a. 32
b. 17
c. 15
d. 9

8. $\frac{3}{4}$, $\frac{8}{16}$

The answer is _____.

a. 2
b. 3
c. 4
d. 12

9. $\frac{4}{12}$, $\frac{5}{15}$

The answer is _____.

a. 0
b. 1
c. 2
d. 3

10. $\frac{1}{2}$, $\frac{1}{8}$

The answer is _____.

a. 2
b. 4
c. 8
d. 16

Find the answers to the following problems and reduce the fractions to the lowest terms.

11. $\frac{7}{16} + \frac{1}{4} =$ _____

a. $\frac{8}{16}$
b. $\frac{11}{16}$
c. $\frac{28}{16}$
d. $\frac{3}{6}$

12. $\frac{3}{8} + \frac{9}{16} =$ _____

a. $\frac{7}{8}$
b. $\frac{15}{16}$
c. $\frac{3}{4}$
d. $\frac{12}{16}$

13. $\frac{11}{32} - \frac{2}{8} =$ _____

a. $\frac{9}{32}$
b. $\frac{19}{32}$
c. $\frac{3}{32}$
d. $\frac{1}{8}$

14. $18 - \frac{7}{8} =$ _____

a. $18\frac{7}{8}$
b. $18\frac{1}{8}$
c. $17\frac{7}{8}$
d. $17\frac{1}{8}$

15. $\frac{9}{16} \times \frac{2}{4} =$ _____

a. $\frac{18}{64}$
b. $\frac{9}{32}$
c. $\frac{11}{20}$
d. $\frac{9}{16}$

16. $\frac{1}{8}$ of 72 is _____

a. 8
b. $8\frac{1}{2}$
c. 9
d. 12

17. On a scale drawing, if $\frac{1}{2}$ inch represents a distance of 1 foot, then a line on the drawing measuring $5\frac{1}{2}$ inches long represents _____ feet.

a. $5\frac{1}{2}$
b. 7
c. 10
d. 11

Find the answers to the following problems and reduce the fractions to the lowest terms.

18. $\frac{9}{16} \div \frac{7}{8} =$ _____

a. $\frac{63}{128}$
b. $\frac{7}{16}$
c. $\frac{23}{16}$
d. $\frac{9}{14}$

19. $\frac{7}{16} \div 7 =$ _____

a. $\frac{1}{64}$
b. $\frac{1}{16}$
c. $2\frac{17}{16}$
d. $\frac{1}{7}$

20. You can cut _____ $\frac{3}{16}$-inch lengths from a 9-inch strip.

a. 16
b. 33
c. 48
d. 64

MODULE 00102-04 ◆ INTRODUCTION TO CONSTRUCTION MATH 2.23

Show Transparency 6 (Figure 14). Explain how to read a machinist's rule. Ensure that trainees can locate 1.3 inches on the machinist's rule.

5.0.0 ◆ DECIMALS

Decimals represent values less than one whole unit. You are already familiar with decimals in the form of money.

25¢ = 0.25 or 25/100
10¢ = 0.10 or 10/100
50¢ = 0.50 or 50/100

Now you will learn how to use decimals in mathematical operations.

5.1.0 Reading a Machinist's Rule

On the job, you may need to use decimals to read instruments or calculate flow rates. Look at the scale on a typical **machinist's rule,** as shown in *Figure 13.* Each number shows the distance, in inches, from the squared end of the rule. The marks between the numbers divide each inch into ten equal parts. Each of these ten parts is referred to as a tenth.

In *Figure 14,* the nail spans one whole inch plus three-tenths of a second inch. It is one and three-tenths of an inch long. This is written as 1.3 inches.

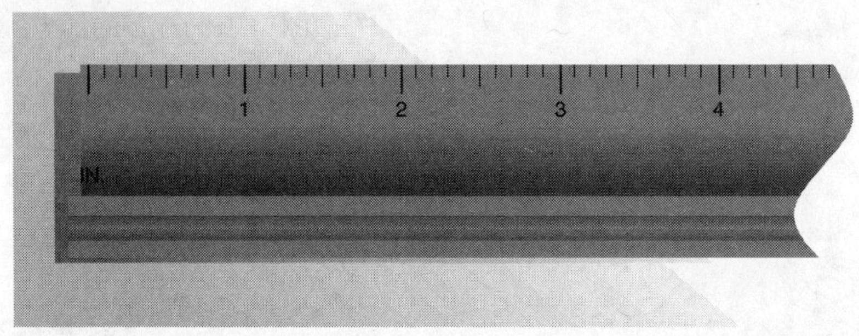

102F13.EPS

Figure 13 ◆ The machinist's rule (divided into tenths).

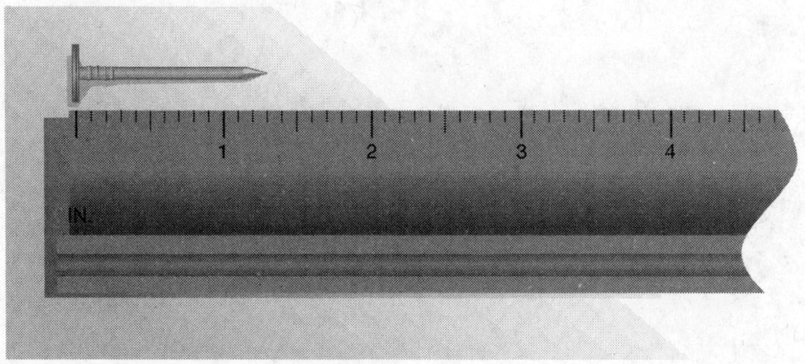

102F14.EPS

Figure 14 ◆ Showing 1.3 inches on a machinist's rule.

Instructor's Notes:

2.24 CORE CURRICULUM ◆ INTRODUCTORY CRAFT SKILLS

5.1.1 Study Problems: Reading a Machinist's Rule

Use the machinist's rule in *Figure 15* to find the answers to the following questions.

1. P is at the _____ tenths of an inch mark.
a. .09
b. 1.9
c. 9
d. 9.9

2. Q is at the _____-inch mark.
a. 1.2
b. 1.3
c. 1.4
d. 1.5

3. R is at the _____ tenths of an inch mark.
a. 24
b. 26
c. 28
d. 30

4. S is at the _____-inch mark.
a. 4.0
b. 4.2
c. 4.4
d. 4.6

5. T is at the _____-inch mark.
a. 5
b. $\frac{4}{5}$
c. $\frac{45}{10}$
d. $\frac{55}{10}$

Ask trainees to complete the study problems for reading a machinist's rule. Go over any questions trainees may have.

Compare whole number place values with decimal place values.

Demonstrate how measurements are read in the field. Show, for example, that a measurement will not read 26 tenths of an inch but, rather, two and six-tenths of an inch.

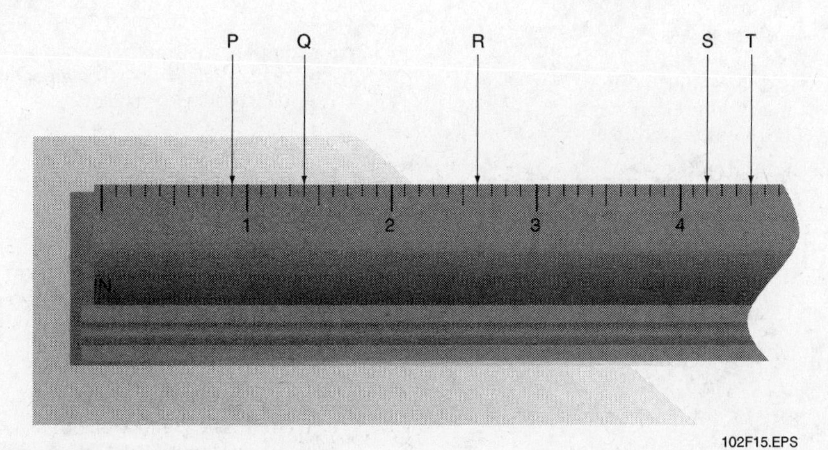

Figure 15 ◆ Study problems machinist's rule.

DID YOU KNOW?
Why Base-Ten?

When we write a number, we use a symbol for that number. This symbol is called a *numeral* or *digit.* Most of the world uses a decimal system with the numerals or digits 1, 2, 3, 4, 5, 6, 7, 8, 9, 0. The decimal system is a counting system based on the number 10 and groups of ten. The word *decimal* is from the Latin word meaning ten. That is why it is called a *base-ten* system. The word *digit* is from the Latin word for finger. Early people naturally used their fingers as a means of counting—a base-ten system.

5.2.0 Comparing Whole Numbers with Decimals

The following chart compares whole number place values with decimal place values:

Whole Numbers		Decimals	
1	ones	1.0	
10	tens	.1	tenths
100	hundreds	.01	hundredths
1000	thousands	.001	thousandths

To read a decimal, say the number as it is written and then the name of its place value. For example, read 0.56 as "fifty-six hundredths."

Mixed numbers also appear in decimals. You read 15.7 as "fifteen and seven-tenths." Notice the use of the word "and" to separate the whole number from the decimal.

MODULE 00102-04 ◆ INTRODUCTION TO CONSTRUCTION MATH 2.25

Have trainees complete the study problems for comparing whole numbers with decimals. Answer any questions trainees may have.

Review the procedures to compare decimals with decimals.

Ask trainees to complete the study problems for comparing decimals. Ask whether trainees have any questions.

5.2.1 Study Problems: Comparing Whole Numbers with Decimals

For the following problems, find the words that mean the same as the decimal or the decimal equivalent of the words.

1. 0.4 = _____

a. four
b. four-tenths
c. four-hundredths
d. four-thousandths

2. 0.05 = _____

a. five
b. five-tenths
c. five-hundredths
d. five-thousandths

3. 2.5 = _____

a. two and five-tenths
b. two and five-hundredths
c. two and five-thousandths
d. twenty-five-hundredths

4. Eighteen-hundredths = _____

a. 1.8
b. 0.18
c. 0.018
d. 0.0018

5. Five and eight-tenths = _____

a. 5.0
b. 5.8
c. 5.08
d. 5.008

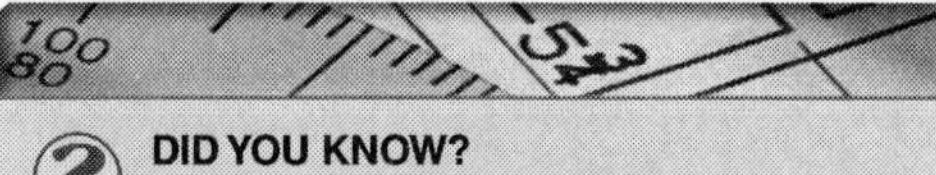

DID YOU KNOW?

Place-Value Systems

As mathematics advanced, counting systems became more efficient for performing calculations and solving problems. These systems made it easier to represent large numbers and to simplify the process of computing. These were called place-value systems. In a place-value system, the value of a particular symbol depends not only on the symbol but also on its position or place in the number.

In our decimal system, each place value is ten times greater than the place to the right. Place values in the decimal system are:

1,000	100	10	1	
5	3	4	9	= 5,349

5.3.0 Comparing Decimals with Decimals

Which decimal is the larger of the two?

0.4 or 0.42?

Here's how to compare decimals:

Step 1 Line up the decimal points of all the numbers.

0.4
0.42

Step 2 Place zeros to the right of each number until all numbers end with the same place value.

0.40
0.42

Step 3 Compare the numbers.

0.42 (42 hundredths) is larger than 0.40 (40 hundredths).

5.3.1 Study Problems: Comparing Decimals

For the following problems, put the decimals in order from *smallest* to *largest*.

1. 0.400, 0.004, 0.044, 0.404

The answer is _____.

a. 0.400, 0.004, 0.044, 0.404
b. 0.004, 0.044, 0.404, 0.400
c. 0.004, 0.044, 0.400, 0.404
d. 0.404, 0.044, 0.400, 0.004

2. 0.567, 0.059, 0.56, 0.508

The answer is _____.

a. 0.508, 0.56, 0.567, 0.059
b. 0.059, 0.56, 0.508, 0.567
c. 0.567, 0.059, 0.56, 0.508
d. 0.059, 0.508, 0.56, 0.567

3. 0.320, 0.032, 0.302, 0.003

The answer is _____.

a. 0.003, 0.032, 0.302, 0.320
b. 0.320, 0.302, 0.032, 0.003
c. 0.302, 0.320, 0.003, 0.032
d. 0.003, 0.032, 0.320, 0.302

4. 0.867, 0.086, 0.008, 0.870

The answer is _____.

a. 0.870, 0.867, 0.086, 0.008
b. 0.008, 0.086, 0.867, 0.870
c. 0.086, 0.008, 0.867, 0.870
d. 0.008, 0.870, 0.867, 0.086

Instructor's Notes:

____

____

____

2.26 CORE CURRICULUM ◆ INTRODUCTORY CRAFT SKILLS

5. 0.626, 0.630, 0.616, 0.641

The answer is ____.

a. 0.616, 0.641, 0.630, 0.626
b. 0.616, 0.626, 0.630, 0.641
c. 0.061, 0.616, 0.626, 0.630
d. 0.630, 0.616, 0.626, 0.641

5.4.0 Adding and Subtracting Decimals

There is only one major rule to remember when adding and subtracting decimals:

Keep your decimal points lined up!

Suppose you want to add 4.76 and 0.834. Line up the problem like this:

$$\begin{array}{r} 4.760 \\ +0.834 \\ \hline 5.594 \end{array}$$

You can add a 0 to help keep the numbers lined up.

The same thing is true for subtraction of decimals. Line up the decimal points.

$$\begin{array}{r} 5.6 \\ -2.724 \\ \hline \end{array} \qquad \begin{array}{r} 5.600 \\ -2.724 \\ \hline 2.876 \end{array}$$

Notice that two zeros were added to the end of the first number to make it easier to see where you need to borrow.

Explain how to add and subtract decimals. Ensure that trainees understand how important it is to keep the numbers lined up.

DID YOU KNOW?

Other Counting Systems

Ancient Babylonians developed one of the first place-value systems—a base-sixty system. It was based on the number 60, and numbers were grouped by sixties. At the beginning, there was no symbol for zero in the Babylonian system. This made it difficult to perform calculations. It was not always possible to determine if a number represented 24, 204, or 240. Although this system is no longer used today, we still use a base-sixty system for measuring time: 60 seconds in a minute and 60 minutes in an hour.

A base-twelve system requires 12 digits. This system is called a *duodecimal* system, from the Latin word *duodecim,* meaning 12. Since the decimal system has only ten digits, two new digits must be added to the base-twelve system. In a duodecimal system, each place value is 12 times greater than the place to the right. Although this is not a common system, a base-twelve system is used to count objects by the dozen (12) or by the gross ($144 = 12 \times 12$).

A hexadecimal system groups numbers by sixteens. The word *hexadecimal* comes from the Greek word for six and the Latin word for ten. Just as the duodecimal system required two new place numerals, the hexadecimal system requires six additional digits. In a hexadecimal system, each place value is 16 times greater than the place to the right. Computers often use the hexadecimal system to store information. Common configurations of RAM come in multiples of 16.

$$64 = 4 \times 16$$
$$96 = 6 \times 16$$
$$128 = 8 \times 16$$
$$256 = 16 \times 16$$

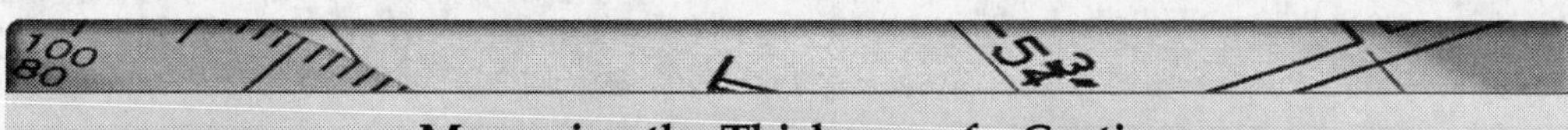

ON-SITE

Measuring the Thickness of a Coating

Coating thickness is important because either too little or too much can cause problems. A coating such as paint needs a minimum thickness to prevent corrosion, withstand abrasion, and look good. A coating that is too thick may crack, flake, blister, or not cure properly.

Many jobs have requirements that specify the thickness of the coating applied to an object or surface. To ensure that the specifications are met, periodic checks of the wet-film thickness can be made using a wet-film thickness gauge. Typically these gauges use measurements in mils or microns. For example, the required wet-film thickness for a coat of paint may be 10.25 mils.

This makes understanding decimals and properly reading gauges an essential job skill.

Ask trainees to complete the study problems for adding and subtracting decimals. Go over any questions trainees may have.

See the Teaching Tip for Sections 5.4.0–5.4.1 at the end of this module.

On the whiteboard/chalkboard, demonstrate how to multiply decimals. Ensure that trainees understand how to properly place the decimal point.

Ask trainees to complete the study problems for multiplying decimals. Ask whether trainees have any questions.

5.4.1 Study Problems: Adding and Subtracting Decimals

Find the answers to these addition and subtraction problems without using a calculator, and don't forget to line up the decimal points.

1. 2.50
   4.20
   +5.00

2. 1.82 + 3.41 + 5.25 = _____

3. 6.43 plus 86.4 = _____

4. The combined thickness of a piece of sheet metal 0.078 inch thick and a piece of band iron 0.25 inch thick is _____.
   a. 0.308
   b. 0.328
   c. 3.08
   d. 32.8

5. Yesterday, a lumberyard contained 6.7 tons of wood. Since then, 2.3 tons were removed. The lumberyard now contains _____ tons of wood.
   a. 3.4
   b. 4.4
   c. 5.4
   d. 6.4

5.5.0 Multiplying Decimals

While unloading wood panels, you measure one panel as 4.5 feet wide. You have seven panels the same width. What is the total width if you put the panels side-by-side?

Step 1 Set up the problem just like the multiplication of whole numbers.

```
 4.5
× 7
```

Step 2 Proceed to multiply.

```
 4.5
× 7
 315
```

Step 3 Once you have the answer, count the number of digits to the right of the decimal point in both numbers being multiplied. (In this example, there is only one number with a decimal point [4.5] and only one number to the right of it.)

Step 4 In the answer, count over the same number of digits (from right to left) and place the decimal point there.

```
 4.5   (There is one total digit to the right of
× 7    the decimal point in the two numbers;
31.5   count in one digit from right to left in
       the answer, and place the decimal point
       there.)
```

NOTE

You may have to add a zero if there are more digits to the right of the decimal points than there are in the answer, as shown in the following example.

```
  0.507   (Add the total digits to the right
× 0.022   of the decimal point in the two
  1014    numbers. There are six.)
  1014
+ 000
 11154 = .011154 (Count six digits from right to
                 left in the product. In this case,
                 you'll need to add a zero.)
```

5.5.1 Study Problems: Multiplying Decimals

Use the following to answer questions 1 and 2: You are machining a part. The starting thickness of the part is 6.18 inches. You take three cuts. Each cut is three-tenths of an inch.

1. You have removed _____ inches of material.
   a. 0.6
   b. 0.8
   c. 0.9
   d. 1.09

2. The remaining thickness of the part is _____ inches.
   a. 5.28
   b. 6.08
   c. 6.10
   d. 6.15

Instructor's Notes:

2.28 CORE CURRICULUM ◆ INTRODUCTORY CRAFT SKILLS

Use the following to answer questions 3 and 4: An electrician wants to know if a light circuit is overloaded. The circuit supplies two different machines.

3. The first machine has 11 bulbs lit. Each bulb uses 4.68 watts. The lights on the first machine need ____ watts.
   a. 0.5148
   b. 5.148
   c. 51.48
   d. 514.8
4. The second machine has seven bulbs lit. Each of these bulbs uses 5.14 watts. The lights on both machines need a total of ____ watts.
   a. 35.98
   b. 76.76
   c. 87.46
   d. 874.6
5. Ceramic tile weighs 4.75 pounds per square foot. Therefore, 128 square feet of ceramic tile weighs ____ pounds.
   a. 598
   b. 608
   c. 908
   d. 1108

5.6.0 Dividing with Decimals

Review the three types of division problems involving decimals.

When would you divide with decimals? Perhaps you need to cut a 44.5-inch pipe into as many 22-inch pieces as possible. How many 22-inch pieces will you be able to cut? How much will be left over?

There are three types of division problems involving decimals:

- Those that have a decimal point in the number being divided (the dividend)

$$22\overline{)44.5}$$

- Those that have a decimal point in the number you are dividing by (the divisor)

$$0.22\overline{)4{,}450}$$

- Those that have decimal points in both numbers (the dividend and the divisor)

$$0.22\overline{)44.5}$$

ON-SITE

Decimals at Work

When are you going to use decimals on the job? Here are two examples of using decimals to get your work done:

- You are installing a boiler to specifications on its concrete pad. When measuring with your level, you find that one corner of the boiler is level within 0.003 inch. Another corner is 0.005 too high, and a third corner is 0.001 too high. The concrete pad is not as even as it should be. To adjust for this, you must place shims (thin, tapered pieces of material) between the boiler base and the concrete pad. Shims come in 0.001- and 0.002-inch widths. How many shims of each size will it take to make the boiler level?
- You are working on a conveyor system. You check the lubricant by taking an oil sample and measuring the metal particles with a micrometer. (A micrometer is a precision tool that can measure to the nearest 1/1000 of an inch.) You find metal particles that are 0.0001 inch in size. The system currently has a 20-micron filter installed to filter pieces bigger than 0.0002 inch. What size filter do you need to filter the 0.0001-inch metal particles out of the oil?

Equivalents

Forty-four and one-half is the same as forty-four point five. If you measure a piece of pipe and the tape measure says it is 44½ inches, another way to say this is "forty-four point five," which is the decimal equivalent. The chart below shows some other decimal equivalents.

Chart of Equivalents

½ = 0.5 ¼ = 0.25 ⅛ = 0.125 1/16 = 0.0625

MODULE 00102-04 ◆ INTRODUCTION TO CONSTRUCTION MATH 2.29

Discuss the three procedures for dividing with decimals.

Have trainees complete the three sections of study problems for dividing with decimals. Answer any questions trainees may have.

5.6.1 Dividing with a Decimal in the Number Being Divided

For the first type of problem, let's use 44.5 ÷ 22 as our example.

Step 1 Place a decimal point directly above the decimal point in the dividend.

$$22\,\overline{)\,44.5}$$

Step 2 Divide as usual.

```
     2.0
22 ) 44.5
    -44
     00.5r
```

How many 22-inch pieces of pipe will you have? The answer: two, with a little (0.5 inch) left over.

5.6.2 Study Problems: Dividing with Decimals, Part 1

Find the answers to the following division problems without using a calculator, and don't go any further than the hundredths (0.01) place, unless otherwise noted.

1. 45.36 ÷ 18 = ____
2. 4.536 ÷ 18 = ____
3. 0.4536 ÷ 18 = ____ [to nearest thousandths (0.001) place]
4. 25) 10.20
5. 6) 31.2

5.6.3 Dividing with a Decimal in the Number You Are Dividing By

For the second type of problem, let's use 4,450 ÷ 0.22 as our example.

Step 1 Move the decimal point in the divisor to the right until you have a whole number.

0.22) 4,450. *(The decimal point in the divisor will be moved two places to the right.)*

Step 2 Move the decimal point in the dividend the same number of places to the right. You may have to add zeros first. Then divide as usual.

```
     20227.2
22 ) 4450.00,0
    -44
     0050
    -0044
     00060
    -00044
     000160
    -000154
     0000060
    -0000044
     0000016r
```

(After adding zeros, move decimal in dividend two places to the right so number becomes 445,000.)

5.6.4 Study Problems: Dividing with Decimals, Part 2

Perform the following division problems on a separate piece of paper without using a calculator. Don't go any further than the hundredths (0.01) place in your answer.

1. 282 ÷ 14.1 = ____
2. 694 ÷ 3.2 = ____
3. 99 ÷ 0.45 = ____
4. 2.5) 102
5. 0.6) 312

5.6.5 Dividing with Decimals in Both Numbers

For the third type of problem, let's use 44.5 ÷ 0.20 as our example.

Step 1 Move the decimal point in the divisor to the right until you have a whole number.

0.20) 44.50 *(Moving the decimal point in the divisor to the right, 0.20 becomes 20.)*

Step 2 Move the decimal point in the dividend the same number of places to the right. Then divide as usual.

```
     222.5
20 ) 4450.
```

(Moving the decimal point in the dividend two places to the right, 44.5 becomes 4,450.)

Now you can see that 44.5 ÷ 0.20 is 222.5.

Instructor's Notes:

2.30 CORE CURRICULUM ◆ INTRODUCTORY CRAFT SKILLS

5.6.6 *Study Problems: Dividing with Decimals, Part 3*

Find the answers to the following division problems without using a calculator. Don't go any further than the hundredths (0.01) place.

1. 20.82 ÷ 4.24 = _____
2. 38.9 ÷ 3.7 = _____
3. 9.9 ÷ 0.45 = _____
4. $0.25\overline{)10.20}$
5. $0.6\overline{)31.2}$

5.7.0 Rounding Decimals

Sometimes the answer is a bit more precise than you require. For example, if tubing costs $3.76 per foot and you spend $800, how much tubing will you buy?

The precise answer is 212.7659574 feet. But you probably only need to measure it to the nearest tenth. What would you do? For this exercise, you will round 212.7659574 to the nearest tenth (0.1).

Step 1 Underline the place to which you are rounding.

$212.\underline{7}659574$

Step 2 Look at the digit one place to its right.

$212.\underline{7}659574$

Step 3 If the digit to the right is 5 or more, you will round up by adding 1 to the underlined digit. If the digit is 4 or less, leave the underlined digit the same. In this example, the digit to the right is 6, which is more than 5, so you round up by adding 1 to the underlined digit.

$212.\underline{8}659574$

Step 4 Drop all other digits to the right.

212.8

DID YOU KNOW?

When calculating a division problem by hand, it is usually only necessary to divide an answer out to one more digit than you want in your final answer. Then you can round the final number out to the required number of places. Rounding is covered in more detail in Rounding Decimals.

5.7.1 *Study Problems: Rounding Decimals*

Solve these problems to practice rounding decimals.

1. You need to cut a 90.5-inch pipe into as many 3.75-inch pieces as possible. You will be able to cut _____ 3.75-inch pieces.
   a. 14
   b. 24
   c. 34
   d. 44
2. If you drove your car 622 miles on 40.1 gallons of gas, you got _____ miles per gallon. (Round your answer to the nearest tenth.)
   a. 15.5
   b. 15.6
   c. 155.1
   d. 156
3. If wire costs $4.30 per pound and you pay a total of $120.95, then you have purchased _____ pounds of wire. (Round your answer to the nearest tenth.)
   a. 0.28
   b. 2.8
   c. 28.0
   d. 28.1
4. Vent pipe is on sale at XYZ Supply Company this week for $0.37 per linear foot. If you spend $115.38, you will purchase _____ linear feet of pipe. (Round your answer to the nearest tenth.)
   a. 308.11
   b. 310.8
   c. 311.8
   d. 311.9
5. Vent pipe at XYZ Supply normally costs $0.48 per linear foot. If you spend the same amount of money ($115.38) when vent pipe is not on sale, you will purchase _____ linear feet of pipe. (Round your answer to the nearest tenth.)
   a. 240
   b. 240.4
   c. 241
   d. 241.4

Review the procedures to round decimals to the nearest tenth.

Ask trainees to complete the study problems for rounding decimals. Ask whether trainees have any questions.

MODULE 00102-04 ◆ INTRODUCTION TO CONSTRUCTION MATH 2.31

Ask trainees to take out their calculators and follow along as you demonstrate how to use the calculator to perform decimal operations.

Ask trainees to complete the study problems for using decimals on the calculator. Go over any questions trainees may have.

Ask trainees to complete the Review Questions for Section 5.0.0. Have trainees review Sections 6.0.0–6.4.1 for the next session.

5.8.0 Using the Calculator to Add, Subtract, Multiply, and Divide Decimals

Performing operations on the calculator using decimals is very much like performing the operations on whole numbers. Follow these steps using the problem 45.6 + 5.7 as an example.

Step 1 Turn the calculator on. A zero (0) appears in the display.

Step 2 Press 4, 5, . (decimal point), and 6. The number 45.6 appears in the display.

Step 3 Press the + key. The 45.6 is still displayed.

> **NOTE**
> For this step, press whichever operation key the problem calls for: + to add, − to subtract, × to multiply, ÷ to divide.

Step 4 Press 5, . (decimal point), and 7. The number 5.7 is displayed.

Step 5 Press the = key. After you press the = key, whether you are adding, subtracting, multiplying, or dividing, the answer will appear on your display.

45.6 + 5.7 = 51.3
45.6 − 5.7 = 39.9
45.6 × 5.7 = 259.92
45.6 ÷ 5.7 = 8

Step 6 Press the ON/C key to clear the calculator.

5.8.1 *Study Problems: Using Decimals on the Calculator*

Use your calculator to find the answers to the following problems, and round your answers to the nearest hundredth (0.01).

1. 45.89 + 7.85

2. 7.6 × 0.12

3. 685.79 − 56.266

4. 6.45 ÷ 3.25 =

5. 34.76 + 3.64

Review Questions

Section 5.0.0

Use the machinist's rule in *Figure 16* to find the answers to questions 1 and 2.

1. Bolt A is _____ inches long.

a. 2.4
b. 2.5
c. 2.6
d. 2.7

2. Bolt B is _____ inches long.

a. 1.5
b. 1.6
c. 1.7
d. 1.8

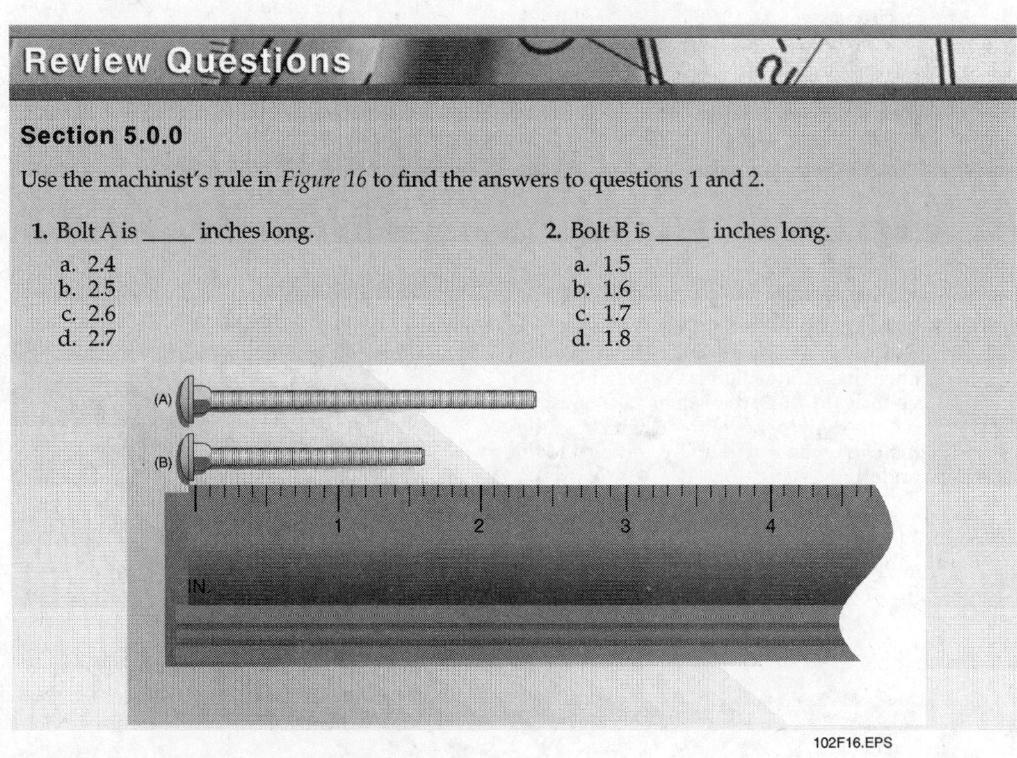

Figure 16 ◆ Review questions machinist's rule.

Instructor's Notes:

2.32 CORE CURRICULUM ◆ INTRODUCTORY CRAFT SKILLS

Review Questions

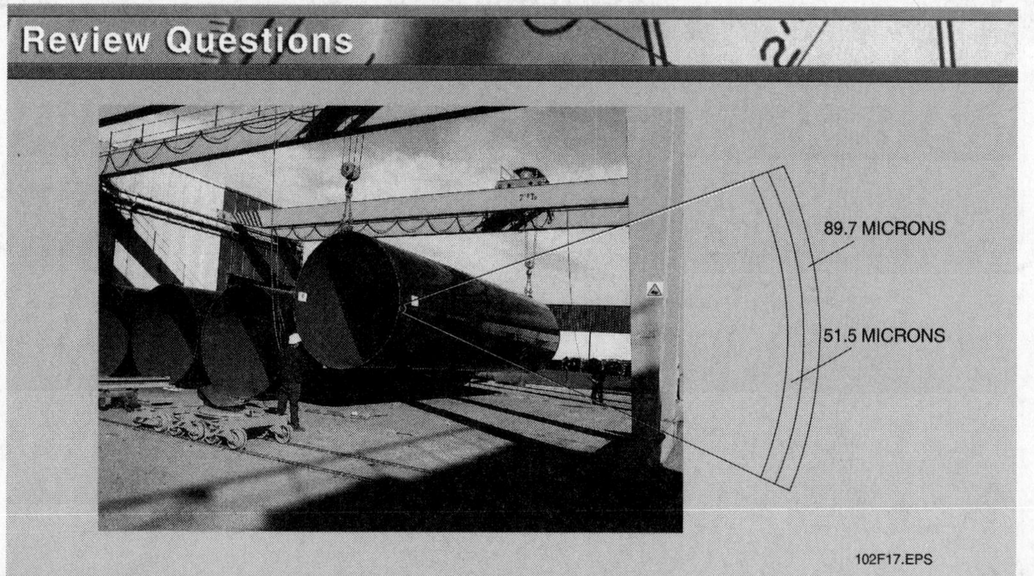

Figure 17 ◆ Pipe coating.

3. Find the words that mean the same as 7.5.

a. seventy and five-tenths
b. seven and five-tenths
c. seven and five-hundredths
d. seven and five-thousandths

4. Find the decimal equivalent of ninety and four-hundredths.

a. 9.04
b. 9.14
c. 90.04
d. 90.004

5. Put the following decimals in order from smallest to largest: 0.402, 0.420, 0.042, 0.442

a. 0.420, 0.042, 0.442, 0.402
b. 0.042, 0.402, 0.420, 0.442
c. 0.442, 0.420, 0.402, 0.042
d. 0.042, 0.402, 0.442, 0.420

6. 9.78 − 4.9 = ____

a. 4.68
b. 4.78
c. 4.88
d. 4.98

7. 6.001 + 9.003 + 13.3 = ____

a. 15.214
b. 15.412
c. 28.138
d. 28.304

8. Two coatings have been applied to the pipe in *Figure 17*. The first coating is 51.5 microns thick; the second coating is 89.7 microns thick. How thick is the combined coating?

a. 141.2 microns
b. 142.12 microns
c. 144.2 microns
d. 145.02 microns

9. The tank in *Figure 18* held 675.57 gallons of fuel last week. Yesterday 356.9 gallons were removed. How many gallons of fuel remain?

a. 3.18
b. 31.86
c. 318.67
d. 3186.7

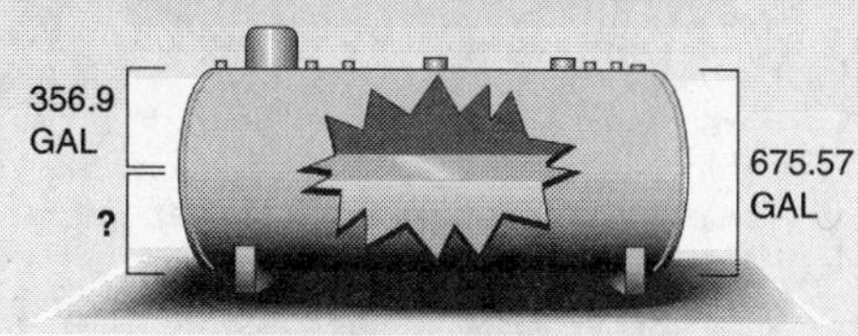

Figure 18 ◆ Fuel tank.

MODULE 00102-04 ◆ INTRODUCTION TO CONSTRUCTION MATH 2.33

Ensure that you have everything required for the demonstration and laboratories during this session.

Go over the Review Questions for Section 5.0.0. Answer any questions trainees may have.

Ensure that trainees understand the relationship among decimals, percentages, and fractions.

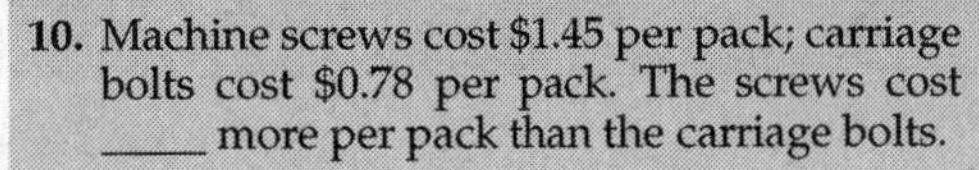

10. Machine screws cost $1.45 per pack; carriage bolts cost $0.78 per pack. The screws cost _____ more per pack than the carriage bolts.

a. $0.67
b. $1.37
c. $1.67
d. $1.87

11. 57.80 × 40 = _____

a. 2,312
b. 2,316
c. 2,318
d. 2,320

12. 3.53 × 9.75 = _____

a. 12.28
b. 34.4175
c. 36.1387
d. 48.13

13. It costs $2.37 to paint one square foot of wall. You need to paint a wall that measures 864.5 square feet. It will cost _____ to paint that wall. (Round your answer to the nearest hundredth.)

a. $204.88
b. $2,048.87
c. $2,848.88
d. $2,888.86

14. 57.80 ÷ 40 = _____

a. 1.445
b. 14.45
c. 145.5
d. 1455

15. 89.435 ÷ 0.05 = _____

a. 1788.7
b. 17.887
c. 4.47175
d. 447.175

Show Transparency 7 (Figure 19). Ask a trainee to determine the percentage of the tank that is filled in Figure 19.

Review the procedures to convert decimals to percentages.

Explain how to convert percentages to decimals.

6.0.0 ◆ CONVERSION PROCESSES

Sometimes you will need to convert some of the numbers you want to work with so that all your numbers appear in the same form. For example, some numbers may appear as decimals, some as **percentages,** and some as fractions. Decimals, percentages, and fractions are all just different ways of expressing the same thing. The decimal 0.25, the percent 25%, and the fraction 1/4 all mean the same thing. In order to work with the different forms of numbers like these, you will need to know how to convert them from one form into another.

6.1.0 Converting Decimals to Percentages and Percentages to Decimals

What are percentages? Think of a whole number divided into 100 parts. You can express any part of the whole as a percentage. Let's look at an example: The tank shown in *Figure 19* has a capacity of 100 gallons. It is now filled with 50 gallons. What percentage of the tank is filled?

If you answered 50 percent (50%), you are correct. Percentage means out of 100. How many gallons out of 100 does the tank contain? It contains 50 out of 100, or 50 percent. Percentages are an easy way to express parts of a whole. Decimals and fractions also express parts of a whole. Let's look at the relationship among percentages, decimals, and fractions.

The tank in *Figure 19* is 50 percent full. If you expressed this as a fraction you would say it was ½ full. You could also express this as a decimal and say it's 0.50 full.

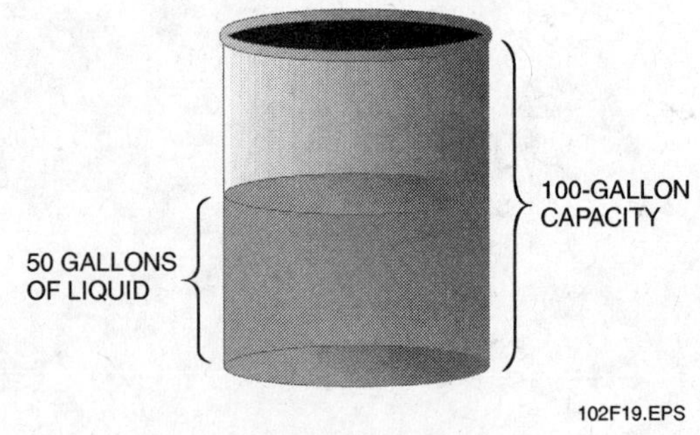

Figure 19 ◆ 100-gallon-capacity tank.

Instructor's Notes:

2.34 CORE CURRICULUM ◆ INTRODUCTORY CRAFT SKILLS

Sometimes you may need to express decimals as percentages or percentages as decimals. Suppose you are preparing a gallon of cleaning solution. The mixture should contain 10 to 15 percent of cleaning agent. The rest should be water. You have 0.12 gallon of cleaning agent. Will you have enough to prepare a gallon of the solution? To answer the question, you must convert a decimal (0.12) to a percentage. You will change 0.12 to a percentage for this exercise:

Step 1 Multiply the decimal by 100. (Tip: When multiplying by 100, simply move the decimal point two places to the right.)

$$0.12 \times 100 = 12$$

Step 2 Add a % sign.

12%

Recall that the mixture should be from 10 to 15 percent cleaning agent. You have 12 percent of a gallon, enough cleaning agent to make the solution.

You may also need to convert percentages to decimals. Let's say that another mixture should contain 22 percent of a certain chemical by weight. You're making 1 pound of the mixture. You weigh the ingredients on a digital scale. How much of the chemical should you add? To answer this, you must convert a percentage (22%) to a decimal. You will change 22 percent to a decimal in the following exercise:

Step 1 Drop the % sign.

22

Step 2 Divide the number by 100. (Tip: When dividing by 100 or more, move the decimal point two places to the left.)

$$22 \div 100 = 0.22$$

The answer to the problem is that you would add 0.22 pound of the chemical to 0.78 pound of the other ingredient to make a 22 percent mixture.

6.1.1 Study Problems: Converting Decimals to Percentages and Percentages to Decimals

Find the answers to the following conversion problems.

1. 0.62 = ____

2. 0.475 = ____

3. 0.7 = ____

4. 72% = ____

5. 12.5% = ____

6.2.0 Converting Fractions to Decimals

You will often need to change a fraction to a decimal. For example, you need ¾ of a dollar. How do you convert ¾ to its decimal equivalent?

Step 1 Divide the numerator of the fraction by the denominator.

$$4\overline{)3.0}$$

In this example, you need to put the decimal point and the zero after the number 3, because you need a number large enough to divide by 4.

Step 2 Put the decimal point directly above its location within the division symbol.

$$\begin{array}{r} .? \\ 4\overline{)3.0} \end{array}$$

Step 3 Once the decimal point is in its proper place above the line, you can divide as you normally would. The decimal point holds everything in place.

$$\begin{array}{r} .75 \\ 4\overline{)3.00} \\ \underline{-2.8} \\ 0.20 \\ \underline{-0.20} \\ 0.00 \end{array}$$

Step 4 Read the answer. The fraction ¾ converted to a decimal is 0.75. So ¾ of a dollar is the same as $0.75.

To save time, you can refer to a chart to convert inches to decimals or decimals to inches. See *Appendix B.*

Ask trainees to complete the study problems for converting decimals to percentages and percentages to decimals. Answer any questions trainees may have.

Review the steps to change fractions to decimals.

See the Teaching Tip for Section 6.2.0 at the end of this module.

MODULE 00102-04 ◆ INTRODUCTION TO CONSTRUCTION MATH 2.35

Have trainees complete the study problems for converting fractions to decimals. Ask whether trainees have any questions.

6.2.1 Study Problems: Converting Fractions to Decimals

Convert the following fractions to their decimal equivalents without using a calculator.

1. $\frac{1}{4}$ = ____
2. $\frac{3}{4}$ = ____
3. $\frac{1}{8}$ = ____
4. $\frac{5}{16}$ = ____
5. $\frac{20}{64}$ = ____

Explain how to convert decimals to fractions. Ensure that trainees understand how to express decimals in words.

6.3.0 Converting Decimals to Fractions

Let's say you have 0.25 of a dollar. What fraction of a dollar is that? Follow these steps to find out:

Step 1 Say the decimal in words.
0.25 is expressed as "twenty-five hundredths"

Step 2 Write the decimal as a fraction.
0.25 is written as a fraction as $\frac{25}{100}$

Step 3 Reduce it to its lowest terms.

$$\frac{25}{100} = \frac{25 \div 25}{100 \div 25} = \frac{1}{4}$$

Step 4 So 0.25 converted to a fraction is $\frac{1}{4}$. If you have 0.25 of a dollar, you have $\frac{1}{4}$ of a dollar.

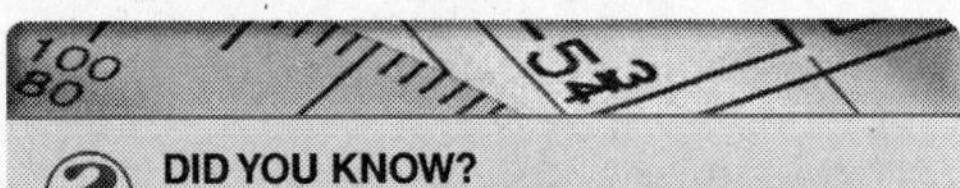

DID YOU KNOW?

HVAC Technicians

The heating, ventilating, and air conditioning (HVAC) trade is really many trades. It requires electrical, plumbing, carpentry, welding, and some insulation and sheet metal work. HVAC technicians install, maintain, diagnose, and correct problems in heating and cooling systems. To do this, they work with many mechanical, electrical, and electronic components. HVAC systems can involve electricity, chemicals and gases, oil, water, or coal. Technicians may specialize in new installations or maintenance of existing climate-control systems. HVAC technicians must have good mathematical skills so they can perform accurate installations and operate precision testing equipment to ensure that the systems function properly.

Ask trainees to complete the study problems for converting decimals to fractions. Go over any questions trainees may have.

6.3.1 Study Problems: Converting Decimals to Fractions

Convert the following decimals to fractions without using a calculator and express them in lowest terms.

1. 0.5 = ____
2. 0.12 = ____
3. 0.125 = ____
4. 0.8 = ____
5. 0.45 = ____

On the whiteboard/chalkboard, demonstrate how to convert inches to decimal equivalents in feet. Explain that, first, trainees should express the inches as a fraction using 12 as the denominator.

6.4.0 Converting Inches to Decimal Equivalents in Feet

What happens if you need to convert inches to their decimal equivalents in feet? For example: 3 inches equals what decimal equivalent in feet?

Here's a hint: First, express the inches as a fraction that has 12 as the denominator. You use 12 because there are 12 inches in a foot. Then reduce the fraction and convert it to a decimal.

In this example, the fraction $\frac{3}{12}$ reduces to $\frac{1}{4}$.

You convert the fraction $\frac{1}{4}$ to a decimal by dividing the 4 into 1.00:

$$\begin{array}{r} 0.25 \\ 4\,\overline{)\,1.00} \\ \underline{-0.8} \\ 0.20 \\ \underline{-0.20} \\ 0 \end{array}$$

Thus, 3 inches converts to 0.25 feet.

Have trainees complete the study problems for converting inches to decimals. Answer any questions trainees may have.

6.4.1 Study Problems: Converting Inches to Decimals

Find the answers to the following conversion problems and round them to the nearest hundredth.

1. 9 inches = ____ feet
2. 10 inches = ____ feet
3. 2 inches = ____ feet
4. 4 inches = ____ feet
5. 8 inches = ____ feet

Instructor's Notes:

____

____

____

2.36 CORE CURRICULUM ◆ INTRODUCTORY CRAFT SKILLS

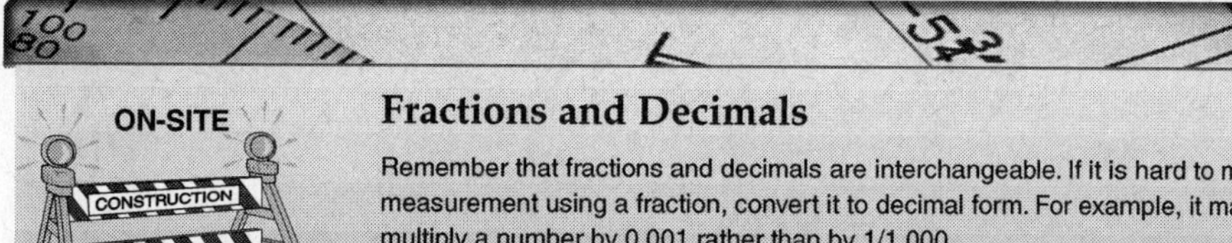

ON-SITE

Fractions and Decimals

Remember that fractions and decimals are interchangeable. If it is hard to multiply a measurement using a fraction, convert it to decimal form. For example, it may be easier to multiply a number by 0.001 rather than by 1/1,000.

When using a calculator, you will have to convert fractions into decimals to perform your calculations.

Review Questions

Section 6.0.0

Solve the following problems without using a calculator.

1. 0.78 = _____ %

a. 0.0078
b. 0.78
c. 7.8
d. 78

2. 13.9% = _____

a. 0.009
b. 0.013
c. 0.139
d. 1.39

3. $\frac{3}{16}$ = _____

a. 0.0087
b. 0.087
c. 0.185
d. 0.1875

4. $\frac{7}{32}$ = _____

a. 0.21875
b. 2.1875
c. 21.875
d. 218.75

5. Convert the following decimal to its equivalent fraction and express it in lowest terms: 0.6 = _____

a. $\frac{1}{3}$
b. $\frac{1}{4}$
c. $\frac{2}{3}$
d. $\frac{3}{5}$

6. Convert the following decimal to its equivalent fraction and express it in lowest terms: 0.875 = _____

a. $\frac{8}{7}$
b. $\frac{875}{1000}$
c. $\frac{7}{8}$
d. $\frac{35}{40}$

7. Convert the following mixed decimal to its equivalent improper fraction expressed in lowest terms: 14.75 = _____

a. $14\frac{75}{100}$
b. $14\frac{3}{4}$
c. $14\frac{3}{5}$
d. $14\frac{15}{20}$

8. Convert the following mixed decimal to its equivalent improper fraction expressed in lowest terms: 2.04 = _____

a. $2\frac{4}{10}$
b. $2\frac{2}{5}$
c. $2\frac{4}{100}$
d. $2\frac{1}{25}$

9. Find the answer to the following conversion problem and round it to the nearest hundredth: 6 inches = _____ feet

a. 0.4
b. 0.5
c. 0.6
d. 0.8

10. Find the answer to the following conversion problem and round it to the nearest hundredth: 11 inches = _____ feet

a. 0.74
b. 0.827
c. 0.96
d. 0.92

Go over the Review Questions for Section 6.0.0. Answer any questions trainees may have.

Have trainees review Sections 7.0.0–7.3.1 for the next session.

MODULE 00102-04 ◆ INTRODUCTION TO CONSTRUCTION MATH 2.37

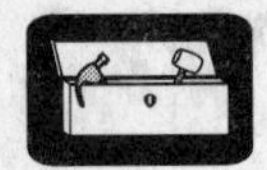

Ensure that you have everything required for the demonstration and laboratories during this session.

Ask trainees to provide examples of common measurements using the metric system.

Show Transparency 8 (Figure 20). Review the information that can be learned from the name of each metric measurement.

Refer to Figure 21. Discuss the relationship among the different metric units.

7.0.0 ◆ INTRODUCTION TO THE METRIC SYSTEM

The metric system is a system of measurement that uses a base-ten method of determining weight, length, **volume,** and temperature. That means that all measurements are counted in tens. Much of the world uses the metric system, and the company you work for may do business in places where the metric system is in use. You need to learn the metric system in case the projects you work on rely on metric system measures.

You may be surprised to find you are already familiar with some of the common metric units. For example, have you purchased a 2-liter bottle of soda? Have you run a 10K (kilometer) race lately? These are metric measures.

7.1.0 Units of Weight, Length, Volume, and Temperature

The name of each metric measurement (see *Figure 20*) tells you two things:

- What type of measurement it is (the basic unit):
  grams = weight   meters = length
  liters = volume   Celsius = temperature
- Its size (in relation to the basic unit, such as the meter):
  deka (da) = 10   deci (d) = 0.1
  hecto (h) = 100   centi (c) = 0.01
  kilo (k) = 1,000   milli (m) = 0.001
  mega (M) = 1,000,000   micro (μ) = 0.000001

The most common prefixes are kilo, milli, and centi. Hecto is used mainly with meters in calculating land size. Mega and micro are used in scientific and engineering measurements.

See *Figure 21* to learn the relationship between the different metric units.

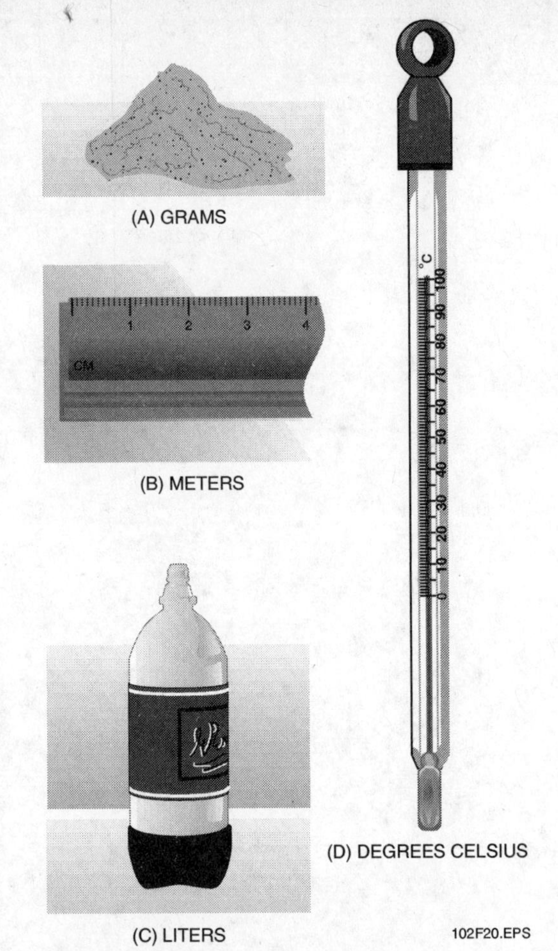

Figure 20 ◆ What type of measurement is it?

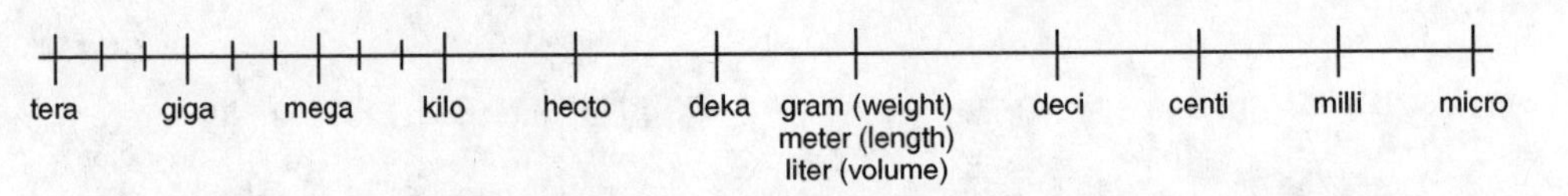

10^0 = 1	BASE UNIT: GRAM (WEIGHT), METER (LENGTH), LITER (VOLUME)				10^0 = 1
MULTIPLIER	**PREFIX**	**MEANING**	**PREFIX**	**MEANING**	**MULTIPLIER**
10^1 = 10	deka-	ten	deci-	tenth	10^{-1} = 10
10^2 = 100	hecto-	hundred	centi-	hundredth	10^{-2} = 100
10^3 = 1,000	kilo-	thousand	milli-	thousandth	10^{-3} = 1,000
10^6 = 1,000,000	mega-	million	micro-	millionth	10^{-6} = 1,000,000
10^9 = 1,000,000,000	giga-	billion	nano-	billionth	10^{-9} = 1,000,000,000
10^{12} = 1,000,000,000,000	tera-	trillion	pico-	trillionth	10^{-12} = 1,000,000,000,000

102F21.EPS

Figure 21 ◆ The decimal scale.

Instructor's Notes:

2.38 CORE CURRICULUM ◆ INTRODUCTORY CRAFT SKILLS

The scale in *Figure 21* is mostly related to metric measurement and solid, liquid, and linear measurements. For every jump on the number line, move the decimal one place in the same direction. Add zeros if necessary. For example, if you have 100 centiliters and want to convert to hectoliters, you move the decimal four places to the left, adding zeros (or divide by 1,000). So 100 centiliters equals 0.01 hectoliter.

7.1.1 Study Problems: Converting within the Metric System

1. A dekagram is _____ gram(s).

2. A hectogram is _____ gram(s).

3. A megagram is _____ gram(s).

4. A decimeter is _____ meter(s).

5. A millimeter is _____ meter(s).

7.2.0 Using a Metric Ruler

Metric rulers are often used with blueprints, in which most measurements are given either in centimeters or in millimeters. In metalworking, for example, it is most common to take measurements in millimeters.

The metric ruler in *Figure 22* is divided into centimeters (cm) and millimeters (mm).

DID YOU KNOW?

Metric System as Modern Standard

As scientific thought developed during the 1500s and later, scholars and scientists had trouble explaining their measurements to one another. The measuring standards varied among countries and sometimes even within one country. By the 1700s, scientists were debating how to establish a uniform system for measurement.

In France in the 1790s, scientists created a standard length called a **meter** (based on the Latin word for measure). This became the basis for the metric system that is still in use throughout most of the world. Not until the 1970s did both the United States and Canada begin to switch over to the metric system. There is still resistance to the metric system in the United States, even though virtually every other country in the world uses it.

The original international standard bar measuring a meter was a platinum-iridium bar kept in the International Bureau of Weights and Measures near Paris, France. The bar was made from a platinum-iridium mixture because it would not rust or change over time. That ensured the accuracy of the standard.

In more modern times, scientists have found that a natural standard measure was more accurate than anything made out of metal. In 1983 the speed of light was calculated as exactly 299,792,458 meters per second. So the distance that light travels in 1/299,792,458 second is the definition of the length of a meter. This may be a more precise measure than you will need on the construction site, but it illustrates the accuracy of modern techniques.

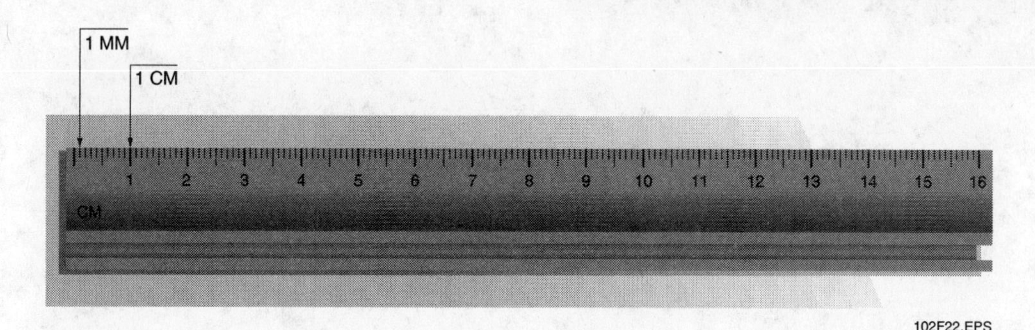

Figure 22 ◆ Metric ruler.

Ask trainees to complete the study problems for converting within the metric system. Go over any questions trainees may have.

See the Teaching Tip for Sections 7.1.0–7.1.1 at the end of this module.

Show Transparency 9 (Figure 22). Explain how to read measurements on a metric ruler.

MODULE 00102-04 ◆ INTRODUCTION TO CONSTRUCTION MATH 2.39

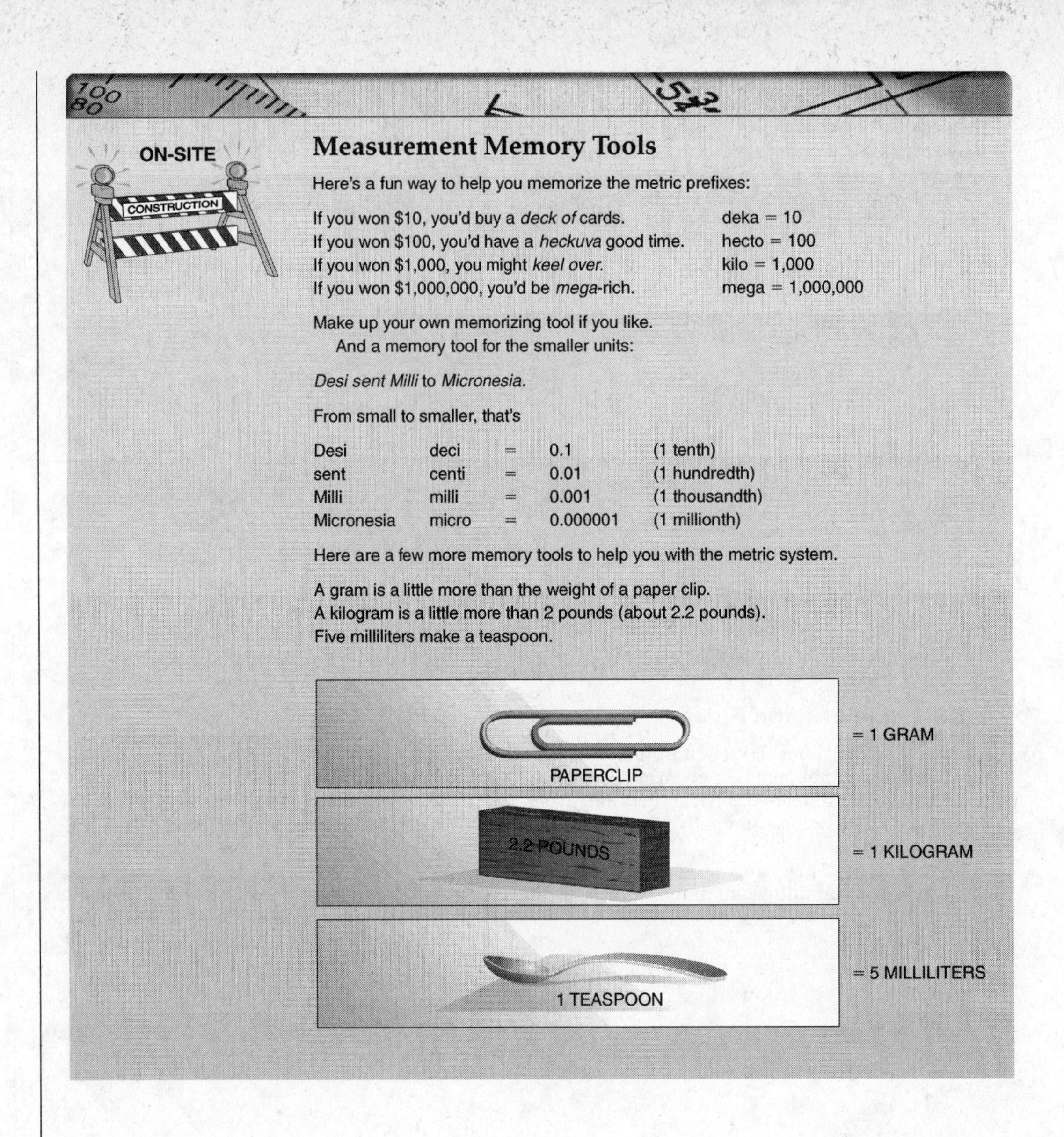

Measurement Memory Tools

Here's a fun way to help you memorize the metric prefixes:

If you won $10, you'd buy a *deck of* cards.	deka = 10
If you won $100, you'd have a *heckuva* good time.	hecto = 100
If you won $1,000, you might *keel over.*	kilo = 1,000
If you won $1,000,000, you'd be *mega*-rich.	mega = 1,000,000

Make up your own memorizing tool if you like.

And a memory tool for the smaller units:

Desi sent Milli to *Micronesia.*

From small to smaller, that's

Desi	deci	=	0.1	(1 tenth)
sent	centi	=	0.01	(1 hundredth)
Milli	milli	=	0.001	(1 thousandth)
Micronesia	micro	=	0.000001	(1 millionth)

Here are a few more memory tools to help you with the metric system.

A gram is a little more than the weight of a paper clip.
A kilogram is a little more than 2 pounds (about 2.2 pounds).
Five milliliters make a teaspoon.

Instructor's Notes:

2.40 CORE CURRICULUM ◆ INTRODUCTORY CRAFT SKILLS

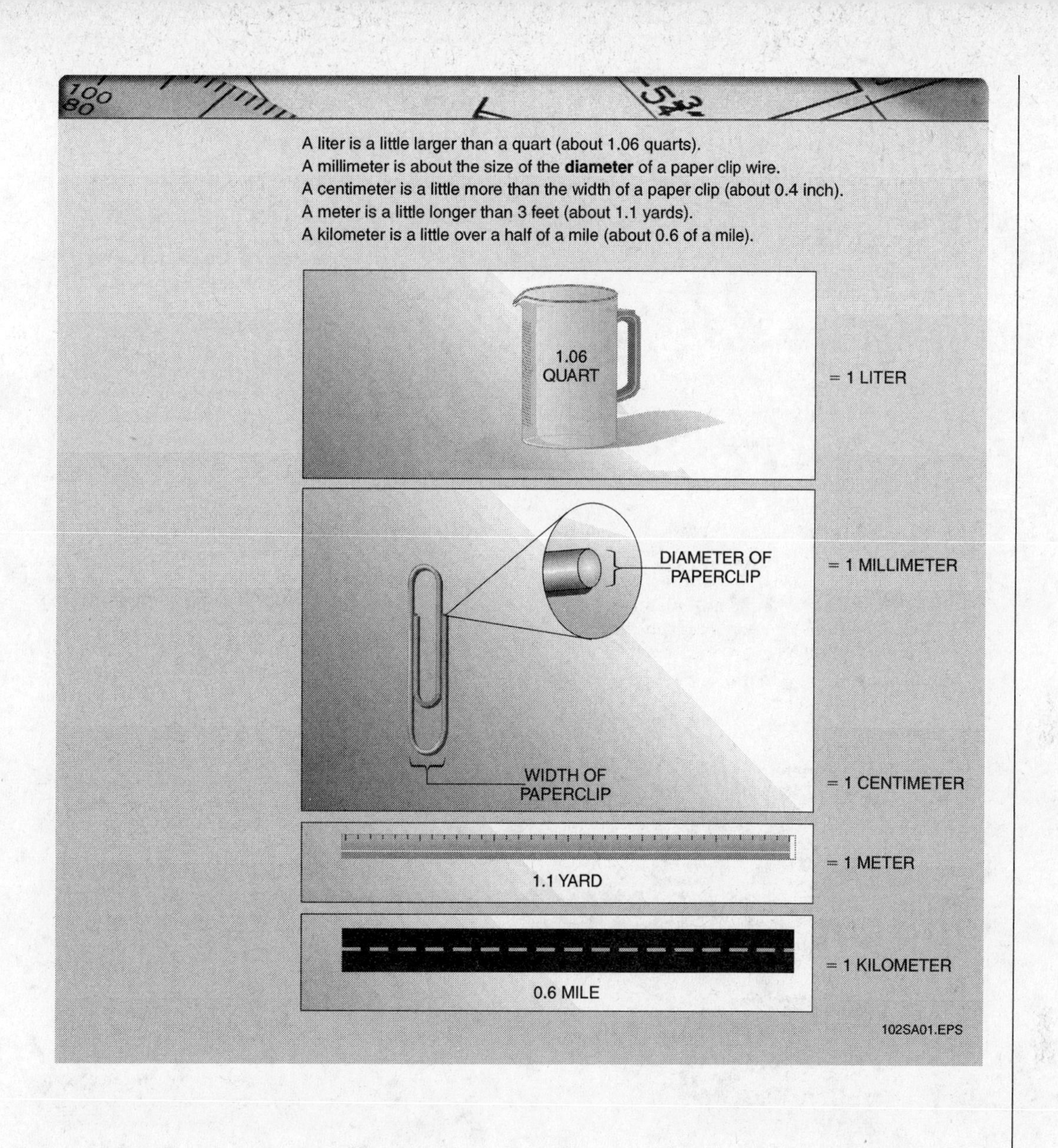
A liter is a little larger than a quart (about 1.06 quarts).
A millimeter is about the size of the diameter of a paper clip wire.
A centimeter is a little more than the width of a paper clip (about 0.4 inch).
A meter is a little longer than 3 feet (about 1.1 yards).
A kilometer is a little over a half of a mile (about 0.6 of a mile).
1.06 QUART
= 1 LITER
DIAMETER OF PAPERCLIP
= 1 MILLIMETER
WIDTH OF PAPERCLIP
= 1 CENTIMETER
1.1 YARD
= 1 METER
0.6 MILE
= 1 KILOMETER
102SA01.EPS

MODULE 00102-04 ◆ INTRODUCTION TO CONSTRUCTION MATH 2.41

Have trainees complete the study problems for using a metric ruler. Answer any questions trainees may have.

7.2.1 *Study Problems: Using a Metric Ruler*

Use the metric ruler in *Figure 23* to find the answers to the following questions.

1. P is at the _____ mark.
  a. 5 centimeter
  b. 5.5 centimeter
  c. 50 centimeter
  d. 5 centimeter

2. Q is at the _____ mark.
  a. 5 centimeter
  b. 55 centimeter
  c. 5 millimeter
  d. 55 millimeter

3. R is at the _____ mark.
  a. 2.5 millimeter
  b. 2.5 centimeter
  c. 250 millimeter
  d. 250 centimeter

4. S is at the _____ mark.
  a. 38 centimeter
  b. 3 centimeter and 8 millimeter
  c. 30 millimeter and 80 centimeter
  d. 83 millimeter

5. T is at the _____ mark.
  a. 1.3 millimeter
  b. 133 millimeter
  c. 13 centimeter
  d. 1.3 centimeter

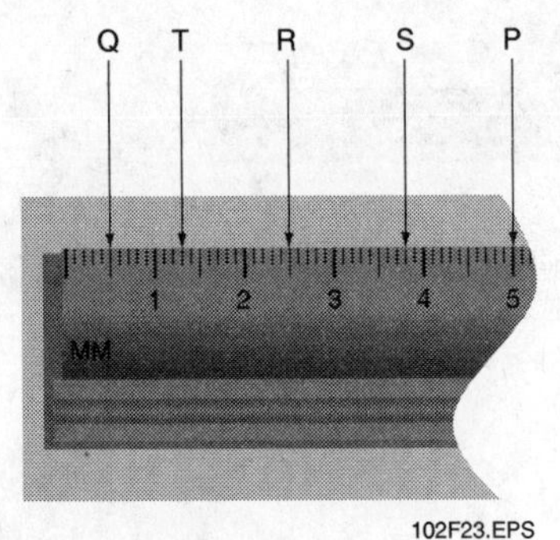

Figure 23 ◆ Study problems metric ruler.

DID YOU KNOW?

The Standard Measure

Determining the length of an object or the distance to a neighbor's farm may have been the earliest measurement of interest to people. Imagine that you wanted to measure the distance to a favorite fishing pond. You could walk—pace off—the distance and count the number of steps you took. You might find that it was 70 paces between your home and the pond. You could then pace off the distance to another pond. By comparing the paces of one distance to the other, you could tell which distance was longer and by how much. Pace became a unit of measure.

Feet, arms, hands, and fingers were useful for measuring all sorts of things. In fact, the body was such a common basis for measuring that we still have traces of that system in our measuring. Lengths are measured in feet. Horses are said to be so many hands high. An inch roughly matches the width of someone's thumb.

The problem develops when you have to decide whose legs, arms, or hands to use for measuring. If you wanted to mark off a piece of land, you might choose a tall person with long legs in hopes of getting more than your money's worth. If you owed several lengths of rope, you would want someone with short arms to measure the quantity and possibly save you some rope.

The solution is to arrive at a measurement that everyone agrees on. Legend says that the foot measurement used by the early French was the length of Charlemagne's foot. The yard, which is three feet long, is supposed to have been the distance from King Henry I of England's nose to the fingertips of his outstretched arm. Kings were not going to travel around their countries or to foreign countries to make measurements for the people, so the solution was to transfer the measurement to a stick. The distance marked on the stick became the standard measurement, and the government could send out duplicate sticks to each of the towns and cities as secondary standards. Civil officials could then check local merchants' measurements using the secondary standards.

During the Middle Ages, associations of craftspeople enforced strict adherence to the established standards. The success and reputation of the craft depended on providing accurately measured goods. Violators who used faulty measurements were often punished.

Instructor's Notes:

2.42 CORE CURRICULUM ◆ INTRODUCTORY CRAFT SKILLS

DID YOU KNOW?

Simple System

The metric system is very easy to use. The basic unit of length is a meter. Anything longer than 1 meter is a multiple of the meter. For example, *kilo-* is a prefix meaning thousand. So a kilometer is 1,000 meters. The prefix *mega-* means one million. So a megameter is 1,000,000 meters (although you rarely hear this term used).

Anything shorter than a meter is a part of a meter and can be calculated using division. The most common smaller units of a meter used in measuring are the centimeter and the millimeter. The *centi-* prefix means one hundredth ($\frac{1}{100}$) and *milli-* means one thousandth ($\frac{1}{1,000}$). A centimeter is $\frac{1}{100}$ of a meter. A millimeter is $\frac{1}{1,000}$ of a meter.

All the units of measure in the metric system differ by multiples of 10. This makes it easy to convert measurements within the metric system.

Centimeter = 1 × 0.01
Millimeter = 1 × 0.001
Meter = 1
Kilometer = 1 × 1,000
Megameter = 1 × 1,000,000

To convert 1 kilometer to meters, you multiply by 1,000. Thus, 1 kilometer equals 1,000 meters (1 × 1000). To convert 1 meter to centimeters, you divide by 0.01. Thus, 1 meter equals 100 centimeters (1 ÷ 0.01 = 100).

7.3.0 Converting Measurements

Sometimes you may need to change from one unit of measurement to another—say, from inches to yards or from centimeters to meters.

In the standard measurement system, also called the English system, this may involve several steps. For example, to change from inches to yards, you must first divide the number of inches by 12 (the number of inches in a foot) and then divide that number by 3 (the number of feet in a yard), or just divide by 36 (the number of inches in a yard).

How many yards are in 72 inches? See *Figure 24.*

There are 2 yards in 72 inches.

The metric system makes the conversion much simpler. You can simply move the decimal point, because the system is built on multiples of 10.

How many meters are there in 72 centimeters? See *Figure 25.*

Because 1 centimeter = 0.01 meter (move decimal two places to the left), 72 centimeters = 0.72 meter.

There are 0.72 meter in 72 centimeters.

This problem is similar to working with money. If you had 72 cents in your pocket, how much of a dollar would you have? You'd have $0.72, or 72 hundredths of a dollar.

Converting measurements from the English system to the metric system, and vice versa, is more complicated. At this stage in your training, you will not be responsible for making such conversions, but it is important that you at least be aware of them. Many dictionaries and other reference books contain simple comparison charts that show some basic equivalents between English system measurements and metric system measurements. See *Appendix C* for a sample comparison chart.

Refer to Fig
Explain how
vert inches

Show how you
move the decir
point to find ho
many meters ar
specific numbe
centimeters.

See the Teaching
for Section 7.3.0 a
the end of this
module.

Review the basic prefixes used in the metric system. Ensure that trainees understand how to convert metric measurements within the metric system.

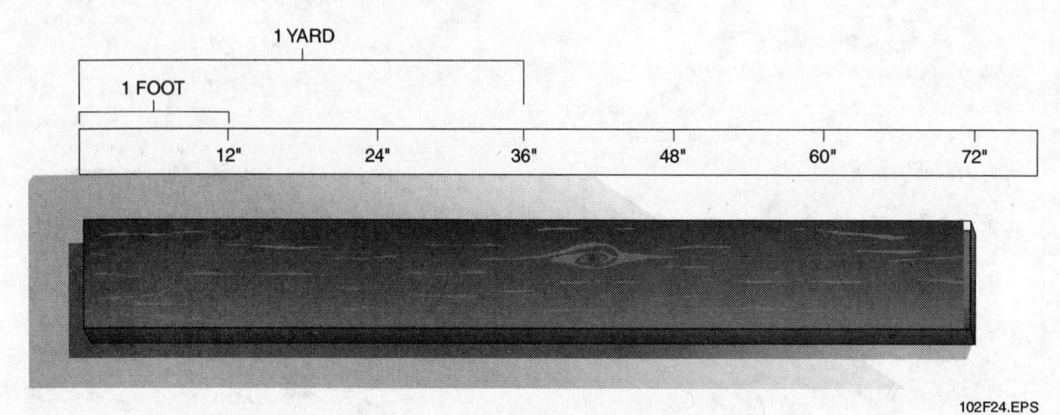

Figure 24 ◆ 72 inches of material.

MODULE 00102-04 ◆ INTRODUCTION TO CONSTRUCTION MATH 2.43

Ask trainees to complete the study problems for converting measurements. Go over any questions trainees may have.

Go over the Review Questions for Section 7.0.0. Answer any questions trainees may have.

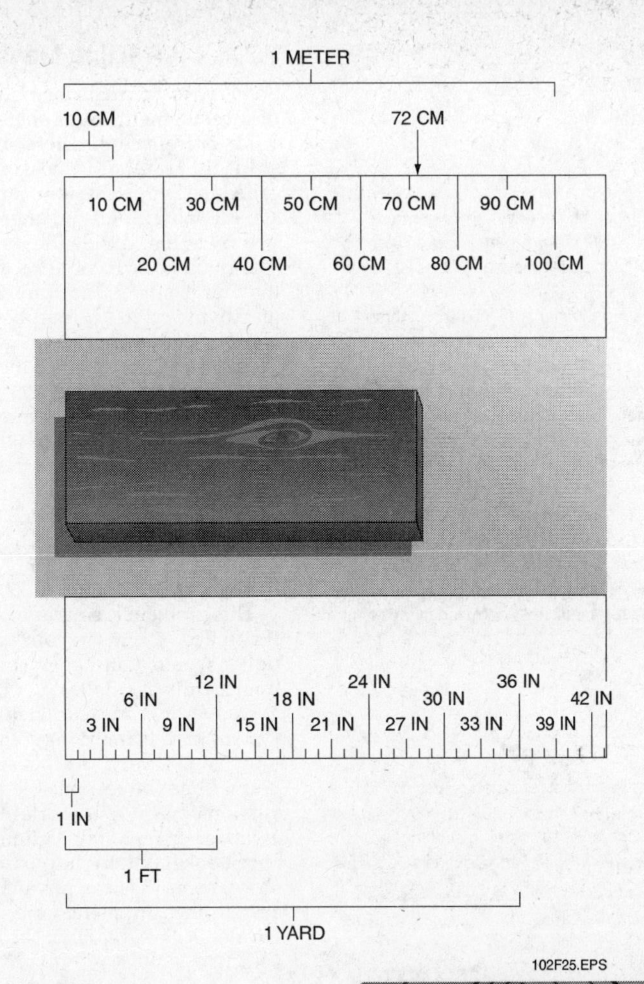

Figure 25 ◆ 72 centimeters of material.

7.3.1 *Study Problems: Converting Measurements*

Find the answers to the following conversion problems without using a calculator.

1. 0.45 meter = _____ centimeter(s)

2. 3 yards = ______ inches

3. 36 inches = _____ yard(s)

4. 90 inches = _____ yards

5. 1 centimeter = _____ meters

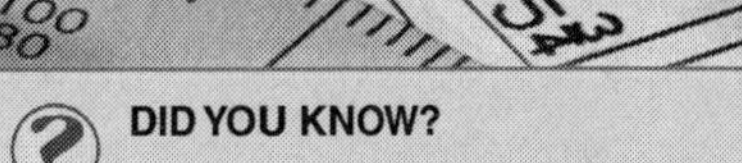

DID YOU KNOW?

Common Terms

The most common units of measure in the metric system are the meter, kilometer, millimeter, and centimeter. Unit names for measurements larger than a kilometer are not common. For example, the term *hectometer* (100 meters) is not often used. In the Olympics you hear about the 200-meter and the 400-meter races, not the 2-hectometer and the 4-hectometer races.

Measurements smaller than the millimeter (0.001 meter) are usually used by scientists. The micrometer (0.000001 meter) and nanometer (0.000000001) are simply not used in everyday measuring.

Instructor's Notes:

____

____

____

2.44 CORE CURRICULUM ◆ INTRODUCTORY CRAFT SKILLS

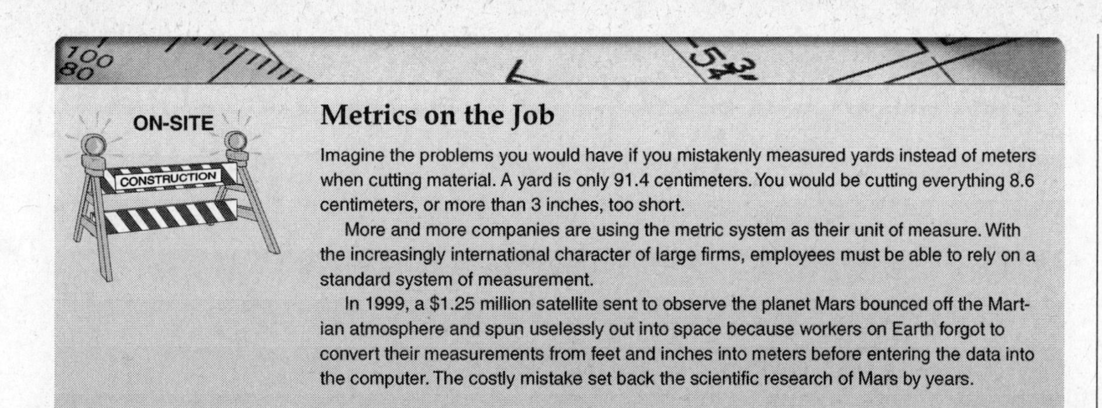

ON-SITE

Metrics on the Job

Imagine the problems you would have if you mistakenly measured yards instead of meters when cutting material. A yard is only 91.4 centimeters. You would be cutting everything 8.6 centimeters, or more than 3 inches, too short.

More and more companies are using the metric system as their unit of measure. With the increasingly international character of large firms, employees must be able to rely on a standard system of measurement.

In 1999, a $1.25 million satellite sent to observe the planet Mars bounced off the Martian atmosphere and spun uselessly out into space because workers on Earth forgot to convert their measurements from feet and inches into meters before entering the data into the computer. The costly mistake set back the scientific research of Mars by years.

Have trainees review Sections 8.0.0–8.4.1 for the next session.

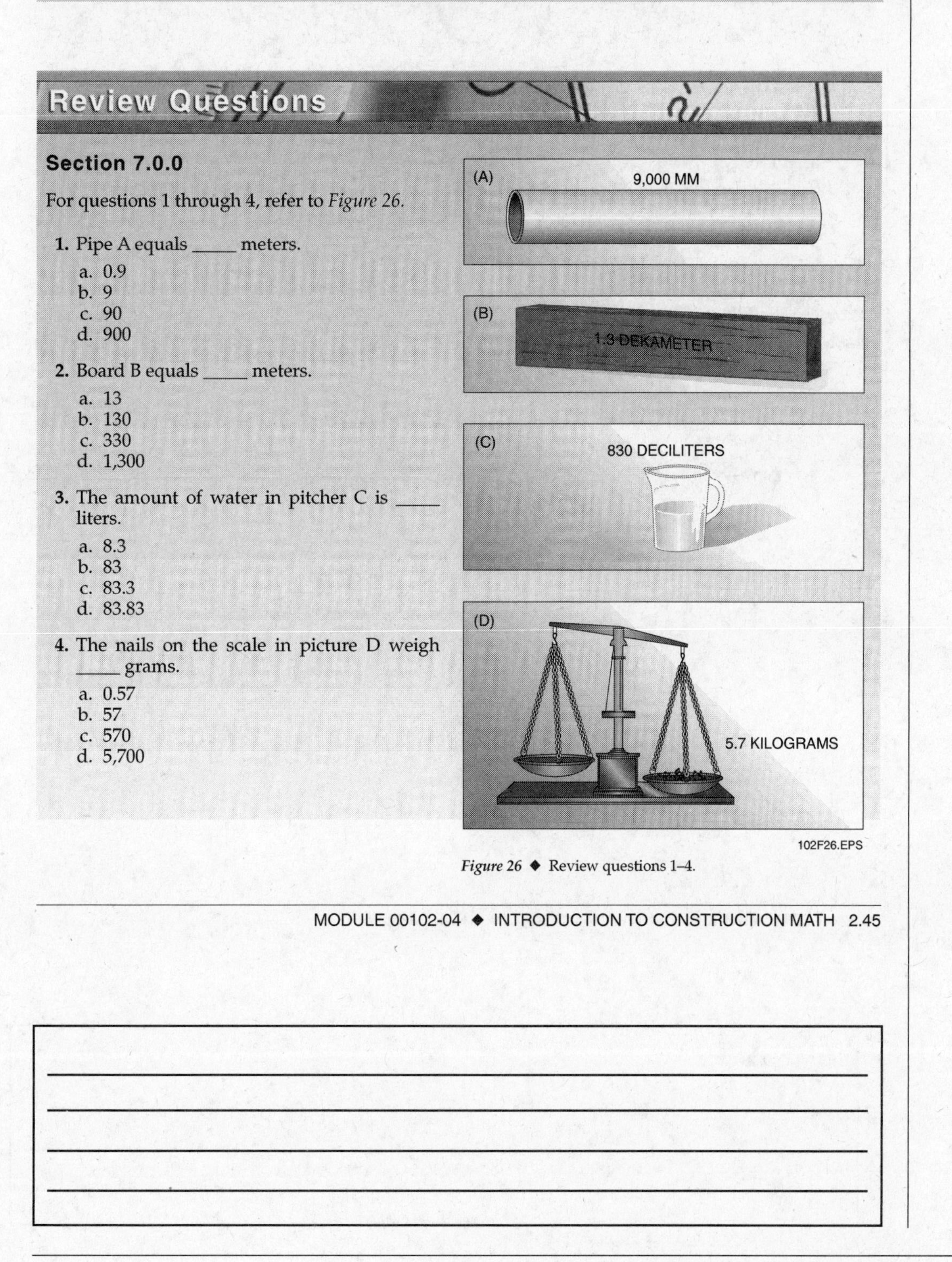

Review Questions

Section 7.0.0

For questions 1 through 4, refer to *Figure 26.*

1. Pipe A equals _____ meters.
   a. 0.9
   b. 9
   c. 90
   d. 900

2. Board B equals _____ meters.
   a. 13
   b. 130
   c. 330
   d. 1,300

3. The amount of water in pitcher C is _____ liters.
   a. 8.3
   b. 83
   c. 83.3
   d. 83.83

4. The nails on the scale in picture D weigh _____ grams.
   a. 0.57
   b. 57
   c. 570
   d. 5,700

Figure 26 ◆ Review questions 1–4.

Review Questions

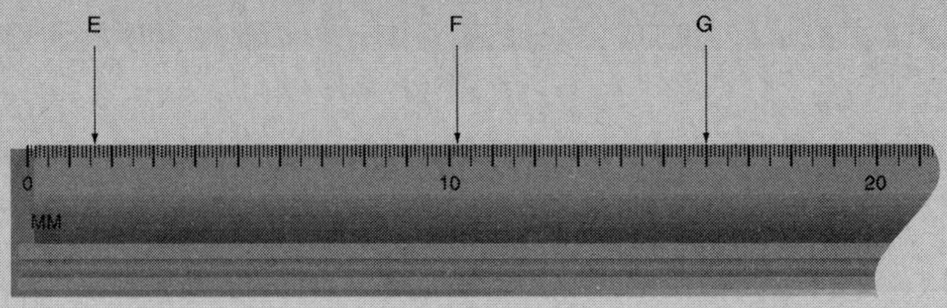

E
F
G
0
10
20
MM

102F27.EPS

Figure 27 ◆ Review questions 5–7.

For questions 5 through 7, refer to *Figure 27*.

5. E is at the _____ mark.

a. 2.6 cm
b. 2.6 mm
c. 16 mm
d. 26 cm

6. F is at the _____ mark.

a. 10.2 cm
b. 10.2 mm
c. 1,020 mm
d. 1,020 cm

7. G is at the _____ mark.

a. 16 mm
b. 16 cm
c. 1,600 mm
d. 1,600 cm

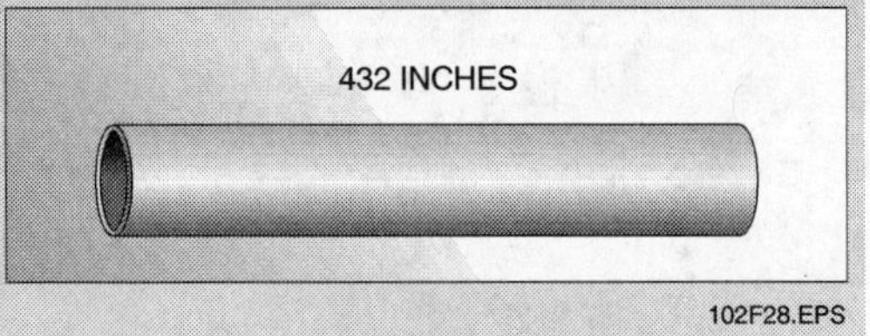

432 INCHES

102F28.EPS

Figure 28 ◆ Review questions 8 and 9.

For questions 8 and 9, refer to *Figure 28*.

8. The pipe measures _____ feet.

a. 32
b. 34
c. 36
d. 38

9. The pipe measures _____ yards.

a. 6
b. 9
c. 12
d. 15

10. 17 centimeters = _____ mm

a. 170
b. 1.7
c. 0.17
d. 0.017

Instructor's Notes:

8.0.0 ◆ INTRODUCTION TO CONSTRUCTION GEOMETRY

Geometry might sound scary, but it is really made up of everyday things you already know—**circles, triangles, squares,** and **rectangles!** The construction industry exists in a world of measurements. You should recognize the basic shapes and measurements that make your work possible.

8.1.0 Angles

An angle is an important term in the construction trades. It is used by all building trades to describe the shape made by two straight lines that meet in a point or **vertex.** Angles are measured in **degrees.** To measure angles, you use a tool called a protractor. The following are the typical angles (see *Figure 29*) you will measure in construction:

- **Acute angle** – An angle that measures between 0 and 90 degrees is an acute angle. The most common acute angles are 30, 45, and 60 degrees.
- **Right angle** – An angle that measures 90 degrees is called a right angle. The two lines that form the right angle are perpendicular to each other. Imagine the shape of a capital letter L. This is a right angle. The sides of the L are perpendicular to one another. This is the angle used most often in the construction trade. A right angle is indicated in plans or drawings with this symbol:

∟

- **Obtuse angle** – An angle that measures between 90 and 180 degrees is obtuse.
- **Straight angle** – A straight angle measures 180 degrees (a flat line).

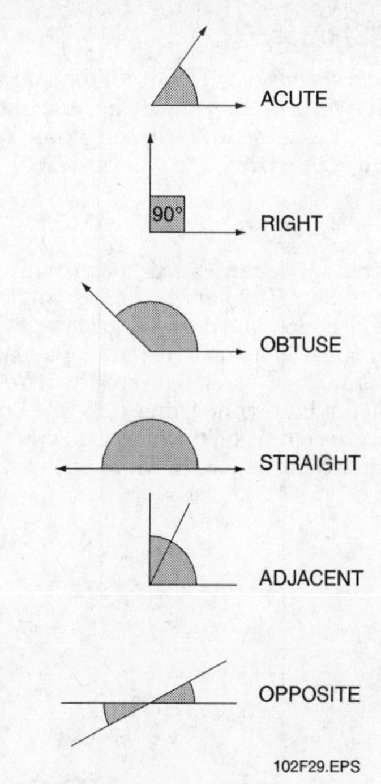

Figure 29 ◆ Typical angles.

- **Adjacent angles** – These angles have the same vertex and one side in common. Adjacent refers to objects that are next to each other.
- **Opposite angles** – Angles formed by two straight lines that cross are opposite. Opposite angles are always equal.

DID YOU KNOW?

Rope Stretchers

The word *geometry* comes from two Greek words: *geos,* meaning land, and *metrein,* meaning to measure.

In ancient Egypt, most farms were located beside the Nile River. Every year during the flood season, the Nile overflowed its banks and deposited mineral-rich silt over the farmland. The floodwaters destroyed the markers used to establish property lines between farms. When this happened, the farms had to be measured again. Men called rope stretchers re-marked the property lines.

The rope stretchers calculated distances and directions using ropes that had equally spaced knots tied along the length of the rope. Stretching the ropes to measure distances on level land was easy. In many cases, however, the rope stretchers had to measure property lines from one side of a hill to the opposite side or across a pond.

The hills and ponds made this measurement difficult. To adjust for the uneven land or ponds that stood in the way, rope stretchers determined new ways of measuring. Such discoveries became the foundation of geometry.

Ensure that you have everything required for demonstrations, laboratories, and testing during this session.

Show Transparency 10 (Figure 29). Review the various types of angles. Ask a trainee to identify the vertex in one of the angles in Figure 29.

On the whiteboard/ chalkboard, draw a variety of angles, and demonstrate how to measure angles using a protractor.

Show Transparency 11 (Figure 30). Review common shapes.

See Figure 31. Ask trainees to identify the common shapes in Figure 31.

Show Transparency 12 (Figure 32). Explain how to cut a rectangle on the diagonal to produce two right triangles.

8.2.0 Shapes

Common shapes (see *Figure 30*) that are essential to your work in the trades include rectangles, squares, triangles, and circles. Look for these shapes in *Figure 31.*

8.2.1 Rectangle

A rectangle is a four-sided shape with four 90-degree angles. (The sum of all four angles in any rectangle is 360 degrees.) A rectangle has two pairs of equal sides that are parallel to each other. The **diagonals** of a rectangle are always equal. Diagonals are lines connecting opposite corners. If you cut a rectangle on the diagonal, you will have two **right triangles,** as shown in *Figure 32.*

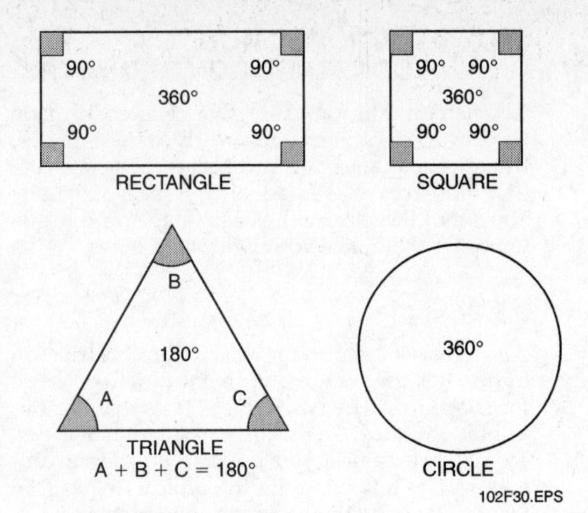

Figure 30 ◆ Common shapes.

Figure 31 ◆ What common shapes can you find at this construction site?

Instructor's Notes:

8.2.2 Square

A square is a special type of rectangle with four equal sides and four 90-degree angles. (The sum of all four angles in all squares is 360 degrees.) If you cut a square on the diagonal, you will also have two right triangles. Each right triangle will have two 45-degree angles and one 90-degree angle, as shown in *Figure 33.*

When measuring the outside lines of a rectangle or a square, you are determining the **perimeter.** The perimeter is the sum of all four sides of a rectangle or square. You may need to calculate the perimeter of a shape so that you can measure, mark, and cut the right amount of material. For example, if you need to install shoe molding along all four walls of a room, you must know the perimeter measurement. If the room is 14 feet by 12 feet, you would calculate: $14 + 12 + 14 + 12 =$ 52 feet of shoe molding. Another way to calculate would be $(2 \times 14 \text{ feet}) + (2 \times 12 \text{ feet}) = 52$ feet.

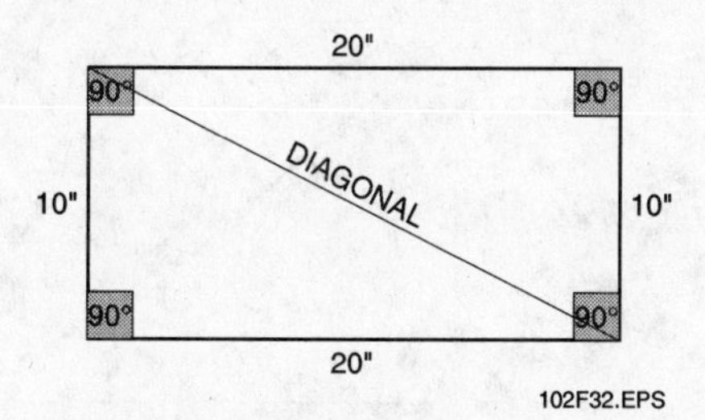

Figure 32 ◆ Cutting a rectangle on the diagonal produces two right triangles.

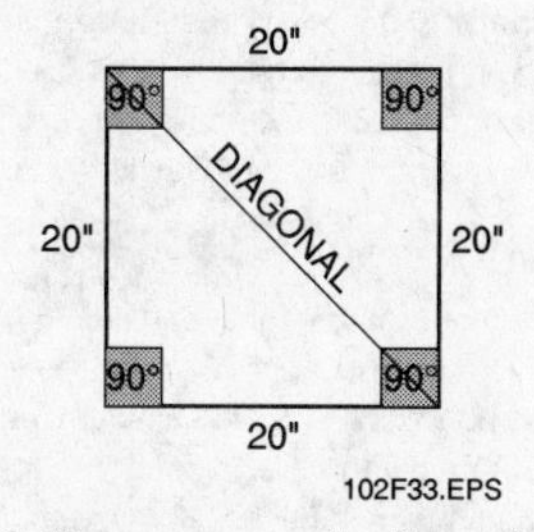

Figure 33 ◆ Cutting a square on the diagonal produces two right triangles.

DID YOU KNOW?

Millwrights

Millwrights install, repair, replace, and dismantle the machinery and heavy equipment used in almost every industry. They may be responsible for placement and installation of machines in a plant or shop. They use hoists, pulleys, jacks, and come-alongs to perform tasks. They also use mechanical trade hand tools such as micrometers and calipers.

Millwrights fit bearings, align gears and wheels, attach motors, and connect belts according to manufacturers' specifications. They may be in charge of preventive maintenance such as lubrication and fixing or replacing worn parts. Precision leveling and alignment are important in the assembly process. Millwrights must have good mathematical skills so they can measure **angles,** material thickness, and small distances.

Sheet Metal Workers

Sheet metal workers make, install, and maintain air conditioning, heating, ventilation, and pollution control duct systems; roofs; siding; rain gutters and downspouts; skylights; restaurant equipment; outdoor signs; and many other building parts and products made from metal sheets. They may also work with fiberglass and plastic materials. They use math and geometry to calculate angles for fabrication and installation of mechanical systems. Some workers may specialize in testing, balancing, adjusting, and servicing existing air conditioning and ventilation systems.

Explain how to determine whether a piece of sheathing is a true rectangle.

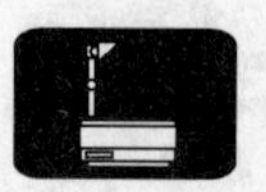

Show Transparency 13 (Figure 33). Explain how cutting a square on the diagonal produces two right triangles.

On the whiteboard/chalkboard, draw a rectangle or a square. Demonstrate how to determine the perimeter by measuring and adding the outside lines.

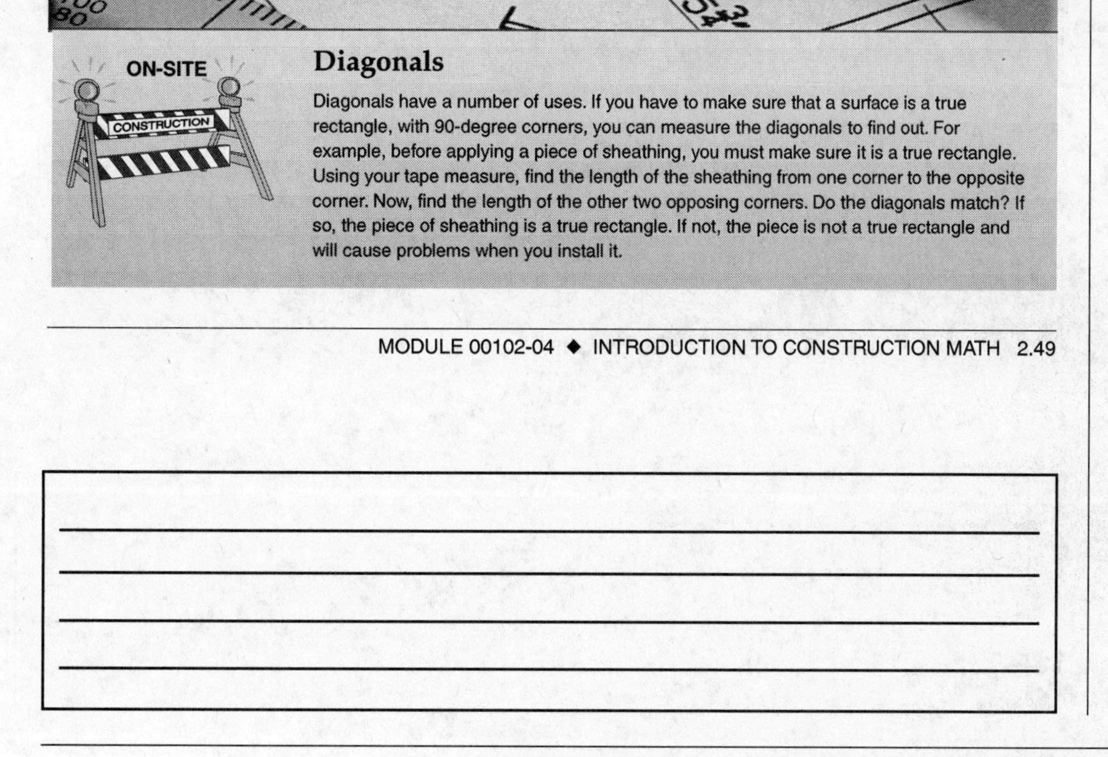

Diagonals

Diagonals have a number of uses. If you have to make sure that a surface is a true rectangle, with 90-degree corners, you can measure the diagonals to find out. For example, before applying a piece of sheathing, you must make sure it is a true rectangle. Using your tape measure, find the length of the sheathing from one corner to the opposite corner. Now, find the length of the other two opposing corners. Do the diagonals match? If so, the piece of sheathing is a true rectangle. If not, the piece is not a true rectangle and will cause problems when you install it.

MODULE 00102-04 ◆ INTRODUCTION TO CONSTRUCTION MATH 2.49

On the whiteboard/chalkboard, draw a right triangle, and assign values to two of the three sides. Have trainees practice solving the Pythagorean formula for a right triangle.

Show Transparency 14 (Figure 34). Measure the three angles in the triangles in Figure 34. Verify that the sum of the three angles equals 180 degrees.

8.2.3 Triangle

A triangle is a closed shape that has three sides and three angles. Although the angles in a triangle can vary, the sum of the three angles is always 180 degrees (see *Figure 34*). The following are different types of triangles you will use in construction (see *Figure 34*):

- *Right triangle* – A right triangle has one 90-degree angle.
- ***Equilateral triangle*** – An equilateral triangle has three equal angles and three equal sides.
- ***Isosceles triangle*** – An isosceles triangle has two equal angles and two equal sides. A line that bisects (runs from the center of the base of the triangle to the highest point) an isosceles triangle creates two adjacent right angles.
- ***Scalene triangle*** – A scalene triangle has three sides of unequal lengths.

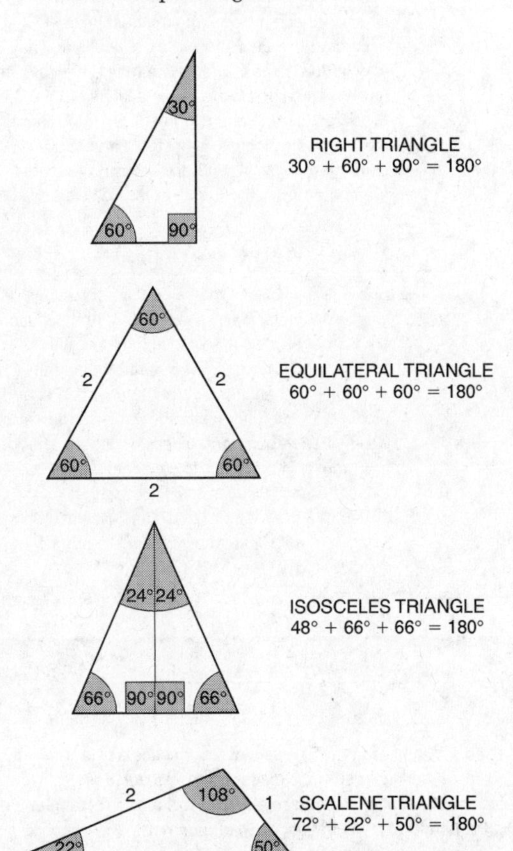

Figure 34 ◆ The sum of a triangle's three angles always equals 180 degrees.

DID YOU KNOW?

Pythagorean Theorem

A formula (a mathematical process used to solve a problem) developed by a Greek mathematician named Pythagoras helps you find the side lengths for any right triangle. The Pythagorean theorem states that the sum of the squares of the two shorter sides is equal to the square of the longest side, or the hypotenuse. The hypotenuse is the longest side and is always opposite the 90-degree angle of the triangle. The mathematical equation for this theorem looks like this:

$$A^2 + B^2 = C^2$$

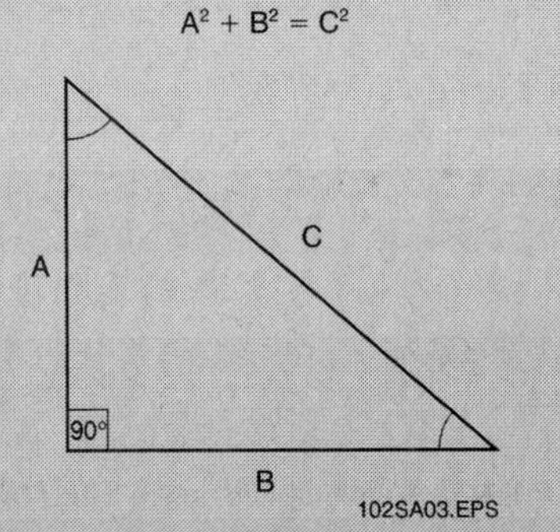

You know that the word *square* refers to shape, but in mathematical terms, it also refers to the product of a number multiplied by itself. For example, 25 is the square of 5, and 16 is the square of 4. Another way to say this is that 25 is 5 squared, or 5 times itself, and 16 is 4 squared, or 4 times itself.

In a mathematical equation this might appear as $5^2 = 25$ or $4^2 = 16$. In these examples, the numbers 5 and 4 are called the square roots, because you have to square them—or multiply them by themselves—to arrive at the squares.

Instructor's Notes:

2.50 CORE CURRICULUM ◆ INTRODUCTORY CRAFT SKILLS

8.2.4 *Circle*

A circle is a closed curved line around a center point. Every point on the curved line is exactly the same distance from the center point. A circle measures 360 degrees. The following measurements apply to circles (see *Figure 35*):

- *Circumference* – The circumference of a circle is the length of the closed curved line that forms the circle. The **formula** for finding circumference is **pi** (3.14) × diameter.
- *Diameter* – The diameter of a circle is the length of a straight line that crosses from one side of the circle through the center point to a point on the opposite side. The diameter is the longest straight line you can draw inside a circle.
- *pi* or π – pi is a mathematical constant value of approximately 3.14 (or $^{22}/_{7}$) used to determine the **area** and circumference of circles.
- *Radius* – The radius of a circle is the length of a straight line from the center point of the circle to any point on the closed curved line that forms the circle. It is equal to half the diameter.

8.3.0 Area of Shapes

Area is the measurement of the surface of an object. You must calculate the area of a shape, such as a floor or a wall, to order the proper amount of material, such as carpeting or paint. Square units of measure describe the amount of surface area. Measurements in the English system are in square inches (sq in.), square feet (sq ft), and square yards (sq yd). Measurements in the metric system include square centimeters (sq cm) and square meters (sq m).

- Square inch = 1 inch × 1 inch = inch^2
- Square foot = 1 foot × 1 foot = foot^2
- Square yard = 1 yard × 1 yard = yard^2
- Square centimeter = 1 cm × 1 cm = cm^2
- Square meter = 1 m × 1 m = m^2

You must be able to calculate the area of basic shapes. Mathematical formulas make this very easy to do. In *Appendix D* you will find a list of the formulas for calculating the areas of these shapes. You need to become familiar with these formulas at this stage in your training.

- *Rectangle:* Area = length × width. For example, you have to paint a wall that is 20 feet long and 8 feet high. To calculate the area, multiply 20 ft × 8 ft = 160 sq ft.

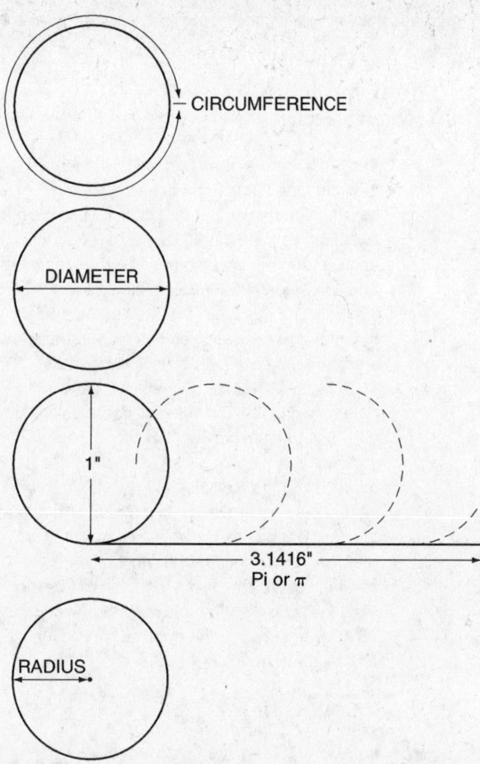
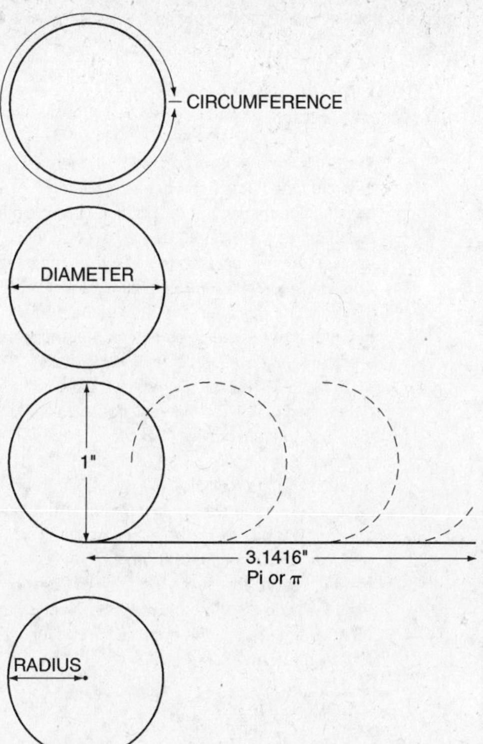

Figure 35 ◆ Measurements that apply to circles.

- *Square:* Area = length × width (but remember that all sides of a square are equal, so the formula can also be area = side × side, or side^2). For example, you have to tile a 12-meter square room. The area is 12 m × 12 m = 144 sq m, or (12 m)^2 = 144 sq m.
- *Circle:* Area = pi × radius^2. In this formula, you must use the mathematical constant pi, which has an approximate value of 3.14. You multiply pi by the radius of the circle squared. For example, to find the area of a circular driveway to be sealed, you must first find the radius, which is 20 feet. The calculation is 3.14 × (20 ft)^2 or 3.14 × 400 sq ft = 1,256 sq ft.
- *Triangle:* Area = 0.5 × base × height. The base is the side the triangle sits on. The height is the length of the triangle from its base to the highest point. For example, you have to install siding on a triangular section of a building. You find the triangle has a base of 2 feet and a height of 4 feet. The calculation is 0.5 × 2 ft × 4 ft = 4 sq ft.

Show Transparency 15 (Figure 35). Identify the circumference, diameter, and radius of a circle.

Ask trainees to use the formula for finding the circumference of a variety of circles.

Explain when and how to measure the surface of an object by calculating the area of a shape.

MODULE 00102-04 ◆ INTRODUCTION TO CONSTRUCTION MATH 2.51

See the Teaching Tip for Section 8.3.0 at the end of the module.

Ask trainees to complete the study problems for calculating the area of shapes. Answer any questions trainees may have.

DID YOU KNOW?

3/4/5 Rule

The 3/4/5 Rule is based on the Pythagorean theorem, and it has been used in building construction for centuries. This simple method for laying out or checking 90-degree angles (right angles) requires only the use of a tape measure. The numbers 3/4/5 represent dimensions in feet that describe the sides of a right triangle. Right triangles that are multiples of the 3/4/5 triangle are commonly used, such as 9/12/15, 12/16/20, and 15/20/25. The specific multiple used is determined by the relative distances involved in the job being laid out or checked.

Refer to the figure for an example of the 3/4/5 theory using the multiples of 15/20/25. In order to square or check a corner, first measure and mark 15'-0" down the line in one direction, then measure and mark 20'-0" down the line in the other direction. The distance measured between the 15'-0" and 20'-0" points must be exactly 25'-0" to ensure that the angle is a perfect right (90 degree) angle.

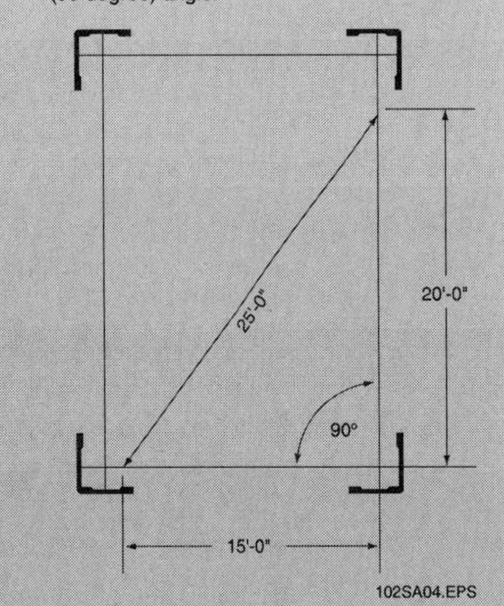

102SA04.EPS

8.3.1 Study Problems: Calculating the Area of Shape

1. The area of a rectangle that is 8 feet long and 4 feet wide is _____.
   a. 12 sq ft
   b. 22 sq ft
   c. 32 sq ft
   d. 36 sq ft
2. The area of a 16-cm square is _____.
   a. 256 sq cm
   b. 265 sq cm
   c. 276 sq cm
   d. 278 sq cm
3. The area of a circle with a 14-foot diameter is _____.
   a. 15.44 sq ft
   b. 43.96 sq ft
   c. 153.86 sq ft
   d. 196 sq ft
4. The area of a triangle with a base of 4 feet and a height of 6 feet is _____.
   a. 12 sq ft
   b. 24 sq ft
   c. 32 sq ft
   d. 36 sq ft
5. The area of a rectangle that is 14 meters long and 5 meters wide is _____.
   a. 60 sq m
   b. 65 sq m
   c. 70 sq m
   d. 75 sq m

2.52 CORE CURRICULUM ◆ INTRODUCTORY CRAFT SKILLS

Instructor's Notes:

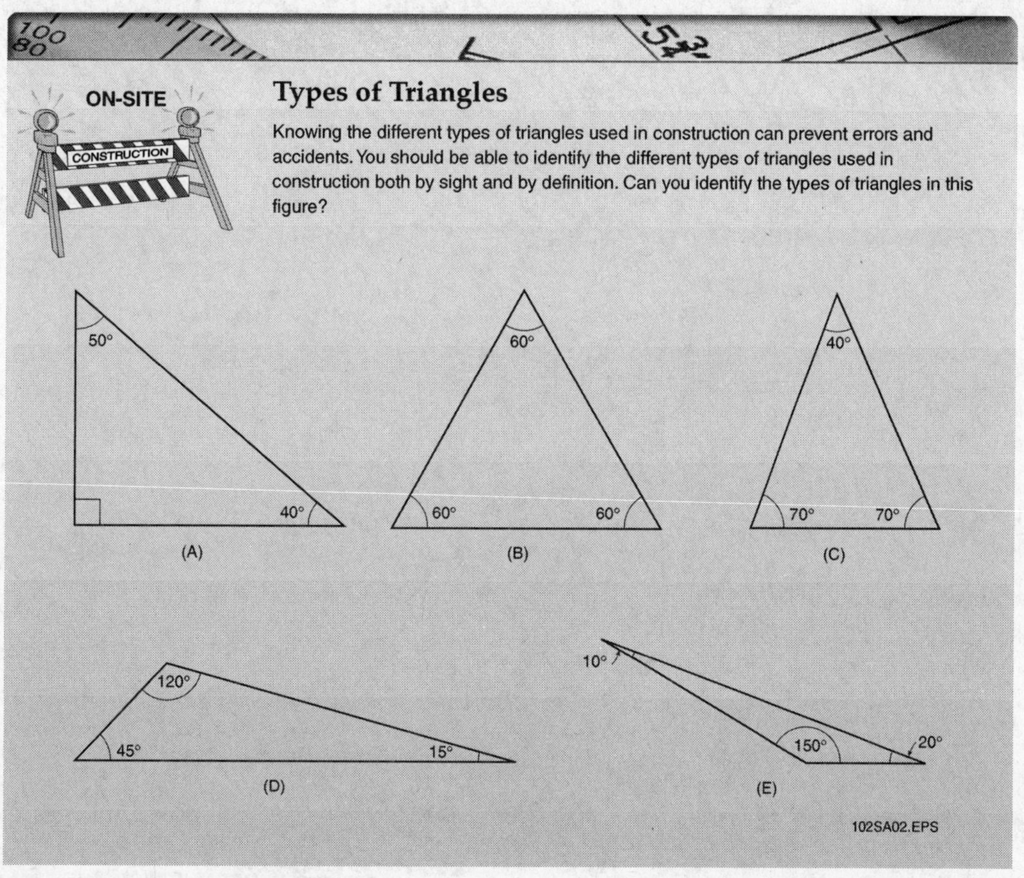
ON-SITE
CONSTRUCTION
Types of Triangles
Knowing the different types of triangles used in construction can prevent errors and accidents. You should be able to identify the different types of triangles used in construction both by sight and by definition. Can you identify the types of triangles in this figure?
50°
40°
(A)
60°
60°
60°
(B)
40°
70°
70°
(C)
120°
45°
15°
(D)
10°
150°
20°
(E)
102SA02.EPS

Show Transparency 16. Ask trainees to identify the right, equilateral, isosceles, and scalene triangles used in construction.

Review the cubic units of measure used to describe the volume of different spaces.

8.4.0 Volume of Shapes

Volume is the amount of space occupied in three dimensions. To measure volume, you must use three measurements: length, width, and height (depth or thickness). **Cubic** units of measure describe the volume of different spaces. Measurements in the English system are in cubic inches (cu in.), cubic feet (cu ft), and cubic yards (cu yd). Metric measurements include cubic centimeters (cu cm) and cubic meters (cu m).

- Cubic inch = 1 inch × 1 inch × 1 inch = inch^3
- Cubic foot = 1 foot × 1 foot × 1 foot = foot^3
- Cubic yard = 1 yard × 1 yard × 1 yard = yard^3
- Cubic centimeter = 1 centimeter × 1 centimeter × 1 centimeter = cm^3
- Cubic meter = 1 meter × 1 meter × 1 meter = m^3

MODULE 00102-04 ◆ INTRODUCTION TO CONSTRUCTION MATH 2.53

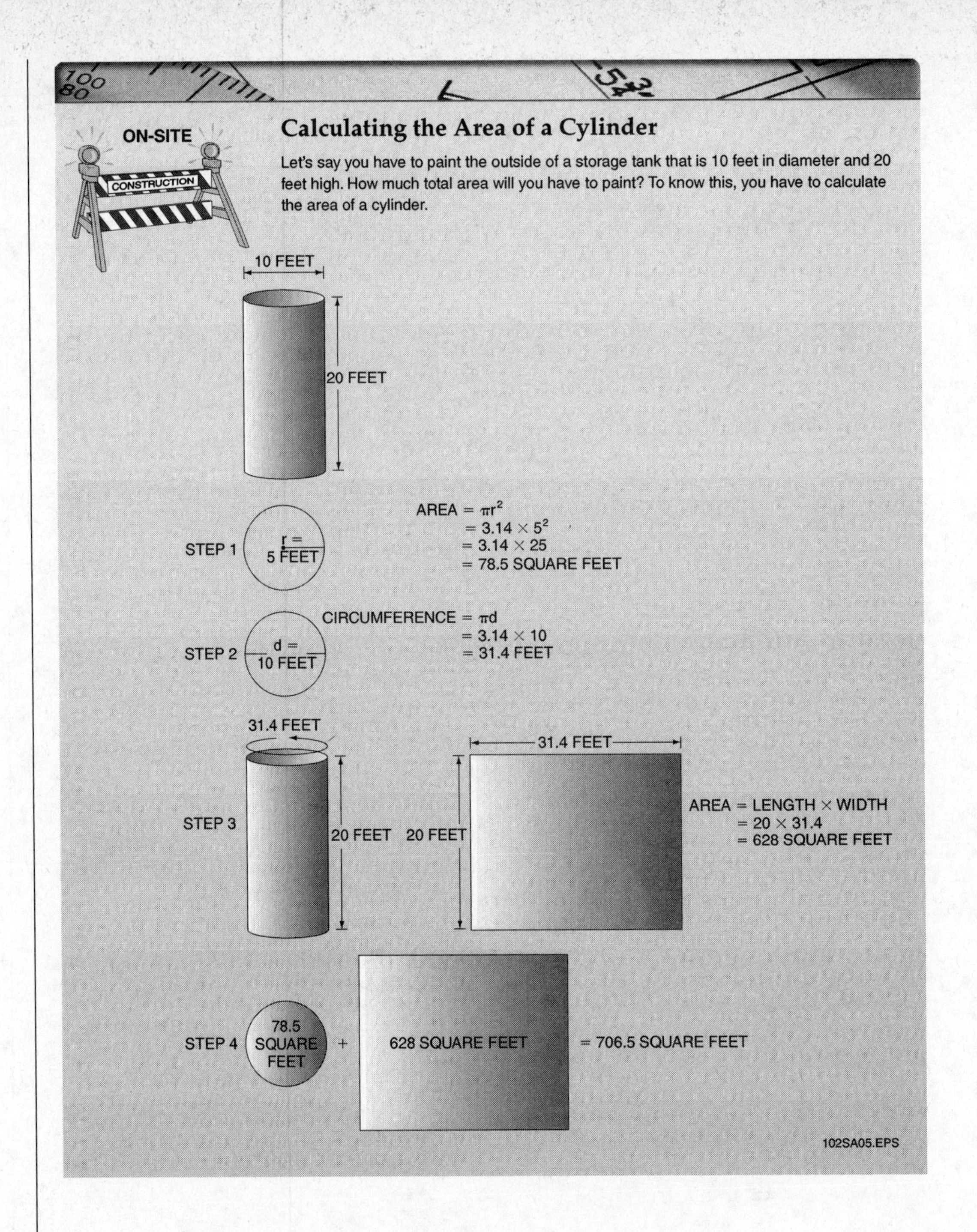

Instructor's Notes:

2.54 CORE CURRICULUM ◆ INTRODUCTORY CRAFT SKILLS

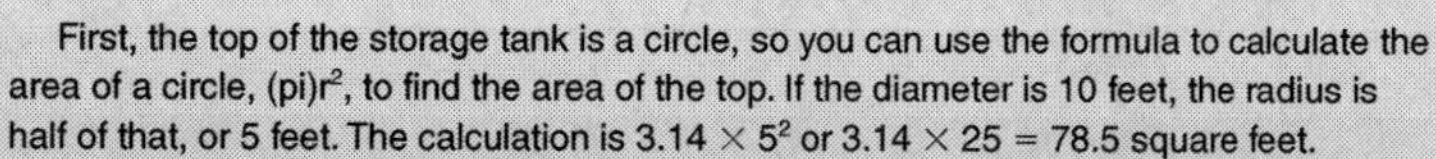

First, the top of the storage tank is a circle, so you can use the formula to calculate the area of a circle, $(pi)r^2$, to find the area of the top. If the diameter is 10 feet, the radius is half of that, or 5 feet. The calculation is 3.14×5^2 or $3.14 \times 25 = 78.5$ square feet.

Then, what do you do about the sides? Imagine that you could unroll the tank—you would see a rectangle shape! You know that the height of the tank (which is also the width of the rectangle) is 20 feet. To calculate the length (remember, although you're visualizing a rectangle, it's still a circle), you must find the circumference of the top (pi, or 3.14, × diameter). Therefore, the length is 31.4 feet. You now know both the length and the width. To find the area (area = length × width), calculate 20 ft × 31.4 ft = 628 sq ft.

Now add the two areas together to find out how much area you must paint: 78.5 sq ft + 628 sq ft = 706.5 sq ft of tank surface.

Show Transparency 17 (Figure 36). Review the procedures to determine the volume of a rectangle and the volume of a square.

Show Transparency 18 (Figure 37). Ask a trainee to explain how to calculate the volume of a cube.

You must be able to calculate the volume of the shapes discussed before. The following are mathematical formulas that make this very easy to do. In *Appendix D* you will find a list of the formulas for calculating the volumes of these shapes. You will need to become familiar with these formulas at this stage in your training. Remember to convert dimensions before multiplying (see On-Site: Unit Conversion).

- *Rectangle:* Volume = length × width × depth. For example, you have to order the right amount of cubic yards of concrete for a slab that is 20 feet long and 8 feet wide and 4 inches thick (see *Figure 36*). You must know the total volume of the slab. To calculate that, perform the following steps:

  Step 1 Convert inches to feet.

  20 ft × 8 ft × (4 in ÷ 12) =

8 FEET

20 FEET

4 INCHES

102F36.EPS

Figure 36 ◆ Volume of a rectangle.

  Step 2 Multiply length × width × depth.

  20 ft × 8 ft × 0.33 ft = 52.8 cu ft

  Step 3 Convert cubic feet to cubic yards.

  52.8 cu ft ÷ 27 (cu ft per cu yd) = 1.96 cu yd of concrete

- *Square:* Volume = length × width × depth. For example, you have to order more concrete for a slab that is 12 feet square and 5 inches thick. The calculation for volume is as follows:

  Step 1 Convert inches to feet.

  12 ft × 12 ft × (5 in ÷ 12) =

  Step 2 Multiply length × width × depth.

  12 ft × 12 ft × 0.42 ft = 60.5 cu ft

  Step 3 Convert cubic feet to cubic yards.

  60.5 cu ft ÷ 27 (cu ft per cu yd) = 2.24 cu yd of concrete

- *Cube (square):* A cube is a special type of three-dimensional rectangle; its length, width, and height are equal (see *Figure 37*). To find the volume of a cube, you can cube one dimension (multiply the number by itself three times). Perform the following steps to find how much concrete to order for a cube to be used as a support member of a structure:

  Step 1 Determine the volume of a cube that is 8 feet cubed.

  8 ft × 8 ft × 8 ft = 512 cu ft

  Step 2 Convert cubic feet to cubic yards.

  512 cu ft ÷ 27 = 18.96 cu yd of concrete

MODULE 00102-04 ◆ INTRODUCTION TO CONSTRUCTION MATH 2.55

Show Transparency 19 (Figure 38). Explain how to find the volume of a cylinder.

Have trainees complete the study problems for calculating the volume of shapes. Ask whether trainees have any questions.

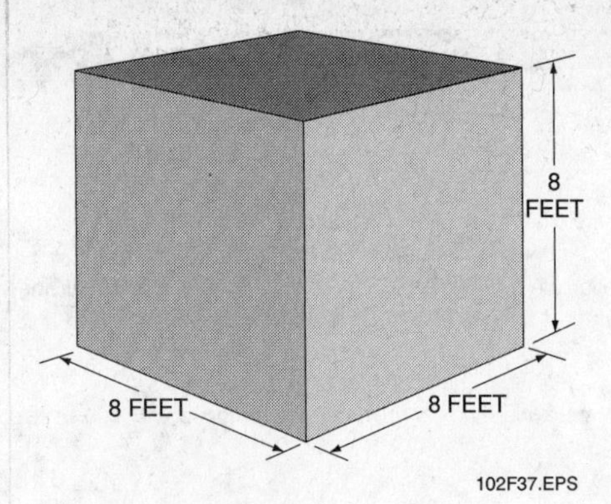

Figure 37 ◆ Volume of a cube.

- *Cylinder (circle):* Volume = pi × radius^2 × height. In this formula, you can use a shortcut: the area of a circle × height. For example, you must fill a cylinder that is 22 feet in diameter and 10 feet high (see *Figure 38*):

  Step 1 Calculate the area of the circle [(pi)r^2].

  $3.14 \times 11^2 = 379.94$ sq ft

  Step 2 Calculate volume (area × height).

  379.94 sq ft × 10 ft = 3,799.4 cu ft to fill

- *Triangle:* Volume = 0.5 × base × height × depth (thickness). In this formula you can use a shortcut: the area of a triangle × depth. For example, you must fill a triangular shape that has a base of 6 inches, a height of 12 inches, and a depth of 11 inches:

  Step 1 Calculate the area of the triangle.

  0.5 × 6 in. × 12 in. = 36 sq in.

  Step 2 Calculate volume.

  36 sq in. × 11 in. = 396 cu in. to fill

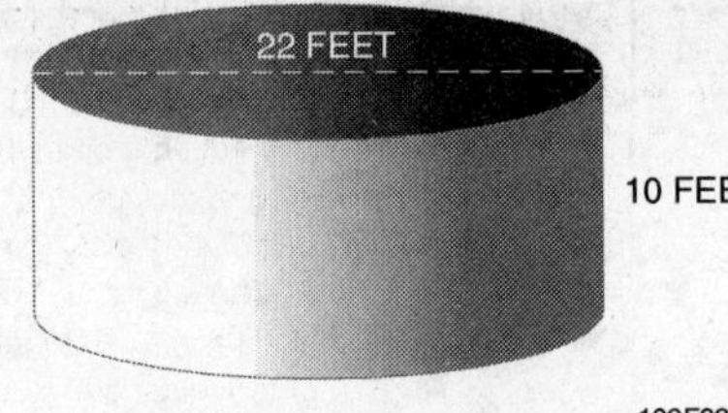

Figure 38 ◆ Volume of a cylinder.

8.4.1 Study Problems: Calculating the Volume of Shapes

1. The volume of a rectangular shape 5 feet high, 6 feet thick, and 13 feet long is _____.
   a. 24 cu ft
   b. 390 cu ft
   c. 43 ft
   d. 95 ft
2. The volume of a 3-cm cube is _____.
   a. 6 cu cm
   b. 9 cu cm
   c. 12 cu cm
   d. 27 cu cm
3. The volume of a triangular shape that has a 6-inch base, a 2-inch height, and a 4-inch depth is _____.
   a. 12 sq in.
   b. 24 cu in.
   c. 48 sq in.
   d. 36 cu in.
4. The volume of a cylinder that is 6 meters in diameter and 60 centimeters high is _____.
   a. 1130.4 cu m
   b. 18 cu m
   c. 16.956 cu m
   d. 6782.4 cu m
5. If a concrete block measures 17 feet square and is 6 inches thick, its volume is _____ cubic yards.
   a. 144.5
   b. 5.35
   c. 102
   d. 3.77

Instructor's Notes:

2.56 CORE CURRICULUM ◆ INTRODUCTORY CRAFT SKILLS

Unit Conversion

When calculating areas or volumes of shapes, you must often work with different units of measure (inch, foot, yard) for dimensions. You cannot use different units together in a calculation. Before doing your calculation, you must convert the measurements to the same units. Use the following conversion table to change dimensions to all inches, feet, or yards. (Also remember to convert metric measurements to the same units.)

Unit Conversion Table

Inches	Feet	Yards
1	1/12	1/36
12	1	1/3
36	3	1

For example, you may have lumber that is sized 2 inches × 4 inches × 12 feet. You cannot multiply these different units together to determine the number of board feet (one board foot equals a board that is 12 inches long, 12 inches wide, and 1 inch thick, or any board that has the same volume) in this piece of lumber. You cannot multiply 2 inches by 4 inches by 12 feet and get a correct answer. You must convert all the dimensions to inches using the conversion table:

2 inches × 1 = 2 inches (no change)
4 inches × 1 = 4 inches (no change)
12 feet × 12 inches per foot = 144 inches (changed feet to inches)

Now you can calculate the total cubic inches in this piece of lumber: 2 inches × 4 inches × 144 inches = 1,152 cubic inches. You must now perform a final conversion to determine how many board feet this is. There are 144 cubic inches in 1 board foot. Use this number to convert the cubic inches to board feet.

1,152 cu. in. ÷ 144 cu. in. = 8 board feet

In another example, you must determine the area of a hallway that is 30 feet long and 36 inches wide. First, convert the dimensions to the same units. In this case, you calculate: 36 inches ÷ 12 (inches in 1 ft) = 3 feet. Now perform the calculation for area (length × width = area): 30 ft × 3 ft = 90 sq ft. The area of the hallway is 90 square feet.

See the Teaching Tip for Section 8.4.0 at the end of this module.

MODULE 00102-04 ◆ INTRODUCTION TO CONSTRUCTION MATH 2.57

Go over the Review Questions for Section 8.0.0. Answer any questions trainees may have.

Review Questions

Section 8.0.0

1. An angle that measures between 0 and 90 degrees is called a(n) _____ angle.
   a. acute
   b. lateral
   c. obtuse
   d. right

2. You can use _____ to check if material is cut in a true rectangle.
   a. diagonals
   b. diameters
   c. perimeters
   d. right angles

3. The _____ is the sum of all four sides of a rectangle or a square.
   a. circumference
   b. diagonal
   c. perimeter
   d. square

4. A triangle with two equal angles and two equal sides is a(n) _____ triangle.
   a. equilateral
   b. hypotenuse
   c. isosceles
   d. right

5. A closed curved shape in which every point on the line is an equal distance from a center point is a _____.
   a. circle
   b. circumference
   c. diameter
   d. radius

6. The area of a rectangular window that measures 7 meters long and 3 meters wide is _____.
   a. 10 sq m
   b. 21 sq m
   c. 21 cu m
   d. 10 cu m

7. The area of a triangular roof section with a base of 7 feet and a height of 3.5 feet is _____.
   a. 7 cu ft
   b. 12 sq ft
   c. 12.25 sq ft
   d. 21 sq ft

8. The volume of a 9-inch cube is _____.
   a. 243 cu yd
   b. 729 cu in
   c. 818 sq in
   d. 1243 cu in

You may use a calculator to answer questions 9 and 10.

9. The volume of a cylindrical tank that is 23 feet high with a radius of 6.25 feet is _____. (Round your answer to the nearest hundredth.)
   a. 143.75 sq ft
   b. 898.44 cu ft
   c. 902.75 sq ft
   d. 2,821.09 cu ft

10. The volume of a triangular shape that has a 7-cm base, a 4-cm height, and a 30-mm depth is _____. (Remember to convert all measurements to the same unit.)
    a. 42 cu cm
    b. 42 cu mm
    c. 82 cu mm
    d. 82 cu cm

Instructor's Notes:

Summary

Mathematics is not just something you need when you are in school. The construction site requires math every day to get the work done. Whether you are cutting stock, painting a wall, or installing cable, you need math skills on the job. Basic operations such as addition, subtraction, multiplication, and division are the keys to completing your tasks. Knowing how to measure, mark, and use materials and supplies increases your value to your employer, which helps ensure your job security.

Notes

MODULE 00102-04 ◆ INTRODUCTION TO CONSTRUCTION MATH 2.59

Ask trainees to complete the Key Terms Quiz. Review trainees' answers and go over any questions trainees may have.

Administer the Module Examination. Be sure to record the results of the Exam on Craft Training Report Form 200, and submit the results to the Training Program Sponsor.

Key Terms Quiz

Fill in the blank with the correct key term that you learned from your study of this module.

1. The _________ is equal to half the diameter of a circle.
2. When you solve the problem 46 − 9, you must _________ from the tens column.
3. A(n) _________ measures between 0 and 90 degrees.
4. When you _________ an angle, you divide it into equal parts.
5. When you add larger, 2-digit numbers in a column you may need to _________ one amount to the next column.
6. _________ can be measured in sq ft, sq in., and sq cm.
7. When multiplying 3 inches by 7 feet, you must _________ the measurements to the same units.
8. A(n) _________ is the shape made by two straight lines coming together at a point.
9. Volume can be measured in _________ feet or meters.
10. In the number 3.45, .45 is the _________ part of the number.
11. A(n) _________ is a closed curved line around a central point.
12. _________ is a unit of measurement for angles.
13. In the fraction ¾, 3 is the _________ and 4 is the _____.
14. A line drawn from one corner of a rectangle to the opposite corner is called a(n) _________.
15. The _________ is the longest straight line you can draw inside a circle.
16. In the problem 17 − 8 = 9, the number 9 is the _________.
17. The numerical symbols from 0 to 9 are called _________.
18. A(n) _________, also called a(n) _____, is an instrument that measures English measurements.
19. In the number 7,890,342, the _________ of the 3 is three hundred.
20. Fractions having different numerators and denominators but equal values, such as ¼ and 2⁄8, are called _________.
21. The _________ for finding the area of a triangle is ½ b × h.
22. The formula for finding _________ is pi × diameter.
23. ¼ and ⅛ are examples of _________.
24. The _________ is the base unit of length in the metric system.
25. 9⁄5 and 5⁄4 are examples of _________.
26. To solve the problem ¾ ÷ ½, you must _________ the second fraction and multiply.
27. To solve 8,994 ÷ 45 without a calculator, you would use _________.
28. In the problem 25 + 25 = 50, the numeral 50 is the _________.
29. A(n) _________ divides each inch into ten equal parts.
30. In the problem 86 ÷ 4, the answer is 21 with a(n) _________ of 2.
31. A(n) _________ is an instrument that measures metric lengths.
32. A(n) _________ measures between 90 and 180 degrees.
33. A(n) _________ is a combination of a whole number with a fraction or decimal.
34. _________ are formed by two straight lines that cross, while _____ are angles that have the same vertex and one side in common.
35. A(n) _________ is a part of one hundred.
36. To determine the _________, measure the distance around the outside of any closed shape.
37. Complete units without fractions or decimals are called _________.
38. Numbers less than zero are _________, and numbers greater than zero are _____.

Instructor's Notes:

2.60 CORE CURRICULUM ◆ INTRODUCTORY CRAFT SKILLS

39. A(n) ________ and a(n) _____ have two pairs of equal sides that are parallel to each other.

40. A(n) ________ has sides of unequal lengths, a(n) _____ has one 90-degree angle, a(n) _____ has three equal sides and three equal angles, and a(n) _____ has two equal sides and two equal angles.

41. A(n)________ measures 90 degrees.

42. A(n) ________ measures 180 degrees.

43. A(n) ________ has three sides and three angles.

44. Two or more lines or curves come together at the ________.

45. ________ equals 3.14 or 22/7.

46. To calculate the ________ of a cube, multiply the length by the width by the height.

Key Terms

Acute angle
Adjacent angles
Angle
Area
Bisect
Borrow
Carry
Circle
Circumference
Convert
Cubic
Decimal
Degree
Denominator
Diagonal
Diameter
Difference
Digit
English ruler
Equilateral triangle
Equivalent fractions
Formula
Fraction
Improper fraction
Invert
Isosceles triangle
Long division
Machinist's rule
Meter
Metric ruler
Mixed number
Negative numbers
Numerator
Obtuse angle
Opposite angles
Percent
Perimeter
Pi
Place value
Positive numbers
Radius
Rectangle
Remainder
Right angle
Right triangle
Scalene triangle
Square
Standard ruler
Straight angle
Sum
Triangle
Vertex
Volume
Whole numbers

MODULE 00102-04 ◆ INTRODUCTION TO CONSTRUCTION MATH 2.61

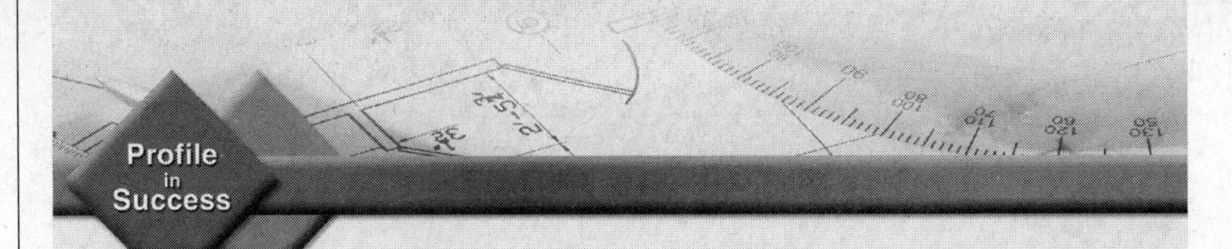

Joe Beyer

Building Trades and Carpentry Instructor
Monroe #1 BOCES
Penfield, New York

Joe was born in Rochester, New York and developed his love of carpentry and teaching from his parents at an early age. Joe earned his BA from the School for American Craftsmen of the Rochester Institute of Technology and went on to earn a master's degree in educational administration from the University of New Hampshire. He also holds a certificate of advanced study from Oswego State University, where he completed five certifications. Joe then taught math, science, and shop classes at area high schools. He is also an OSHA-certified instructor.

The BOCES program (Board of Cooperative Educational Services) allows teenage and adult students to explore alternatives to a traditional academic education. Students may work toward NCCER certification, union apprenticeship, or college credits in their chosen area.

How did you become interested in the construction industry?
I've been interested in furniture and teaching since I was very young. My mother had an antique furniture collection, and my father's family owned sawmills and also manufactured furniture. I started toward a career in carpentry but decided to go more into teaching when I became involved in a project to disassemble and move a special barn built around 1815 or 1820 and reassemble it on a museum site. I got very involved with running the crew and doing the work. I loved working with a timber frame building, including pulling the pins, labeling all the parts, putting the foundation in, and building the barn to last for another 200 years. I saw the tremendous workmanship that had gone into the building, and I decided that this type of skill should continue to be taught.

How did you decide on your current position?
I taught in high schools for 18 years. I began to feel that the construction education programs there were being watered down, so I went into career education. I believed that I could integrate math, science, business, work ethic, and other important skills into the 2½-hour construction class. And I have been able to do that.

What are some of the things you do in your job?
The program provides students with training in hands-on and social skills as they work their way through projects. They perform a variety of smaller projects and, during their junior and senior years, they build a house. In the past, the completed homes were auctioned off, but recently I have worked to find a buyer up front. It's less complicated than an auction, and it adds to the students' knowledge base. They not only have to learn to work together in teams, but they also have to interact with an owner, which is a learning experience they'll find useful in the real world. Students learn how to work through the same problems they're going to face on the job.

The modular format of the soft skills course (Contren®'s *Tools for Success*) works really well for me. The students are in 2½-hour classes five days a week, and I break up the material into what I call profiles. They combine soft skills with hands-on projects. For example, an electrical skills instructor and I build and wire a 4' × 6' cube to demonstrate team skills and the importance of learning more than one craft. The students then work in teams to build and wire their own cubes. Everything in the soft skills, from communication skills to getting along with others to solving problems, gets pulled into this training.

The courses demand a high level of energy from students and from me as well. The students have to do quite a bit of remedial work in the 3 Rs, and that makes the teaching more challenging. But when you see a kid who has been struggling to find success, when you see that light go on and you know you had a role in that experience, that's all the reward you

Instructor's Notes:

2.62 CORE CURRICULUM ◆ INTRODUCTORY CRAFT SKILLS

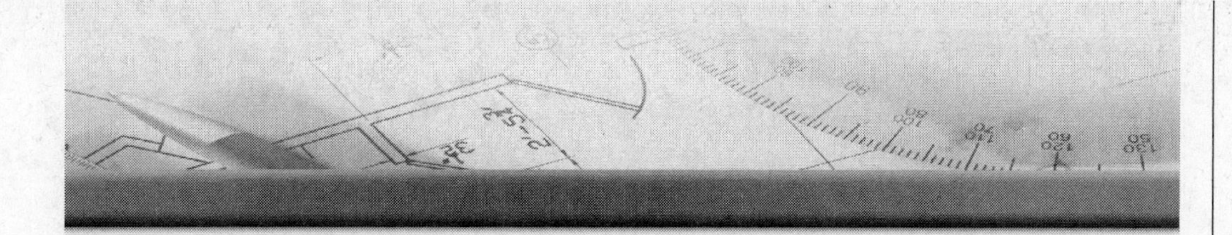

need. Students seem to learn faster when you can relate training to a practical application. Usually they'll catch on at some point that I've actually been teaching them math or English right along with their craft skills, but by then they've learned something.

What do you like most about your job?

I really enjoy the passing on of information and skills to the next generation, and I also enjoy working with the variety of students that comes through our program, whether they are adults or high school students. I love carpentry, but I'd wanted to be a teacher since I was in the ninth grade. Now I have the best of both worlds: making furniture over the summer and teaching carpentry and cabinetmaking the rest of the year.

Approximately 20 to 30 percent of my students go on to a two-year vocational college, and 30 to 40 percent go straight to work. That number has been higher in recent years because the demand for skilled construction workers is high and the pay and benefits are excellent. Many former students return for a visit, and some talk about how my teaching has affected their lives. And some past students even enroll their own children in the course. It does make you feel a little bit old, but it's a nice feeling to see the results of my work.

What do you think it takes to be a success in the construction trades?

The first thing you've got to have is knowledge of the subject matter, and that includes not only book knowledge but also working knowledge. You need to know how to hold a hammer so that it doesn't hurt your body when you use it. If you have only the working knowledge, then as codes change and you decide to keep doing things the way you want, you will cause major problems. And that's why I like the Contren® *Core Curriculum, Tools for Success,* and the multi-level arrangement of the different curricula. If you do the activities and follow the book, you get both types of knowledge.

What would you say to someone entering the trades today?

Be honest and hard working. Meet the customer's needs, and even exceed them when you can, because word of mouth is what's going to keep you employed. You may have the smarts and the ability, but you need good people skills to meet the customer's needs. That way, they get the good job they deserve and they will refer you to the next person. Construction is tougher to do than a lot of jobs, but if you're a bright person, it is not that difficult because it's very concrete, you can visualize it, and you get to see what you're doing. You get instant gratification.

Safety is very important. I invite other OSHA-certified instructors in to talk to students about how their clothing and attitudes can affect workplace safety. In my 29 years as a trainer, I have had only two minor accidents on student projects. One student, an adult, shot a finishing nail through his fingernail. The other accident involved a visitor to a project worksite. This fellow was wearing baggy pants and started to climb a ladder before anyone could stop him. Both he and the students got a quick lesson in wearing appropriate clothing when he fell off the second rung.

I believe that you must demonstrate the proper work ethic. If you don't, then you will not have a chance. And that goes for me, too, as an instructor. Students will use you as either an excuse or an example. It reminds me of songs dating back to when I was a kid: "If you're not part of the solution, you're part of the problem," and "Fix it or get out of the way." It takes a while to pick those words out, though, because a lot of times they're hard to understand!

MODULE 00102-04 ◆ INTRODUCTION TO CONSTRUCTION MATH 2.63

Trade Terms Introduced in This Module

Acute angle: Any angle between 0 degrees and 90 degrees.

Adjacent angles: Angles that have the same vertex and one side in common.

Angle: The shape made by two straight lines coming together at a point. The space between those two lines is measured in degrees.

Area: The surface or amount of space occupied by a two-dimensional object such as a rectangle, circle, or square. To calculate the area for rectangles and squares, multiply the length and width. To calculate the area for circles, multiply the radius squared and pi.

Bisect: To divide into equal parts.

Borrow: To move numbers from one value column (such as the tens column) to another value column (such as units) to perform subtraction problems.

Carry: To transfer an amount from one column to another column.

Circle: A closed curved line around a central point. A circle measures 360 degrees.

Circumference: The distance around the curved line that forms a circle.

Convert: To change from one unit of expression to another. For example, convert a decimal to a percentage: 0.25 to 25%; or convert a fraction to an equivalent: $\frac{3}{4}$ to $\frac{6}{8}$.

Cubic: Measurement found by multiplying a number by itself three times; it describes volume measurement.

Decimal: Part of a number represented by digits to the right of a point, called a decimal point. For example, in the number 1.25, .25 is the decimal part of the number.

Degree: A unit of measurement for angles. For example, a right angle is 90 degrees, an acute angle is between 0 and 90 degrees, and an obtuse angle is between 90 and 180 degrees.

Denominator: The part of a fraction below the dividing line. For example, the 2 in $\frac{1}{2}$ is the denominator.

Diagonal: Line drawn from one corner of a rectangle or square to the farthest opposite corner.

Diameter: The length of a straight line that crosses from one side of a circle, through the center point, to a point on the opposite side. The diameter is the longest straight line you can draw inside a circle.

Difference: The result you get when you subtract one number from another. For example, in the problem $8 - 3 = 5$, the number 5 is the difference.

Digit: Any of the numerical symbols 0 to 9.

English ruler: Instrument that measures English measurements; also called the standard ruler. Units of English measure include inches, feet, and yards.

Equilateral triangle: A triangle that has three equal sides and three equal angles.

Equivalent fractions: Fractions having different numerators and denominators, but equal values, such as $\frac{1}{2}$ and $\frac{2}{4}$.

Formula: A mathematical process used to solve a problem. For example, the formula for finding the area of a rectangle is side A times side B = Area, or $A \times B$ = Area.

Fraction: A number represented by a numerator and a denominator, such as $\frac{1}{2}$.

Improper fraction: A fraction whose numerator is larger than its denominator. For example, $\frac{8}{4}$ and $\frac{6}{3}$ are improper fractions.

Invert: To reverse the order or position of numbers. In fractions, to turn upside down, such as $\frac{3}{4}$ to $\frac{4}{3}$. When you are dividing by fractions, one fraction is inverted.

Isosceles triangle: A triangle that has two equal sides and two equal angles.

Long division: Process of writing out each step of a division problem until you reach the answer and identify any remainder that can no longer be divided by the divisor.

Machinist's rule: A ruler that is marked so that the inches are divided into 10 equal parts, or tenths.

Instructor's Notes:

Meter: The base unit of length in the metric system; approximately 39.37 inches.

Metric ruler: Instrument that measures metric lengths. Units of measure include millimeters, centimeters, and meters.

Mixed number: A combination of a whole number with a fraction or decimal. For example, mixed numbers are 3⁷⁄₁₆, 5.75, and 1¼.

Negative numbers: Numbers less than zero. For example, −1, −2, and −3 are negative numbers.

Numerator: The part of a fraction above the dividing line. For example, the 1 in ½ is the numerator.

Obtuse angle: Any angle between 90 degrees and 180 degrees.

Opposite angles: Two angles that are formed by two straight lines crossing. They are always equal.

Percent: Of or out of one hundred. For example, 8 is 8 percent (%) of 100.

Perimeter: The distance around the outside of any closed shape, such as a rectangle, circle, or square.

Pi: A mathematical value of approximately 3.14 (or ²²⁄₇) used to determine the area and circumference of circles. It is sometimes symbolized by π.

Place value: The exact quantity of a digit, determined by its place within the whole number or by its relationship to the decimal point.

Positive numbers: Numbers greater than zero. For example, 1, 2, and 3 are positive numbers.

Radius: The distance from a center point of a circle to any point on the curved line, or half the width (diameter) of a circle.

Rectangle: A four-sided shape with four 90-degree angles. Opposite sides of a rectangle are always parallel and the same length. Adjacent sides are perpendicular and are not equal in length.

Remainder: The leftover amount in a division problem. For example, in the problem 34 ÷ 8, 8 goes into 34 four times (8 × 4 = 32) and 2 is left over, or, in other words, it is the remainder.

Right angle: An angle that measures 90 degrees. The two lines that form a right angle are perpendicular to each other. This is the angle used most in the trades.

Right triangle: A triangle that includes one 90-degree angle.

Scalene triangle: A triangle with sides of unequal lengths.

Square: (1) A special type of rectangle with four equal sides and four 90-degree angles. (2) The product of a number multiplied by itself. For example, 25 is the square of 5; 16 is the square of 4.

Standard ruler: An instrument that measures English lengths (inches, feet, and yards). See English ruler.

Straight angle: A 180-degree angle or flat line.

Sum: The total in an addition problem. For example, in the problem 7 + 8 = 15, 15 is the sum.

Triangle: A closed shape that has three sides and three angles.

Vertex: A point at which two or more lines or curves come together.

Volume: The amount of space occupied in three dimensions (length, width, and height/depth/thickness).

Whole numbers: Complete units without fractions or decimals.

MODULE 00102-04 ◆ INTRODUCTION TO CONSTRUCTION MATH 2.65

Appendix A

Multiplication Table

Trace across and down from the numbers that you want to multiply, and find the answer. In the example, $6 \times 7 = 42$.

	2	**3**	**4**	**5**	**6**	**7**	**8**	**9**	**10**	**11**	**12**
2	4	6	8	10	12	14	16	18	20	22	24
3	6	9	12	15	18	21	24	27	30	33	36
4	8	12	16	20	24	28	32	36	40	44	48
5	10	15	20	25	30	35	40	45	50	55	60
6	12	18	24	30	36	42	48	54	60	66	72
7	14	21	28	35	42	49	56	63	70	77	84
8	16	24	32	40	48	56	64	72	80	88	96
9	18	27	36	45	54	63	72	81	90	99	108
10	20	30	40	50	60	70	80	90	100	110	120
11	22	33	44	55	66	77	88	99	110	121	132
12	24	36	48	60	72	84	96	108	120	132	144

MODULE 00102-04 ◆ INTRODUCTION TO CONSTRUCTION MATH 2.67

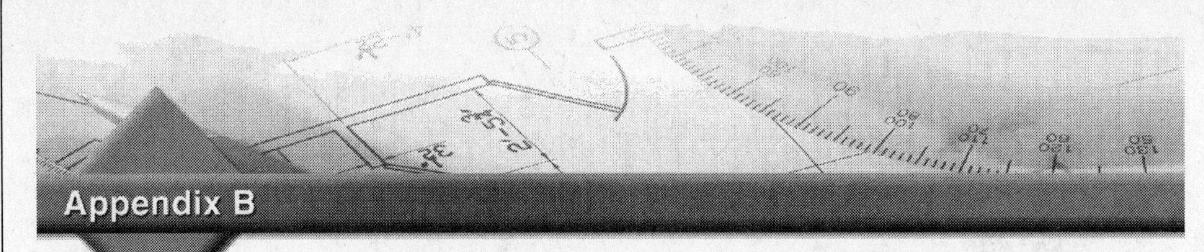

Appendix B

Inches Converted to Decimals of a Foot

Inches	Decimals of a Foot	Inches	Decimals of a Foot	Inches	Decimals of a Foot
1/16	0.005	2 1/16	0.172	4 1/16	0.339
1/8	0.010	2 1/8	0.177	4 1/8	0.344
3/16	0.016	2 3/16	0.182	4 3/16	0.349
1/4	0.021	2 1/4	0.188	4 1/4	0.354
5/16	0.026	2 5/16	0.193	4 5/16	0.359
3/8	0.031	2 3/8	0.198	4 3/8	0.365
7/16	0.036	2 7/16	0.203	4 7/16	0.370
1/2	0.042	2 1/2	0.208	4 1/2	0.374
9/16	0.047	2 9/16	0.214	4 9/16	0.380
5/8	0.052	2 5/8	0.219	4 5/8	0.385
11/16	0.057	2 11/16	0.224	4 11/16	0.391
3/4	0.063	2 3/4	0.229	4 3/4	0.396
13/16	0.068	2 13/16	0.234	4 13/16	0.401
7/8	0.073	2 7/8	0.240	4 7/8	0.406
15/16	0.078	2 15/16	0.245	4 15/16	0.411
1	0.083	3	0.250	5	0.417
1 1/16	0.089	3 1/16	0.255	5 1/16	0.422
1 1/8	0.094	3 1/8	0.260	5 1/8	0.427
1 3/16	0.099	3 3/16	0.266	5 3/16	0.432
1 1/4	0.104	3 1/4	0.271	5 1/4	0.438
1 5/16	0.109	3 5/16	0.276	5 5/16	0.443
1 3/8	0.115	3 3/8	0.281	5 3/8	0.448
1 7/16	0.120	3 7/16	0.286	5 7/16	0.453
1 1/2	0.125	3 1/2	0.292	5 1/2	0.458
1 9/16	0.130	3 9/16	0.297	5 9/16	0.464
1 5/8	0.135	3 5/8	0.302	5 5/8	0.469
1 11/16	0.141	3 11/16	0.307	5 11/16	0.474
1 3/4	0.146	3 3/4	0.313	5 3/4	0.479
1 13/16	0.151	3 13/16	0.318	5 13/16	0.484
1 7/8	0.156	3 7/8	0.323	5 7/8	0.490
1 15/16	0.161	3 15/16	0.328	5 15/16	0.495
2	0.167	4	0.333	6	0.500

Instructor's Notes:

2.68 CORE CURRICULUM ◆ INTRODUCTORY CRAFT SKILLS

Inches	Decimals of a Foot	Inches	Decimals of a Foot	Inches	Decimals of a Foot
6 1/6	0.505	8 1/6	0.672	10 1/6	0.839
6 1/8	0.510	8 1/8	0.677	10 1/8	0.844
6 3/16	0.516	8 3/16	0.682	10 3/16	0.849
6 1/4	0.521	8 1/4	0.688	10 1/4	0.854
6 5/16	0.526	8 5/16	0.693	10 5/16	0.859
6 3/8	0.531	8 3/8	0.698	10 3/8	0.865
6 7/16	0.536	8 7/16	0.703	10 7/16	0.870
6 1/2	0.542	8 1/2	0.708	10 1/2	0.875
6 9/16	0.547	8 9/16	0.714	10 9/16	0.880
6 5/8	0.552	8 5/8	0.719	10 5/8	0.885
6 11/16	0.557	8 11/16	0.724	10 11/16	0.891
6 3/4	0.563	8 3/4	0.729	10 3/4	0.896
6 13/16	0.568	8 13/16	0.734	10 13/16	0.901
6 7/8	0.573	8 7/8	0.740	10 7/8	0.906
6 15/16	0.578	8 15/16	0.745	10 15/16	0.911
7	0.583	9	0.750	11	0.917
7 1/6	0.589	9 1/6	0.755	11 1/6	0.922
7 1/8	0.594	9 1/8	0.760	11 1/8	0.927
7 3/16	0.599	9 3/16	0.766	11 3/16	0.932
7 1/4	0.604	9 1/4	0.771	11 1/4	0.938
7 5/16	0.609	9 5/16	0.776	11 5/16	0.943
7 3/8	0.615	9 3/8	0.781	11 3/8	0.948
7 7/16	0.620	9 7/16	0.786	11 7/16	0.953
7 1/2	0.625	9 1/2	0.792	11 1/2	0.958
7 9/16	0.630	9 9/16	0.797	11 9/16	0.964
7 5/8	0.635	9 5/8	0.802	11 5/8	0.969
7 11/16	0.641	9 11/16	0.807	11 11/16	0.974
7 3/4	0.646	9 3/4	0.813	11 3/4	0.979
7 13/16	0.651	9 13/16	0.818	11 13/16	0.984
7 7/8	0.656	9 7/8	0.823	11 7/8	0.990
7 15/16	0.661	9 15/16	0.828	11 15/16	0.995
8	0.667	10	0.833	12	1.000

MODULE 00102-04 ◆ INTRODUCTION TO CONSTRUCTION MATH 2.69

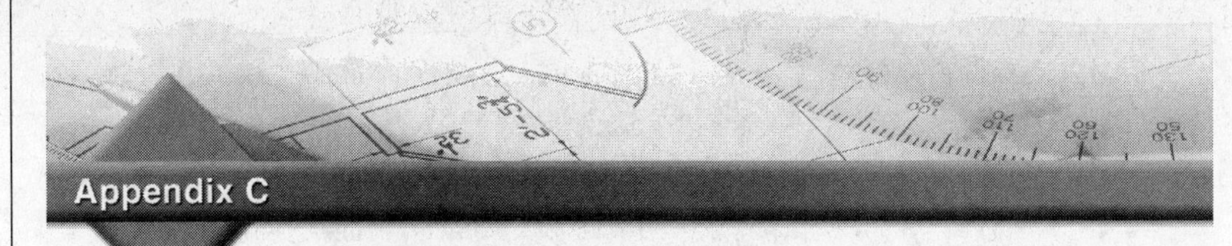

Conversion Tables

How to Convert Units of Volume

	Metric to English			**English to Metric**	
From	**Multiply By**	**To Obtain**	**From**	**Multiply By**	**To Obtain**
Liters	1.0567	Quarts	Quarts	0.946	Liters
Liters	2.1134	Pints	Pints	0.473	Liters
Liters	0.2642	Gallons	Gallons	3.785	Liters

How to Convert Units of Weight

	Metric to English			**English to Metric**	
From	**Multiply By**	**To Obtain**	**From**	**Multiply By**	**To Obtain**
Grams	0.0353	Ounces	Pounds	0.4536	Kilograms
Grams	15.4321	Grains	Pounds	453.6	Grams
Kilograms	2.2046	Pounds	Ounces	28.35	Grams
Kilograms	0.0011	Tons (short)	Grains	0.0648	Grams
Tons (metric)	1.1023	Tons (short)	Tons (short)	0.9072	Tons (metric)

How to Convert Units of Length

	Metric to English			**English to Metric**	
From	**Multiply By**	**To Obtain**	**From**	**Multiply By**	**To Obtain**
Meters	39.37	Inches	Inches	2.54	Centimeters
Meters	3.2808	Feet	Inches	0.0254	Meters
Meters	1.0936	Yards	Inches	25.4	Millimeters
			Miles	1,609,344	Millimeters
Centimeters	0.3937	Inches	Feet	0.3048	Meters
Millimeters	0.03937	Inches	Feet	30.48	Centimeters
Kilometers	0.6214	Miles	Yards	0.9144	Meters
			Yards	91.44	Centimeters
			Miles	1.6093	Kilometers

Instructor's Notes:

2.70 CORE CURRICULUM ◆ INTRODUCTORY CRAFT SKILLS

Appendix D

Area and Volume Formulas

Area Formulas

Rectangle:	area = length × width
Square:	area = length × width or area = side^2
Circle:	area = pi × radius^2
Triangle:	area = 0.5 × base × height

Volume Formulas

Rectangle:	volume = length × width × depth
Square:	volume = length × width × depth
Cube (square):	volume = side^3
Cylinder (circle):	volume = pi × radius^2 × height or volume = area of a circle × height
Triangle:	volume = 0.5 × base × height × depth (thickness) or volume = area of a triangle × depth

MODULE 00102-04 ◆ INTRODUCTION TO CONSTRUCTION MATH 2.71

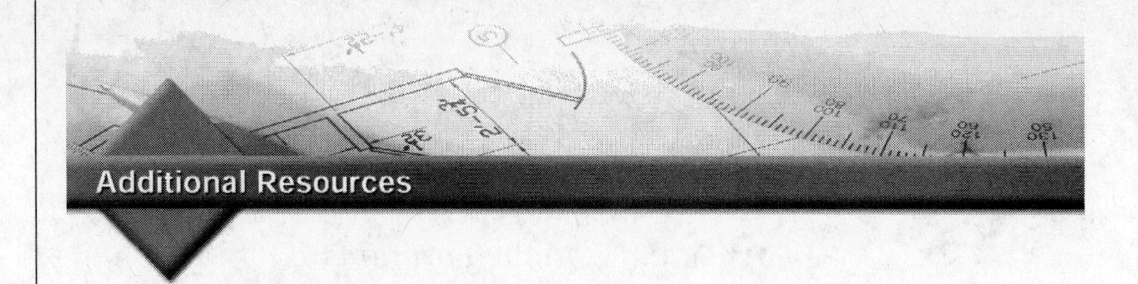

Additional Resources

This module is intended to present thorough resources for task training. The following reference works are suggested for further study. These are optional materials for continued education rather than for task training.

All the Math You'll Ever Need, 1999. Stephen Slavin. New York: John Wiley & Sons.

Basic Construction Math Review: A Manual of Basic Construction Mathematics for Contractor and Tradesman License Exams. Printcorp Business Printing. Construction Book Express.

Mastering Math for the Building Trades, 2000. James Gerhart. McGraw-Hill/TAB Electronics.

Math to Build On: A Book for Those Who Build, 1997. Johnny and Margaret Hamilton. Clinton, NC: Construction Trades Press.

Mathematics for the Million, 1993. Lancelot Thomas Hogben. New York: W. W. Norton & Company.

Instructor's Notes:

2.72 CORE CURRICULUM ◆ INTRODUCTORY CRAFT SKILLS

The following are suggested activities or instructional methods to help you teach the material in this AIG.

General

When you call on someone to answer a question, the rest of the class relaxes or even tunes out because they expect that the question and answer will take place only between you and the trainee you called on. Instead, use this technique to involve more trainees in answering questions and to keep them on their toes.

1. Ask trainees to define a term or explain a concept.
2. After one trainee has answered, ask a trainee seated nearby if the answer is right. Then ask whether a trainee in the back of the room agrees.
3. Ask trainees to explain why they think an answer is right or wrong.
4. Use the session to clear up incorrect ideas, and encourage trainees to learn from their mistakes.

Section 2.4.0

Multiplying Simple Whole Numbers

This exercise allows trainees to practice their multiplication tables. Trainees will need their Trainee Guides and pencil and paper to take notes for discussion. Allow between 20 and 30 minutes for this exercise.

1. Refer to *Appendix A*, and divide the class into small groups. Ask trainees to study the multiplication table.
2. After allowing the trainees sufficient time to study, conduct a timed multiplication drill.
3. Starting with the 1s, ask each group to answer a different multiplication problem.
4. If the group answers their problem correctly, they stay in the drill, and you move onto the next group. If the group gets the answer incorrectly, they are out of the drill.
5. As the tables become more challenging, allow more time for each group to answer.

Sections 2.0.0–2.6.9

Whole Numbers

This exercise will familiarize trainees with on-site mathematical functions. You will need sample work orders that require trainees to perform mathematical functions. Trainees will need their Trainee Guides and pencil and paper to take notes for discussion. Allow 30 minutes for this exercise.

1. Bring in sample work orders that require math.
2. Have trainees review the orders and perform the mathematical functions required to complete the tasks.
3. Ask trainees to share examples of mathematical functions they have had to perform on site.
4. Answer any questions trainees may have.

Sections 5.4.0–5.4.1

Adding and Subtracting Decimals

This exercise will give trainees the opportunity to practice adding and subtracting decimals. Trainees will need their Trainee Guides and pencil and paper to take notes for discussion. Allow 20 minutes for this exercise.

1. Have trainees review the problems in "On-Site: Decimals at Work."
2. Ask trainees to calculate the number of shims needed to install a boiler to specifications on its concrete pad.
3. Ask trainees to determine the size filter needed to remove metal particles from the oil in a conveyor system.
4. Review trainees' answers, and discuss other examples of using decimals on the job.

Section 6.2.0

Converting Fractions to Decimals

This exercise allows trainees to practice converting inches to decimals. Trainees will need their Trainee Guides and pencil and paper to take notes for discussion. Allow 20 minutes for this exercise.

1. Ask trainees to refer to the chart in *Appendix B: Inches Converted to Decimals of a Foot.*
2. On the whiteboard/chalkboard, write a variety of conversion problems.
3. Have trainees convert inches to decimals using the table in *Appendix B.* Next, have trainees double-check their work by converting inches to decimals using the steps outlined in Section 6.2.0.
4. Ask trainees to compare their answers. Review trainees' answers for accuracy, and go over any questions trainees may have.

Sections 7.1.0–7.1.1

Units of Weight, Length, Volume, and Temperature

This exercise demonstrates the necessity for standard units of measurement. Trainees will need standard and metric rulers, their Trainee Guides, and pencil and paper to take notes for discussion. Allow 20 minutes for this exercise.

1. Have trainees walk off a distance of 10 paces.
2. Ask trainees to mark where they start and end and then measure the distance using a standard and metric ruler.
3. Ask trainees to mark off 5 thumb lengths on a piece of paper and then measure the length using a standard and metric ruler.
4. Compare the various lengths, and discuss the importance of following standards for measurements.

Section 7.3.0

Converting Measurements

This exercise enables trainees to convert measurements from the English system to the metric system. Trainees will need their Trainee Guides and pencil and paper to take notes for discussion. Allow 20 minutes for this exercise.

1. Ask trainees to review *Appendix C: Conversion Tables.*
2. On the whiteboard/chalkboard, write a variety of metric conversion problems.
3. Have trainees convert a variety of measurements from the English system to the metric system using values that you specify.
4. Provide examples of converting measurements on the job.

Section 8.3.0 *Area of Shapes*

This exercise allows trainees to practice calculating area. Trainees will need their Trainee Guides and pencil and paper to take notes for discussion. Allow 30 minutes for this exercise.

1. Have trainees review the formulas for calculating area.
2. On the whiteboard/chalkboard, draw a variety of rectangles, squares, circles, and triangles. Assign the necessary values to calculate area.
3. Ask trainees to calculate the area of the shapes you have drawn.
4. If time permits, have trainees practice calculating the area of a cylinder.
5. Review trainees' answers, and go over any questions trainees may have.

Section 8.4.0 *Volume of Shapes*

This exercise allows trainees to practice calculating volume. Trainees will need their Trainee Guides and pencil and paper to take notes for discussion. Allow 30 minutes for this exercise.

1. Have trainees review the formulas for calculating volume.
2. On the whiteboard/chalkboard, draw a rectangle, square, cube, cylinder, and triangle. Assign the necessary values to calculate volume.
3. Ask trainees to calculate the volume of the shapes you have drawn.
4. Ensure that trainees follow the unit conversion table in On-Site: Unit Conversion to convert different units of measure to the same units.
5. Review trainees' answers, and go over any questions trainees may have.

MODULE 00102-04 — ANSWERS TO STUDY PROBLEMS

Section 2.1.1

1. a
2. b
3. b
4. a
5. c

Section 2.2.2

1. 32
+ 75
107 pounds of nails

2. 73
+ 45
118 in. of molding

3. 83
+ 53
136 ft of cable

4. 1
452
+ 74
526 ft of baseboard

5. 1
323
+ 758
1,081 yds of binding

Section 2.3.1

1. 7 17
~~87~~
− 38
49 sockets

2. 1 16
~~26~~
− 17
9 connectors

3. 8 12
~~92~~
− 34
58 hours

4. 3 16
2~~46~~
− 18
228 ft of cable

5. 1 16
8~~26~~
− 717
109 bricks

Section 2.4.1

1. a

2. 9
× 8
72 doors

3. 9
× 6
54 bolts

4. 7
× 2
14 fixtures

5. 8
× 4
32 beams

Section 2.4.3

1. 12
× 21
12
+ 24
252 wheelbarrows

2. 11
× 15
55
+ 11
165 ladders

3. 30
× 25
150
+ 60
750 barrels

4. 2
452
× 4
1,808 sq. ft.

5. 3 1
1
162
× 52
324
+ 810
8,424 bricks

Section 2.5.1

1. $$\begin{array}{r} 5 \\ 3\overline{)15} \\ -15 \\ \hline 0 \end{array}$$

2. $$\begin{array}{r} 9 \\ 4\overline{)36} \\ -36 \\ \hline 0 \end{array}$$

3. $$\begin{array}{r} 10 \\ 5\overline{)54} \\ -5 \\ \hline 04 \\ -0 \\ \hline 4r \end{array}$$

4. $$\begin{array}{r} 2 \\ 6\overline{)17} \\ -12 \\ \hline 5r \end{array}$$

5. $$\begin{array}{r} 5 \\ 7\overline{)39} \\ -35 \\ \hline 4r \end{array}$$

Section 2.5.3

1. $$\begin{array}{r} 21 \\ 12\overline{)263} \\ -24 \\ \hline 23 \\ -12 \\ \hline 11r \end{array}$$

2. $$\begin{array}{r} 263 \\ 16\overline{)4218} \\ -32 \\ \hline 101 \\ -96 \\ \hline 58 \\ -48 \\ \hline 10r \end{array}$$

3. $$\begin{array}{r} 302 \\ 15\overline{)4532} \\ -45 \\ \hline 03 \\ -00 \\ \hline 32 \\ -30 \\ \hline 2r \end{array}$$

4. c $$\begin{array}{r} 10 \\ 15\overline{)150} \\ -15 \\ \hline 00 \\ -00 \\ \hline 0 \end{array}$$

5. a $$\begin{array}{r} 5 \\ 20\overline{)100} \\ -100 \\ \hline 0 \end{array}$$

Section 2.6.2

1. 69
2. 209
3. 168
4. 1,059
5. 2,356

Section 2.6.4

1. 92
2. 38
3. 697
4. 346
5. $$\begin{array}{r} 99 \\ -75 \\ \hline 24 \end{array}$$ (Use calculator to solve.)

Section 2.6.6

1. 71,478
2. 87,699
3. 215,871

4. b $$\begin{array}{r} 465 \\ \times\ 16 \\ \hline 7{,}440 \end{array}$$ (Use calculator to solve.)

5. b $$\begin{array}{r} 12 \\ \times\ 18 \\ \hline 216 \end{array}$$ (Use calculator to solve.)

Section 2.6.9

1. 4
2. 6
3. 5.33

4. a $100 \div 20 = 5$ (Use calculator to solve.)

5. a $8 \div 3 = 2.67$
You can cut two 3-foot lengths with 0.67 of a length left over.
$0.67 \times 3 = 2.01$, so 2 feet are left over.

Section 4.1.1

1. b $\dfrac{1 \times 4}{4 \times 4} = \dfrac{4}{16}$

2. c $\dfrac{2 \times 2}{16 \times 2} = \dfrac{4}{32}$

3. d $\dfrac{3 \times 2}{4 \times 2} = \dfrac{6}{8}$

4. a $\dfrac{3 \times 16}{4 \times 16} = \dfrac{48}{64}$

5. c $\dfrac{3 \times 2}{16 \times 2} = \dfrac{6}{32}$

Section 4.2.1

1. $\dfrac{2 \div 2}{16 \div 2} = \dfrac{1}{8}$

2. $\dfrac{2 \div 2}{8 \div 2} = \dfrac{1}{4}$

3. $\dfrac{12 \div 4}{32 \div 4} = \dfrac{3}{8}$

4. $\dfrac{4 \div 4}{8 \div 4} = \dfrac{1}{2}$

5. $\dfrac{4 \div 4}{64 \div 4} = \dfrac{1}{16}$

Section 4.3.1

1. c $\frac{2}{6}, \frac{3}{4}$

$\frac{2 \div 2}{6 \div 2} = \frac{1}{3}$

$\frac{1 \times 4}{3 \times 4} = \frac{4}{12}$

$\frac{3 \times 3}{4 \times 3} = \frac{9}{12}$

$\frac{4}{12}, \frac{9}{12}$

12

2. b $\frac{1}{4}, \frac{3}{8}$

$\frac{1 \times 2}{4 \times 2} = \frac{2}{8}$

$\frac{2}{8}, \frac{3}{8}$

8

3. d $\frac{1}{8}, \frac{1}{2}$

$\frac{1 \times 4}{2 \times 4} = \frac{4}{8}$

$\frac{1}{8}, \frac{4}{8}$

8

4. b $\frac{1}{4}, \frac{3}{16}$

$\frac{1 \times 4}{4 \times 4} = \frac{4}{16}$

$\frac{4}{16}, \frac{3}{16}$

16

5. a $\frac{4}{32}, \frac{5}{8}$

$\frac{4 \div 4}{32 \div 4} = \frac{1}{8}$

$\frac{1}{8}, \frac{5}{8}$

8

Section 4.4.1

1. $\frac{1}{8} + \frac{4}{16}$

$\frac{4 \div 2}{16 \div 2} = \frac{2}{8}$

$\frac{1}{8} + \frac{2}{8} = \frac{3}{8}$

2. $\frac{4}{8} + \frac{6}{16}$

$\frac{6 \div 2}{16 \div 2} = \frac{3}{8}$

$\frac{4}{8} + \frac{3}{8} = \frac{7}{8}$

3. $\frac{2}{4} + \frac{1}{4} = \frac{3}{4}$

4. $\frac{3}{4} + \frac{2}{8}$

$\frac{2 \div 2}{8 \div 2} = \frac{1}{4}$

$\frac{3}{4} + \frac{1}{4} = \frac{4}{4} = 1$

5. $\frac{14}{16} + \frac{3}{8}$

$\frac{14 \div 2}{16 \div 2} = \frac{7}{8}$

$\frac{7}{8} + \frac{3}{8} = \frac{10}{8} = 1\frac{2}{8}$

$\frac{2 \div 2}{8 \div 2} = \frac{1}{4}$

$1\frac{1}{4}$

Section 4.5.1

1. $\frac{3}{8} - \frac{5}{16}$

$\frac{3 \times 2}{8 \times 2} = \frac{6}{16}$

$\frac{6}{16} - \frac{5}{16} = \frac{1}{16}$

2. $\frac{11}{16} - \frac{5}{8}$

$\frac{5 \times 2}{8 \times 2} = \frac{10}{16}$

$\frac{11}{16} - \frac{10}{16} = \frac{1}{16}$

3. $\frac{3}{4} - \frac{2}{6}$

$\frac{3 \times 3}{4 \times 3} = \frac{9}{12}$

$\frac{2 \times 2}{6 \times 2} = \frac{4}{12}$

$\frac{9}{12} - \frac{4}{12} = \frac{5}{12}$

4. $\frac{11}{12} - \frac{4}{8}$

$\frac{11 \times 2}{12 \times 2} = \frac{22}{24}$

$\frac{4 \times 3}{8 \times 3} = \frac{12}{24}$

$\frac{22}{24} - \frac{12}{24} = \frac{10}{24}$

$\frac{10 \div 2}{24 \div 2} = \frac{5}{12}$

5. $\frac{11}{16} - \frac{1}{2}$

$\frac{1 \times 8}{2 \times 8} = \frac{8}{16}$

$\frac{11}{16} - \frac{8}{16} = \frac{3}{16}$

Section 4.5.3

1. $8 - \frac{3}{4}$

$7 + 1 - \frac{3}{4}$

$7 + \frac{4}{4} - \frac{3}{4}$

$7 + \frac{1}{4} = 7\frac{1}{4}$

2. $12 - \frac{5}{8}$

$11 + 1 - \frac{5}{8}$

$11 + \frac{8}{8} - \frac{5}{8}$

$11 + \frac{3}{8} = 11\frac{3}{8}$

3. d Two punches $= 4\frac{1}{64} + 4\frac{3}{32}$

$\frac{3}{32} \times \frac{2}{2} = \frac{6}{64}$

$4\frac{1}{64} + 4\frac{6}{64} = 8\frac{7}{64}$

$9\frac{7}{16} - 8\frac{7}{64}$

$\frac{7}{16} \times \frac{4}{4} = \frac{28}{64}$

$9\frac{28}{64} - 8\frac{7}{64} = 1\frac{21}{64}$

4. b $20\frac{3}{4} - 12\frac{1}{16}$

$\frac{3}{4} \times \frac{4}{4} = \frac{12}{16}$

$20\frac{12}{16} - 12\frac{1}{16} = 8\frac{11}{16}$

5. c $36\frac{3}{8} - 35\frac{15}{16}$

$\frac{3}{8} \times \frac{2}{2} = \frac{6}{16}$

$36\frac{6}{16} - 35\frac{15}{16}$

$35 + 1 + \frac{6}{16}$

$35 + \frac{16}{16} + \frac{6}{16}$

$35 + \frac{22}{16}$

$35\frac{22}{16} - 35\frac{15}{16} = \frac{7}{16}$

Section 4.6.1

1. $\frac{4}{16} \times \frac{5}{8} = \frac{20}{128} \div \frac{4}{4} = \frac{5}{32}$

2. $\frac{3}{4} \times \frac{7}{8} = \frac{21}{32}$

3. $\frac{2}{8} \times \frac{15}{1} = \frac{30}{8} \div \frac{2}{2} = \frac{15}{4}$

4. $\frac{3}{7} \times \frac{49}{1} = \frac{147}{7} \div \frac{7}{7} = \frac{21}{1} = 21$

5. $\frac{8}{16} \times \frac{32}{64}$

$\frac{8}{16} \div \frac{8}{8} = \frac{1}{2}$ $\frac{32}{64} \div \frac{32}{32} = \frac{1}{2}$

$\frac{1}{2} \times \frac{1}{2} = \frac{1}{4}$

or $\frac{8}{16} \times \frac{32}{64} = \frac{256}{1024} \div \frac{256}{256} = \frac{1}{4}$

Section 4.7.1

1. $\frac{3}{8} \div 3 = \frac{3}{8} \times \frac{1}{3} = \frac{3}{24} \div \frac{3}{3} = \frac{1}{8}$

2. $\frac{5}{8} \div \frac{1}{2} = \frac{5}{8} \times \frac{2}{1} = \frac{10}{8} \div \frac{2}{2} = \frac{5}{4}$ or $1\frac{1}{4}$

3. $\frac{3}{4} \div \frac{3}{8} = \frac{3}{4} \times \frac{8}{3} = \frac{24}{12} \div \frac{12}{12} = \frac{2}{1} = 2$

4. a $8\frac{1}{2} \div \frac{1}{4} = \frac{17}{2} \div \frac{1}{4} = \frac{17}{2} \times \frac{4}{1} = \frac{68}{2} \div \frac{2}{2} =$

$\frac{34}{1} = 34$

5. d $7 \div \frac{7}{8} = \frac{7}{1} \times \frac{8}{7} = \frac{56}{7} \div \frac{7}{7} = \frac{8}{1} = 8$

Section 5.1.1

1. c
2. c
3. b
4. b
5. c

Section 5.2.1

1. b
2. c
3. a
4. b
5. b

Section 5.3.1

1. c
2. d
3. a
4. b
5. b

Section 5.4.1

1.
$$\begin{array}{r} 2.50 \\ 4.20 \\ +5.00 \\ \hline 11.70 \end{array}$$

2.
$$\begin{array}{r} {}^{1} \\ 1.82 \\ 3.41 \\ +5.25 \\ \hline 10.48 \end{array}$$

3.
$$\begin{array}{r} {}^{1} \\ 6.43 \\ +86.40 \\ \hline 92.83 \end{array}$$

4. b
$$\begin{array}{r} {}^{1} \\ 0.078 \\ +0.250 \\ \hline 0.328 \end{array}$$

5. b
$$\begin{array}{r} 6.7 \\ -2.3 \\ \hline 4.4 \end{array}$$

Section 5.5.1

1. c

3
× 0.3
0.9

2. a

5 11
6.18
− 0.90
5.28

3. c

4.68
× 11
468
+ 468
51.48

4. c

2
5.14
× 7
35.98
(Watts on second machine)

First machine + second machine =

1 1
51.48
+35.98
87.46

5. b

1 3
1 5
1 4
128
× 4.75
6 40
89 6
+ 512
608.00

Section 5.6.2

1. 18) 45.36 = 2.52

−36
093
−090
0036
−0036
0000

2. 18) 4.536 = 0.25

−36
093
−090
003*r*

3. 18) 0.4536 = 0.025

−00
045
−036
0093
−0090
0003*r*

4. 25) 10.20 = 0.40 (0.41 is also acceptable)

−100
0020
−0000
0020*r*

5. 6) 31.2 = 5.2

−30
012
−012
000

Section 5.6.4

1. $14.1\overline{)282} = \begin{array}{r} 20 \\ 141\overline{)2820} \\ \underline{-282} \\ 0000 \\ \underline{-0000} \\ 0000 \end{array}$

2. $3.2\overline{)694} = \begin{array}{r} 216.87 \\ 32\overline{)6940.00} \\ \underline{-64} \\ 054 \\ \underline{-032} \\ 0220 \\ \underline{-0192} \\ 00280 \\ \underline{-00256} \\ 000240 \\ \underline{-000224} \\ 000016r \end{array}$

3. $0.45\overline{)99} = \begin{array}{r} 220 \\ 45\overline{)9900} \\ \underline{-90} \\ 090 \\ \underline{-090} \\ 0000 \\ \underline{-0000} \\ 0000 \end{array}$

4. $2.5\overline{)102} = \begin{array}{r} 40.8 \\ 25\overline{)1020.0} \\ \underline{-100} \\ 0020 \\ \underline{-0000} \\ 00200 \\ \underline{-00200} \\ 00000 \end{array}$

5. $0.6\overline{)312} = \begin{array}{r} 520 \\ 6\overline{)3120} \\ \underline{-30} \\ 012 \\ \underline{-012} \\ 0000 \\ \underline{-0000} \\ 0000 \end{array}$

Section 5.6.6

1. $4.24\overline{)20.82} = \begin{array}{r} 4.91 \\ 424\overline{)2082.00} \\ \underline{-1696} \\ 3860 \\ \underline{-3816} \\ 440 \\ \underline{-424} \\ 16r \end{array}$

2. $3.7\overline{)38.9} = \begin{array}{r} 10.51 \\ 37\overline{)389.00} \\ \underline{-37} \\ 19 \\ \underline{-0} \\ 190 \\ \underline{-185} \\ 50 \\ \underline{-37} \\ 13r \end{array}$

3. $0.45\overline{)9.9} = \begin{array}{r} 22 \\ 45\overline{)990} \\ \underline{-90} \\ 90 \\ \underline{-90} \\ 0 \end{array}$

4. $0.25\overline{)10.20} = \begin{array}{r} 40.8 \\ 25\overline{)1020.0} \\ \underline{-100} \\ 20 \\ \underline{-0} \\ 200 \\ \underline{-200} \\ 0 \end{array}$

5. $0.6\overline{)31.2} = \begin{array}{r} 52 \\ 6\overline{)312} \\ \underline{-30} \\ 12 \\ \underline{-12} \\ 0 \end{array}$

Section 5.7.1

```
                               24.1
1. b  3.75 ) 90.5  =  375 ) 9050.00
                            -750
                             1550
                            -1500
                              500
                             -375
                              125r
```

```
                             15.51
2. a  40.1 ) 622  =  401 ) 6220.00
                          -401
                           2210
                          -2005
                            2050
                           -2005
                             450
                            -401
                              49r
```

Rounded to the nearest tenth = 15.5

```
                                 28.12
3. d  4.30 ) 120.95  =  430 ) 12095.00
                              -860
                               3495
                              -3440
                                 550
                                -430
                                 1200
                                 -860
                                  340r
```

Rounded to the nearest tenth = 28.1

```
                                311.83
4. c  0.37 ) 115.38  =  37 ) 11538.00
                            -111
                              43
                             -37
                               68
                              -37
                               310
                              -296
                                140
                               -111
                                 29r
```

Rounded to the nearest tenth = 311.8

```
                                240.37
5. b  0.48 ) 115.38  =  48 ) 11538.00
                             -96
                              193
                             -192
                                18
                                -0
                               180
                              -144
                                360
                               -336
                                 24r
```

Rounded to the nearest tenth = 240.4

Section 5.8.1

1. 53.74
2. 0.91
3. 629.52
4. 1.98
5. 38.40

Section 6.1.1

1. $0.62 = {}^{62}/_{100} = 62\%$

2. $0.475 = {}^{475}/_{1000} = 47.5\%$

3. $0.7 = {}^{7}/_{10} \times {}^{10}/_{10} = {}^{70}/_{100} = 70\%$

4. $72\% = {}^{72}/_{100} = 0.72$

5. $12.5\% = {}^{125}/_{1000} = 0.125$

Section 6.2.1

```
       0.25
1. 4 ) 1.00
      -8
        20
       -20
         0
```

```
       0.75
2. 4 ) 3.00
      -28
        20
       -20
         0
```

```
       0.125
3. 8 ) 1.000
       -8
        20
       -16
         40
        -40
          0
```

4. $16\overline{)5.000}$ = 0.3125

−48
20
−16
40
−32
80
−80
0

5. $64\overline{)20.0000}$ = 0.3125

−192
80
−64
160
−128
320
−320
0

Section 6.3.1

1. $0.5 = \frac{5 \div 5}{10 \div 5} = \frac{1}{2}$

2. $0.12 = \frac{12 \div 4}{100 \div 4} = \frac{3}{25}$

3. $0.125 = \frac{125 \div 125}{1000 \div 125} = \frac{1}{8}$

4. $0.8 = \frac{8 \div 2}{10 \div 2} = \frac{4}{5}$

5. $0.45 = \frac{45 \div 5}{100 \div 5} = \frac{9}{20}$

Section 6.4.1

1. $\frac{9 \div 3}{12 \div 3} = \frac{3}{4}$

$4\overline{)3.00}$ = 0.75

−28
20
−20
0

2. $\frac{10 \div 2}{12 \div 2} = \frac{5}{6}$

$6\overline{)5.00}$ = 0.833

−48
20
−18
20
−18
2*r*

Rounded to the nearest hundredth = 0.83

3. $\frac{2 \div 2}{12 \div 2} = \frac{1}{6}$

$6\overline{)1.000}$ = 0.166

−6
40
−36
40
−36
6*r*

Rounded to the nearest hundredth = 0.17

4. $\frac{4 \div 4}{12 \div 4} = \frac{1}{3}$

$3\overline{)1.000}$ = 0.333

−9
10
−9
10
−9
1*r*

Rounded to the nearest hundredth = 0.33

5. $\frac{8 \div 4}{12 \div 4} = \frac{2}{3}$

$3\overline{)2.000}$ = 0.666

−18
20
−18
20
−18
2*r*

Rounded to the nearest hundredth = 0.67

Section 7.1.1

1. 10
2. 100
3. 1,000,000
4. 0.1
5. 0.001

Section 7.2.1

1. a
2. c
3. b
4. b
5. d

Section 7.3.1

1. $2.45 \text{ m} \times \dfrac{100 \text{ cm}}{1 \text{ m}} = 45 \text{ cm}$

2. $3 \text{ yards} \times \dfrac{3 \text{ ft}}{1 \text{ yd}} \times \dfrac{12 \text{ in.}}{1 \text{ ft}} = 108 \text{ in.}$

3. $36 \text{ in.} \times \dfrac{1 \text{ yd}}{36 \text{ in.}} = \dfrac{36}{36} = 1 \text{ yd}$

4. $90 \text{ in.} \times \dfrac{1 \text{ ft}}{12 \text{ in.}} \times \dfrac{1 \text{ yd}}{3 \text{ ft}} = \dfrac{90 \div 18}{36 \div 18} = \dfrac{5}{2}$ or $2\dfrac{1}{2}$ or 2.5 yds

5. $1 \text{ cm} \times \dfrac{1 \text{ m}}{100 \text{ cm}} = \dfrac{1}{100} \text{ m}$

$$\begin{array}{r} 0.01 \\ 100\,\overline{)\,1.00} \\ \underline{-0} \\ 100 \\ \underline{-100} \\ 0 \end{array}$$

Section 8.3.1

1. c Area = length × width = 8 × 4 = 32 sq ft

2. a Area = length × width = 16 × 16 = 256 sq cm

3. c radius = ½ diameter

radius = ½ × 14 = 7 ft

Area = pi × radius^2 = 3.14 × 7 × 7 = 3.14 × 49

$$\begin{array}{r} {}^{1}_{1\ 3} \\ 3.14 \\ \underline{\times\ 49} \\ 2826 \\ \underline{+\ 1256\ } \\ 153.86 \text{ sq ft} \end{array}$$

4. a Area = 0.5 × base × height = 0.5 × 4 × 6 = 0.5 × 24 = 12 sq ft

$$\begin{array}{r} {}^{2} \\ 24 \\ \underline{\times 0.5} \\ 12.0 \end{array}$$

5. c Area = base × height = 14 × 5 = 70 sq m

Section 8.4.1

1. b Volume = length × width × depth = 5 × 6 × 13 = 30 × 13 = 390 cu ft

$$\begin{array}{r} 30 \\ \underline{\times 13} \\ 90 \\ \underline{+30\ } \\ 390 \end{array}$$

2. d Volume = length × width × depth = 3 × 3 × 3 = 27 cu cm

3. b Volume = 0.5 × base × height × depth = 0.5 × 6 × 2 × 4 = 24 cu in.

4. c Radius = ½ diameter = ½ × 6 = 6/2 ÷ 2/2 = 3/1 = 3 m

Height = 60 cm = 0.6 m

Volume = pi × radius^2 × h = 3.14 × 3 × 3 × 0.6 = 3.14 × 9 × 0.6

$$\begin{array}{r} 9 \\ \underline{\times 0.6} \\ 5.4 \end{array}$$

Volume = 3.14 × 5.4

$$\begin{array}{r} {}^{2}_{1} \\ 3.14 \\ \underline{\times 5.4} \\ 1256 \\ \underline{+1570\ } \\ 16.956 \end{array}$$

Volume = 16.956 cu m

5. b Depth = 6 in.
6/12 = ½ = 0.5 ft

Volume = length × width × depth = 17 × 17 × 0.5 =

$$\begin{array}{r} {}^{4} \\ 17 \\ \underline{\times 17} \\ 119 \\ \underline{+17\ } \\ 289 \end{array}$$

Volume = 289 × 0.5

$$\begin{array}{r} {}^{4\,4} \\ 289 \\ \underline{\times 0.5} \\ 144.5 \end{array}$$

Volume = 144.5 cu ft

1 yd = 3 ft

1 cu yd = 3 × 3 × 3 = 27 cu ft

$144.5 \text{ cu ft} \times \dfrac{1 \text{ cu yd}}{27 \text{ cu ft}} = 144.5 \div 27$

$$\begin{array}{r} 5.35 \\ 27\,\overline{)\,144.50} \\ \underline{-135} \\ 95 \\ \underline{-81} \\ 140 \\ \underline{-135} \\ 5r \end{array}$$

5.35 cu yds

Section 2.0.0

1. d

2. b

3. b
$$\begin{array}{r} {}^{2}\\ 35\\ 18\\ \underline{+8}\\ 61 \end{array}$$

4. a
$$\begin{array}{r} {}^{1\,1}\\ 649\\ 632\\ \underline{+478}\\ 1{,}759 \end{array}$$

5. b
$$\begin{array}{r} {}^{2}\\ 31\\ 46\\ 49\\ \underline{+16}\\ 142 \end{array}$$

6. b
$$\begin{array}{r} {}^{4\,4}\\ 142\\ 57\\ 35\\ 79\\ 32\\ 79\\ 53\\ \underline{+95}\\ 572 \end{array}$$

7. d
$$\begin{array}{r} 36{,}000\\ \underline{+1{,}500}\\ 37{,}500 \end{array}$$

8. b
$$\begin{array}{r} 86\\ \underline{-12}\\ 74 \end{array}$$

9. a
$$\begin{array}{r} {}^{8\,13}\\ \not{9}\not{3}\\ \underline{-27}\\ 66 \end{array}$$

10. c
$$\begin{array}{r} {}^{0\;13\,16\,18}\\ \not{1}{,}\not{4}\not{7}\not{8}\\ \underline{-489}\\ 989 \end{array}$$

11. c
$$\begin{array}{r} 11\\ \underline{\times 8}\\ 88 \end{array}$$

12. c
$$\begin{array}{r} {}^{1}\\ {}^{3}\\ 15\\ \underline{\times 26}\\ 90\\ \underline{+30\;}\\ 390 \end{array}$$

13. d
$$\begin{array}{r} {}^{4}\\ 65\\ \underline{\times 8}\\ 520 \end{array}$$

14. a
215r
9
$$\begin{array}{r} 215\\ 27\,\overline{)\,5814}\\ \underline{-54}\;\;\\ 41\;\\ \underline{-27}\;\\ 144\\ \underline{-135}\\ 9 \end{array}$$

15. c \$448 ÷ \$4 = \$112
$$\begin{array}{r} 112\\ 4\,\overline{)\,448}\\ \underline{-4}\;\;\\ 04\;\\ \underline{-4}\;\\ 08\\ \underline{-8}\\ 0 \end{array}$$

16. c $96 \div 4 = 24$

17. a $72 + 456 + 125 = 653$

18. b $17 \times 5 = 85$

19. d $560 + 978 + 234 = 1{,}772$
$8{,}000 - 1{,}772 = 6{,}228$

20. d $628 \div 8 = 78.5$
78 pieces with ½ of a piece (½ of 8 inches = 4 inches) left over

Section 3.0.0

1. c

2. a

3. b

4. c

5. d

Section 4.0.0

1. c $\frac{3}{8}$ inch $\times \frac{8}{8} = \frac{24}{64}$ inch

2. b $\frac{5}{16}$ inch $\times \frac{2}{2} = \frac{10}{32}$ inch

3. d $\frac{1}{2}$ inch $\times \frac{8}{8} = \frac{8}{16}$ inch

4. c $\dfrac{16}{64} \div \dfrac{16}{16}$

$$\begin{array}{r} 1\\ 16\,\overline{)\,16}\\ \underline{-16}\\ 0 \end{array} \qquad \begin{array}{r} 4\\ 16\,\overline{)\,64}\\ \underline{-64}\\ 0 \end{array}$$

$$\frac{16 \div 16}{64 \div 16} = \frac{1}{4}$$

5. c $\frac{8 \div 8}{16 \div 8}$

$$\begin{array}{r} 1 \\ 8\overline{)8} \\ -8 \\ \hline 0 \end{array} \qquad \begin{array}{r} 2 \\ 8\overline{)16} \\ -16 \\ \hline 0 \end{array}$$

$\frac{8 \div 8}{16 \div 8} = \frac{1}{2}$

6. c $\frac{2 \div 2}{4 \div 2}$

$$\begin{array}{r} 1 \\ 2\overline{)2} \\ -2 \\ \hline 0 \end{array} \qquad \begin{array}{r} 2 \\ 2\overline{)4} \\ -4 \\ \hline 0 \end{array}$$

$\frac{2 \div 2}{4 \div 2} = \frac{1}{2}$

7. a Step 1. Reduce fractions to lowest terms.

$\frac{8 \div 8}{64 \div 8} = \frac{1}{8}$

$\frac{1}{32} = \frac{1}{32}$

Step 2. Find the lowest common multiple.

$\frac{1 \times 4}{8 \times 4} = \frac{4}{32}$

$\frac{1}{32}$

Answer: 32 is the lowest common denominator.

8. c Step 1. Reduce fractions to lowest terms.

$\frac{3}{4} = \frac{3}{4}$

$\frac{8 \div 8}{16 \div 8} = \frac{1}{2}$

Step 2. Find the lowest common multiple.

$\frac{1 \times 2}{2 \times 2} = \frac{2}{4}$

$\frac{3}{4}$

Answer: 4 is the lowest common denominator.

9. d Step 1. Reduce fractions to lowest terms.

$\frac{4 \div 4}{12 \div 4} = \frac{1}{3}$

$\frac{5 \div 5}{15 \div 5} = \frac{1}{3}$

Answer: Both fractions reduce to ⅓, so 3 is the lowest common denominator.

10. c Step 1. Reduce fractions to lowest terms.

$\frac{1}{2} = \frac{1}{2}$

$\frac{1}{8} = \frac{1}{8}$

Both fractions are already in lowest terms.

Step 2. Find the lowest common multiple.

$\frac{1 \times 4}{2 \times 4} = \frac{4}{8}$

$\frac{1}{8}$

Answer: 8 is the lowest common denominator.

11. b $\frac{7}{16} + \frac{1}{4}$

$\frac{1 \times 4}{4 \times 4} = \frac{4}{16}$

$\frac{7}{16} + \frac{4}{16} = \frac{11}{16}$

12. b $\frac{3}{8} + \frac{9}{16}$

$\frac{3 \times 2}{8 \times 2} = \frac{6}{16}$

$\frac{6}{16} + \frac{9}{16} = \frac{15}{16}$

13. c $\frac{11}{32} - \frac{2}{8}$

$\frac{2 \times 4}{8 \times 4} = \frac{8}{32}$

$\frac{11}{32} - \frac{8}{32} = \frac{3}{32}$

14. d $18 - \frac{7}{8}$

$18 = 17 + 1 = 17 + \frac{8}{8}$

$17 + \frac{8}{8} - \frac{7}{8} = 17\frac{1}{8}$

15. b $\frac{9}{16} \times \frac{2}{4} = \frac{18 \div 2}{64 \div 2} = \frac{9}{32}$

16. c $\frac{1}{8} \times 72 = 9$

17. d $5\frac{1}{2} \div \frac{1}{2}$

$$\frac{11}{2} \div \frac{1}{2} = \frac{11}{2} \times \frac{2}{1} = \frac{22}{2} = 11$$

18. d $\frac{9}{16} \div \frac{7}{8}$

$$\frac{9}{16} \times \frac{8}{7} = \frac{72 \div 8}{112 \div 8} = \frac{9}{14}$$

19. b $\frac{7}{16} \div 7$

$$\frac{7}{16} \times \frac{1}{7} = \frac{7 \div 7}{112 \div 7} = \frac{1}{16}$$

20. c $9 \div \frac{3}{16}$

$$\frac{9}{1} \times \frac{16}{3} = \frac{144 \div 3}{3 \div 3} = \frac{48}{1} = 48$$

Section 5.0.0

1. a
2. b
3. b
4. c
5. b

6. c

$$\begin{array}{r} {\scriptstyle 8\ 17} \\ 9.78 \\ -4.90 \\ \hline 4.88 \end{array}$$

7. d

$$\begin{array}{r} {\scriptstyle 1} \\ 6.001 \\ 9.003 \\ +13.300 \\ \hline 28.304 \end{array}$$

8. a

$$\begin{array}{r} {\scriptstyle 1\ 1} \\ 51.5 \\ +89.7 \\ \hline 141.2 \end{array}$$ microns

9. c

$$\begin{array}{r} {\scriptstyle 6\ 14\ 15} \\ 675.57 \\ -356.90 \\ \hline 318.67 \end{array}$$

10. a

$$\begin{array}{r} {\scriptstyle 0\ 13\ 15} \\ 1.45 \\ -0.78 \\ \hline \$0.67 \end{array}$$

11. a

$$\begin{array}{r} {\scriptstyle 3\ 3} \\ 57.80 \\ \times 40 \\ \hline 0000 \\ +23120 \\ \hline 2{,}312.00 \end{array}$$

12. b

$$\begin{array}{r} {\scriptstyle 4\ 2} \\ {\scriptstyle 3\ 2} \\ {\scriptstyle 2\ 1} \\ 3.53 \\ \times 9.75 \\ \hline 17\ 65 \\ 247\ 1 \\ +3177 \\ \hline 34.41\ 75 \end{array}$$

13. b $864.5 \times \$2.37$

$$\begin{array}{r} {\scriptstyle 1\ \ \ 1} \\ {\scriptstyle 1\ 1\ 1} \\ {\scriptstyle 4\ 3\ 3} \\ 864.5 \\ \times 2.37 \\ \hline 60515 \\ 25935 \\ +17290 \\ \hline \$2{,}048.865 \end{array}$$

Round up to $2,048.87

14. a $57.80 \div 40$

$$\begin{array}{r} 1.445 \\ 40\,\overline{)\,57.800} \\ -40 \\ \hline 178 \\ -160 \\ \hline 180 \\ -160 \\ \hline 200 \\ -200 \\ \hline 0 \end{array}$$

15. a $89.435 \div 0.05$

$$\begin{array}{r} 1788.7 \\ 0.05.\,\overline{)\,89.43.5} \\ -5 \\ \hline 39 \\ -35 \\ \hline 44 \\ -40 \\ \hline 43 \\ -40 \\ \hline 35 \\ -35 \\ \hline 0 \end{array}$$

Section 6.0.0

1. d $0.78 \times 100 = 78\%$

$$\begin{array}{r} 100 \\ \underline{\times 0.78} \\ 800 \\ \underline{+700} \\ 78.00 \end{array}$$

78%

2. c 13.9%

$13.9 \div 100 =$

$$\begin{array}{r} 0.139 \\ 100\overline{)13.900} \\ \underline{-100} \\ 390 \\ \underline{-300} \\ 900 \\ \underline{-900} \\ 0 \end{array}$$

3. d $\frac{3}{16} = 3 \div 16$

$$\begin{array}{r} 0.1875 \\ 16\overline{)3.0000} \\ \underline{-16} \\ 140 \\ \underline{-128} \\ 120 \\ \underline{-112} \\ 80 \\ \underline{-80} \\ 0 \end{array}$$

4. a $\frac{7}{32} = 7 \div 32$

$$\begin{array}{r} 0.21875 \\ 32\overline{)7.00000} \\ \underline{-64} \\ 60 \\ \underline{-32} \\ 280 \\ \underline{-256} \\ 240 \\ \underline{-224} \\ 160 \\ \underline{-160} \\ 0 \end{array}$$

5. d $0.6 = \frac{6}{10}$

$\frac{6 \div 2}{10 \div 2} = \frac{3}{5}$

6. c $0.875 = \frac{875}{1000}$

$\frac{875 \div 125}{1000 \div 125} = \frac{7}{8}$

7. b $14.75 = 14\frac{75}{100}$

$\frac{75 \div 25}{100 \div 25} = \frac{3}{4}$

$14\frac{3}{4}$

8. d $2.04 = 2\frac{4}{100}$

$\frac{4 \div 4}{100 \div 4} = \frac{1}{25}$

$2\frac{1}{25}$

9. b $\frac{6}{12}$ inches

$$\begin{array}{r} 0.5 \\ 12\overline{)6.0} \\ \underline{-60} \\ 0 \end{array}$$

10. d $\frac{11}{12}$

$$\begin{array}{r} 0.916 \\ 12\overline{)11.000} \\ \underline{-108} \\ 20 \\ \underline{-12} \\ 80 \\ \underline{-72} \\ 8 \end{array}$$

0.916 rounds up to 0.92

Section 7.0.0

1. b
2. a
3. b
4. d
5. c
6. a
7. b
8. c
9. c
10. a 17 cm = 170 mm

Section 8.0.0

1. a
2. a
3. c
4. c
5. a
6. b $7\text{ m} \times 3\text{ m} = 21$ sq m
7. c $0.5\text{ b} \times \text{h} = 0.5 \times 7\text{ ft} \times 3.5\text{ ft} = 12.25$ sq ft
8. b $9\text{ in.} \times 9\text{ in.} \times 9\text{ in.} = 729$ cu in.
9. d $\text{pi} \times r^2 \times h = 3.14 \times 6.25\text{ ft} \times 6.25\text{ ft} \times 23\text{ ft} = 2821.09$ cu ft
10. a $0.5 \times b \times h \times d = 0.5 \times 7\text{ cm} \times 4\text{ cm} \times 30\text{ mm}$

    $30\text{ mm} = 3.0\text{ cm}$

    $0.5 \times 7\text{ cm} \times 4\text{ cm} \times 3\text{ cm} = 42$ cu cm

MODULE 00102-04 — ANSWERS TO KEY TERMS QUIZ

1. radius
2. borrow
3. acute angle
4. bisect
5. carry
6. Area
7. convert
8. angle
9. cubic
10. decimal
11. circle
12. Degree
13. numerator, denominator
14. diagonal
15. diameter
16. difference
17. digits
18. English ruler, standard ruler (Note: Terms are acceptable in either order.)
19. place value
20. equivalent fractions
21. formula
22. circumference
23. fractions
24. meter
25. improper fractions
26. invert
27. long division
28. sum
29. machinist's rule
30. remainder
31. metric ruler
32. obtuse angle
33. mixed number
34. Opposite angles, adjacent angles
35. percent
36. perimeter
37. whole numbers
38. negative numbers, positive numbers
39. rectangle, square (Note: Both terms are acceptable.)
40. scalene triangle, right triangle, equilateral triangle, isosceles triangle
41. right angle
42. straight angle
43. triangle
44. vertex
45. Pi
46. volume

CONTREN® LEARNING SERIES — USER FEEDBACK

The NCCER makes every effort to keep these textbooks up-to-date and free of technical errors. We appreciate your help in this process. If you have an idea for improving this textbook, or if you find an error, a typographical mistake, or an inaccuracy in NCCER's *Contren®* textbooks, please write us, using this form or a photocopy. Be sure to include the exact module number, page number, a detailed description, and the correction, if applicable. Your input will be brought to the attention of the Technical Review Committee. Thank you for your assistance.

Instructors – If you found that additional materials were necessary in order to teach this module effectively, please let us know so that we may include them in the Equipment/Materials list in the Annotated Instructor's Guide.

Write: Product Development
National Center for Construction Education and Research
P.O. Box 141104, Gainesville, FL 32614-1104

Fax: 352-334-0932

E-mail: curriculum@nccer.org

Craft ______________ Module Name ______________

Copyright Date ______________ Module Number ______________ Page Number(s) ______________

Description

____

____

____

____

(Optional) Correction

____

____

____

(Optional) Your Name and Address

____

____

____

Introduction to Hand Tools

00103-04

NCCER STANDARDIZED CRAFT TRAINING PROGRAM

The National Center for Construction Education and Research (NCCER) provides a standardized national program of accredited craft training. Key features of the program include instructor certification, competency-based training, and performance testing. The program provides trainees, instructors, and companies with a standard form of recognition through a National Craft Training Registry. The program is described in full in the *Guidelines for Accreditation*, published by the NCCER. For more information on standardized craft training, contact the NCCER by writing us at P.O. Box 141104, Gainesville, FL 32614-1104; calling 352-334-0911; or e-mailing info@nccer.org. More information may be found at our Web site, www.nccer.org.

HOW TO USE THIS ANNOTATED INSTRUCTOR'S GUIDE

Each page presents two sections of information. The larger section displays each page exactly as it appears in the Trainee Module. The narrow column ties suggested trainee and instructor actions to each page and provides icons (detailed below) to call your attention to material, safety, audiovisual, or testing requirements. The bottom of each page includes space for your notes.

The **Audiovisual** icon indicates an appropriate time to show a transparency or other audiovisual aid.

The **Classroom** icon prompts you to define a term, stress a point, ask trainees to explain a concept, or give examples.

The **Demonstration** icon directs you to show trainees how to perform tasks.

The **Examination** icon tells you to administer the written module examination.

The **Homework** icon is placed where you may wish to assign reading for the next class, to assign a project, or to advise trainees to prepare for an examination.

The **Laboratory** icon is used when trainees are to practice performing tasks.

The **Materials** icon is a reminder for you to gather materials needed for classes, labs, and testing.

The **Performance Testing** icon tells you to administer a performance test or a portion thereof.

The **Safety** icon is used to emphasize safety issues. It is often keyed to *Caution* and *Warning* statements in the Trainee Module.

The **Teaching Tip** icon indicates additional guidance is available, such as how to conduct an exercise, get the most educational value from a field trip, or encourage class participation. Teaching Tips may expand on a feature (*Think About It, Did You Know?*) or provide Quick Quizzes or similar exercises. You will be referred to the Teaching Tips section at the back of the module if there is additional material.

The **Combination** icon indicates that the laboratory listed corresponds with a performance task. If desired, you can note the proficiency of the trainees during the laboratory and use it to satisfy performance testing requirements.

PREPARATION

Before teaching this module, you should review the Objectives, Performance Tasks, Materials and Equipment List, and the Module Outline. Be sure to allow ample time to prepare your own training or lesson plan and gather all required materials and equipment.

MODULE OVERVIEW

This module explains how to inspect and properly use hand tools. Trainees will learn how to identify and take care of basic hand tools.

PREREQUISITES

Prior to training with this module, it is recommended that the trainee shall have successfully completed the following: *Core Curriculum: Introductory Craft Skills,* Modules 00101-04 and 00102-04

OBJECTIVES

Upon completion of this module, the trainee will be able to:

1. Recognize and identify some of the basic hand tools used in the construction trade.
2. Use hand tools safely.
3. Describe the basic procedures for taking care of hand tools.

PERFORMANCE TASKS

Under the supervision of the instructor, the trainee should be able to:

1. Visually inspect the following tools to determine if they are safe to use:
   - Hammer
   - Screwdriver
   - Saw
2. Make a straight, square cut using a crosscut saw. To prepare for sawing, the trainee should mark the wood with a combination square and support the work. The first stroke should be toward the trainee's body. The saw should be kept vertical to the work.
3. Safely and properly use the following tools:
   - Hammer and cat's paw (to drive and pull nails)
   - Screwdriver (slotted or Phillips)
   - Adjustable wrench
   - Tongue-and-groove pliers and adjustable wrench
   - Spirit level
   - Carpenter's square and steel tape
   - Saw

MATERIALS AND EQUIPMENT LIST

Transparencies
Markers/chalk
Blank acetate sheets
Transparency pens
Pencils and scratch paper
Overhead projector and screen
Whiteboard/chalkboard
Appropriate personal protective equipment
Copies of your local code
Claw hammer
Wood board with nails to practice using hammers and mallets
Ball peen hammer
Mallet
Screwdriver
Wood board with screws to practice using screwdrivers
Sledgehammer
Stake
Nail pullers, including:
- Cat's paws
- Chisel bars
- Flat bars

Wood boards with nails to practice using nail pullers
Slip-joint pliers
Pliers, including:
- Slip-joint
- Long-nose
- Lineman
- Tongue-and-groove
- Vise-Grip®

Boards with wire and soft metals to practice using pliers

Rulers and measuring tools, including:
- Steel rules
- Measuring tapes
- Wooden folding rules

Spirit level
Combination square
Try square
Square wood frames to practice using the measuring tools
Plumb bob
Self-chalker/plumb bob
Clamps, including:
- C-clamp
- Locking clamp
- Spring clamp
- Bar clamp
- Pipe clamp
- Hand-screw clamp
- Web clamp

Crosscut saws
Ripsaws
Sections of wood suitable for sawing
Files and rasps, including:
- Veneer knife file
- Square file
- Triangle file
- Flat file
- Rat-tail file

Materials to be filed
Wood chisel
Cold chisel
Wood and metal to practice using the chisels
Nonadjustable wrenches
Adjustable wrenches
Torque wrench
Wedges
Utility knife
Cardboard box to practice cutting with a utility knife
Wire brushes
Dirty tools to clean using wire brushes
Module Examinations*
Performance Profile Sheets*

*Located in the Test Booklet.

SAFETY CONSIDERATIONS

Ensure that the trainees are equipped with appropriate personal protective equipment. Always work in a clean, well-lit, appropriate work area.

ADDITIONAL RESOURCES

This module is intended to present thorough resources for task training. The following reference works are suggested for both instructors and motivated trainees interested in further study. These are optional materials for continued education rather than for task training.

Field Safety, 2003. NCCER. Upper Saddle River, NJ: Prentice Hall.

Hand Tools & Techniques, 1999. Minneapolis, MN: Handyman Club of America.

The Long and Short of It: How to Take Measurements. Video. Charleston, WV: Cambridge Vocational & Technical, 800-468-4227.

Reader's Digest Book of Skills and Tools, 1993. Pleasantville, NY: Reader's Digest.

TEACHING TIME FOR THIS MODULE

An outline for use in developing your lesson plan is presented below. Note that each Roman numeral in the outline equates to one session of instruction. Each session has a suggested time period of 2½ hours. This includes 10 minutes at the beginning of each session for administrative tasks and one 10-minute break during the session. Approximately 10 hours are suggested to cover *Introduction to Hand Tools.* You will need to adjust the time required for hands-on activity and testing based on your class size and resources. Because laboratories often correspond to Performance Tasks, the proficiency of the trainees may be noted during these exercises for Performance Testing purposes.

Topic	Planned Time
Session I. Hand Tools, Part One	
A. Hammers	________
B. Screwdrivers	________
C. Sledgehammers	________
D. Ripping Bars and Nail Pullers	________
E. Pliers and Wire Cutters	________
Session II. Hand Tools, Part Two	
A. Rulers and Other Measuring Tools	________
B. Levels	________
C. Squares	________
D. Plumb Bob	________
E. Chalk Lines	________
F. Bench Vises	________
Session III. Hand Tools, Part Three	
A. Clamps	________
B. Saws	________
C. Files and Rasps	________
D. Chisels and Punches	________
E. Wrenches	________
F. Performance Testing (Tasks 1 and 2)	________
Session IV. Hand Tools, Part Four	
A. Sockets and Ratchets	________
B. Torque Wrenches	________
C. Wedges	________
D. Utility Knives	________
E. Chain Falls and Come-Alongs	________
F. Wire Brushes	________
G. Shovels	________
H. Review	________
I. Module Examination	________
1. Trainees must score 70% or higher to receive recognition from NCCER.	
2. Record the testing results on Craft Training Report Form 200, and submit the results to the Training Program Sponsor.	
J. Performance Testing (Task 3)	________
1. Trainees must perform each task to the satisfaction of the instructor to receive recognition from NCCER. If applicable, proficiency noted during laboratory exercises can be used to satisfy the Performance Testing requirements.	
2. Record the testing results on Craft Training Report Form 200, and submit the results to the Training Program Sponsor.	

Introduction to Hand Tools
00103-04

Build America Winner—
Environmental

The Genzyme Center in Cambridge, Massachusetts, was built by Turner Construction Company. This building began as a radical concept. The Center is a sustainable "living" building that will set new standards in environmental responsibility, design aesthetics, and workplace well-being. Turner Construction made the Genzyme's vision a reality, creating an office building that will serve as a prototype for sustainable construction for years to come.

00103-04
Introduction to Hand Tools

Topics to be presented in this unit include:

Overview

Hand tools are the backbone of the construction trades. Hammers, wrenches, pliers, saws, shovels, and squares are just a few of the hand tools that are used every day on construction sites. Hand tools must be properly used and maintained at all times. Tools that are damaged or don't work are dangerous.

Safety is an important part of using hand tools. This means that tools must always be clean, dry, well-maintained, and used only for the job they were designed to do. Workers must be thoroughly trained in the proper use and maintenance of hand tools. This helps prevent accidents, damages, and injuries.

Take the time to learn how to properly maintain and use tools. It can make the difference between a job well done and the damage of equipment or loss of lives.

Instructor's Notes:

Objectives

When you have completed this module, you will be able to do the following:

1. Recognize and identify some of the basic hand tools used in the construction trade.
2. Use hand tools safely.
3. Describe the basic procedures for taking care of hand tools.

Key Trade Terms

Allen wrench
Ball peen hammer
Bell-faced hammer
Bevel
Box-end wrench
Carpenter's square
Cat's paw
Chisel
Chisel bar
Claw hammer
Combination square
Combination wrench
Crescent wrench
Dowel
Fastener
Flat bar
Flats
Foot-pounds
Inch-pounds
Joint
Kerf
Level
Miter joint
Nail puller
Open-end wrench
Peening
Pipe wrench
Planed
Pliers
Plumb
Points
Punch
Rafter angle square
Ripping bar
Round off
Spud wrench
Square
Striking (or slugging) wrench
Strip
Tang
Tempered
Tenon
Torque
Try square
Vise
Weld

Required Trainee Materials

1. Appropriate personal protective equipment
2. Sharpened pencils and paper

Prerequisites

Before you begin this module, it is recommended that you successfully complete the following: *Core Curriculum: Introductory Craft Skills*, Modules 00101-04 and 00102-04.

This course map shows all of the modules in *Core Curriculum: Introductory Craft Skills.* The suggested training order begins at the bottom and proceeds up. Skill levels increase as you advance on the course map. The local Training Program Sponsor may adjust the training order.

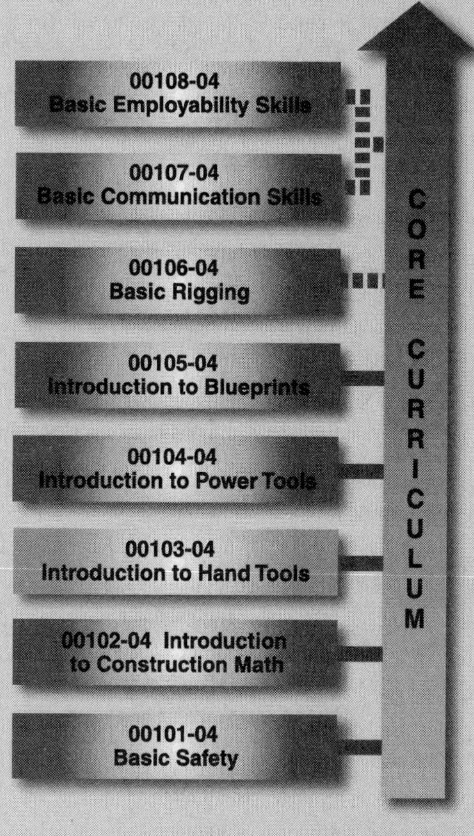

103CMAP.EPS

Ensure that you have everything required to teach the course. Check the Materials and Equipment List at the front of this module.

See the general Teaching Tip at the end of this module.

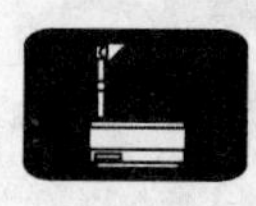

Show Transparency 1, Course Objectives.

Show Transparency 2, Performance Tasks.

Explain that terms shown in bold (blue) are defined in the Glossary at the back of this module.

Emphasize the importance of wearing appropriate personal protective equipment when using hand tools.

Discuss basic safety procedures for using hand tools.

Discuss the consequences of failing to focus when using a hammer.

Review the uses of a claw hammer. Demonstrate how to use a claw hammer to drive a nail and to pull a nail.

1.0.0 ◆ INTRODUCTION

Every profession has its tools. A surgeon uses a scalpel, a teacher uses a chalkboard, and an accountant uses a calculator. The construction trade has a whole collection of hand tools, such as hammers, screwdrivers, and **pliers,** that everyone uses. Even if you are already familiar with some of these tools, you need to learn to maintain them and use them safely. The better you use and maintain your tools, the better you will be in your craft.

This module shows you how to safely use and maintain some of the most common hand tools of the construction trade. It also highlights some specialized crafts and uses of hand tools.

1.1.0 Safety

To work safely, you must think about safety. Before you use any tool, you should know how it works and some of the possible dangers of using it the wrong way. Always read and understand the procedures and safety tips in the manufacturer's guide for every tool you use. Make sure every tool you use is in good condition. Never use worn or damaged tools.

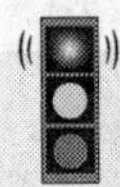

WARNING!

Always protect yourself when you are using tools by wearing appropriate personal protective equipment (PPE), such as safety gloves and eye protection.

2.0.0 ◆ HAMMERS

Hammers are made in different sizes and weights for specific types of work. Two of the most common hammers are the **claw hammer** and the **ball peen hammer** (see *Figure 1*).

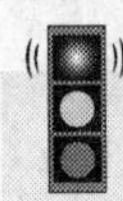

WARNING!

The most important safety consideration when using a hammer is focusing on the work. If you look away from the work while using a hammer, you may accidentally strike yourself or damage the work.

2.1.0 The Claw Hammer

The claw hammer has a steel head and a handle made of wood, steel, or fiberglass. You use the head to drive nails, wedges, and **dowels.** You use the claw to pull nails out of wood. The face of the hammer may be flat or rounded. It's easier to drive nails with the flat face (plain) claw hammer, but the flat face may leave hammer marks when you drive the head of the nail flush (even) with the surface of the work.

A claw hammer with a slightly rounded (or convex) face is called a **bell-faced hammer.** A skilled worker can use it to drive the nail head flush without damaging the surface of the work.

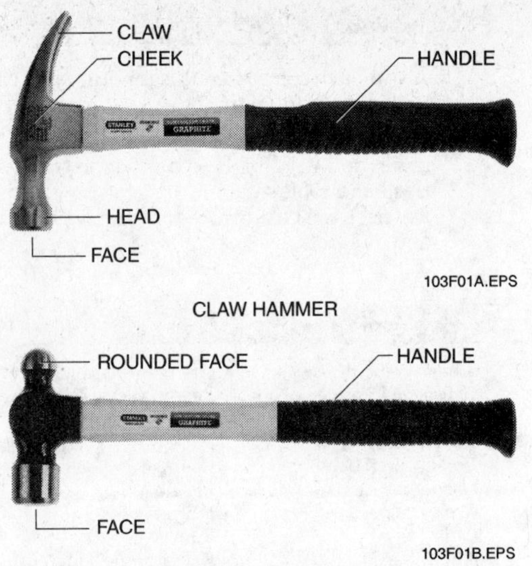

Figure 1 ◆ Claw and ball peen hammers.

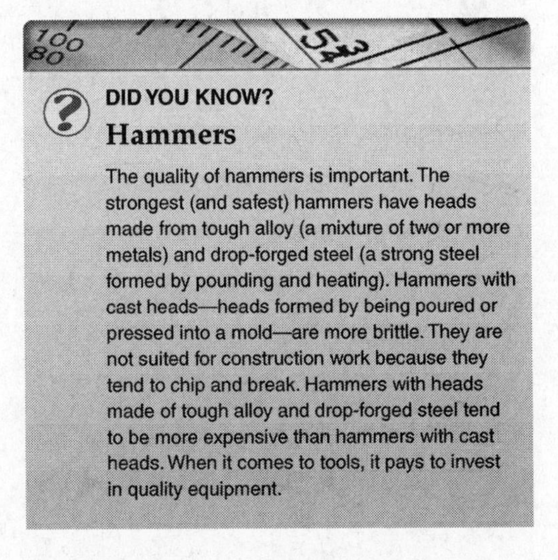

DID YOU KNOW?

Hammers

The quality of hammers is important. The strongest (and safest) hammers have heads made from tough alloy (a mixture of two or more metals) and drop-forged steel (a strong steel formed by pounding and heating). Hammers with cast heads—heads formed by being poured or pressed into a mold—are more brittle. They are not suited for construction work because they tend to chip and break. Hammers with heads made of tough alloy and drop-forged steel tend to be more expensive than hammers with cast heads. When it comes to tools, it pays to invest in quality equipment.

Instructor's Notes:

3.2 CORE CURRICULUM ◆ INTRODUCTORY CRAFT SKILLS

2.1.1 *How to Use a Claw Hammer to Drive a Nail*

Follow these simple steps to use a claw hammer properly when driving a nail:

Step 1 Hold the nail straight, at a 90-degree angle to the surface being nailed.

Step 2 Grip the handle of the hammer. Hold the end of the handle even with the lower edge of your palm.

Step 3 Rest the face of the hammer on the nail.

Step 4 Draw the hammer back and give the nail a few light taps to start it.

Step 5 Move your fingers away from the nail and hit the nail firmly with the center of the hammer face. Hold the hammer **level** with the head of the nail and strike the face squarely (see *Figure 2*). Deliver the blow through your wrist, your elbow, and your shoulder.

2.1.2 *How to Use a Claw Hammer to Pull a Nail*

Pulling a nail with a claw hammer is as easy as driving one. Follow these steps:

Step 1 Slip the claw of the hammer under the nail head and pull until the handle is nearly straight up (vertical) and the nail is partly drawn out of the wood.

Step 2 Pull the nail straight up from the wood.

2.2.0 The Ball Peen Hammer

A ball peen hammer has a flat face for striking and a rounded face that is used to align brackets. You use this hammer with **chisels** and **punches** (discussed later in this module). In welding operations, the ball peen hammer is used to reduce stress in the **weld** by **peening** or striking the **joint** as it cools. Ball peen hammers are classified by weight. They weigh from 6 ounces to 2½ pounds.

WARNING!

Do not use a hammer with a cast head. A chip could easily break off and injure you or a co-worker.

Never use a hammer to strike the head of another hammer. Flying fragments from drop-forged alloy steel are dangerous.

2.2.1 *How to Use a Ball Peen Hammer*

Using a ball peen hammer is not that much different from using a claw hammer. Follow these steps:

Step 1 Grip the handle. Keep the end of the handle flush with the lower edge of your palm. Keep the face of the hammer parallel to the work.

Step 2 Use the face for hammering. Use the ball peen for rounding off (peening) rivets and similar jobs.

Explain why a ball peen hammer with a cast head should not be used.

Discuss the consequences of striking the head of a hammer with another hammer.

Review the procedures to use a ball peen hammer. Ensure that trainees understand how to properly grip the handle. Discuss the relationship between distance and force when using a hammer.

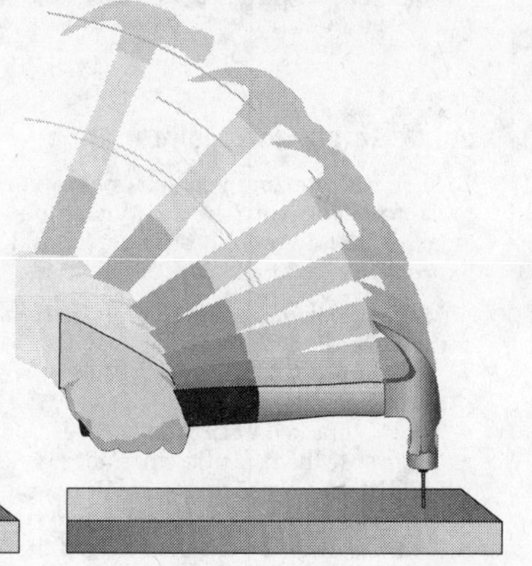

103F02.EPS

Figure 2 ◆ Proper use of a claw hammer.

MODULE 00103-04 ◆ INTRODUCTION TO HAND TOOLS 3.3

Review tips for maintaining and safely using hammers.

Bring in a claw hammer, a ball peen hammer, and a mallet. Ask trainees to demonstrate how to properly use each tool. Have trainees comment on what was done correctly and/or incorrectly.

Go over the Review Questions for Section 2.0.0. Answer any questions trainees may have.

Show Transparency 3 (Figure 3). Compare the common types of screw heads.

Discuss methods to select the appropriate screwdriver for a particular screw. Explain how to ensure the correct fit.

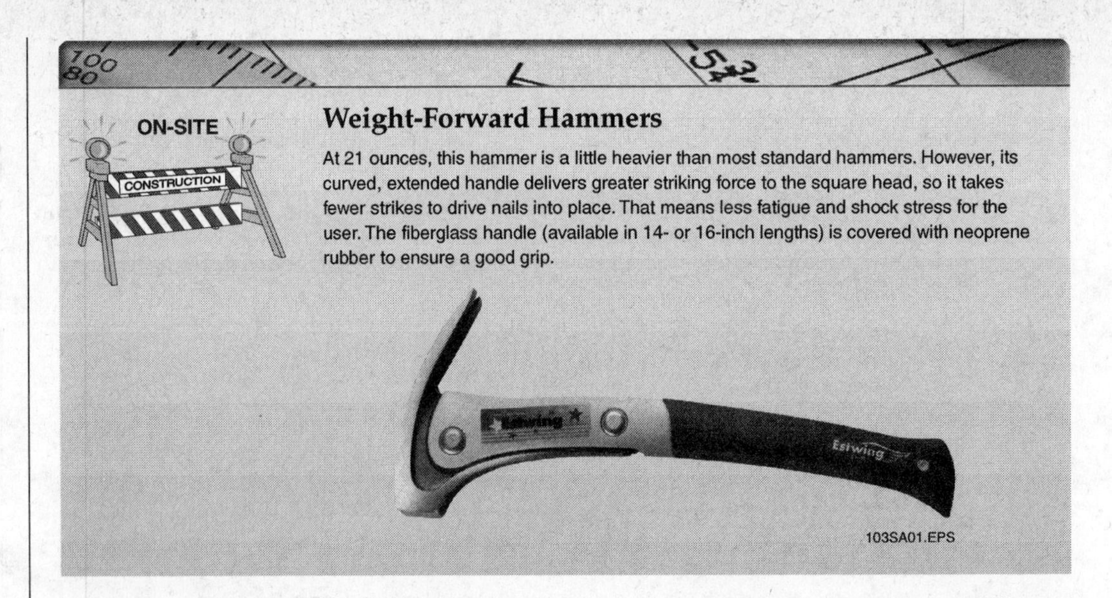

Weight-Forward Hammers

At 21 ounces, this hammer is a little heavier than most standard hammers. However, its curved, extended handle delivers greater striking force to the square head, so it takes fewer strikes to drive nails into place. That means less fatigue and shock stress for the user. The fiberglass handle (available in 14- or 16-inch lengths) is covered with neoprene rubber to ensure a good grip.

103SA01.EPS

Physics and the Hammer

The hammer is designed to produce a certain amount of force on the object it strikes. If you hold the hammer incorrectly, you cancel out the design factor. Always remember to hold the end of the handle even with the lower edge of your palm. The distance between your hand and the head of the hammer affects the force you use to drive a nail. The closer you hold the hammer to the head, the harder you will need to swing to achieve the desired force. Make it easier on yourself by holding the hammer properly; it takes less effort to drive the nail.

2.3.0 Safety and Maintenance

To keep from hurting yourself or a co-worker, you must focus on your work. Make sure you are aware of these guidelines for safety and maintenance when using all types of hammers:

- Make sure there are no splinters in the handle of the hammer.
- Make sure the handle is set securely in the head of the hammer.
- Replace cracked or broken handles.
- Make sure the face of the hammer is clean.
- Hold the hammer properly. Grasp the handle firmly near the end and hit the nail squarely.
- Don't hit with the cheek or side of the hammer head.
- Don't use hammers with chipped, mushroomed (overly flattened by use), or otherwise damaged heads.
- Don't use a hammer with a cast head. A chip could easily break off and injure you or a co-worker.
- Don't use one hammer to strike another.

3.0.0 ◆ SCREWDRIVERS

A screwdriver is used to tighten or remove screws. It is identified by the type of screw it fits. *Figure 3* shows six common types of screw heads.

The most common screwdrivers are slotted (also known as straight-blade, flat, or standard tip) and Phillips head screwdrivers. You will also use more specialized screwdrivers such as a

Instructor's Notes:

3.4 CORE CURRICULUM ◆ INTRODUCTORY CRAFT SKILLS

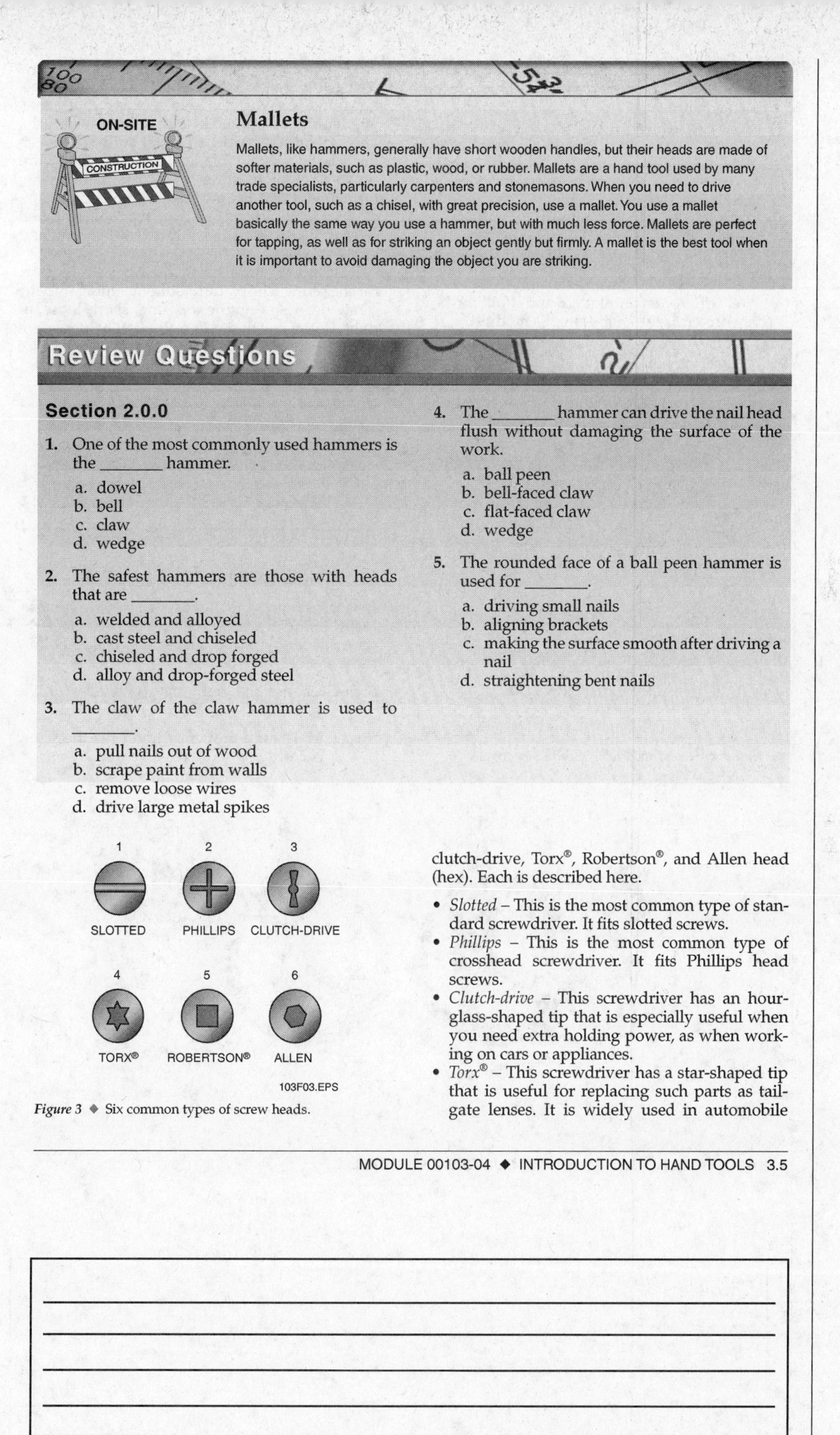

ON-SITE: Mallets

Mallets, like hammers, generally have short wooden handles, but their heads are made of softer materials, such as plastic, wood, or rubber. Mallets are a hand tool used by many trade specialists, particularly carpenters and stonemasons. When you need to drive another tool, such as a chisel, with great precision, use a mallet. You use a mallet basically the same way you use a hammer, but with much less force. Mallets are perfect for tapping, as well as for striking an object gently but firmly. A mallet is the best tool when it is important to avoid damaging the object you are striking.

Review Questions

Section 2.0.0

1. One of the most commonly used hammers is the _______ hammer.

a. dowel
b. bell
c. claw
d. wedge

2. The safest hammers are those with heads that are _______.

a. welded and alloyed
b. cast steel and chiseled
c. chiseled and drop forged
d. alloy and drop-forged steel

3. The claw of the claw hammer is used to _______.

a. pull nails out of wood
b. scrape paint from walls
c. remove loose wires
d. drive large metal spikes

4. The _______ hammer can drive the nail head flush without damaging the surface of the work.

a. ball peen
b. bell-faced claw
c. flat-faced claw
d. wedge

5. The rounded face of a ball peen hammer is used for _______.

a. driving small nails
b. aligning brackets
c. making the surface smooth after driving a nail
d. straightening bent nails

Figure 3 ◆ Six common types of screw heads.

clutch-drive, Torx®, Robertson®, and Allen head (hex). Each is described here.

- *Slotted* – This is the most common type of standard screwdriver. It fits slotted screws.
- *Phillips* – This is the most common type of crosshead screwdriver. It fits Phillips head screws.
- *Clutch-drive* – This screwdriver has an hourglass-shaped tip that is especially useful when you need extra holding power, as when working on cars or appliances.
- *Torx®* – This screwdriver has a star-shaped tip that is useful for replacing such parts as tailgate lenses. It is widely used in automobile

MODULE 00103-04 ◆ INTRODUCTION TO HAND TOOLS 3.5

Have a trainee explain problems that could result from using a dirty screwdriver.

Have trainees practice using a screwdriver without stripping the screw head.

repair work. Torx® screws are also used in household appliances, as well as lawn and garden equipment.

- *Robertson® (square)* – This screwdriver has a square drive that provides high **torque** power. Usually color coded according to size, it can reach screws that are sunk below the surface.
- *Allen (hex)* – This screwdriver works with screws that can also be operated with hex keys. It is suitable for socket-head screws that are recessed.

We will focus on slotted and Phillips head screwdrivers (see *Figure 4*) in this module.

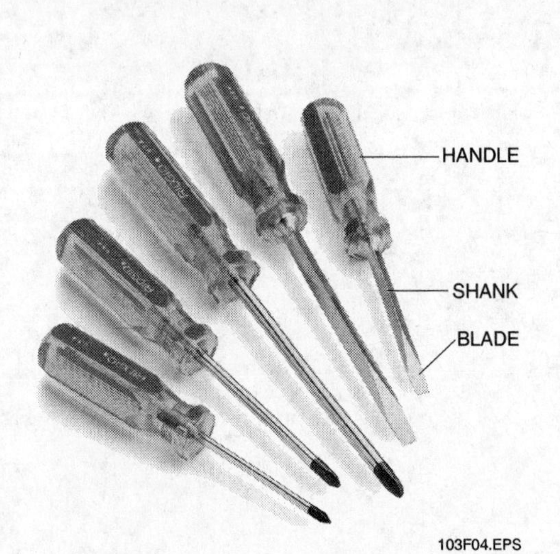

Figure 4 ◆ Slotted and Phillips head screwdrivers.

To choose the right screwdriver and use it correctly, you have to know a little bit about the sections of a screwdriver. Each section has a name. The handle is designed to give you a firm grip. The shank is the hardened metal portion between the handle and the blade. The shank can withstand a lot of twisting force. The blade is the formed end that fits into the head of a screw. Industrial screwdriver blades are made of **tempered** steel to resist wear and to prevent bending and breaking.

It is important to choose the right screwdriver for the screw. The blade should fit snugly into the screw head and not be too long, short, loose, or tight. If you use the wrong size blade, you might damage the screwdriver or the screw head (see *Figure 5*).

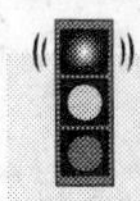

WARNING!

Keep the screwdriver clean. A dirty or greasy screwdriver can slip out of your hand or out of the screw head and possibly cause injury or equipment damage.

3.1.0 How to Use a Screwdriver

It is very important to use a screwdriver correctly. Using one the wrong way can damage the screwdriver or **strip** the screw head. Follow these steps:

Step 1 Choose the right type of blade for the screw head (see *Figure 4*).

Step 2 Make sure the screwdriver fits the screw correctly, as shown in *Figure 5*.

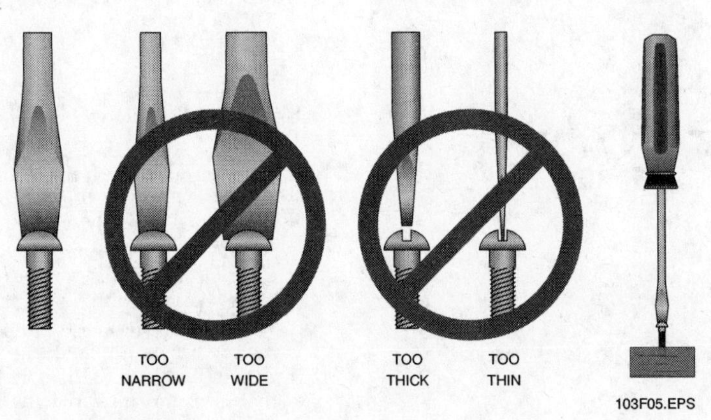

Figure 5 ◆ Proper use of a screwdriver.

Instructor's Notes:

3.6 CORE CURRICULUM ◆ INTRODUCTORY CRAFT SKILLS

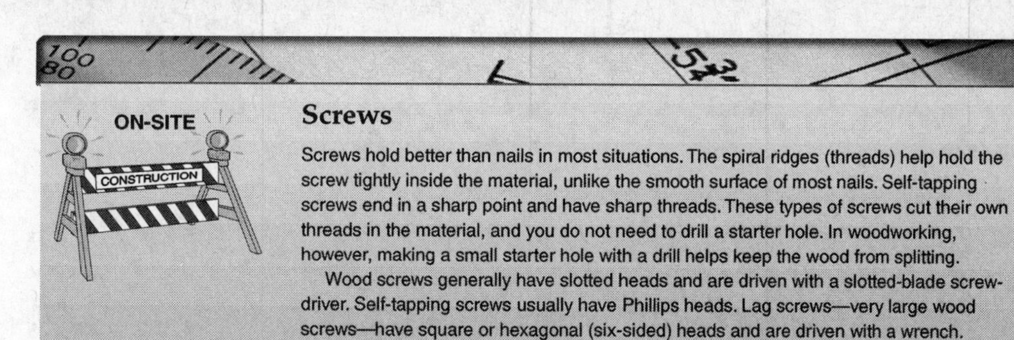

Screws

Screws hold better than nails in most situations. The spiral ridges (threads) help hold the screw tightly inside the material, unlike the smooth surface of most nails. Self-tapping screws end in a sharp point and have sharp threads. These types of screws cut their own threads in the material, and you do not need to drill a starter hole. In woodworking, however, making a small starter hole with a drill helps keep the wood from splitting.

Wood screws generally have slotted heads and are driven with a slotted-blade screwdriver. Self-tapping screws usually have Phillips heads. Lag screws—very large wood screws—have square or hexagonal (six-sided) heads and are driven with a wrench.

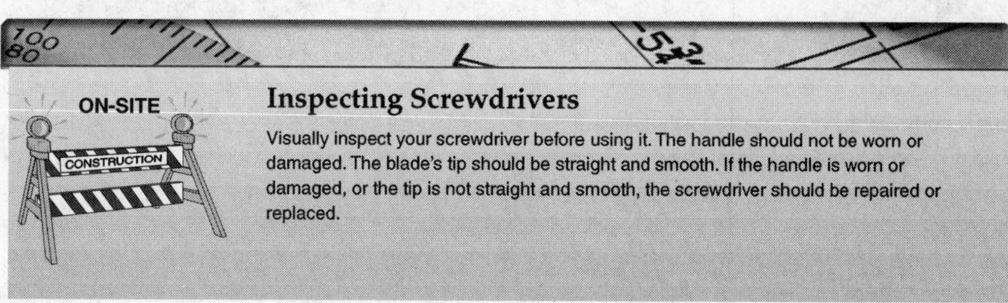

Inspecting Screwdrivers

Visually inspect your screwdriver before using it. The handle should not be worn or damaged. The blade's tip should be straight and smooth. If the handle is worn or damaged, or the tip is not straight and smooth, the screwdriver should be repaired or replaced.

Explain how to prevent the blade from slipping when starting a screw.

Discuss procedures for maintaining and safely using screwdrivers. Emphasize that a screwdriver should not be used as an electrical tester.

Step 3 Position the shank perpendicular (at a right angle) to your work.

Step 4 Apply firm, steady pressure to the screw head and turn: clockwise to tighten (right is tight); counterclockwise to loosen (left is loose).

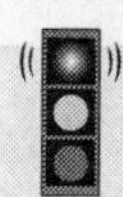

WARNING!

When you're starting the screw, it's easy to hurt your fingers if the blade slips. Work with caution.

3.2.0 Safety and Maintenance

If you follow the steps in the section on how to use a screwdriver, you will be effective with a screwdriver. You also want to be safe, though. There are many guidelines you must follow for your own safety and the safety of others, as well as for maintaining your tool, when using a screwdriver. Learn the following usage guidelines for screwdrivers:

- Keep the screwdriver free of dirt, grease, and grit so the blade will not slip out of the screwhead slot.
- File the blade tip to restore a worn straight edge.
- Don't ever use the screwdriver as a punch, chisel, or pry bar.
- Don't ever use a screwdriver near live wires or as an electrical tester.
- Don't expose a screwdriver to excessive heat.
- Don't use a screwdriver that has a worn or broken handle.
- Don't point the screwdriver blade toward yourself or anyone else.

MODULE 00103-04 ◆ INTRODUCTION TO HAND TOOLS 3.7

Go over the Review Questions for Section 3.0.0. Go over any questions trainees may have.

Show Transparency 4 (Figure 6). Discuss the difference between double-face and crosspeen sledgehammers.

Demonstrate the correct way to secure an object before hitting it with a sledgehammer.

Refer to Figure 7. Explain how to use a long-handled sledgehammer.

Review Questions

Section 3.0.0

1. A screwdriver is identified by _______.
   a. the length of its handle
   b. its torque
   c. the type of screw it fits
   d. the width of its tip

2. The most common standard screwdriver is the _______ screwdriver.
   a. Phillips
   b. slotted
   c. Robertson®
   d. Allen

3. The most common crosshead screwdriver is the _______ screwdriver.
   a. slotted
   b. Torx®
   c. Robertson®
   d. Phillips

4. For safety's sake, industrial screwdriver blades are made of _______.
   a. tempered steel
   b. Torx®
   c. clutch-driven steel
   d. fiberglass

5. If you use the wrong screwdriver head for the job, you might _______.
   a. strike a live wire
   b. turn the screw counterclockwise
   c. twist the shank of the screwdriver
   d. damage the screw head

DID YOU KNOW?

Drywall Workers and Lathers

Drywall workers can perform either installation or finishing work. Installers measure, cut, fit, and fasten drywall panels to the inside framework of buildings. They also install the metal or vinyl-beaded edge around the corners. Finishers prepare the panels for painting by taping and finishing joints and imperfections using drywall mud or spackling. They also sand the material lightly using a sanding pole and sanding block. Finishers are respected for their skills in handling the spackling tools, trowels, mud pans, and taping knives.

Lathers apply metal or gypsum lathe to walls, ceilings, or ornamental frameworks to form the support base for plaster coatings. Lathers nail, screw, staple, or wire-tie the lathe directly to the structural framework. Accuracy and precision must be practiced at all times to ensure that the base is prepared properly. The base must be prepared properly; otherwise the plaster will crack and fall off the surface.

4.0.0 ◆ SLEDGEHAMMERS

A sledgehammer is a heavy-duty tool used to drive posts or other large stakes. You can also use it to break up cast iron or concrete. The head of the sledgehammer is made of high-carbon steel and weighs 2 to 20 pounds. The shape of the head depends on the job the sledgehammer is designed to do. Sledgehammers can be either long-handled or short-handled, depending on the jobs for which they are designed.

Figure 6 shows two types of sledgehammers: the double-face and the crosspeen.

4.1.0 How to Use a Sledgehammer

Obviously, a sledgehammer can cause injury to you or to anyone working near you. You must use a sledgehammer the right way, and you must focus on what you are doing the entire time you use one. Follow these steps (see *Figure 7*):

Step 1 Wear appropriate personal protective equipment.

Step 2 Inspect the sledgehammer to ensure that there are no defects.

Step 3 Be sure that no co-workers are standing in the surrounding area.

Step 4 Hold the sledgehammer with both hands (hand over hand).

Instructor's Notes:

3.8 CORE CURRICULUM ◆ INTRODUCTORY CRAFT SKILLS

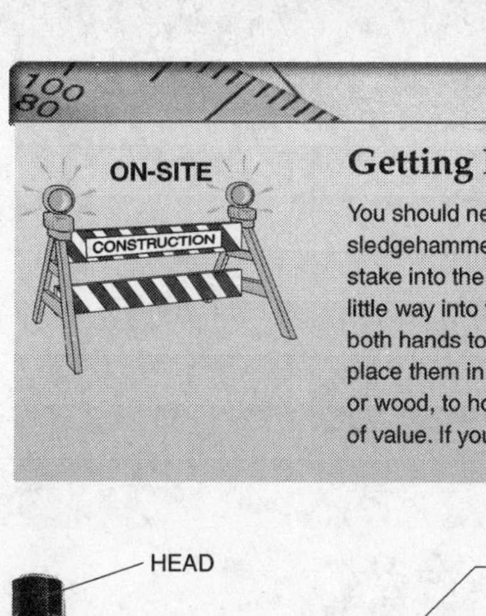

Getting It Started

You should never use your hand, or anyone else's, to hold an object before you hit it with a sledgehammer. So how do you hold the object you want to strike? If you are driving a stake into the ground, for example, use a mallet to get the stake started. Tap the stake a little way into the ground so that it stands on its own. Then you can step back and use both hands to wield the sledgehammer. For objects that need to be broken up, you can place them in between other objects, such as cinder blocks or broken pieces of concrete or wood, to hold them. If you do this, make sure that you never use anything that might be of value. If you are ever in doubt, ask your instructor or immediate supervisor.

Stress the importance of using two hands when holding a sledgehammer.

Discuss the consequences of swinging a sledgehammer behind your body or head.

Ask trainees to review considerations for maintaining and safely using sledgehammers. Explain how to use the right amount of force for the job.

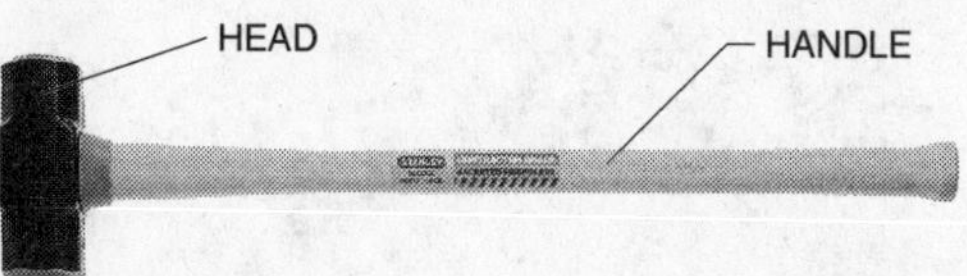

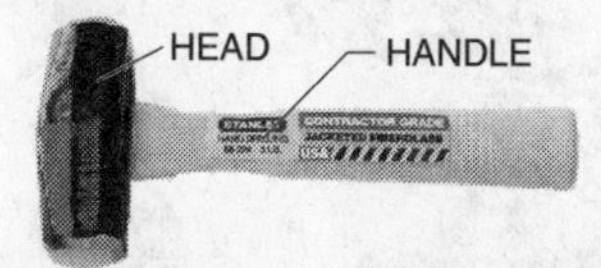

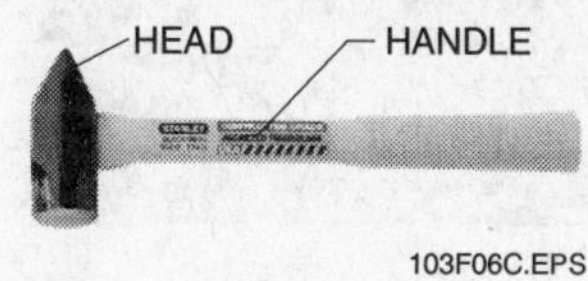

Figure 6 ◆ Types of sledgehammers.

Step 5 Stand directly in front of the object you want to drive.

Step 6 Lift the sledgehammer straight up above the target.

Step 7 Set the head of the sledgehammer on the target.

Step 8 Begin delivering short blows to the target and gradually increase the length and force of the stroke.

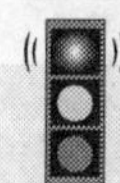

WARNING!

Hold the sledgehammer with both hands. Never use your hands to hold the object as either you or someone else drives with a sledgehammer. Doing so could result in serious injury, such as crushed or broken bones.

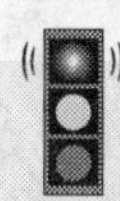

WARNING!

Avoid swinging the sledgehammer behind you or beyond your head. Doing so may cause injury to your back and could limit the control you have directing the blow to the target.

4.2.0 Safety and Maintenance

Remember that using a sledgehammer the correct way not only gets the job done right, but also keeps everyone in the area, including you, safe. Of course, as with all tools, there are other considerations for safety and maintenance when working with this tool. Here are the guidelines for working with a sledgehammer:

- Wear eye protection when you are using a sledgehammer. It's also a good idea to wear safety gloves.
- Replace cracked or broken handles before you use the sledgehammer.
- Make sure the handle is secured firmly at the head.
- Use the right amount of force for the job.
- Keep your hands away from the object you are driving.
- Don't swing until you have checked behind you to make sure you have enough room and no one is behind you.

Figure 7 ◆ Proper use of a long-handled sledgehammer.

Instructor's Notes:

Review Questions

Section 4.0.0

1. You use a sledgehammer to break up concrete and _______.
   a. remove overhead wires
   b. drive a post or stake
   c. hammer nails
   d. drive bolts

2. Two types of sledgehammers are the double-face and _______.
   a. ball peen
   b. double-edged
   c. crosspeen
   d. round-head

3. The head of a sledgehammer is made of _______.
   a. high-carbon steel
   b. heavy rubber
   c. fiberglass
   d. welded iron

4. The shape of a sledgehammer head depends on _______.
   a. the torque of the sledgehammer
   b. whether the sledgehammer is single-face or double-face
   c. the composition of the head
   d. the job the sledgehammer is intended to do

5. When using a sledgehammer, it is important to _______.
   a. swing from between your knees
   b. wear appropriate personal protective equipment
   c. hold tightly to the object you are driving
   d. place a protective covering over the sledgehammer head

5.0.0 ◆ RIPPING BARS AND NAIL PULLERS

A number of tools are made to rip and pry apart woodwork as well as to pull nails. In this section, you will learn about **ripping bars** and **nail pullers** (see *Figure 8*). These tools are necessary in the construction trade because often the job involves building where something else already exists. The existing structure needs to be torn apart before the building can begin. That is where the ripping bar and the nail puller come in.

5.1.0 Ripping Bars

The ripping bar—also called a pinch, pry, or wrecking bar—can be 12 to 36 inches long. This bar is used for heavy-duty dismantling of woodwork, such as tearing apart building frames or concrete forms. The ripping bar has an octagonal (eight-sided) shaft and two specialized ends. A deeply curved nail claw at one end is used as a nail puller. An angled, wedge-shaped face at the other end is used as a prying tool to pull apart materials that are nailed together.

5.1.1 How to Use a Ripping Bar

Take the following steps when using a ripping bar:

Step 1 Wear appropriate personal protective equipment.

Step 2 Use the angled prying end to force apart pieces of wood or use the heavy claw to pull large nails and spikes.

5.2.0 Nail Pullers

There are three main types of nail-pulling tools: **cat's paw** (also called nail claws and carpenter's pincers), **chisel bars,** and **flat bars.**

The cat's paw is a straight steel rod with a curved claw at one end. It is used to pull nails that have been driven flush with the surface of the wood or slightly below it. You use the cat's paw to pull nails to just above the surface of the wood so they can be pulled completely out with the claw of a hammer or a pry bar.

The chisel bar has a claw at each end and is ground to a chisel-like **bevel** (slant) on both ends. You can use it like a claw hammer to pull nails.

Go over the Review Questions for Section 4.0.0. Ask whether trainees have any questions.

Show Transparency 5 (Figure 8). Discuss the tools that are used to rip and pry apart woodwork.

Review the procedures for safely using a ripping bar.

Set up stations with cat's paws, chisel bars, and flat bars. Have trainees practice pulling nails using each of the nail pullers.

Discuss the importance of protecting yourself from debris when using a ripping bar or nail puller.

Explain how most accidents occur when using nail pullers. Review safety procedures.

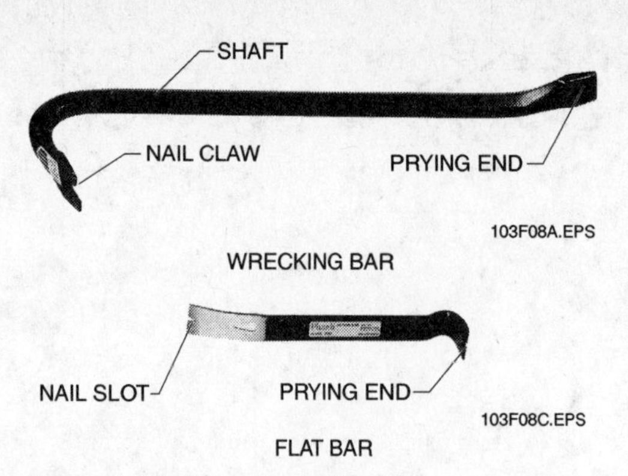

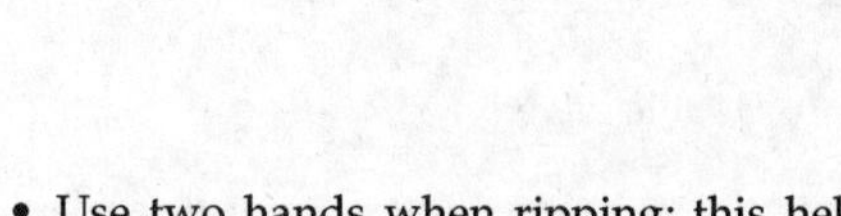

Figure 8 ◆ Ripping bars and nail pullers.

You can also drive it into wood to split and rip apart the pieces.

The flat bar (ripping chisel, wonder bar, action bar) has a nail slot at the end to pull nails out from tightly enclosed areas. It can also be used as a small pry bar. The flat bar is usually 2 inches wide and 15 inches long.

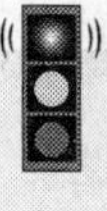

WARNING!

A piece of material can break off and fly through the air when you are using a ripping bar or a nail puller. Wear a hard hat, safety glasses, and gloves to protect yourself from flying debris. Make sure others around you are similarly protected.

5.2.1 How to Use a Nail Puller (Cat's Paw)

Take the following steps when using a nail puller:

Step 1 Wear appropriate personal protective equipment.

Step 2 Drive the claw into the wood, grabbing the nail head.

Step 3 Pull the handle of the bar to lift the nail out of the wood.

5.3.0 Safety and Maintenance

Here are the guidelines for ripping and nail pulling:

- Wear appropriate personal protective equipment.
- Use two hands when ripping; this helps ensure you keep even pressure on your back as you pull.
- When nail pulling, be sure the material holding the nail is braced securely before you pull the nail, to keep it from hitting you in the face.

Most accidents with prying tools occur when a pry bar slips and the craftworker falls to the ground. Be sure to keep a balanced footing and a firm grip on the tool. This technique also helps reduce damage to materials that must be reused, such as concrete forms.

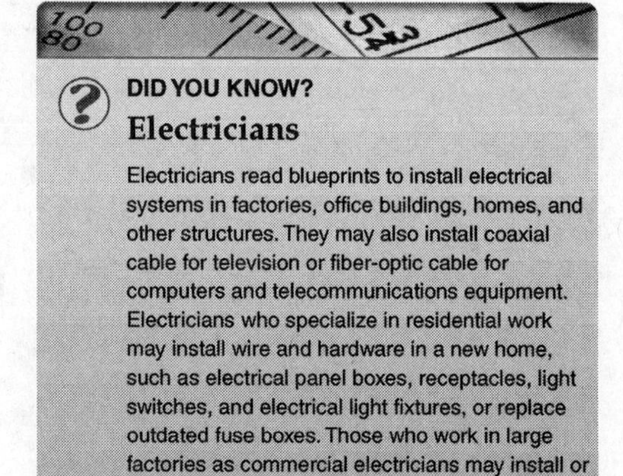

DID YOU KNOW?

Electricians

Electricians read blueprints to install electrical systems in factories, office buildings, homes, and other structures. They may also install coaxial cable for television or fiber-optic cable for computers and telecommunications equipment. Electricians who specialize in residential work may install wire and hardware in a new home, such as electrical panel boxes, receptacles, light switches, and electrical light fixtures, or replace outdated fuse boxes. Those who work in large factories as commercial electricians may install or repair motors, transformers, generators, or electronic controllers on machine tools and industrial robots. They use many hand tools, including pliers, wrenches, screwdrivers, hammers, and saws.

Instructor's Notes:

3.12 CORE CURRICULUM ◆ INTRODUCTORY CRAFT SKILLS

Review Questions

Section 5.0.0

1. The ripping bar is used for _______.
   a. gripping large metal objects for demolition
   b. heavy-duty dismantling of woodwork
   c. hammering nails
   d. breaking up concrete

2. The angled prying end of the ripping bar is used to _______.
   a. rip out nails
   b. drive the claw into the wood
   c. bevel wood
   d. force apart pieces of wood

3. A cat's paw is a kind of _______.
   a. sledgehammer
   b. ripping tool
   c. nail-pulling tool
   d. crushing tool

4. A chisel bar can be used to _______.
   a. pry apart steel beams
   b. split and rip apart pieces of wood
   c. break apart concrete
   d. make ridges in wood beams

5. When using prying tools, be sure to _______.
   a. keep a balanced footing
   b. hold the tool loosely
   c. swing firmly from above
   d. keep the material loosely braced

Go over the Review Questions for Section 5.0.0. Answer any questions trainees may have.

Show Transparency 6 (Figure 9). Review the common types of pliers. Explain why pliers cannot be used on nuts or bolt heads.

6.0.0 ◆ PLIERS AND WIRE CUTTERS

Pliers are a special type of adjustable wrench. They are scissor-shaped tools with jaws. The jaws usually have teeth to help grip objects. The jaws are adjustable because the two legs (or handles) move on a pivot. You will generally use pliers to hold, cut, and bend wire and soft metals. Do not use pliers on nuts or bolt heads. They will **round off** the edges of the hex (six-sided) head, and wrenches will no longer fit properly.

High-quality pliers are made of hardened steel. Pliers come in many different head styles, depending on their use. The following types of pliers are the most commonly used (see *Figure 9*):

- Slip-joint (combination) pliers
- Long-nose (needle-nose) pliers
- Lineman pliers (side cutters)
- Tongue-and-groove (or water pump) pliers
- Vise-Grip® (locking) pliers

You will learn about each type of plier in this module.

6.1.0 Slip-Joint (Combination) Pliers

You use slip-joint (or combination) pliers to hold and bend wire and to grip and hold objects during assembly operations. They have adjustable jaws. There are two jaw settings: one for small materials and one for larger materials.

6.1.1 How to Use Slip-Joint Pliers

When using slip-joint pliers, be sure to wear appropriate personal protective equipment. Take the following steps to use slip-joint pliers properly:

Step 1 Place the jaws on the object to be held.

Step 2 Squeeze the handles until the pliers grip the object.

Review the procedures to use slip-joint, long-nose, lineman, tongue-and-groove, and Vise-Grip® pliers. Discuss the type of tasks for which each of the pliers is used.

6.2.0 Long-Nose (Needle-Nose) Pliers

Long-nose (or needle-nose) pliers are used to get into tight places where other pliers won't reach or to grip parts that are too small to hold with your fingers. These pliers are useful for bending angles in wire or narrow metal strips. They have a sharp wire cutter near the pivot. Long-nose pliers, like many other types of pliers, are available with spring openers (see *Figure 9*), which are spring-like devices between the handles that keep the handles apart—and therefore the jaws open—unless you purposely close them. This device can make long-nose pliers easier to use.

Demonstrate how to place the jaws of the different pliers on the objects to be held.

Have trainees practice cutting, holding, and bending wire using the different pliers.

6.2.1 How to Use Long-Nose Pliers

Be sure to wear appropriate personal protective equipment when working with long-nose pliers.

MODULE 00103-04 ◆ INTRODUCTION TO HAND TOOLS 3.13

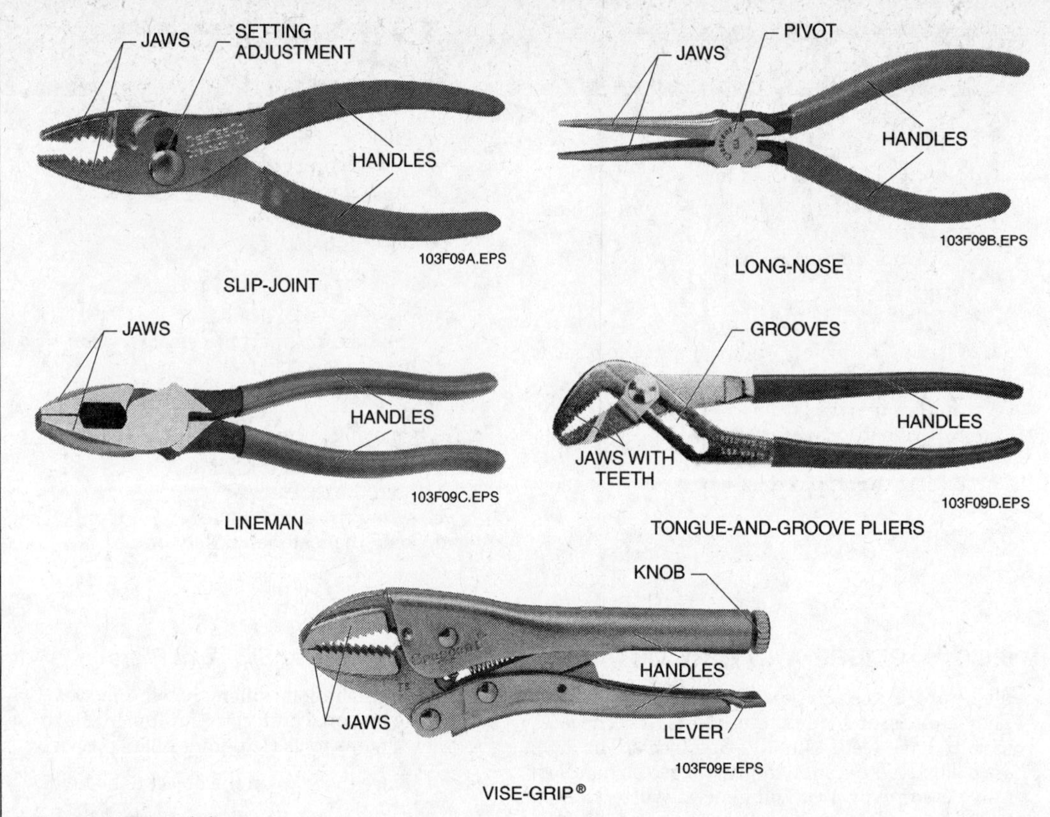

Figure 9 ◆ Types of pliers.

Note the following points in using long-nose pliers properly:

- If the pliers do not have a spring between the handles to keep them open, place your third or little finger inside the handles to keep them open.
- Use the sharp cutter near the pivot for cutting wire.

6.3.0 Lineman Pliers (Side Cutters)

Lineman pliers (or side cutters) have wider jaws than slip-joint pliers do. You use them to cut heavy or large-gauge wire and to hold work. The wedged jaws reduce the chance that wires will slip, and the hook bend in both handles gives you a better grip.

6.3.1 How to Use Lineman Pliers

Be sure to wear appropriate personal protective equipment when working with lineman pliers. Take the following steps to use lineman pliers properly:

Step 1 When you cut wire, always point the loose end of the wire down.

Step 2 Cut at a right angle to the wire.

6.4.0 Tongue-and-Groove Pliers

Tongue-and-groove pliers, manufactured by CHANNELLOCK®, Inc., have serrated teeth that grip flat, square, round, or hexagonal objects. You can set the jaws in up to five positions by slipping the curved ridge into the desired groove (see *Figure 10*). Large tongue-and-groove pliers are often used to hold pipes because the longer handles give more leverage. The jaws stay parallel and give a better grip than slip-joint pliers.

6.4.1 How to Use Tongue-and-Groove Pliers

Be sure to wear appropriate personal protective equipment when using tongue-and-groove pliers. Take the following steps to use tongue-and-groove pliers properly:

Instructor's Notes:

3.14 CORE CURRICULUM ◆ INTRODUCTORY CRAFT SKILLS

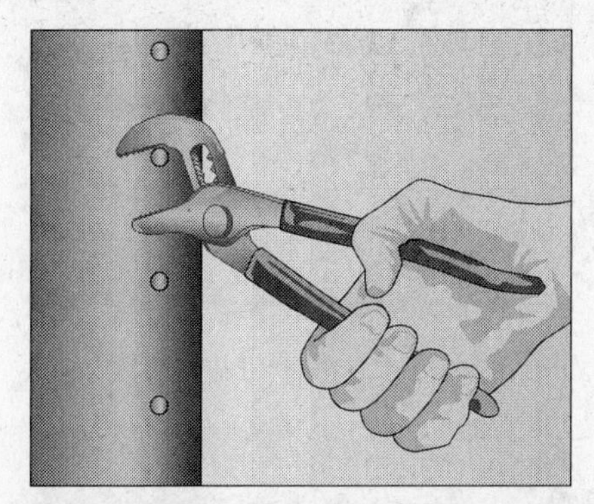

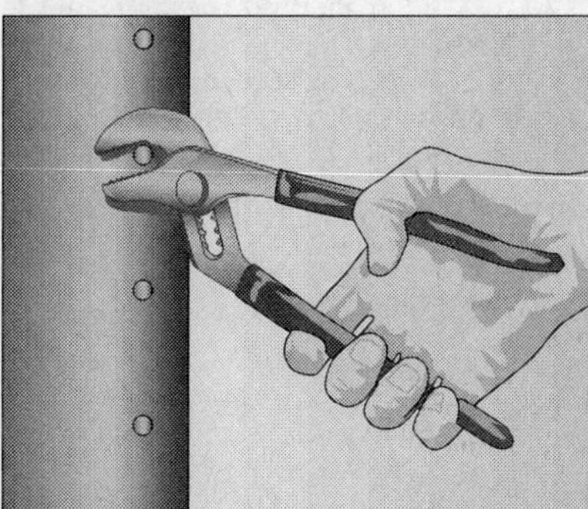

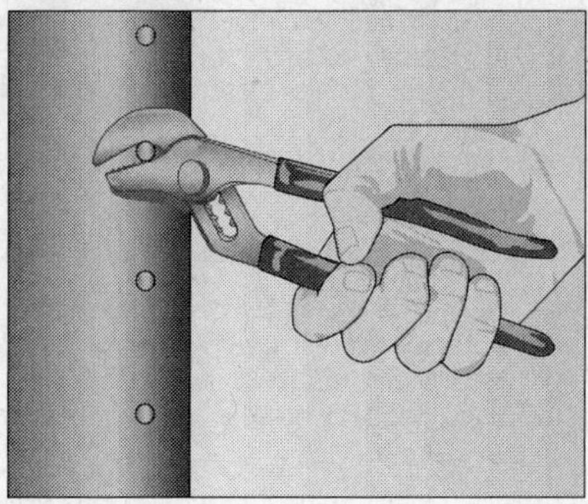

103F10.EPS

Figure 10 ◆ Proper use of tongue-and-groove pliers.

Step 1 With pliers open to the widest position, place the jaws on the object to be held.

Step 2 Determine which groove provides the proper position.

Step 3 Squeeze the handles until the pliers grip the object (see *Figure 10*).

6.5.0 Vise-Grip® (Locking) Pliers

Vise-Grip® (locking) pliers clamp firmly onto objects the way a **vise** does. (You will learn about vises later in this module.) A knob in the handle controls the width and tension of the jaws. You close the handles to lock the pliers. You release the pliers by pressing the lever to open the jaws.

6.5.1 How to Use Vise-Grip® Pliers

As with all the types of pliers, be sure to use appropriate personal protective equipment when working with Vise-Grip® pliers. Take the following steps to use Vise-Grip® pliers properly:

Step 1 Place the jaws on the object to be held.

Step 2 Turn the adjusting screw in the handle until the pliers grip the object.

Step 3 Squeeze the handles together to lock the pliers.

Step 4 Squeeze the release lever when you want to remove the pliers (see *Figure 11*).

6.6.0 Safety and Maintenance

You might not think that misusing pliers could cause injury, but it can. Proper safety precautions when using pliers and proper maintenance of your pliers are very important. Here are some guidelines to remember when using pliers:

- Hold pliers close to the end of the handles to avoid pinching your fingers in the hinge.
- Don't extend the length of the handles for greater leverage. Use a larger pair of pliers instead.
- Wear appropriate personal protective equipment, especially when you cut wire.

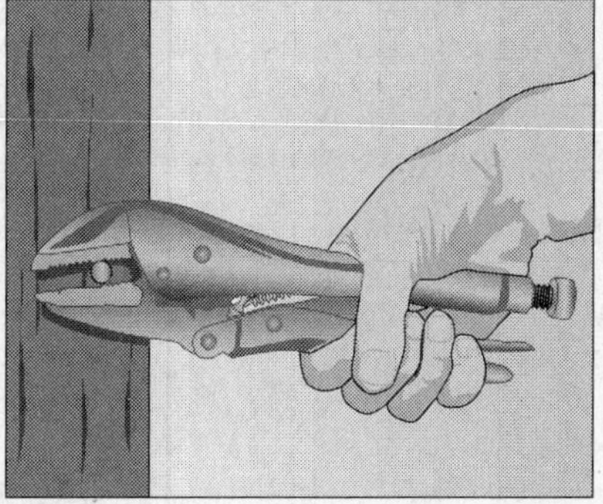

103F11.EPS

Figure 11 ◆ Proper use of Vise-Grip® pliers.

Discuss safety procedures for avoiding injuries when using pliers.

MODULE 00103-04 ◆ INTRODUCTION TO HAND TOOLS 3.15

Explain why trainees should never rock pliers from side to side when cutting.

Go over the Review Questions for Section 6.0.0. Go over any questions trainees may have.

See the Teaching Tip for Sections 1.0.0–6.6.0 at the end of this module.

Have trainees review Sections 7.0.0–12.2.0 for the next session.

- Hold the short ends of wires to avoid flying metal bits when you cut.
- Always cut at right angles. Don't rock the pliers from side to side or bend the wire back and forth against the cutting blades. Loose wire can fly up and injure you or someone else.
- Oil pliers regularly to prevent rust and to keep them working smoothly.
- Don't use pliers around energized electrical wires. Although the handles may be plastic-coated, they are not insulated against electrical shock.
- Don't expose pliers to extreme heat.
- Don't use pliers to turn nuts or bolts; they are not wrenches.
- Don't use pliers as hammers.

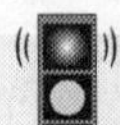

WARNING!
Never rock pliers from side to side when you are cutting. The object you are cutting could fly in your face.

DID YOU KNOW?
Stonemasons

Stone used to be one of our primary building materials. Because of its strength, stone was often used for dams, bridges, fortresses, foundations, and important buildings. Today, steel and concrete have replaced stone as a basic construction material. Stone is used primarily as sheathing for buildings, for flooring in high-traffic areas, and for decorative uses.

A stonemason's job requires precision. Stones have uneven, rough edges that must be trimmed and finished before each stone can be set. The process of trimming projections and jagged edges is called dressing the stone. This requires skill and experience using specialized hand tools. Many craftworkers consider stonework an art.

Stonemasons build stone walls as well as set stone exteriors and floors, working with natural cut and artificial stones. These include marble, granite, limestone, cast concrete, marble chips, or other masonry materials. Stonemasons usually work on structures such as houses, churches, hotels, and office buildings. Special projects include zoos, theme parks, and movie sets.

Review Questions

Section 6.0.0

1. Pliers are generally used to _______.
   a. pry open objects that are stuck together
   b. turn nuts or bolt heads
   c. hold, cut, and bend wire and soft metals
   d. substitute for wrenches in tight spaces

2. Pliers should not be used on a nut or bolt because _______.
   a. they will round off the edges of the hex head
   b. they are not strong enough
   c. they are designed only for tightening
   d. their jaws will not open wide enough

3. The best-quality pliers are made of _______.
   a. fiberglass
   b. hardened steel
   c. high-carbon steel
   d. alloys

4. Long-nose pliers are used to _______.
   a. clamp objects loosely
   b. hold pipes because the long handles give more leverage
   c. get into tight places where other pliers can't fit
   d. cut heavy or large-gauge wire

5. When using pliers, you should always _______.
   a. extend the ends for better leverage
   b. detach the spring between the handles
   c. rock the pliers from side to side
   d. wear appropriate personal protective equipment

Instructor's Notes:

____

____

____

3.16 CORE CURRICULUM ◆ INTRODUCTORY CRAFT SKILLS

7.0.0 ◆ RULERS AND OTHER MEASURING TOOLS

Craftworkers use four basic types of measuring tools:

- Flat steel rule
- Measuring tape
- Wooden folding rule
- Digital measuring device

When you choose a measuring tool, keep the following in mind:

- It must be accurate.
- It should be easy to use.
- It should be durable.
- The numbers should be easy to read (black on yellow or off-white are good).

Ensure that you have everything required for laboratories and demonstrations during this session.

Review basic measuring tools and how to select the proper tool for the job.

7.1.0 Steel Rule

Show Transparency 7 (Figure 12). Explain why thin rulers measure more accurately. Identify the four sets of marks on a steel rule.

The flat steel rule (see *Figure 12*) is the simplest and most common measuring tool. The flat steel rule is usually 6 or 12 inches long, but longer sizes are available.

Steel rules can be flexible or nonflexible, thin or wide. The thinner the rule, the more accurately it measures, because the division marks are closer to the work.

Generally, a steel rule has four sets of marks, two on each side of the rule. On one side are the inch marks. The longest lines are for 1-inch increments. On one edge of that side, each inch is divided into eight equal spaces of 1/8 inch each. On the other edge of that side, each inch is divided into 1/16-inch spaces. To make counting easier, the 1/4-inch and the 1/2-inch marks are normally longer than the smaller division marks. The other side of the steel rule is divided into 32 and 64 spaces to the inch. Each fourth division in the inch is usually numbered for easier reading.

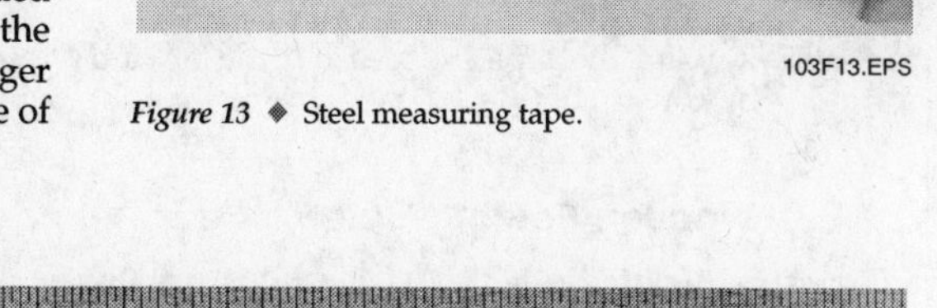

103F12.EPS

Figure 12 ◆ Steel rule.

7.2.0 Measuring Tape

Measuring tapes are available in different lengths. The shorter tapes are usually made with a curved cross section, so they are flexible enough to roll up but stay rigid when they are extended. When using a long, flat tape, lay it along a surface to keep it from sagging in the middle.

Steel measuring tapes (see *Figure 13*) are usually wound into metal cases. A hook at the end of the tape hooks over the object you are measuring. The tape may have a lock that holds the blade open and a rewind spring that returns the blade to the case. Look for a tape that is easy to read. Good-quality tapes have a polyester film bonded to the steel blade to guard against wear.

Refer to Figure 13. Review the procedures for using a steel measuring tape.

103F13.EPS

Figure 13 ◆ Steel measuring tape.

MODULE 00103-04 ◆ INTRODUCTION TO HAND TOOLS 3.17

Show Transparency 8 (Figure 14). Discuss the advantages of using a wooden folding rule.

Bring in a steel rule, measuring tape, and wooden folding rule. Set up one station for each. Include an item of equal measure at each of the three stations. Ask trainees to measure the items at each station. See which measuring tool provides the most accurate measurement.

Review the guidelines for using rulers and measuring tools. Discuss the importance of taking accurate measurements.

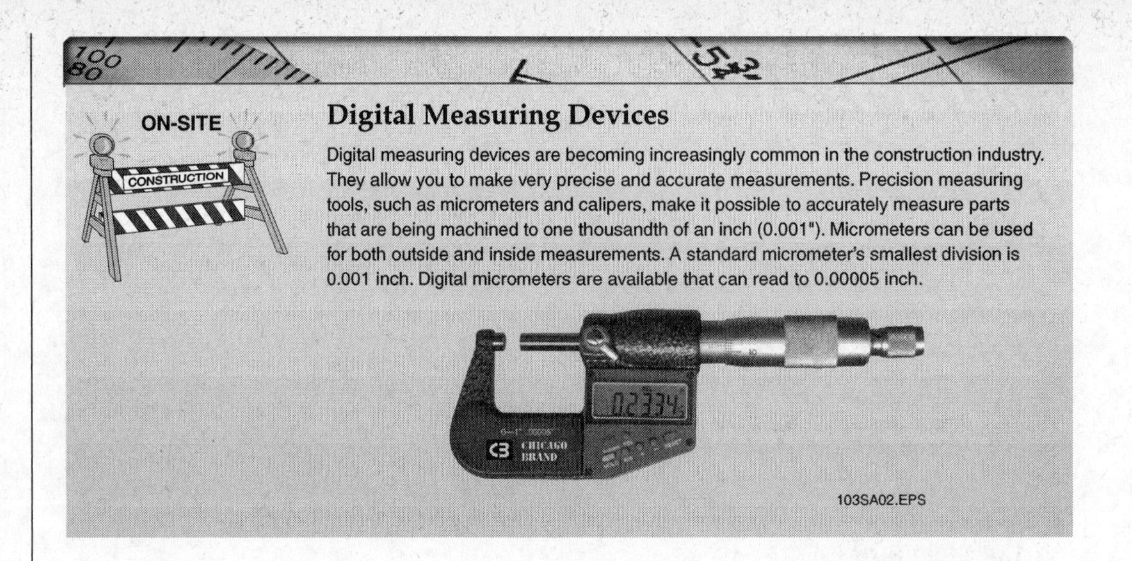

Digital Measuring Devices

Digital measuring devices are becoming increasingly common in the construction industry. They allow you to make very precise and accurate measurements. Precision measuring tools, such as micrometers and calipers, make it possible to accurately measure parts that are being machined to one thousandth of an inch (0.001"). Micrometers can be used for both outside and inside measurements. A standard micrometer's smallest division is 0.001 inch. Digital micrometers are available that can read to 0.00005 inch.

103SA02.EPS

7.2.1 How to Use a Steel Tape

Measuring with a steel tape is simple. Follow these steps to use a steel tape properly:

Step 1 Pull the tape out to the desired length.

Step 2 Place the hook over the edge of the material you are measuring. Lock the tape if necessary (use the lock button on the holder).

Step 3 Mark or record the measurement.

Step 4 Unhook the tape from the edge.

Step 5 Rewind the tape by pressing the rewind button.

7.3.0 Wooden Folding Rule

A wooden folding rule (see *Figure 14*) is usually marked in sixteenths of an inch on both edges of each side. Folding rules come in 6- and 8-foot lengths. Because of its stiffness, a folding rule is better than a cloth or steel tape for measuring vertical distance. This is because, unlike tape, it holds itself straight up. This makes it easier to measure some distances, such as those where you might need a ladder to reach one end.

103F14.EPS

Figure 14 ◆ Wooden folding rule.

7.4.0 Safety and Maintenance

There are some safety concerns, as well as some needed maintenance, for rulers and measuring tools. Here are the guidelines to remember:

- Occasionally apply a few drops of light oil on the spring joints of a wooden folding rule and steel tape.
- Wipe moisture off steel tape to keep it from rusting.
- Don't kink or twist steel tape, because this could cause it to break.
- Don't use steel tape near exposed electrical parts.
- Don't let digital measuring devices get wet.

Instructor's Notes:

3.18 CORE CURRICULUM ◆ INTRODUCTORY CRAFT SKILLS

Review Questions

Section 7.0.0

1. A measuring tool must be accurate and _______.
   a. easy to read
   b. made of strong metal
   c. flat
   d. equipped with a carrying case

2. The simplest and most common measuring tool is the _______.
   a. measuring tape
   b. wooden folding rule
   c. digital measuring device
   d. flat steel rule

3. One side of a steel rule is divided into 32 and 64 spaces to the inch, and the other side shows _______.
   a. vertical distance
   b. digital measurements
   c. inch marks
   d. a curved cross section

4. Steel measuring tapes are usually wound into metal cases and have _______.
   a. inch marks on the cases
   b. a hook at the end of the tape
   c. conversion tables on the cases
   d. markings on both sides of the tape

5. Because of its stiffness, a folding rule is better than a cloth or steel tape for measuring _______.
   a. vertical distance
   b. horizontal distance
   c. plumb
   d. level

8.0.0 ◆ LEVELS

A level is a tool used to determine both how level a horizontal surface is and how **plumb** a vertical surface is. If a surface is described as level, that means it is exactly horizontal. If a surface is described as plumb, that means it is exactly vertical. Levels are used to determine how near to exactly horizontal or exactly vertical a surface is.

Types of levels range from simple spirit levels to electronic and laser instruments. The spirit level (see *Figure 15*) is the most commonly used level in the construction trade.

8.1.0 Spirit Levels

Most levels are made of tough, lightweight metals such as magnesium or aluminum. The spirit level has three vials filled with alcohol that measure either plumb (vertical) or level (horizontal).

The amount of liquid in each vial is not enough to fill it, so there is always a bubble in the vial. When the bubble is centered between the lines on the vial, the surface is either level or plumb, depending on the vial (see *Figure 16*).

Spirit levels come in a variety of sizes. The longer the level, the greater its accuracy.

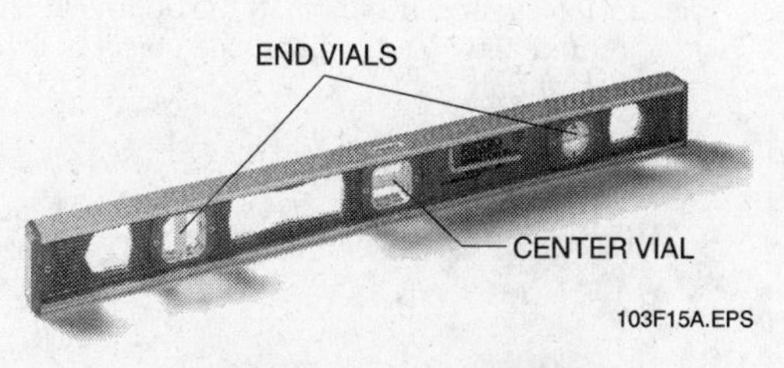

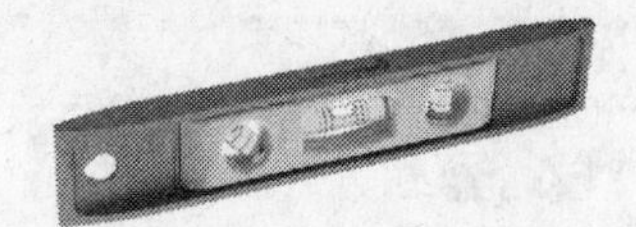

Figure 15 ◆ Spirit levels.

DID YOU KNOW?

The Spirit Level

The spirit level got its name because the vials in it are filled with alcohol. Alcohol used to be called spirits. Thus, the spirit level.

Go over the Review Questions for Section 7.0.0. Ask whether trainees have any questions.

Show Transparency 9 (Figure 15). Explain how levels are used to determine whether a surface is level and/or plumb.

MODULE 00103-04 ◆ INTRODUCTION TO HAND TOOLS 3.19

Demonstrate how to use and read a spirit level to determine whether a surface is either level and/or plumb.

Review procedures for preventing damage to precision instruments.

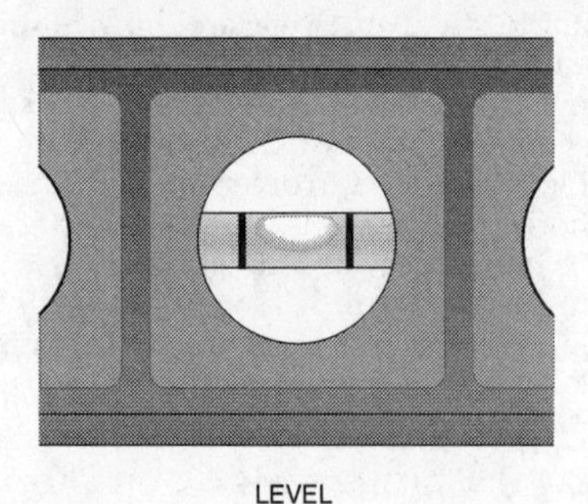

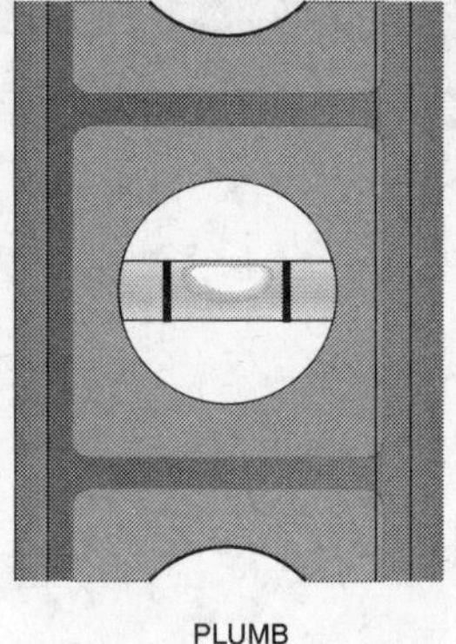

Figure 16 ◆ An air bubble centered between the lines shows level or plumb.

8.1.1 How to Use a Spirit Level

Using a spirit level is very simple. It just requires a careful eye to be sure you are reading it correctly. Follow these steps to use a spirit level properly:

Step 1 Put the spirit level on the object you are checking.

Step 2 Look at the air bubble. If the bubble is centered between the lines, the object is level or plumb.

8.2.0 Safety and Maintenance

Levels are precision instruments that must be handled with care. Although there is little chance of personal injury when working with levels, there is a chance of damaging or breaking the level. Here are your guidelines to remember when working with levels:

- Replace a level if a crack or break appears in any of the vials.
- Keep levels clean and dry.
- Don't bend or apply too much pressure on your level.
- Don't drop or bump your level.

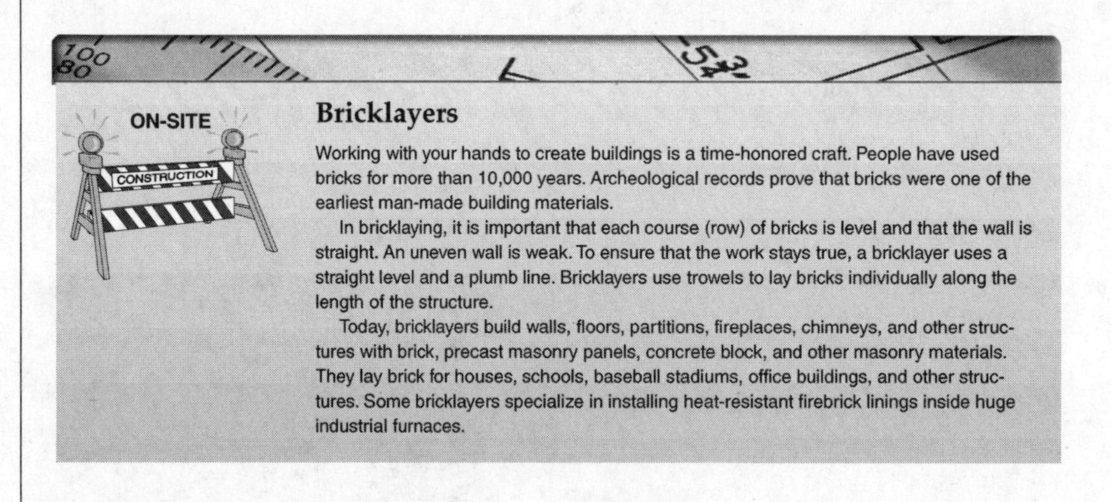

Bricklayers

Working with your hands to create buildings is a time-honored craft. People have used bricks for more than 10,000 years. Archeological records prove that bricks were one of the earliest man-made building materials.

In bricklaying, it is important that each course (row) of bricks is level and that the wall is straight. An uneven wall is weak. To ensure that the work stays true, a bricklayer uses a straight level and a plumb line. Bricklayers use trowels to lay bricks individually along the length of the structure.

Today, bricklayers build walls, floors, partitions, fireplaces, chimneys, and other structures with brick, precast masonry panels, concrete block, and other masonry materials. They lay brick for houses, schools, baseball stadiums, office buildings, and other structures. Some bricklayers specialize in installing heat-resistant firebrick linings inside huge industrial furnaces.

Instructor's Notes:

3.20 CORE CURRICULUM ◆ INTRODUCTORY CRAFT SKILLS

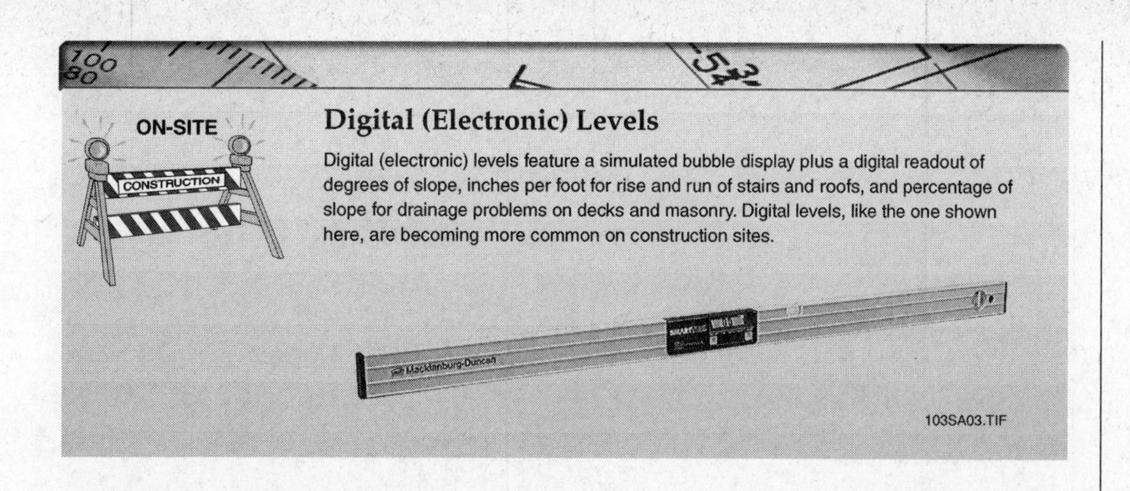

Digital (Electronic) Levels

Digital (electronic) levels feature a simulated bubble display plus a digital readout of degrees of slope, inches per foot for rise and run of stairs and roofs, and percentage of slope for drainage problems on decks and masonry. Digital levels, like the one shown here, are becoming more common on construction sites.

103SA03.TIF

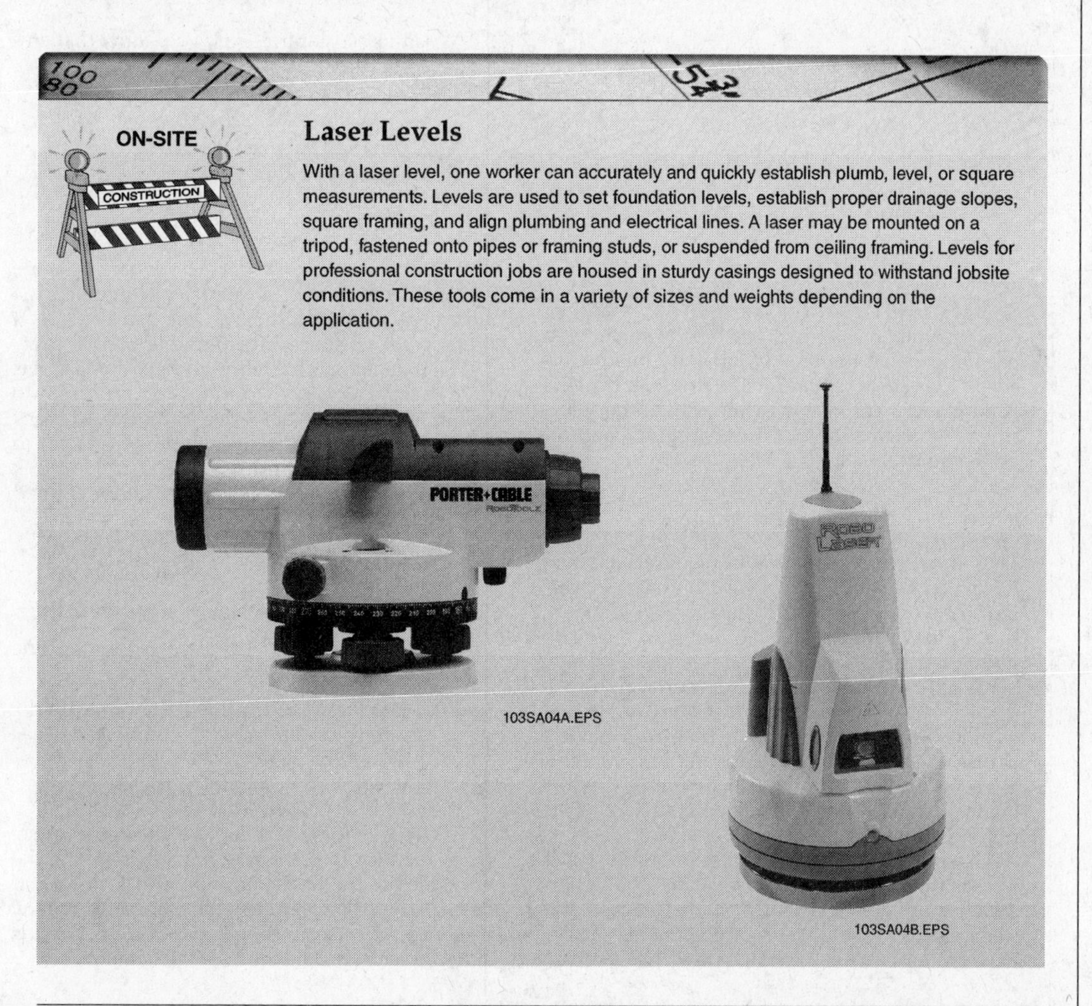

Laser Levels

With a laser level, one worker can accurately and quickly establish plumb, level, or square measurements. Levels are used to set foundation levels, establish proper drainage slopes, square framing, and align plumbing and electrical lines. A laser may be mounted on a tripod, fastened onto pipes or framing studs, or suspended from ceiling framing. Levels for professional construction jobs are housed in sturdy casings designed to withstand jobsite conditions. These tools come in a variety of sizes and weights depending on the application.

103SA04A.EPS

103SA04B.EPS

MODULE 00103-04 ◆ INTRODUCTION TO HAND TOOLS 3.21

Go over the Review Questions for Section 8.0.0. Answer any questions trainees may have.

Show Transparency 10 (Figure 17). Discuss the types of squares used for marking, checking, and measuring.

Refer to Figure 18. Explain how squares are used to check the squareness of adjoining surfaces.

Demonstrate how to use a try square to lay out cutting lines, check a joint to make sure it is square, and check a planed piece of lumber to see whether it is warped or cupped.

Review Questions

Section 8.0.0

1. When determining whether a surface is level, you gauge the _______.
   a. vertical surface
   b. spirit
   c. horizontal surface
   d. amount of bubbles

2. When determining whether a surface is plumb, you gauge the _______.
   a. horizontal surface
   b. vertical surface
   c. spirit
   d. degrees of slope

3. The instrument that has three vials and is used to find out if a surface is level or plumb is called a _______.
   a. plumb bob
   b. spirit level
   c. line level
   d. horizontal level

4. The two end vials in a spirit level measure _______.
   a. plumb
   b. level
   c. spirits
   d. horizontal slope

5. When using a spirit level, you know that an object is level if the air bubble _______.
   a. settles at the bottom
   b. disappears
   c. is centered between the lines
   d. sits on the bottom

9.0.0 ◆ SQUARES

Squares (see *Figure 17*) are used for marking, checking, and measuring. The type of square you use depends on the type of job and your preference. Common squares are the **carpenter's square, rafter angle square** (also called the speed square or magic square), **try square,** and **combination square.**

9.1.0 The Carpenter's Square

The carpenter's square (framing square) is shaped like an L and is used mainly for squaring up sections of work such as wall studs and sole plates, that is, to ensure that they are at right angles to each other. The carpenter's square has a 24-inch blade and a 16-inch tongue, forming a right (90-degree) angle. The blade and tongue are marked with inches and fractions of an inch. You can use the blade and the tongue as a rule or a straightedge. Tables and formulas are printed on the blade for making quick calculations such as determining area and volume.

The rafter angle square (also called a speed square or magic square) is another type of carpenter's square, frequently made of cast aluminum. It is a combination protractor, try square, and framing square. It is marked with degree gradations for fast, easy layout. The square is small, so it's easy to store and carry. By positioning the square on a piece of lumber, you can use it as a guide when cutting with a portable circular saw.

The try square is a fixed, 90-degree angle and is used mainly for woodworking. You can use it to lay out cutting lines at 90-degree angles, to check (or try) the squareness of adjoining surfaces, to check a joint to make sure it is square, and to check if a **planed** piece of lumber is warped or cupped (bowed).

9.1.1 *How to Use a Carpenter's Square*

Take the following steps to mark a line for cutting:

Step 1 Find and mark the place where the line will be drawn.

Step 2 Place the square so that it lines up with the bottom of the object to be marked (see *Figure 18*).

Step 3 Mark the line and cut off excess material.

To check that joints meet at a 90-degree angle, place the blades of the framing square along the two sides of the angle, as shown in *Figure 19.* If both blades fit there tightly, the material is square. If there is any space between either blade and the side closest to it, the material is not square.

Instructor's Notes:

3.22 CORE CURRICULUM ◆ INTRODUCTORY CRAFT SKILLS

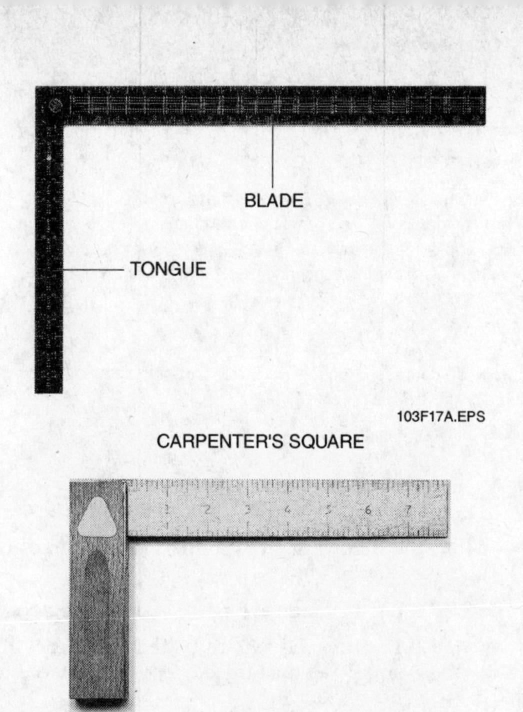

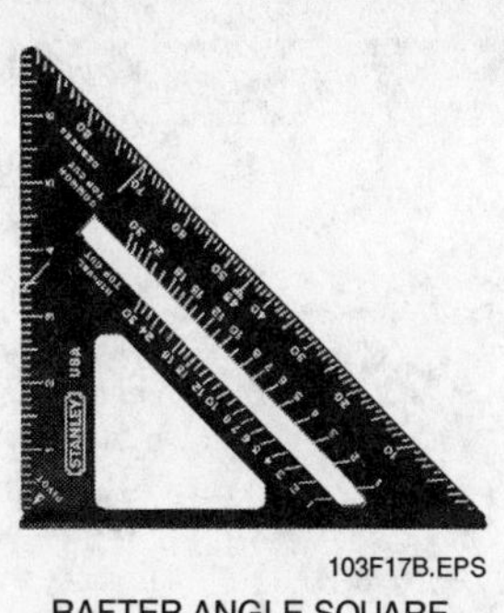

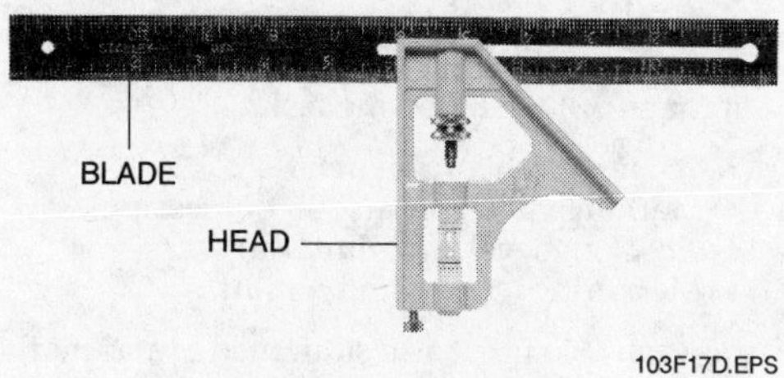

Figure 17 ◆ Types of squares.

Figure 18 ◆ Marking a line for cutting.

Figure 19 ◆ Checking squareness.

Take the following steps to check the flatness of material:

Step 1 Place the edge of the blade on the surface to be checked.

Step 2 Look to see if there is light between the square and the surface of the material. If you can see light, the surface is not flat.

9.2.0 The Combination Square

The combination square has a 12-inch blade that moves through a head. The head is marked with 45-degree and 90-degree angle measures. Some squares also contain a small spirit level and a carbide scriber, which is a sharp, pointed tool for marking metal. The combination square is one of

Discuss the tasks for which a combination square can be used.

MODULE 00103-04 ◆ INTRODUCTION TO HAND TOOLS 3.23

Have trainees practice using a combination square to mark 45- and 90-degree angles.

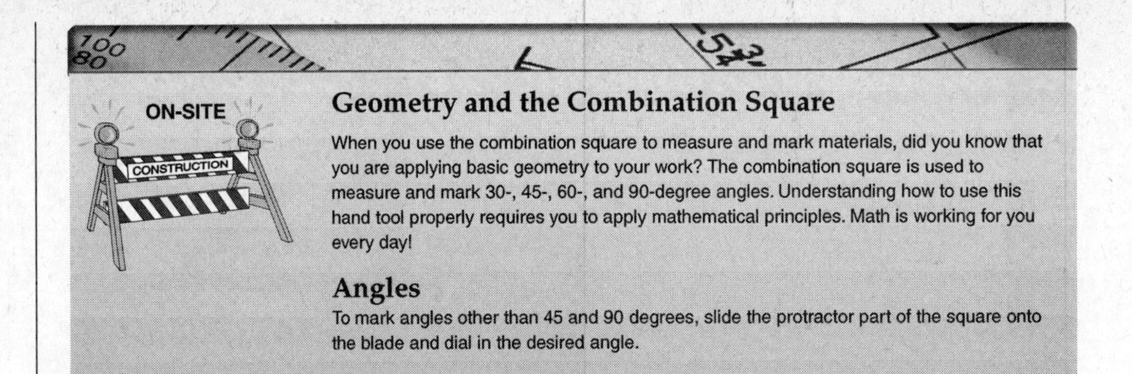

Geometry and the Combination Square

When you use the combination square to measure and mark materials, did you know that you are applying basic geometry to your work? The combination square is used to measure and mark 30-, 45-, 60-, and 90-degree angles. Understanding how to use this hand tool properly requires you to apply mathematical principles. Math is working for you every day!

Angles

To mark angles other than 45 and 90 degrees, slide the protractor part of the square onto the blade and dial in the desired angle.

the most useful tools for layout work. You can use it for any of the following tasks:

- Testing work for squareness
- Marking 90-degree and 45-degree angles
- Checking level and plumb surfaces
- Measuring lengths and widths

You can also use it as a straightedge and marking tool.

Good combination squares have all-metal parts, a blade that slides freely but can be clamped securely in position, and a glass tube spirit level that is truly level and tightly fastened.

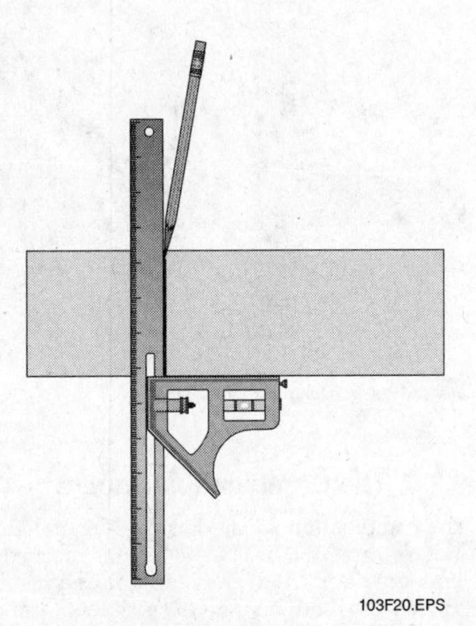

103F20.EPS

Figure 20 ◆ Using a combination square to mark a 90-degree angle.

9.2.1 How to Use a Combination Square

Take the following steps to mark a 90-degree angle (see *Figure 20*):

Step 1 Set the blade at a right angle (90 degrees).

Step 2 Position the square so that the head fits snugly against the edge of the material to be marked.

Step 3 Starting at the edge of the material, use the blade as a straightedge to guide the mark.

Take the following steps to mark a 45-degree angle (see *Figure 21*):

Step 1 Set the blade at a 45-degree angle.

Step 2 Position the square so that the head fits snugly against the edge of the material to be marked.

Step 3 Starting at the edge of the material, use the blade as a straightedge to guide the mark.

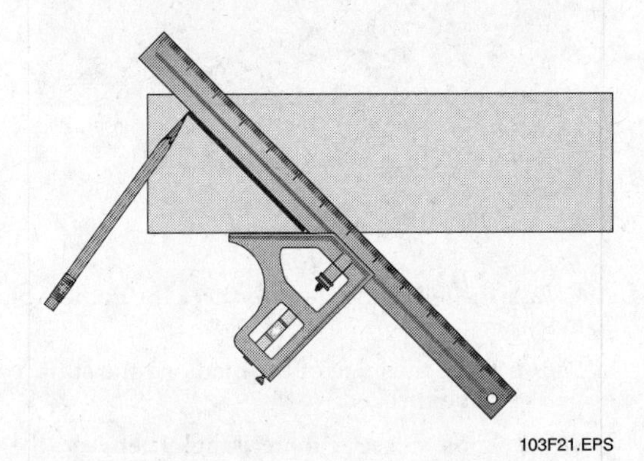

103F21.EPS

Figure 21 ◆ Using a combination square to mark a 45-degree angle.

Instructor's Notes:

9.3.0 Safety and Maintenance

Here are the guidelines to remember when using squares:

- Keep the square dry to prevent it from rusting.
- Use a light coat of oil on the blade, and occasionally clean the blade's grooves and the setscrew.
- Don't use a square for something it wasn't designed for, especially prying or hammering.
- Don't bend a square or use one for any kind of horseplay. They are expensive!
- Don't drop or strike the square hard enough to change the angle between the blade and the head.

Review Questions

Section 9.0.0

1. Squares are used for marking, checking, and _______.
   a. cutting
   b. bending
   c. twisting
   d. measuring
2. The carpenter's square is used mainly for _______.
   a. squaring up sections of work
   b. placing nails along a beam
   c. measuring 360-degree angles
   d. reaching areas where hammers won't fit
3. The try square is a fixed _______ angle.
   a. 45-degree
   b. 180-degree
   c. 90-degree
   d. 360-degree
4. A combination square can be used to test work for squareness; measure lengths, widths, and angles; and _______.
   a. turn nuts and bolts
   b. check level and plumb surfaces
   c. hold material in place
   d. pry apart objects that are stuck together
5. All of the following are appropriate when using squares *except*_______.
   a. using the square for prying or hammering
   b. wearing the appropriate PPE
   c. keeping the square dry
   d. using a light coat of oil on the blade

10.0.0 ◆ PLUMB BOB

The plumb bob (see *Figure 22*), which is a pointed weight attached to a string, uses the force of gravity to make the line hang vertical, or plumb. Plumb bobs come in different weights: 12 ounces, 8 ounces, and 6 ounces are the most common.

When the weight is allowed to hang freely, the string is plumb (see *Figure 23*). You can use a plumb bob to make sure a wall or a doorjamb is vertical. Or, suppose you want to install a post under a beam. A plumb bob can show what point on the floor is directly under the section of the beam you need to support.

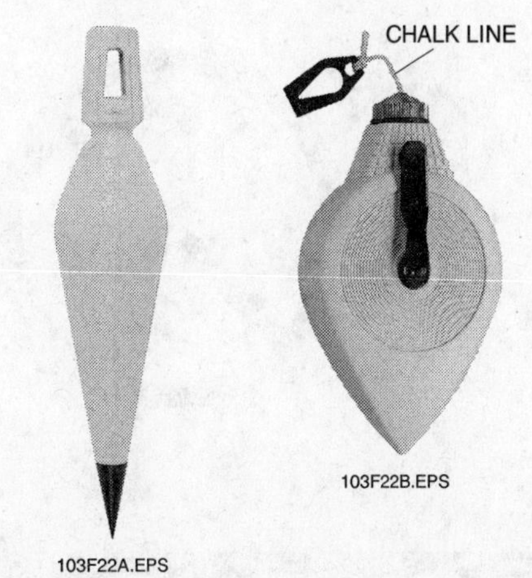

Figure 22 ◆ Plumb bobs.

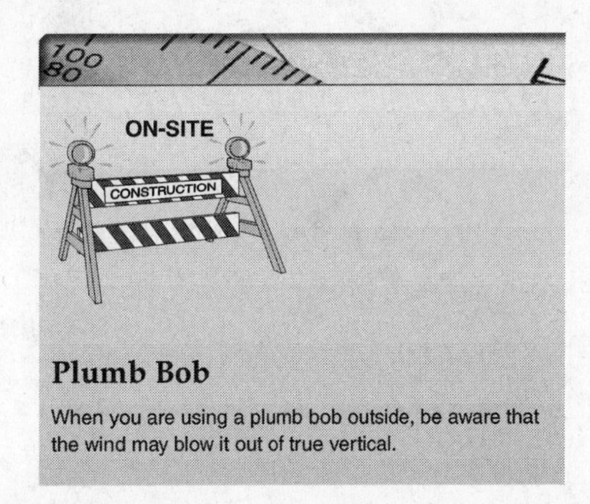

Plumb Bob

When you are using a plumb bob outside, be aware that the wind may blow it out of true vertical.

Review the guidelines for using squares. Ensure that trainees take care not to change the angle between the blade and the head.

Go over the Review Questions for Section 9.0.0. Go over any questions trainees may have.

See the Teaching Tip for Sections 7.0.0–9.3.0 at the end of this module.

Show Transparency 11 (Figure 22). Explain how a plumb bob uses gravity. Ask a trainee how to determine whether a line is vertical.

MODULE 00103-04 ◆ INTRODUCTION TO HAND TOOLS 3.25

Demonstrate how to hang a plumb bob and accurately mark the point below the tip of the bob.

Discuss the consequences of dropping the plumb bob on its point.

Go over the Review Questions for Section 10.0.0. Ask whether trainees have any questions.

10.1.0 How to Use a Plumb Bob

Follow these steps to use a plumb bob properly:

Step 1 Make sure the line is attached at the exact top center of the plumb bob.

Step 2 Hang the bob from a horizontal member, such as a doorjamb, joist, or beam.

Step 3 When the weight is allowed to hang freely and stops swinging, the string is plumb (vertical).

Step 4 Mark the point directly below the tip of the plumb bob. This point is precisely below the point where you attached the bob.

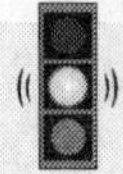

CAUTION

Do not drop the plumb bob on its point. A bent or rounded point causes inaccurate readings.

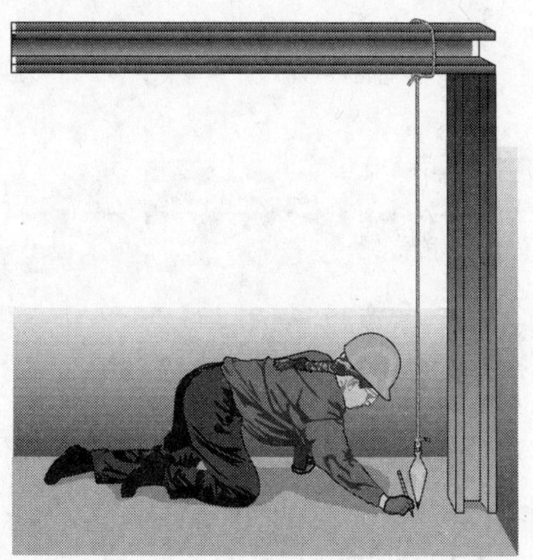

103F23.EPS

Figure 23 ◆ Proper use of a plumb bob.

Review Questions

Section 10.0.0

1. A plumb bob uses _______ to make a line hang vertical.
   a. magnetic forces
   b. a point on the floor
   c. the force of gravity
   d. a spirit level

2. When something is plumb, it is _______.
   a. vertical
   b. horizontal
   c. at a 30-degree angle
   d. bobbed

3. When using a plumb bob outside, remember that _______ may affect the plumb bob.
   a. magnetic forces
   b. wind
   c. noise
   d. the point of suspension

4. When a plumb bob hangs freely, its string is _______.
   a. level
   b. horizontal
   c. vertical
   d. at a 45-degree angle

5. A plumb bob will be damaged if you drop it on its _______.
   a. line
   b. head
   c. joist
   d. point

Instructor's Notes:

11.0.0 ◆ CHALK LINES

A chalk line is a piece of string or cord that is coated with chalk. You stretch the line tightly between two points and then snap it to release a chalky line to the surface. You can use a piece of string rubbed with chalk if you need to snap only a couple of lines. But for frequent use, a mechanical self-chalking line is much handier.

A mechanical self-chalking line is a metal box (see *Figure 24*) containing a line on a reel. The box is filled with colored chalk powder. The line is automatically chalked each time you pull it out of the box. Some models have a point on the end of the box so it can be used as a plumb bob also.

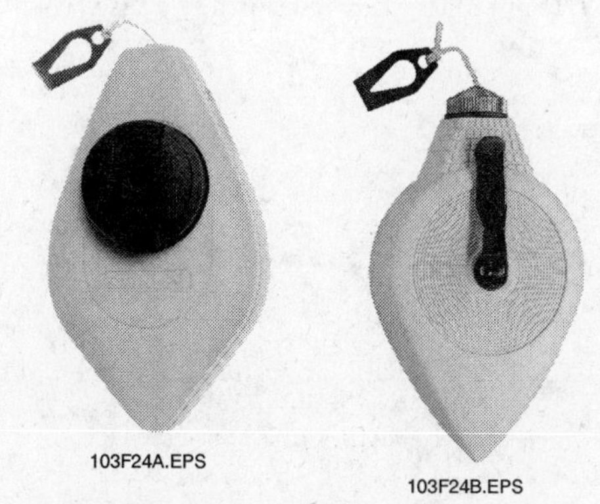

Figure 24 ◆ Mechanical self-chalkers.

11.1.0 How to Use a Chalk Line

Follow these steps to use a chalk line properly:

Step 1 Pull the line from the case. Have a partner hold one end.

Step 2 Stretch the line between the two points to be connected.

Step 3 After the line has been pulled tight, pull straight away from the work and then release. This marks the surface underneath with a straight line of chalk (see *Figure 25*).

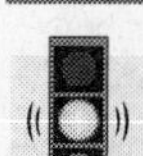

CAUTION

Store the chalk line in a dry place. Damp or wet chalk is unusable.

Refer to Figure 25. Explain how to use a chalk line to mark a line.

Have trainees practice using a self-chalker/plumb bob to mark the line between two points.

Stress that damp or wet chalk is unusable.

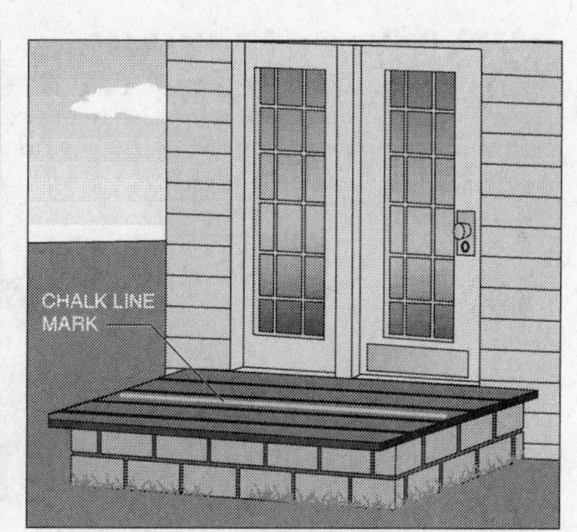

Figure 25 ◆ Proper use of a chalk line.

MODULE 00103-04 ◆ INTRODUCTION TO HAND TOOLS 3.27

Refer to Figure 26. Explain how vises enable one person to do work that would normally require two people.

Discuss the consequence of using a hammer to tighten the handle of a bench vise.

Explain how to properly use a vise to avoid damage to the vise and the object it is holding.

Go over the Review Questions for Sections 11.0.0–12.0.0. Answer any questions trainees may have.

Have trainees review Sections 13.0.0–17.3.0 for the next session.

12.0.0 ◆ BENCH VISES

Vises are gripping and holding tools. By using a vise, you can do work that would otherwise require two people. Vises are used to secure an object while you work on it. Vises can be portable or fixed, which means they stay in one place.

The bench vise (see *Figure 26*) is a stationary vise with two sets of jaws: one to hold flat work and another to hold round work, such as pipe. Some bench vises have swivel bases so that you can turn the vise in any horizontal direction.

12.1.0 How to Use a Bench Vise

Follow these steps to use a bench vise properly:

Step 1 Place the object in the open clamp of the bench vise.

Step 2 To clamp the object, turn the sliding T-handle screw clockwise.

Step 3 To release the object, turn the T-handle screw counterclockwise.

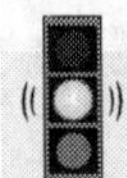

CAUTION

Never use a hammer to tighten the handle, and never use a piece of pipe for leverage. Doing so may damage both the vise and the object being clamped.

12.2.0 Safety and Maintenance

You can damage a vise or the object it is holding by not using it properly. You can also injure yourself. For example, if the object in the vise is not clamped tightly enough, you could slip while using a saw on the object. Remember these guidelines for using a vise properly:

- Fasten the vise securely to the bench.
- Clamp work evenly in the vise.
- Support the ends of any long piece of wood or other material that is being held in the vise.
- Saw as close as possible to the jaws of the vise when you are sawing an object.
- Keep threaded parts clean.
- Don't use the jaws of the vise as a pounding surface.
- Don't place your hand inside a vise when adjusting it.

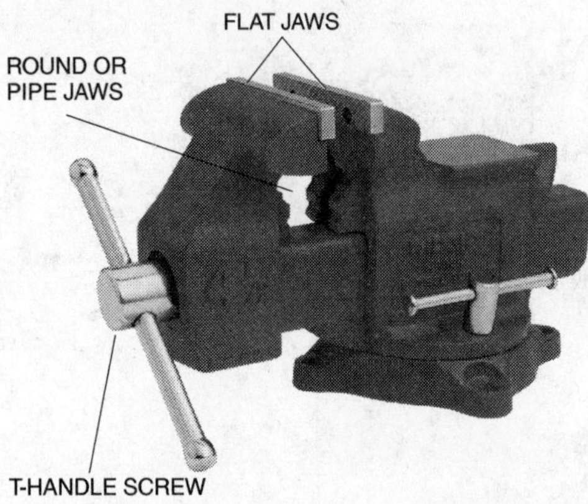

Figure 26 ◆ Bench vise.

Review Questions

Sections 11.0.0 and 12.0.0

1. To use a chalk line, stretch the line tightly between two points and then _______ to transfer a chalky line to the surface.
   a. walk over it
   b. snap it
   c. moisten it
   d. press it
2. A vise is a _______ tool.
   a. measuring and gauging
   b. holding and gripping
   c. slicing and sawing
   d. ripping and prying
3. One set of vise jaws is for holding flat work; the other set is for holding _______ work.
   a. slippery
   b. extra
   c. large
   d. round
4. When using a bench vise, turn the sliding T-handle screw _______ to clamp the object.
   a. clockwise
   b. counterclockwise
   c. downward
   d. upward
5. When sawing an object, you should saw as close as possible to the jaws of the vise.
   a. True
   b. False

Instructor's Notes:

3.28 CORE CURRICULUM ◆ INTRODUCTORY CRAFT SKILLS

13.0.0 ◆ CLAMPS

There are many types and sizes of clamps, each designed to solve a different holding problem. Clamps (see *Figure 27*) are sized by the maximum opening of the jaw. They come in sizes from 1 inch to 24 inches. The depth (or throat) of the clamp determines how far from the edge of the work the clamp can be placed. The following are common types of clamps:

- *C-clamp* – This multipurpose clamp has a C-shaped frame. It is used primarily for clamping metalwork. The clamp has a metal shoe at the end of a screw. Using a T-bar, you tighten the clamp so that it holds material between the metal jaw of the frame and the shoe.
- *Locking C-clamp* – This clamp works like vise-grip pliers. A knob in the handle controls the width and tension of the jaws. You close the handles to lock the clamp. You release the clamp by pressing the lever to open the jaws.
- *Spring clamp* – You use your hand to open the spring-operated clamp. When you release the handles, the spring holds the clamp tightly shut, applying even pressure to the material. The jaws are usually made of steel, some with plastic coating to protect the material's surface against scarring.
- *Bar clamp* – A rectangular piece of steel or aluminum is the spine of the bar clamp. It has a fixed jaw at one end and a sliding jaw (tail slide) with a spring-locking device that moves along the bar. You position the fixed jaw against the object you want to hold and then move the sliding jaw into place. The screw set is tightened as with a C-clamp.
- *Pipe clamp* – Although this clamp looks like a bar clamp, the spine is actually a length of pipe. It has a fixed jaw and a movable jaw that work the same way as the bar clamp. The movable jaw has a lever mechanism that you squeeze when sliding the movable jaw along the spine.
- *Hand-screw clamp* – This clamp has wooden jaws. It can spread pressure over a wider area than other clamps can. Each jaw works independently. You can angle the jaws toward or away from each other or keep them parallel. You tighten the clamp using spindles that screw through the jaws.
- *Web (strap, band) clamp* – This clamp uses a belt-like canvas or nylon strap or band to apply even pressure around a piece of material. After looping the band around the work, you use the clamp head to secure the band. Using a wrench or screwdriver, you ratchet (tighten by degrees) the bolt tight in the clamp head. A quick-release device loosens the band after you are finished.

Ensure that you have everything required for demonstrations, laboratories, and testing during this session.

Show Transparencies 12 and 13 (Figure 27). Review the various types and sizes of clamps.

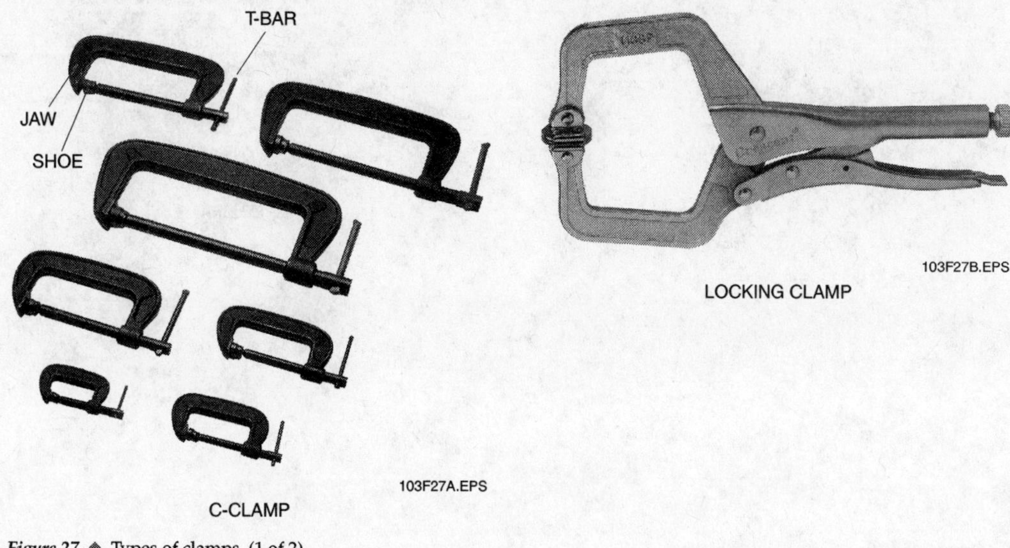

Figure 27 ◆ Types of clamps. (1 of 2)

MODULE 00103-04 ◆ INTRODUCTION TO HAND TOOLS 3.29

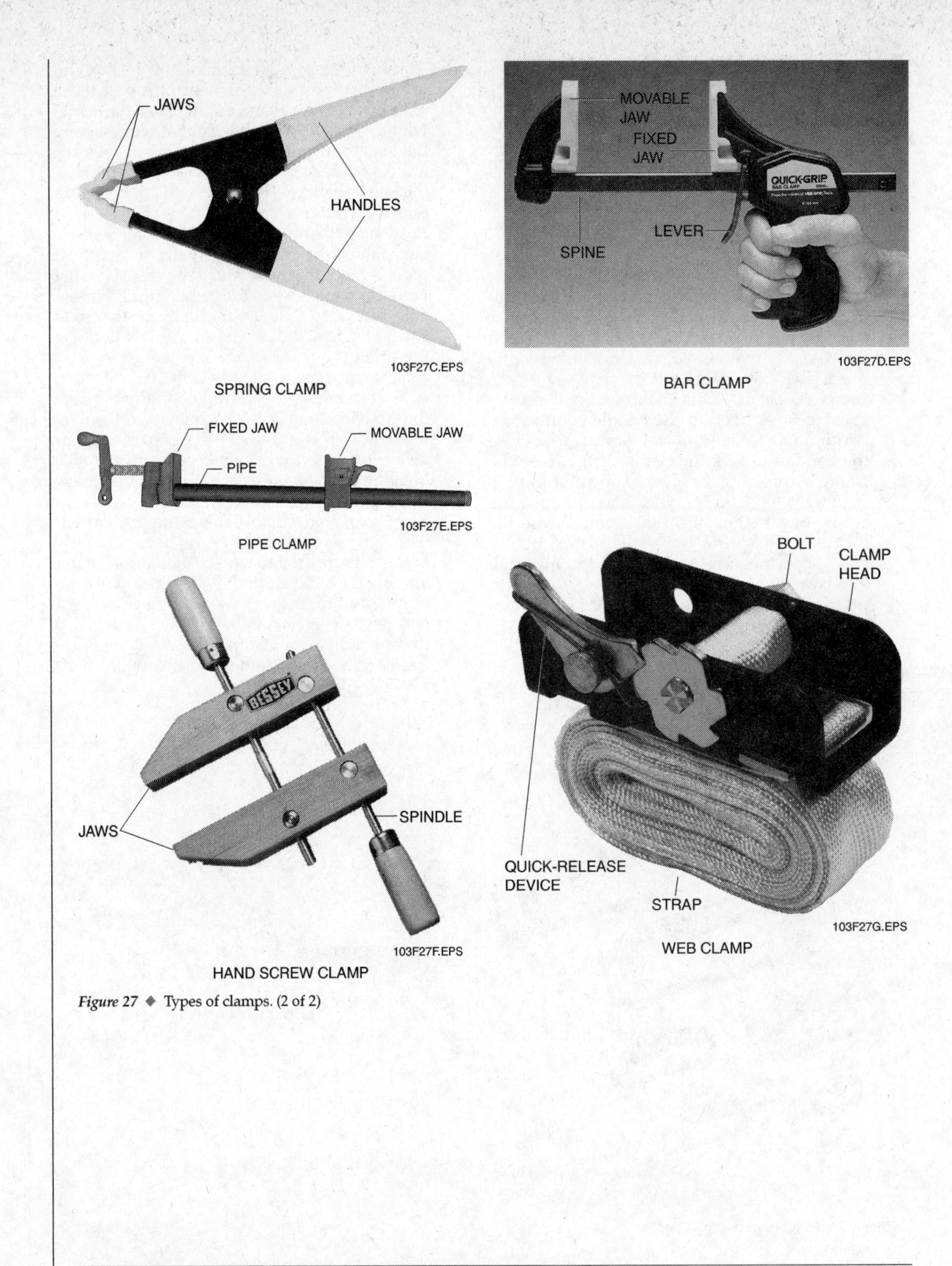

Figure 27 ◆ Types of clamps. (2 of 2)

Instructor's Notes:

13.1.0 How to Use a Clamp

Take the following steps to use a clamp properly:

Step 1 When clamping wood or other soft material, place pads or thin blocks of wood between the work piece and the clamp to protect the work (see *Figure 28*).

Step 2 Tighten the clamp's pressure mechanism, such as the T-bar handle. Don't force it.

13.2.0 Safety and Maintenance

Here are some guidelines to remember when using clamps:

- Store clamps by clamping them to a rack.
- Use pads or thin wood blocks when clamping wood or other soft materials.

CAUTION

When tightening a clamp, do not use pliers or a section of pipe on the handle to extend your grip or gain more leverage. Doing so means you will have less, not more, control over the clamp's tightening mechanism.

If you are clamping work that has been glued, do not tighten the clamps so much that all the glue is squeezed out of the joint.

- Discard clamps with bent frames.
- Clean and oil threads.
- Check the swivel at the end of the screw to make sure it turns freely.
- Don't use a clamp for hoisting (pulling up) work.
- Don't overtighten clamps.

Bring in a variety of clamps. Demonstrate how to protect your work when clamping wood or soft materials.

Review procedures for maintaining and safely using clamps.

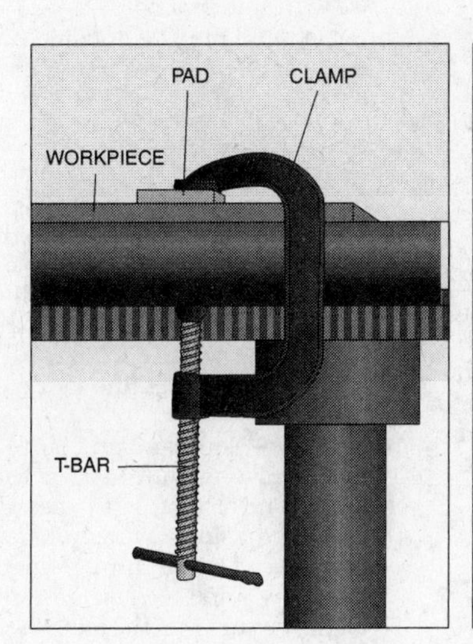

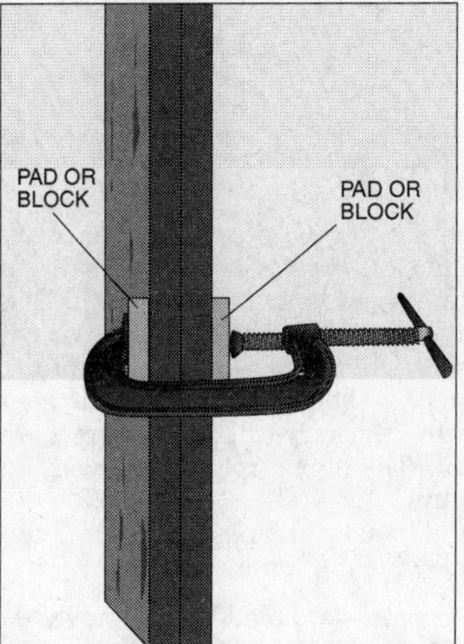

103F28.EPS

Figure 28 ◆ Placing pads and wood blocks.

MODULE 00103-04 ◆ INTRODUCTION TO HAND TOOLS 3.31

Go over the Review Questions for Section 13.0.0. Answer any questions trainees may have.

Show Transparency 14 (Figure 29). Review the differences among the types of saws. Explain how shape, number, and pitch of teeth make it possible to cut different materials.

Discuss the classifications used for handsaws. Explain that the fewer teeth per inch (tpi), the coarser and faster the cut.

Review Questions

Section 13.0.0

1. Clamps are sized by the _______.
   a. length of the handles
   b. distance between the metal bar and the shoe
   c. maximum opening of the jaw
   d. kind of pad you insert to protect the work

2. A C-clamp is used mainly for clamping _______.
   a. wood
   b. nylon
   c. T-bars
   d. metal

3. A hand-screw clamp has _______ jaws.
   a. metal
   b. nylon
   c. wooden
   d. fiberglass

4. When clamping soft materials, you need to use _______ to protect the work.
   a. a web
   b. pads
   c. a rack
   d. pipes

5. Using pliers or pipe on the handle of a clamp _______.
   a. gives you less control
   b. gives you more control
   c. is a safe practice
   d. should only be done with woodwork

14.0.0 ◆ SAWS

Using the right saw for the job makes cutting easy. The main differences between types of saws are the shape, number, and pitch of their teeth. These differences make it possible to cut across or with the grain of wood, along curved lines, or through metal, plastic, or wallboard. Generally, the fewer **points** or teeth per inch (tpi), the coarser and faster the cut. The more tpi, the slower and smoother the cut.

Figure 29 shows several types of saws and their parts. The following are common types of saws:

- *Backsaw* – The standard blade of this saw is 8 to 14 inches long with 11 to 14 tpi. A backsaw has a broad, flat blade and a reinforced back edge. It is used for cutting joints, especially **miter joints** and **tenons.**
- *Compass (keyhole) saw* – The standard blade of this saw is 12 to 14 inches long with 7 or 8 tpi. This saw cuts curves quickly in wood, plywood, or wallboard. It is also used to cut holes for large-diameter pipes, vents, and plugs or switch boxes. It can fit in tight places where a larger handsaw will not.
- *Coping saw* – This saw has a narrow, flexible 6¾-inch blade attached to a U-shaped frame. Holders at each end of the frame can be rotated so you can cut at angles. Standard blades range from 10 to 20 tpi. The coping saw is used for making irregular-shaped moldings fit together cleanly.
- *Dovetail saw* – This is a small backsaw with a straight handle. The standard blade is 10 inches long with 16 to 20 tpi. The dovetail saw is used for cutting fine work, especially dovetail joints.
- *Hacksaw* – The standard blade of this saw is 8 to 16 inches long with 14 to 32 tpi. It has a sturdy frame and a pistol-grip handle. The blade is tightened using a wing nut and bolt. The hacksaw is used to cut through metal, such as nails, bolts, or pipe. When installing a hacksaw blade, be sure that the teeth face away from, not toward, the saw handle. Hacksaws are designed to cut on the push stroke, not on the pull stroke.
- *Handsaw (crosscut saw or ripsaw)* – The standard blade of this saw is 26 inches long with 8 to 14 tpi for a crosscut saw and 5 to 9 tpi for a ripsaw. You will learn how to use handsaws in this module.

14.1.0 Handsaws

The handsaw's blade is made of tempered steel so it will stay sharp and will not bend or buckle. Handsaws are classified mainly by the number, shape, size, slant, and direction of the teeth. Saw

Instructor's Notes:

3.32 CORE CURRICULUM ◆ INTRODUCTORY CRAFT SKILLS

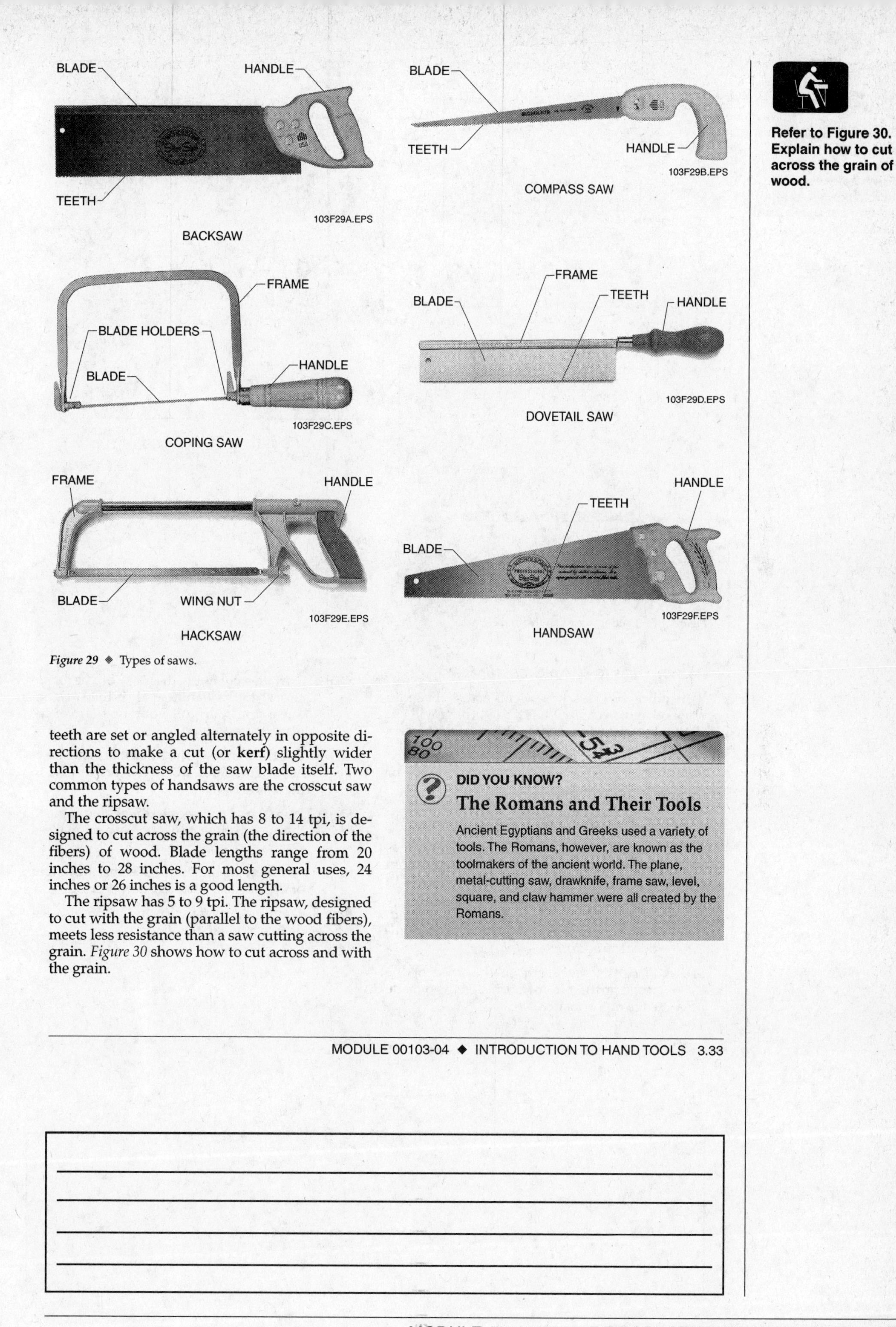

Figure 29 ◆ Types of saws.

Refer to Figure 30. Explain how to cut across the grain of wood.

teeth are set or angled alternately in opposite directions to make a cut (or **kerf**) slightly wider than the thickness of the saw blade itself. Two common types of handsaws are the crosscut saw and the ripsaw.

The crosscut saw, which has 8 to 14 tpi, is designed to cut across the grain (the direction of the fibers) of wood. Blade lengths range from 20 inches to 28 inches. For most general uses, 24 inches or 26 inches is a good length.

The ripsaw has 5 to 9 tpi. The ripsaw, designed to cut with the grain (parallel to the wood fibers), meets less resistance than a saw cutting across the grain. *Figure 30* shows how to cut across and with the grain.

DID YOU KNOW?

The Romans and Their Tools

Ancient Egyptians and Greeks used a variety of tools. The Romans, however, are known as the toolmakers of the ancient world. The plane, metal-cutting saw, drawknife, frame saw, level, square, and claw hammer were all created by the Romans.

MODULE 00103-04 ◆ INTRODUCTION TO HAND TOOLS 3.33

Set up stations with a variety of crosscut saws and ripsaws. Have trainees practice making a straight square cut using both saws.

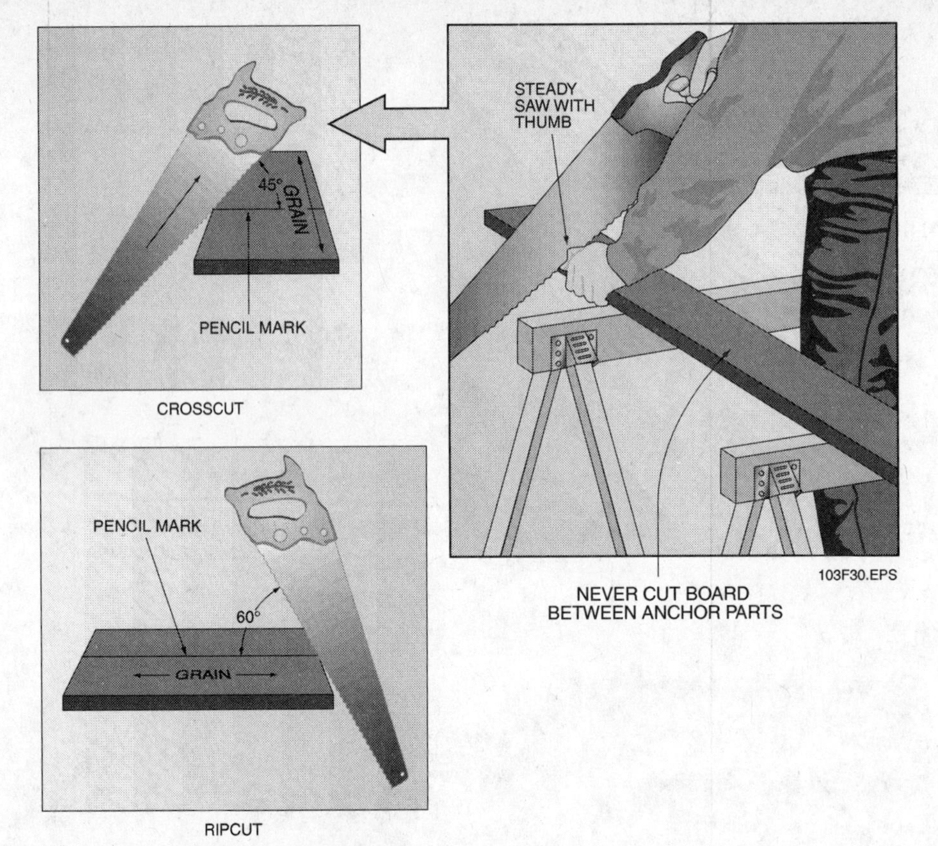

Figure 30 ◆ Cutting across and with the grain.

14.1.1 How to Use a Crosscut Saw

Remember, the crosscut saw cuts across the grain of wood and, because it has 8 to 14 tpi, it will cut slowly but smoothly. Follow these steps to use a crosscut saw properly:

Step 1 Mark the cut to be made with a square or other measuring tool.

Step 2 Make sure the piece to be cut is well-supported (on a sawhorse, jack, or other support). Support the scrap end as well as the main part of the wood to keep it from splitting as the kerf nears the edge. With short pieces of wood, you can support the scrap end of the piece with your free hand. With longer pieces, you will need additional support.

Step 3 Place the saw teeth on the edge of the wood farthest from you, just at the outside edge of the mark.

Step 4 Start the cut with the part of the blade closest to the handle-end of the saw, because you will pull your first stroke toward your body.

Step 5 Use the thumb of the hand that is not sawing to guide the saw so it stays vertical to the work.

Step 6 Place the saw at about a 45-degree angle to the wood, then pull the saw to make a small groove (see *Figure 30*).

Step 7 Start sawing slowly, increasing the length of the stroke as the kerf deepens.

Step 8 Continue to saw with the blade at a 45-degree angle to the wood.

Instructor's Notes:

3.34 CORE CURRICULUM ◆ INTRODUCTORY CRAFT SKILLS

Emphasize that saws can be dangerous if used incorrectly. Review safety procedures for working with handsaws.

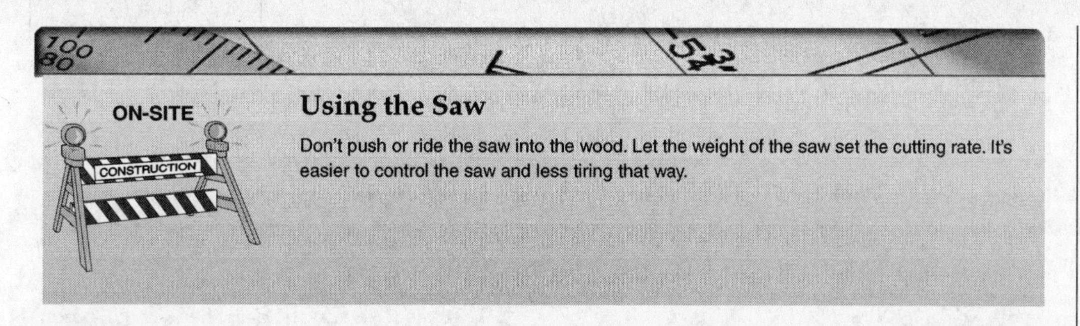

ON-SITE

Using the Saw

Don't push or ride the saw into the wood. Let the weight of the saw set the cutting rate. It's easier to control the saw and less tiring that way.

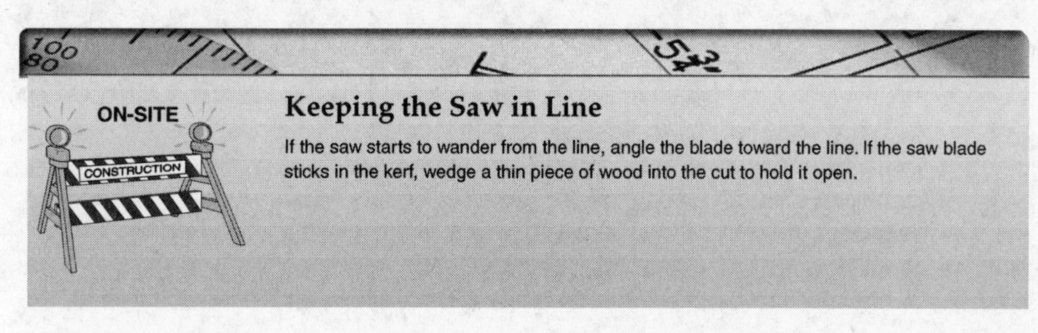

ON-SITE

Keeping the Saw in Line

If the saw starts to wander from the line, angle the blade toward the line. If the saw blade sticks in the kerf, wedge a thin piece of wood into the cut to hold it open.

14.1.2 How to Use a Ripsaw

The ripsaw cuts along the grain of wood. Because it has fewer points (5 to 9 tpi) than the crosscut saw, it will make a coarser, but faster, cut. Follow these steps to use a ripsaw properly:

Step 1 Mark and start a ripping cut the same way you would start cutting with a crosscut saw.

Step 2 Once you've started the kerf, saw with the blade at a steeper angle to the wood—about 60 degrees.

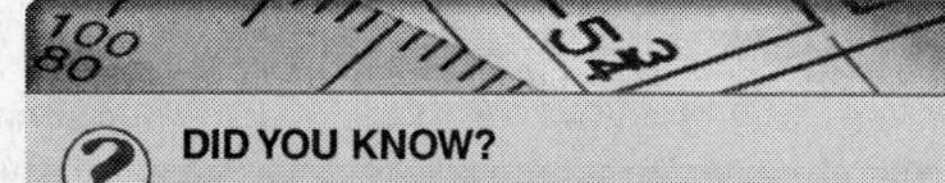

DID YOU KNOW?

Emery Cloth

Emery cloth is a maintenance and cleaning tool often used in the construction industry. It is usually used for cleaning tools made of metal, like handsaws. It may be used wet or dry in a manner very similar to sandpaper. Emery cloth is coated with a substance called powdered emery, which is a granular form of pure carborundum.

14.2.0 Safety and Maintenance

You must maintain your saws for them to work properly. Also, it is very important to focus on your work when you are sawing—saws can be dangerous if used incorrectly or if you are not paying attention. Here are the guidelines for working with handsaws:

- Clean your saw blade with a fine emery cloth and apply a coat of light machine oil if it starts to rust—rust ruins the saw blade.
- Always lay a saw down gently.
- Have your saw sharpened by an experienced sharpener.
- Brace yourself when sawing so you are not thrown off balance on the last stroke.
- Don't let saw teeth come in contact with stone, concrete, or metal.

MODULE 00103-04 ◆ INTRODUCTION TO HAND TOOLS 3.35

Go over the Review Questions for Section 14.0.0. Ask whether trainees have any questions.

Refer to Figure 31. Discuss the various types of files and rasps, and explain how they are used to shape metal parts.

Bring in a variety of files and rasps. Demonstrate how to choose a file or rasp to fit the area and material you are filing.

Refer to Table 1. Discuss the various file classifications and their uses.

Set up several stations with materials to be filed. Have trainees mount the work in a vise and practice their filing techniques.

Review Questions

Section 14.0.0

1. The main differences between types of saws relate to their _______.
   a. handles
   b. blades
   c. frames
   d. teeth

2. If you want to cut metal, you would choose the _______.
   a. ripsaw
   b. compass saw
   c. hacksaw
   d. crosscut saw

3. A type of saw that cuts *across* the grain of the wood is the _______.
   a. ripsaw
   b. crosscut saw
   c. hacksaw
   d. coping saw

4. A type of saw that cuts *with* the grain of the wood is the _______.
   a. dovetail saw
   b. ripsaw
   c. crosscut saw
   d. hacksaw

5. If a saw blade sticks in the kerf, you should _______.
   a. wedge a piece of wood in the kerf to keep it open
   b. apply a coat of light machine oil to the kerf
   c. sharpen the saw
   d. switch to a saw with fewer teeth per inch

15.0.0 ◆ FILES AND RASPS

You use files and rasps to cut, smooth, or shape metal parts. You can also use them to finish and shape all metals except hardened steel, and to sharpen many tools.

Files have slanting rows of teeth. Rasps have individual teeth. Files and rasps are usually made from a hardened piece of high-grade steel (see *Figure 31*). Both are sized by the length of the body. The size does not include the handle because the handle is generally separate from the file or rasp. For most sharpening jobs, files and rasps range from 4 inches to 14 inches (see *Figure 32*).

Choose a file or rasp whose shape fits the area you are filing. Files and rasps are available in round, square, flat, half-round, and triangular shapes. For filing large concave (curved inward) or flat surfaces, you might use a half-round shape. For filing small curves or for enlarging and smoothing holes, you might use a round shape with a tapered end, called a rat-tail file. For filing angles, you might use a triangular file.

There is a specific type of file for each of the common soft metals, hard metals, plastics, and wood. In general, the teeth of files for soft materials are very sharp and widely spaced. Those for hard materials are blunter and closer together. The shape of the teeth also depends on the material to be worked.

If you use a file designed for soft material on hard material, the teeth will quickly chip and dull. If you use a file designed for hard material on soft material, the teeth will clog.

Most files are sold without a handle. You can use a single handle for different files. The sharp metal point at the end of the file, the **tang**, fits into the handle. You can tighten the handle to prevent the tang from coming loose.

Files are classified by the cut of their teeth. File classifications include the following:

- Single-cut and double-cut
- Rasp-cut
- Curved-tooth

Table 1 lists types of files and some uses for each.

Rasps are also classified by the size of their teeth: coarse, medium, and fine.

15.1.0 How to Use a File

Trying to use a file the wrong way will only frustrate you. Follow these steps to use a file properly:

Step 1 Mount the work you are filing in a vise at about elbow height.

Instructor's Notes:

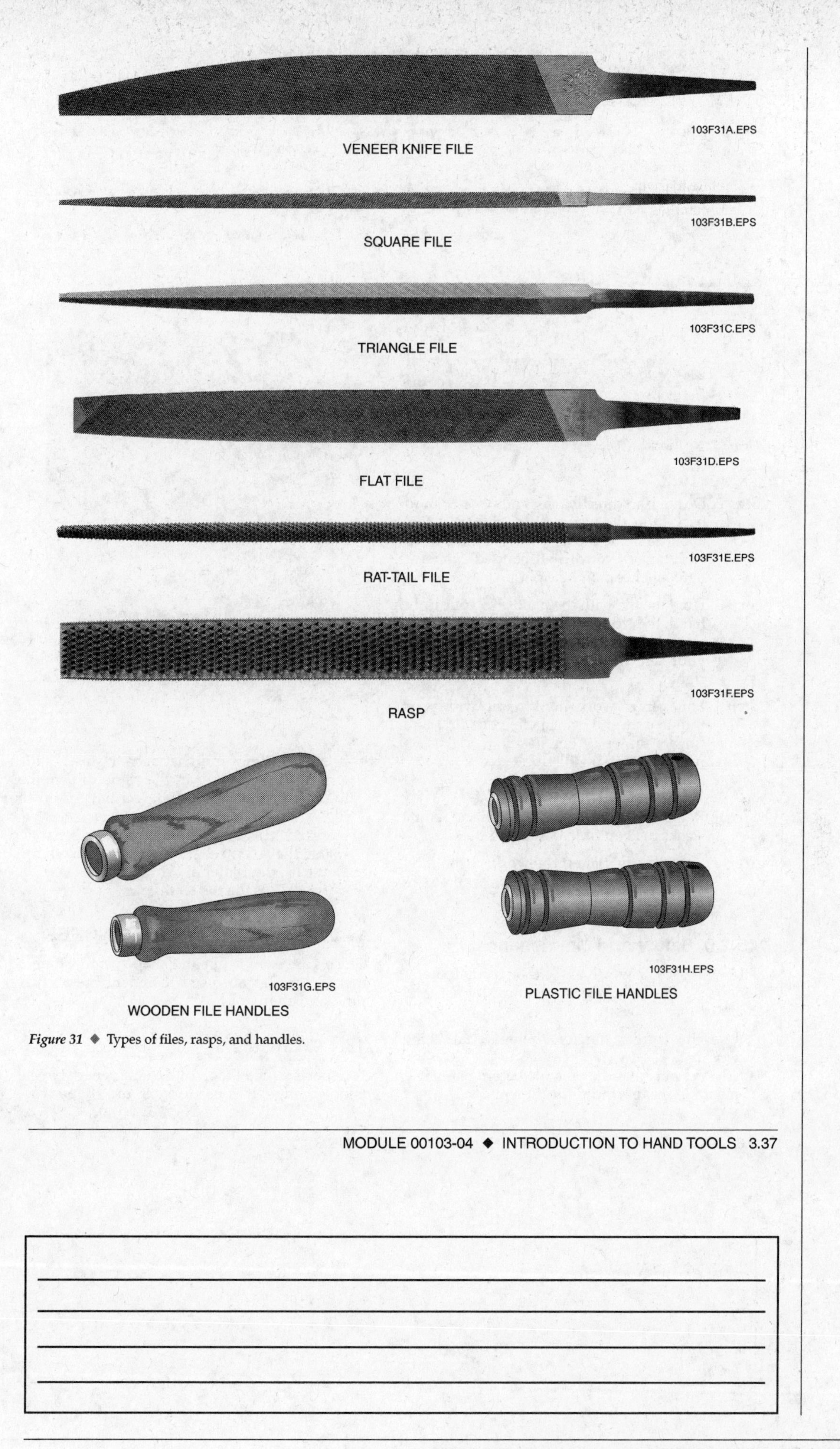

Figure 31 ◆ Types of files, rasps, and handles.

MODULE 00103-04 ◆ INTRODUCTION TO HAND TOOLS 3.37

Explain that, if not cared for properly, files will become unusable. Review maintenance procedures for working with files.

Go over the Review Questions for Section 15.0.0. Answer any questions trainees may have.

Table 1 Types and Uses of Files

Type	Decription	Uses
Rasp-cut file	The teeth are individually cut; they are not connected to each other.	Gives a very rough surface. Used mostly on aluminum, lead, and other soft metals to remove waste materials. Also used on wood.
Single-cut file	Has a single set of straight-edged teeth running across the file at an angle.	Used to sharpen edges, such as rotary mower blades.
Double-cut file	Two sets of teeth crisscross each other. Types are bastard (roughest cut), second cut, and smooth.	Used for fast cutting.

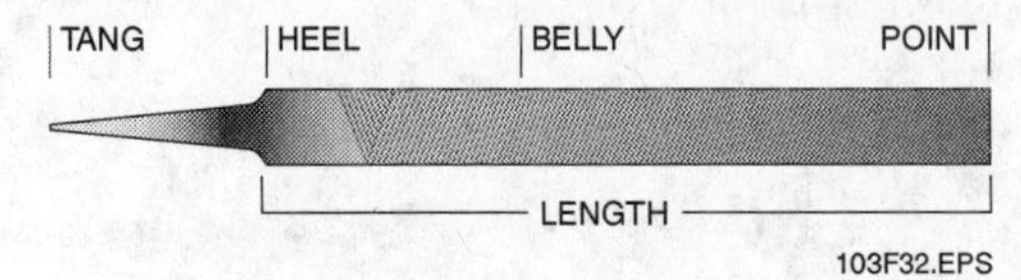

Figure 32 ◆ Parts of a file.

Step 2 Do not lean directly over your work. Stand back from the vise a little with your feet about 24 inches apart, the right foot ahead of the left. (If you are left-handed, put your left foot ahead of the right.)

Step 3 Hold the file with the handle in your right hand, the tip of the blade in your left. (If you are left-handed, hold the handle in your left hand and the blade in your right.)

Step 4 For average work, hold the tip with your thumb on top of the blade, your first two fingers under it. For heavy work, use a full-hand grip on the tip.

Step 5 Apply pressure only on the forward stroke.

Step 6 Raise the file from the work on the return stroke to keep from damaging the file.

Step 7 Keep the file flat on the work. Clean it by tapping lightly at the end of each stroke (see *Figure 33*).

15.2.0 Safety and Maintenance

Files will become worthless without proper maintenance. Here are some guidelines for use and maintenance of files:

- Use the correct file for the material being worked.
- Always put a handle on a file before using it—most files have handle attachments.

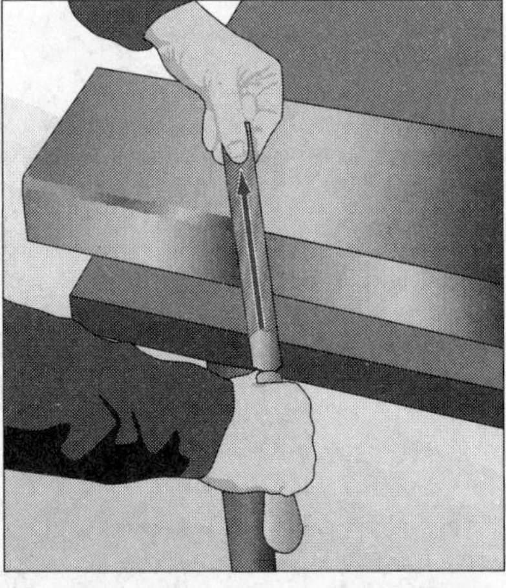

Figure 33 ◆ Proper use of a file.

- Brush the filings from between the teeth with a wire brush, pushing in the same direction as the line of the teeth, after you have used the file.
- Store files in a dry place and keep them separated so that they won't chip or damage each other.
- Don't let the material vibrate in the vise, because it dulls the file teeth.

16.0.0 ◆ CHISELS AND PUNCHES

Chisels are used to cut and shape wood, stone, or metal. Punches are used to indent metal, drive pins, and align holes.

16.1.0 Chisels

A chisel is a metal tool with a sharpened, beveled (sloped) edge. It is used to cut and shape wood,

Instructor's Notes:

3.38 CORE CURRICULUM ◆ INTRODUCTORY CRAFT SKILLS

Review Questions

Section 15.0.0

1. Files and rasps are usually made from _______.
   a. high-grade steel
   b. high-carbon steel
   c. cast aluminum
   d. lead
2. Files are classified by the _______ of their teeth.
   a. width
   b. angle
   c. taper
   d. cut
3. The teeth of files for soft materials are _______.
   a. soft
   b. sharp
   c. loose
   d. dull
4. Files have slanting rows of teeth and rasps have _______ teeth.
   a. smooth
   b. individual
   c. coarse
   d. wire
5. When cleaning files, _______.
   a. brush in the opposite direction of the line of teeth
   b. use an old toothbrush
   c. brush in the same direction as the line of teeth
   d. use soap and water

Refer to Figure 34. Explain how chisels are used to cut and shape wood, stone, or metal.

Demonstrate how to use a wood chisel to make a notch in wooden material.

ON-SITE

Caring for Hand Tools

You have to use power tools to care for some types of hand tools, such as chisels, screwdrivers, hammers, and punches. If the edge or striking surface of a hand tool is damaged or worn, it should be ground back to its desired shape using a grinder. Grinding a hammer face or a punch point will remove unwanted burrs or mushrooming. For a chisel to cut well, the blade needs to be beveled (sloped) at a precise angle. A grinder can be used to remove nicks. The cutting edge must then be sharpened on an oilstone to produce a keen, precise edge. Screwdriver blades can also be cleaned up using a grinder.

stone, or metal. You will learn about two kinds of chisels in this section: the wood chisel and the cold chisel (see *Figure 34*). Both chisels are made from steel that is heat-treated to make it harder. A chisel can cut any material that is softer than the steel of the chisel.

16.1.1 How to Use a Wood Chisel

You use the wood chisel to make openings or notches in wooden material. For instance, you can use it to make a recess for butt-type hinges, such as the hinges in a door. Follow these steps to use a wood chisel properly:

Step 1 Wear appropriate personal protective equipment.

Step 2 Outline the opening (recess) to be chiseled.

Step 3 Set the chisel at one end of the outline, with its edge on the cross-grain line and the bevel facing the recess to be made.

Step 4 Strike the chisel head lightly with a mallet.

Step 5 Repeat this process at the other end of the outline, again with the bevel of the chisel blade toward the recess. Then make a series of cuts about ¼ inch apart from one end of the recess to the other.

Have trainees practice using a cold chisel to cut metal rivets.

Show Transparency 15 (Figure 36). Compare center, prick, and tapered punches.

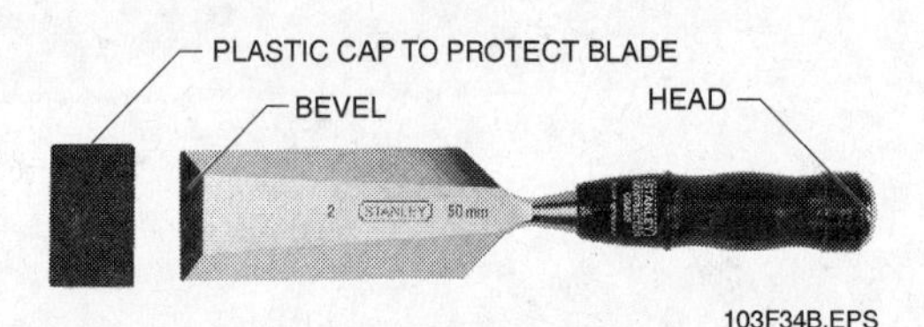

Figure 34 ◆ Cold and wood chisels.

Step 6 To pare (trim) away the notched wood, hold the chisel bevel-side down to slice inward from the end of the recess (see *Figure 35*).

16.1.2 How to Use a Cold Chisel

You use the cold chisel to cut metal. For instance, you can use it to cut rivets, nuts, and bolts made of brass, bronze, copper, or iron. Follow these steps to use a cold chisel properly:

Step 1 Wear appropriate personal protective equipment.

Step 2 Secure the object you want to cut in a vise, if possible.

Step 3 Use a holding tool to hold the chisel in the spot where you want to cut the metal.

Step 4 Using a holding tool, place the blade of the chisel at the spot where you want to cut the material.

Step 5 Hit the chisel handle with a ball peen hammer to force the chisel into and through the material. Use a holding tool to hold material in place. Repeat if necessary.

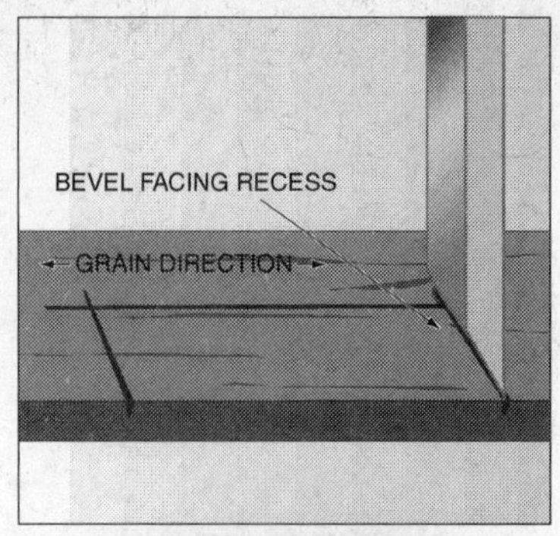

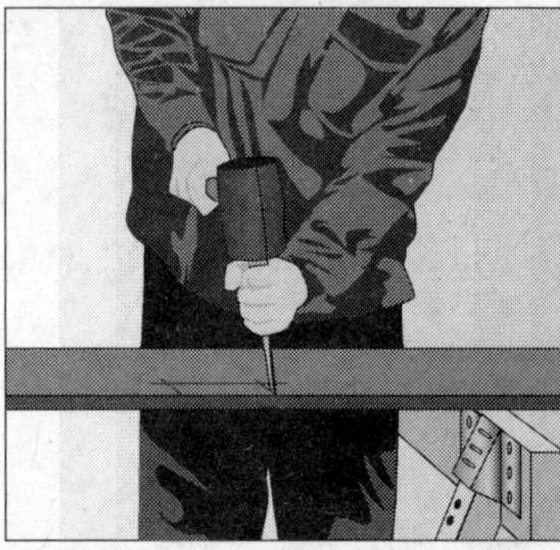

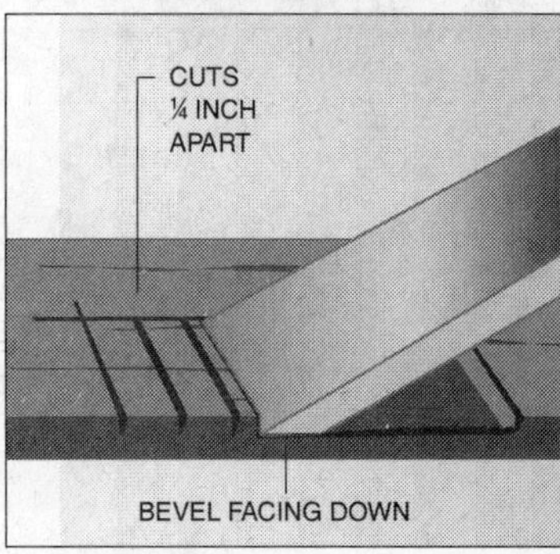

Figure 35 ◆ Proper use of a wood chisel.

16.2.0 Punches

A punch (see *Figure 36*) is used to indent metal (from the impact of a hammer) before you drill a hole, to drive pins, and to align holes in two parts that are mates. Punches are made of hardened and tempered steel. They come in various sizes.

Three common types of punches are the center punch, the prick punch, and the straight punch. The center and prick punches are used to make small locating points for drilling holes. The straight punch is used to punch holes in thin sheets of metal.

Instructor's Notes:

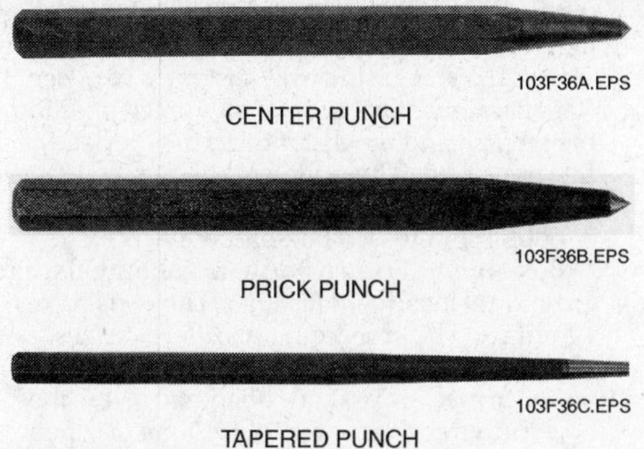

Figure 36 ◆ Punches.

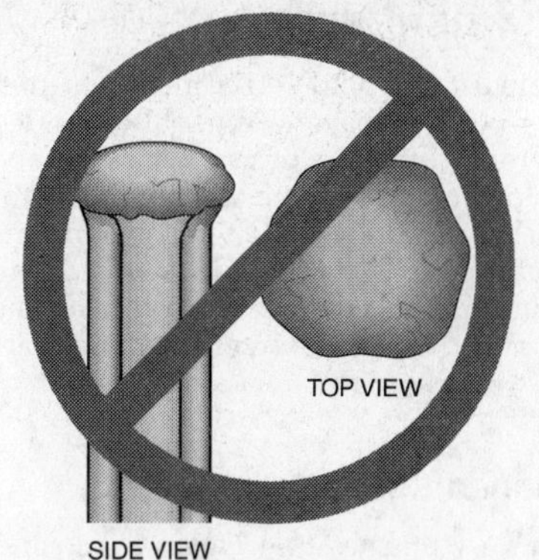

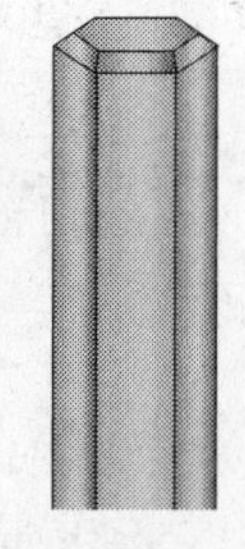

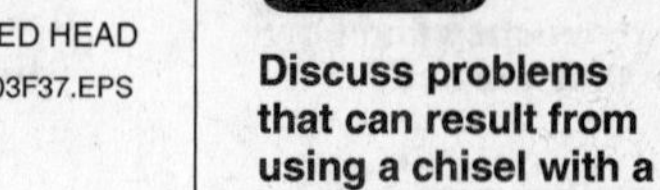

Figure 37 ◆ Chisel damage.

16.3.0 Safety and Maintenance

Here are the guidelines to remember when you're working with punches and chisels:

- Always wear safety goggles.
- Make sure the wood chisel blade is beveled at a precise 25-degree angle so it will cut well.
- Make sure the cold chisel blade is beveled at a 60-degree angle so it will cut well.
- Sharpen the cutting edge of a chisel on an oilstone to produce a keen edge.
- Don't use a chisel head or hammer that has become mushroomed or flattened (see *Figure 37*).
- Don't use a cold chisel to cut or split stone or concrete.

WARNING!

Striking a chisel that has a mushroom-shaped head can cause metal chips to break off. These flying chips can cause serious injury. If a chisel has a mushroom-shaped head, it is damaged. Replace the chisel or have a qualified person repair it.

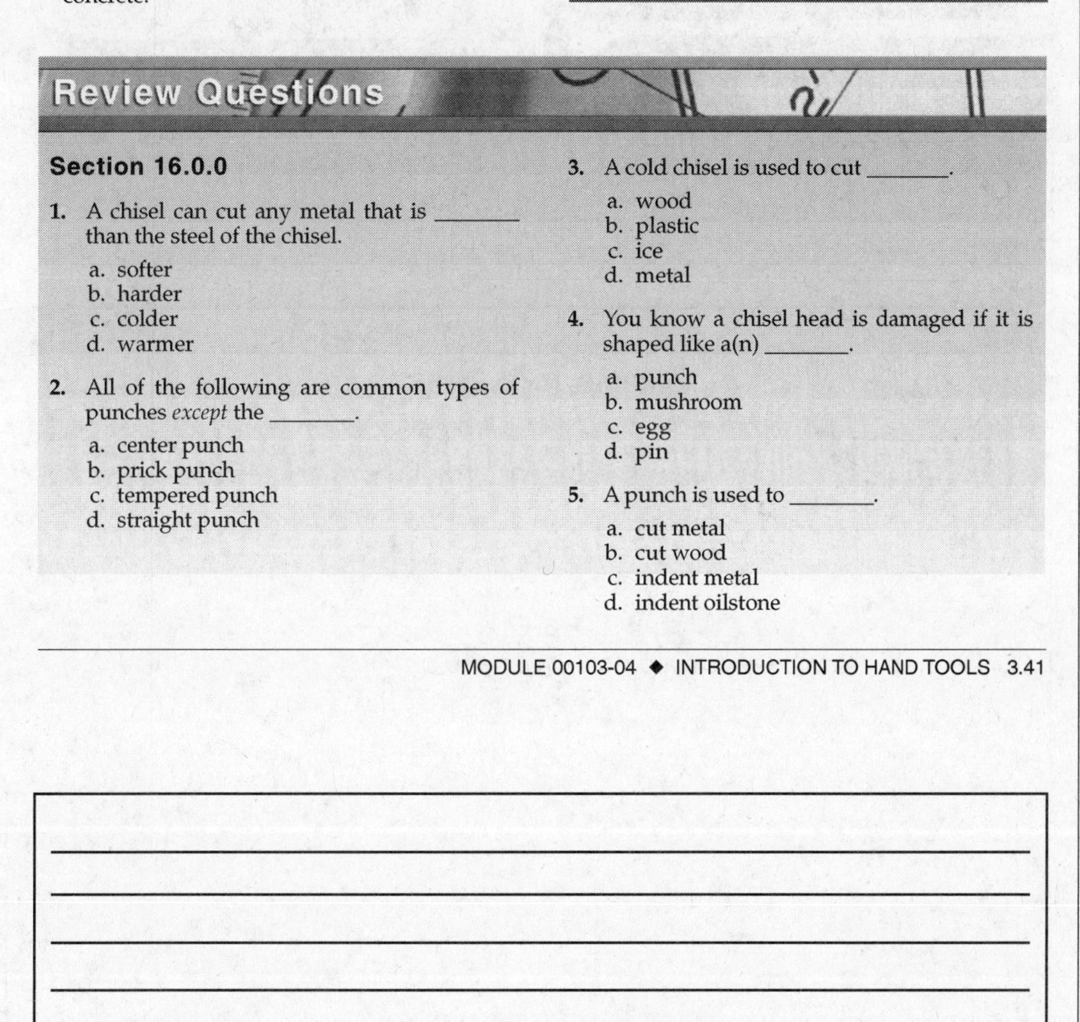

Review Questions

Section 16.0.0

1. A chisel can cut any metal that is _______ than the steel of the chisel.
   a. softer
   b. harder
   c. colder
   d. warmer

2. All of the following are common types of punches *except* the _______.
   a. center punch
   b. prick punch
   c. tempered punch
   d. straight punch

3. A cold chisel is used to cut _______.
   a. wood
   b. plastic
   c. ice
   d. metal

4. You know a chisel head is damaged if it is shaped like a(n) _______.
   a. punch
   b. mushroom
   c. egg
   d. pin

5. A punch is used to _______.
   a. cut metal
   b. cut wood
   c. indent metal
   d. indent oilstone

Review the guidelines for working with punches and chisels. Explain how to verify whether a chisel's blade is beveled at the correct angle.

Discuss problems that can result from using a chisel with a mushroom-shaped head.

Go over the Review Questions for Section 16.0.0. Go over any questions trainees may have.

See the Teaching Tip for Sections 10.0.0–16.3.0 at the end of this module.

Have each trainee perform Performance Tasks 1 and 2 to your satisfaction. Fill out a Performance Profile Sheet for each trainee.

MODULE 00103-04 ◆ INTRODUCTION TO HAND TOOLS 3.41

Show Transparency 16 (Figure 38). Review the different types of nonadjustable wrenches.

Compare the types of wrenches and how they are used.

Show Transparency 17 (Figure 39). Explain when it is appropriate to use a striking wrench and how to avoid damaging screw threads and bolt heads.

17.0.0 ◆ WRENCHES

Wrenches are used to hold and turn screws, nuts, bolts, and pipes that have hexagonal (six-sided) heads. There are many types of wrenches, but they fall into two main categories: nonadjustable and adjustable. Nonadjustable wrenches fit only one size nut or bolt. They come in both standard (English) and metric sizes. Adjustable wrenches can be expanded to fit different-sized nuts and bolts.

17.1.0 Nonadjustable Wrenches

Nonadjustable wrenches (see *Figure 38*) include the **open-end wrench,** the **box-end wrench,** the **Allen** (or hex key) **wrench,** the **striking** (or **slugging**) **wrench,** and the **combination wrench.**

The open-end wrench is one of the easiest nonadjustable wrenches to use. It has an opening at each end that determines the size of the wrench. Often, one wrench has two different-sized openings, such as $\frac{7}{16}$ inch and $\frac{1}{2}$ inch, one on each end. These sizes measure the distance between the **flats** (straight sides or jaws of wrench opening) of the wrench and the distance across the head of the **fastener** used. The open end allows you to slide the tool around the fastener when there is not enough room to fit a box-end wrench.

Box-end wrenches form a continuous circle around the head of a fastener. The ends have 6 or 12 **points.** The ends come in different sizes, ranging from $\frac{3}{8}$ inch to $\frac{15}{16}$ inch. Box-end wrenches offer a firmer grip than open-end wrenches. A box-end wrench is safer to use than an open-end wrench because it will not slip off the sides of certain kinds of bolts. The handles of box-end wrenches are available in a range of lengths.

Striking or slugging wrenches (see *Figure 39*) are similar to box-end wrenches in that they have an enclosed circular opening designed to lock on to the fastener when the wrench is struck. The

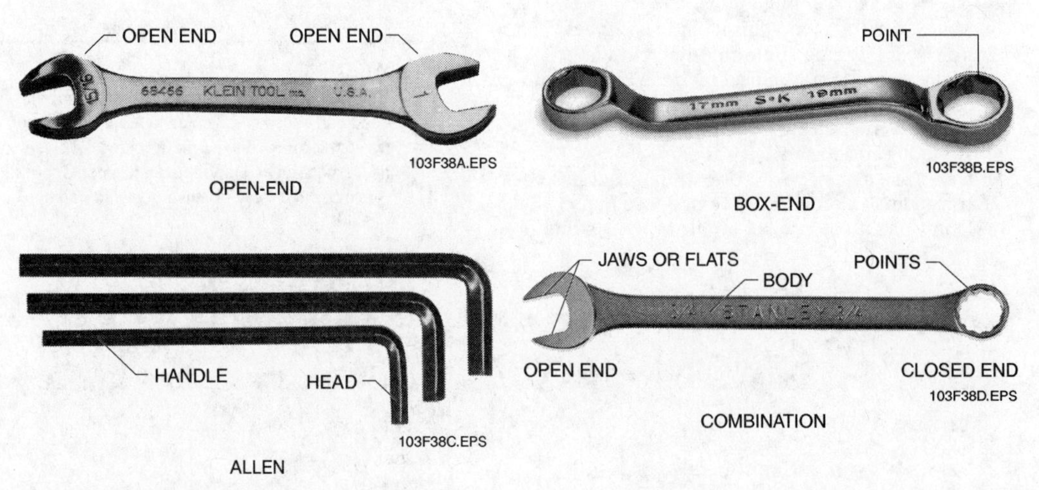

Figure 38 ◆ Nonadjustable wrenches.

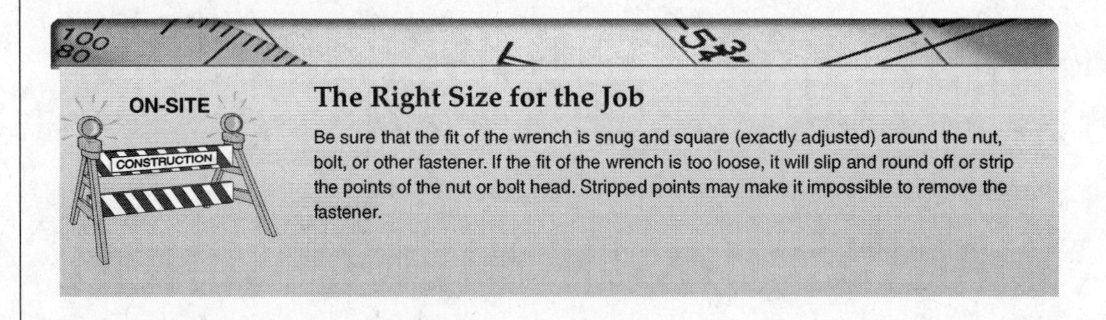

The Right Size for the Job

Be sure that the fit of the wrench is snug and square (exactly adjusted) around the nut, bolt, or other fastener. If the fit of the wrench is too loose, it will slip and round off or strip the points of the nut or bolt head. Stripped points may make it impossible to remove the fastener.

Instructor's Notes:

3.42 CORE CURRICULUM ◆ INTRODUCTORY CRAFT SKILLS

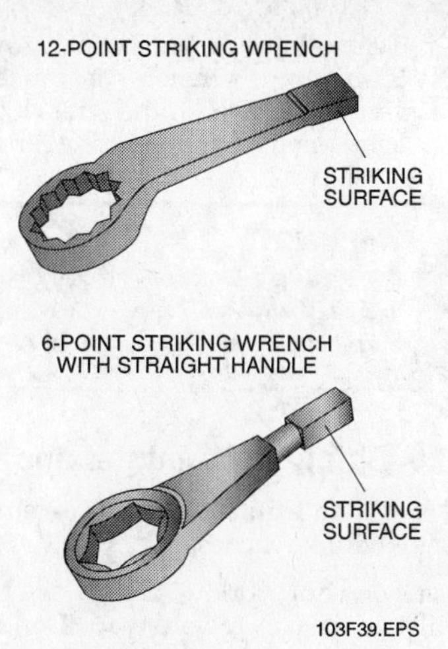

Figure 39 ◆ Striking wrenches.

wrenches have a large striking surface so you can hit them more accurately, usually with a mallet or handheld sledgehammer. The ends have 6 or 12 points. Striking wrenches are used only in certain situations, such as when a bolt has become stuck to another material through rust or corrosion. Striking wrenches can damage screw threads and bolt heads. If you are ever in doubt about whether or not to use a striking wrench, ask your instructor or immediate supervisor.

Allen or hex key wrenches are L-shaped, hexagonal (six-sided) steel bars. Both ends fit the socket of a screw or bolt. The shorter length of the L-shape is called the head, and the longer length is the handle. These wrenches generally have a 1⁄16-inch to 3⁄4-inch diameter. You might use them with setscrews. Setscrews are used in tools and machinery to set two parts tightly together so they don't move from the set position.

Combination wrenches are, as the name implies, a combination of two types of wrenches. One end of the combination wrench is open and the other is closed, or box-end. Combination wrenches can speed up your work because you don't have to keep changing wrenches.

17.1.1 How to Use a Nonadjustable Wrench

Follow these steps when using a nonadjustable wrench:

Step 1 Always use the correct size wrench for the nut or bolt.

Step 2 Pull the wrench toward you. Pushing the wrench can cause injury.

17.2.0 Adjustable Wrenches

Adjustable wrenches are used to tighten or remove nuts and bolts and all types and sizes of pipes. They have one fixed jaw and one movable jaw. The adjusting nut on the wrench joins the teeth in the body of the wrench and moves the adjustable jaw. These wrenches come in lengths from 4 to 24 inches and open as wide as 27⁄16 inches. Common types of adjustable wrenches include **pipe wrenches, spud wrenches,** and **crescent wrenches** (see *Figure 40*). Using an adjustable wrench may save time when you're working with different sizes of nuts and bolts.

Pipe wrenches (often called monkey wrenches) are used to tighten and loosen all types and sizes of threaded pipe. You adjust the upper jaw of the wrench by turning the adjusting nut (see

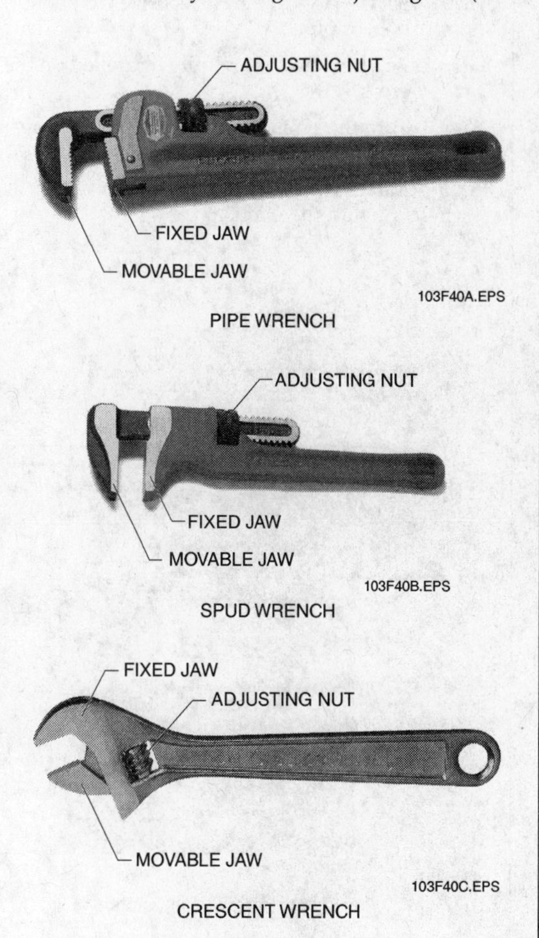

Figure 40 ◆ Adjustable wrenches.

Demonstrate how to select the correct size wrench according to the nut or bolt size.

Show Transparency 18 (Figure 40). Explain how pipe, spud, and crescent wrenches can be adjusted.

MODULE 00103-04 ◆ INTRODUCTION TO HAND TOOLS 3.43

Set up stations with a variety of adjustable wrenches. Ask trainees to practice selecting and using adjustable wrenches.

Explain the types of problems that can occur if the jaws are improperly adjusted.

Discuss the procedures for maintaining and safely using wrenches.

Figure 41). Both jaws have serrated teeth for gripping power. The jaw is spring-loaded and slightly angled so you can release the grip and reposition the wrench without having to readjust the jaw.

Spud wrenches loosen and tighten fittings on drain traps, sink strainers, toilet connections, and large, odd-shaped nuts. Spud wrenches have narrow jaws to fit into tight places.

Crescent wrenches are smooth-jawed for turning nuts, bolts, small pipe fittings, and chrome-plated pipe fittings.

17.2.1 How to Use an Adjustable Wrench

To use an adjustable wrench properly, follow these steps:

Step 1 Set the jaws to the correct size for the nut, bolt, or pipe.

Step 2 Be sure the jaws are fully tightened on the work.

Step 3 Turn the wrench so you are putting pressure on the fixed jaw (see *Figure 41*).

Step 4 Make sure there is room for your fingers as you turn the wrench.

Step 5 Generally, pull the wrench toward you. Pushing the wrench can cause injury. If you must push on the wrench, keep your hand open to avoid getting pinched.

WARNING!

If the jaws are improperly adjusted, you could be injured. The wrench could slip, causing you to hurt your hand or lose your balance.

17.3.0 Safety and Maintenance

Here are some guidelines to remember when working with wrenches:

- Focus on your work.
- Pull the wrench toward you. Don't push the wrench because that can cause injury.
- Keep adjustable wrenches clean. Don't allow mud or grease to clog the adjusting screw and slide; oil these parts frequently.
- Don't use the wrench as a hammer.
- Don't use any wrench beyond its capacity. For example, never add an extension to increase its leverage. This could cause serious injury.

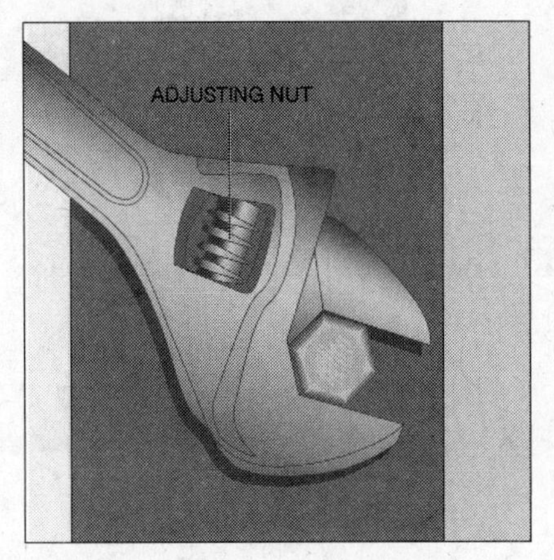

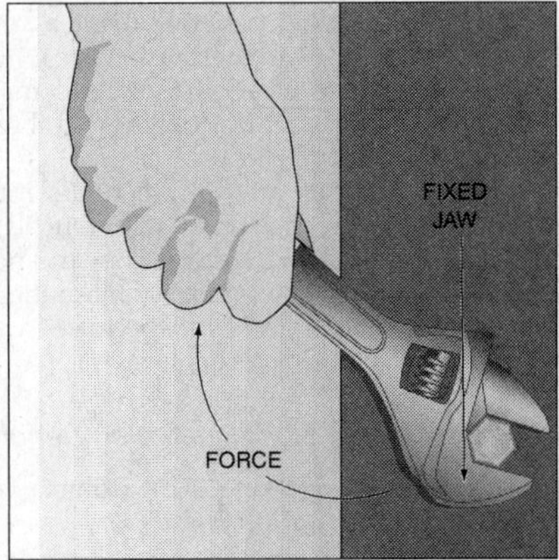

103F41.EPS

Figure 41 ◆ Proper use of an adjustable wrench.

Instructor's Notes:

3.44 CORE CURRICULUM ◆ INTRODUCTORY CRAFT SKILLS

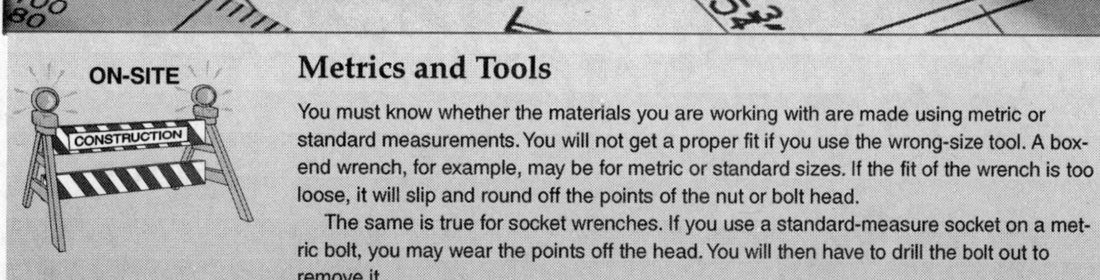

Metrics and Tools

You must know whether the materials you are working with are made using metric or standard measurements. You will not get a proper fit if you use the wrong-size tool. A box-end wrench, for example, may be for metric or standard sizes. If the fit of the wrench is too loose, it will slip and round off the points of the nut or bolt head.

The same is true for socket wrenches. If you use a standard-measure socket on a metric bolt, you may wear the points off the head. You will then have to drill the bolt out to remove it.

Go over the Review Questions for Section 17.0.0. Answer any questions trainees may have.

Have trainees review Sections 18.0.0–24.2.0 for the next session.

Review Questions

Section 17.0.0

1. Wrenches are used to _______.
   a. pound nails into wood
   b. turn screws, nuts, bolts, and pipes
   c. pry open fittings on drains
   d. strip threaded pipe

2. An adjustable wrench _______.
   a. can be used for many purposes, such as hammering or prying
   b. automatically adjusts its jaws
   c. fits only one size of screw, nut, bolt, and pipe
   d. can be used on different sizes of screws, nuts, bolts, and pipes

3. A nonadjustable wrench _______.
   a. fits only one size nut or bolt
   b. can be used on all sizes of screws, nuts, bolts, and pipes
   c. can be used only for hammering
   d. must have its jaws adjusted by hand each time it is used

4. One of the easiest nonadjustable wrenches to use is the _______ wrench.
   a. open-end
   b. box-end
   c. Allen
   d. pipe

5. Using an adjustable wrench can save time when you are working with _______.
   a. nuts and bolts that are all the same size
   b. stripped heads
   c. different sizes of nuts and bolts
   d. nails and plywood

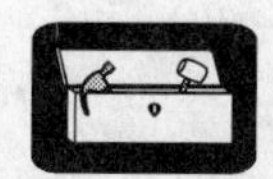

Ensure that you have everything required for demonstrations, laboratories, and testing during this session.

Refer to Figure 42. Discuss the role of sockets and ratchets.

Refer to Figure 43. Identify the lever on a ratchet handle, and explain how it can be used to change the turning direction.

18.0.0 ◆ SOCKETS AND RATCHETS

Socket wrench sets include different combinations of sockets (the part that grips the nut or bolt) and ratchets (handles) that are used to turn the sockets.

Most sockets (see *Figure 42*) have 6 or 12 gripping points. The end of the socket that fits into the handle is square. Sockets also come in different lengths. The long socket is called a deep socket. It is used when normal sockets will not reach down over the end of the bolt to grip the nut.

Socket sets contain different types of handles for different uses. The ratchet handle (see *Figure 43*) has a small lever that you can use to change the turning direction.

MODULE 00103-04 ◆ INTRODUCTION TO HAND TOOLS 3.45

Review the procedures to use sockets and ratchets.

Go over the Review Questions for Section 18.0.0. Answer any questions trainees may have.

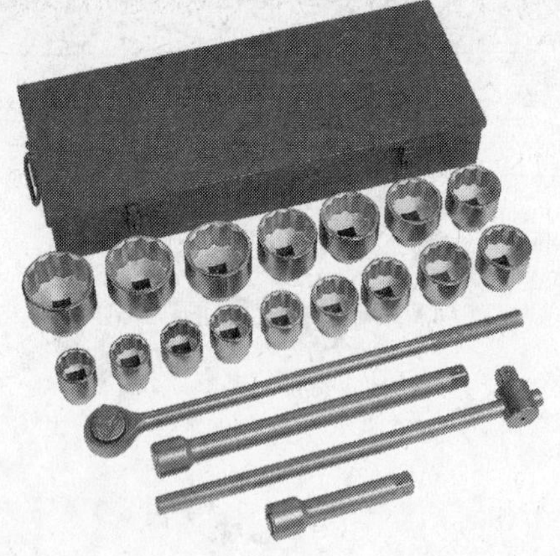

103F42.EPS

Figure 42 ◆ Sockets.

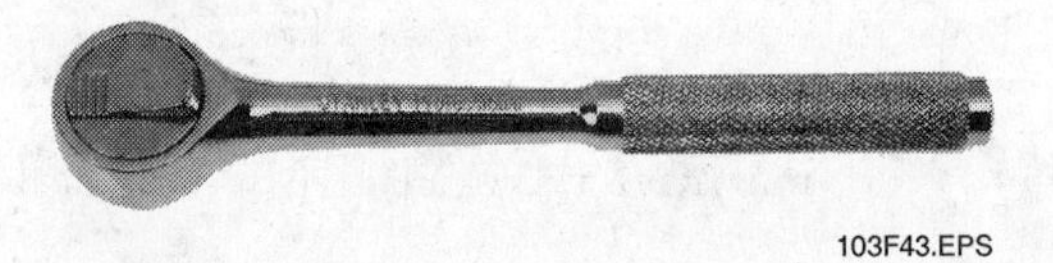

103F43.EPS

Figure 43 ◆ Ratchet handle.

18.1.0 How to Use Sockets and Ratchets

Follow these steps to use sockets and ratchets properly:

Step 1 Select a socket that fits the fastener (such as a nut or bolt) you want to tighten or loosen.

Step 2 Place the square end of the socket over the spring-loaded button on the ratchet shaft.

Step 3 Place the socket over the nut or bolt.

Step 4 Pull on the handle in one direction to turn the nut. (Moving the handle in the other direction has no effect.) To reverse the direction of the socket, use the adjustable lock mechanism.

18.2.0 Safety and Maintenance

Follow these guidelines to maintain your sockets and ratchets in good working order:

- Never force the ratchet handle beyond hand-tight. This could break the head off the fastener.
- Don't use a cheater pipe (a longer piece of pipe slipped over the ratchet handle to provide more leverage). This could snap the tool or break the head off the bolt or nut.

Review Questions

Section 18.0.0

1. A socket is used to _______.

a. turn the ratchet
b. grip a nut or bolt
c. fasten the ratchet
d. adjust the lock

2. A ratchet is used to _______.

a. grip a nut or bolt
b. fasten the socket
c. turn the socket
d. adjust the lock

3. Most sockets have either 6 or 12 _______.

a. ratchets
b. locks
c. handles
d. gripping points

4. When turning the ratchet, _______.

a. stop when it is hand-tight
b. force it just a little past hand-tight
c. use a wrench to make sure it's very tight
d. stop just short of making it hand-tight

5. You should never use a(n) _______ with a ratchet handle.

a. machine-oiled cloth
b. oilstone
c. locking bar
d. cheater pipe

Instructor's Notes:

19.0.0 ◆ TORQUE WRENCHES

Torque wrenches (see *Figure 44*) measure resistance to turning. You need them when you are installing fasteners that must be tightened in sequence without distorting the workpiece. You will use a torque wrench only when a torque setting is specified for a particular bolt.

Torque specifications are usually stated in **inch-pounds** for small fasteners or **foot-pounds** for large fasteners.

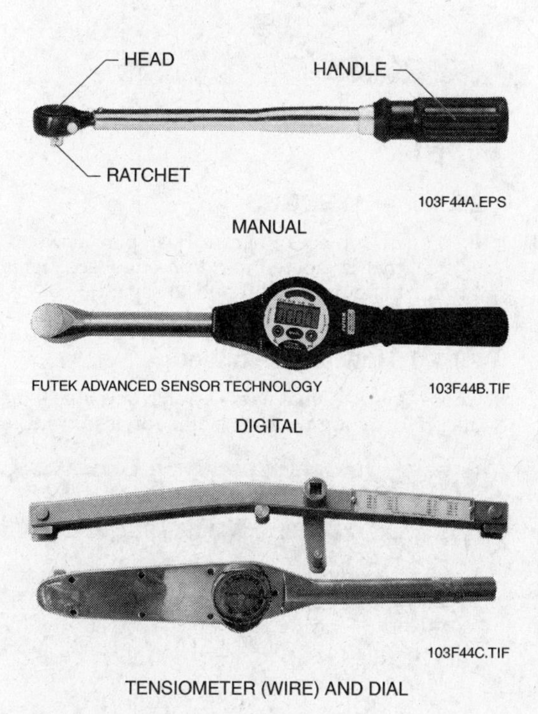

Figure 44 ◆ Torque wrenches.

19.1.0 How to Use a Torque Wrench

Take these steps to use a torque wrench properly:

Step 1 Look on the tool to find out how many inch-pounds or foot-pounds you need to torque to. Set the controls on the wrench to the desired torque level (wrench models vary).

Step 2 Find out the torque sequence (which fastener comes first, second, and so on). If you use the wrong sequence, you could damage what you are fastening

Step 3 Place the torque wrench on the object to be fastened, such as a bolt. Hold the head of the wrench with one hand to support the bolt and to make sure it is properly aligned.

Step 4 Watch the torque indicator or listen for the click (depending on the model of the wrench) as you tighten the bolt (see *Figure 45*).

19.2.0 How to Calculate Torque When Using an Adaptor

When you use an adaptor or extension, the torque wrench becomes longer. The extra length and the applied torque will be greater than the torque indicator. Use this formula to determine the correct torque.

$$\text{Preset torque} = \frac{\text{Length of torque wrench} \times \text{desired torque}}{\text{Length of torque wrench} + \text{length of extension}}$$

To determine the length of the torque wrench, measure the distance from the center of the square drive of the wrench to the center of the handle. To determine the length of the extension, measure the distance from the center of the square drive of the extension to the center of the bolt or nut. Be sure to measure only the length that is parallel to the handle.

Refer to Figure 44. Explain how torque wrenches measure resistance to turning.

Demonstrate how to determine and set the desired number of inch-pounds or foot-pounds of torque.

Have trainees practice determining torque sequence and tightening bolts using a torque wrench.

On the whiteboard/chalkboard, write the formula to determine the correct torque. Demonstrate how to calculate torque when using an adaptor.

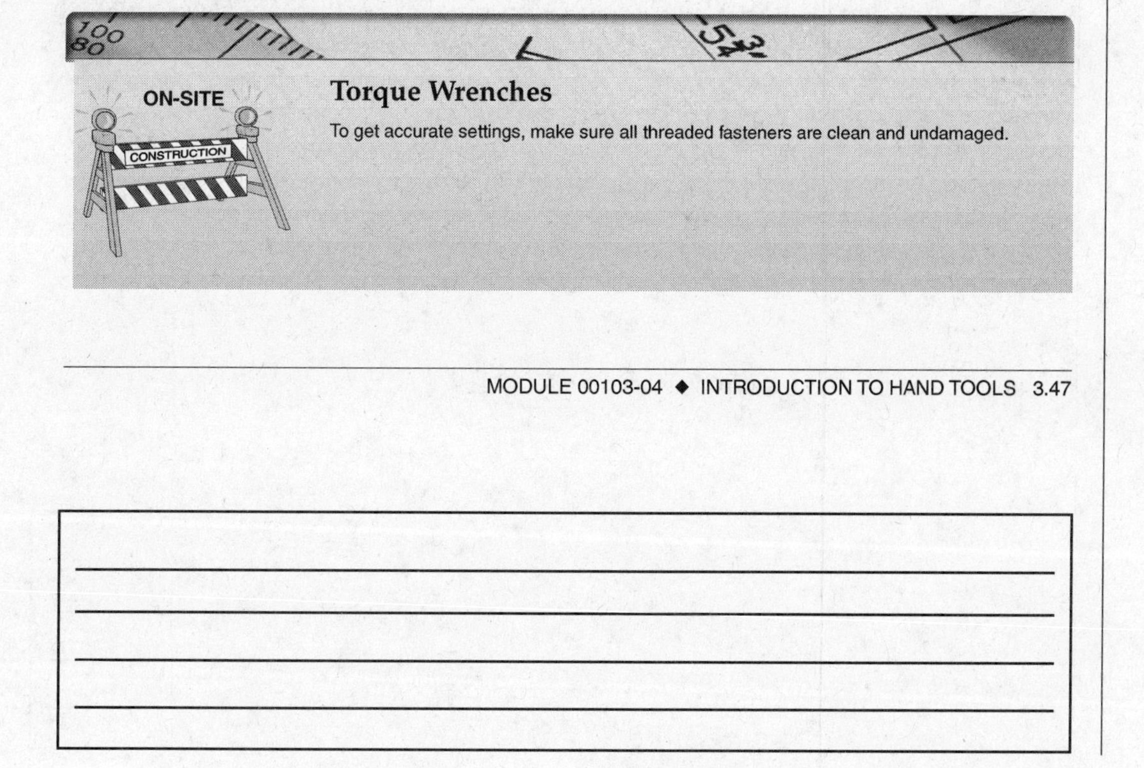

Torque Wrenches

To get accurate settings, make sure all threaded fasteners are clean and undamaged.

MODULE 00103-04 ◆ INTRODUCTION TO HAND TOOLS 3.47

Discuss the importance of following the manufacturer's recommendations when using a torque wrench.

Go over the Review Questions for Section 19.0.0. Go over any questions trainees may have.

Show Transparency 19 (Figure 47). Explain how wedges can be used to lift and separate objects.

Bring in a variety of wedges. Have trainees practice using the wedges to lift and separate a variety of objects in the classroom.

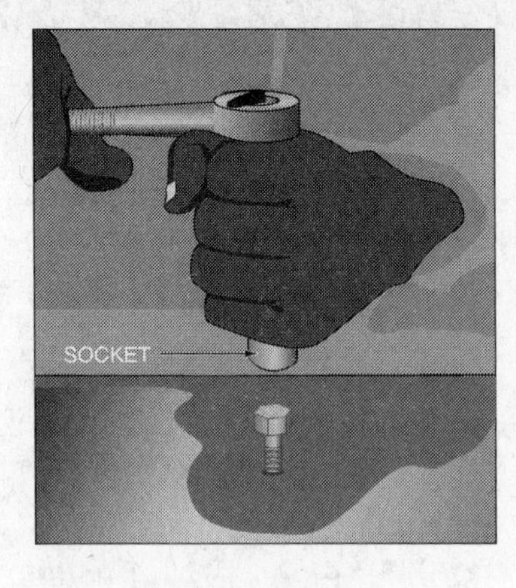

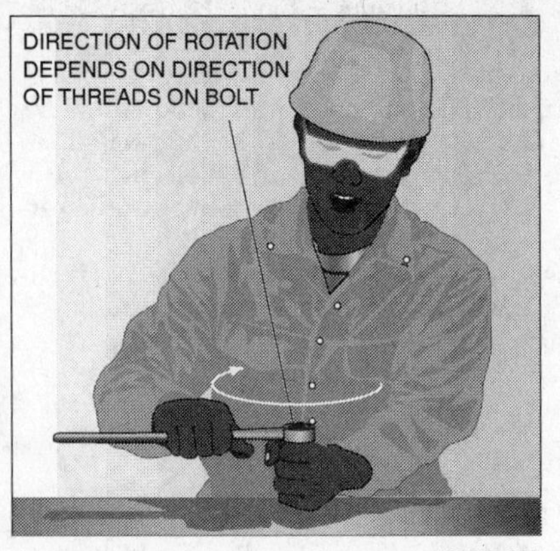

103F45.EPS

Figure 45 ◆ Proper use of a torque wrench.

19.3.0 Safety and Maintenance

A torque wrench can cause property damage and injury if used incorrectly. Follow these guidelines when using your torque wrench:

- Always follow the manufacturer's recommendations for safety, maintenance, and calibration.
- Always store the wrench in its case.
- Never use the wrench as a ratchet or as anything other than its intended purpose.

20.0.0 ◆ WEDGES

A wedge is a piece of hard rubber, plastic, wood, or steel that is tapered to a thin edge (see *Figure 46*). It can be used to lift and to separate objects, among other uses.

20.1.0 How to Use a Wedge

Choose a wedge that won't scratch or damage the material you are working with. You also want to

Review Questions

Section 19.0.0

1. Torque wrenches measure resistance to _______.
   a. accuracy
   b. alignment
   c. turning
   d. plumb

2. Torque specifications for small fasteners are usually specified in _______.
   a. foot-pounds
   b. inch-pounds
   c. wrench widths
   d. threads

3. Torque specifications for large fasteners are usually specified in _______.
   a. foot-pounds
   b. inch-pounds
   c. wrench widths
   d. threads

4. Torque wrenches are used when you need to install _______ that must be tightened in sequence.
   a. torques
   b. clean threads
   c. ratchets
   d. fasteners

5. You should place one hand on the _______ to support the bolt and to make sure it is properly aligned.
   a. vise bench
   b. head of the wrench
   c. plumb bob
   d. head of the bolt

Instructor's Notes:

3.48 CORE CURRICULUM ◆ INTRODUCTORY CRAFT SKILLS

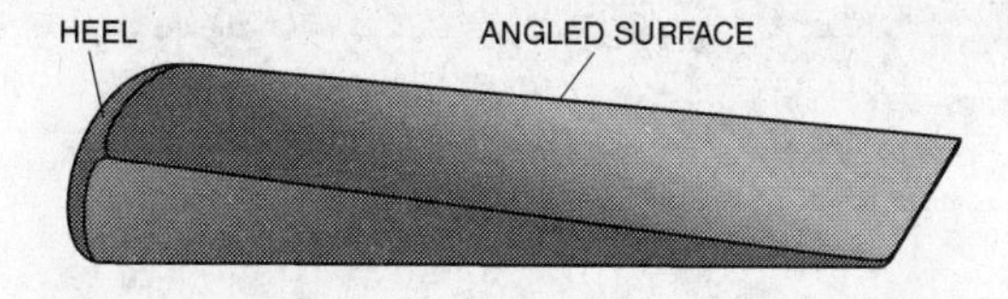

Figure 46 ◆ Wedge.

choose one of the proper size so it will lift or separate the material only as far as you need to. Follow these steps to use a wedge properly (see *Figure 47*).

Follow these steps to properly lift an object using a wedge:

Step 1 Place the wedge at the edge of the object.

Step 2 Check to be sure that the object is well supported.

Step 3 Strike the heel of the wedge with a hammer.

Follow these steps to separate two objects using a wedge:

Step 1 Place the wedge between them.

Step 2 Strike the heel of the wedge with a hammer to force the two objects apart.

20.2.0 Safety and Maintenance

A wedge can be dangerous if used without precautions. Remember these guidelines whenever you use a wedge:

- Wear appropriate personal protective equipment, including safety glasses and a face shield, when using a wedge because a piece of it could fly off.
- Keep your hands away from the heel of the wedge when you are striking it.

Figure 47 ◆ Proper use of a wedge.

Remind trainees to keep their hands away from the heel of the wedge.

MODULE 00103-04 ◆ INTRODUCTION TO HAND TOOLS 3.49

Go over the Review Questions for Section 20.0.0. Ask whether trainees have any questions.

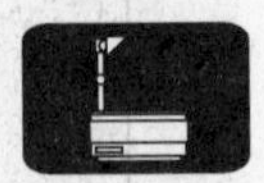

Show Transparency 20 (Figure 48). Discuss the uses of a utility knife.

Demonstrate how to safely make a straight cut with a utility knife.

Emphasize that utility knives should never be used on live wires. Review basic safety procedures.

Review Questions

Section 20.0.0

1. A wedge can be used to lift and _______ objects.
   a. weld
   b. separate
   c. taper
   d. fasten
2. A wedge can be made of _______, plastic, wood, or steel that is tapered to a thin edge.
   a. hard rubber
   b. aluminum
   c. iron
   d. stone
3. When using a wedge, always _______.
   a. wear eye protection
   b. keep your hand close to the heel of the wedge
   c. oil the wedge lightly
   d. sharpen the angled edge first
4. To separate two objects, strike the _______ of the wedge with a hammer.
   a. thin edge
   b. rubber edge
   c. toe
   d. heel
5. Before using a wedge to lift an object, check to make sure the object is _______.
   a. made of the same material as the wedge
   b. well supported
   c. beveled
   d. plumb

21.0.0 ◆ UTILITY KNIVES

A utility knife is used for a variety of purposes including cutting roofing felt, fiberglass or asphalt shingles, vinyl or linoleum floor tiles, fiberboard, and gypsum board. You can also use it for trimming insulation.

The utility knife has a replaceable razor-like blade. It has a handle about 6 inches long, made of cast iron or plastic, to hold the blade. The handle is made in two halves, held together with a screw (see *Figure 48*).

With many utility knives, you can lock the blade in the handle in one, two, or three positions, depending on the type of knife. Models that have a retractable blade (the blade pulls into the handle when not in use) are the safest. Some models have different blades for cutting different materials.

103F48.EPS

Figure 48 ◆ Utility knife.

21.1.0 How to Use a Utility Knife

Follow these steps to use a utility knife properly:

Step 1 Unlock the knife blade. Push the blade out.

Step 2 Lock the blade in the open position.

Step 3 Place some scrap, such as a piece of wood, under the object you are cutting. This will protect the surface under the object.

Step 4 Use the sharp side of the blade (the longer side) to cut straight lines.

Step 5 As soon as you have finished cutting, unlock the blade, pull it back to the closed position, and lock the blade.

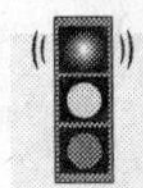

WARNING!
Never use a utility knife on live electrical wires. You could be electrocuted.

21.2.0 Safety and Maintenance

Here are some guidelines to remember about your utility knife:

- Don't bother sharpening your utility knife blade (even though you can) because the blades are so inexpensive. It makes more sense to buy new blades.
- Always keep the blade closed and locked when you are not using the knife.

Instructor's Notes:

Review Questions

Section 21.0.0

1. A utility knife is used to cut such materials as roofing felt, vinyl floor tiles, and _______.
   a. plastic pipes
   b. gypsum board
   c. electrical wires
   d. porcelain floor tiles
2. The safest kind of utility knife is one with _______.
   a. a leather sheath
   b. a blue blade
   c. at least three blades
   d. a retractable blade
3. When using a utility knife, place a scrap under the object you are cutting in order to _______.
   a. see the object more clearly
   b. keep the blade sharp
   c. protect the surface under the object
   d. automatically unlock the knife blade
4. Once you have pushed the utility knife blade out, always _______.
   a. lock the blade in the closed position
   b. lock the blade in the open position
   c. scuff the blade
   d. sharpen the blade
5. A dull utility knife blade should be sharpened rather than replaced because of the cost of the blade.
   a. True
   b. False

Go over the Review Questions for Section 21.0.0. Answer any questions trainees may have.

Explain how chain falls and come-alongs are used to move heavy loads.

Refer to Figure 49. Review the parts of a manual chain fall. Ask trainees to identify each and explain how it is used.

22.0.0 ◆ CHAIN FALLS AND COME-ALONGS

Chain falls and come-alongs are used to move heavy loads safely. A chain fall, also called a chain block or chain hoist, is a tackle device fitted with an endless chain used for hoisting heavy loads by hand. It is usually suspended from an overhead track. A come-along is used to move loads horizontally over the ground for short distances.

22.1.0 Chain Falls

The chain fall (see *Figure 49*) has an automatic brake that holds the load after it is lifted. As the load is lifted, a screw forces fiber discs together to keep the load from slipping. The brake pressure increases as the loads get heavier. The brake holds the load until the lowering chain is pulled. Manual chain falls are operated by hand. Electrical chain falls are operated from an electrical control box.

In a chain fall, the suspension hook is a steel hook used to hang the chain fall. It is one size larger than the load hook. The gear box contains the gears that provide lifting power. The hand chain is a continuous chain used to operate the gearbox. The load chain is attached to the load hook and used to lift loads. A safety latch prevents the load from slipping off the load hook, which is attached to the load.

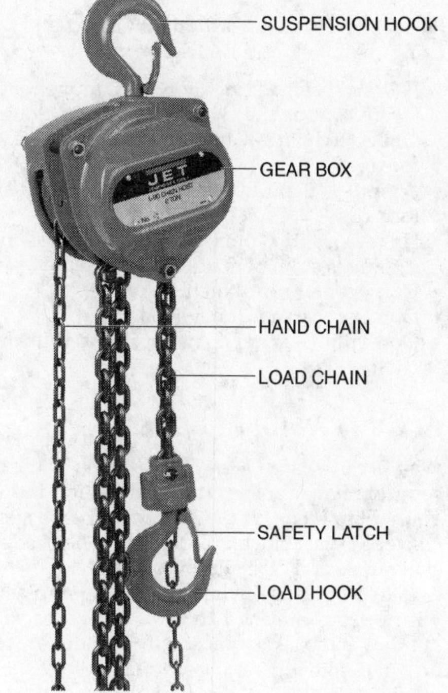

Figure 49 ◆ Parts of a manual chain fall.

MODULE 00103-04 ◆ INTRODUCTION TO HAND TOOLS 3.51

Refer to Figure 50. Explain how to use a ratchet handle to move a load using a come-along.

Discuss the consequences of using a come-along for vertical overhead lifting.

Discuss the guidelines for maintaining and safely using chain falls and come-alongs. Emphasize that a qualified person must ensure that the support rigging can handle the load.

Go over the Review Questions for Section 22.0.0. Answer any questions trainees may have.

Show Transparency 21 (Figure 51). Explain how wire brushes are used to clean tools and metal hardware.

22.2.0 Come-Alongs

Come-alongs, also called cable pullers, use a ratchet handle to position loads or to move heavy loads horizontally over short distances (see *Figure 50*). They can support loads from 1 to 6 tons. Some come-alongs use a chain for moving their loads; others use wire ropes.

When using a chain come-along, you can use the ratchet handle to take up the chain and the ratchet release to allow the chain to be pulled out. You can also use the fast-wind handle to take up or let out slack in the chain without using the ratchet handle.

CAUTION

Never use a come-along for vertical overhead lifting. Use a come-along only to move loads horizontally over the ground for short distances. Come-alongs are not equipped with the safety features, such as a pawl to check the ratchet's motion, that would ensure the safety of anyone underneath a load.

22.3.0 Safety and Maintenance

Here are the guidelines for maintaining and safely using chain falls and come-alongs:

- Follow the manufacturer's recommendations for lubricating the chain fall or come-along.
- Inspect a chain fall and come-along for wear before each use.
- Try out a chain fall or come-along on a small load first.
- Have a qualified person ensure that the support rigging is strong enough to handle the load.
- Don't get lubricant on the clutches.
- Don't ever stand under a load.
- Don't put your hands near pinch-points on the chain.

23.0.0 ◆ WIRE BRUSHES

Wire brushes (see *Figure 51*) are some of the most common hand tools in the construction industry. You will find one on practically every job site. All craft areas use wire brushes to clean objects, especially tools and other hardware made of metal.

Like all brushes, wire brushes are implements with bristles attached to a handle or back. The handles and backs of most wire brushes are made of wood, although some are made of plastic and other materials. The wire that makes up the brush itself is composed of many individual filaments or slender rods of drawn metal. Wire brushes have different types of bristles for cleaning different types of metals.

For example, carpenters use wood-handled brushes with stainless steel bristles to clean rusty tools and to remove paint. Pipefitters and plumbers use wire brushes to clean tools and also to clean welds on soldered pipe. Many wire brushes have brass bristles for use on especially heavy-duty jobs. Other types of wire brushes are used to clean wire rope and chains used for rigging operations.

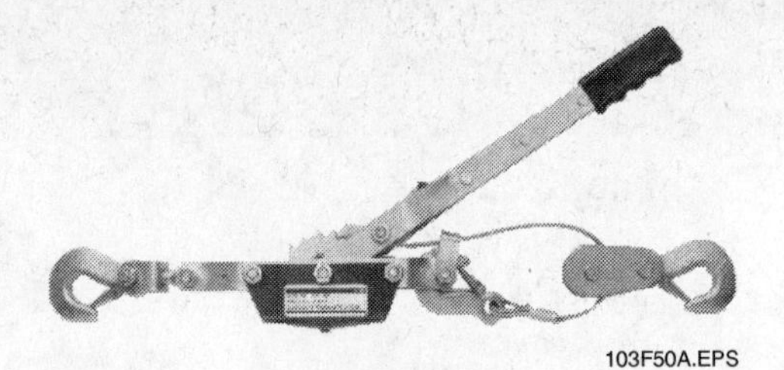

103F50A.EPS

CABLE COME-ALONG

103F50B.EPS

CHAIN COME-ALONG

Figure 50 ◆ Come-alongs.

103F51.EPS

Figure 51 ◆ Wire brush.

Instructor's Notes:

3.52 CORE CURRICULUM ◆ INTRODUCTORY CRAFT SKILLS

Review Questions

Section 22.0.0

1. Chain falls are used to _______.
   a. transport light loads safely
   b. supplement come-along pulls
   c. rig light loads safely
   d. safely move heavy loads vertically

2. An important feature of the chain fall is the _______.
   a. automatic brake that holds the load after it is lifted
   b. automatic lubricating device
   c. fast-wind handle for lifting or lowering
   d. ratchet handle with automatic release

3. A cable come-along is used to _______.
   a. transport light loads
   b. move heavy loads horizontally over short distances
   c. supplement the work done by a chain fall
   d. hold the load after the chain fall has lifted it

4. You can use the fast-wind handle on a come-along to _______.
   a. perform emergency lifts
   b. transfer the load from the chain fall
   c. take up or let out the chain slack without using the ratchet handle
   d. lift loads weighing more than 6 tons

5. When you lubricate a chain fall or come-along, avoid getting lubricant on the _______.
   a. clutches
   b. chain
   c. ratchet release
   d. jack

Set up stations with a variety of tools and wire brushes. Have trainees select the appropriate brush and practice cleaning the hand tools.

Ask a trainee to explain why wire brushes should not be used for finishing work.

Show Transparency 22 (Figure 52). Review the basic shapes of shovel blades and how each is used.

23.1.0 How to Use a Wire Brush

Use a wire brush only for its intended purposes. If you are ever in doubt about whether or not a job calls for using a wire brush, ask your instructor or immediate supervisor. Wire brushes can damage many materials, especially wood and plastic. Also, because they are used to keep other tools clean, wire brushes themselves can become quite dirty. Be sure to clean a wire brush properly each time you use it or your work may become soiled or damaged. At this point in your training, you will most likely be relied upon to clean up periodically during and especially at the end of each workday. You'll need to know how to make good use of a wire brush.

CAUTION

Do not use a wire brush for finishing work. It will scratch the surface. Some wire brushes cannot be used on stainless steel.

24.0.0 ◆ SHOVELS

Shovels are used by many different construction trades. An electrician running underground wiring may dig a trench. A concrete mason may dig footers for a foundation. A carpenter may clear dirt from an area for concrete form-building. A plumber may dig a ditch to lay pipe. A welder may use a shovel to clean up scrap metal and slag after the job is finished.

There are three basic shapes of shovel blades: round, square, and spade (see *Figure 52*). Use a round-bladed shovel to dig holes or remove large amounts of soil. Use a square-bladed shovel to move gravel or clean up construction debris. Use a spade to move large amounts of soil or dig trenches that need smooth, straight sides.

Shovels can have wooden or fiberglass handles. They generally come in two lengths. A long handle is usually 47 to 48 inches long. A short handle is usually 27 inches long.

MODULE 00103-04 ◆ INTRODUCTION TO HAND TOOLS 3.53

Explain how to select the appropriate shovel for a specific task.

24.1.0 How to Use a Shovel

Follow these steps to shovel properly (see *Figure 53*):

Step 1 Select the type of shovel that is best for the job.

Figure 52 ◆ Shapes of shovel blades.

For a round shovel or spade:

Step 2 Place the tip of the shovel blade or spade at the point where you will begin digging or removing soil.

Step 3 With your foot balanced on the turned step (ridge), press down and cut into the soil with the blade.

For a square shovel:

Step 2 Place the leading edge of the shovel blade against the gravel or construction debris and push until the shovel is loaded.

24.2.0 Safety and Maintenance

Here are some guidelines for working with shovels:

- Always check the handle before using a shovel. There should be no cracks or splits.
- Use appropriate personal protective equipment when digging, trenching, or clearing debris. Wear steel-toed boots to protect your feet from dropped materials.
- Don't let dirt or debris build up on the blade. Always rinse off the shovel blade after using it.

Figure 53 ◆ Proper use of a shovel.

Instructor's Notes:

3.54 CORE CURRICULUM ◆ INTRODUCTORY CRAFT SKILLS

Review Questions

Sections 23.0.0 and 24.0.0

1. Wire brushes are used for all of the following jobs *except* _______.
   a. removing paint
   b. cleaning rusty tools
   c. cleaning welds
   d. leveling concrete

2. _______ is easily damaged with wire brushes.
   a. Wood
   b. Metal
   c. Concrete
   d. Wire rope

3. A round-bladed shovel is used to _______.
   a. move gravel or clean up construction debris
   b. move large amounts of soil or dig trenches that need straight sides
   c. dig holes or remove large amounts of soil
   d. spread gravel on driveways

4. A square-bladed shovel is used to _______.
   a. move gravel or clean up construction debris
   b. dig holes or remove large amounts of soil
   c. cut roots of plantings
   d. dig footers for a foundation

5. A spade is used to _______.
   a. dig holes or remove large amounts of soil
   b. move gravel or clean up construction debris
   c. tamp down soil along a building's foundation
   d. move large amounts of soil or dig trenches with straight sides

Go over the Review Questions for Sections 23.0.0–24.0.0. Ask whether trainees have any questions.

See the Teaching Tip for Sections 17.0.0–24.2.0 at the end of this module.

MODULE 00103-04 ◆ INTRODUCTION TO HAND TOOLS 3.55

Go over the Key Terms Quiz. Answer any questions trainees may have.

Administer the Module Examination. Be sure to record the results of the Exam on Craft Training Report Form 200, and submit the results to the Training Program Sponsor.

Have each trainee perform Performance Task 3 to your satisfaction. Fill out a Performance Profile Sheet for each trainee.

Ensure that all Performance Tests have been completed and Performance Profile Sheets for each trainee are filled out. If desired, trainee proficiency noted during laboratory sessions may be used to complete the Performance Test. Be sure to record the results of the testing on Craft Training Report Form 200, and submit the results to the Training Program Sponsor.

Summary

As a craft professional, your tools are essential to your success. In this module, you learned to identify and work with many of the basic hand tools commonly used in construction. Learning to properly use and maintain your tools is an essential skill for every craftworker. Although you may not work with all of the tools introduced in this module, you will use many of them as you progress in your career, regardless of what craft area you choose to work in.

When you use tools properly, you are working safely and efficiently. You are not only preventing accidents that can cause injuries and equipment damage, you are showing your employer that you are a responsible, safe worker.

The same pride you take in using your tools to do a job well is important when it comes to maintaining your tools. When you maintain your tools properly, they last longer, work better, and function more safely. The simple act of maintaining your tools will help you prevent accidents, make your tools last longer, and help you perform your job better. Taking the time to learn to use and maintain these tools properly now will help keep you safe and save you time and money down the road.

Notes

Instructor's Notes:

Key Terms Quiz

Fill in the blank with the correct key term that you learned from your study of this module.

1. Used mainly for woodworking, the _________ is a fixed, 90-degree angle.
2. A(n) _________ is an L-shaped, hexagonal steel bar.
3. The _________ has a flat face for striking and a rounded face that is used to align brackets and drive out bolts.
4. Usually 2 inches wide and 15 inches long, the _________ has a nail slot at the end to pull nails out from tightly enclosed areas.
5. Shaped like an L, the _________ is used to make sure wall studs and sole plates are at right angles to each other.
6. A(n) _________ is a metal tool with a sharpened, beveled edge that is used to cut and shape wood, stone, or metal.
7. The _________ is used to drive nails and to pull nails out of wood.
8. To _________ is to cut on a slant at an angle that is not a right angle.
9. The _________ has a 12-inch blade that moves through a head that is marked with 45-degree and 90-degree angle measures.
10. If you use a screwdriver incorrectly, you can damage the screwdriver or _________ the screw head.
11. Use a(n) _________ to turn nuts, bolts, small pipe fittings, and chrome-plated pipe fittings.
12. To fasten or align two pieces or material, you can use a(n) _________, which is a pin that fits into a corresponding hole.
13. A(n) _________ is a device such as a nut or bolt used to attach one material to another.
14. Use a(n) _________ for heavy-duty dismantling of woodwork.
15. The straight sides or jaws of a wrench opening are called the ____________.
16. A(n) _________ is a claw hammer with a slightly rounded face.
17. _________ is a unit of measure used to describe the torque needed to tighten a large object.
18. _________ is a unit of measure used to describe the torque needed to tighten a small object.
19. The point at which members or the edges of members are joined is called the ____________.
20. The _________ is the cut or channel made by a saw.
21. Using a(n) _________ can speed up your work because it has an open wrench at one end and a box-end at the other.
22. Use a(n) _________ to determine if a surface is exactly horizontal.
23. You make a(n) _________ by fastening together usually perpendicular parts with the ends cut at an angle.
24. A(n) _________ is a tool used to remove nails.
25. A(n) _________ has an opening at each end that determines its size.
26. To reduce stress in a weld, use a special type of hammer for _________ the joint as it cools.
27. Used for marking, checking, and measuring, a(n) _________ comes in several types: carpenter's, rafter angle, try, and combination.
28. A(n) _________ has serrated teeth on both jaws for gripping power.
29. _________, which is the turning force applied to an object, is measured in inch-pounds or foot-pounds.
30. The _________ is used to pull nails that have been driven flush with the surface of the wood or slightly below it.
31. A box-end wrench has 6 or 12 ____________.
32. The _________, a nonadjustable wrench, forms a continuous circle around the head of a fastener.

MODULE 00103-04 ◆ INTRODUCTION TO HAND TOOLS 3.57

33. To indent metal before you drill a hole, to drive pins, or to align holes in two parts that are mates, use a(n) _________.

34. Also called a speed square or magic square, the _________ is a combination protractor, try square, and framing square.

35. Using pliers on nuts or bolt heads may _________ the edges of the hex head and cause wrenches to no longer fit properly.

36. Use a(n) _________ to loosen and tighten fittings on drain traps, sink strainers, toilet connections, and large, odd-shaped nuts.

37. The _________ has a claw at each end that can be used to pull nails or split wood.

38. Use a(n) _________ when a bolt has become stuck to another material through rust or corrosion.

39. A special type of adjustable wrench, _________ are scissor-shaped tools with jaws.

40. The _________ fits into a wooden file handle.

41. Some tools are made of _________ steel so that they resist wear and do not bend or break.

42. A(n) _________ piece of lumber is one that has had its surface made smooth.

43. If a surface is _________, it is exactly vertical.

44. A(n) _________ is a piece that projects out of wood so it can be placed into a hole or groove to form a joint.

45. Use a(n) _________ to secure an object while you work on it.

46. A(n) _________ is a joint that has been created by heating pieces of metal.

Key Terms

Allen wrench
Ball peen hammer
Bell-faced hammer
Bevel
Box-end wrench
Carpenter's square
Cat's paw
Chisel
Chisel bar
Claw hammer
Combination square
Combination wrench
Crescent wrench
Dowel
Fastener
Flat bar
Flats
Foot-pounds
Inch-pounds
Joint
Kerf
Level
Miter joint
Nail puller
Open-end wrench
Peening
Pipe wrench
Planed
Pliers
Plumb
Points
Punch
Rafter angle square
Ripping bar
Round off
Spud wrench
Square
Striking (or slugging) wrench
Strip
Tang
Tempered
Tenon
Torque
Try square
Vise
Weld

Instructor's Notes:

3.58 CORE CURRICULUM ◆ INTRODUCTORY CRAFT SKILLS

Profile in Success

Jim Evans

Supervisor of Maintenance Training
Calvert Cliffs Nuclear Power Plants
Lusby, Maryland

How did you become interested in the construction industry?

I was born and raised in Baltimore, and I guess I was about 9 years old at the time I first became interested. Our radio broke, and my father threw it out. I dug it out of the trash and got it running again. Ever since then, electrical work has pretty much been my main interest.

Because I expressed so much interest in it as I was growing up, one day my father sat me down with the newspaper and together we looked at the Help Wanted ads. There were a lot of ads for electricians, so it looked like a great field to get into!

I took electronics and shop in high school, and then I went into the Navy and became an electrician's mate. That was my first job in construction. After my discharge, I worked as an electrician at many places, including U.S. Gypsum in Baltimore, Esskay Meats, and the Chessie System railroad.

What drew you to the Calvert Cliffs Nuclear Power Plants?

It was the challenge. The environment and the culture are completely different from any of the other places I've worked. Our number one job is public safety. On that, everything else hinges. For example, you could be working on a motor. When you finish working on it and you walk away from it, you have to know that that motor is going to work exactly when it's supposed to, and work exactly the way it is supposed to. In the nuclear power industry, you can't have a bad day. And I really like that kind of challenge.

What do you think it takes to be a success in your trade?

One word: *learn.* Learn something every day. And when you don't feel like learning any more, then learn some more!

The real challenge of this job is learning how and why things work. I think that it's a fascinating field. It can be an extremely interesting one for you too, but only if you want it to be interesting for you. You really have to enjoy learning to be successful in the electrical industry.

What are some of the things you do in your job?

In maintenance training we try to identify trends before they become a problem, and then train people to perform their tasks more safely. The process begins by identifying what we call a low-level trend. Then we document it, pull data about it, and track it. When we have identified what's behind the trend, we develop the appropriate training to correct the trend.

For example, and this is purely hypothetical, say that we start to see an increased incidence of injuries on the job. We look at the situation and find that people are not using their fall protection properly, because the harnesses are fitted with a new lanyard. So we train everyone how to use the harness properly. After that, the number of fall-related injuries goes back down again.

A constant part of my job is analyzing the performance problems that we're having in the field. We ask ourselves questions such as, "Is the problem that people haven't been trained properly? Is it that

MODULE 00103-04 ◆ INTRODUCTION TO HAND TOOLS 3.59

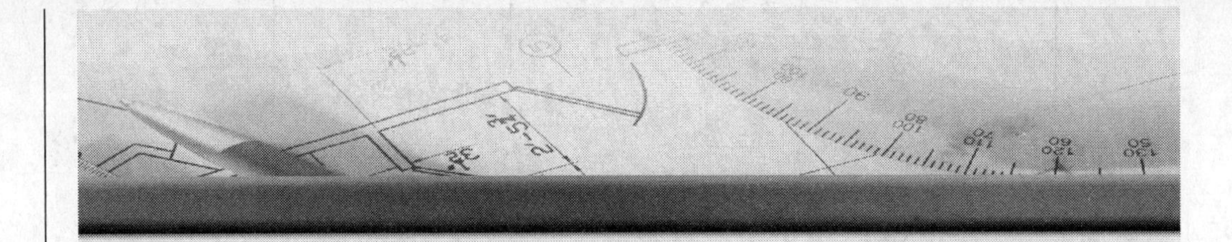

they aren't being rewarded enough? Is it the equipment itself?" It could be a knowledge issue; maybe they never learned how to do it. Or maybe they haven't done the task in two years and they need a refresher. Whatever the question is, we identify it and develop appropriate training in response.

What do you like most about your job?
I learn something new every day. I really do. And I really enjoy working with the people I work with. They are professional and knowledgeable, and I learn from them all the time.

What would you say to someone entering the trades today?
I've got quite a few things to say about that, actually! But basically, it all comes down to this: If you come into the craft willing to learn, and you carry that attitude with you, then there is no end to what you can accomplish in your field. I've worked in a lot of places, and I am where I am today because of my willingness to learn and my attitude. With a good attitude, people want to be around you. And that really makes your learning experience better than it would be otherwise.

Instructor's Notes:

3.60 CORE CURRICULUM ◆ INTRODUCTORY CRAFT SKILLS

Trade Terms Introduced in This Module

Allen wrench: A hexagonal steel bar that is bent to form a right angle. Also called a hex key wrench.

Ball peen hammer: A hammer with a flat face that is used to strike cold chisels and punches. The rounded end—the peen—is used to bend and shape soft metal.

Bell-faced hammer: A claw hammer with a slightly rounded, or convex, face.

Bevel: To cut on a slant at an angle that is not a right angle (90 degrees). The angle or inclination of a line or surface that meets another at any angle but 90 degrees.

Box-end wrench: A wrench, usually double-ended, that has a closed socket that fits over the head of a bolt.

Carpenter's square: A flat, steel square commonly used in carpentry.

Cat's paw: A straight steel rod with a curved claw at one end that is used to pull nails that have been driven flush with the surface of the wood or slightly below it.

Chisel: A metal tool with a sharpened, beveled edge used to cut and shape wood, stone, or metal.

Chisel bar: A tool with a claw at each end, commonly used to pull nails.

Claw hammer: A hammer with a flat striking face. The other end of the head is curved and divided into two claws to remove nails.

Combination square: An adjustable carpenter's tool consisting of a steel rule that slides through an adjustable head.

Combination wrench: A wrench with an open end and a closed end.

Crescent wrench: A smooth-jawed adjustable wrench used for turning nuts, bolts, and pipe fittings.

Dowel: A pin, usually round, that fits into a corresponding hole to fasten or align two pieces.

Fastener: A device such as a bolt, clasp, hook, or lock used to attach or secure one material to another.

Flat bar: A prying tool with a nail slot at the end to pull nails out in tightly enclosed areas. It can also be used as a small pry bar.

Flats: The straight sides or jaws of a wrench opening. Also, the sides on a nut or bolt head.

Foot-pounds: Unit of measure used to describe the amount of pressure exerted (torque) to tighten a large object.

Inch-pounds: Unit of measure used to describe the amount of pressure exerted (torque) to tighten a small object.

Joint: The point where members or the edges of members are joined. The types of welding joints are butt joint, corner joint, and T-joint.

Kerf: A cut or channel made by a saw.

Level: Perfectly horizontal; completely flat; a tool used to determine if an object is level.

Miter joint: A joint made by fastening together usually perpendicular parts with the ends cut at an angle.

Nail puller: A tool used to remove nails.

Open-end wrench: A nonadjustable wrench with an opening at each end that determines the size of the wrench.

Peening: The process of bending, shaping, or cutting material by striking it with a tool.

Pipe wrench: A wrench for gripping and turning a pipe or pipe-shaped object; it tightens when turned in one direction.

Planed: Describing a surface made smooth by using a tool called a plane.

Pliers: A scissor-shaped type of adjustable wrench equipped with jaws and teeth to grip objects.

Plumb: Perfectly vertical; the surface is at a right angle (90 degrees) to the horizon or floor and does not bow out at the top or bottom.

Points: Teeth on the gripping part of a wrench. Also refers to the number of teeth per inch on a handsaw.

Punch: A steel tool used to indent metal.

Rafter angle square: A type of carpenter's square made of cast aluminum that combines a protractor, try square, and framing square.

MODULE 00103-04 ◆ INTRODUCTION TO HAND TOOLS 3.61

Ripping bar: A tool used for heavy-duty dismantling of woodwork, such as tearing apart building frames or concrete forms.

Round off: To smooth out threads or edges on a screw or nut.

Spud wrench: An adjustable wrench used for fittings on drain traps, sink strainers, toilet connections, and odd-shaped nuts.

Square: Exactly adjusted; any piece of material sawed or cut to be rectangular with equal dimensions on all sides; a tool used to check angles.

Striking (or slugging) wrench: A nonadjustable wrench with an enclosed, circular opening designed to lock on to the fastener when the wrench is struck.

Strip: To damage the threads on a nut or bolt.

Tang: Metal handle-end of a file. The tang fits into a wooden or plastic file handle.

Tempered: Treated with heat to create or restore hardness in steel.

Tenon: A piece that projects out of wood or another material for the purpose of being placed into a hole or groove to form a joint.

Torque: The turning or twisting force applied to an object, such as a nut, bolt, or screw, using a socket wrench or screwdriver to tighten it. Torque is measured in inch-pounds or foot-pounds.

Try square: A square whose legs are fixed at a right angle.

Vise: A holding or gripping tool, fixed or portable, used to secure an object while work is performed on it.

Weld: To heat or fuse two or more pieces of metal so that the finished piece is as strong as the original; a welded joint.

Instructor's Notes:

3.62 CORE CURRICULUM ◆ INTRODUCTORY CRAFT SKILLS

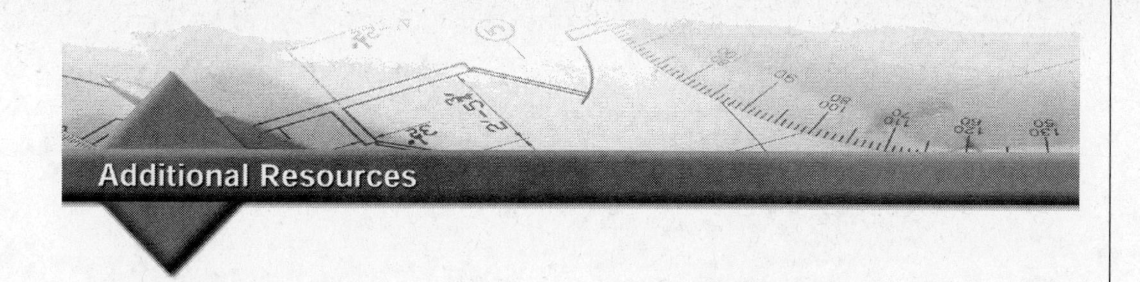

Additional Resources

This module is intended to present thorough resources for task training. The following reference works are suggested for further study. These are optional materials for continued education rather than for task training.

Field Safety, 2003. NCCER. Upper Saddle River, NJ: Prentice Hall.

Hand Tools & Techniques, 1999. Minneapolis, MN: Handyman Club of America.

The Long and Short of It: How to Take Measurements. Video. Charleston, WV: Cambridge Vocational & Technical, 800-468-4227.

Reader's Digest Book of Skills and Tools, 1993. Pleasantville, NY: Reader's Digest.

MODULE 00103-04 ◆ INTRODUCTION TO HAND TOOLS 3.63

MODULE 00103-04 — TEACHING TIPS

The following are suggested activities or instructional methods to help you teach the material in this AIG.

General

When you call on someone to answer a question, the rest of the class relaxes or even tunes out because they expect that the question and answer will take place only between you and the trainee you called on. Instead, use this technique to involve more trainees in answering questions and to keep them on their toes.

1. Ask trainees to define a term or explain a concept.
2. After one trainee has answered, ask a trainee seated nearby if the answer is right. Then ask whether a trainee in the back of the room agrees.
3. Ask trainees to explain why they think an answer is right or wrong.
4. Use the session to clear up incorrect ideas, and encourage trainees to learn from their mistakes.

Sections 1.0.0–6.6.0

Hand Tools, Part One

This exercise will familiarize trainees with the safety and maintenance procedures for a variety of hand tools. You will need hammers, screwdrivers, sledgehammers, ripping bars and nail pullers, and pliers and wire cutters. Trainees will need their personal protective equipment, Trainee Guides, and pencils and paper to take notes for discussion. Allow 30 minutes for this exercise.

1. On the whiteboard/chalkboard, write a column heading for each of the hand tools discussed to this point.
2. Divide the class into small groups, and assign a different set of tools to each group.
3. Have each group properly identify its tools and write a list of safety and maintenance procedures for each.
4. Ask trainees to review each group's list, discuss the procedures, and make any necessary corrections.
5. Go over any questions trainees may have about hand tools discussed to this point.

Sections 7.0.0–9.3.0

Hand Tools, Part Two

This exercise will give trainees the opportunity to practice using a variety of provided measuring tools, levels, squares, plumb bobs, and chalk lines. Trainees will need their personal protective equipment, Trainee Guides, and pencils and paper to take notes for discussion. Allow 30 minutes for this exercise.

1. Set up stations with a variety of square wood frames. Divide the class into small groups according to the number of frames.
2. Using two different measuring devices, have trainees measure the beams to determine whether all four beams are equal.
3. Ask trainees to determine whether the beams are level and/or plumb.
4. Using the square, have trainees check the squareness of the adjoining beams.
5. Answer any questions trainees may have about hand tools discussed to this point.

Sections 10.0.0–16.3.0

Hand Tools, Part Three

This exercise will provide trainees with the opportunity to practice using a variety of hand tools. You will need to provide bench vises and clamps, handsaws, files and rasps, chisels and punches, and wrenches. Trainees will need their personal protective equipment, Trainee Guides, and pencils and paper to take notes for discussion. Allow 30 minutes for this exercise.

1. Set up several work stations for the following hand tools, and assign a small group to each station:
   - Bench vises and clamps
   - Saws
   - Files and rasps
   - Chisels and punches
   - Wrenches
2. Ask each group to visually inspect the tools at its station for damage and demonstrate how to safely use them.
3. After each group demonstrates how to use its tools, have trainees comment on what was done correctly and/or incorrectly.
4. Ask trainees to follow procedures for cleaning, storing, and maintaining the tools at their stations.
5. Ask whether trainees have any questions about hand tools discussed in this module.

Sections 17.0.0–24.2.0

Hand Tools, Part Four

This exercise will allow trainees to practice using a variety of hand tools. You will need to provide sockets and ratchets, torque wrenches, wedges, utility knives, chain falls and come-alongs, wire brushes, and shovels. Trainees will need their personal protective equipment, Trainee Guides, and pencils and paper to take notes for discussion. Allow 30 minutes for this exercise.

1. Set up a variety of stations for each of the hand tools.
2. Ask trainees to visit each station and identify the type of hand tool, how the tool is used, and a specific application for each.
3. Have trainees list at least one safety procedure and one maintenance guideline for each of the tools.
4. Review and discuss trainees' answers. Answer any questions trainees may have about hand tools discussed to this point.

MODULE 00103-04 — ANSWERS TO REVIEW QUESTIONS

Section 2.0.0
1. c
2. d
3. a
4. b
5. b

Section 3.0.0
1. c
2. b
3. d
4. a
5. d

Section 4.0.0
1. b
2. c
3. a
4. d
5. b

Section 5.0.0
1. b
2. d
3. c
4. b
5. a

Section 6.0.0
1. c
2. a
3. b
4. c
5. d

Section 7.0.0
1. a
2. d
3. c
4. b
5. a

Section 8.0.0
1. c
2. b
3. b
4. a
5. c

Section 9.0.0
1. d
2. a
3. c
4. b
5. a

Section 10.0.0
1. c
2. a
3. b
4. c
5. d

Sections 11.0.0 and 12.0.0
1. b
2. b
3. d
4. a
5. a

Section 13.0.0
1. c
2. d
3. c
4. b
5. a

Section 14.0.0
1. d
2. c
3. b
4. b
5. a

Section 15.0.0
1. a
2. d
3. b
4. b
5. c

Section 16.0.0
1. a
2. c
3. d
4. b
5. c

Section 17.0.0
1. b
2. d
3. a
4. a
5. c

Section 18.0.0
1. b
2. c
3. d
4. a
5. d

Section 19.0.0
1. c
2. b
3. a
4. d
5. b

Section 20.0.0
1. b
2. a
3. a
4. d
5. b

Section 21.0.0
1. b
2. d
3. c
4. b
5. b

Section 22.0.0
1. d
2. a
3. b
4. c
5. a

Sections 23.0.0 and 24.0.0
1. d
2. a
3. c
4. a
5. d

MODULE 00103-04 — ANSWERS TO KEY TERMS QUIZ

1. try square
2. Allen wrench
3. ball peen hammer
4. flat bar
5. carpenter's square
6. chisel
7. claw hammer
8. bevel
9. combination square
10. strip
11. crescent wrench
12. dowel
13. fastener
14. ripping bar
15. flats
16. bell-faced hammer
17. Foot-pounds
18. Inch-pounds
19. joint
20. kerf
21. combination wrench
22. level
23. miter joint
24. nail puller
25. open-end wrench
26. peening
27. square
28. pipe wrench
29. Torque
30. cat's paw nail bar
31. points
32. box-end wrench
33. punch
34. rafter angle square
35. round off
36. spud wrench
37. chisel bar
38. striking (or slugging) wrench
39. pliers
40. tang
41. tempered
42. planed
43. plumb
44. tenon
45. vise
46. weld

The NCCER makes every effort to keep these textbooks up-to-date and free of technical errors. We appreciate your help in this process. If you have an idea for improving this textbook, or if you find an error, a typographical mistake, or an inaccuracy in NCCER's *Contren®* textbooks, please write us, using this form or a photocopy. Be sure to include the exact module number, page number, a detailed description, and the correction, if applicable. Your input will be brought to the attention of the Technical Review Committee. Thank you for your assistance.

Instructors – If you found that additional materials were necessary in order to teach this module effectively, please let us know so that we may include them in the Equipment/Materials list in the Annotated Instructor's Guide.

Write: Product Development
National Center for Construction Education and Research
P.O. Box 141104, Gainesville, FL 32614-1104

Fax: 352-334-0932

E-mail: curriculum@nccer.org

Craft ________ Module Name ________

Copyright Date ________ Module Number ________ Page Number(s) ________

Description

____

____

____

____

(Optional) Correction

____

____

____

(Optional) Your Name and Address

____

____

____

Introduction to Power Tools

00104-04

NCCER STANDARDIZED CRAFT TRAINING PROGRAM

The National Center for Construction Education and Research (NCCER) provides a standardized national program of accredited craft training. Key features of the program include instructor certification, competency-based training, and performance testing. The program provides trainees, instructors, and companies with a standard form of recognition through a National Craft Training Registry. The program is described in full in the *Guidelines for Accreditation*, published by the NCCER. For more information on standardized craft training, contact the NCCER by writing us at P.O. Box 141104, Gainesville, FL 32614-1104; calling 352-334-0911; or e-mailing info@nccer.org. More information may be found at our Web site, www.nccer.org.

HOW TO USE THIS ANNOTATED INSTRUCTOR'S GUIDE

Each page presents two sections of information. The larger section displays each page exactly as it appears in the Trainee Module. The narrow column ties suggested trainee and instructor actions to each page and provides icons (detailed below) to call your attention to material, safety, audiovisual, or testing requirements. The bottom of each page includes space for your notes.

The **Audiovisual** icon indicates an appropriate time to show a transparency or other audiovisual aid.

The **Classroom** icon prompts you to define a term, stress a point, ask trainees to explain a concept, or give examples.

The **Demonstration** icon directs you to show trainees how to perform tasks.

The **Examination** icon tells you to administer the written module examination.

The **Homework** icon is placed where you may wish to assign reading for the next class, to assign a project, or to advise trainees to prepare for an examination.

The **Laboratory** icon is used when trainees are to practice performing tasks.

The **Materials** icon is a reminder for you to gather materials needed for classes, labs, and testing.

The **Performance Testing** icon tells you to administer a performance test or a portion thereof.

The **Safety** icon is used to emphasize safety issues. It is often keyed to *Caution* and *Warning* statements in the Trainee Module.

The **Teaching Tip** icon indicates additional guidance is available, such as how to conduct an exercise, get the most educational value from a field trip, or encourage class participation. Teaching Tips may expand on a feature (*Think About It*, *Did You Know?*) or provide Quick Quizzes or similar exercises. You will be referred to the Teaching Tips section at the back of the module if there is additional material.

The **Combination** icon indicates that the laboratory listed corresponds with a performance task. If desired, you can note the proficiency of the trainees during the laboratory and use it to satisfy performance testing requirements.

PREPARATION

Before teaching this module, you should review the Objectives, Performance Tasks, Materials and Equipment List, and the Module Outline. Be sure to allow ample time to prepare your own training or lesson plan and gather all required materials and equipment.

MODULE OVERVIEW

This module introduces power tools commonly used in the construction trade. Trainees will learn how to safely use and properly maintain a variety of power tools.

PREREQUISITES

Prior to training with this module, it is recommended that the trainee shall have successfully completed the following: *Core Curriculum: Introductory Craft Skills,* Modules 00101-04 through 00103-04.

OBJECTIVES

Upon completion of this module, the trainee will be able to:

1. Identify power tools commonly used in the construction trades.
2. Use power tools safely.
3. Explain how to maintain power tools properly.

PERFORMANCE TASKS

Under the supervision of the instructor, the trainee should be able to:

1. Safely and properly operate an electric drill, ensuring the following:
   - The equipment and work area are safe prior to operation, and appropriate personal protective equipment is being worn.
   - The right bit is chosen for the job and is loaded properly into the chuck opening.
   - The drill is held with both hands, and a small indent is made in the material before the hole is drilled.
   - The work that is being drilled is firmly clamped or supported.
   - Two clean holes are produced after drilling.
2. Safely and properly operate a circular saw, ensuring the following:
   - The equipment and work area are safe prior to operation, and appropriate personal protective equipment is being worn.
   - The material to be cut is properly secured.
   - The front edge of the base plate is properly placed on the work, so the guide notch is in line with the cutmark.
   - The blade depth is properly adjusted to the thickness of the wood being cut.
   - The blade has revved to full speed before moving the saw forward.
   - The saw handles are gripped firmly with two hands.
   - The blade is used as a guide during the cutting operation.
   - The trigger switch is released.
3. Safely and properly operate a bench grinder, ensuring the following:
   - The equipment and work area are safe prior to operation, and appropriate personal protective equipment (including a face shield) is being worn.
   - An adjustable tool rest is being used for support if working on metal pieces.
   - A ⅛-inch maximum gap is between the tool rest and the wheel.
   - The wheel comes to full speed before it touches the work.
   - The face of the wheel is used per the manufacturer's recommendations.
4. Safely and properly operate a portable belt sander, ensuring the following:
   - The equipment and work area are safe prior to operation, and appropriate personal protective equipment is being worn.
   - The correct grade of sandpaper is chosen for the job.
   - The sander is resting on its heel when the motor is started.
   - The sander is lowered onto the workpiece once the sander is moving.
   - The sander is kept level as it is being moved across the workpiece.
   - The sander is moving at all times it is in contact with the workpiece.
   - The sander is tipped back onto its heel when the work is finished.

- The wheel comes to full speed before it touches the work.
- The face of the wheel is used whenever possible.

5. Safely and properly operate a pneumatic power nailer, ensuring the following:
   - The equipment and work area are safe prior to operation, and appropriate personal protective equipment is being worn.
   - The manufacturer's instructions are read before operating the nailer.
   - The nails are properly loaded into the nailer, and the correct nails are selected for the job.
   - The hoses are connected properly.
   - The air compressor is checked and adjusted.
   - A test nail is first tried in scrap material, and the pressure is adjusted accordingly.
   - If nailing wall materials, the trainee knows what is in the wall and what is on the other side before nailing.
   - If nailing wall materials, the trainee has located and marked the wall studs before nailing.
   - The nailer is held firmly against the material to be fastened before pressing the trigger.
   - The air hose is disconnected once the job is finished.

MATERIALS AND EQUIPMENT LIST

Transparencies
Markers/chalk
Blank acetate sheets
Transparency pens
Pencils and scratch paper
Overhead projector and screen
Whiteboard/chalkboard
Appropriate personal protective equipment
Copies of your local code
Power drills, including:
  Electric drills
  Cordless drills
  Hammer drills
  Electromagnetic drills
  Pneumatic drills (air hammers)
  Electric screwdrivers
Variety of drill bits
Saws, including:
  Circular saws (Skilsaw®)
  Saber saws
  Reciprocating saws (Sawsall®)
  Portable handheld bandsaws
  Power miter box saws
Variety of saw blades
Changeable blades for saber saws
Boards to practice cutting
Grinders and sanders, including:
  Angle grinders (side grinders), end grinders, and detail grinders
  Bench grinders
  Portable belt sanders
  Random orbital sanders (finishing sanders)
Attachments for detail grinders
Metalwork to practice polishing
Materials to demonstrate a ring test
Miscellaneous power tools, including:
  Pneumatically powered nailers (nail guns)
  Powder-actuated fastening systems
  Air impact wrenches
  Pavement breakers
  Hydraulic jacks (Porta-Power™)
Nails
Air compressor
Nuts and bolts to practice using the air impact wrenches
Module Examinations*
Performance Profile Sheets*

*Located in the Test Booklet.

SAFETY CONSIDERATIONS

Ensure that the trainees are equipped with appropriate personal protective equipment. Always work in a clean, well-lit, appropriate work area.

ADDITIONAL RESOURCES

This module is intended to present thorough resources for task training. The following reference works are suggested for both instructors and motivated trainees interested in further study. These are optional materials for continued education rather than for task training.

29 CFR 1926, OSHA Construction Industry Regulations, latest edition. Washington, DC: Occupational Safety and Health Administration, U.S. Department of Labor, U.S. Government Printing Office.

All About Power Tools, 2002. Des Moines, IA: Meredith Books.

Hand and Power Tool Training. Video. All About OSHA. Surprise, AZ.

Power Tools, 1997. Minnetonka, MN: Handyman Club of America.

Powered Hand Tool Safety: Handle with Care. Video. 20 minutes. Coastal Training Technologies Corp. Virginia Beach, VA.

Reader's Digest Book of Skills and Tools, 1993 edition. Pleasantville, NY: Reader's Digest.

TEACHING TIME FOR THIS MODULE

An outline for use in developing your lesson plan is presented below. Note that each Roman numeral in the outline equates to one session of instruction. Each session has a suggested time period of 2½ hours. This includes 10 minutes at the beginning of each session for administrative tasks and one 10-minute break during the session. Approximately 5 hours are suggested to cover *Introduction to Power Tools.* You will need to adjust the time required for hands-on activity and testing based on your class size and resources. Because laboratories often correspond to Performance Tasks, the proficiency of the trainees may be noted during these exercises for Performance Testing purposes.

Topic	Planned Time
Session I. Power Tools, Part One	
A. Introduction to Electric, Pneumatic, and Hydraulic Tools	____________
B. Power Drills	____________
C. Saws	____________
D. Performance Testing (Tasks 1 and 2)	____________
Session II. Power Tools, Part Two	
A. Grinders and Sanders	____________
B. Miscellaneous Power Tools	____________
C. Review	____________
D. Module Examination	____________
1. Trainees must score 70% or higher to receive recognition from NCCER.	
2. Record the testing results on Craft Training Report Form 200, and submit the results to the Training Program Sponsor.	
E. Performance Testing (Tasks 3 through 5)	____________
1. Trainees must perform each task to the satisfaction of the instructor to receive recognition from NCCER. If applicable, proficiency noted during laboratory exercises can be used to satisfy the Performance Testing requirements.	
2. Record the testing results on Craft Training Report Form 200, and submit the results to the Training Program Sponsor.	

Introduction to
Power Tools
00104-04
Excellence in Construction Winner—
Renovation, $10–99 Million
Manhattan Construction Company and Flintco, Inc. formed a joint venture to work on the Oklahoma State Capital Dome in Oklahoma City. This two-phase, design-build project was the first of its kind awarded by the state of Oklahoma. Manhattan/Flintco completed restoration of the five-million-pound dome with two weeks to spare for its unveiling during Oklahoma Statehood Day.

00104-04
Introduction to Power Tools

Topics to be presented in this unit include:

Overview

Power tools are frequently used throughout construction trades. It is very common to see power tools such as drills, saws, grinders, sanders, and nailers on a construction site. These tools are typically powered by electricity, pressurized air, or pressurized fluids.

When power is added to a tool, there is a significant increase in the risks associated with using the tools. For example, you might sustain a serious cut when using a handsaw. However, the same amount of contact with a circular saw could cause a much more severe injury.

Much like the safety precautions required for nonpowered hand tools, anyone using a power tool must be thoroughly trained in the maintenance and proper use of power tools. Power tools should only be used for the work they were designed to do. They should also be kept dry, clean, and in good working order. Never put yourself at risk by using a power tool you don't know how to use or one that isn't safe.

Instructor's Notes:

Objectives

When you have completed this module, you will be able to do the following:

1. Identify power tools commonly used in the construction trades.
2. Use power tools safely.
3. Explain how to maintain power tools properly.

Required Trainee Materials

1. Appropriate personal protective equipment
2. Sharpened pencils and paper

Prerequisites

Before you begin this module, it is recommended that you successfully complete the following: *Core Curriculum: Introductory Craft Skills*, Modules 00101-04 through 00103-04.

This course map shows all of the modules in *Core Curriculum: Introductory Craft Skills*. The suggested training order begins at the bottom and proceeds up. Skill levels increase as you advance on the course map. The local Training Program Sponsor may adjust the training order.

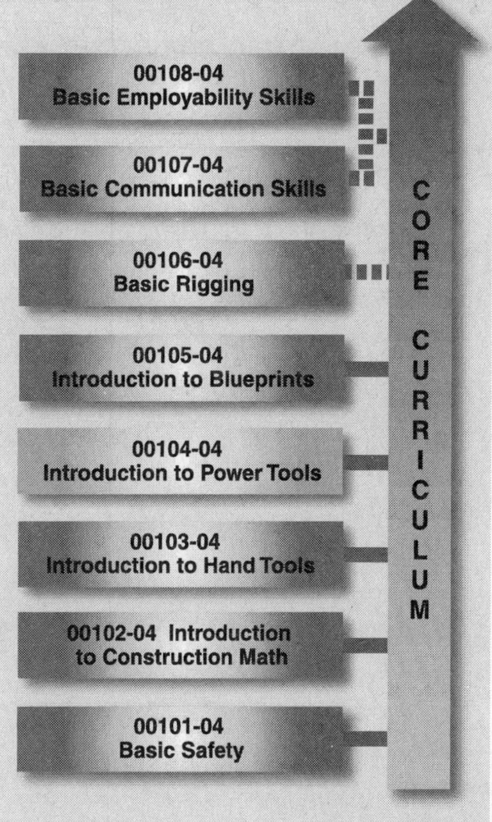

104CMAP.EPS

Key Trade Terms

Abrasive
AC (alternating current)
Auger
Booster
Carbide
Chuck
Chuck key
Countersink
DC (direct current)
Electric tools
Ferromagnetic
Grit
Ground fault circuit interrupter (GFCI)
Ground fault protection
Hazardous materials
Hydraulic tools
Masonry
Pneumatic tools
Reciprocating
Revolutions per minute (rpm)
Ring test
Shank
Trigger lock

Ensure that you have everything required to teach the course. Check the Materials and Equipment List at the front of this module.

See the general Teaching Tip at the end of this module.

Show Transparency 1, Course Objectives.

Show Transparencies 2 through 4, Performance Tasks.

Explain that terms shown in bold (blue) are defined in the Glossary at the back of this module.

Ensure that trainees have completed *Basic Safety* before beginning this module.

Review the three categories of power tools, and explain how each is powered. Provide examples from your experience for each type of power tool.

Discuss general safety issues that trainees have learned to this point. Emphasize the importance of using power tools safely.

1.0.0 ◆ INTRODUCTION

Power tools are used to make holes; cut, smooth, and shape materials; and even demolish pavement in almost every construction industry. As a construction worker, you will probably use power tools on the job. Knowing how to identify and use power tools safely and correctly is very important. This module provides an overview of the various types of power tools and how they work. You will also learn the proper safety techniques required to operate these tools.

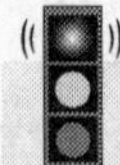

WARNING!

If you have not completed the *Basic Safety* module, stop here! You must complete the *Basic Safety* module first. Also, you must wear appropriate personal protective equipment when you operate any power tool or when you are near someone else who is operating a power tool.

2.0.0 ◆ ELECTRIC, PNEUMATIC, AND HYDRAULIC TOOLS

This module introduces three kinds of power tools: electric, pneumatic, and hydraulic.

- *Electric tools* – These tools are powered by electricity. They are operated from either an **alternating current (AC)** source (such as a wall plug) or a **direct current (DC)** source (such as a battery). Belt sanders and circular saws are examples of electric tools.
- *Pneumatic tools* – These tools are powered by air. Electric or gasoline-powered compressors produce the air pressure. Air hammers and pneumatic nailers are examples of pneumatic tools.
- *Hydraulic tools* – These tools are powered by fluid pressure. Hand pumps or electric pumps are used to produce the fluid pressure. Jackhammers and Porta-Powers™ are examples of hydraulic tools.

2.1.0 Safety

You must complete the *Basic Safety* module before you take this course. It is easy to hurt yourself or others if you use a power tool incorrectly or unsafely. Safety issues for each tool are covered in this module, but general safety issues—such as safety in the work area, safety equipment, and working with electricity—are covered in the *Basic Safety* module in this book. This information is vital for working with power tools.

One of the most important things about working with power tools is to always disconnect the power source for any tool before you replace parts such as bits, blades, or discs. Always disconnect the power source before you perform maintenance on any power tool. Never activate the **trigger lock** on any power tool.

Review Questions

Section 2.0.0

1. Trigger locks should always be activated when a power tool is in use.
   a. True
   b. False

2. Electric tools get their power from _____.
   a. hand pumps
   b. an AC (wall plug) or DC (battery) source
   c. fluid pressure
   d. a gasoline-powered compressor

3. Pneumatic tools get their power from _____.
   a. air pressure
   b. fluid pressure
   c. hand pumps
   d. AC power sources

4. Hydraulic tools get their power from _____.
   a. rotary engines
   b. air pressure
   c. fluid pressure
   d. solar panels

5. Always disconnect the _____ before you perform maintenance on any power tool.
   a. trigger lock
   b. drill bit
   c. power source
   d. belt sander

Instructor's Notes:

4.2 CORE CURRICULUM ◆ INTRODUCTORY CRAFT SKILLS

3.0.0 ◆ POWER DRILLS

The power drill is used often in the construction industry. It is most commonly used to make holes by spinning drill bits into wood, metal, plastic, and other materials. However, with different attachments and accessories, the power drill can be used as a sander, polisher, screwdriver, grinder, or **countersink**—even as a saw.

3.1.0 Types of Power Drills

In this section, you will learn about various types of power drills, including the following:

- Electric drills
- Cordless drills
- Hammer drills
- Electromagnetic drills
- Pneumatic drills (air hammers)
- Electric screwdrivers

Most of these drills are similar, so you will first learn about what they have in common.

Most power drills have a pistol grip with a trigger switch for controlling power (see *Figure 1*). The harder you pull on the trigger of a variable-speed drill, the faster the speed. Drills also have reversing switches that allow you to back the drill bit out if it gets stuck in the material while drilling. Most drills have replaceable bits for use on different kinds of jobs (see *Figure 2*). On most power drills, you can insert a screwdriver bit in place of a drill bit and use the drill as a screwdriver. Be sure to use screwdriver bits that are designed for use in a power drill.

Twist drill bits are used to drill wood and plastics at high speeds or to drill metal at a lower speed. A forstner bit is used on wood and is particularly good for boring any part of a circle. A paddle bit or spade bit is also used in wood. The bit size is measured by the paddle's diameter, which ranges from ½ inch to ¼ inch. A **masonry** bit, which has a **carbide** tip, is used in concrete, stone, slate, and ceramic. The **auger** drill bit is used for drilling wood and other soft materials, but not for drilling metal. As a rule, the point of a bit should be sharper for softer materials than for harder ones. All bits are held in the drill by the drill **chuck** (see *Figure 3*). Chucks can be either keyed or keyless.

Review the different types of power drills. Ask trainees to share their experiences working with these types of drills.

Show Transparency 5 (Figure 1). Identify the parts of a power drill, and explain how each operates.

Bring in a variety of drill bits. Demonstrate how each bit is used, and explain which bits are suitable for different materials.

Refer to Figure 3. Explain the role of a chuck key in a power drill.

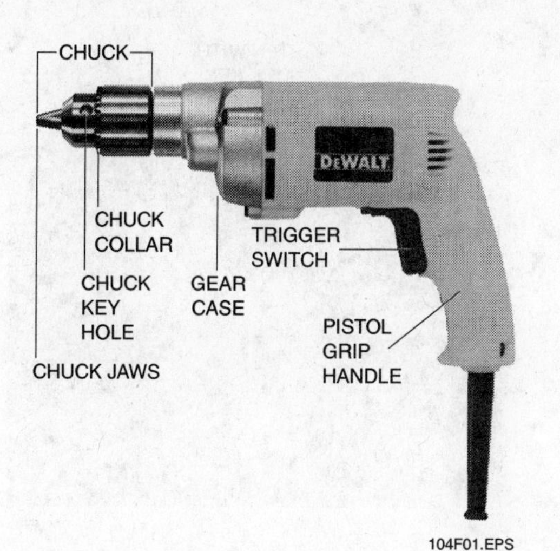

Figure 1 ◆ Parts of the power drill.

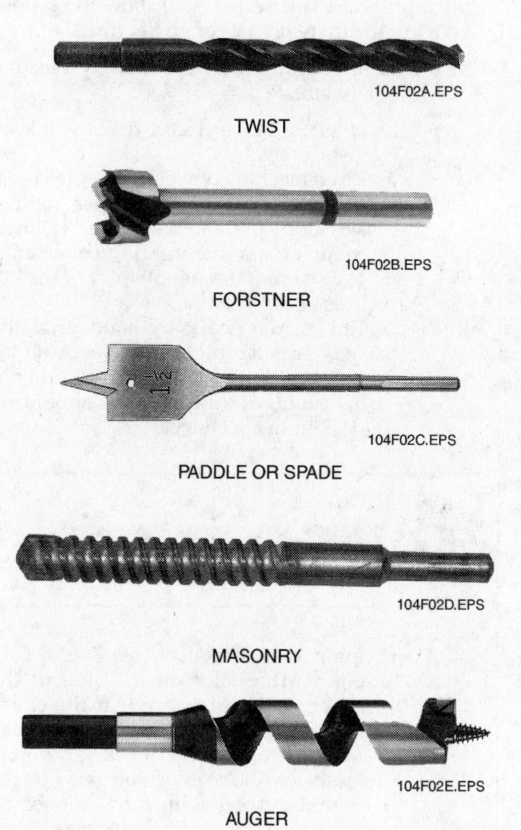

Figure 2 ◆ Drill bits.

MODULE 00104-04 ◆ INTRODUCTION TO POWER TOOLS 4.3

Demonstrate how to load a bit in an electric drill and how to use the power drill safely.

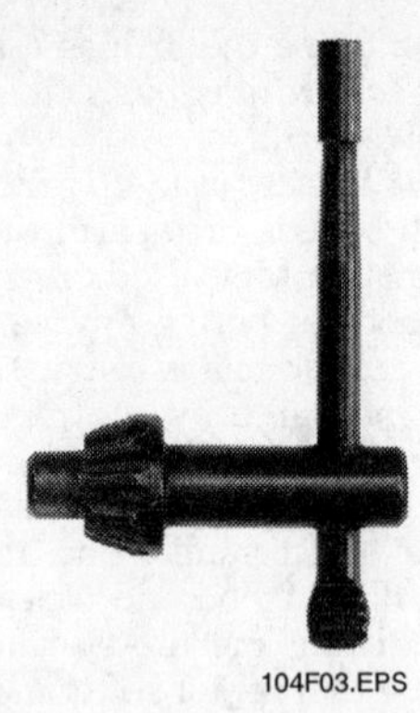

Figure 3 ◆ Chuck key.

3.1.1 How to Use a Power Drill with a Chuck

Power drills can be dangerous if you do not use them properly. This section will show the proper way to use different types of power drills.

Step 1 Wear appropriate personal protective equipment.

Step 2 Load the bit in the electric drill by following these steps:

a. Disconnect the power. Open the chuck and turn it counterclockwise (to the left) until the chuck opening is large enough for you to insert the bit shank. The shank is the smooth part of the bit.
b. Insert the bit shank.
c. Tighten the chuck by hand until the jaws grip the bit shank. Keep the bit centered as you tighten it. It should not be leaning to one side but should be straight in the chuck.

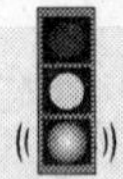

NOTE
Stop here if you are using a keyless chuck.

d. Insert the **chuck key** (see *Figure 4*) in one of the holes on the side of the chuck. You will notice that the chuck key has a grooved ring called a gear. Make sure that the chuck key's gear meshes with the matching gears on the geared end of the chuck. In larger drills, tighten the bit by inserting the chuck key into each of the holes in the three-jawed chuck. This ensures that all the jaws are uniformly tight around the bit.

(A) INSERT THE BIT SHANK INTO THE CHUCK OPENING.

(B) TIGHTEN WITH THE CHUCK KEY.

(C) HOLD THE DRILL PERPENDICULAR TO THE MATERIAL AND START THE DRILL.

Figure 4 ◆ Proper drill use.

Instructor's Notes:

4.4 CORE CURRICULUM ◆ INTRODUCTORY CRAFT SKILLS

Drilling Metal

When you are drilling metal, lubricate the bit to help cool the cutting edges and produce a smoother finished hole. A very small amount of cutting oil that is not combustible (capable of catching fire and burning) makes a good lubricant for drilling softer metals. No lubrication is needed for wood drilling. When you are drilling deep holes, pull the drill bit partly out of the hole every so often. This helps to clear the hole of shavings.

Discuss problems that could result if you do not remove the key from the chuck.

Emphasize the importance of keeping your hands clear of the drill.

Discuss problems that can result if a drill comes in contact with electrical wiring, plumbing, or insulation.

Review the safety guidelines for working with drills. Explain how to determine whether a tool has the proper ground fault protection.

e. Turn the chuck key clockwise (right) to tighten the grip on the bit.
f. Remove the key from the chuck.

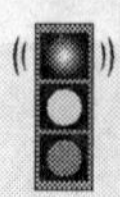

WARNING!

Always remember to remove the key from the chuck. Otherwise, when you start the drill, the key could fly out and injure you or a co-worker.

Step 3 Make a small indent exactly where you want the hole drilled.
a. In wood, use a small punch to make an indent.
b. In metal, first use a center punch.

Step 4 Firmly clamp or support the work that is being drilled.

Step 5 Hold the drill perpendicular (at a right angle) to the material surface and start the drill motor. Be sure the drill is rotating in the right direction (with the bit facing away from you; it should be turning clockwise). Hold the drill with both hands and apply only moderate pressure when drilling. The drill motor should operate at approximately the same **revolutions per minute (rpm)** as it does when it is not drilling through anything. For power drills, the term *rpm* refers to how many times the drill bit completes one full rotation every minute. *Figure 4(C)* shows the proper way to hold the drill when you are operating it.

Step 6 Lessen the pressure when the bit is about to come through the other side of the work, especially when you are drilling metal. If you are still pressing hard when the bit comes out the other side, the drill itself will hit the surface of your material. This could damage or dent the metal surface. If the drill bit gets stuck in the material while you are drilling, release the trigger, use the reversing switch to change the direction of the drill, and back it gently out of the material. When you are finished backing it out, switch back to your original drilling position.

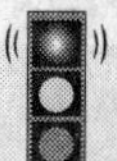

WARNING!

Be sure your hand is not in contact with the drill bit. The spinning bit will cut your hand. Keep an even pressure on the drill to keep the drill from twisting or binding.

WARNING!

Before you start drilling into or through a wall, find out what is on the other side, and take steps to avoid hitting anything that would endanger your safety or cause damage. Spaces between studs (upright pieces in the walls of a building) often contain electrical wiring, plumbing, or insulation, for instance. If you are not careful, you will drill directly into the wiring, pipes, or insulation.

3.1.2 Safety and Maintenance

In addition to the general safety rules you learned in *Basic Safety*, there are some specific safety rules for working with drills:

- Always wear appropriate personal protective equipment, especially safety glasses.
- To prevent an electrical shock, operate only those tools that are double-insulated electric power tools with proper **ground fault protection.** Using a **ground fault circuit interrupter (GFCI)** device protects the equipment from continued electrical current in case of a circuit fault. The GFCI monitors the current flow and opens the circuit (which stops the flow of electricity) if it detects a difference between positive and negative flow. The interruption

Explain why electric drills should not be used around combustible materials.

Refer to Figure 5. Discuss the advantages of using a cordless drill.

Refer to Figure 6. Explain how to load a bit on a cordless drill with a keyless chuck.

Refer to Figure 7. Explain how hammer drills are able to drill faster than regular drills. Discuss the types of drill bits that can be used with hammer drills.

typically takes place in less than one-tenth of a second.

- Before you connect to the power source, make sure the trigger is not turned on. It should be off. Always disconnect the power source before you change bits or work on the drill.
- Find out what is inside the wall or on the other side of the work material before you cut through a wall or partition. Avoid hitting water lines or electrical wiring.
- Ensure that electric tools with two-prong plugs are double insulated. If a tool is not double insulated, its plug must have a third prong to provide grounding.
- Ensure that the switch can be operated with one finger.
- Use the right bit for the job.
- Always use a sharp bit.
- Make sure the drill bit is tightened in the chuck before you start the drill.
- Make sure the chuck key is removed from the chuck before you start the drill.
- Hold the drill with both hands and apply steady pressure. Let the drill do the work.
- Never ram the drill while you are drilling. This chips the cutting edge and damages the bearings.
- Never use the trigger lock. The trigger lock is a small lever, switch, or part that you push or pull to lock the trigger in operating mode.
- Drills do not need much maintenance, but they should be kept clean. Many drills have gears and bearings that are lubricated for life. Some drills have a small hole in the case for lubricating the motor bearings. Apply about three drops of oil occasionally, but don't overdo it. Extra lubricant can leak onto electrical contacts and burn the copper surfaces.
- Keep the drill's air vent clean with a small brush or small stick. Airflow is crucial to the maintenance and safety of a drill.
- Attach the chuck key to the power cord when you are not using the key, so it does not get lost.
- Do not overreach when using a power drill while standing on a ladder. You could fall.

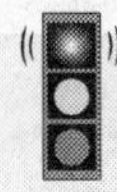

WARNING!
Do not use electric drills around combustible materials. Motors and bits can create sparks, which can cause an explosion.

3.2.0 Cordless Drill

Cordless power drills (see *Figure 5*) are useful for working in awkward spaces or in areas where a power source is hard to find.

Cordless drills usually contain a rechargeable battery pack that runs the motor. The pack can be detached and plugged into a battery charger any time you are not using the drill. Some chargers can recharge the battery pack in an hour, while others require more time. Workers who use cordless drills a lot usually carry an extra battery pack with them. Some cordless drills have adjustable clutches so that the drill motor can also serve as a power screwdriver. Many cordless drills are now available with keyless chucks.

3.2.1 How to Use a Cordless Drill Without a Chuck

When using a cordless drill without a chuck, follow these guidelines:

Step 1 Wear appropriate personal protective equipment.

Step 2 To load the bit on a cordless drill with a keyless chuck (see *Figure 6*), follow these steps:

a. Remove the power pack/battery. Open the chuck by turning it counterclockwise until the jaws are wide enough for you to insert the bit shank.
b. Tighten the chuck by hand until the jaws grip the bit shank. Be sure to keep the bit centered as you tighten it. It should not be leaning to one side, but should be straight in the chuck.
c. Grip the chuck in one hand and apply a small amount of pressure to the trigger. This action spins the drill a little. Resist the drill's spin by holding tightly to the chuck. This locks the bit shank into the chuck.

Step 3 To operate the drill, follow the procedures previously outlined for the power drill.

3.2.2 Safety and Maintenance

Follow the power drill safety practices that you learned earlier in this module.

3.3.0 Hammer Drill

The hammer drill (see *Figure 7*) has a pounding action that lets you drill into concrete, brick, or tile. The bit rotates and hammers at the same time, allowing you to drill much faster than you could with a regular drill. The depth gauge on a hammer drill can be set to the depth of the hole you want to drill.

You need special hammer drill bits that can take the pounding. Some hammer drills use percussion and masonry bits (see *Figure 8*).

Instructor's Notes:

4.6 CORE CURRICULUM ◆ INTRODUCTORY CRAFT SKILLS

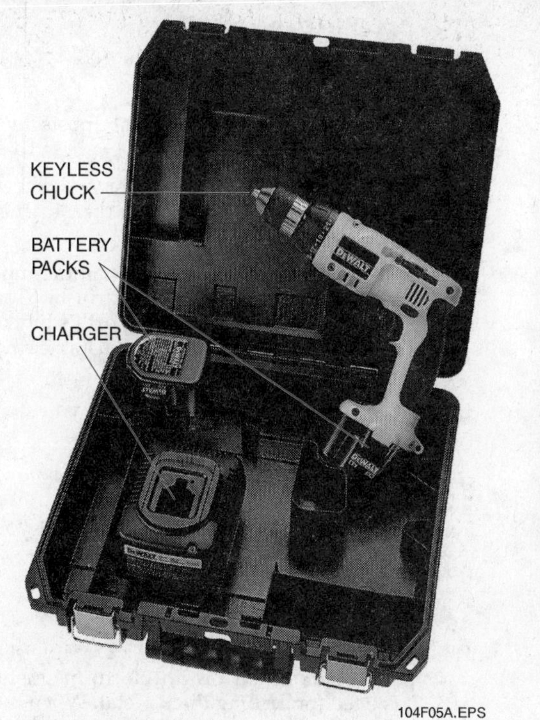

104F05A.EPS

104F05B.EPS

Figure 5 ◆ Cordless drill.

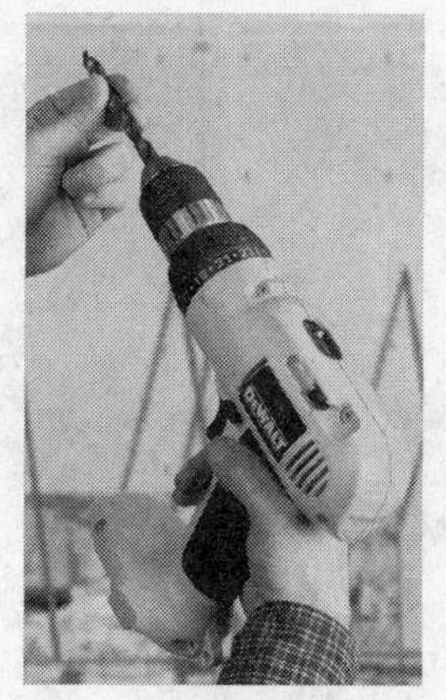

104F06A.EPS

(A) INSERT THE
BIT SHANK.

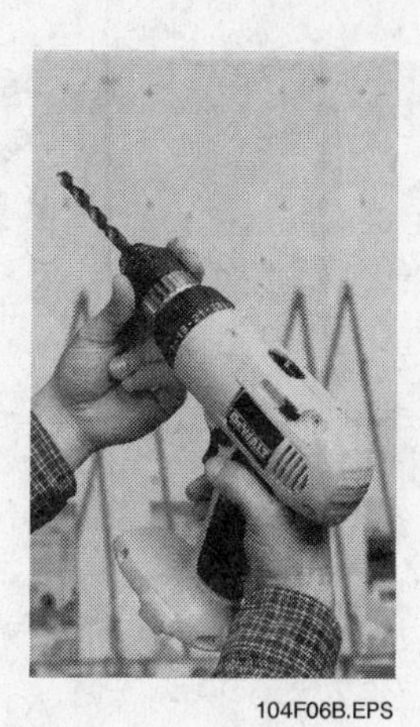

104F06B.EPS

(B) TIGHTEN THE CHUCK.

Figure 6 ◆ Loading the bit on a cordless drill.

MODULE 00104-04 ◆ INTRODUCTION TO POWER TOOLS 4.7

Set up stations with hammer drills and drill bits. Ask trainees to practice loading the bits into the drills. Review the procedures for using a hammer drill.

Refer to Figure 10. Explain how an electromagnetic drill is held in place for drilling.

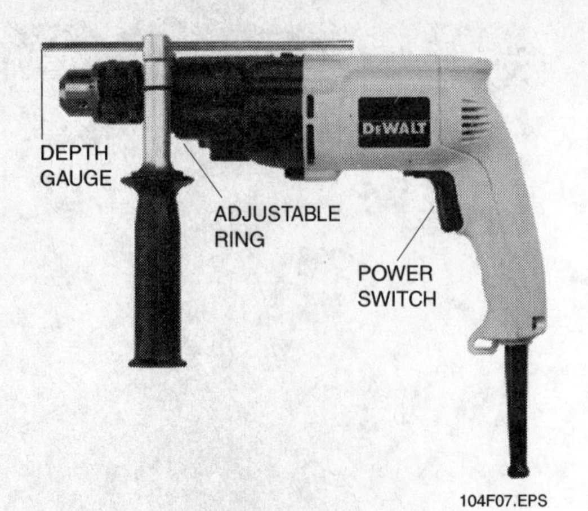

Figure 7 ◆ Hammer drill.

Figure 8 ◆ Hammer drill bits.

3.3.1 How to Use a Hammer Drill

When using a hammer drill, follow these guidelines:

Step 1 Wear appropriate personal protective equipment.

Step 2 Follow the procedures for using a power drill (which you learned earlier in this module).

Step 3 Most hammer drills will not hammer until you put pressure on the drill bit (see *Figure 9*). You can adjust the drill's blows per minute by turning the adjustable ring (refer to *Figure 7*).

Step 4 The hammer action stops when you stop applying pressure to the drill.

3.3.2 Safety and Maintenance

Follow the power drill safety practices that you learned about earlier in this module.

3.4.0 Electromagnetic Drill

The electromagnetic drill (see *Figure 10*) is a portable drill mounted on an electromagnetic base. It is used for drilling thick metal. When the drill is placed on metal and the power is turned on, the magnetic base will hold the drill in place for drilling. The drill can also be rotated on the base.

A switch on the junction box controls the electromagnetic base. When the switch is turned on,

Figure 9 ◆ Proper use of a hammer drill.

4.8 CORE CURRICULUM ◆ INTRODUCTORY CRAFT SKILLS

Instructor's Notes:

4.8 CORE CURRICULUM ◆ INTRODUCTORY CRAFT SKILLS

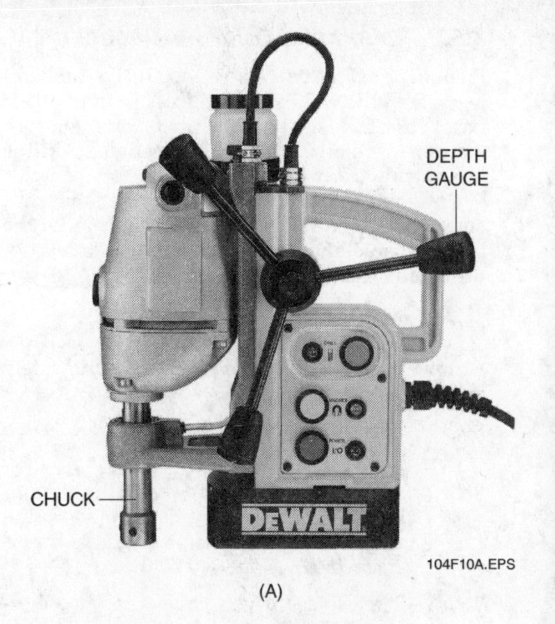

104F10A.EPS

(A)

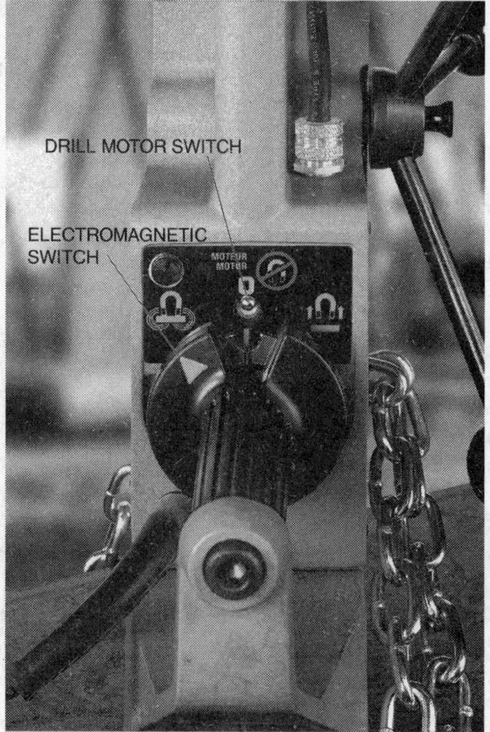

104F10B.EPS

(B)

104F10C.EPS

Figure 10 ◆ Electromagnetic drill.

Bring in and demonstrate how to set up an electromagnetic drill.

the magnet holds the drill in place on a **ferromagnetic** metal surface. (Ferromagnetic refers to substances, especially metals, that have magnetic properties.) The switch on the top of the drill turns the drill on and off. You can also set the depth gauge to the depth of the hole you are drilling.

3.4.1 How to Set Up an Electromagnetic Drill

The use of this tool is explained in detail in the specific craft areas that use it. For now, you will learn only the setup procedures for using the electromagnetic drill.

MODULE 00104-04 ◆ INTRODUCTION TO POWER TOOLS 4.9

Discuss the consequence of failing to use a safety chain to secure an electromagnetic drill.

Refer to Figure 10. Ask a trainee to explain when the DO NOT UNPLUG tag should be placed on an electromagnetic drill's power cord.

Emphasize the importance of using a nonsparking drill when working near combustible materials.

Refer to Figure 11. Explain how pneumatic drills are powered using compressed air.

Set up stations with pneumatic drills, couplers, and whip checks. Have trainees practice attaching the coupler to the connector on the pneumatic drill. Explain how to add a whip check, and ask trainees to practice installing a whip check.

See the Teaching Tip for Sections 3.0.0–3.5.2 at the end of this module.

Step 1 Wear appropriate personal protective equipment.

Step 2 Place the drill face down into the metal holder.

Step 3 Put the electromagnetic switch (not the drill) in the ON position. Doing this holds the drill in place by magnetizing the base of the drill directly onto the metal to be drilled.

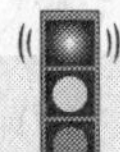

WARNING!

Expect the unexpected. Use a safety chain to secure the electromagnetic drill in case the power is shut off. If there is no power, you lose the electromagnetic field that holds the base to the metal being drilled.

Step 4 Lock the drill in place.

Step 5 Set the depth gauge to the depth of the hole you are going to drill.

Step 6 Fasten the work securely on the drilling surface with clamps.

Step 7 Proceed to drill.

3.4.2 Safety and Maintenance

In addition to the general safety rules you learned in *Basic Safety*, there are some specific safety rules for working with electromagnetic drills:

- Clamp the material securely. Unsecured materials can become deadly flying objects.
- Make sure the electrical power is not interrupted. Put a DO NOT UNPLUG tag on the cord (refer to *Figure 10*).
- Safety attachments, such as shields to block flying objects and safety lines to keep the drill from falling if the power is cut off, are available. In some states, they are required. Ask your instructor or supervisor about requirements for safety attachments in your area.
- Support the drill before you turn it off. It will fall over if you do not hold it when you turn off the power.

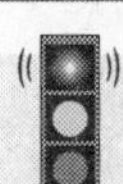

WARNING!

When you are working near combustible materials, be sure to use a nonsparking drill. A drill that gives off sparks could start a fire.

3.5.0 Pneumatic Drill (Air Hammer)

Pneumatic drills (see *Figure 11*), also called air hammers, are powered by compressed air from an air hose. They have many of the same parts, controls, and uses as electric drills. The pneumatic drill is typically used when there is no source of electricity.

Common sizes of pneumatic drills are ¼, ⅜, and ½ inch. The size refers to the diameter of the largest shank that can be gripped in the chuck, not the drilling capacity.

3.5.1 How to Use a Pneumatic Drill

Follow these steps to use a pneumatic drill safely and efficiently (see *Figure 12*):

Step 1 Wear appropriate personal protective equipment.

Step 2 Hold the coupler at the end of the air supply line, slide the ring back, and slip the coupler on the connector or nipple that is attached to the pneumatic drill.

Step 3 Check to see if you have a good connection. You cannot take apart a good coupling without first sliding the ring back.

Step 4 Once you have a good connection, install a whip check as required.

Step 5 Proceed to drill as needed.

Step 6 When your work is completed, disconnect the drill from the hose.

3.5.2 Safety and Maintenance

Follow the power drill safety practices that you learned earlier in this module.

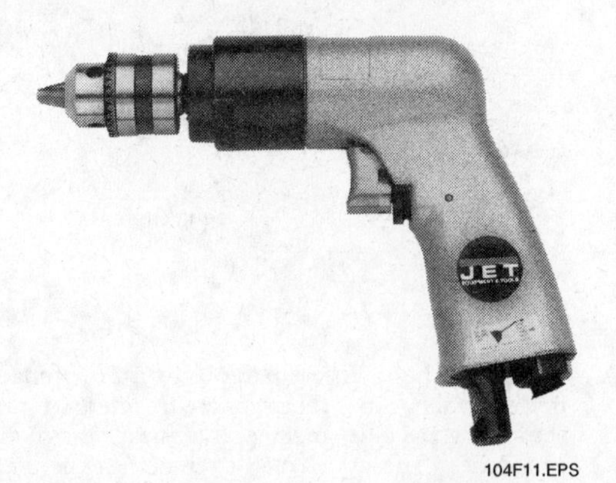

104F11.EPS

Figure 11 ◆ Pneumatic drill.

Instructor's Notes:

4.10 CORE CURRICULUM ◆ INTRODUCTORY CRAFT SKILLS

Go over the Review Questions for Section 3.0.0. Ask whether trainees have any questions.

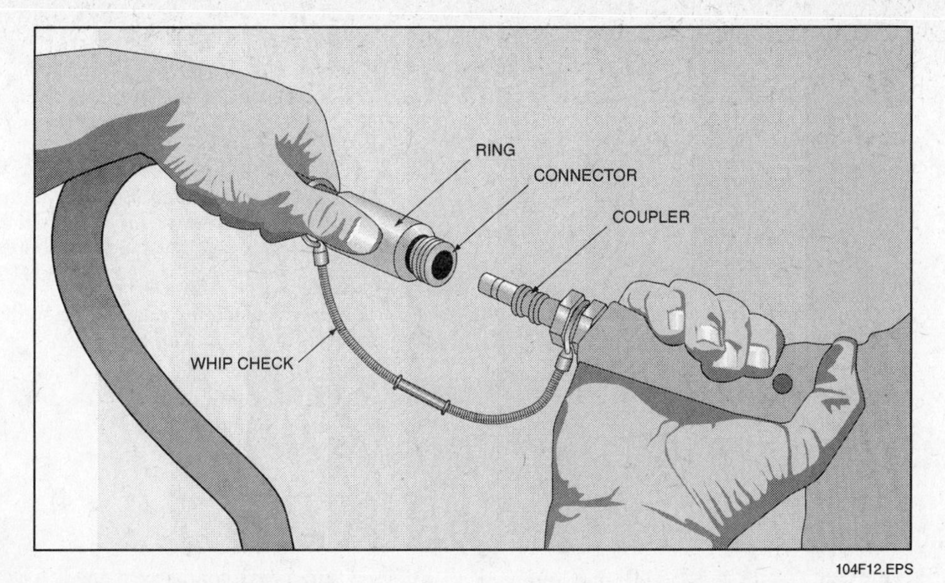

Figure 12 ◆ Proper use of a pneumatic drill.

Review Questions

Section 3.0.0

1. The most common use of the power drill is to _____.
   a. cut wood, metal, and plastic
   b. drive nails into wood, metal, and plastic
   c. make holes in wood, metal, and plastic
   d. carve letters in wood, metal, and plastic

2. The trigger switch on the variable-speed power drill _____.
   a. controls the speed
   b. shoots nails into walls
   c. blows air to clean the work area
   d. tightens the chuck jaws

3. If the drill bit gets stuck in the material while you are drilling, you should first _____.
   a. unplug the drill
   b. release the trigger switch
   c. back the drill out of the material
   d. use the reversing switch to change directions

4. Before you start drilling, be sure _____ is not in contact with the drill bit.
   a. the bit shank
   b. the material being drilled
   c. the chuck
   d. your hand

5. Before you start drilling through a material, always _____.
   a. find out what is on the other side
   b. disconnect the power source
   c. reinsert the chuck key
   d. drive a nail in first

6. When using a power drill, proper ground fault protection will prevent _____.
   a. drilling into electrical wiring
   b. you from losing your grip
   c. electric shock
   d. excess noise

MODULE 00104-04 ◆ INTRODUCTION TO POWER TOOLS 4.11

Bring in a variety of saw blades. Demonstrate the differences among each type of blade, and provide examples of suitable jobs for each type of blade.

Show Transparency 6 (Figure 13). Explain how circular saw size is measured.

Review Questions

7. When drilling, hold the drill with _____.
   a. one hand and push it into the material
   b. both hands and apply steady pressure
   c. one hand and the power cord with the other
   d. your arms fully extended
8. Hammer drills are designed to drill into _____.
   a. wood, metal, and plastic
   b. concrete, brick, and tile
   c. drywall
   d. roofing shingles
9. The electromagnetic drill is a _____.
   a. handheld drill used on wood
   b. cordless drill used on masonry and tile
   c. portable drill used on thick metal
   d. pneumatic drill that has a pounding action
10. When you are drilling near combustible materials, be sure to use a(n) _____.
    a. cordless drill
    b. electromagnetic drill
    c. nonsparking drill
    d. masonry bit

4.0.0 ◆ SAWS

Using the right saw for the job will make your work much easier. Always make sure that the blade is right for the material being cut. In this section, you will learn about the following types of power saws:

- Circular saws (Skilsaw®)
- Saber saws
- **Reciprocating** saws (Sawsall®)
- Portable handheld bandsaws
- Power miter box saws

4.1.0 Circular Saw

Many years ago a company named Skil® made power-tool history by introducing the portable circular saw. Today many different companies make dozens of models, but a lot of people still call any portable circular saw a skilsaw. Other names you might hear are utility saw, electric handsaw, and builder's saw. The portable circular saw (see *Figure 13*) is designed to cut lumber and boards to size for a project.

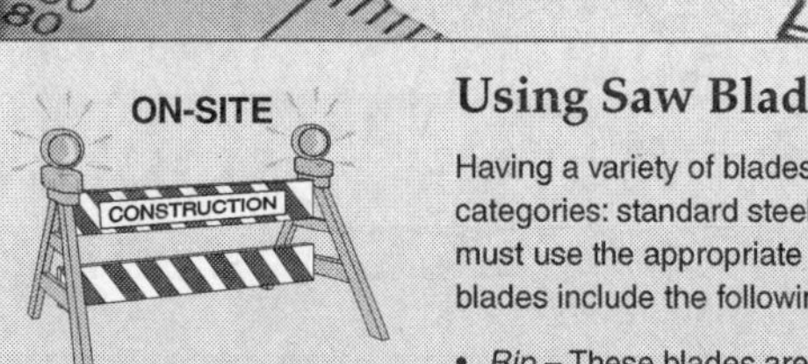

Using Saw Blades

Having a variety of blades will allow you to work on different projects. Blades fall into two categories: standard steel, which must be sharpened regularly, and carbide-tipped. You must use the appropriate type of saw blade for the job. Some common types of saw blades include the following:

- *Rip* – These blades are designed to cut with the grain of the wood. The square chisel teeth cut parallel with the grain and are generally larger than other types of blade teeth.
- *Crosscut* – These blades are designed to cut against the grain of the wood (at a 90-degree angle). Crosscut teeth cut at an angle and are finer than rip blade teeth.
- *Combination* – These blades are designed to cut hard or soft wood, either with or across the grain. The combination blade features both rip and crosscut teeth with deep troughs (gullets) between the teeth.
- *Nail cutter* – This blade has large carbide-tipped teeth that can make rough cuts through nails that may be embedded in the work.
- *Nonferrous metal cutter* – This blade has carbide-tipped teeth for cutting aluminum, copper, lead, and brass. It should be lubricated with oil or wax before each use.

Always follow the manufacturer's instructions when using saw blades.

Instructor's Notes:

4.12 CORE CURRICULUM ◆ INTRODUCTORY CRAFT SKILLS

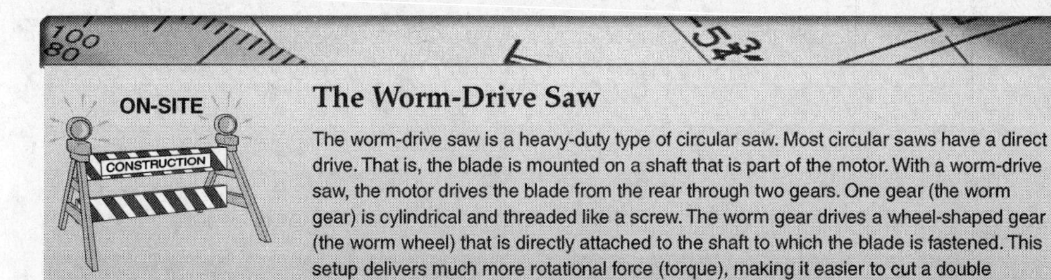
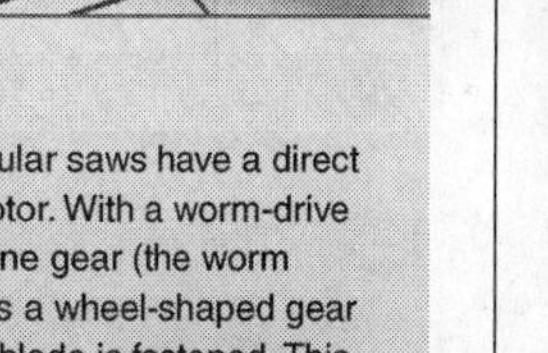

ON-SITE

The Worm-Drive Saw

The worm-drive saw is a heavy-duty type of circular saw. Most circular saws have a direct drive. That is, the blade is mounted on a shaft that is part of the motor. With a worm-drive saw, the motor drives the blade from the rear through two gears. One gear (the worm gear) is cylindrical and threaded like a screw. The worm gear drives a wheel-shaped gear (the worm wheel) that is directly attached to the shaft to which the blade is fastened. This setup delivers much more rotational force (torque), making it easier to cut a double thickness of lumber. The worm-drive saw is almost twice as heavy as a conventional circular saw. This saw should be used only by an experienced craftworker.

Explain why the lower blade guard must be properly attached to a saw.

Review the procedures for using a circular saw.

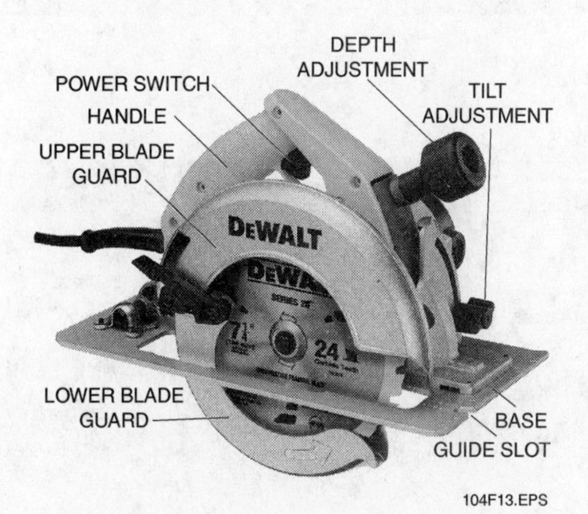

Figure 13 ◆ Circular saw.

Saw size is measured by the diameter of the circular blade. Saw blade diameters range from $3\frac{3}{8}$ to $16\frac{1}{4}$ inches. The $7\frac{1}{4}$-inch size is the most popular. A typical circular saw weighs between 9 and 12 pounds. The handle of the circular saw has a trigger switch that starts the saw. The motor is protected by a rigid plastic housing. Blade speed when the blade is not engaged in cutting is given in rpm. The teeth of the blade point in the direction of the rotation. The blade is protected by two guards. On top, a rigid plastic guard protects you from flying debris and from touching the spinning blade if you lean forward accidentally. The lower guard is spring-loaded—as you push the saw forward, it retracts up and under the top guard to allow the saw to cut.

WARNING!

Never use the saw unless the lower blade guard is properly attached. The guard protects you from the blade and from flying particles.

4.1.1 How to Use a Circular Saw

Follow these steps to use a circular saw safely and efficiently:

Step 1 Wear appropriate personal protective equipment.

Step 2 Properly secure the material to be cut. If the work isn't heavy enough to stay in position without moving, weight or clamp it down.

Step 3 Make your cut mark with a pencil or other marking tool.

Step 4 Place the front edge of the baseplate on the work so the guide notch is in line with the cut mark.

Step 5 Adjust the blade depth to the thickness of the wood you are cutting plus $\frac{1}{4}$ inch.

Step 6 Start the saw. After the blade has revved up to full speed, move the saw forward to start cutting. The lower blade guard will automatically rotate up and under the top guard when you push the saw forward.

Step 7 While cutting with the saw, grip the saw handles firmly with two hands, as shown in *Figure 14.*

MODULE 00104-04 ◆ INTRODUCTION TO POWER TOOLS 4.13

Demonstrate the proper way to use a circular saw. Show how to support the material being cut. Discuss using a circular saw on hand-held materials.

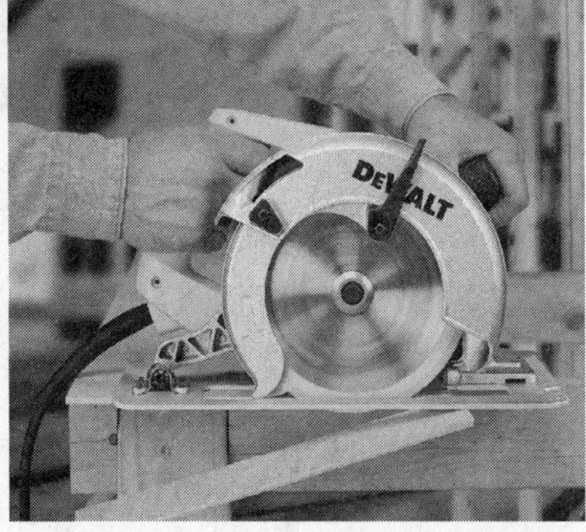

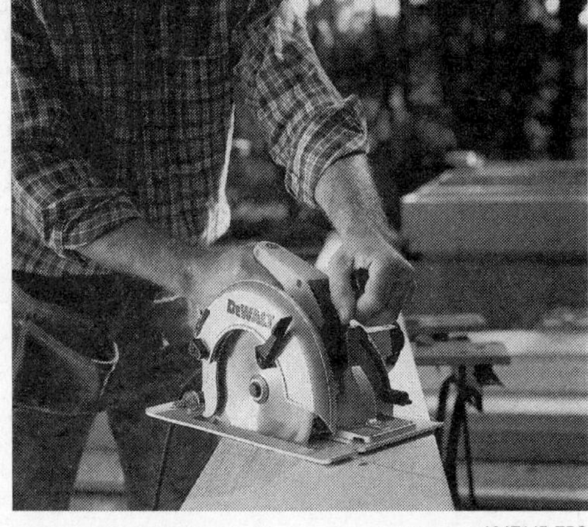

104F14A.EPS 104F14B.EPS

Figure 14 ◆ Proper use of a circular saw.

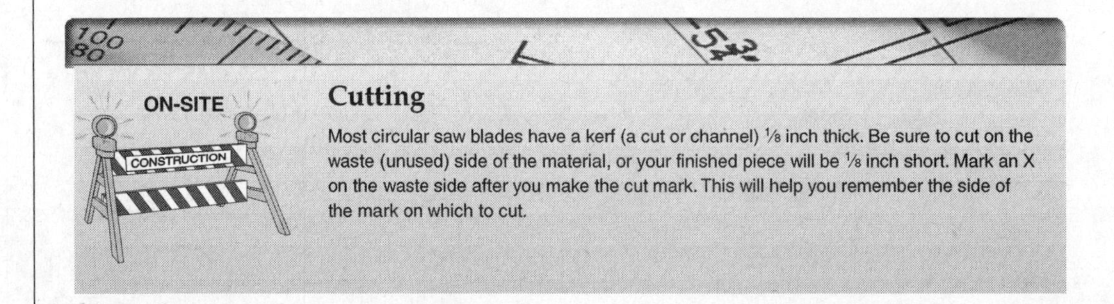

ON-SITE

Cutting

Most circular saw blades have a kerf (a cut or channel) ⅛ inch thick. Be sure to cut on the waste (unused) side of the material, or your finished piece will be ⅛ inch short. Mark an X on the waste side after you make the cut mark. This will help you remember the side of the mark on which to cut.

Step 8 If the saw cuts off the line, stop, back out, and restart the cut. Do not force the saw.

Step 9 As you get to the end of the cut, the guide notch on the baseplate will move off the end of the work. Use the blade as your guide.

Step 10 Release the trigger switch. The blade will stop rotating.

Step 11 Ensure that the work is properly supported.

CAUTION

Make sure the blade is appropriate for the material being cut.

4.1.2 Safety and Maintenance

To use a circular saw safely and ensure its long life, follow these guidelines:

- Wear appropriate personal protective equipment.
- Ensure that the blade is tight.
- Check that the blade guard is working correctly before you connect the saw to the power source.
- Before you cut through a wall or partition, find out what is inside the wall or on the other side of the partition. Avoid hitting water lines or electrical wiring.
- Whenever possible, keep both hands on the saw grips while you are operating the saw.
- Never force the saw through the work. This causes binding and overheating and may cause injury.
- Never reach underneath the work while you are operating the saw.

Instructor's Notes:

4.14 CORE CURRICULUM ◆ INTRODUCTORY CRAFT SKILLS

- Never stand directly behind the work. Always stand to one side of it.
- Do not use your hands to try to secure small pieces of material to be cut. Use a clamp instead.
- Know where the power cord is located. You don't want to cut through the power cord by accident and electrocute yourself!
- The most important maintenance on a circular saw is at the lower blade guard. Sawdust builds up and causes the guard to stick. If the guard sticks and does not move quickly over the blade after it makes a cut, the bare blade may still be turning when you set the saw down and may cause damage. Remove sawdust from the blade guard area. Remember to always disconnect the power source before you do maintenance.
- To avoid injury to you, your materials, and your co-workers, check often to make sure the guard snaps shut quickly and smoothly. To ensure smooth operation of the guard, disconnect the saw from its power source, allow it to cool, and clean foreign material from the track. Be aware of fire hazards when using cleaning liquids such as isopropyl alcohol.
- Do not lubricate the guard with oil or grease. This could cause sawdust to stick in the mechanism.
- Always keep blades clean and sharp to reduce friction and kickback. Blades can be cleaned with hot water or mineral spirits. Be careful with mineral spirits; they are very flammable.
- Do not hold material to be ripped with your hands.

4.2.0 Saber Saw

Saber saws have very fine blades, which makes the saws great tools for doing delicate and intricate work, such as cutting out patterns or irregular shapes from wood or thin, soft metals. They are also some of the best tools for cutting circles.

The saber saw (see *Figure 15*) is a very useful portable power tool. It can make straight or curved cuts in wood, metal, plastic, wallboard, and other materials. It can also make its own starting hole if a cut must begin in the middle of a board. The saber saw cuts with a blade that moves up and down, unlike the spinning circular saw blade. This means that each cutting stroke (upward) is followed by a return stroke (downward), so the saw is cutting only half the time it is in operation. This is called up-cutting or clean-cutting.

WARNING!
Never use a circular saw when holding material to be cut with your hands. Doing so violates OSHA regulations.

Many models are available with tilting baseplates for cutting beveled edges. Models come with a top handle or a barrel handle. Some cordless models are available.

The saber saw has changeable blades that let it cut many different materials, from wood and metal to wallboard and ceramic tile. Most saber saws can be operated at various blade speeds. Types of saber saws include single-speed, two-speed, and variable-speed. The variable-speed saber saw can cut at low and high speeds. The low-speed setting is for cutting hard materials, and the high-speed setting is for soft materials. An important part of the saber saw is the baseplate (shoeplate or footplate). Its broad surface helps to

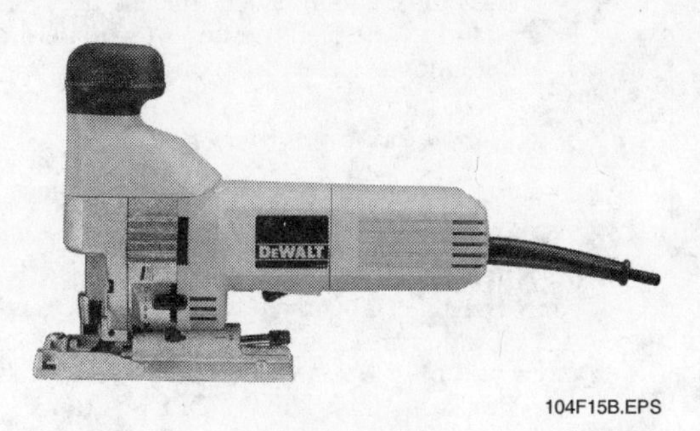

104F15A.EPS

TOP HANDLE

104F15B.EPS

BARREL HANDLE

Figure 15 ◆ Saber saws.

Refer to Figure 15. Review the uses of a saber saw. Ask a trainee to explain how a saber saw can make its own starting hole.

Emphasize the importance of following OSHA regulations when using circular saws.

Bring in a variety of changeable blades for saber saws. Demonstrate the different materials that can be cut using the blades.

On the whiteboard/chalkboard, write a list of materials to be cut using a saber saw. Have trainees select the appropriate saw blade and speed for each of the materials.

MODULE 00104-04 ◆ INTRODUCTION TO POWER TOOLS 4.15

Discuss the consequences of lifting a saw's blade out of the work while the saw is still running.

Discuss safety and maintenance guidelines for using saber saws.

keep the blade lined up. It keeps the work from vibrating and allows the blade teeth to bite into the material.

4.2.1 How to Use a Saber Saw

Follow these steps to ensure that you use a saber saw safely and efficiently:

Step 1 Wear appropriate personal protective equipment.

Step 2 To avoid vibration, clamp the work to a pair of sawhorses or hold the work in a vise.

Step 3 Check the blade to see that it is the right blade for the job and that it is sharp and undamaged.

Step 4 Measure and mark the work.

Step 5 When you cut from the edge of a board or panel, be sure the front of the baseplate is resting firmly on the surface of the work before you start the saw. The blade should not be touching the work at this stage.

Step 6 Start the saw (pull the trigger) and move the blade gently but firmly into the work. Continue feeding the saw into the work as fast as possible without forcing it. Do not push the blade into the work. *Figure 16* shows the proper way to use a saber saw.

CAUTION

Do not lift the blade out of the work while the saw is still running. If you do, the tip of the blade may hit the wood surface, marring the work and possibly breaking the blade.

Step 7 When the cut is finished, release the trigger and let the blade come to a stop before you remove it from the work.

4.2.2 Safety and Maintenance

When using a saber saw, follow these guidelines:

- Always wear appropriate personal protective equipment.
- Secure the material you are working with to reduce vibration and ensure safety.
- Before you plug the saw into a power source, make sure the switch is in the OFF position.
- Before you cut through a wall or partition, find out what is inside the wall or on the other side of the partition. Avoid hitting water lines or electrical wiring.

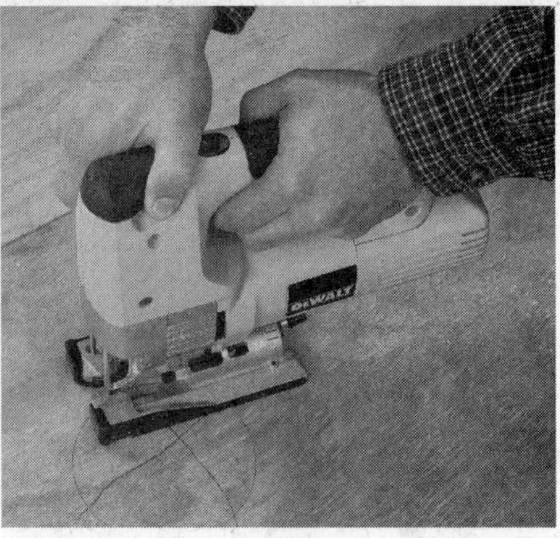

104F16.EPS

Figure 16 ◆ Proper use of a saber saw.

- Always use a sharp blade and never force the blade through the work.
- Do not force or lean into the blade. You could lose your balance and fall forward, or your hands could slip onto the work surface and you could cut yourself.
- When cutting metal pieces, use a metal-cutting blade. Lubricate the blade with an agent such as beeswax to help make tight turns and to reduce the chance of breaking the blade.
- When you are replacing a broken blade, look for any pieces of the blade that may be stuck inside the collar.
- When you install a blade in the saw, make sure it is in as far as it will go, and tighten the setscrew securely. Always disconnect the power source before you change blades or perform maintenance.

4.3.0 Reciprocating Saw

Both the saber saw and the reciprocating saw (also called a Sawsall™) can make straight and curved cuts. They are used to cut irregular shapes and holes in plaster, plasterboard, plywood, studs, metal, and most other materials that can be cut with a saw.

Both saws have straight blades that move as you guide them in the direction of the cut. But here's the difference: The saber saw's blade moves up and down, whereas the reciprocating saw's blade moves back and forth. The reciprocating saw is designed for more heavy-duty jobs than the saber saw is. It can use longer and tougher blades

Instructor's Notes:

4.16 CORE CURRICULUM ◆ INTRODUCTORY CRAFT SKILLS

Blade Safety

If you are cutting into the middle of a piece with a saber saw, first make a starter hole using a power drill. Once you have a drilled hole, tip the saber saw forward on the front of its baseplate, positioning the blade over your drilled hole. Press the trigger and slowly tip the baseplate and the blade down toward the surface. When it strikes the surface, the blade may jump. Keep a steady hand, and the gentle pressure will eventually push the blade through the workpiece. Plunging the blade into the work with sudden force is one of the most common causes of broken blades. The other cause of broken blades is pushing a saber saw too fast. The common result of too much pressure too fast is a snapped blade.

Ask a trainee to practice making a starter hole using a power drill. Have trainees comment on what was done correctly or incorrectly.

Refer to Figure 17. Compare a reciprocating saw with a saber saw.

Explain how to operate a reciprocating saw at the different speed settings.

Discuss the importance of using both hands to grip a saw.

than a saber saw. Also, because of its design, you can get into more places with it. The reciprocating saw (see *Figure 17*) is used for jobs that require brute strength. It can saw through walls or ceilings and create openings for windows, plumbing lines, and more. It is a basic tool in any demolition work.

Like the saber saw, reciprocating saws come in single-speed, two-speed, and variable-speed models. The two-speed reciprocating saw can cut at low and high speeds. The low-speed setting is best for metal work. The high-speed setting is for sawing wood and other soft materials.

The baseplate (shoeplate or footplate) may have a swiveling action, or it may be fixed. Whatever the design, the baseplate is there to provide a brace or support point for the sawing operation.

4.3.1 How to Use a Reciprocating Saw

Follow these steps to use a reciprocating saw safely and efficiently (see *Figure 18*):

Step 1 Wear appropriate personal protective equipment.

Step 2 To avoid vibration, clamp the work to a pair of sawhorses or secure it in a vise.

Step 3 Set the saw to the desired speed. Remember these guidelines:
- Use lower speeds for sawing metal.
- Use higher speeds for sawing wood and other soft materials.

Step 4 Grip the saw with both hands. Place the baseplate firmly against the workpiece.

Step 5 Squeeze the trigger ON switch. The blade moves back and forth, cutting on the backstroke.

CAUTION

Use both hands to grip the saw firmly. Otherwise, the pull created by the blade's grip might jerk the saw out of your grasp.

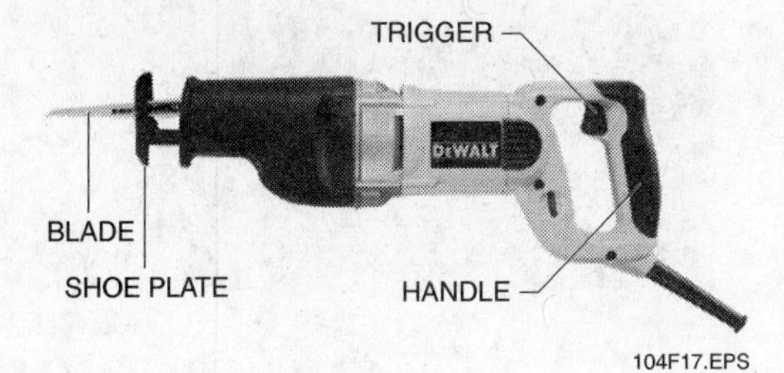

Figure 17 ◆ Reciprocating saw.

Figure 18 ◆ Proper use of a reciprocating saw.

MODULE 00104-04 ◆ INTRODUCTION TO POWER TOOLS 4.17

Review safety guidelines for using reciprocating saws.

Refer to Figure 19. Explain when it is ideal to use a portable handheld bandsaw.

Review the procedures for using a portable bandsaw.

Emphasize that bandsaws cut on the pull, not the push.

4.3.2 Safety and Maintenance

Follow these guidelines to ensure safety for yourself and your nearby co-workers, and a long life for the saw:

- Always wear appropriate personal protective equipment.
- Before you cut through a wall or partition, find out what is inside the wall or on the other side of the partition. Avoid hitting water lines or electrical wiring.
- Always disconnect the power source before you change blades or perform maintenance.

4.4.0 Portable Handheld Bandsaw

The portable handheld bandsaw (see *Figure 19*) is used when it is better to move the saw to the work than to move the work to the saw. The bandsaw can cut pipe, metal, plastics, wood, and irregularly shaped materials. It is especially good for cutting heavy metal, but it will also do fine cutting work.

The bandsaw has a one-piece blade that runs in one direction around guides at either end of the saw. The blade is a thin, flat piece of steel. It is sized according to the diameter of the revolving pulleys that drive and support the blade. The saw often works at various speeds.

4.4.1 How to Use a Portable Bandsaw

Follow these steps to use a bandsaw safely and efficiently:

Step 1 Wear appropriate personal protective equipment.

Step 2 Place the stop firmly against the object to be cut. This will keep the saw from bouncing against the object and breaking the band.

Step 3 Gently pull the trigger. Only a little pressure is needed to make a good clean cut because the weight of the saw gives you more leverage for cutting. *Figure 20* shows the proper way to use a portable bandsaw.

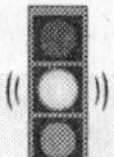

CAUTION

The portable bandsaw cuts on the pull, not the push. You must be especially careful because, in some situations, the saw blade might be moving directly toward your body. Always wear appropriate personal protective equipment and keep your mind focused on the work in front of you.

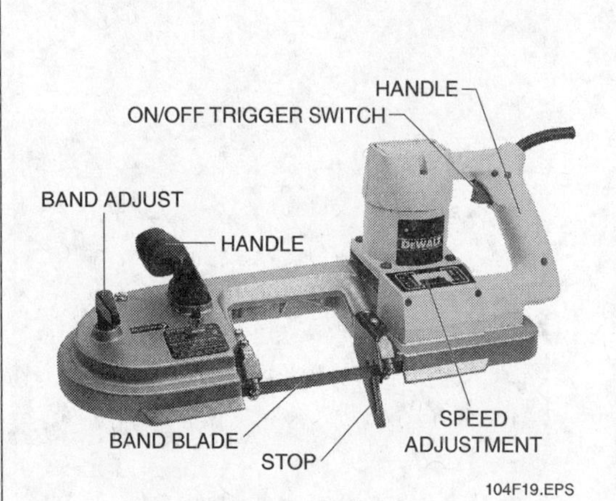

104F19.EPS

Figure 19 ◆ Portable handheld bandsaw.

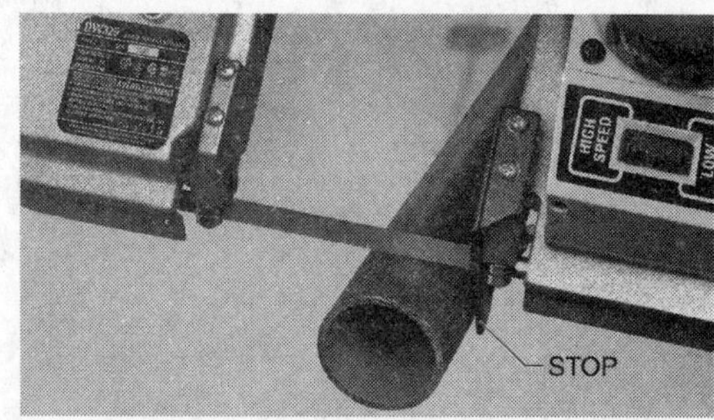

104F20A.EPS

(A) PLACE THE STOP FIRMLY AGAINST THE OBJECT.

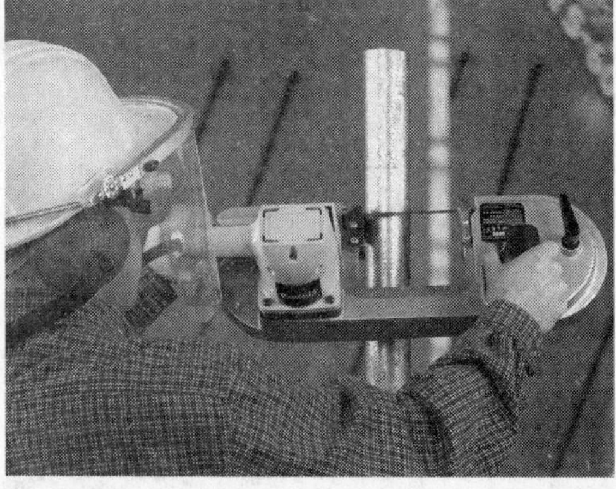

104F20B.EPS

(B) APPLY ONLY A LITTLE PRESSURE TO MAKE A CUT.

Figure 20 ◆ Proper use of a portable bandsaw.

Instructor's Notes:

4.18 CORE CURRICULUM ◆ INTRODUCTORY CRAFT SKILLS

4.4.2 Safety and Maintenance

Follow these guidelines to ensure safety for yourself and your co-workers, and a long life for the saw:

- Always wear appropriate personal protective equipment.
- Use only a bandsaw that has a stop.
- Before you cut through a wall or partition, find out what is inside the wall or on the other side of the partition. Avoid hitting water lines or electrical wiring.
- The blade of a portable bandsaw gets stuck very easily. Never force a portable bandsaw. Let the saw do the cutting.
- The blades should be waxed with an appropriate lubricant, such as the one recommended by the blade's manufacturer. Always disconnect the power source before you do maintenance.

4.5.0 Power Miter Box

The power miter saw combines a miter box with a circular saw, allowing it to make straight and miter cuts. There are two types of power miter boxes: power miter saws and compound miter saws.

In a power miter box (see *Figure 21*), the saw blade pivots horizontally from the rear of the table and locks in position to cut angles from 0 degrees to 45 degrees right and left. Stops are set for common angles. The difference between the power miter saw and the compound miter saw is that the blade on the compound miter saw can be tilted vertically, allowing the saw to be used to make a compound cut (combined bevel and miter cut).

4.5.1 How to Use a Power Miter Box

Follow these steps to use a power miter box safely and efficiently:

Step 1 Wear appropriate personal protective equipment.

Step 2 Be sure the saw blade has reached its maximum speed before starting the cut.

Step 3 Hold the workpiece firmly against the fence when making the cut.

Step 4 Turn off the saw immediately after making the cut and use the brake to stop the blade.

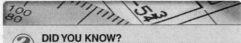

DID YOU KNOW?

For Portable Bandsaws, Low Speed Works Best

The portable bandsaw cuts best at a low speed. Using a high speed will cause the blade's teeth to rub rather than cut. This can create heat through friction, which will cause the blade to wear out quickly.

104F21.EPS

Figure 21 ◆ Power miter box saw.

4.5.2 Safety and Maintenance

Follow these guidelines to ensure safety for yourself and your co-workers, and a long life for the miter box:

- Always check the condition of the blade and be sure the blade is secure before starting the saw.
- Keep your fingers clear of the blade.
- Be sure the blade guards are in place and working properly.
- Never make adjustments while the saw is running.
- Never leave a saw until the blade stops.
- Be sure the saw is sitting on a firm base and is properly fastened to the base.
- Be sure the saw is securely locked at the correct angle.
- If working on long stock, have a helper support the end of the stock.

Refer to Figure 21. Explain how the saw blade pivots in a power miter box. Compare a power miter saw with a compound miter saw.

Refer again to Figure 21. Explain how to hold the workpiece when making a cut using a power miter box.

Review the guidelines for maintaining and safely using a power miter box.

Go over the Review Questions for Section 4.0.0. Answer any questions trainees may have.

MODULE 00104-04 ◆ INTRODUCTION TO POWER TOOLS 4.19

See the Teaching Tip for Sections 4.0.0–4.5.2 at the end of this module.

Have each trainee perform Performance Tasks 1 and 2 to your satisfaction. Fill out a Performance Profile Sheet for each trainee.

Have trainees review Sections 5.0.0–6.5.2 for the next session.

Ensure that you have everything required for demonstrations, laboratories, and testing during this session.

Explain that grinding tools can be used to power a variety of abrasive wheels.

Explain how sanders are used to shape workpieces.

Review Questions

Section 4.0.0

1. Never use a circular saw that doesn't have a lower blade guard because the _____.

a. trigger lock won't work properly
b. saw blade may fall off
c. guard protects you from the blade and from flying particles
d. wood will split

2. For circular saws, saw size is measured by the _____.

a. diameter of the circular blade
b. depth of the upper blade guard
c. area of the base
d. length of the guide slot

3. When cutting with a circular saw, grip the saw handles _____.

a. and pull the saw toward you
b. firmly with one hand
c. loosely with one hand
d. firmly with two hands

4. If the material you are cutting with a circular saw isn't heavy enough to stay in position without moving, _____.

a. use a heavier saw
b. weight or clamp it down
c. hold it down with your free hand
d. use a lighter saw

5. To secure small pieces of material when you are cutting them, always use _____.

a. a clamp
b. your hands
c. nails or screws
d. the safety guard

6. Saber saws are good for _____.

a. heavy-duty materials
b. oversized pieces of lumber
c. delicate and intricate work
d. masonry and concrete

7. When using a saber saw, avoid vibration by _____.

a. holding the workpiece down with your free hand.
b. setting a heavy object on the workpiece
c. using a low-speed setting
d. using a clamp or vise to hold the work

8. Before you cut through a wall or partition, always _____.

a. remove the lower blade guard
b. find out what is on the other side
c. increase the revolutions per minute
d. lubricate the guard with oil or grease

9. Before you plug any saw into a power source, make sure the _____.

a. power switch is in the OFF position
b. blade has been loosened
c. power switch is in the ON position
d. lower blade guard has been removed

10. Use only a bandsaw that has a _____.

a. breastplate with a broad surface
b. battery pack
c. thick, three-piece blade
d. stop

5.0.0 ◆ GRINDERS AND SANDERS

Grinding tools can power all kinds of **abrasive** wheels, brushes, buffs, drums, bits, saws, and discs. These wheels come in a variety of materials and **grits.** They can drill, cut, smooth, and polish; shape or sand wood or metal; mark steel and glass; and sharpen or engrave. They can even be used on plastics.

Sanders can shape workpieces, remove imperfections in wood and metal, and create the smooth surfaces needed before finishing work can begin.

Sanding is an essential part of all finish carpentry. Sanding gives a smooth, professional look to the completed work regardless of whether it will be painted. You will learn about the following types of grinders and sanders in this section:

- Angle grinders (side grinders), end grinders, detail grinders
- Bench grinders
- Portable belt sanders
- Random orbital sanders (finishing sanders)

Instructor's Notes:

4.20 CORE CURRICULUM ◆ INTRODUCTORY CRAFT SKILLS

5.1.0 Angle Grinders, End Grinders, and Detail Grinders

These types of grinders are grouped together because they are all handheld.

The angle grinder (also called a side grinder) is used to grind away hard, heavy materials and to grind surfaces such as pipes, plates, or welds (see *Figure 22*).

End grinders (see *Figure 23*) are also called horizontal grinders or pencil grinders. These smaller grinders are used to smooth the inside of materials such as pipe.

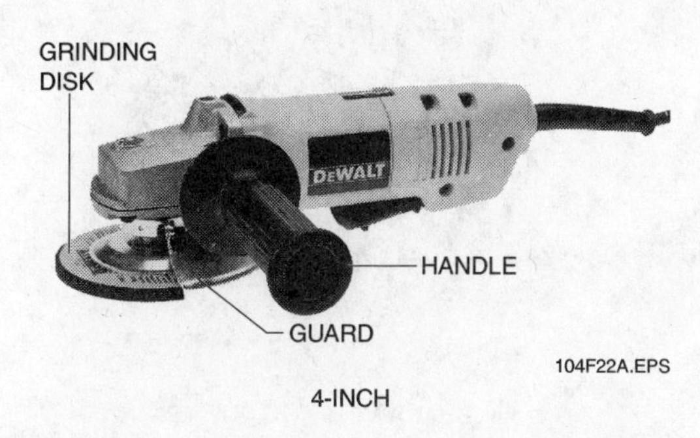

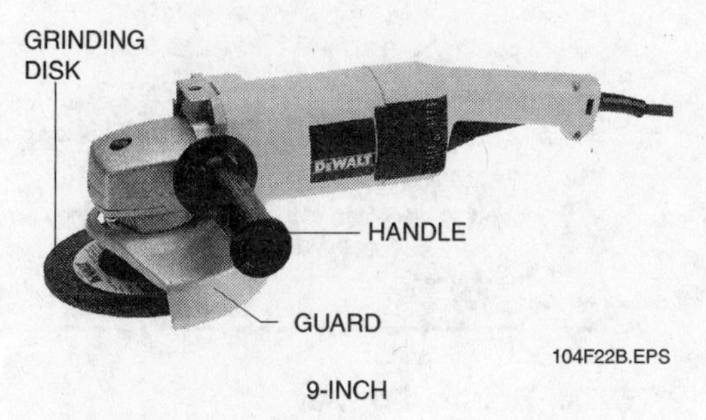

Figure 22 ◆ Angle grinders.

Detail grinders (see *Figure 24*) use small attachments, also called points, to smooth and polish intricate metallic work. These attachments, some of which are shown in *Figure 24*, are commonly made in sizes ranging from 1⁄16 to 1⁄4 inch.

The angle grinder has a rotating grinding disc set at a right angle to the motor shaft. The grinding disc on the end grinder rotates in line with the motor shaft. Grinding is also done with the outside of the grinding disc. The detail grinder has a shank that extends from the motor shaft; points of different sizes and shapes can be mounted on the shank.

5.1.1 How to Use an Angle Grinder, End Grinder, or Detail Grinder

Follow these steps to use an angle, end, or detail grinder safely and efficiently (see *Figure 25*):

Step 1 Wear appropriate personal protective equipment.

Step 2 If it is not already secured, secure the material in a vise or clamp it to the bench.

Step 3 To use an angle grinder, place one hand on the handle of the grinder and one on the trigger. To use an end grinder or detail grinder, grip the grinder at the shaft end with one hand and cradle the opposite end of the tool in your other hand.

Step 4 Finish the work by removing any loose material with a wire brush.

5.1.2 Safety and Maintenance

Follow these guidelines to ensure safety for yourself and your co-workers, and a long life for the grinder:

- Always wear appropriate personal protective equipment.
- Never use an angle grinder, end grinder, or detail grinder unless it is equipped with the guard that surrounds the grinding wheel.
- Choose a grinding disc that is appropriate for the type of work you are doing.

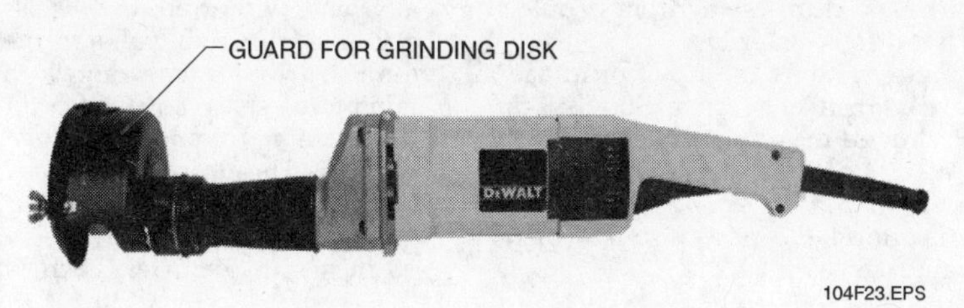

Figure 23 ◆ End grinder.

Refer to Figure 22. Explain how angle grinders can be used to grind hard, heavy materials.

Refer to Figure 23. Explain how end grinders can be used to smooth the inside of pipes.

Bring in a variety of attachments for detail grinders. Demonstrate how these attachments are used to polish metallic work.

Bring in a variety of grinders. Ask trainees to practice using the different grinders to smooth and polish a variety of metal objects.

Review the safety and maintenance guidelines for using handheld grinders.

MODULE 00104-04 ◆ INTRODUCTION TO POWER TOOLS 4.21

Discuss problems that can result from using a cracked grinding disc.

Refer to Figure 26. Explain how to use a bench grinder to smooth mushroomed chisel heads.

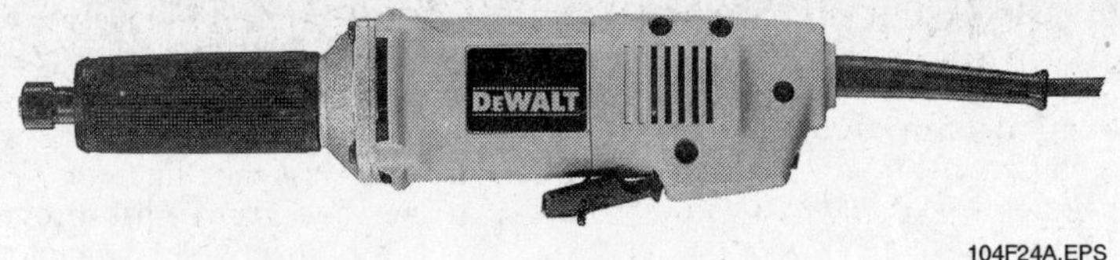

104F24A.EPS

DETAIL GRINDER

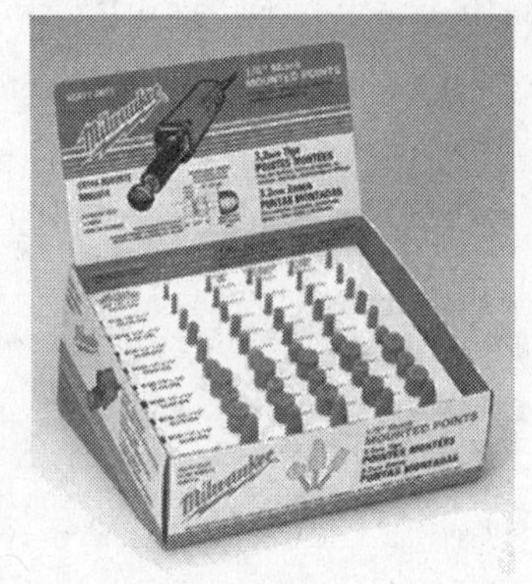

104F24B.EPS

1/8-INCH
SHANK-MOUNTED POINTS

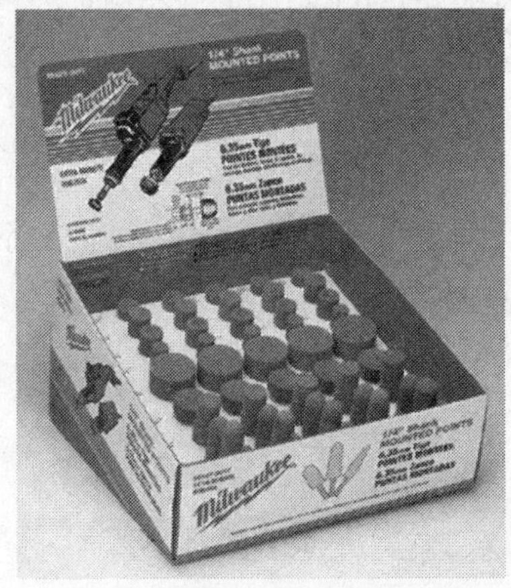

104F24C.EPS

1/4-INCH
SHANK-MOUNTED POINTS

Figure 24 ◆ Detail grinder and points.

- Make sure that you are using a disc that is properly sized for the grinder.
- Before you start the grinder, make sure the grinding disc is secured and is in good condition.
- Make sure all guards are in place.
- Be sure to have firm footing and a firm grip before you use a grinder. Grinders have a tendency to pull you off balance.
- Always hold the grinder with both hands.
- Always use a spark deflector (shield) as well as proper eye protection.
- Direct sparks and debris away from people or from any **hazardous materials.**
- When you are grinding on a platform, use a flame-retardant blanket to catch falling sparks.
- When you shut off the power, do not leave the tool until the grinding disc has come to a complete stop.
- Always disconnect the power source before you do maintenance.

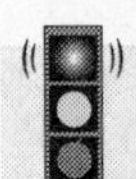

WARNING!

Grinding discs can explode if used when they are cracked. Inspect the disc for cracks before using the grinder.

5.2.0 Bench Grinder

Bench grinders (see *Figure 26*) are electrically powered stationary grinding machines. They usually have two grinding wheels that are used for grinding, rust removal, and metal buffing. They are also great for renewing worn edges and maintaining the sharp edges of cutting tools. Remember learning about the danger of mushroomed cold chisel heads in *Introduction to Hand Tools?* The bench grinder can smooth these heads.

Heavy-duty grinder wheels range from 6¾ to 10 inches in diameter. Each wheel's maximum

Instructor's Notes:

4.22 CORE CURRICULUM ◆ INTRODUCTORY CRAFT SKILLS

104F25.EPS

Figure 25 ◆ Proper use of a handheld grinder.

speed is given in rpm. Never use a grinding wheel above its rated maximum speed. Bench grinders come with an adjustable tool rest. This is the surface on which you position the material you are grinding, such as cold chisel heads. There should be a distance of only ⅛ inch between the tool rest and the wheel. Attachments for the bench grinder include knot-wire brushes for removing rust, scale, and file marks from metal surfaces, and cloth buffing wheels for polishing and buffing metal surfaces.

Explain how to position a grinding wheel.

Explain problems that could result if adjustments are made when the grinder is on.

5.2.1 How to Use a Bench Grinder

Follow these steps to use a bench grinder safely and efficiently:

Step 1 Wear appropriate personal protective equipment. A face shield is essential.

Step 2 Always use the adjustable tool rest as a support when you are grinding or beveling metal pieces. There should be a maximum gap of ⅛ of an inch between the tool rest and the wheel and ¼ inch between the top guard and wheel. Make sure the bench grinder is placed on a secure surface.

CAUTION

Never change the adjustment of tool rests when the grinder is on or when the grinding wheels are spinning. Doing so may damage the work.

104F26.EPS

Figure 26 ◆ Bench grinder.

MODULE 00104-04 ◆ INTRODUCTION TO POWER TOOLS 4.23

Demonstrate how to perform a ring test before you mount a grinding wheel.

Refer to Figure 28. Explain that you use a portable belt sander to remove rough areas from sections of wood.

Refer to Figure 29. Explain how to use a portable belt sander.

Step 3 Let the wheel come up to full speed before you touch the work.

Step 4 Keep the metal you are grinding cool. If the metal gets too hot, it can destroy the temper (hardness) of the material you are grinding, such as a metal chisel head.

Step 5 Whenever possible, work on the face of the wheel. For many jobs you must work on the side of the wheel, but inspect the wheel frequently to be sure you do not reduce the thickness so much that it can break. *Figure 27* shows the proper way to use a bench grinder.

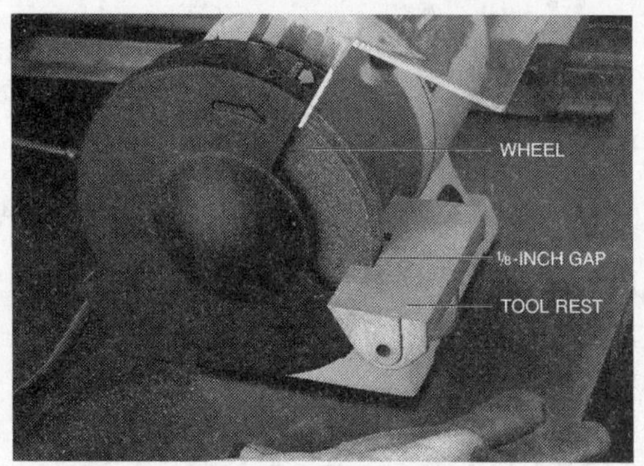

104F27A.EPS

THERE SHOULD BE A 1/8-INCH GAP BETWEEN THE TOOL REST AND THE WHEEL AND A 1/4-INCH GAP BETWEEN THE TOP GUARD AND THE WHEEL.

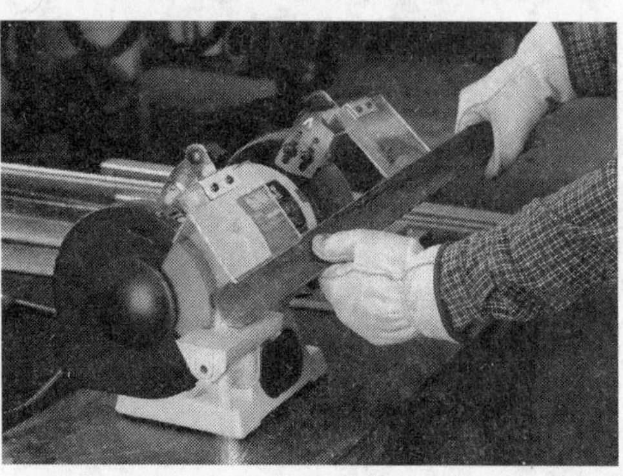

104F27B.EPS

WHENEVER POSSIBLE, WORK ON THE FACE OF THE WHEEL.

Figure 27 ◆ Proper use of a bench grinder.

5.2.2 Safety and Maintenance

Follow these guidelines to ensure safety for yourself and your co-workers, and a long life for the grinder:

- Always wear appropriate personal protective equipment.
- Never wear loose clothing or jewelry when you are grinding. It can get caught in the wheels.
- Grinding metal creates sparks, so keep the area around the grinder clean.
- Always adjust the tool rests so they are within 1/8 inch of the wheel. This reduces the chance of getting the work wedged between the rest and the wheel.
- Keep your hands away from the grinding wheels.
- Let the wheel come up to full speed before you touch the work.
- Never use a grinding wheel above its rated maximum speed.
- When you are finished using the bench grinder, shut it off.
- Always make sure the bench grinder is disconnected before you change grinding wheels.
- Perform a **ring test** before you mount a wheel. After you look for chipped edges and cracks, mount the wheel on a rod that you pass through the wheel hole. Tap the wheel gently on the side with a piece of wood. The wheel will ring clearly if it is in good condition. A dull thud may mean that there is a crack that you can't see. Get rid of the wheel if this happens.

5.3.0 Portable Belt Sander

The portable belt sander (see *Figure 28*) is used to remove rough areas from large, flat sections of wood; for trimming off excess wood; and for stripping old finishes such as paint and varnish.

A portable belt sander uses a continuous-loop abrasive belt that is stretched between two drums. The back roller is powered by the motor; the front roller is spring-loaded to correct belt tension. The longer and wider the belt, the heavier and more powerful the sander. The size of the belt sander refers to the width of the belt. Heavy-duty models usually have belt widths of 3 to 4 inches. All models should have an integrated dust-collection bag.

5.3.1 How to Use a Belt Sander

Follow these steps to use a belt sander safely and efficiently. *Figure 29* shows the proper way to use a portable belt sander.

Instructor's Notes:

4.24 CORE CURRICULUM ◆ INTRODUCTORY CRAFT SKILLS

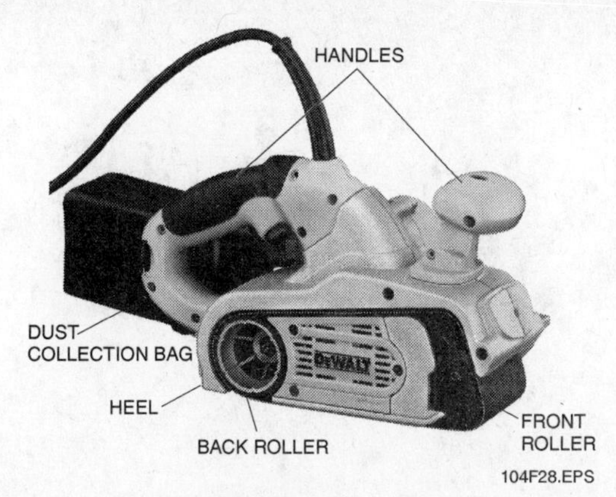

Figure 28 ◆ Portable belt sander.

Step 1 Wear appropriate personal protective equipment.

Step 2 Rest the sander on its heel when starting the motor.

Step 3 Once the sander is moving, lower it onto the workpiece.

Step 4 Keep the sander level while you move it across the workpiece.

Step 5 Keep the sander moving whenever it is in contact with the workpiece.

CAUTION

Do not allow the sander to remain stationary. This will create a groove, gouge, or depression in the surface of the workpiece.

Step 6 Tip the sander back on its heel when you finish. Do not set the sander down until the belt stops moving.

5.3.2 Safety and Maintenance

Follow these guidelines to ensure safety for yourself and your co-workers, and a long life for the sander:

- Always wear appropriate personal protective equipment. Use a dust mask, even with models that have a dust-collection system.
- Never wear loose clothing or jewelry when you are sanding. It could get caught in the belt.
- Keep your hands away from the sanding belt.
- Keep the power cord away from the area being sanded.
- Let the belt come up to full speed before you apply it to the workpiece.
- When you are finished using the belt sander, shut it off and unplug it.
- Always make sure the belt sander is disconnected before you change belts.
- Do not light matches when you are sanding in a confined area. Dust from sanding can be explosive.

Explain why the sander cannot remain stationary on the workpiece.

Emphasize the importance of wearing a dust mask when working with sanders.

Figure 29 ◆ Proper use of a portable belt sander.

MODULE 00104-04 ◆ INTRODUCTION TO POWER TOOLS 4.25

Refer to Figure 30. Explain how to use finishing sanders to smooth surfaces.

Demonstrate how to use a back-and-forth hand motion with an orbital sander's circular motion.

Discuss the consequences of leaving a sander in one place.

5.4.0 Random Orbital Sander (Finishing Sander)

Finishing sanders are used to create an even, smooth surface, such as one needed for painting. Finishing sander models used to have either orbital (circular) or oscillating (back and forth) movement. The random orbital sander (see *Figure 30*) combines both types of movement.

The random orbital sander's circular motion works together with your own back-and-forth hand motion to eliminate any telltale patterns that orbital sanders tend to leave on the work. Using a fine-grit abrasive paper and moving the sanding disc in a smooth motion over the workpiece maintains a swirl-free finish. Random orbital sanders come in single- and variable-speed models. Many have dust-collection bags attached to the body of the sander.

104F30.EPS

Figure 30 ◆ Random orbital sander.

5.4.1 How to Use a Random Orbital Sander

Follow these steps to use a random orbital sander safely and efficiently:

Step 1 Wear appropriate personal protective equipment.

Step 2 Maintain a firm hold when you start the sander.

Step 3 Let the sander come up to full speed before applying it to the surface of the workpiece.

Step 4 Keep the sander level while you move it across the workpiece.

Step 5 Keep the sander moving whenever it is in contact with the workpiece.

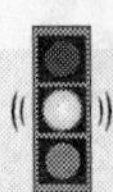

CAUTION

Do not let the sander remain in one place. This will create a groove, gouge, or depression in the surface of the workpiece.

Step 6 Lift the sander off the workpiece when you finish. Do not set the sander down until it stops moving. *Figure 31* shows the proper way to use a random orbital sander.

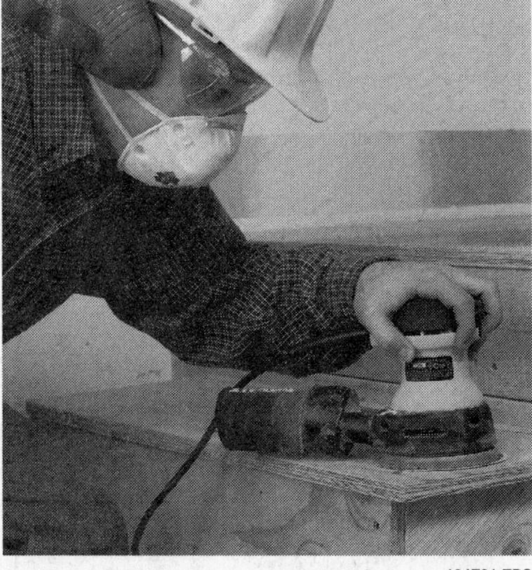

104F31.EPS

Figure 31 ◆ Proper use of a random orbital sander.

5.4.2 Safety and Maintenance

Follow the belt-sander safety practices that you learned earlier in this module when using a finishing sander.

Instructor's Notes:

4.26 CORE CURRICULUM ◆ INTRODUCTORY CRAFT SKILLS

Explain why you should always sand with rougher grits of paper first.

Go over the Review Questions for Section 5.0.0. Ask whether trainees have any questions.

Sanding Tips

Always sand with rougher grits of paper first. As you work, change to progressively finer grits for a smoother finish. Before giving wood its final sanding, wipe it down with a damp rag to remove sanding dust from between the grain lines.

Review Questions

Section 5.0.0

1. The angle grinder is used to grind _____.
   a. soft, porous materials
   b. imperfections in wood
   c. hard, heavy materials
   d. nonmetals only

2. The end grinder is used to _____.
   a. polish intricate work
   b. grind surfaces
   c. smooth the work before painting
   d. smooth the inside of materials, such as pipe

3. The _____ is an electrically powered stationary grinding machine.
   a. angle grinder
   b. bench grinder
   c. end grinder
   d. detail grinder

4. The adjustable tool rest on a bench grinder should be positioned a maximum of _____ of an inch from the wheel.
   a. ⅛
   b. ¼
   c. ½
   d. ¾

5. When using a grinder, try to work on the _____ of the wheel whenever possible.
   a. face
   b. edge
   c. bottom
   d. side

6. The adjustable tool rest on a bench grinder should be positioned _____ inch from the wheel.
   a. ⅛
   b. ¼
   c. ½
   d. ¾

7. Never use a grinding wheel _____.
   a. to remove rust
   b. above its rated maximum speed
   c. to bevel chiseling tools
   d. on cool metal surfaces

8. The portable belt sander is used to _____.
   a. remove rough areas from wood
   b. maintain sharp edges on cutting tools
   c. sand heavy-duty materials only
   d. sand portable items only

9. When you are sanding, wear _____.
   a. loose clothing
   b. a full face guard
   c. a dust mask
   d. a belt

10. The random orbital sander is used to _____.
    a. strip off paint
    b. bevel the edges of chisel heads
    c. eliminate circular swirls on finished work
    d. remove rust from heavily corroded materials

MODULE 00104-04 ◆ INTRODUCTION TO POWER TOOLS 4.27

Refer to Figure 32. Discuss the common uses of nail guns on the construction site.

6.0.0 ◆ MISCELLANEOUS POWER TOOLS

It is very common to see several different types of power tools on a construction site. In this section you will learn about some of the commonly used power tools we haven't yet covered. It is likely you will see them used on your site or use them yourself.

- Pneumatically powered nailers (nail guns)
- Powder-actuated fastening systems
- Air impact wrenches
- Pavement breakers
- Hydraulic jacks (Porta-Power™)

6.1.0 Pneumatically Powered Nailer (Nail Gun)

Pneumatically powered nailers (see *Figure 32*), or nail guns, are commonplace on construction jobs. They greatly speed up the installation of materials such as wallboard, molding, framing members, and shingles.

Nail guns are driven by compressed air traveling through air lines connected to an air compressor. Nailers are designed for specific purposes, such as roofing, framing, siding, flooring, sheathing, trim, and finishing. Nailers use specific types of nails depending on the material to be fastened. The nails come in coils and in strips and are loaded into the nail gun.

104F32.EPS

Figure 32 ◆ Pneumatic nailer.

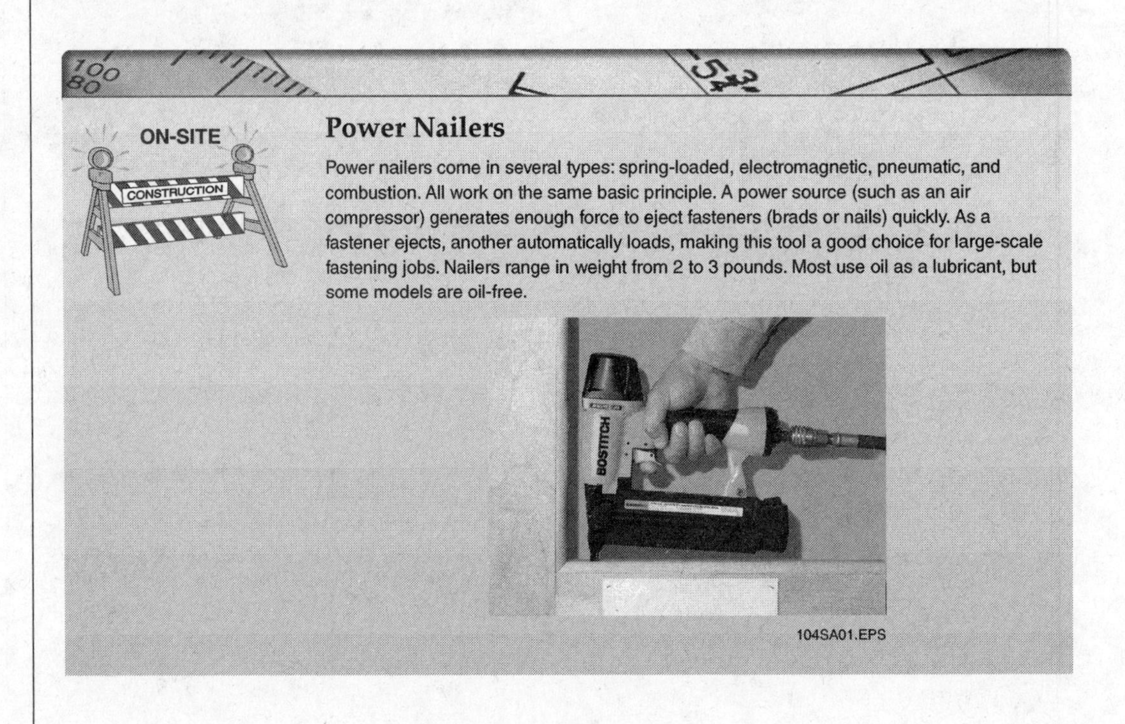

Power Nailers

Power nailers come in several types: spring-loaded, electromagnetic, pneumatic, and combustion. All work on the same basic principle. A power source (such as an air compressor) generates enough force to eject fasteners (brads or nails) quickly. As a fastener ejects, another automatically loads, making this tool a good choice for large-scale fastening jobs. Nailers range in weight from 2 to 3 pounds. Most use oil as a lubricant, but some models are oil-free.

104SA01.EPS

Instructor's Notes:

4.28 CORE CURRICULUM ◆ INTRODUCTORY CRAFT SKILLS

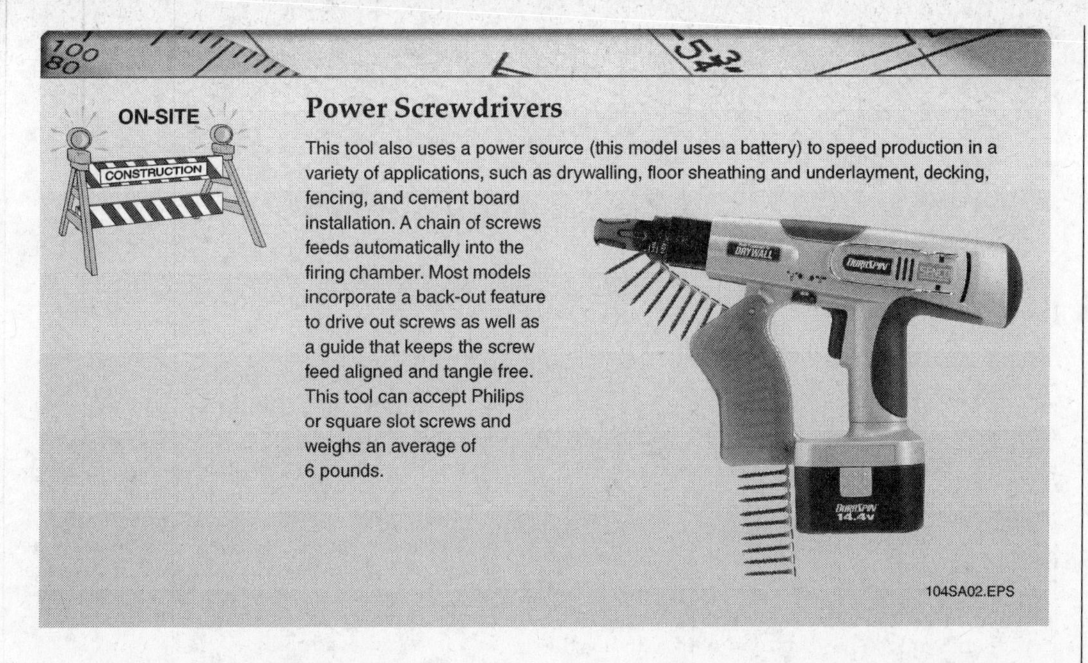

Power Screwdrivers

This tool also uses a power source (this model uses a battery) to speed production in a variety of applications, such as drywalling, floor sheathing and underlayment, decking, fencing, and cement board installation. A chain of screws feeds automatically into the firing chamber. Most models incorporate a back-out feature to drive out screws as well as a guide that keeps the screw feed aligned and tangle free. This tool can accept Philips or square slot screws and weighs an average of 6 pounds.

104SA02.EPS

Bring in a variety of nail guns. Have trainees inspect them for damage and loose connections. Ask trainees to practice using the nail guns safely.

Ask a trainee to explain why a pneumatic nailer should never be carried by its trigger.

6.1.1 How to Use a Power Nailer

Follow these steps to use a power nailer safely and efficiently:

Step 1 Read the manufacturer's instructions before using a pneumatically powered nailer. Wear all appropriate personal protective equipment.

Step 2 Inspect the nailer for damage and loose connections.

Step 3 Load the nails into the nailer. Be sure to use the correct type of nail for the job.

Step 4 Ensure that hoses are connected properly.

Step 5 Check the air compressor and adjust the pressure.

Step 6 Try a test nail in scrap material. Most nailers operate at pressures of 70 to 120 pounds per square inch (psi).
  a. If nail penetration is too deep, adjust the regulator on the air compressor to get a lower reading on the pressure gauge.
  b. If the nailhead sticks up above the surface, adjust the compressor for higher pressure.

Step 7 When nailing wall materials, locate and mark wall studs before nailing. Otherwise, you won't be able to feel a missed nail that penetrates the wallboard but misses the stud.

Step 8 Hold the nailer firmly against the material to be fastened, then press the trigger (see *Figure 33*).

Step 9 Disconnect the air hose as soon as you finish.

104F33.EPS

Figure 33 ◆ Proper use of a nailer.

Emphasize the importance of adhering to the maximum specified operating pressure of a pneumatic nailer.

Explain how to keep a nailer properly oiled.

Ask a trainee to explain problems that can occur if a nail gun is used as a toy.

Refer to Figure 34. Explain the role of boosters in a powder-actuated fastening system.

Demonstrate how to set up a powder-actuated fastening tool. Explain how to use fastening tools.

Review the guidelines for maintaining and safely using powder-actuated tools.

Power Nailer Safety

Pneumatic nailers are designed to fire when the trigger is pressed and the tool is pressed against the material being fastened. An important safety feature of all pneumatic nailers is that they will not fire unless pressed against the material.

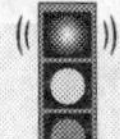

WARNING!
Never exceed the maximum specified operating pressure of a pneumatic nailer. Doing so will damage the pneumatic nailer and cause injury.

6.1.2 Safety and Maintenance

Follow these guidelines to ensure safety for yourself and your co-workers, and a long life for the nail gun:

- Always wear appropriate personal protective equipment.
- Review the operating manual before using any nailer.
- Keep the nailer oiled. Add a few drops to the air inlet before each use, according to the manufacturer's recommendations.
- Use the correct nailer for the job. Use the correct size and type of nail for the job.
- Never load the nailer with the compressor hose attached.
- Never leave the nailer unattended.
- If the nailer is not firing, disconnect the air hose before you attempt repairs.
- Keep all body parts and co-workers away from the nail path to avoid serious injury. Nails can go through paneling and strike someone on the other side.
- Check for pipes, electrical wiring, vents, and other materials behind wallboard before nailing.

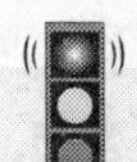

WARNING!
A nail gun is not a toy. Playing with a nail gun can cause serious injury. Nails can pierce a hand, leg, or eye easily. Never point a nail gun at anyone or carry one with your finger on the trigger. Use the nail gun only as directed.

6.2.0 Powder-Actuated Fastening Systems

The use of powder-actuated anchor or fastening systems has been increasing rapidly in recent years. They are used for anchoring static loads to steel and concrete beams, walls, and so forth.

A powder-actuated tool is a low-velocity fastening system powered by gunpowder cartridges called **boosters.** The tools are used to drive steel pins or threaded steel studs directly into masonry and steel (see *Figure 34*).

6.2.1 How to Set Up and Use a Powder-Actuated Fastening Tool

Follow these steps to use a powder-actuated tool safely and efficiently:

Step 1 Wear the appropriate personal protective equipment, including safety goggles, ear protection, and a hard hat.

Step 2 Feed the pin or stud into the piston.

Step 3 Feed the gunpowder cartridge (booster or charge) into position.

Step 4 Position the tool in front of the item to be fastened and press it against the mounting surface. This pressure releases the safety lock.

Step 5 Pull the trigger handle to fire the booster charge (see *Figure 34*).

6.2.2 Safety and Maintenance

Follow these guidelines to ensure safety for yourself and your co-workers, and a long life for the powder-actuated tool:

- Always wear appropriate personal protective equipment, including ear protection, safety goggles, and a hard hat.
- Do not use a powder-actuated tool until you are certified on the model you will be using.

Instructor's Notes:

Emphasize that all powder-actuated tools must be operated by qualified and certified personnel.

Discuss the consequence of using a powder-actuated tool on easily penetrated materials.

Refer to Figure 35. Explain how to adjust the speed and strength of an air impact wrench.

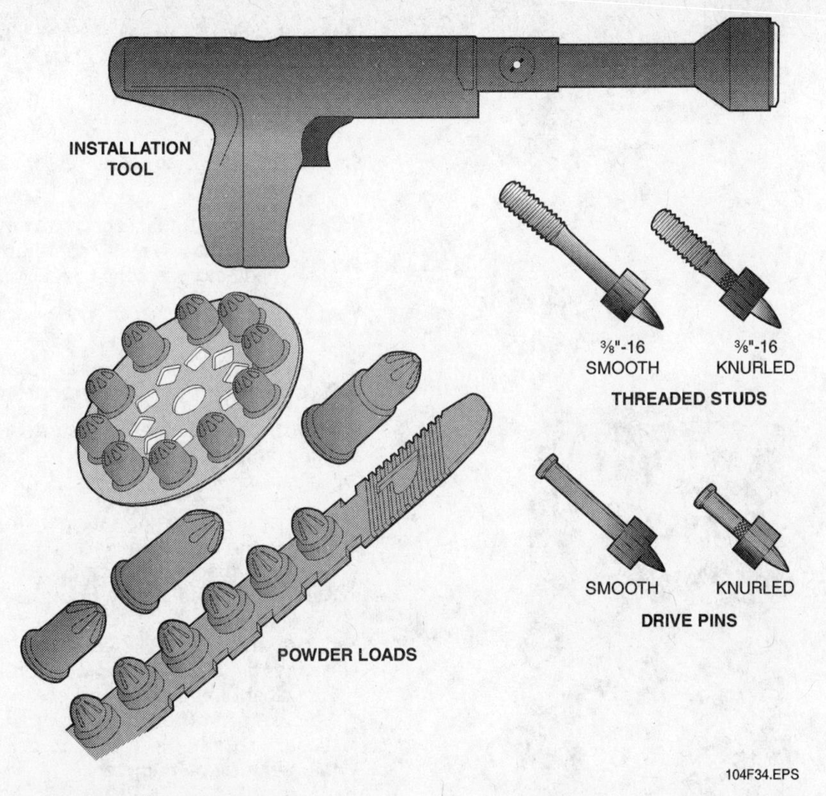

Figure 34 ◆ Powder-actuated fastening system.

- Follow all safety precautions in the manufacturer's instruction manual.
- Always install the charge until just prior to charging.
- Use the proper size pin for the job you are doing.
- When loading the driver, put the pin in before you put in the charge.
- Use the correct booster charge according to the manufacturer's instructions for the tool being used.
- Never initiate a charge until just prior to firing.
- Never hold the end of the barrel against any part of your body or cock the tool against your hand.
- Never hold your hand behind the material you are fastening.
- Do not fire the tool close to the edge of concrete. Pieces of concrete may chip off and strike someone, or the projectile could continue past the concrete and strike a co-worker.
- Never try to pry the booster out of the magazine with a sharp instrument.

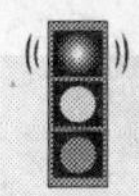

WARNING!
OSHA requires that all operators of powder-actuated tools be qualified and certified by the manufacturer of the tool. You must carry a certification card when using the tool.

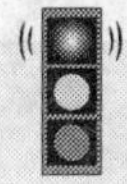

WARNING!
Avoid firing a powder-actuated tool into easily penetrated materials. The fastener may pass through the material and become a flying missile on the other side.

6.3.0 Air Impact Wrench

Air impact wrenches (see *Figure 35*) are power tools that are used to fasten, tighten, and loosen nuts and bolts. The speed and strength (torque) of

MODULE 00104-04 ◆ INTRODUCTION TO POWER TOOLS 4.31

Set up stations with a variety of air impact wrenches. Have trainees practice connecting the wrench to the air hose and adjusting the pressure on the air compressor.

Review safety and maintenance guidelines for air impact wrenches.

Explain damage that can be caused by a handheld socket.

Refer to Figure 36. Explain that, unlike hammer drills, which rotate, demolition tools such as pavement breakers typically reciprocate.

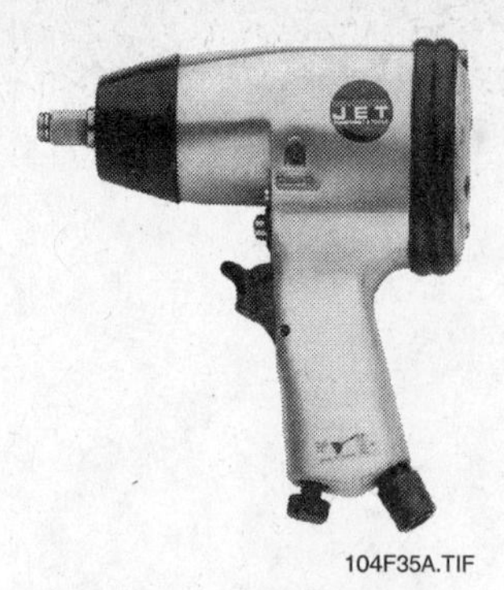

104F35A.TIF

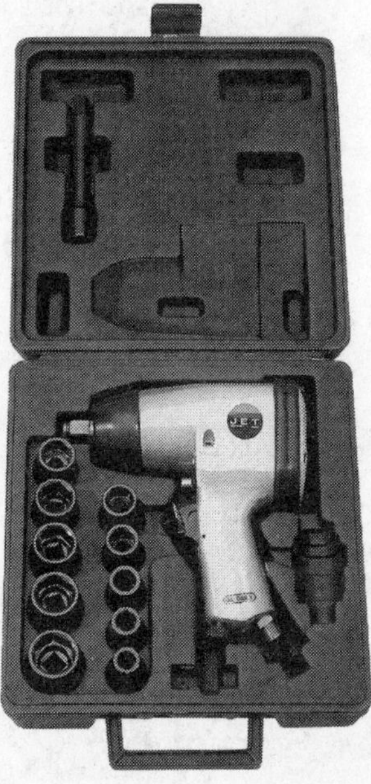

104F35B.TIF

Figure 35 ◆ Air impact wrench.

these wrenches can easily be adjusted depending on the type of job. Air impact wrenches are powered pneumatically (with compressed air). In order to operate an air impact wrench, it must be attached with a hose to an air compressor.

6.3.1 How to Set Up and Use an Air Impact Wrench

Follow these steps to use an air impact wrench safely and efficiently:

Step 1 Read the manufacturer's instructions before using an air impact wrench. Wear all appropriate personal protective equipment.

Step 2 Inspect the wrench for damage.

Step 3 Select the appropriate impact socket.

Step 4 Connect the wrench to the appropriate air hose.

Step 5 Turn on the air compressor and adjust the pressure.

Step 6 Place the impact socket firmly against the material to be fastened, removed, or loosened, and press the trigger.

Step 7 Disconnect the air hose as soon as you finish.

6.3.2 Safety and Maintenance

Follow these guidelines to ensure safety for yourself and your nearby co-workers, and a long life for the air impact wrench:

- Always wear appropriate personal protective equipment, including eye and ear protection.
- Keep your body stance balanced.
- Keep your hands away from the working end of the wrench.
- Ensure that the workpiece is secure.
- Always use clean, dry air at the proper pressure.
- Stay clear of whipping air hoses.
- Always turn off the air supply and disconnect the air supply hose before performing any maintenance on the wrench.

WARNING!

Using handheld sockets can damage property and cause injury. Use only impact sockets made for air impact wrenches.

6.4.0 Pavement Breaker

Several large-scale demolition tools are frequently used in construction. They include pavement breakers, clay spades, and rock drills (see *Figure 36*). These tools do not rotate like hammer drills. They reciprocate (move back and forth). The name *jackhammer* comes from a trade name, but has come to refer to almost any of the handheld impact tools. There are differences in the tools and their uses, however. In this section, we will look at the pavement breaker.

The pavement breaker is used for large-scale demolition work, such as tearing down brick and concrete walls and breaking up concrete or pavement.

A pavement breaker weighs from 50 to 90 pounds. On most pavement breakers, a throttle is

Instructor's Notes:

4.32 CORE CURRICULUM ◆ INTRODUCTORY CRAFT SKILLS

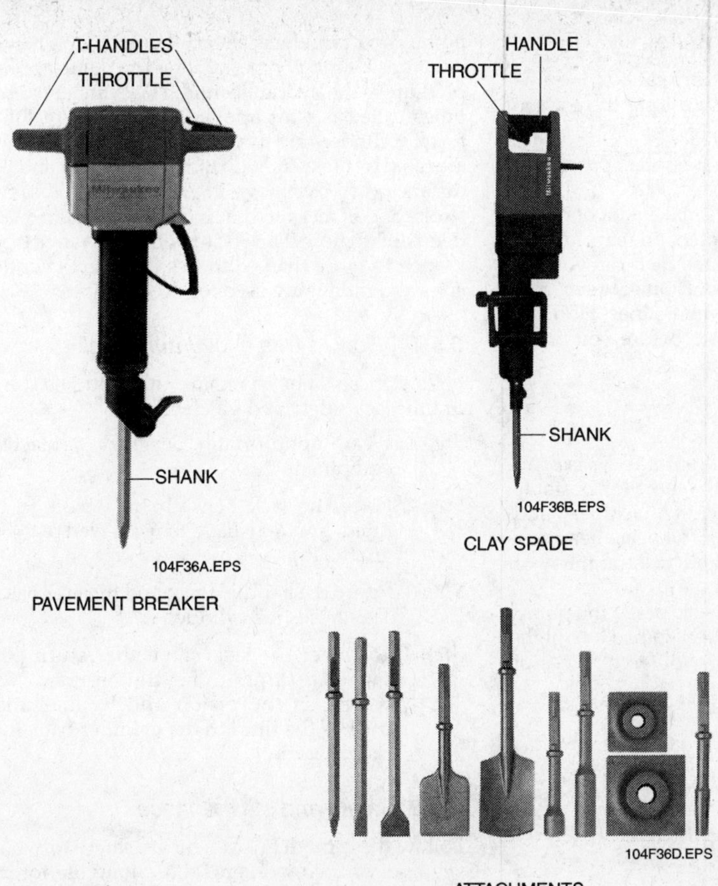

Figure 36 ◆ Three typical demolition tools and attachments.

Explain how to operate a pavement breaker. Ensure that trainees understand how to determine whether they have a good connection.

Ask a trainee to explain the consequences of failing to properly connect and secure an air hose.

located on the T-handle. When you push the throttle, compressed air operates a piston inside the tool. The piston drives the steel-cutting shank into the material you want to break up. You can use attachments, such as spades or chisels, for different tasks.

6.4.1 How to Set Up and Use a Pavement Breaker

Follow these steps to use a pavement breaker safely and efficiently:

Step 1 Wear appropriate personal protective equipment.

Step 2 Make sure that the air pressure is shut off at the main air outlet.

Step 3 Hold the coupler at the end of the air supply line, slide the ring back, and slip the coupler on the connector or nipple that is attached to the air drill.

Step 4 Check to see if you have a good connection. (A good coupling cannot be taken apart without first sliding the ring back.)

Step 5 Add a whip check.

Step 6 Once you have a good connection, turn on the air supply valve. The pavement breaker is now ready to use.

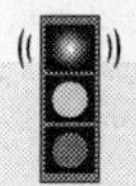

WARNING!

The air hose must be connected properly and securely. An unsecured air hose can come loose and whip around violently, causing serious injury. Some fittings require the use of whip checks to keep them from coming loose.

MODULE 00104-04 ◆ INTRODUCTION TO POWER TOOLS 4.33

Emphasize the importance of locating utilities before using a pavement breaker.

Refer to Figure 37. Review the components in a hydraulic jack. Explain how each operates.

Demonstrate how to properly position a hydraulic jack beneath the object to be lifted. Explain how to operate the jack.

Explain how to prevent a jack from kicking out.

6.4.2 Safety and Maintenance

Follow these guidelines to ensure safety for yourself and your co-workers, and a long life for the demolition tool:

- Always wear appropriate personal protective equipment.
- Because some of these tools make a lot of noise, you must wear hearing protection (earplugs).
- Be aware of what is under the material you are about to break. Know the location of water, gas, electricity, sewer, and telephone lines. Find out what is there and where it is before you break the pavement!

6.5.0 Hydraulic Jack

The portable hydraulic jack (see *Figure 37*) is commonly called a Porta-Power™ (short for portable power). It is used to push heavy machinery and other heavy objects; to pull wheels, bearings, gears, and cylinder liners; and to straighten or bend frames.

Hydraulic jacks have two basic parts: the pump and the cylinder. The two parts are joined by a high-pressure hydraulic hose. Many different types of pumps and cylinders can be used in different combinations for many types of jobs. The pump applies pressure to the hydraulic fluid. The cylinder (sometimes called a ram) applies a lifting or pushing force. Cylinders are available in many sizes. They are rated by the weight (in tons) they can lift and the distance they can move it. This distance is called stroke and is measured in inches. Hydraulic cylinders can lift more than 500 tons. Strokes range from ¼ inch to more than 48 inches. Different cylinder sizes and ratings are used for different jobs.

104F37.EPS

Figure 37 ◆ Portable hydraulic jack.

6.5.1 How to Use a Hydraulic Jack

Follow these steps to ensure that you use a hydraulic jack safely and efficiently.

Step 1 Wear appropriate personal protective equipment.

Step 2 Place the jack beneath the object to be lifted. You may have to use a wedge to begin the lift.

Step 3 Pump the handle down and then release it. This raises the cylinder.

Step 4 To lower the jack, open the return passage by turning the thumbscrew. The weight of the piston and its load then pushes the fluid in the cylinder back into the reservoir.

6.5.2 Safety and Maintenance

Follow these guidelines to ensure safety for yourself and your co-workers, and a long life for the hydraulic jack:

- Always wear appropriate personal protective equipment.
- Check the fluid level in the pump before using it.
- Make sure the hydraulic hose is not twisted.
- Do not move the pump if the hose is under pressure.
- Clear the work area when you are making a lift.
- When you are lifting, make sure the cylinder is on a secure, level surface to prevent the jack from kicking out.
- Do not use a cheater bar (extension) on the pump handle.
- Watch for leaks.

Instructor's Notes:

____

____

____

4.34 CORE CURRICULUM ◆ INTRODUCTORY CRAFT SKILLS

Review Questions

Section 6.0.0

1. Most pneumatic nailers operate at pressures of _____ per square inch (psi).
   a. 70 to 120 grams
   b. 7 to 120 ounces
   c. 7 to 12 pounds
   d. 70 to 120 pounds

2. When loading nails into a pneumatic nailer, _____.
   a. never leave the compressor hose attached
   b. always connect the compressor hose
   c. uncoil the strips
   d. reattach the air lines

3. If a pneumatic nailer is not firing properly, _____ before you attempt repairs.
   a. adjust the compressor
   b. disconnect the air hose
   c. adjust the regulator
   d. load a new coil of nails

4. According to manufacturers, it's a good idea to add a _____ to the air inlet each time you use a pneumatic nailer.
   a. few drops of oil
   b. few drops of water
   c. quart of oil
   d. quart of water

5. Powder-actuated fastening systems are used to _____.
   a. penetrate drywall
   b. hammer nails
   c. anchor static loads to steel beams
   d. remove nails

6. Before you begin setting up a pavement breaker for use, make sure that the air pressure is _____.
   a. shut off at the coupler
   b. turned on only halfway
   c. turned on full
   d. shut off at the main air outlet

7. Before you begin using a demolition tool, make sure you _____.
   a. rotate the hammer drill
   b. disconnect the coupler
   c. disconnect the air supply
   d. know what is underneath the material you are breaking up

8. The two basic parts of a hydraulic jack are the _____.
   a. pump and cylinder
   b. reservoir and relief pump
   c. cylinder and hose
   d. hose and pump

9. Porta-Power™ cylinders are rated by how much weight they can lift and by _____.
   a. their torque
   b. the amount of electromagnetic material they have
   c. how much they weigh
   d. the distance they can move the weight

10. Before using a Porta-Power™, make sure the hose _____.
    a. reaches to the wall outlet
    b. is connected to the reservoir
    c. is disconnected from the hydraulic jack
    d. is not twisted

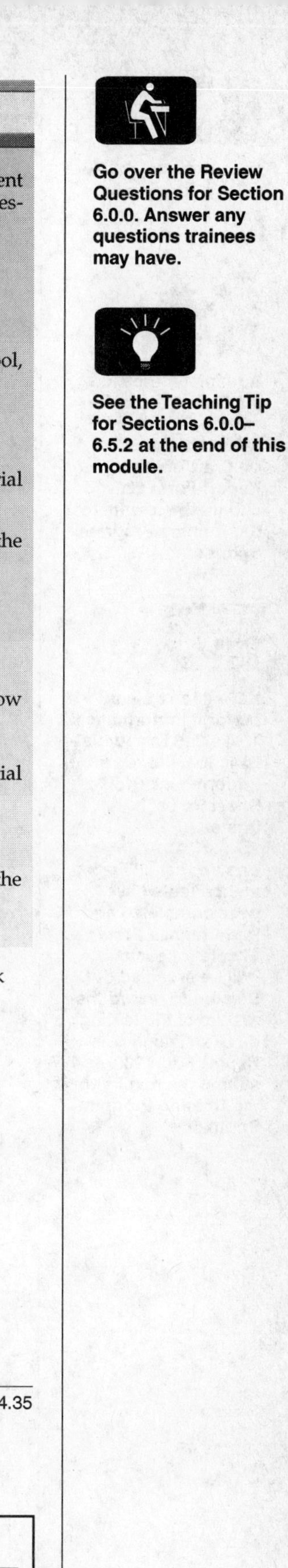

Go over the Review Questions for Section 6.0.0. Answer any questions trainees may have.

See the Teaching Tip for Sections 6.0.0–6.5.2 at the end of this module.

Review the Key Terms Quiz. Ask whether trainees have any questions.

Administer the Module Examination. Be sure to record the results of the Exam on Craft Training Report Form 200, and submit the results to the Training Program Sponsor.

Have each trainee perform Performance Tasks 3–5 to your satisfaction. Fill out a Performance Profile Sheet for each trainee.

Ensure that all Performance Tests have been completed and Performance Profile Sheets for each trainee are filled out. Be sure to record the results of the testing on Craft Training Report Form 200, and submit the results to the Training Program Sponsor.

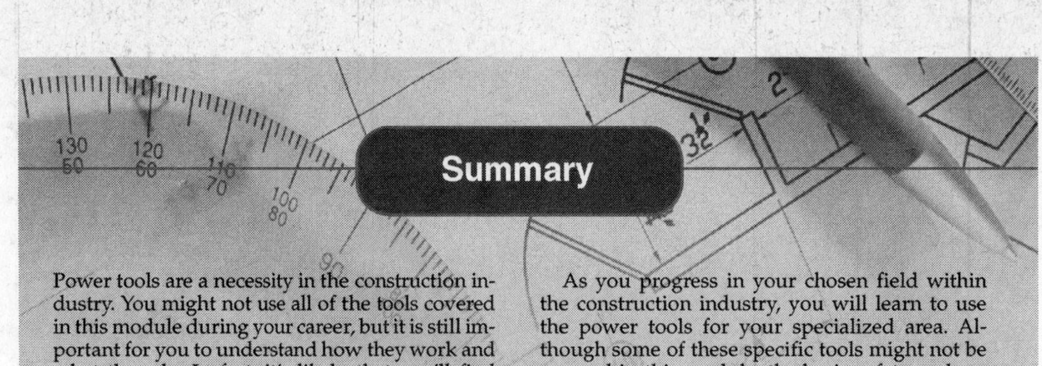

Summary

Power tools are a necessity in the construction industry. You might not use all of the tools covered in this module during your career, but it is still important for you to understand how they work and what they do. In fact, it's likely that you'll find yourself working around other craftworkers who use them. You and your co-workers will be safer if everyone is familiar with the tools being used on the job site.

You must also learn how to maintain your power tools properly, whether they belong to you or your employer. The better care you take of your tools, the better and more safely they will function, and the longer they'll last. Proper maintenance of power tools saves you and your employer time and money.

As you progress in your chosen field within the construction industry, you will learn to use the power tools for your specialized area. Although some of these specific tools might not be covered in this module, the basic safety and usage concepts are always applicable. Remember to always read the manufacturer's manual for any new power tool you use and never to use a tool on which you have not been properly trained. Following the basic use and safety guidelines explained in this module, maintaining your tools well, and educating yourself before using any new equipment will help you progress in your career, work efficiently, and stay safe.

Notes

Instructor's Notes:

Key Terms Quiz

Fill in the blank with the correct key term that you learned from your study of this module.

1. Activate the _________ to make the trigger stay in operating mode even without your finger on the trigger.
2. _________ reverses its direction at regularly recurring intervals; this type of current is delivered through wall plugs.
3. A(n) _________ saw's straight blades move back and forth.
4. A(n) _________ powers a powder-actuated tool.
5. _________ must be accompanied by material safety data sheets.
6. Masonry bits and nail cutter saw blades have a(n) _________ tip.
7. A(n) _________ is a substance, such as sandpaper, that is used to wear away material.
8. The _________ is a clamping device that holds an attachment.
9. _________ is the number of times a drill bit completes one full rotation in a minute.
10. A(n) _________ is used to set the head of a screw at or below the surface of the material.
11. _________ flows in one direction, from the negative to the positive terminal of the source.
12. Belt sanders and circular saws are examples of _________.
13. Use a(n) _________ to bore holes in wood and other materials.
14. An electromagnet holds an electromagnetic drill in place on a(n) _________ metal surface.
15. _________ is applied to the surface of a grinding wheel to give it a nonslip finish.
16. The _________ of the drill holds the drill bit.
17. To prevent an electrical shock, do not operate electric power tools without proper _________.
18. Jackhammers and Porta-Powers™ are examples of _________.
19. _________ refers to building material, including stone, brick, or concrete block.
20. Air hammers and pneumatic nailers are examples of _________.
21. Perform a(n) _________ to check the condition of a grinding wheel.
22. The _________ is the smooth part of a drill bit that fits into the chuck.
23. A(n) _________ protects people from electric shock and protects equipment from damage by interrupting the flow of electricity if a circuit fault occurs.

Key Terms

Abrasive
AC (alternating current)
Auger
Booster
Carbide
Chuck
Chuck key
Countersink
DC (direct current)
Electric tools
Ferromagnetic
Grit
Ground fault circuit interrupter (GFCI)
Ground fault protection
Hazardous materials
Hydraulic tools
Masonry
Pneumatic tools
Reciprocating
Revolutions per minute (rpm)
Ring test
Shank
Trigger lock

MODULE 00104-04 ◆ INTRODUCTION TO POWER TOOLS 4.37

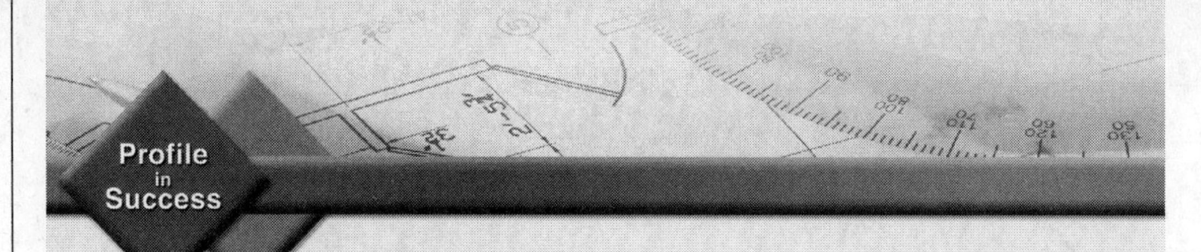

R. P. Hughes

Department Chair for Carpentry and Construction Management
Guilford Technical Community College
Jamestown, North Carolina

R. P. was born in North Carolina and has lived there all his life. After graduating from high school in 1962, R. P. worked in a mill for a year before accepting a job with a residential building company. He enrolled in a construction and apprenticeship program at a local community college, where he quickly came to appreciate the value of education and where he picked up many valuable skills. R. P. worked with the residential builder for 12 years, becoming a superintendent and then becoming a teacher at Guilford Technical Community College.

How did you become interested in the construction industry?
When I graduated from high school, I had neither the money nor the desire to go to college. I was 17 years old, and the only job I could find at that time was in a mill. I was very unhappy about it, and when I had a chance to go into the construction profession with a residential builder, I took it. I started out as a helper.

I saw that construction had the potential to be an excellent career and that it had a promising future. I liked seeing a house going up, and I enjoyed working with my hands. Later, when we finished a house, I enjoyed driving by it and knowing that I had helped to build it.

How did you decide to become an educator?
I got into teaching strictly by accident. At the time, I was working with the residential firm. One day I went to sign up for classes at Guilford. The fellow standing in line ahead of me was applying for the carpentry class, but the person at the desk told him that they weren't going to be able to offer the class because there was no qualified instructor. So I came back the next day and put in my application for carpentry instructor, and I got the job!

What do you think it takes to be a success in your trade?
You need to be able to think problems through and to visualize. You need excellent math skills. You also need strong communication skills to be able to work as part of a team. A lot of people think that it's enough just to know how to drive a nail, but there's so much more involved. You need to be able to deal with subcontractors, with suppliers, and with architects every day. You must have a broad range of skills to be successful in construction.

What are some of the things you do in your job?
A large part of my time is spent teaching. This semester, for example, I am teaching a blueprint class, a safety class, a carpentry class, and an estimating class. Currently in the carpentry class I have 40 students. We established a nonprofit corporation associated with the carpentry program, for which students build a house throughout the year. I supervise the construction of those houses. I try to help individual students determine where their strengths are and then guide them in that direction.

We teach not just the hands-on skills, but also safety skills, estimating techniques, and how to read building codes. If I can teach them to think, then they're well on their way to a successful career in construction. And I think that critical thinking is a skill we are gradually losing. You can punch numbers into a calculator, but you also need to understand what goes on behind those numbers, and how those numbers work the way they do.

I am responsible for two full-time instructors and three or four part-time instructors. I act as a general contractor for the school. I even do the paperwork to see that the bills are paid every month, which is probably one of the least favorite things I have to do!

What do you like most about your job?
I really like the fact that every day is different. You may end up building the same house twice, but you may encounter a unique set of problems each time.

Instructor's Notes:

4.38 CORE CURRICULUM ◆ INTRODUCTORY CRAFT SKILLS

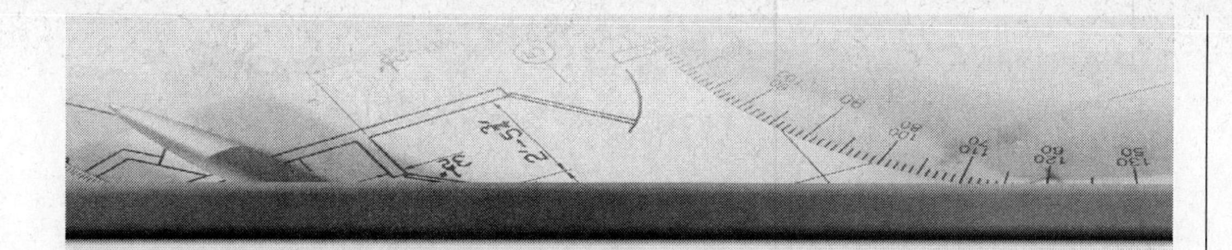

There is a lot of satisfaction that comes from working with your hands. I enjoy going by houses that I built 20 or 30 years ago and seeing them still there. I enjoy giving homeowners a good product and seeing them happy in their homes.

As a teacher and a carpenter, I get both aspects of the profession. I get to build a house, and I get to shape and change some lives and give them some skills that they can work with. Students really develop a lot of skills in a year's time. That's very rewarding to me.

I run into students in the field all the time, and they often come back and ask for my advice on career moves. Often they tell me that, of all the classes they took, one of mine was their favorite and it really meant a lot to them. That makes my job worthwhile. That's why I'm here, why I teach. I could make more money as a builder, but I wouldn't get that reward.

What would you say to someone entering the trades today?

There are so many avenues that you can go into. It depends on your skills and interests. Some students don't mind taking appropriate risks, and they're good at time management. These students could be excellent contractors. Others have more patience and are skilled at detail work. They could do finishing work like trim or cabinets. A lot of students go into framing work and become framing contractors. Whatever you do, you have to want that feeling of accomplishment every day.

There have been some real changes in the industry since I first started. Codes are a lot tougher today, more technical. If you are going to be a success in your chosen field, you need to stay up on your training.

Students today need to be aware of the shortage of skilled workers in the construction industry. A lot of the workers out there are unskilled. The average age of a skilled carpenter today is around 56 years old, I've heard. Someone has to take these people's place when they retire. We see plenty of young, sharp people enter our program, and they need formal training to become skilled workers. You're not going to see skilled jobs like these shipped overseas. People are always going to need a house.

MODULE 00104-04 ◆ INTRODUCTION TO POWER TOOLS 4.39

Trade Terms Introduced in This Module

Abrasive: A substance—such as sandpaper—that is used to wear away material.

AC *(alternating current):* An electrical current that reverses its direction at regularly recurring intervals; the current delivered through wall plugs.

Auger: A tool with a spiral cutting edge for boring holes in wood and other materials.

Booster: Gunpowder cartridge used to power powder-actuated fastening tools.

Carbide: A very hard material made of carbon and one or more heavy metals. Commonly used in one type of saw blade.

Chuck: A clamping device that holds an attachment; for example, the chuck of the drill holds the drill bit.

Chuck key: A small, T-shaped steel piece used to open and close the chuck on power drills.

Countersink: A bit or drill used to set the head of a screw at or below the surface of the material.

DC *(direct current):* Electrical current that flows in one direction, from the negative (−) to the positive (+) terminal of the source, such as a battery.

Electric tools: Tools powered by electricity. The electricity is supplied by either an AC source (wall plug) or a DC source (battery).

Ferromagnetic: Having magnetic properties. Substances such as iron, nickel, cobalt, and various alloys are ferromagnetic.

Grit: A granular abrasive used to make sandpaper or applied to the surface of a grinding wheel to give it a nonslip finish. Grit is graded according to its texture. The grit number indicates the number of abrasive granules in a standard size (per inch or per cm). The higher the grit number, the finer the abrasive material.

Ground fault circuit interrupter (GFCI): A circuit breaker designed to protect people from electric shock and to protect equipment from damage by interrupting the flow of electricity if a circuit fault occurs.

Ground fault protection: Protection against short circuits; a safety device cuts power off as soon as it senses any imbalance between incoming and outgoing current.

Hazardous materials: Materials (such as chemicals) that must be transported, stored, applied, handled, and identified according to federal, state, or local regulations. Hazardous materials must be accompanied by material safety data sheets (MSDSs).

Hydraulic tools: Tools powered by fluid pressure. The pressure is produced by hand pumps or electric pumps.

Masonry: Building material, including stone, brick, or concrete block.

Pneumatic tools: Air-powered tools. The power is produced by electric or fuel-powered compressors.

Reciprocating: Moving back and forth.

Revolutions per minute (rpm): The number of times (or rate) a motor component or accessory (drill bit) completes one full rotation every minute.

Ring test: A method of testing the condition of a grinding wheel. The wheel is mounted on a rod and tapped. A clear ring means the wheel is in good condition; a dull thud means the wheel is in poor condition and should be disposed of.

Shank: The smooth part of a drill bit that fits into the chuck.

Trigger lock: A small lever, switch, or part that you push or pull to activate a locking catch or spring. Activating the trigger lock causes the trigger to stay in the operating mode even without your finger on the trigger.

Instructor's Notes:

4.40 CORE CURRICULUM ◆ INTRODUCTORY CRAFT SKILLS

Additional Resources

This module is intended to present thorough resources for task training. The following reference works are suggested for further study. These are optional materials for continued education rather than for task training.

29 CFR 1926, OSHA Construction Industry Regulations, latest edition. Washington, DC: Occupational Safety and Health Administration, U.S. Department of Labor, U.S. Government Printing Office.

All About Power Tools. 2002. Des Moines, IA: Meredith Books.

Hand & Power Tool Training. Video. All About OSHA. Surprise, AZ.

Power Tools. 1997. Minnetonka, MN: Handyman Club of America.

Powered Hand Tool Safety: Handle With Care. Video. 20 minutes. Coastal Training Technologies Corp. Virginia Beach, VA.

Reader's Digest Book of Skills and Tools, 1993 edition. Pleasantville, NY: Reader's Digest.

MODULE 00104-04 ◆ INTRODUCTION TO POWER TOOLS 4.41

The following are suggested activities or instructional methods to help you teach the material in this AIG.

General

When you call on someone to answer a question, the rest of the class relaxes or even tunes out because they expect that the question and answer will take place only between you and the trainee you called on. Instead, use this technique to involve more trainees in answering questions and to keep them on their toes.

1. Ask trainees to define a term or explain a concept.
2. After one trainee has answered, ask a trainee seated nearby if the answer is right. Then ask whether a trainee in the back of the room agrees.
3. Ask trainees to explain why they think an answer is right or wrong.
4. Use the session to clear up incorrect ideas, and encourage trainees to learn from their mistakes.

Sections 3.0.0–3.5.2

Power Drills

This exercise will familiarize trainees with power drills and drill bits, which you will provide. Trainees will need their personal protective equipment, Trainee Guides, and pencils and paper to take notes for discussion. Allow 30 minutes for this exercise.

1. Set up stations with a variety of power drills and drill bits.
2. Have trainees visit each station and identify each type of drill.
3. On the whiteboard/chalkboard, write a description of a number of hypothetical tasks. Ask trainees to select the right drill and bit for each task.
4. Ask trainees to practice loading the drill bits into the chuck opening. Have trainees demonstrate how to use each drill.
5. Go over any questions trainees may have about power drills.

Sections 4.0.0–4.5.2

Saws

This exercise allows trainees to practice using a variety of saws. You will need a variety of saws and the appropriate materials for cutting. Trainees will need their personal protective equipment, Trainee Guides, and pencils and paper to take notes for discussion. Allow 30 minutes for this exercise.

1. Set up stations with a variety of saws and materials appropriate for cutting with each saw.
2. Have trainees select the appropriate blade, adjust the blade depth, and properly secure the material to be cut.
3. Ask trainees to practice using the saws to cut each of the materials.
4. Answer any questions trainees may have about saws.

Sections 6.0.0–6.5.2

Miscellaneous Power Tools

This exercise will give trainees practice using miscellaneous power tools. You will provide nail guns, powder-actuated fastening systems, air impact wrenches, pavement breakers, and hydraulic jacks. Trainees will need their personal protective equipment, Trainee Guides, and pencils and paper to take notes for discussion. Allow 30 minutes for this exercise.

1. Set up one station for each of the different power tools.
2. Ask trainees to visit each station, identify each tool, and briefly explain how each is used.
3. Assign one trainee to each station, and ask the trainee to demonstrate how to operate the tool.
4. Have trainees observe the demonstrations and explain what was done correctly or incorrectly.
5. Go over any questions trainees may have about power tools.

MODULE 00104-04 — ANSWERS TO REVIEW QUESTIONS

Section 2.0.0

1. b
2. b
3. a
4. c
5. c

Section 3.0.0

1. c
2. a
3. b
4. d
5. a
6. c
7. b
8. b
9. c
10. c

Section 4.0.0

1. c
2. a
3. d
4. b
5. a
6. c
7. d
8. b
9. a
10. d

Section 5.0.0

1. c
2. d
3. b
4. a
5. a
6. a
7. b
8. a
9. c
10. c

Section 6.0.0

1. d
2. a
3. b
4. a
5. c
6. d
7. d
8. a
9. d
10. d

MODULE 00104-04 — ANSWERS TO KEY TERMS QUIZ

1. trigger lock
2. AC (alternating current)
3. reciprocating
4. booster
5. Hazardous materials
6. carbide
7. abrasive
8. chuck key
9. Revolutions per minute (rpm)
10. countersink
11. DC (direct current)
12. electric tools
13. auger
14. ferromagnetic
15. Grit
16. chuck
17. ground fault protection
18. hydraulic tools
19. Masonry
20. pneumatic tools
21. ring test
22. shank
23. ground fault circuit interrupter (GFCI)

CONTREN® LEARNING SERIES — USER FEEDBACK

The NCCER makes every effort to keep these textbooks up-to-date and free of technical errors. We appreciate your help in this process. If you have an idea for improving this textbook, or if you find an error, a typographical mistake, or an inaccuracy in NCCER's *Contren®* textbooks, please write us, using this form or a photocopy. Be sure to include the exact module number, page number, a detailed description, and the correction, if applicable. Your input will be brought to the attention of the Technical Review Committee. Thank you for your assistance.

Instructors – If you found that additional materials were necessary in order to teach this module effectively, please let us know so that we may include them in the Equipment/Materials list in the Annotated Instructor's Guide.

Write: Product Development
National Center for Construction Education and Research
P.O. Box 141104, Gainesville, FL 32614-1104

Fax: 352-334-0932

E-mail: curriculum@nccer.org

Craft ______ Module Name ______

Copyright Date ______ Module Number ______ Page Number(s) ______

Description

______

______

______

(Optional) Correction

______

______

(Optional) Your Name and Address

______

______

Introduction to Blueprints

00105-04

NCCER STANDARDIZED CRAFT TRAINING PROGRAM

The National Center for Construction Education and Research (NCCER) provides a standardized national program of accredited craft training. Key features of the program include instructor certification, competency-based training, and performance testing. The program provides trainees, instructors, and companies with a standard form of recognition through a National Craft Training Registry. The program is described in full in the *Guidelines for Accreditation*, published by the NCCER. For more information on standardized craft training, contact the NCCER by writing us at P.O. Box 141104, Gainesville, FL 32614-1104; calling 352-334-0911; or e-mailing info@nccer.org. More information may be found at our Web site, www.nccer.org.

HOW TO USE THIS ANNOTATED INSTRUCTOR'S GUIDE

Each page presents two sections of information. The larger section displays each page exactly as it appears in the Trainee Module. The narrow column ties suggested trainee and instructor actions to each page and provides icons (detailed below) to call your attention to material, safety, audiovisual, or testing requirements. The bottom of each page includes space for your notes.

The **Audiovisual** icon indicates an appropriate time to show a transparency or other audiovisual aid.

The **Classroom** icon prompts you to define a term, stress a point, ask trainees to explain a concept, or give examples.

The **Demonstration** icon directs you to show trainees how to perform tasks.

The **Examination** icon tells you to administer the written module examination.

The **Homework** icon is placed where you may wish to assign reading for the next class, to assign a project, or to advise trainees to prepare for an examination.

The **Laboratory** icon is used when trainees are to practice performing tasks.

The **Materials** icon is a reminder for you to gather materials needed for classes, labs, and testing.

The **Performance Testing** icon tells you to administer a performance test or a portion thereof.

The **Safety** icon is used to emphasize safety issues. It is often keyed to *Caution* and *Warning* statements in the Trainee Module.

The **Teaching Tip** icon indicates additional guidance is available, such as how to conduct an exercise, get the most educational value from a field trip, or encourage class participation. Teaching Tips may expand on a feature (*Think About It, Did You Know?*) or provide Quick Quizzes or similar exercises. You will be referred to the Teaching Tips section at the back of the module if there is additional material.

The **Combination** icon indicates that the laboratory listed corresponds with a performance task. If desired, you can note the proficiency of the trainees during the laboratory and use it to satisfy performance testing requirements.

PREPARATION

Before teaching this module, you should review the Objectives, Performance Tasks, Materials and Equipment List, and the Module Outline. Be sure to allow ample time to prepare your own training or lesson plan and gather all required materials and equipment.

MODULE OVERVIEW

This module discusses blueprint terms, components, and symbols. Trainees will learn how to interpret blueprints, recognize classifications of drawings, and use drawing dimensions.

PREREQUISITES

Prior to training with this module, it is recommended that the trainee shall have successfully completed the following: *Core Curriculum: Introductory Craft Skills,* Modules 00101-04 through 00104-04.

OBJECTIVES

Upon completion of this module, the trainee will be able to:

1. Recognize and identify basic blueprint terms, components, and symbols.
2. Relate information on blueprints to actual locations on the print.
3. Recognize different classifications of drawings.
4. Interpret and use drawing dimensions.

PERFORMANCE TASKS

Under the supervision of the instructor, the trainee should be able to:

1. Using the floor plan supplied at the back of this module, locate the game room interior wall.
2. Using the floor plan supplied at the back of this module, give the distance from gridline 1 to gridline 3.
3. Using the floor plan supplied at the back of this module, determine the distance from the edge of the storage room doors across the lobby to the edge of the east set of double doors.

MATERIALS AND EQUIPMENT LIST

Transparencies
Markers/chalk
Blank acetate sheets
Transparency pens
Pencils and scratch paper
Overhead projector and screen
Whiteboard/chalkboard
Appropriate personal protective equipment
Copies of your local code
Door, window, and hardware schedules
A complete set of plans, including:
- Architectural
- Civil
- Structural
- Mechanical
- Plumbing
- Electrical

Specifications
Blueprints with title block
Blueprints with a legend
Measuring tools, including:
- Engineer's scale
- Architect's scale
- Metric scale

Blueprints with a gridline system
Blueprints with interior and exterior measurements
Module Examinations*
Performance Profile Sheets*

*Located in the Test Booklet.

ADDITIONAL RESOURCES

This module is intended to present thorough resources for task training. The following reference works are suggested for both instructors and motivated trainees interested in further study. These are optional materials for continued education rather than for task training.

Blueprint Reading for the Building Trades, 1989. John Traister. Carlsbad, CA: Craftsman Book Co.

Blueprint Reading for Construction, 1997. James Fatzinger. Upper Saddle River, NJ: Prentice Hall.

Construction Blueprint Reading, 1985. Robert Putnam. Englewood Cliffs, NJ: Prentice Hall.

Print Reading for Construction: Residential and Commercial, 1997. Walter C. Brown. Tinley Park, IL: Goodheart-Willcox Co.

Reading Architectural Plans for Residential and Commercial Construction, 2001. Ernest R. Weidhaas. Upper Saddle River, NJ: Prentice Hall.

TEACHING TIME FOR THIS MODULE

An outline for use in developing your lesson plan is presented below. Note that each Roman numeral in the outline equates to one session of instruction. Each session has a suggested time period of 2½ hours. This includes 10 minutes at the beginning of each session for administrative tasks and one 10-minute break during the session. Approximately 7½ hours are suggested to cover *Introduction to Blueprints.* You will need to adjust the time required for hands-on activity and testing based on your class size and resources. Because laboratories often correspond to Performance Tasks, the proficiency of the trainees may be noted during these exercises for Performance Testing purposes.

Topic	Planned Time
Session I. Introduction to Blueprints, Part One	
A. Plans	________
B. Specifications	________
C. Requests for Information	________
Session II. Introduction to Blueprints, Part Two	
A. Components of the Blueprint	________
B. Scale	________
C. Lines of Construction	________
Session III. Introduction to Blueprints, Part Three	
A. Abbreviations, Symbols, and Keynotes	________
B. Using Gridlines to Identify Plan Locations	________
C. Dimensions	________
D. Review	________
E. Module Examination	________
1. Trainees must score 70% or higher to receive recognition from NCCER.	
2. Record the testing results on Craft Training Report Form 200, and submit the results to the Training Program Sponsor.	
F. Performance Testing	________
1. Trainees must perform each task to the satisfaction of the instructor to receive recognition from NCCER. If applicable, proficiency noted during laboratory exercises can be used to satisfy the Performance Testing requirements.	
2. Record the testing results on Craft Training Report Form 200, and submit the results to the Training Program Sponsor.	

Build America Winner— Building New

The General Motors Vehicle Engineering Center in Warren, Michigan, was built by Clark Construction Group, Inc. It's a flexible, high-tech environment designed to stimulate interaction, innovation, productivity, and efficiency and at the same time, serves as an elaborate showroom. This massive facility has silver metal-paneled cylindrical columns, colored tile walls, a glass-enclosed eight-story staircase, and a lake with an events island in the center.

00105-04

Introduction to Blueprints

Topics to be presented in this unit include:

Overview

Blueprints are sets of documents that provide detailed information about structures being built and the materials needed for construction. They are used to plan and design industrial, commercial, and residential structures. They are also used in landscaping.

Several different trade groups have input in the planning, supplying, and building of structures. The plans developed by each group are combined to create a project blueprint. For example, electricians must have a plan for wiring the structure, plumbers must have a plan for the drain and waste system, and architects must have a plan for the overall look and feel of the project. All of these groups must join forces to make the arrangements and agreements related to building the structure.

All blueprints contain standard components such as a title block, border, drawing area, revision block, and legend. These components standardize the blueprint format and ensure that everyone using the blueprint has the same, accurate information. Other standardized components of blueprints include the tools used to create the scale of a drawing, the lines used in a drawing, and the abbreviations, keynotes, and symbols used in a drawing.

Understanding the basics of reading blueprints is a significant skill for all members of the construction trade. They serve not only as a guide but as a method to verify that the structure is being built accurately.

Instructor's Notes:

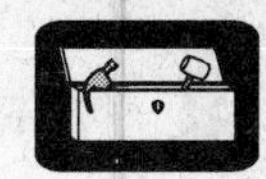

Ensure that you have everything required to teach the course. Check the Materials and Equipment List at the front of this module.

See the general Teaching Tip at the end of this module.

Show Transparency 1, Course Objectives.

Show Transparency 2, Performance Tasks.

Explain that terms shown in bold (blue) are defined in the Glossary at the back of this module.

Objectives

When you have completed this module, you will be able to do the following:

1. Recognize and identify basic blueprint terms, components, and symbols.
2. Relate information on blueprints to actual locations on the print.
3. Recognize different classifications of drawings.
4. Interpret and use drawing dimensions.

Key Trade Terms

Architect	Hidden line
Architect's scale	Isometric drawing
Architectural plans	Leader
Beam	Legend
Blueprints	Mechanical plans
Civil plans	Metric scale
Computer-aided drafting (CAD)	Not to scale (NTS)
Contour lines	Piping and instrumentation drawings (P&IDs)
Detail drawings	Plumbing
Dimension line	Plumbing plans
Dimensions	Request for information (RFI)
Electrical plans	Roof plan
Elevation (EL)	Scale
Elevation drawing	Schematic
Engineer	Section drawing
Engineer's scale	Specifications
Floor plan	Structural plans
Foundation plan	Symbol
Heating, ventilating, and air conditioning (HVAC)	Title block

Required Trainee Materials

1. Appropriate personal protective equipment
2. Sharpened pencils and paper

Prerequisites

Before you begin this module, it is recommended that you successfully complete the following: *Core Curriculum: Introductory Craft Skills,* Modules 00101-04 through 00104-04.

This course map shows all of the modules in *Core Curriculum: Introductory Craft Skills.* The suggested training order begins at the bottom and proceeds up. Skill levels increase as you advance on the course map. The local Training Program Sponsor may adjust the training order.

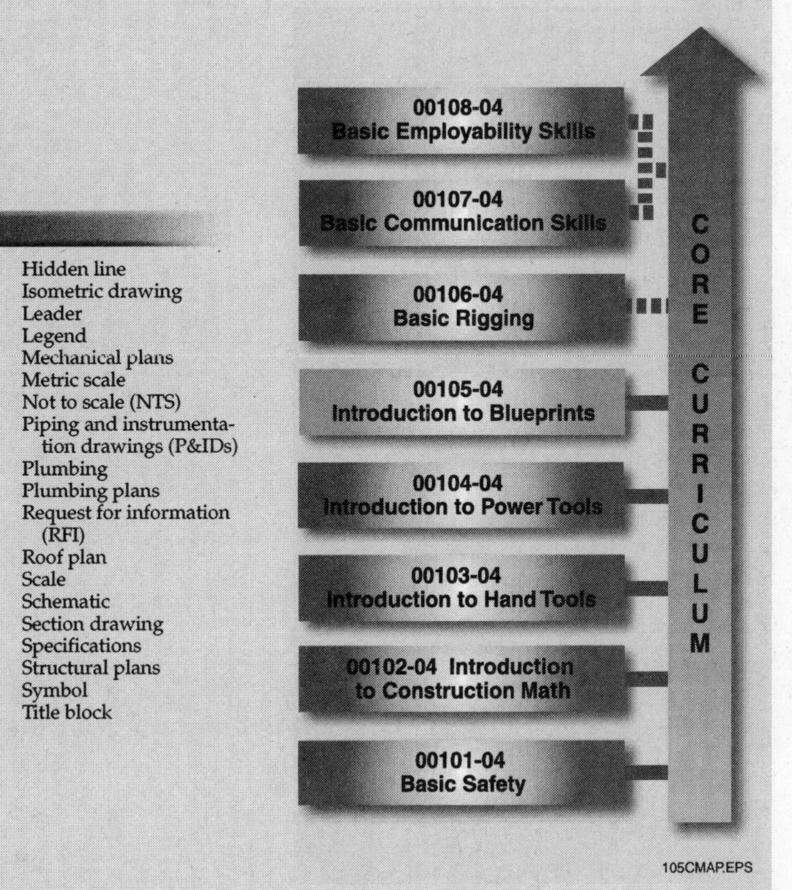

105CMAP.EPS

MODULE 00105-04 ◆ INTRODUCTION TO BLUEPRINTS 5.1

Show Transparencies 3 and 4 (Figure 1). Compare early blueprint designs with a blueprint created by computer-aided drafting (CAD). Discuss the advantages of CAD-generated drawings over hand-drawn blueprints.

Explain that blueprints and specifications dictate what is to be built and what materials are to be used. Discuss the importance of being able to read blueprints.

Discuss the importance of being able to read blueprints.

Show Transparency 5 (Figure 2). Explain how contour lines on a civil plan show the contours of the earth. Ask trainees to identify construction features illustrated on the drawing.

Show Transparency 6 (Figure 3). Review the different types of information that are included in floor and roof plans.

1.0.0 ◆ INTRODUCTION

What is a **blueprint?** A blueprint is a copy of a drawing. In the past, the lines on a blueprint were white and the background was blue. That's where the word *blueprint* comes from. Blueprints are also called prints. Today, however, most prints are created by **computer-aided drafting (CAD),** and they have blue or black lines on a white background. However, they are still called blueprints. *Figure 1* shows examples of an old and a new blueprint. Also see the large-scale blueprints on the insert in this module.

Various kinds of blueprints, including residential blueprints, commercial blueprints, landscaping blueprints, shop drawing blueprints, and industrial blueprints, are used in construction. In this module, you will learn about some basic types of blueprints.

Blueprints, together with the set of **specifications** (specs), detail what is to be built and what materials are to be used.

1.1.0 Set of Blueprints

The set of blueprints forms the basis of agreement and understanding that a building will be built as detailed in the drawings. Therefore, everyone involved in planning, supplying, and building any structure should be able to read blueprints. For any building project, also consult the civil engineering plans for that location, including sewer, highway, and water installation plans.

1.2.0 Plans

A complete set of plans includes six major plan groups:

- Civil
- Architectural
- Structural
- Mechanical
- **Plumbing**
- Electrical

1.2.1 Civil Plans

Civil plans are used for work that has to do with construction in or on the earth. Civil plans are also called site plans, survey plans, or plot plans. They show the location of the building on the site from an aerial view (see *Figure 2*). A civil plan also shows the natural contours of the earth, represented on the plan by **contour lines.** The civil plan also shows any trees on the property; construction features such as walks, driveways, or utilities; the **dimensions** of the property; and possibly a legal description of the property.

This is where it all starts. If the site is not acceptable, there is no reason to continue building!

1.2.2 Architectural Plans

Architectural plans (also called architectural drawings) show the design of the project. One part of an architectural plan is a **floor plan,** also known as a plan view. Any drawing made looking down on an object is commonly called a plan view. The floor plan is an aerial view of the layout of each room. It provides the most information about the project (refer to *Figure 1*). It shows exterior and interior walls, doors, stairways, and mechanical equipment. The floor plan shows the floor as you would see it from above if the upper part of the building were removed.

An architectural plan also includes a **roof plan,** which is a view of the roof from above the building. It shows the shape of the roof and the materials that will be used to finish it. In *Figure 3,* the arrows show that the roof slopes toward the drains.

Elevation (EL) is another element of architectural drawings. **Elevation drawings** are side views. They are called elevations because they show height. On a building drawing, there are standard names for different elevations. For example, the side of a building that faces south is called the south elevation. Exterior elevations show the size of the building; the style of the building; and the placement of doors, windows, chimneys, and decorative trim.

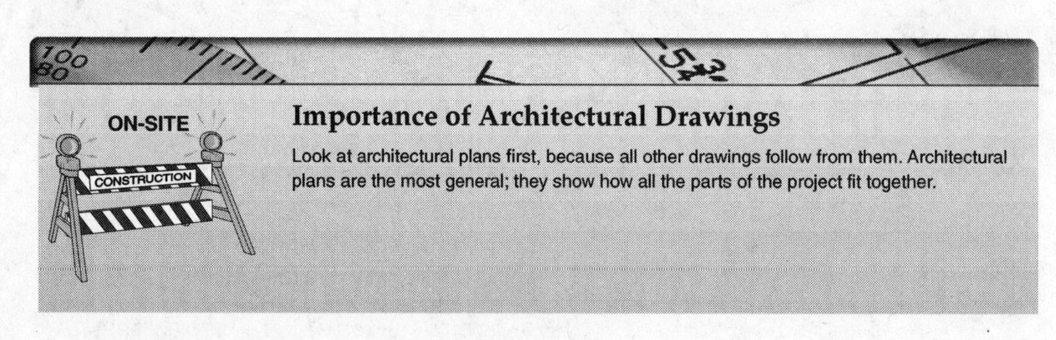

Importance of Architectural Drawings

Look at architectural plans first, because all other drawings follow from them. Architectural plans are the most general; they show how all the parts of the project fit together.

5.2 CORE CURRICULUM ◆ INTRODUCTORY CRAFT SKILLS

Instructor's Notes:

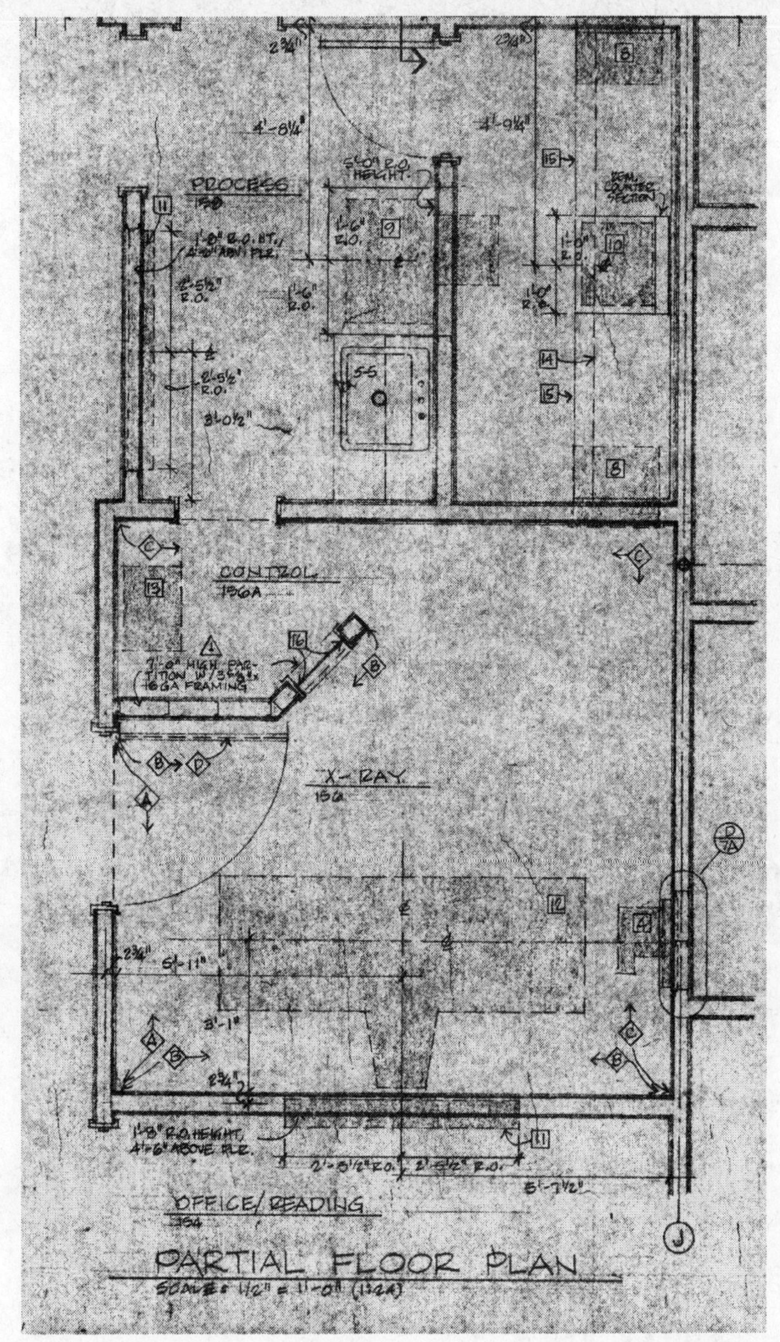

Figure 1 ◆ (A) Old blueprint. (1 of 2)

MODULE 00105-04 ◆ INTRODUCTION TO BLUEPRINTS 5.3

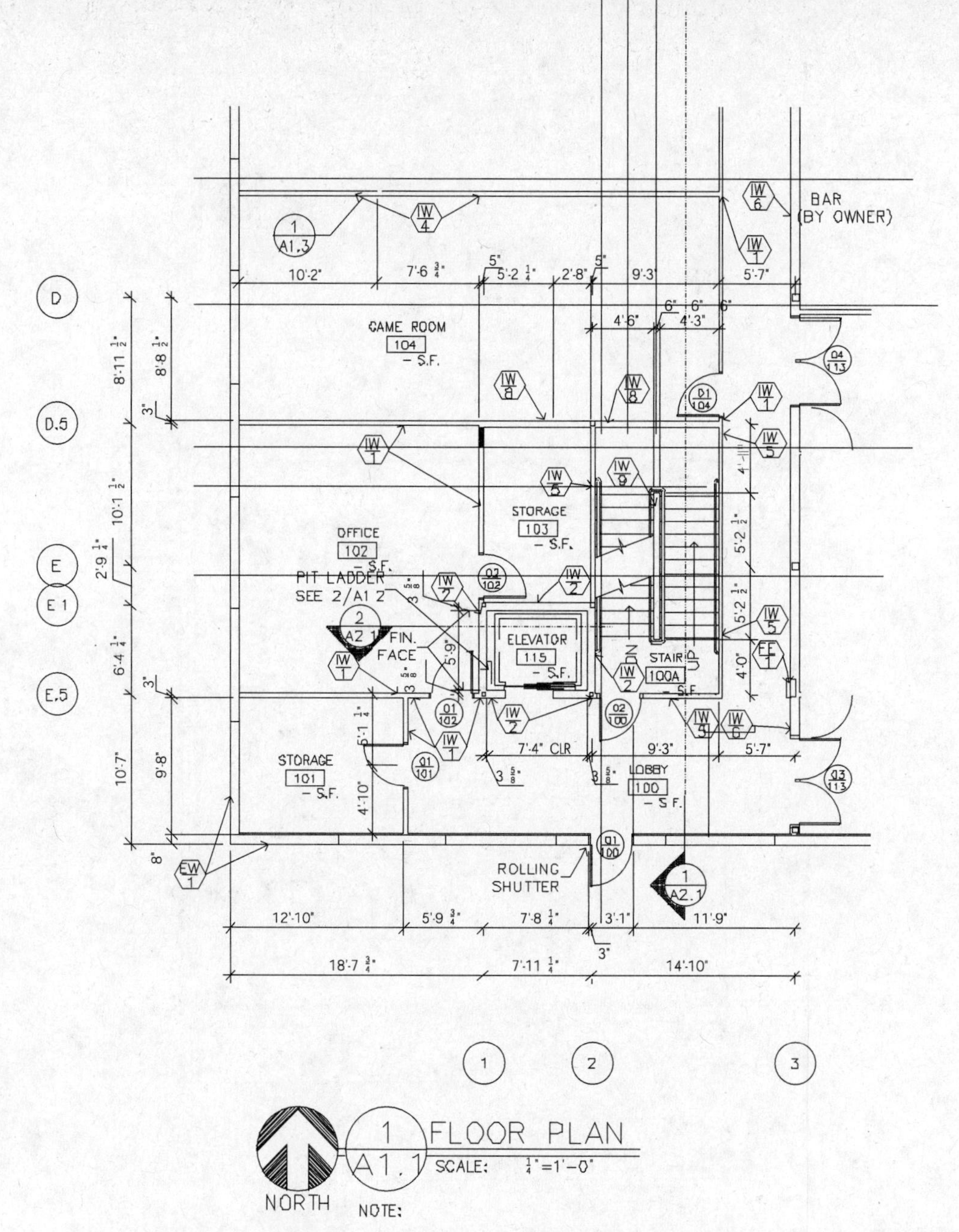

105F01B.EPS

Figure 1 ◆ (B) Blueprint of a floor plan created by computer-aided drafting. (2 of 2)

Instructor's Notes:

5.4 CORE CURRICULUM ◆ INTRODUCTORY CRAFT SKILLS

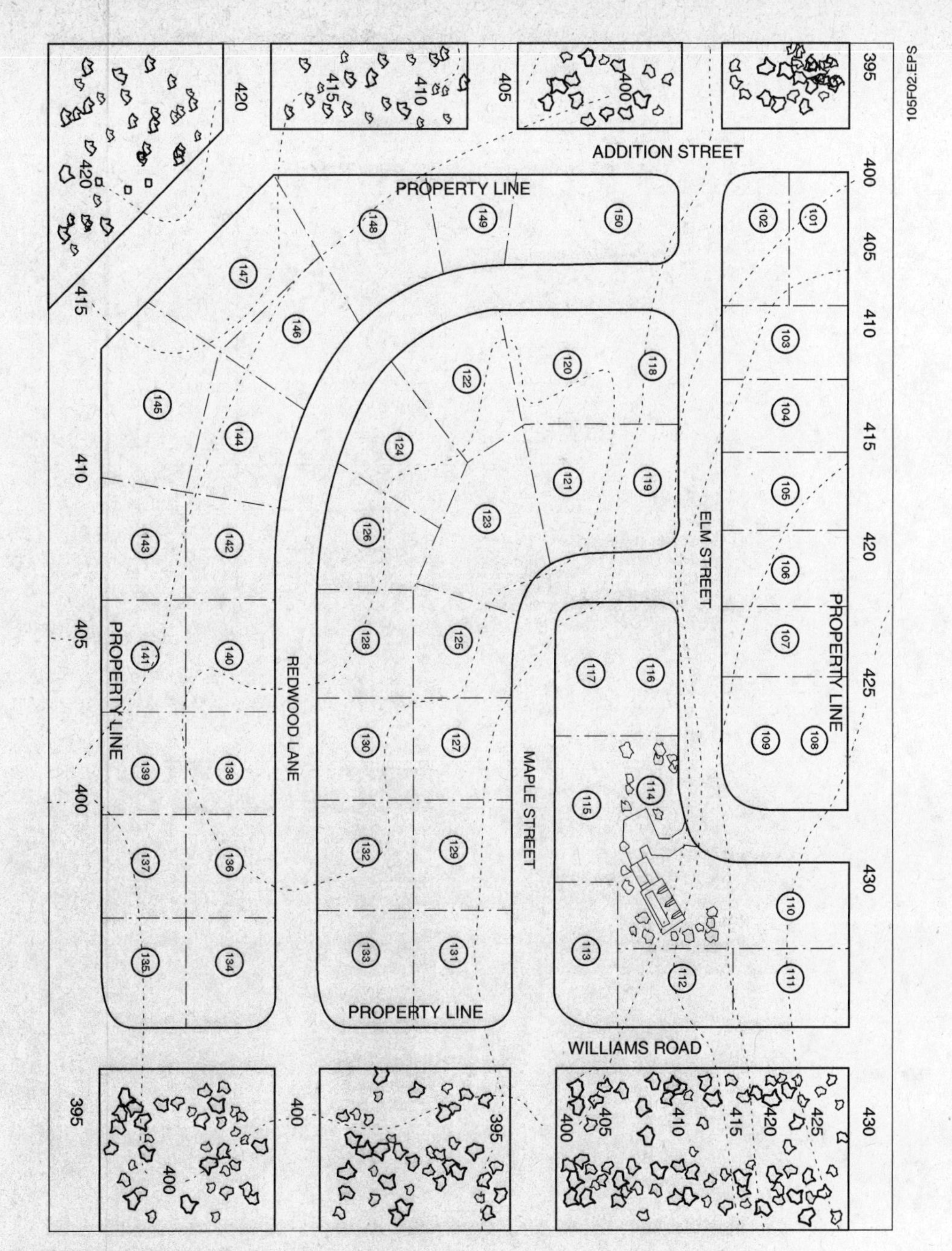

Figure 2 ◆ Civil plan, aerial view.

MODULE 00105-04 ◆ INTRODUCTION TO BLUEPRINTS 5.5

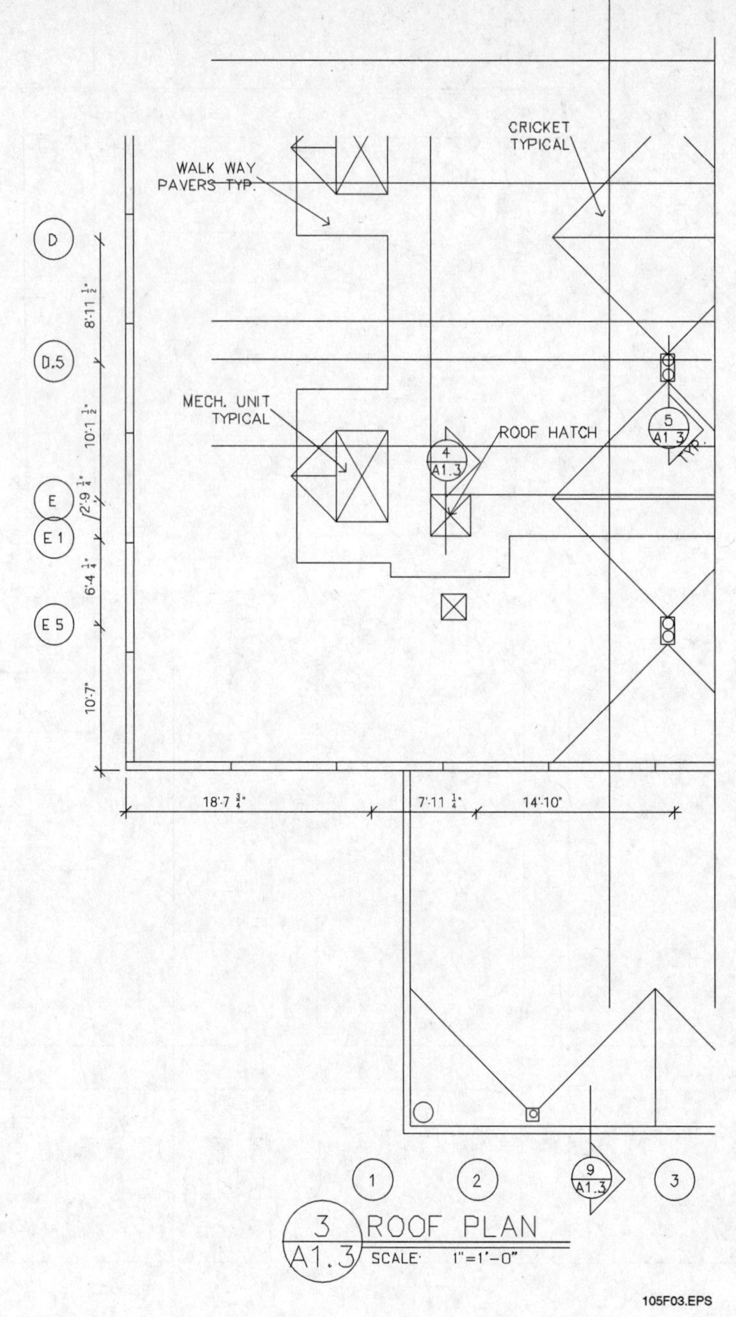

Figure 3 ◆ Roof plan.

Instructor's Notes:

5.6 CORE CURRICULUM ◆ INTRODUCTORY CRAFT SKILLS

What Is Computer-Aided Drafting?

The use of computers is a cost-effective way to increase drafting productivity because the computer program automates much of the repetitive work.

A CAD system generates drawings from computer programs. Using CAD has the following advantages over hand drawing blueprints:

- It is automated.
- The computer performs calculations quickly and easily.
- Changes can be made quickly and easily.
- Commonly used symbols can be easily retrieved.
- CAD can include three-dimensional modeling of the structure.

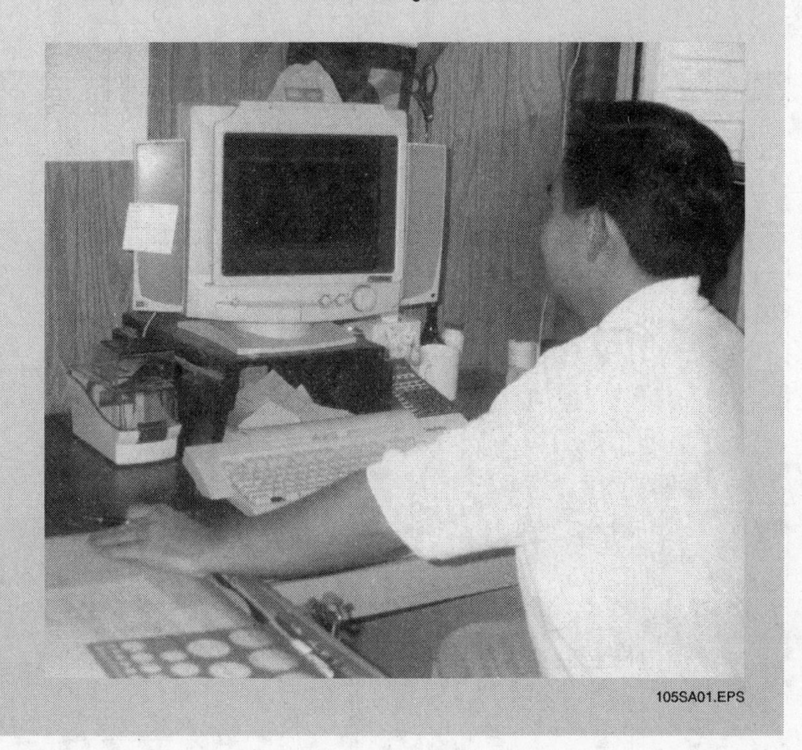

105SA01.EPS

Refer to Figures 4 and 5. Explain how elevation drawings are used to illustrate side views and building interiors.

Refer to Figures 6 and 7. Discuss the types of details that are included in section and detail drawings. Explain how to locate these drawings in the blueprint plan.

Bring in sample door, window, and hardware schedules. Demonstrate how to interpret the information on these schedules.

Figure 4 shows typical elevation drawings. Interior elevations (see *Figure 5*) show details of finishes and designs of the individual rooms in the building. Note that *Figure 5* does not match the elevation drawing exactly. It is included to help you visualize the finished product.

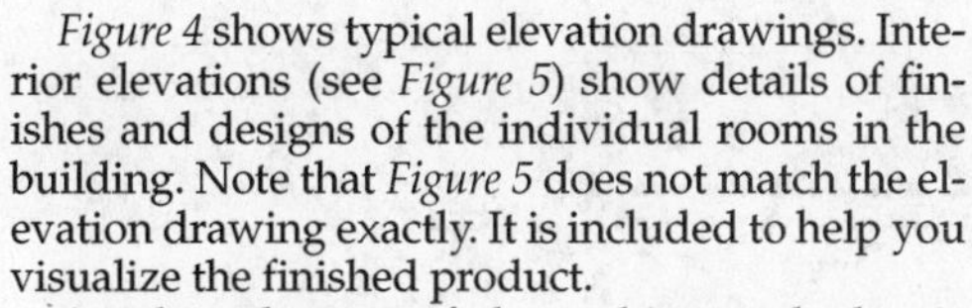

Another element of the architectural plan is **section drawings,** which show how the structure is to be built (*Figure 6*). Section drawings are cross-sectional views that show the inside of an object or building. They show what construction materials to use and how the parts of the object or building fit together. They normally show more detail than plan views. The section drawing in *Figure 6* shows an EL from *Figure 1B*.

Even more detail is shown in **detail drawings** (*Figure 7*), which are enlarged views of some special features of a building, such as floors and walls. They are enlarged to make the details clearer. Often the detail drawings are placed on the same sheet where the feature appears in the plan, but sometimes they are placed on separate sheets and referred to by a number on the plan view.

MODULE 00105-04 ◆ INTRODUCTION TO BLUEPRINTS 5.7

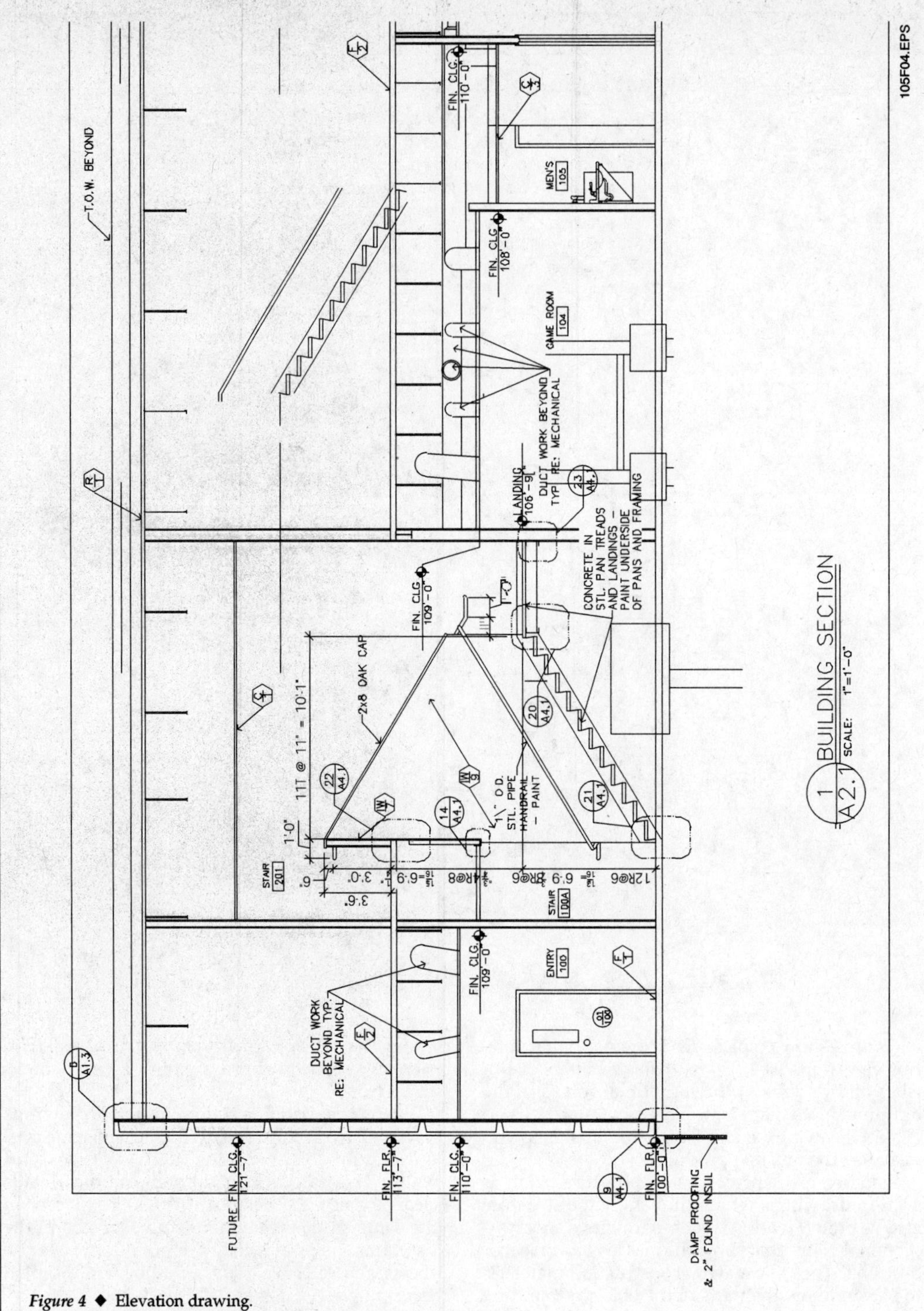

Figure 4 ◆ Elevation drawing.

Instructor's Notes:

5.8 CORE CURRICULUM ◆ INTRODUCTORY CRAFT SKILLS

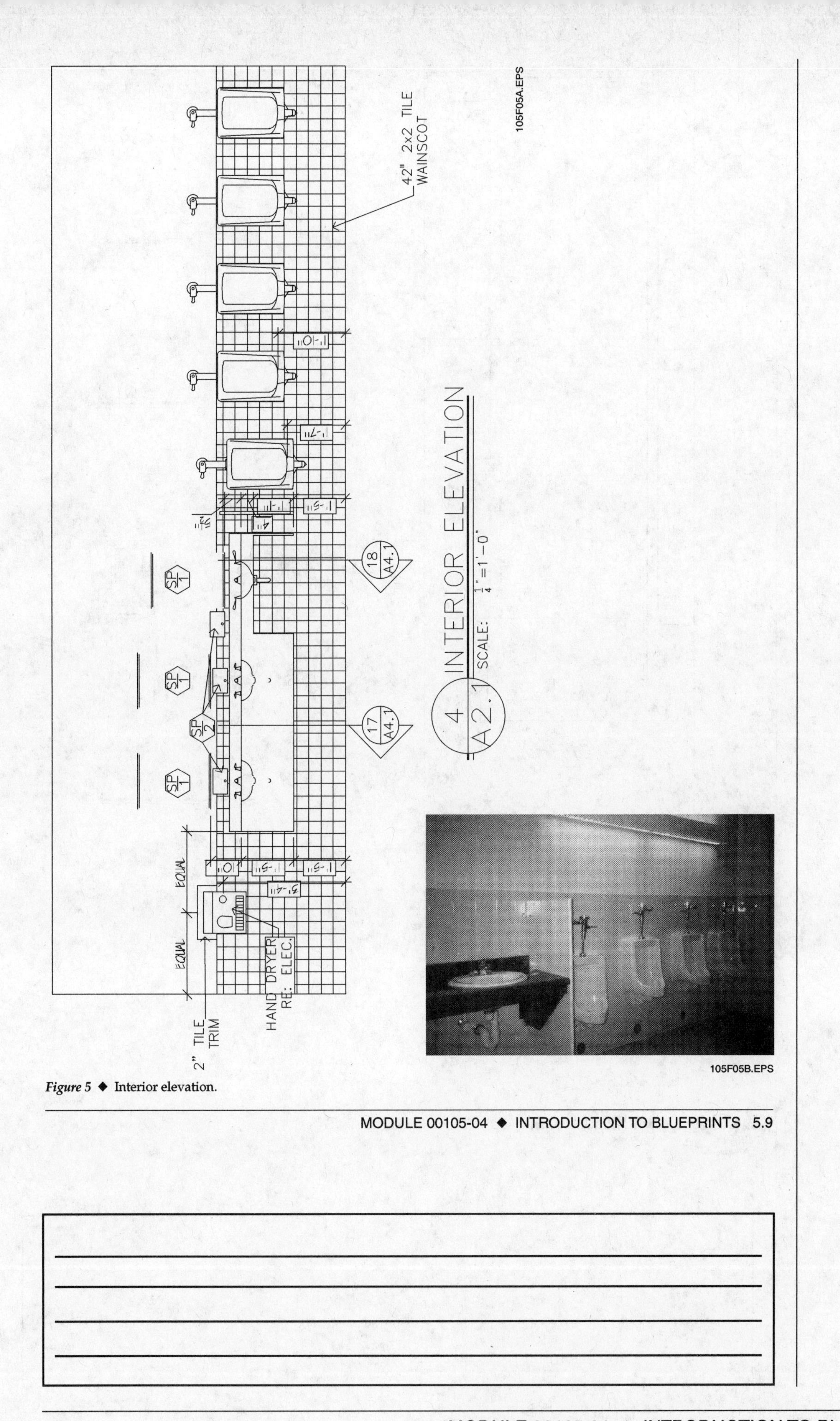

Figure 5 ◆ Interior elevation.

MODULE 00105-04 ◆ INTRODUCTION TO BLUEPRINTS 5.9

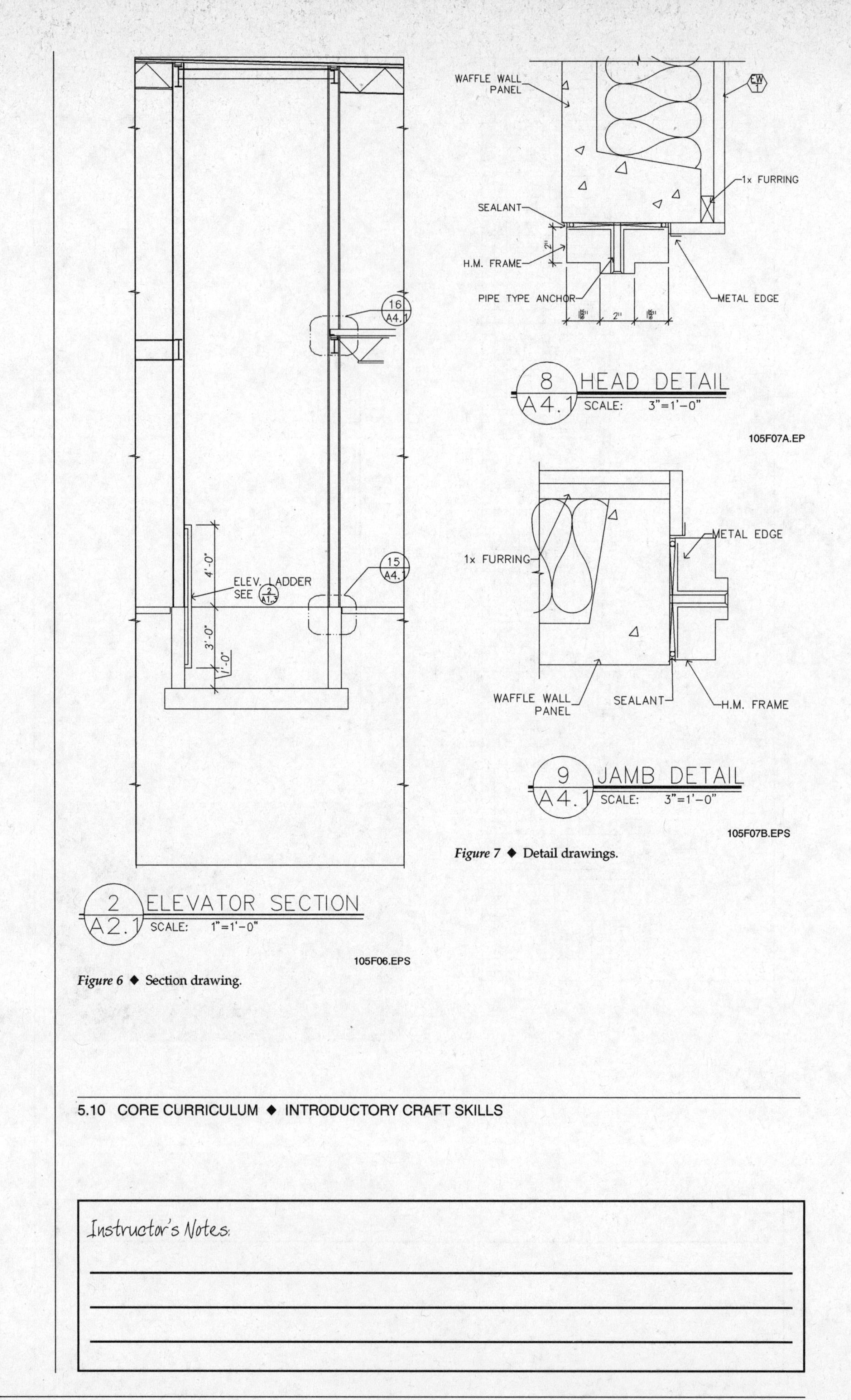

Figure 6 ◆ Section drawing.

Figure 7 ◆ Detail drawings.

5.10 CORE CURRICULUM ◆ INTRODUCTORY CRAFT SKILLS

Instructor's Notes:

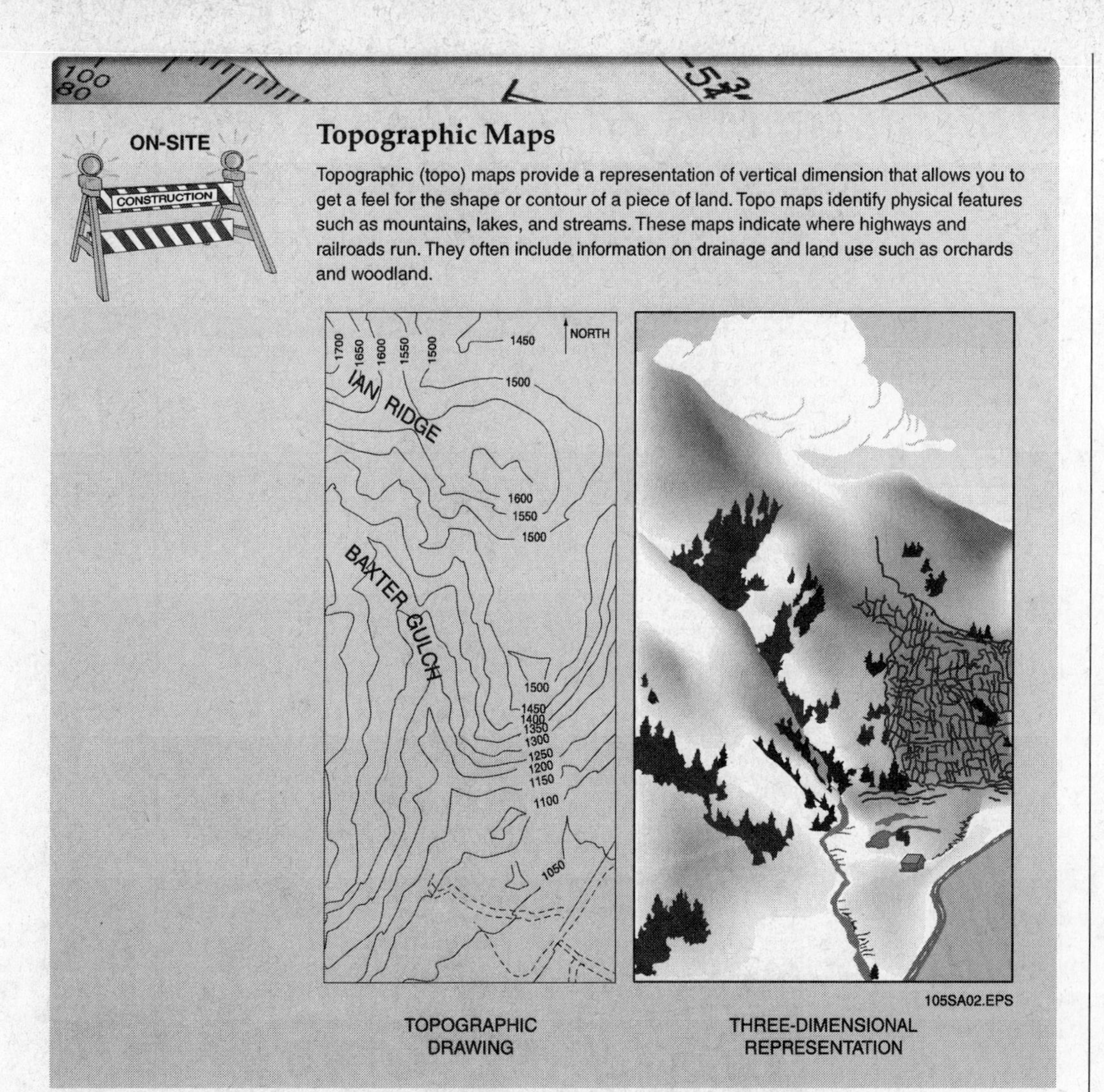

Topographic Maps

Topographic (topo) maps provide a representation of vertical dimension that allows you to get a feel for the shape or contour of a piece of land. Topo maps identify physical features such as mountains, lakes, and streams. These maps indicate where highways and railroads run. They often include information on drainage and land use such as orchards and woodland.

TOPOGRAPHIC DRAWING

THREE-DIMENSIONAL REPRESENTATION

Show Transparency 7 (Figure 8). Explain how to translate the information in the general notes on a structural plan.

The architectural plan also shows the finish schedules to be used for the doors and windows of the building. Door and window schedules, for example, are tables that list the sizes and other information about the various types of doors and windows used in the project. Schedules also may be included for finish hardware and fixtures. These schedules are not drawings, but they are usually included in a set of working drawings.

1.2.3 Structural Plans

The **structural plans** are a set of engineered drawings used to support the architectural design. The first part of the structural plans is the general notes (*Figure 8*). These notes give details of the materials to be used and the requirements to be followed in order to build the structure that the architectural plan depicts. The notes, for instance, might specify the type and strength of concrete

MODULE 00105-04 ◆ INTRODUCTION TO BLUEPRINTS 5.11

GENERAL NOTES

1. LIVE LOADS USED IN DESIGN:
   A. ROOF (SNOW) ---------- 30 PSF
   B. 1st AND 2nd LEVEL ---------- 100 PSF
   C. STAIRS ---------- 100 PSF
   D. WIND ---------- 70 MPH, EXPOSURE C
   E. SEISMIC ---------- ZONE 1

2. CONCRETE:
   A. ALL CONCRETE SHALL DEVELOP 3,000 PSI COMPRESSIVE STRENGTH IN 28 DAYS, WITH THE EXCEPTION OF INTERIOR SLABS ON GRADE, WHICH SHALL DEVELOP 3,500 PSI COMPRESSIVE STRENGTH IN 28 DAYS, AND "WAFFLE-WALL" PANELS WHICH SHALL DEVELOP 4,500 PSI COMPRESSIVE STRENGTH IN 28 DAYS.
   B. ALL REINFORCING SHALL CONFORM TO ASTM A615, GRADE 60, EXCEPT PILASTER TIES, GRADE BEAM STIRRUPS, AND DOWELS TO SLABS ON GRADE MAY BE GRADE 40.
   C. NO SPLICES OF REINFORCEMENT SHALL BE MADE EXCEPT AS DETAILED OR AUTHORIZED BY THE STRUCTURAL ENGINEER. LAP SPLICES, WHERE PERMITTED, SHALL BE A MINIMUM OF 36 BAR DIAMETERS. MAKE ALL BARS CONTINUOUS AROUND CORNERS.
   D. CONTINUOUS BARS IN GRADE BEAMS SHALL BE SPLICED AS FOLLOWS: TOP BARS AT MIDSPAN; BOTTOM BARS OVER SUPPORTS.
   E. DETAIL BARS IN ACCORDANCE WITH A.C.I DETAILING MANUAL AND A.C.I. BUILDING CODE REQUIREMENTS FOR REINFORCED CONCRETE, LATEST EDITIONS.
   F. PROVIDE ALL ACCESSORIES NECESSARY TO SUPPORT REINFORCING AT POSITIONS SHOWN ON THE DRAWINGS.
   G. REINFORCEMENT PROTECTION SHALL BE AS FOLLOWS:
      (1) CONCRETE POURED AGAINST EARTH ---------- 3"
      (2) FORMED CONCRETE EXPOSED TO EARTH OR WEATHER ---------- 2"
      (3) FORMED STAIRS OR WALLS NOT EXPOSED TO WEATHER ---------- 3/4"
   H. PLACE (2)-#5 (ONE EACH FACE) WITH 2'-0 PROJECTION AROUND ALL OPENINGS IN CONCRETE UNLESS OTHERWISE SHOWN OR NOTED.
   I. GRADE BEAMS SHALL NOT HAVE JOINTS IN A HORIZONTAL PLANE. ANY STOP IN CONCRETE WORK MUST BE MADE AT MIDDLE OF SPAN WITH VERTICAL BULKHEADS AND HORIZONTAL KEYS SPACED 8" ON CENTER, UNLESS OTHERWISE SHOWN. ALL CONSTRUCTION JOINTS SHALL BE AS DETAILED OR AS APPROVED BY THE ARCHITECT AND THE STRUCTURAL ENGINEER.
   J. WIRE FABRIC REINFORCEMENT MUST LAP ONE FULL MESH +2" AT SIDE AND END LAPS, AND SHALL BE TIED TOGETHER.

3. STEEL:
   A. ALL STRUCTURAL STEEL SHALL CONFORM TO ASTM A38, EXCEPT TUBE COLUMNS WHICH SHALL CONFORM TO ASTM A500 (GRADE B), LATEST EDITIONS.
   B. STRUCTURAL STEEL SHALL BE DETAILED AND FABRICATED IN ACCORDANCE WITH LATEST PROVISIONS OF THE A.I.S.C. MANUAL OF STEEL CONSTRUCTION.
   C. USE FRAMED BEAM CONNECTIONS WITH 3/4" DIAMETER ASTM A325 BOLTS, OR WELDED EQUIVALENT, UNLESS OTHERWISE SHOWN OR NOTED. FOR BEAMS WITHOUT DESIGNATED LOADS ON DRAWING, SELECT CONNECTIONS TO SUPPORT 50% OF TOTAL UNIFORM LOAD CAPACITY IN BENDING FOR EACH GIVEN BEAM AND SPAN, PLUS THE REACTION DUE TO ANY CONCENTRATED LOADS, MINIMUM OF (2) BOLTS PER CONNECTION.
   D. STEEL JOISTS SHALL BE DESIGNED, FABRICATED, AND ERECTED IN ACCORDANCE WITH STEEL JOISTS INSTITUTE SPECIFICATIONS.
   E. STEEL ROOF DECK:
      (1) STEEL DECK SHALL BE ERECTED IN ACCORDANCE WITH MANUFACTURER'S SUGGESTED SPECIFICATIONS.
      (2) STEEL DECK SHALL BE 1-1/2"x22 GAUGE, TYPE B, SHOP PAINTED.
      (3) DECK TO BE CONTINUOUS OVER A MINIMUM OF 3 SUPPORTS.
      (4) WELD DECK TO ALL SUPPORTS WITH 5/8" DIAMETER PUDDLE WELDS AT 12" ON CENTER (36/4 PATTERN). PROVIDE #10 TEK SCREWS (2 PER SPAN) AT SIDE LAPS. DECK MUST BE CAPABLE OF WITHSTANDING A DIAPHRAGM SHEAR OF 250 PLF. SUBMIT TEST DATA FROM DECK MANUFACTURER FOR DECK SELECTED TO SUBSTANTIATE THAT DECK WILL MEET OR EXCEED REQUIRED DIAPHRAGM SHEAR.
      (5) PROVIDE L3x3x1/4 FRAMING AROUND ALL OPENINGS LARGER THAN 5". ALL ROOF DRAINS SHALL HAVE ANGLE FRAMING.
   F. ALL WELDERS SHALL HAVE EVIDENCE OF PASSING THE A.W.S. STANDARD QUALIFICATION TESTS.
   G. SEE ARCHITECTURAL DRAWINGS FOR NAILER HOLES OR OTHER HOLES REQUIRED IN STEEL MEMBERS.

4. FOUNDATIONS:
   FOUNDATION DESIGN IS BASED ON RECOMMENDATIONS BY LAMBERT AND ASSOCIATES, JOB NO. M97063GE. RECOMMENDATIONS IN THIS REPORT SHOULD BE FOLLOWED.
   A. STEEL H PILES:
      (1) H PILES ARE TO BE HP10x42, WITH A 3/4" TOP PLATE. H PILES SHALL HAVE A BEARING CAPACITY OF 50 TONS.
      (2) ALL PILES ARE TO BE DRIVEN TO REFUSAL INTO THE SHALE LAYER. PILES TO BE A MINIMUM OF 5'-0 IN LENGTH. FOR BID PURPOSES, PILES SHOULD SET UP ABOUT 20'-0 BELOW TOP GRADE.
      (3) SPLICES ARE TO BE FULL PENETRATION FIELD WELDING TO DEVELOP FULL DESIGN LOAD.
      (4) PILE SET SHOULD BE DETERMINED BY THE JANBU FORMULA.
   B. SOILS ENGINEER SHALL BE PRESENT DURING PILE DRIVING TO VERIFY SET AND LENGTHS.

5. ALL DIMENSIONS ON STRUCTURAL DRAWINGS TO BE CHECKED AGAINST ARCHITECTURAL. NOTIFY ARCHITECT AND STRUCTURAL ENGINEER OF ANY DISCREPANCIES BEFORE PROCEEDING WITH CONSTRUCTION.

6. VERIFY ALL OPENINGS THROUGH FLOORS, ROOF, AND WALLS WITH MECHANICAL AND ELECTRICAL REQUIREMENTS.

105F08.EPS

Figure 8 ◆ Structural plan general notes.

Instructor's Notes:

required for the foundation, the loads that the roof and stairs must be built to accommodate, and codes that contractors must follow. General notes may be on a separate general notes sheet or may be part of individual plan sheets.

The structural plans also include a **foundation plan,** which shows the lowest level of the building, including concrete footings, slabs, and foundation walls (*Figure 9*). They also may show steel girders, columns, or **beams,** as well as detail drawings to show where and how the foundation must be reinforced. A related element is the structural floor plan, which depicts the framing, made of either wood or metal joists, and the underlayment of each floor of the structure.

The structural plans show the materials to be used for the walls, whether concrete or masonry, and whether the framing is wood or steel. They include a roof framing plan (*Figure 10*), showing what kinds of ceiling joists and roof rafters are to be used and where trusses are to be placed.

The structural plans include structural section drawings (*Figure 11*), which are similar to the architectural section drawings but show only the structural requirements. Miscellaneous structural details may also be shown in these sections to provide a better understanding of such things as connections and attachments of accessories.

DID YOU KNOW?

How Blueprints Started

The process for making blueprints was developed in 1842 by an English astronomer named Sir John F. Herschel. The method involved coating a paper with a special chemical. After the coating dried, an original hand drawing was placed on top of the paper. Both papers were then covered with a piece of glass and set in the sunlight for about an hour. The coated paper was developed much like a photograph. After a cold-water wash, the coated paper turned blue, and the lines of the drawing remained white.

Refer to Figures 9 and 10. Review the types of information that are included on foundation and structural plans.

Show Transparency 8 (Figure 11). Explain how structural section drawings illustrate connections and attachments of accessories.

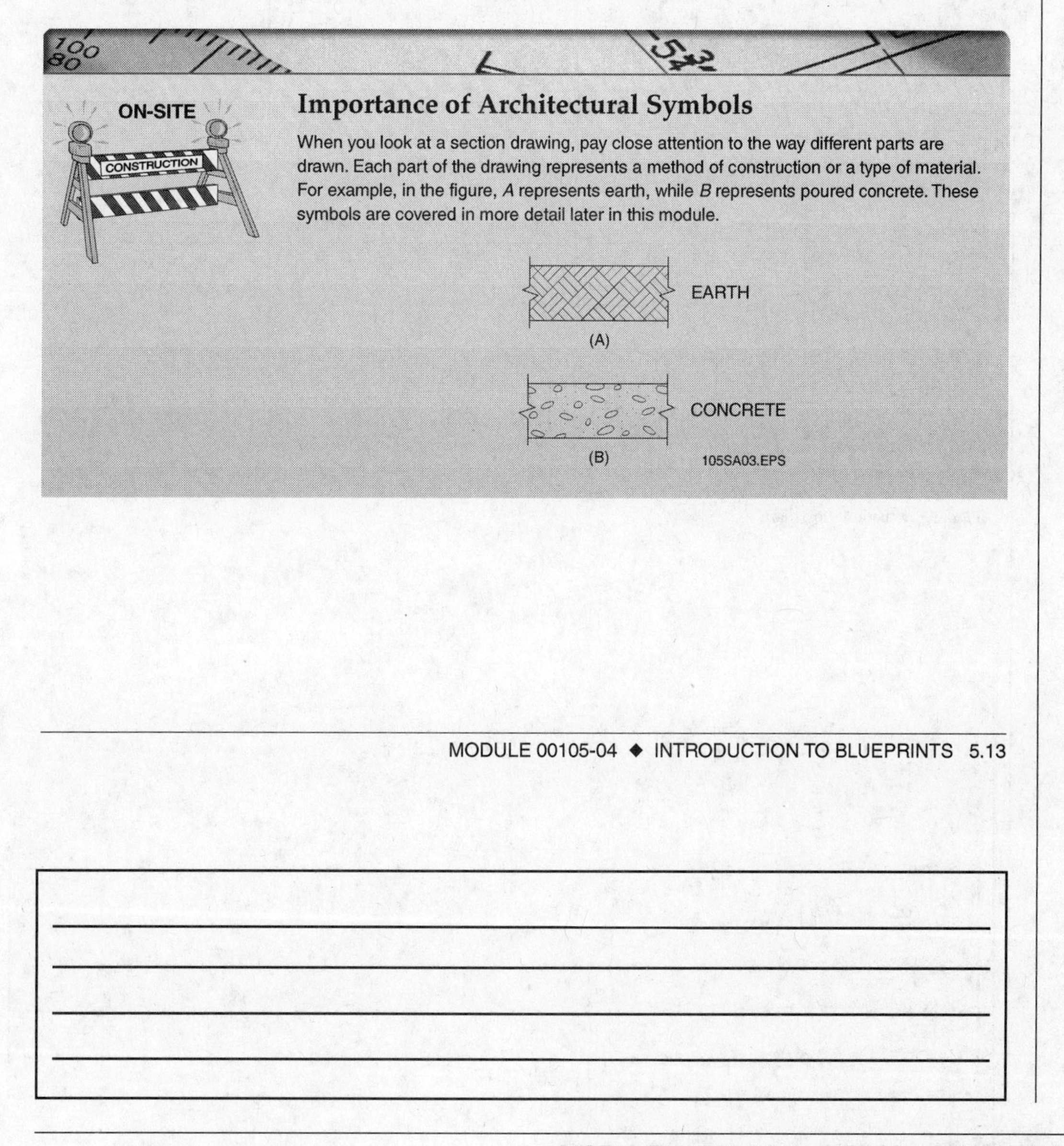

Importance of Architectural Symbols

When you look at a section drawing, pay close attention to the way different parts are drawn. Each part of the drawing represents a method of construction or a type of material. For example, in the figure, *A* represents earth, while *B* represents poured concrete. These symbols are covered in more detail later in this module.

MODULE 00105-04 ◆ INTRODUCTION TO BLUEPRINTS 5.13

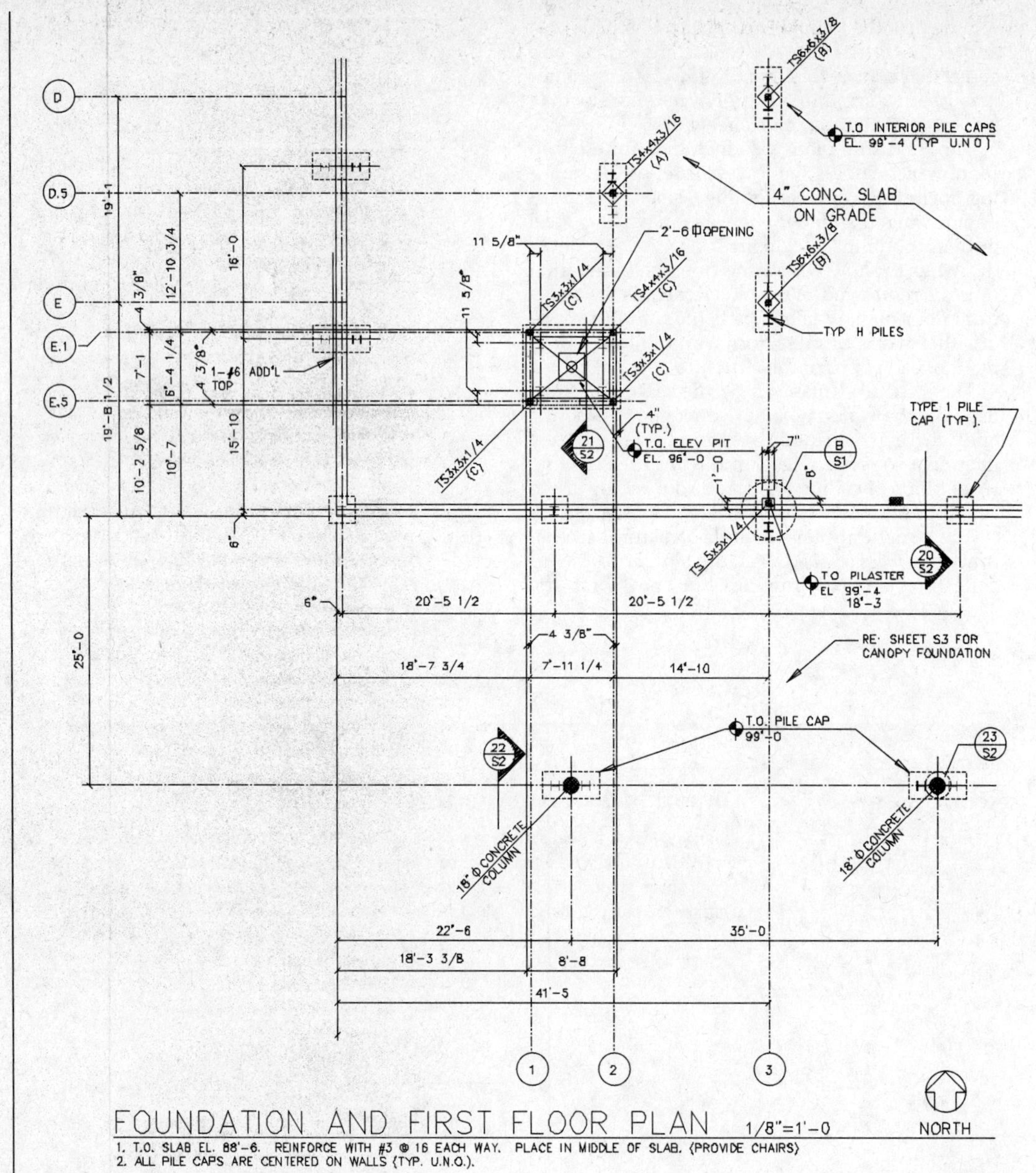

Figure 9 ◆ Foundation plan.

Instructor's Notes:

5.14 CORE CURRICULUM ◆ INTRODUCTORY CRAFT SKILLS

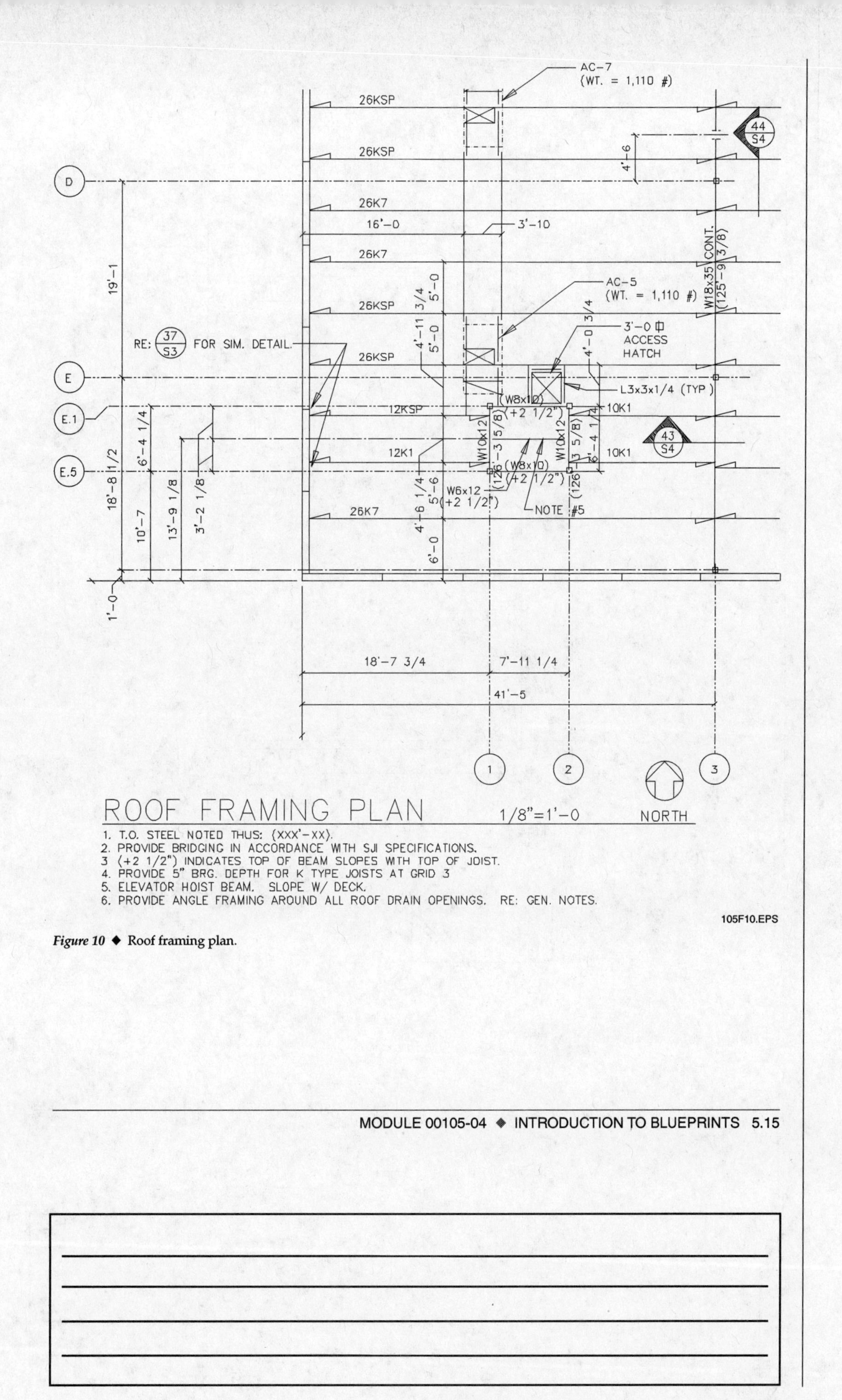

Figure 10 ◆ Roof framing plan.

MODULE 00105-04 ◆ INTRODUCTION TO BLUEPRINTS 5.15

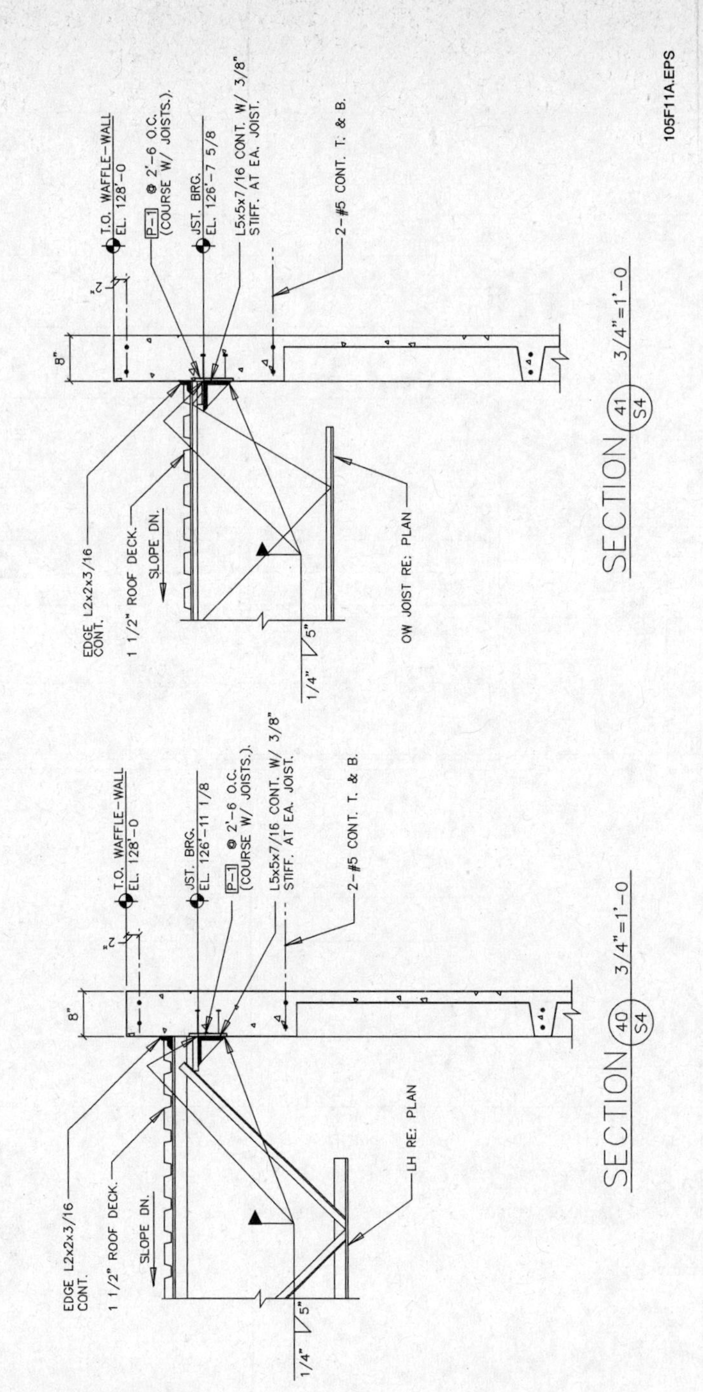

Figure 11 ◆ Structural section drawing.

Instructor's Notes:

5.16 CORE CURRICULUM ◆ INTRODUCTORY CRAFT SKILLS

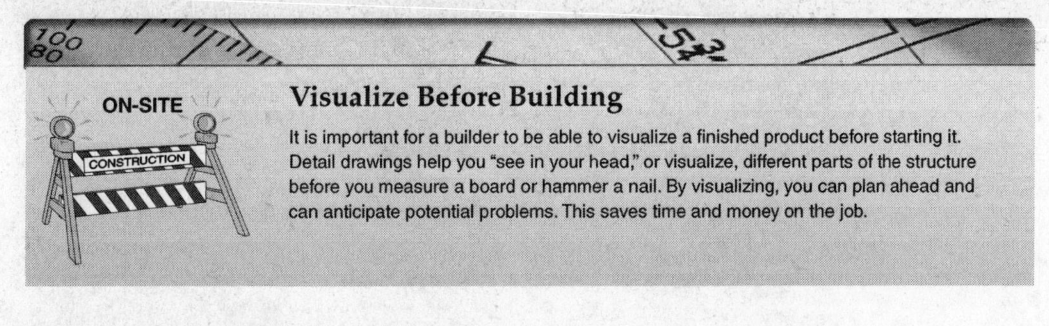

Visualize Before Building

It is important for a builder to be able to visualize a finished product before starting it. Detail drawings help you "see in your head," or visualize, different parts of the structure before you measure a board or hammer a nail. By visualizing, you can plan ahead and can anticipate potential problems. This saves time and money on the job.

Show Transparency 9 (Figure 12). Review the general notes that are typically included on a mechanical plan. Explain how these notes are used to determine the location of grilles and registers.

1.2.4 Mechanical Plans

Mechanical plans are engineered plans for motors, pumps, piping systems, and piping equipment. These plans incorporate general notes (*Figure 12*) containing specifications ranging from what the contractor is to provide to how the contractor determines the location of grilles and registers. A mechanical **legend** (*Figure 13*) defines the **symbols** used on the mechanical plans. A list of abbreviations (*Figure 14*) spells out abbreviations found on the plans.

Piping and instrumentation drawings, or **P&IDs** (*Figure 15*), are **schematic** diagrams of a complete piping system that show the process flow. They also show all the equipment, pipelines, valves, instruments, and controls needed to operate the system. P&IDs are not drawn to **scale** because they are meant only to give a representation, or a general idea, of the work to be done.

For more complex jobs, a separate **heating, ventilating, and air conditioning (HVAC)** plan is added to the set of plans. Piping system plans for gas, oil, or steam heat may be included in the HVAC plan. The mechanical plans include the layout of the HVAC system, showing specific requirements and elements for that system, including a floor, a reflected ceiling, or a roof. HVAC drawings (*Figure 16*) include an electrical schematic that shows the electrical circuitry for the HVAC system. HVAC plans are both mechanical and electrical drawings in one plan.

On the whiteboard/chalkboard, draw or write a variety of symbols and abbreviations used on mechanical plans. Ask trainees to identify the symbols and abbreviations. Discuss the importance of being able to accurately interpret symbols and abbreviations.

Refer to Figures 15 and 16. Explain why piping and instrument drawings are not drawn to scale. Review the need for separate heating, ventilating, and air conditioning plans.

GENERAL NOTES (FOR ALL MECHANICAL DRAWINGS)

1. CONTRACTOR IS TO PROVIDE COMPLETE CONNECTIONS TO ALL NEW AND RELOCATED OWNER FURNISHED EQUIPMENT.
2. CONTRACTOR TO COORDINATE THE LOCATION OF ALL DUCTWORK AND DIFFUSERS WITH REFLECTED CEILING PLAN AND STRUCTURE PRIOR TO BEGINNING WORK.
3. DIMENSIONS FOR INSULATED OR NON-INSULATED DUCT ARE OUTSIDE SHEET METAL DIMENSIONS.
4. DRAWINGS ARE NOT TO BE SCALED FOR DIMENSIONS. TAKE ALL DIMENSIONS FROM ARCHITECTURAL DRAWINGS, CERTIFIED EQUIPMENT DRAWINGS AND FROM THE STRUCTURE ITSELF BEFORE FABRICATING ANY WORK. VERIFY ALL SPACE REQUIREMENTS COORDINATING WITH OTHER TRADES, AND INSTALL THE SYSTEMS IN THE SPACE PROVIDED WITHOUT EXTRA CHARGES TO THE OWNER.
5. LOCATION OF ALL GRILLES, REGISTERS, DIFFUSERS AND CEILING DEVICES SHALL BE DETERMINED FROM THE ARCHITECTURAL REFLECTED CEILING PLANS.
6. THE OWNER AND DESIGN ENGINEER ARE NOT RESPONSIBLE FOR THE CONTRACTOR'S SAFETY PRECAUTIONS OR TO MEANS, METHODS, TECHNIQUES, CONSTRUCTION SEQUENCES, OR PROCEDURES REQUIRED TO PERFORM HIS WORK.
7. ALL WORK SHALL BE INSTALLED IN ACCORDANCE WITH PLRC'S SAFETY PLAN AND ALL APPLICABLE STATE AND LOCAL CODES.
8. ALL EXTERIOR WALL AND ROOF PENETRATIONS SHALL BE SEALED WEATHERPROOF. REFERENCE SPECIFICATION SECTION 15050.
9. ALL MECHANICAL WORK UNDER THIS CONTRACT IS TO FIVE (5) FEET OUTSIDE THE BUILDING.

105F12.EPS

Figure 12 ◆ Mechanical plan general notes.

MODULE 00105-04 ◆ INTRODUCTION TO BLUEPRINTS 5.17

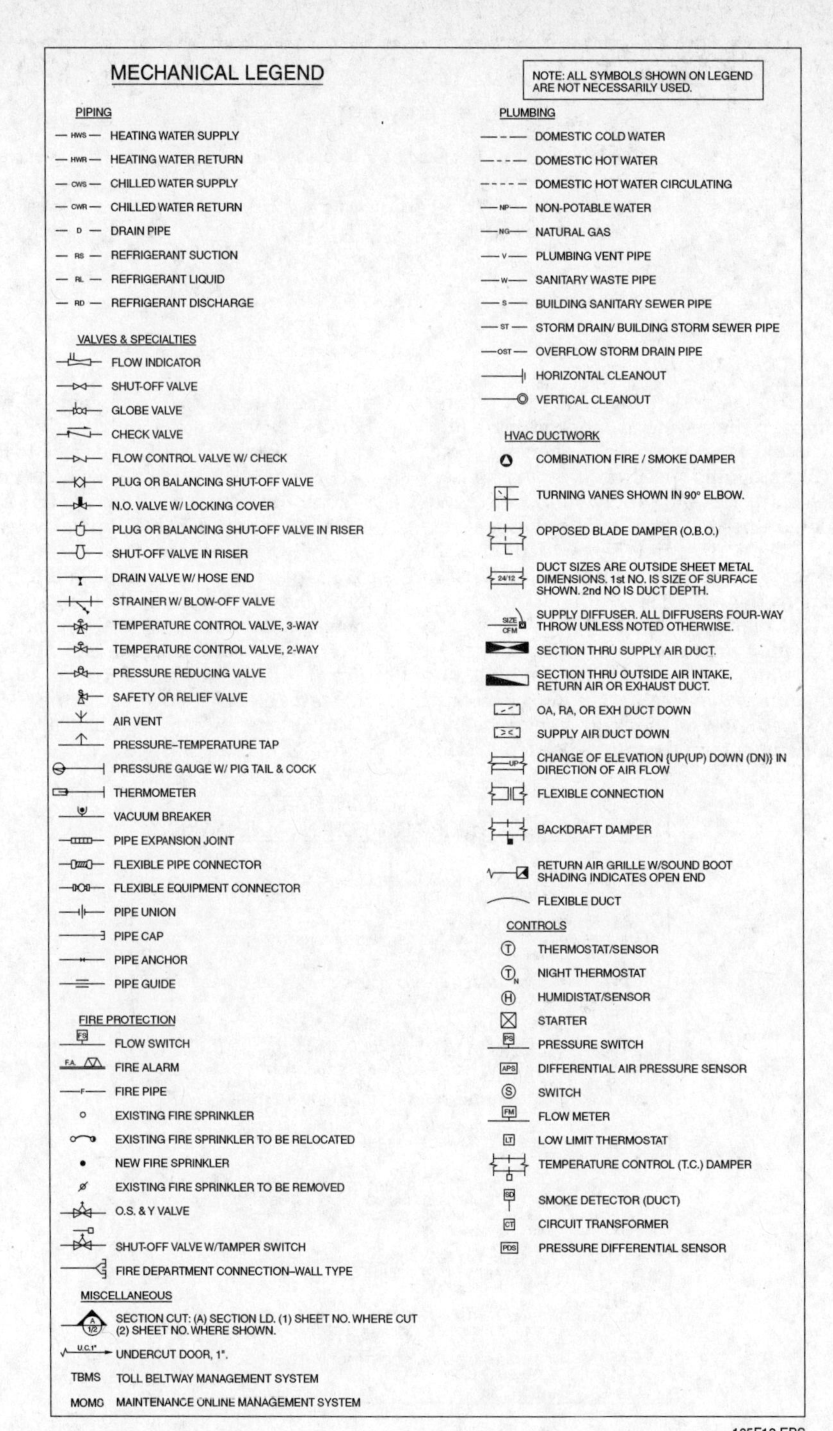

Figure 13 ◆ Mechanical plan legend.

Instructor's Notes:

5.18 CORE CURRICULUM ◆ INTRODUCTORY CRAFT SKILLS

ABBREVIATIONS

AFF	ABOVE FINISHED FLOOR
ALT	ALTITUDE
BHP	BRAKE HORSEPOWER
BTU	BRITISH THERMAL UNIT
Cv	COEFFICIENT, VALVE FLOW
CU FT	CUBIC FEET
CU IN	CUBIC INCH
CFM	CUBIC FEET PER MINUTE
SCFM	CFM, STANDARD CONDITIONS
dB	DECIBEL
DCW	DOMESTIC COLD WATER
DEG OR °	DEGREE
DHW	DOMESTIC HOT WATER
DIA	DIAMETER
DB	DRY-BULB
EAT	ENTERING AIR TEMPERATURE
EFF	EFFICIENCY
ELEV or EL	ELEVATION
ESP	EXTERNAL STATIC PRESSURE
EWT	ENTERING WATER TEMPERATURE
EXH	EXHAUST
F	FAHRENHEIT
FLA	FULL LOAD AMPS
FPM	FEET PER MINUTE
FPS	FEET PER SECOND
FT	FOOT OR FEET
FU	FIXTURE UNITS
GA	GAUGE
GAL	GALLONS
GPH	GALLONS PER HOUR
GPM	GALLONS PER MINUTE
HD	HEAD
HG	MERCURY
HGT	HEIGHT
HORZ	HORIZONTAL
HP	HORSEPOWER
HR	HOUR(S)
HWC	HOT WATER CIRCULATING (DOMESTIC)
HZ	HERTZ
ID	INSIDE DIAMETER
IE	INVERT ELEVATION
IN	INCHES
IN W.C.	INCHES WATER COLUMN
KW	KILOWATT
KWH	KILOWATT HOUR
LAT	LEAVING AIR TEMPERATURE
LBS OR #	POUNDS
LF	LINEAR FEET
LRA	LOCKED ROTOR AMPS
LWT	LEAVING WATER TEMPERATURE
MAX	MAXIMUM
MCA	MINIMUM CIRCUIT AMPS
MBH	BTU PER HOUR (THOUSAND)
MIN	MINIMUM
NC	NOISE CRITERIA
N.O.	NORMALLY OPEN
N.C.	NORMALLY CLOSED
N/A	NOT APPLICABLE
NIC	NOT IN CONTRACT
NTS	NOT TO SCALE
NO	NUMBER
OA	OUTSIDE AIR
OD	OUTSIDE DIAMETER
PPM	PARTS PER MILLION
%	PERCENT
PH OR f	PHASE (ELECTRICAL)
PSF	POUNDS PER SQUARE FOOT
PSI	POUNDS PER SQUARE INCH
PSIA	PSI ABSOLUTE
PSIG	PSI GAUGE
PRESS	PRESSURE
RA	RETURN AIR
RECIRC	RECIRCULATE
RH	RELATIVE HUMIDITY
RLA	RUNNING LOAD AMPS
RPM	REVOLUTIONS PER MINUTE
SL	SEA LEVEL
SENS	SENSIBLE
SPEC	SPECIFICATION
SQ	SQUARE
STD	STANDARD
SP	STATIC PRESSURE
SA	SUPPLY AIR
TEMP	TEMPERATURE
TD	TEMPERATURE DIFFERENCE
TSP	TOTAL STATIC PRESSURE
TSTAT	THERMOSTAT
TONS	TONS OF REFRIGERATION
VAV	VARIABLE AIR VOLUME
VEL	VELOCITY
VERT	VERTICAL
V	VOLT
VOL	VOLUME
W	WATT
WT	WEIGHT
WB	WET-BULB

105F14.EPS

Figure 14 ◆ Mechanical plan list of abbreviations.

MODULE 00105-04 ◆ INTRODUCTION TO BLUEPRINTS 5.19

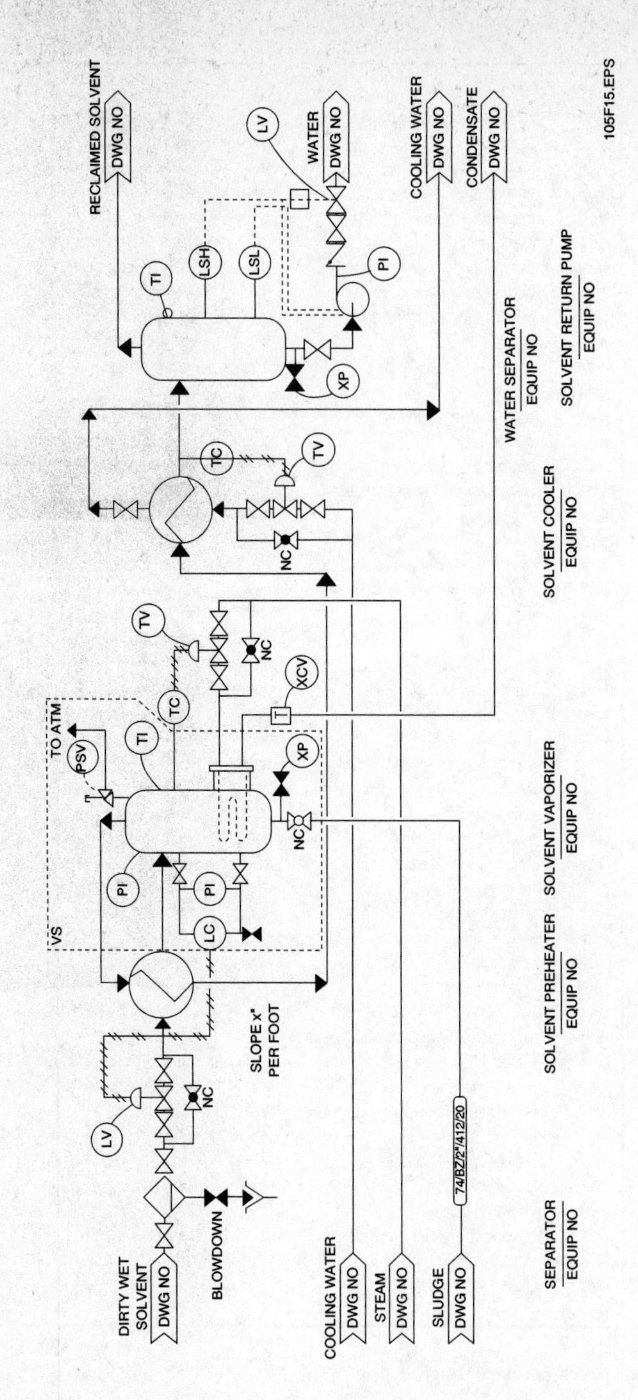

Figure 15 ◆ P&ID.

Instructor's Notes:

5.20 CORE CURRICULUM ◆ INTRODUCTORY CRAFT SKILLS

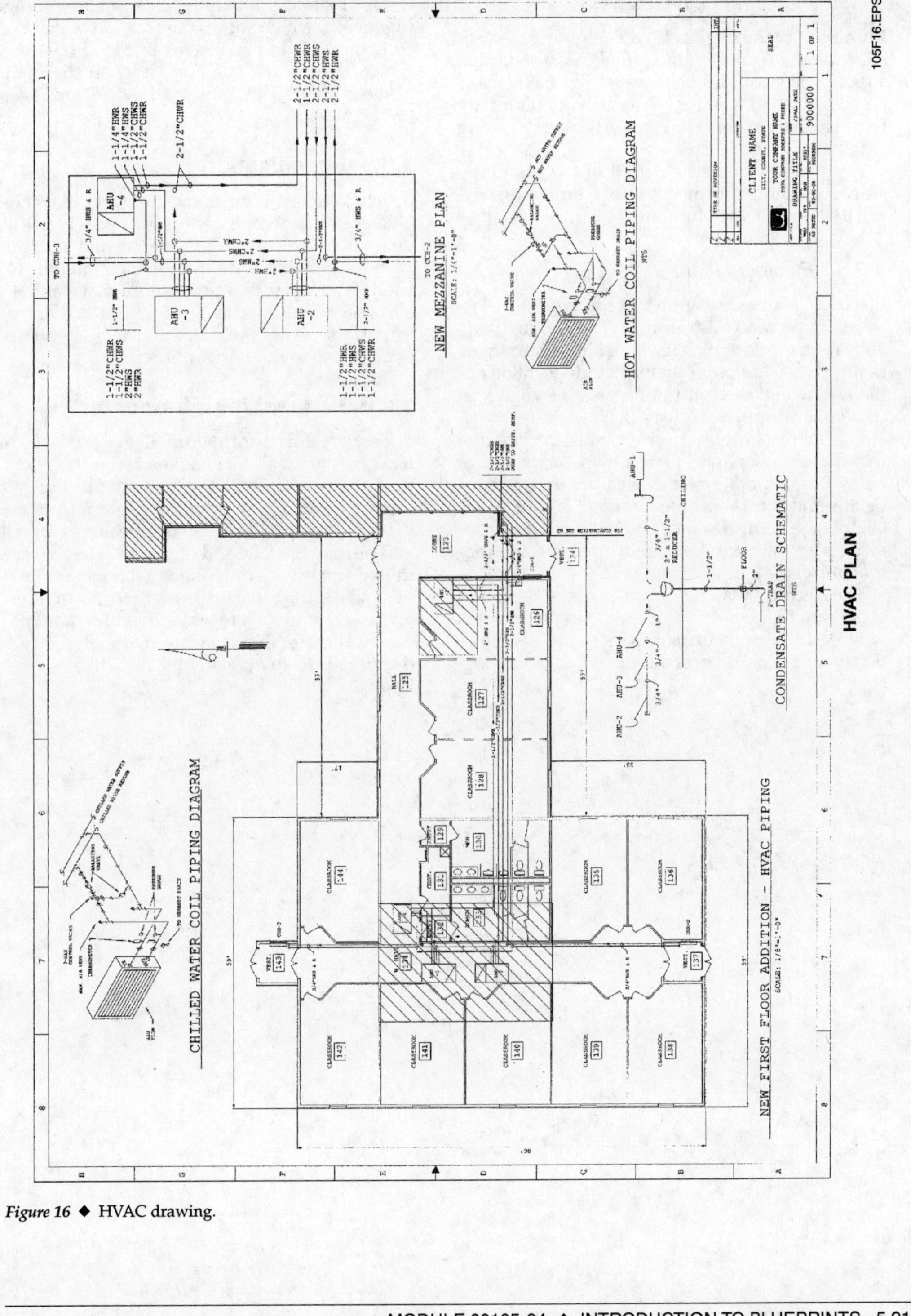

Figure 16 ◆ HVAC drawing.

MODULE 00105-04 ◆ INTRODUCTION TO BLUEPRINTS 5.21

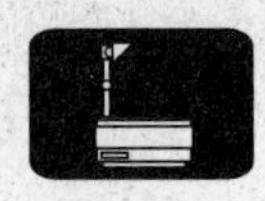

Show Transparency 10 (Figure 17). Review the layouts that are illustrated on plumbing/piping plans. Explain how isometric drawings are used to depict plumbing systems.

Show Transparency 11 (Figure 18). Discuss the advantage of using a separate electrical plan rather than including the information on the floor plan. Ask a trainee to specify the type of information included in an electrical plan's general notes.

On the whiteboard/ chalkboard, draw and write a variety of symbols and abbreviations used on an electrical plan. Ask trainees to identify the symbols and abbreviations. Compare the electrical legend and abbreviations with the mechanical legend and abbreviations.

See the Teaching Tip for Sections 1.2.0–1.2.6 at the end of this module.

Bring in a set of specifications. Distribute the specifications, and ask trainees to locate information that you specify.

Refer to Figure 22. Review the information that is included in a request for information. Provide examples of discrepancies that should be reported to the foreman.

1.2.5 Plumbing/Piping Plans

Plumbing plans (*Figure 17*) are engineered plans showing the layout for the plumbing system that supplies the hot and cold water, for the sewage disposal system, and for the location of plumbing fixtures. For commercial projects, each system may be on a separate plan.

A plumbing isometric is part of the plumbing plan. A plumbing isometric is an **isometric drawing** that depicts the plumbing system.

1.2.6 Electrical Plans

Electrical plans are engineered drawings for electrical supply and distribution. These plans may appear on the floor plan itself for simple construction projects. Electrical plans include locations of the electric meter, distribution panel, switchgear, convenience outlets, and special outlets.

For more complex projects, the information may be on a separate plan added to the set of plans. This separate plan leaves out unnecessary details and shows just the electrical layout. More complex electrical plans include locations of switchgear, transformers, main breakers, and motor control centers.

The electrical plan (*Figure 18*) starts with a set of general notes (*Figure 19*). These notes cover items ranging from main transformers to the coordination of underground penetrations into the building.

This plan also incorporates a power and lighting layout, which shows the location of lights and receptacles. It has an electrical legend (*Figure 20*), which defines the symbols used on the plan. It also has a key to the abbreviations (*Figure 21*) used on the plan.

1.3.0 Specifications

Specifications are written statements that the architectural and engineering firm provides to the general contractors. They define the quality of work to be done and describe the materials to be used. They clarify information that cannot be shown on the drawings. Specifications are very important to the **architect** and owner to ensure compliance to the standards set.

1.4.0 Request for Information

A **request for information (RFI)** (*Figure 22*) is used to clarify any discrepancies in the plans. If you notice a discrepancy, you should notify the foreman. The foreman will write up an RFI, explaining the problem as specifically as possible and putting the date and time on it. The RFI is submitted to the superintendent, who passes it to the general contractor, who passes it to the architect or **engineer,** who then resolves the discrepancy.

Always refer to specifications and the RFI when deciding how to interpret the drawings.

Instructor's Notes:

5.22 CORE CURRICULUM ◆ INTRODUCTORY CRAFT SKILLS

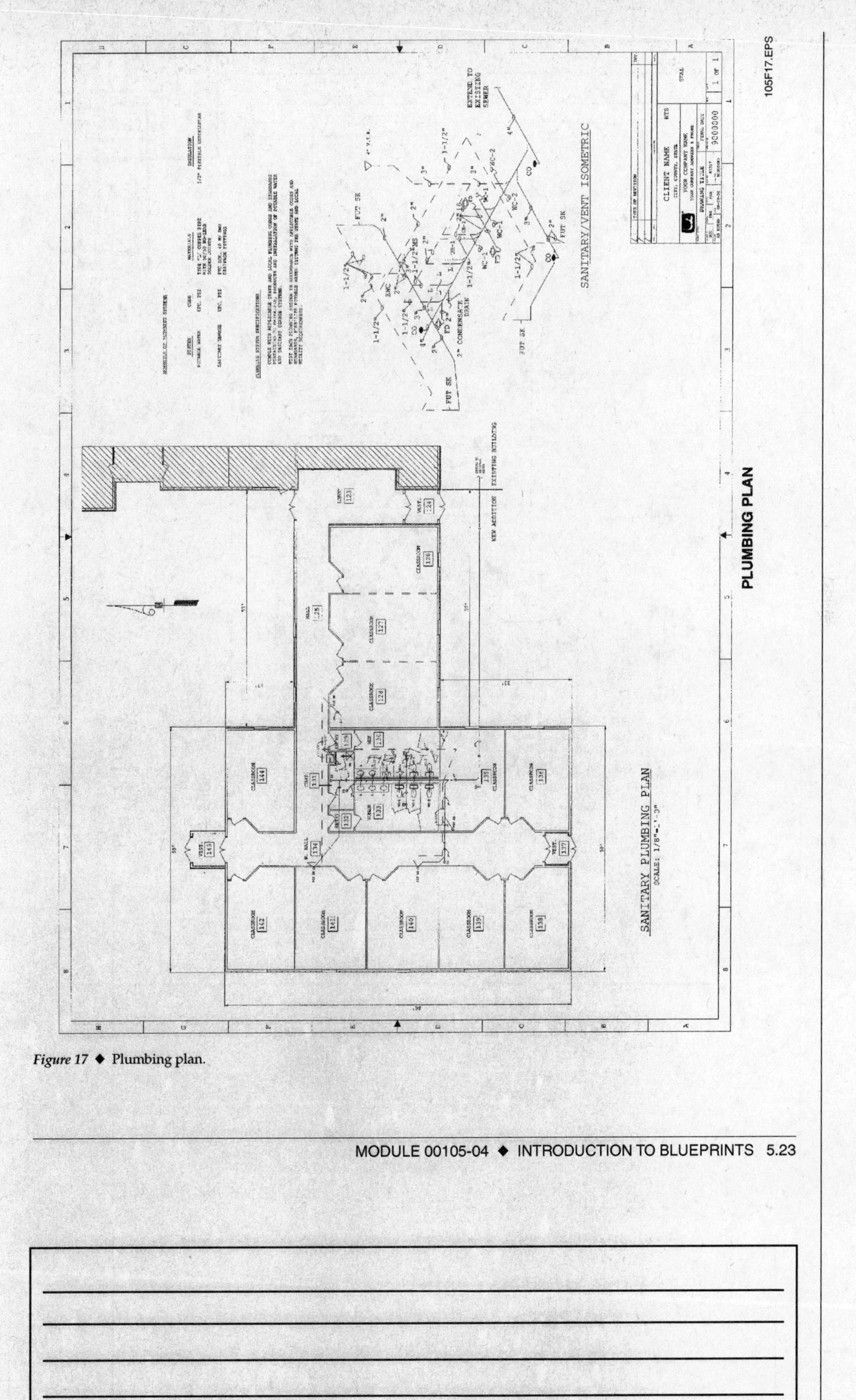

Figure 17 ◆ Plumbing plan.

MODULE 00105-04 ◆ INTRODUCTION TO BLUEPRINTS 5.23

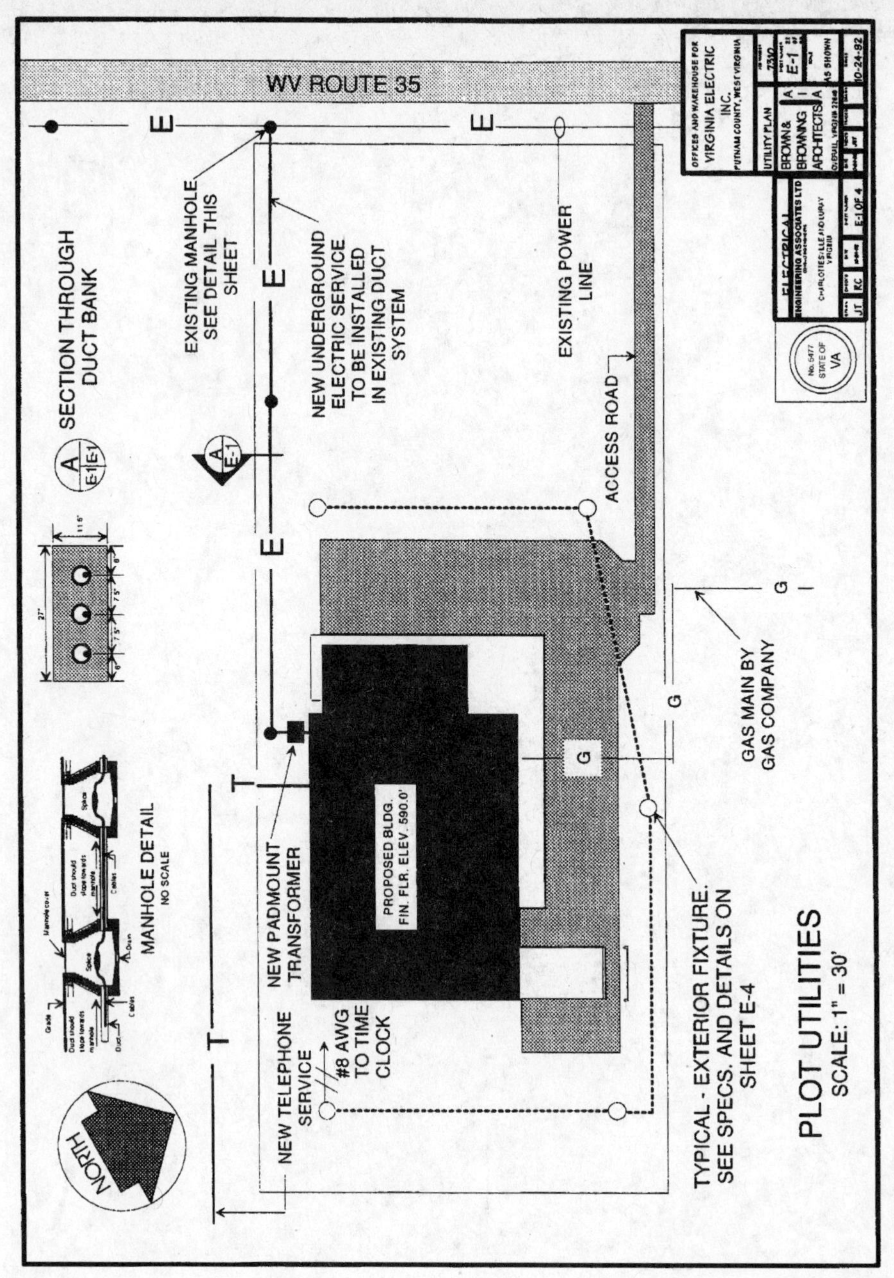

105F18.EPS

Figure 18 ◆ Electrical plan.

Instructor's Notes:

5.24 CORE CURRICULUM ◆ INTRODUCTORY CRAFT SKILLS

GENERAL NOTES (FOR ALL ELECTRICAL SHEETS)

1. COORDINATE LOCATION OF LUMINARIES WITH ARCHITECTURAL REFLECTED CEILING PLANS.
2. COORDINATE LOCATION OF ALL OUTLETS WITH ARCHITECTURAL ELEVATIONS, CASEWORK SHOP DRAWINGS AND EQUIPMENT INSTALLATION DRAWINGS.
3. COORDINATE LOCATION OF MECHANICAL EQUIPMENT WITH MECHANICAL PLANS AND MECHANICAL CONTRACTOR PRIOR TO ROUGH-IN.
4. PROVIDE (1) 3/4"C WITH PULL WIRE FROM EACH TELEPHONE, DATA OR COMMUNICATION OUTLET SHOWN, TO ABOVE ACCESSIBLE CEILING, AND CAP.
5. 3-LAMP FIXTURES SHOWN HALF SHADED HAVE INBOARD SINGLE LAMP CONNECTED TO EMERGENCY BATTERY PACK FOR FULL LUMEN OUTPUT. SEE SPECIFICATIONS.
6. SITE PLAN DOES NOT INDICATE ALL OF THE UG UTILITY LINES, RE: CIVIL DRAWINGS FOR ADDITIONAL INFORMATION. CONTRACTOR TO FIELD VERIFY EXACT LOCATION OF ALL EXISTING UNDERGROUND UTILITY LINES OF ALL TRADES PRIOR TO ANY SITE WORK.
7. THE LOCATIONS OF ALL SMOKE DETECTORS SHOWN ARE CONSIDERED TO BE SCHEMATIC ONLY. THE ACTUAL LOCATIONS (SPACING TO ADJACENT DETECTORS, WALLS, ETC.) ARE REQUIRED TO MEET NFPA 72.
8. ANY ITEMS DAMAGED BY THE CONTRACTOR SHALL BE REPLACED BY THE CONTRACTOR.
9. "CLEAN POWER" AND COMMUNICATION/COMPUTER SYSTEM REQUIREMENTS SHALL BE COORDINATED WITH COMMUNICATION/COMPUTER SYSTEMS CONTRACTOR.
10. REFER TO ARCHITECTURAL PLANS, ELEVATIONS AND DIAGRAMS FOR LOCATIONS OF FLOOR DEVICES AND WALL DEVICES. LOCATION WILL INDICATE VERTICAL AND/ OR HORIZONTAL MOUNTING. IF DEVICES ARE NOT NOTED OTHERWISE THEY SHALL BE MOUNTED LONG AXIS HORIZONTAL AT +16" TO CENTER.
11. ALL PLUGMOLD SHOWN SHALL BE WIREMOLD SERIES V2000 (IVORY FINISH) WITH SNAPICOIL #V20GB06 (OUTLETS 6" ON CENTER). PROVIDE ALL NECESSARY MOUNTING HARDWARE, ELBOWS, CORNERS, ENDS, ETC. REQUIRED FOR A COMPLETE SYSTEM.
12. ALL EMERGENCY RECEPTACLE DEVICES SHALL BE RED IN COLOR.
13. ALL BRANCH CIRCUITS SHALL BE 3-WIRE (HOT,NEUTRAL,GROUND).
14. COORDINATE EXACT EQUIPMENT LOCATIONS AND POWER REQUIREMENTS WITH OWNER AND ARCHITECT PRIOR TO ROUGH-INS.
15. ADA COMPLIANCE: ALL ADA HORN/STROBE UNITS SHALL BE MOUNTED +90" AFF OR 6" BELOW FINISHED CEILING, WHICH EVER IS LOWER. ELECTRICAL DEVICES PROJECTING FROM WALLS WITH THEIR LEADING EDGES BETWEEN 27" AND 80" AFF SHALL PROTRUDE NO MORE THAN 4" INTO WALKS OR CORRIDORS. ELECTRICAL AND COMMUNICATIONS SYSTEMS RECEPTACLES ON WALLS SHALL BE 15" MINIMUM AFF TO BOTTOM OF COVERPLATE.
16. COORDINATE ALL UNDERGROUND PENETRATIONS INTO THE BUILDING AND TUNNEL WITH STRUCTURAL ENGINEER, DUE TO EXPANSIVE SOILS.
17. ELECTRONIC STRIKES, MOTION DETECTORS AND ALARM SHUNTS ARE PROVIDED BY OTHERS. PROVIDE ALL NECESSARY ROUGH-INS FOR THESE ITEMS. COORDINATE WORK WITH SECURITY SYSTEM PROVIDER.

105F19.EPS

Figure 19 ◆ Electrical plan general notes.

MODULE 00105-04 ◆ INTRODUCTION TO BLUEPRINTS 5.25

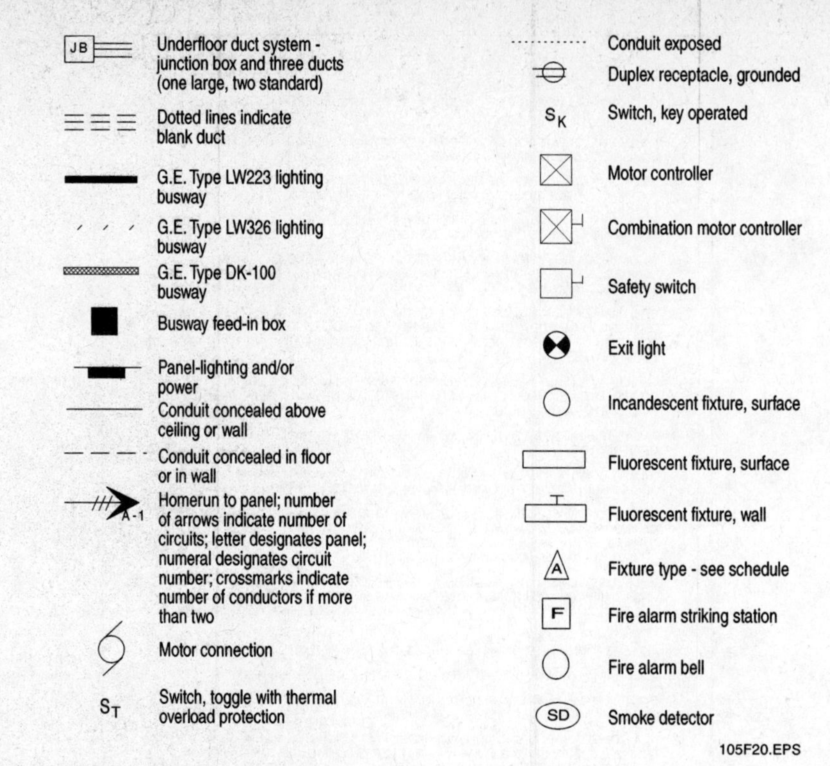

Figure 20 ◆ Electrical legend.

ABBREVIATIONS

NOTE: ALL ABBREVIATIONS SHOWN ARE NOT NECESSARILY USED.

AC	— 6" ABOVE COUNTER	MLO	— MAIN LUGS ONLY
AFF	— ABOVE FINISHED FLOOR	N	— NEUTRAL
AIC	— AMPS INTERRUPTING CURRENT	NIC	— NOT IN CONTRACT
AL	— ALUMINUM	NL	— NIGHT LIGHT
BFG	— BELOW FINISHED GRADE	NTS	— NOT TO SCALE
CB	— CIRCUIT BREAKER	OC	— OVER CURRENT
CU	— COPPER	RECP	— RECEPTACLE
EM	— EMERGENCY	TTB	— TELEPHONE TERMINAL BOARD
EWC	— ELECTRIC WATER COOLER	UG	— UNDERGROUND
FLA	— FULL LOAD AMPS	UNO	— UNLESS NOTED OTHERWISE
G	— GROUND	V	— VOLTS
GFCI	— GROUND FAULT CIRCUIT INTERRUPTER	W	— WATTS
HP	— HORSEPOWER	WG	— WIREGUARD
IG	— ISOLATED GROUND	WP	— WEATHERPROOF
KVA	— KILOVOLT-AMPS	XFMR	— TRANSFORMER
KW	— KILOWATTS	XP	— EXPLOSION PROOF
MCA	— MINIMUM CIRCUIT AMPS	'	— FEET
MCB	— MAIN CIRCUIT BREAKER	"	— INCHES

105F21.EPS

Figure 21 ◆ Electrical abbreviations.

Instructor's Notes:

5.26 CORE CURRICULUM ◆ INTRODUCTORY CRAFT SKILLS

DATE ______________ RFI NO. ______________

PROJECT NAME ______________________________ PROJECT NO.

REQUEST: REF D.W.G.NO. __________ REV. __________ OTHER

BY: ______________________________ REPLY BY (DATE):

REPLY:

DATE: ______________

105F22.EPS

Figure 22 ◆ Sample RFI.

MODULE 00105-04 ◆ INTRODUCTION TO BLUEPRINTS 5.27

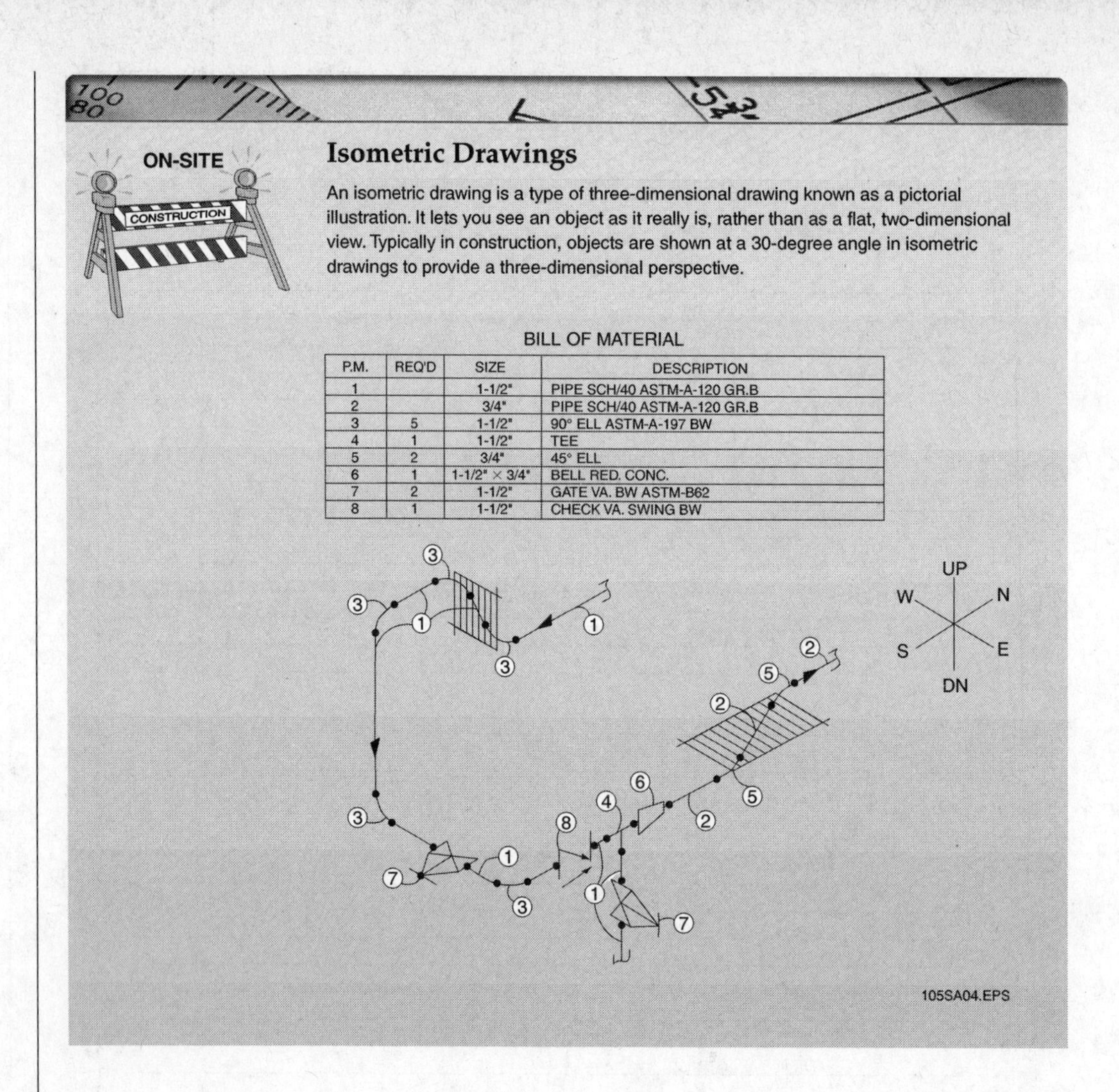

ON-SITE

Isometric Drawings

An isometric drawing is a type of three-dimensional drawing known as a pictorial illustration. It lets you see an object as it really is, rather than as a flat, two-dimensional view. Typically in construction, objects are shown at a 30-degree angle in isometric drawings to provide a three-dimensional perspective.

BILL OF MATERIAL

P.M.	REQ'D	SIZE	DESCRIPTION
1		1-1/2"	PIPE SCH/40 ASTM-A-120 GR.B
2		3/4"	PIPE SCH/40 ASTM-A-120 GR.B
3	5	1-1/2"	90° ELL ASTM-A-197 BW
4	1	1-1/2"	TEE
5	2	3/4"	45° ELL
6	1	1-1/2" × 3/4"	BELL RED. CONC.
7	2	1-1/2"	GATE VA. BW ASTM-B62
8	1	1-1/2"	CHECK VA. SWING BW

105SA04.EPS

Instructor's Notes:

5.28 CORE CURRICULUM ◆ INTRODUCTORY CRAFT SKILLS

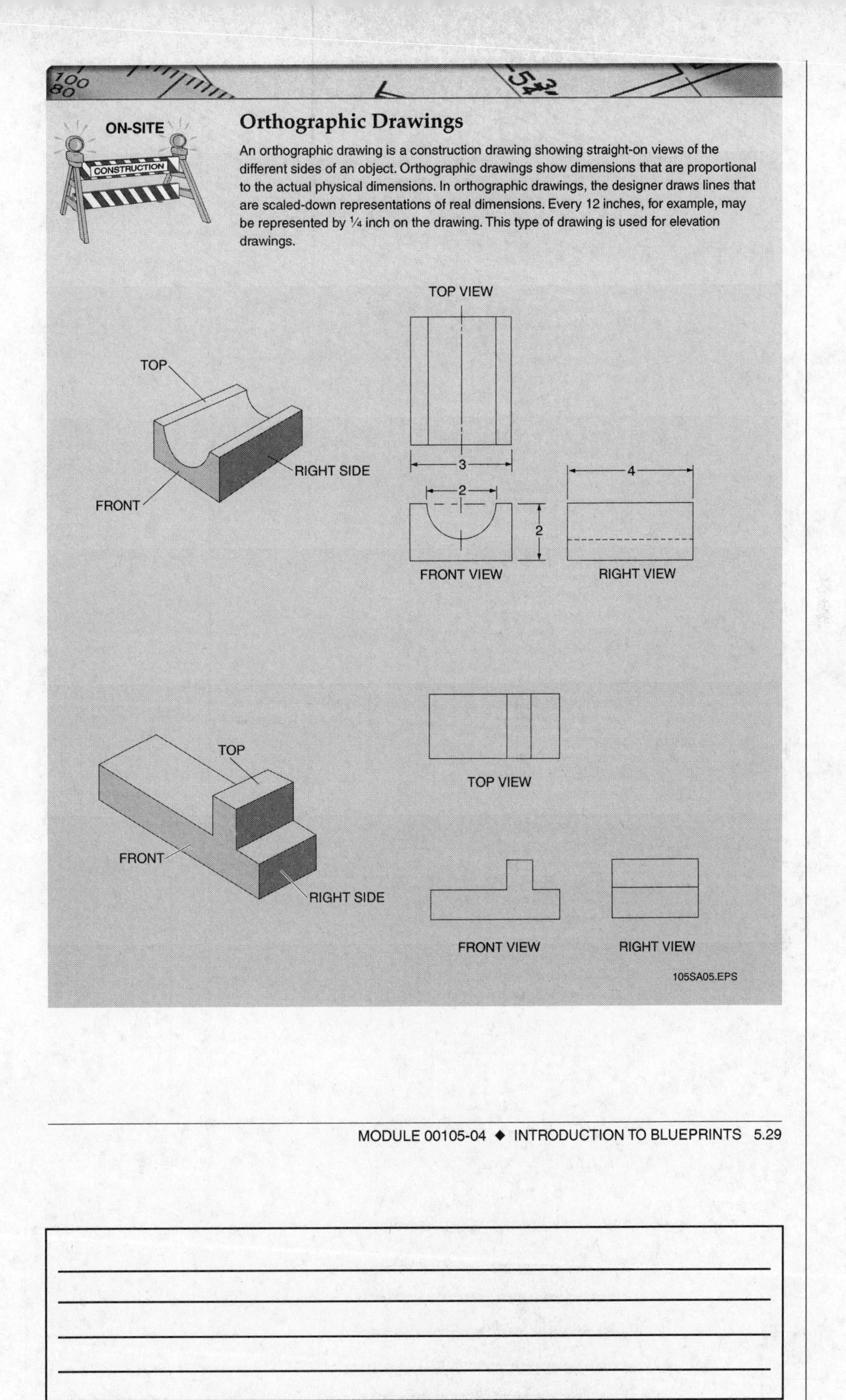

Orthographic Drawings

An orthographic drawing is a construction drawing showing straight-on views of the different sides of an object. Orthographic drawings show dimensions that are proportional to the actual physical dimensions. In orthographic drawings, the designer draws lines that are scaled-down representations of real dimensions. Every 12 inches, for example, may be represented by ¼ inch on the drawing. This type of drawing is used for elevation drawings.

MODULE 00105-04 ◆ INTRODUCTION TO BLUEPRINTS 5.29

ON-SITE

Schematic Drawings

Most plumbing and electrical sketches are single-line drawings or schematic drawings. These drawings illustrate the scale and relationship of the project's components. In a single-line or schematic plumbing drawing, the line represents the centerline of the pipe. In a single-line or schematic electrical drawing, the line represents electrical wiring routing or circuit.

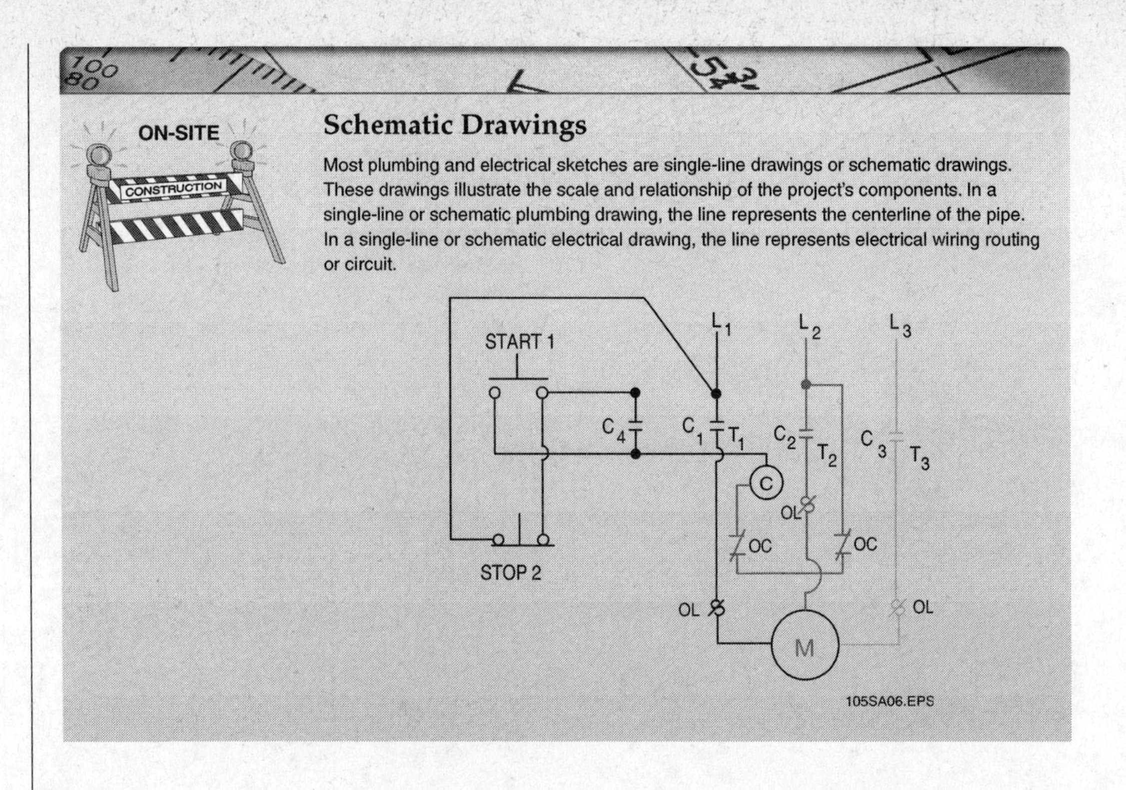

105SA06.EPS

Instructor's Notes:

5.30 CORE CURRICULUM ◆ INTRODUCTORY CRAFT SKILLS

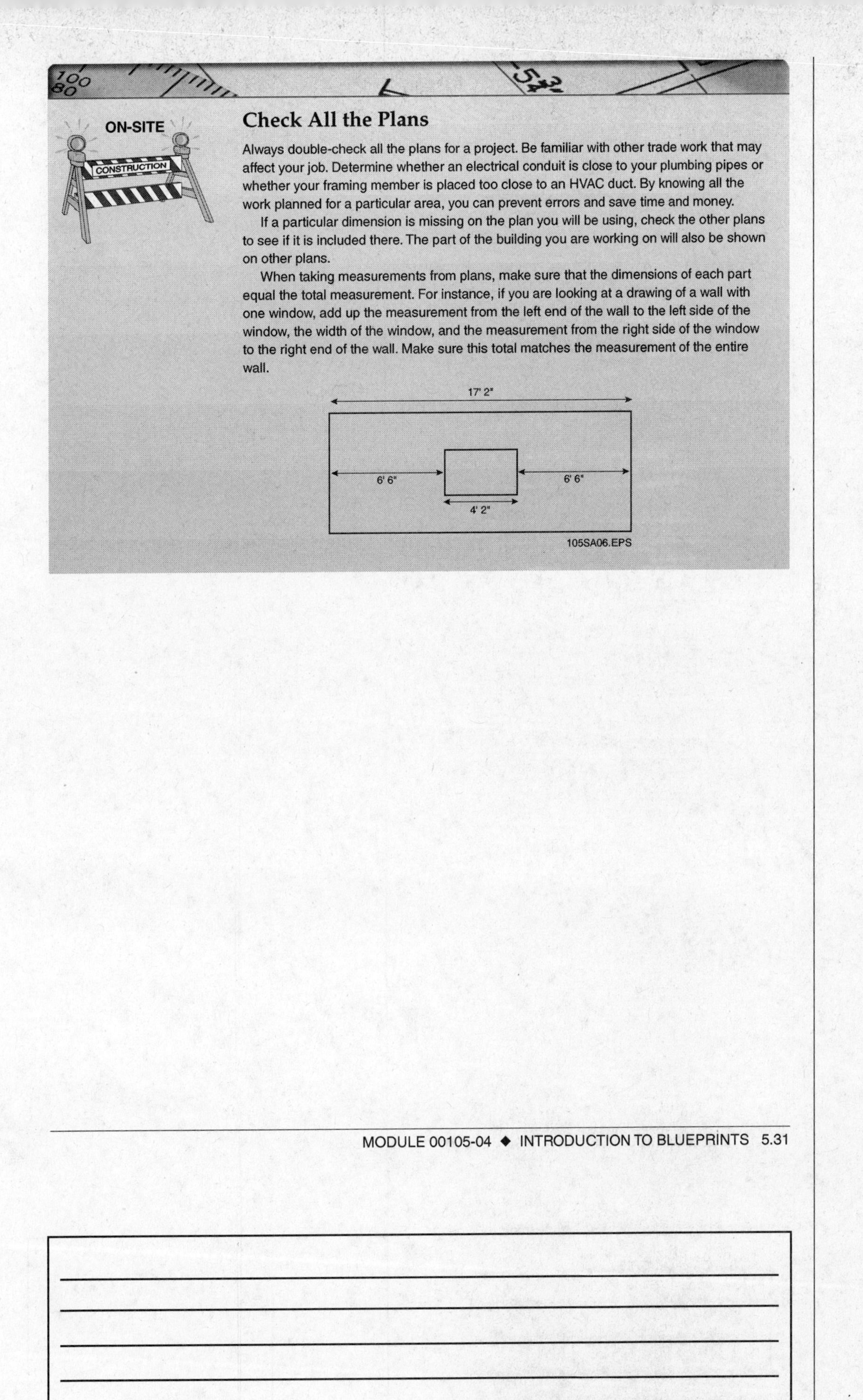

Check All the Plans

Always double-check all the plans for a project. Be familiar with other trade work that may affect your job. Determine whether an electrical conduit is close to your plumbing pipes or whether your framing member is placed too close to an HVAC duct. By knowing all the work planned for a particular area, you can prevent errors and save time and money.

If a particular dimension is missing on the plan you will be using, check the other plans to see if it is included there. The part of the building you are working on will also be shown on other plans.

When taking measurements from plans, make sure that the dimensions of each part equal the total measurement. For instance, if you are looking at a drawing of a wall with one window, add up the measurement from the left end of the wall to the left side of the window, the width of the window, and the measurement from the right side of the window to the right end of the wall. Make sure this total matches the measurement of the entire wall.

MODULE 00105-04 ◆ INTRODUCTION TO BLUEPRINTS 5.31

Go over the Review Questions for Section 1.0.0. Answer any questions trainees may have.

Have trainees review Sections 2.0.0–4.0.0 for the next session.

Review Questions

Section 1.0.0

1. A site plan _____.
   a. does not show features such as trees and driveways
   b. shows the location of the building from an aerial view
   c. is drawn after the floor plan is drawn, before the HVAC is designed
   d. includes information about plumbing fixtures and equipment

2. Elevation drawings _____.
   a. are views taken from a location above the building
   b. show a three-dimensional view of the building
   c. show how the elevator will be installed in the building
   d. are side views of the building or object

3. Detail drawings _____.
   a. are enlarged views of some special features of a building, such as floors and walls
   b. show the details of the site, such as elevation, water tables, and contour lines
   c. illustrate and describe how the roof is constructed
   d. specify the type and strength of concrete required for the foundation

4. The structural plans show the _____.
   a. contours of the site
   b. materials to be used
   c. electrical system
   d. door and window schedules

5. _____ plans show the layout of the HVAC system.
   a. Plumbing
   b. Structural
   c. Mechanical
   d. Foundation

Instructor's Notes:

2.0.0 ◆ COMPONENTS OF THE BLUEPRINT

All blueprints are laid out in a fairly standardized format. In this section, you will learn about the following five parts of the blueprint:

- **Title block**
- Border
- Drawing area
- Revision block
- Legend

2.1.0 Title Block

When you look at any blueprint, the first thing to look at is the title block. The title block is normally in the lower right-hand corner of the drawing or across the right edge of the paper (*Figure 23*).

The title block has two purposes. First, it gives information about the structure or assembly. Second, it is numbered so the print can be filed easily.

Different companies put different information in the title block. Generally, it contains the following:

- *Company logo* – Usually preprinted on the drawing.
- *Sheet title* – Identifies the project.
- *Date* – Date the drawing was checked and readied for seal, or issued for construction.
- *Drawn* – Initials of the person who drafted the drawing.
- *Drawing number* – Code numbers assigned to a project.
- *Scale* – The ratio of the size of the object as drawn to the object's actual size.
- *Revision blocks* – Information on revisions, including (at a minimum) the date and the initials of the person making the revision. Other information may include descriptions of the revision and a revision number.

Every company has its own system for such things as project numbers and departments. Every company also has its own placement locations for the title and revision blocks. Your supervisor should explain your company's system to you.

Ensure that you have everything required for the laboratories and demonstrations during this session.

Review the five parts of a blueprint.

Show Transparency 12 (Figure 23). Discuss the information included in a title block. Explain the two main purposes of a title block.

Display a sample blueprint. Ask trainees to locate on the title block information that you specify.

TYPE OF REVISION OR ENGINEERING STATUS | INT.
REV. | DATE | DESCRIPTION | APP'D.
SEAL
A
YOUR COMPANY NAME
YOUR COMPANY ADDRESS & PHONE
SHEET TITLE DRAWING TITLE
ISSUED FINAL DATE
DESIGNED | DRAWN | CX'D. | JOB NO. | DRAWING NO. | REV. | SHEET OF
SCALE | DATE | FILE NO.
2
1
105F23.EPS

Figure 23 ◆ The title block of a blueprint.

MODULE 00105-04 ◆ INTRODUCTION TO BLUEPRINTS 5.33

Distinguish between a blueprint's border and its drawing area. Explain why a border is necessary.

Show Transparency 13 (Figure 25). Ask a trainee to explain where and how revisions should be included on a blueprint.

Emphasize the importance of using the latest version of a blueprint.

2.2.0 Border

The border is a clear area of approximately half an inch around the edge of the drawing area. It is there so that everything in the drawing area can be printed or reproduced on printing machines with no loss of information.

2.3.0 Drawing Area

The drawing area (*Figure 24*) presents the information for constructing the project: the floor plan, elevations of building, sections, and details.

2.4.0 Revision Block

A revision block is located in the drawing area, usually in the lower right corner inside the title block or near it. Different companies put the revision block in different places. This block is used to record any changes (revisions) to the drawing. It typically contains the revision number, a brief description, the date, and the initials of the person who made the revisions (*Figure 25*). All revisions must be noted in this block and dated and identified by a letter or number.

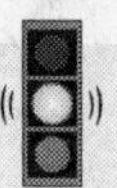

CAUTION

It is essential to note the revision designation on a blueprint and to use only the latest version. Otherwise, costly mistakes may result.

DID YOU KNOW?

Legality of Blueprints

Blueprints are incorporated into building contracts "by reference," making them part of the legal documents associated with a project. When describing the project to be completed, the legal contract between the builder and the owner refers to the accompanying construction drawings for details that would be too lengthy to write out. That makes tracking changes or revisions to blueprints over the course of a project vitally important. If an error is made along the way, either the owner or the builder must be able to find the discrepancy by reading the blueprints. Taking care of blueprints over the course of the project makes good business sense.

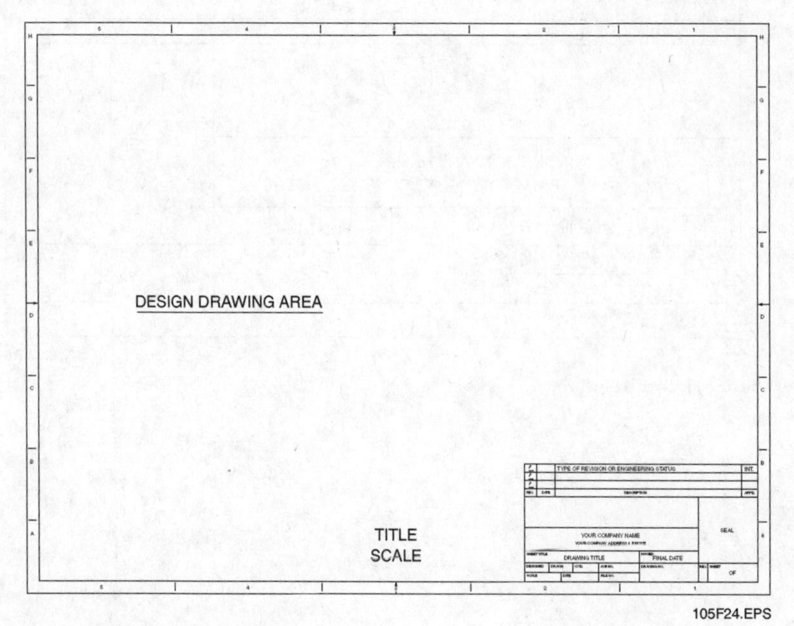

Figure 24 ◆ Drawing area of a blueprint.

Instructor's Notes:

5.34 CORE CURRICULUM ◆ INTRODUCTORY CRAFT SKILLS

PROJ	NO	REVISION	RVSD	CHKD	APPD	DATE
3483	01	RELEASED FOR CONSTRUCTION		APD	NWS	JULY 92
3483	02	DELETED PART OF LINE 12037		APD	NWS	AUG 92
3483	03	⚠ ADDED WELDING SYMBOL		APD	NWS	AUG 92

105F25.EPS

Figure 25 ◆ The revision block of a blueprint.

Bring in a blueprint with a legend. Demonstrate how to interpret the legend and how to use it to locate information about the plan's design.

2.5.0 Legend

Each line on a blueprint has a specific design and thickness that identifies it. (Some of the lines may be used to identify off-site utilities. There may be an identification on the cover sheet or on the civil plans where they are used.) The identification of these lines and other symbols is called the legend. Although a legend doesn't automatically appear on every blueprint, when it does, it explains or defines symbols or special marks used in the drawing (*Figure 26*).

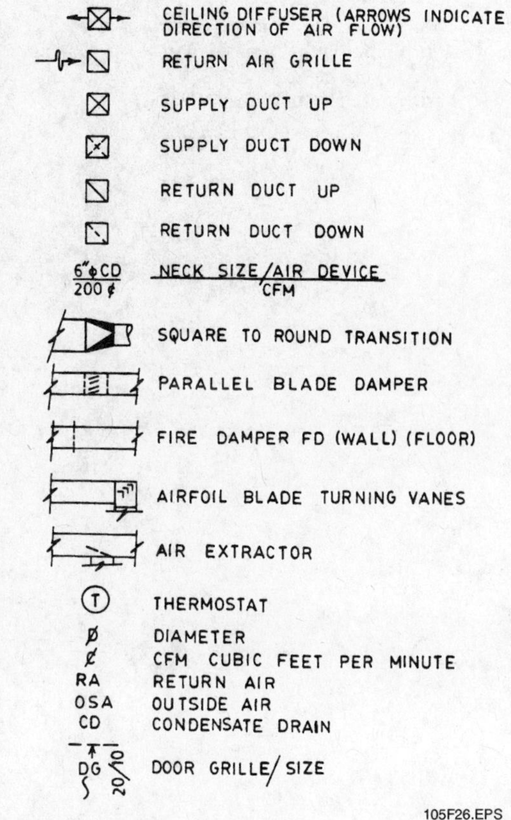

105F26.EPS

Figure 26 ◆ Sample legend.

DID YOU KNOW?

The North Arrow

All blueprints have an arrow indicating north. It may be located near the title block or in some other conspicuous place. The north arrow allows you to describe the locations of objects, walls, and building parts shown on the blueprint using a common reference point.

MODULE 00105-04 ◆ INTRODUCTION TO BLUEPRINTS 5.35

Go over the Review Questions for Section 2.0.0. Ask whether trainees have any questions.

Review Questions

Section 2.0.0

1. The title block generally contains _____.
   a. revision blocks
   b. special marks used in the drawing
   c. the mechanical plans
   d. the legend

2. The latest revision date on a set of blueprints can be found _____.
   a. inside the detail
   b. in the schedule
   c. inside the title block
   d. in the legend

3. The border around the edge of the blueprints _____.
   a. is used for making notes
   b. contains the revision dates
   c. is kept clear
   d. contains the legend

4. Revisions to the drawing are entered in the revision block and must include _____.
   a. the date the drawing was approved
   b. the project description, engineering approvals, and intended date of completion
   c. the date and the initials of the person who made the revision
   d. customer approval

5. The legend is used to _____.
   a. tell the history of the building
   b. explain symbols on the drawing
   c. define terms that are used only by CAD programs
   d. show the latest revision date

Instructor's Notes:

3.0.0 ◆ SCALE

The scale of a drawing tells the size of the object drawn compared with the actual size of the object represented. The scale is shown in one of the spaces in the title block, beneath the drawing itself, or in both places. The type of scale used on a drawing depends on the size of the objects being shown, the space available on the paper, and the type of plan.

On a site plan, the scale may read SCALE: 1" = 20' -0". This means that every 1 inch on the drawing represents 20 feet, 0 inches. The scale used to develop site plans is an **engineer's scale,** described later in this section.

On a floor plan, the scale may read SCALE: 1/4" = 1'-0". This means that every ¼ inch on the drawing represents 1 foot, 0 inches. Floor plans are developed using an **architect's scale,** described later in this section. This scale is divided into fractions of an inch.

Some drawings are *not* drawn to scale. A note on such drawings reads **not to scale (NTS).**

CAUTION

When a plan is marked NTS, you *cannot* measure dimensions on the drawing and use those measurements to build the project. Not-to-scale drawings give relative positions and sizes. The sizes are approximate and are not accurate enough for construction.

CAUTION

Always take your measurements from the drawing's written dimensions. Written dimensions are more accurate than those measured from the drawings because the drawing could have been shrunk or stretched without your knowledge.

3.1.0 Measuring Tools

The term *scale* also can describe a measuring tool used to draw or measure the lines of a blueprint. Three types of scales are used to draw or measure the lines on a blueprint:

- Engineer's scale
- Architect's scale
- **Metric scale**

3.1.1 Engineer's Scale

The engineer's scale is divided into decimal graduations (10, 20, 30, 40, 50, and 60 divisions to the inch). The engineer's scale is used when an area is too large to be represented by the usual scale. For example, the usual scale has a single foot represented by a portion of an inch, and an engineer's scale might be used to represent a larger number of feet per inch. The engineer's scale is used for plotting and map drawing and in the graphic solution of problems, such as survey and site plans (*Figure 27*).

Discuss the factors that affect the type of scale used on a drawing.

Explain why approximate sizes are not accurate enough for construction.

Discuss the consequences of failing to use a drawing's written dimensions.

Show Transparency 14 (Figure 27). Explain when and how to use an engineer's scale.

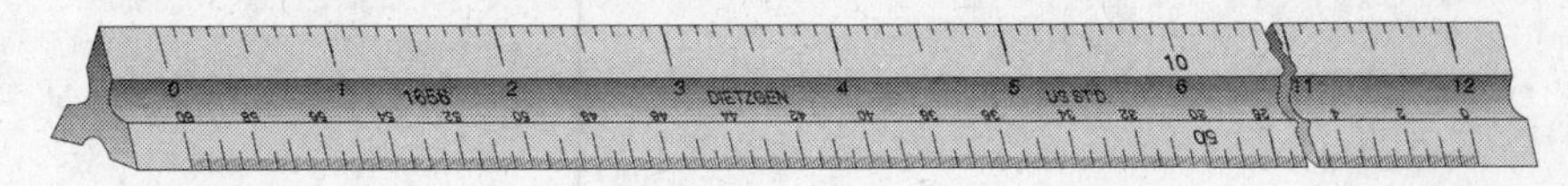

Figure 27 ◆ The engineer's scale.

MODULE 00105-04 ◆ INTRODUCTION TO BLUEPRINTS 5.37

Bring in an architect's scale. Demonstrate how to read the scale from left to right or from right to left.

Show Transparency 15 (Figure 29). Compare the metric scale with the architect's and engineer's scales.

Bring in a variety of scales and sample blueprints. Divide the class into small groups according to the number of blueprints. Have trainees practice using the scales to draw and measure the lines on the blueprints.

3.1.2 Architect's Scale

The architect's scale is used on all plans other than site plans. It is divided into feet and inches. The triangular form is commonly used because it contains a variety of scales on a single tool. It can be read either from left to right or from right to left (*Figure 28*).

Scales developed using an architect's scale include the following:

- ½" × 1'-0"
- ¾" × 1'-0"
- 1½" × 1'-0"
- 1" × 1'-0"

3.1.3 Metric Scale

The metric scale (*Figure 29*) is divided into centimeters (cm), with each centimeter divided into 10 millimeters (mm) or 20 half-millimeters. Some scales are made with metric divisions on one edge and inch divisions on the opposite edge. Many companies express measurements in both metric and English symbols.

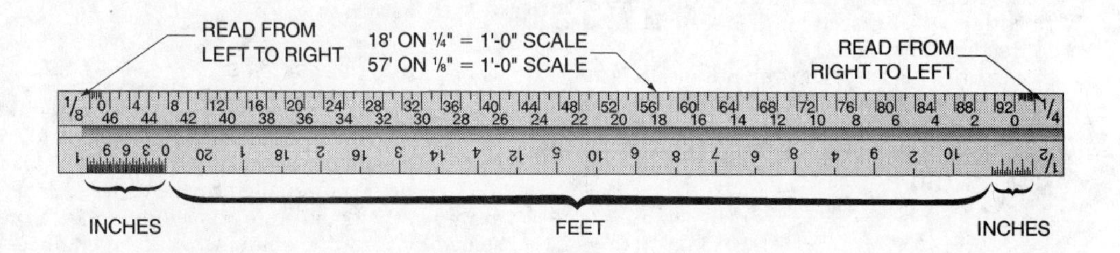

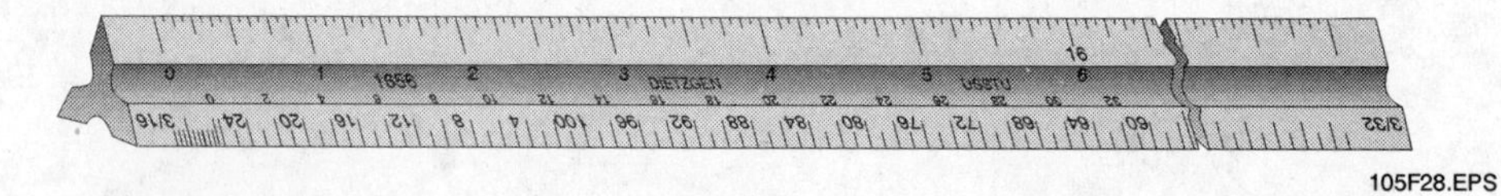

105F28.EPS

Figure 28 ◆ The architect's scale.

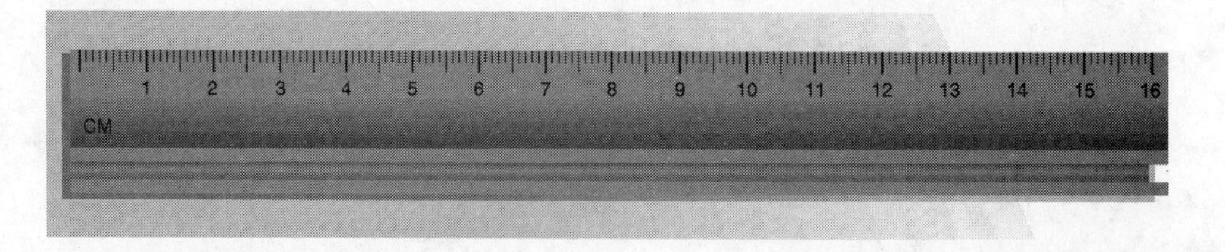

105F29.EPS

Figure 29 ◆ The metric scale.

Instructor's Notes:

5.38 CORE CURRICULUM ◆ INTRODUCTORY CRAFT SKILLS

Go over the Review Questions for Section 3.0.0. Answer any questions trainees may have.

Review Questions

Section 3.0.0

1. If the scale on a site plan reads SCALE: 1" × 20'-0", then every _____.
   a. 1⁄20th of an inch on the drawing represents 20 feet, 0 inches
   b. 20 inches on the drawing represents 1 foot, 0 inches
   c. inch on the drawing represents 20 feet, 0 inches
   d. 20 inches on the drawing represents 20 feet, 0 inches

2. A plan marked NTS _____.
   a. gives relative positions and sizes
   b. should be used only by plumbers
   c. gives accurate dimensions using the metric system
   d. uses all three types of scales

3. A(n) _____ scale is divided into decimal graduations.
   a. architect's
   b. engineer's
   c. electrician's
   d. metric

4. A(n) _____ scale is divided into proportional feet and inches.
   a. architect's
   b. engineer's
   c. electrician's
   d. metric

5. On a(n) _____ scale, each centimeter is divided into 10 millimeters.
   a. architect's
   b. engineer's
   c. electrician's
   d. metric

MODULE 00105-04 ◆ INTRODUCTION TO BLUEPRINTS 5.39

Show Transparency 16 (Figure 30). Review the lines commonly used on blueprints.

Distribute a variety of sample blueprints. Ask trainees to review the drawings and to identify and label each of the lines of construction used on the drawings.

Go over the Review Questions for Section 4.0.0. Ask whether trainees have any questions.

4.0.0 ◆ LINES OF CONSTRUCTION

It is very important to understand the meanings of lines on a drawing. The lines commonly used on a blueprint are sometimes called the Alphabet of Lines. Here are some of the more common types of lines (*Figure 30*):

- ***Dimension lines*** – Dimension lines establish the dimensions (sizes) of parts of a structure. These lines end with arrows (open or closed), dots, or slashes at a termination line drawn perpendicular to the dimension line.
- ***Leaders*** and *arrowheads* – Leaders and arrowheads identify the location of a specific part of the drawing. They are used with words, abbreviations, symbols, or keynotes.
- *Property lines* – Property lines indicate land boundaries.
- *Cut lines* – Cut lines are lines around part of a drawing that is to be shown in a separate cross-sectional view.
- *Section cuts* – Section cuts show areas not included in the cutting line view.
- *Break lines* – Break lines show where an object has been broken off to save space on the drawing.
- ***Hidden lines*** – Hidden lines identify part of a structure that is not visible on the drawing. You may have to look at another drawing to see the part referred to by the lines.
- *Centerlines* – Centerlines show the measured center of an object such as a column or fixture.
- *Object lines* – Object lines identify the object of primary interest or the closest object.

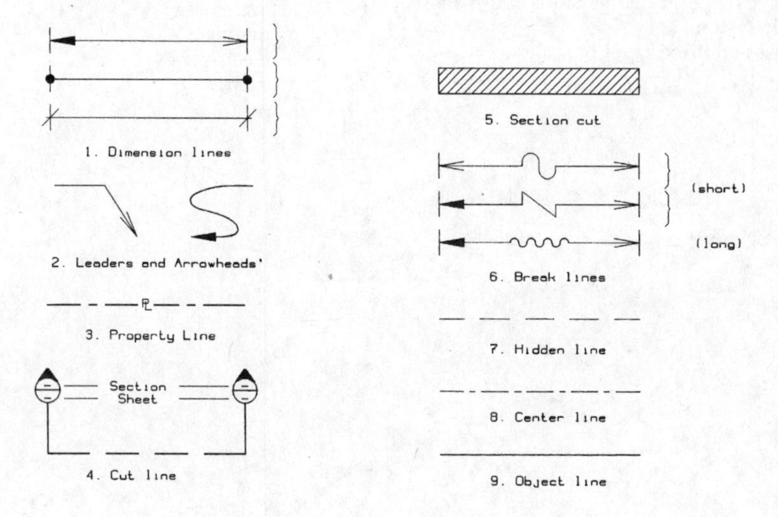

Figure 30 ◆ Lines of construction (Alphabet of Lines).

Instructor's Notes:

5.40 CORE CURRICULUM ◆ INTRODUCTORY CRAFT SKILLS

Review Questions

Section 4.0.0

1. Break lines are used to show that _____.
   a. an object needs to be repaired
   b. an object is hidden
   c. only part of an object is represented to save space
   d. an object is not included on the cutting line view

2. The Alphabet of Lines consists of _____.
   a. the line types commonly used on blueprints
   b. lines that are indicated using letters of the alphabet
   c. lines that are used to show where pages match up when a large page has been broken down to several smaller pages
   d. lines that indicate land boundaries on the site plan

3. _____ lines establish the sizes of parts of a structure.
   a. Object
   b. Dimension
   c. Property
   d. Hidden

4. _____ lines identify parts of the structure that are not visible on the drawing.
   a. Broken
   b. Object
   c. Cutting
   d. Hidden

5. Section cuts _____.
   a. show where an object has been broken off to save space on the drawing
   b. show the center of an object such as a column or fixture
   c. show areas not included in the cutting line view
   d. identify the location of a specific part of the drawing

5.0.0 ◆ ABBREVIATIONS, SYMBOLS, AND KEYNOTES

Architects and engineers use systems of abbreviations, symbols, and keynotes to keep plans uncluttered, making them easier to read and understand.

Each trade has its own symbols, and you should learn to recognize the symbols used by other trades. For example, if you are an electrician, you should understand a carpenter's symbols. If you are a carpenter, you should understand a plumber's symbols, and so on. Then, no matter what symbols you see when you are working on a project, you will understand what they mean.

Abbreviations used in blueprints are short forms of common construction terms. For example, the term *face of wall* is abbreviated *F.O.W.* Some common abbreviations are listed in *Figure 31.*

Abbreviations should always be written in capital letters. Abbreviations for each project should be noted on the title sheet or other introductory drawing page. Books that list construction abbreviations and their meanings are available. You do not need to memorize these abbreviations. You will start to remember them as you use them.

Symbols are used on a drawing to tell what material is required for that part of the project. A combination of these symbols, expanded and drawn to the same size, makes up the pictorial view of the plan. There are architectural symbols (*Figure 32*), civil and structural engineering symbols (*Figure 33*), mechanical symbols (*Figure 34*), plumbing symbols (*Figure 35*), and electrical symbols (*Figure 36*). Slightly different symbols may be used in different parts of the country. The symbols used for each set of plans should be indicated on the title sheet or other introductory drawing. Many code books, manufacturers' brochures, and specifications include symbols and their meanings.

Some plans use keynotes (*Figure 37*) instead of symbols. A keynote is a number or letter (usually in a square or circle) with a leader and arrowhead that is used to identify a specific object. Part of the drawing sheet (usually on the right-hand side) lists the keynotes with their numbers or letters. The keynote descriptions normally use abbreviations.

Have trainees review Sections 5.0.0–7.0.0 for the next session.

Ensure that you have everything required for the laboratories and testing during this session.

Show Transparency 17 (Figure 31). Review the abbreviations commonly used on blueprints. Emphasize that abbreviations should always be written in capital letters.

Ask trainees to practice writing common abbreviations for a hypothetical set of plans. Have trainees write a list of abbreviations as they might appear on a title sheet.

Show Transparency 18 (Figure 32). Explain that there are unique sets of symbols for architectural, civil and structural, mechanical, plumbing, and electrical plans.

Show Transparency 19 (Figure 37). Compare keynotes with symbols. Discuss the advantages and disadvantages of using keynotes.

MODULE 00105-04 ◆ INTRODUCTION TO BLUEPRINTS 5.41

ABBREVIATIONS

- A.B. – ANCHOR BOLT
- ADD'L – ADDITIONAL
- ADJ. – ADJACENT
- A.I.S.C. – AMERICAN INSTITUTE OF STEEL CONSTRUCTION
- ALT. – ALTERNATE
- ARCH. – ARCHITECTURAL
- A.S.T.M. – AMERICAN SOCIETY FOR TESTING & MATERIALS
- BLDG. – BUILDING
- BM. – BEAM
- B.O. – BOTTOM OF
- BOT. – BOTTOM
- BSMT. – BASEMENT
- BTWN. – BETWEEN
- CANT. – CANTILEVER
- CB. – CARDBOARD
- CH. – CHAMFER
- C.J. – CONTROL/CONSTRUCTION JOINT
- CLR. – CLEAR, CLEARANCE
- C.M.U. – CONCRETE MASONRY UNIT
- COL. – COLUMN
- CONC. – CONCRETE
- CONN. – CONNECTION
- CONST. – CONSTRUCTION
- CONT. – CONTINUOUS
- CONTR. – CONTRACTOR
- CTRD. – CENTERED
- DET. – DETAIL
- DIAG. – DIAGONAL
- DIAM. – DIAMETER
- DIM. – DIMENSION
- DISCONT. – DISCONTINUOUS
- DWG. – DRAWING
- EA. – EACH
- E.F. – EACH FACE
- EL. – ELEVATION
- ELECT. – ELECTRICAL
- ELEV. – ELEVATOR
- EQ. – EQUAL
- E.W.B. – END WALL BARS
- E.W. – EACH WAY
- EXIST. – EXISTING
- EXP. JNT. – EXPANSION JOINT
- EXT. – EXTERIOR
- F.D. – FLOOR DRAIN
- FDN. – FOUNDATION
- FIN. – FINISH
- FLR. – FLOOR
- F.O.B. – FACE OF BRICK
- F.O.CONC. – FACE OF CONCRETE
- F.O.W. – FACE OF WALL
- FS. – FLAT SLAB
- FT. – FOOT
- FTG. – FOOTING
- F.W. – FILLET WELD
- GA. – GAUGE
- GAL. – GALVANIZED
- G.L. – GLU-LAM BEAM
- GR. – GRADE
- GR. BM. – GRADE BEAM
- H.A.S. – HEADED ANCHOR STUD
- HORIZ. – HORIZONTAL
- H.S.B. – HIGH STRENGTH BOLT
- I.D. – INSIDE DIAMETER
- IN. – INCH
- INT. – INTERIOR
- JNT. – JOINT
- LB. – POUND
- LIN. FT. – LINEAL FEET
- L.L.V. – LONG LEG VERTICAL
- MAT'L. – MATERIAL
- MAX. – MAXIMUM
- MECH. – MECHANICAL
- MID. – MIDDLE
- MIN. – MINIMUM
- MISC. – MISCELLANEOUS
- MTL. – METAL
- N.I.C. – NOT IN CONTRACT
- NO. – NUMBER
- NOM – NOMINAL
- N.T.S. – NOT TO SCALE
- O.C. – ON CENTER
- O.D. – OUTSIDE DIAMETER
- O.H. – OPPOSITE HAND
- OPNG. – OPENING
- PL – PLATE
- P.S.F. – POUND PER SQUARE FOOT
- P.S.I. – POUND PER SQUARE INCH
- R. – RADIUS
- REINF. – REINFORCEMENT
- REQ'D. – REQUIRED
- RM. – ROOM
- SCHED. – SCHEDULE
- SECT. – SECTION
- SHT. – SHEET
- SIM. – SIMILAR
- S.L.V. – SHORT LEG VERTICAL
- SPC. – SPACE
- SPEC. – SPECIFICATION
- SQ. – SQUARE
- STD. – STANDARD
- STIFF. – STIFFENER
- STL. – STEEL
- STOR. – STORAGE
- SYM. – SYMMETRICAL
- T.&B. – TOP AND BOTTOM
- THK. – THICKNESS
- T.O. – TOP OF
- TYP. – TYPICAL
- U.N.O. – UNLESS NOTED OTHERWISE
- VAR. – VARIES
- VERT. – VERTICAL
- V.I.F. – VERIFY IN FIELD
- WT. – WEIGHT

SYMBOLS

℄	CENTER LINE
ϕ	DIAMETER
 ⊕	ELEVATION
&	AND
W/	WITH
PL	PLATE
X	BY
#	NUMBER
@	AT
⌷	SQUARE
L	ANGLE

105F31.EPS

Figure 31 ◆ Abbreviations.

Instructor's Notes:

5.42 CORE CURRICULUM ◆ INTRODUCTORY CRAFT SKILLS

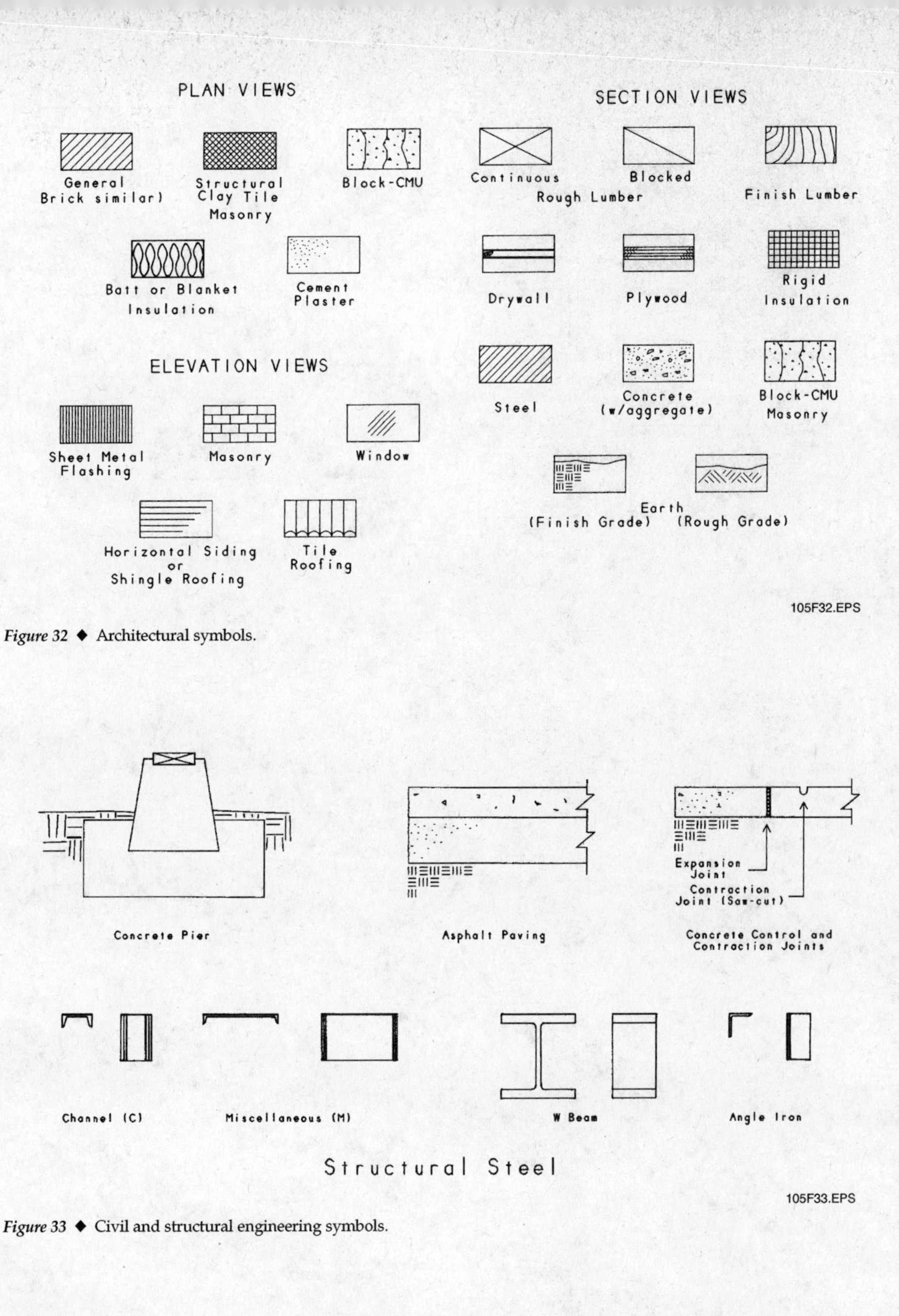

Figure 32 ◆ Architectural symbols.

Figure 33 ◆ Civil and structural engineering symbols.

MODULE 00105-04 ◆ INTRODUCTION TO BLUEPRINTS 5.43

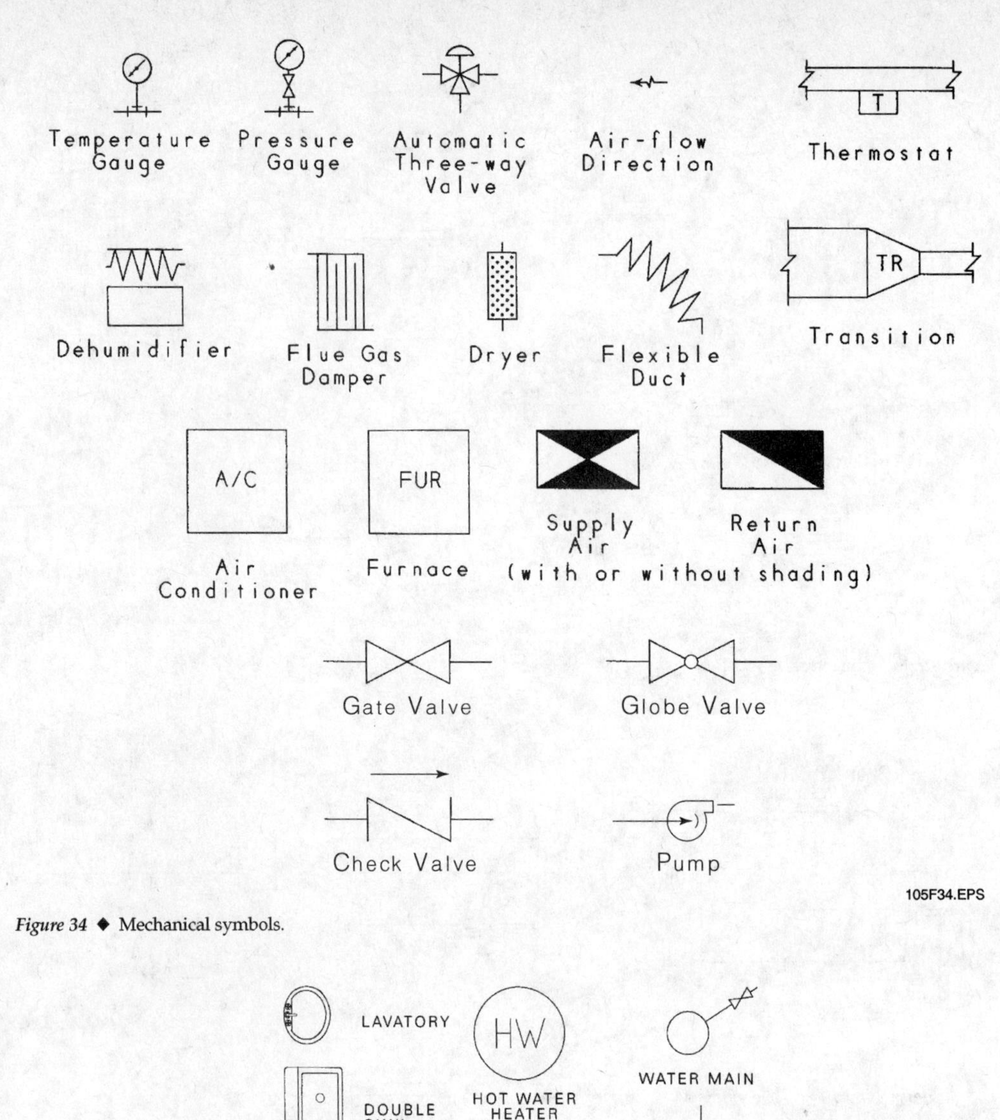

Figure 34 ◆ Mechanical symbols.

Figure 35 ◆ Plumbing symbols.

Instructor's Notes:

5.44 CORE CURRICULUM ◆ INTRODUCTORY CRAFT SKILLS

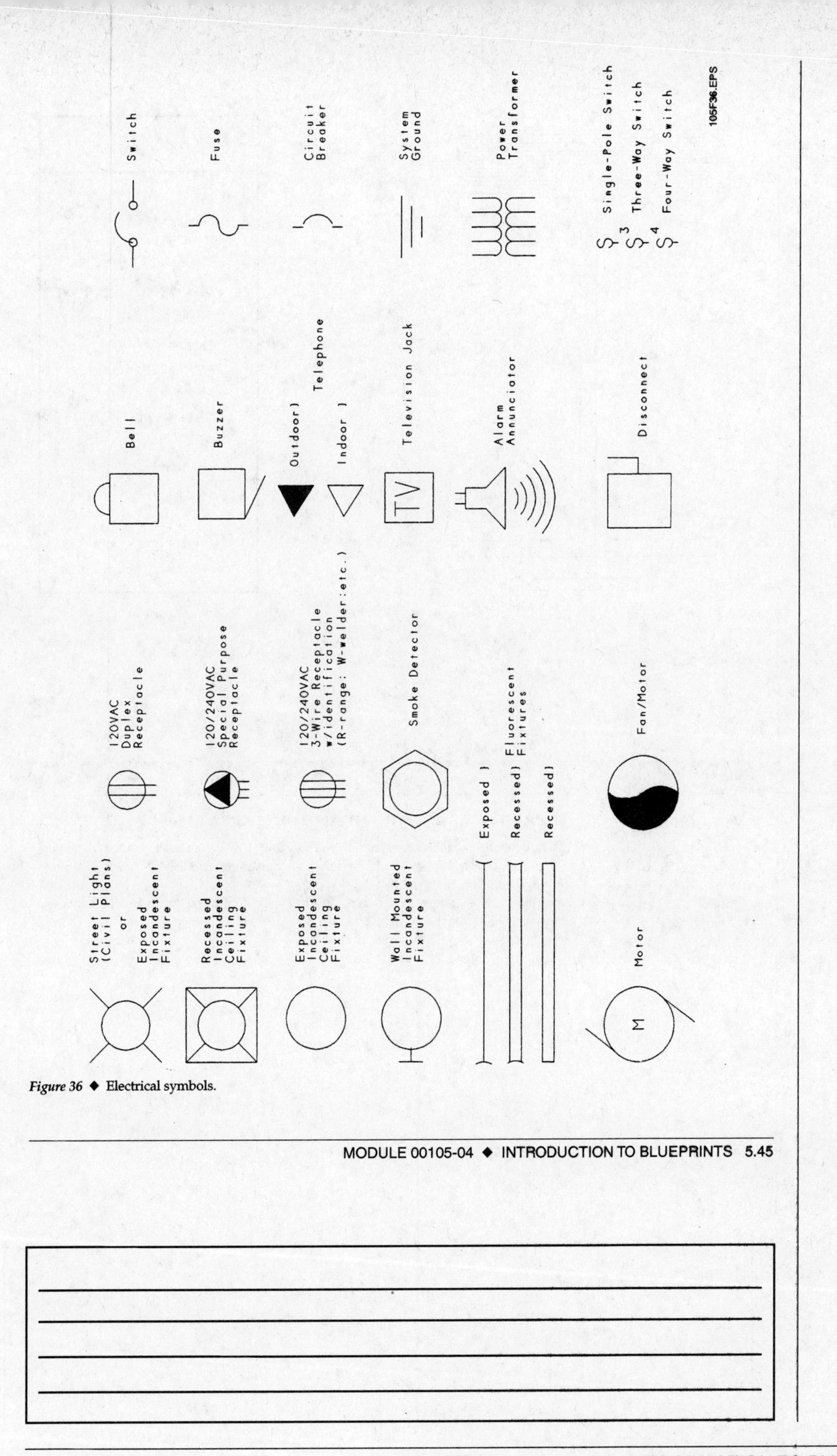

Figure 36 ◆ Electrical symbols.

MODULE 00105-04 ◆ INTRODUCTION TO BLUEPRINTS 5.45

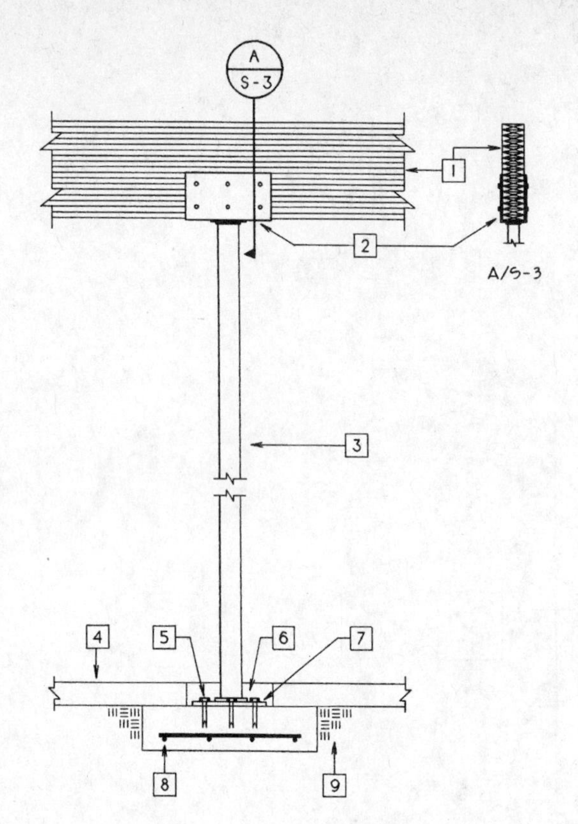

KEYNOTES

1. 6⅛" x 18" glulam beam
2. 6⅛" end cap
3. 6" DIA pipe column
4. 6" concrete slab w/6x6-W2.9 x W2.9 WWF
5. ⅝" DIA expansion bolt (typical of 6)
6. Seismic opening (to be filled with concrete)
7. ½" steel plate column base
8. 1'-0" x 4'-0" concrete footing w/4-#5 rebar each way
9. Natural or compacted grade

105F37.EPS

Figure 37 ◆ Keynotes.

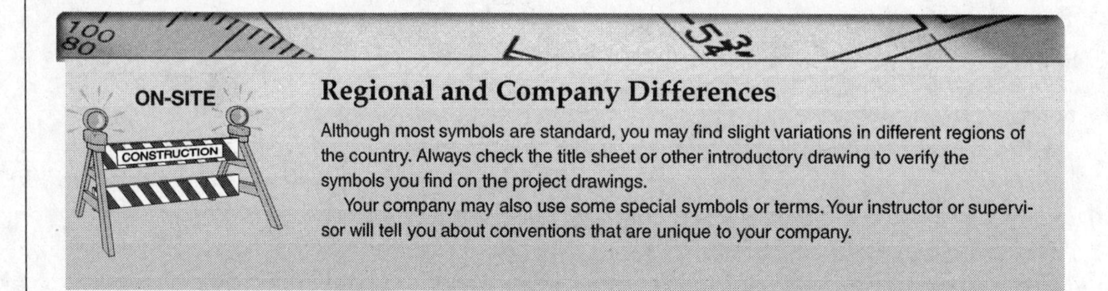

Regional and Company Differences

Although most symbols are standard, you may find slight variations in different regions of the country. Always check the title sheet or other introductory drawing to verify the symbols you find on the project drawings.

Your company may also use some special symbols or terms. Your instructor or supervisor will tell you about conventions that are unique to your company.

Instructor's Notes:

5.46 CORE CURRICULUM ◆ INTRODUCTORY CRAFT SKILLS

6.0.0 ◆ USING GRIDLINES TO IDENTIFY PLAN LOCATIONS

Have you ever used a map to find a street? The map probably used a grid to make locating a detailed area easier. The index might have referred you to section E-7. You located E along the side of the map and 7 along the top. Then you located the intersection of the two and found your street.

The gridline system shown on a plan (*Figure 38*) is used like the grid on a map. On a drawing such as a floor plan, a grid divides the area into small parts called bays.

The numbering and lettering system begins in the upper left-hand corner of the floor plan. The numbers are normally across the top and the letters are along the side. To avoid confusion, certain letters and the symbol for zero are not used. Omitted from the gridline system are:

- Letters I, O, and Q
- Numbers 1 and 0

A gridline system makes it easy to refer to specific locations on a plan. Suppose you want to refer to one outlet, but there are a dozen on a plan. Simply refer to the outlet in bay A-9.

See the Teaching Tip for Section 5.0.0 at the end of this module.

Show Transparency 20 (Figure 38). Explain how to use a gridline to identify plan locations.

Distribute a set of blueprints that uses a gridline system. Ask trainees to locate items on the plan in bays that you specify.

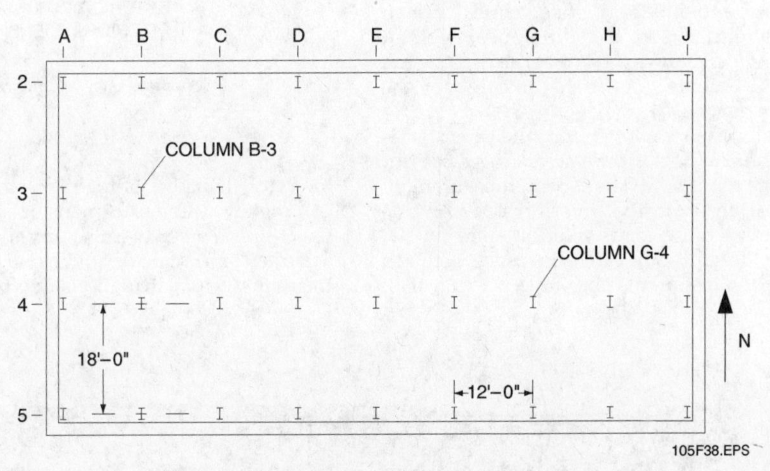

Figure 38 ◆ Grid.

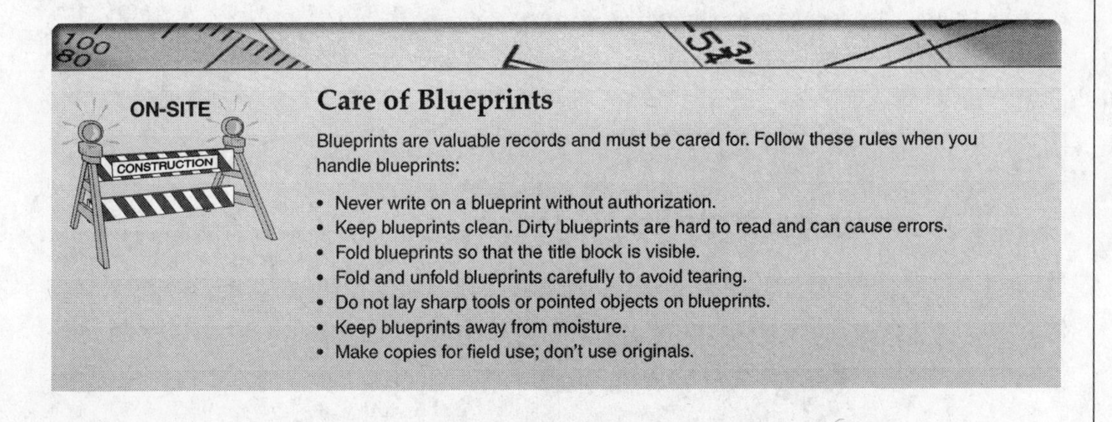

Care of Blueprints

Blueprints are valuable records and must be cared for. Follow these rules when you handle blueprints:

- Never write on a blueprint without authorization.
- Keep blueprints clean. Dirty blueprints are hard to read and can cause errors.
- Fold blueprints so that the title block is visible.
- Fold and unfold blueprints carefully to avoid tearing.
- Do not lay sharp tools or pointed objects on blueprints.
- Keep blueprints away from moisture.
- Make copies for field use; don't use originals.

MODULE 00105-04 ◆ INTRODUCTION TO BLUEPRINTS 5.47

Show Transparency 21 (Figure 39). Compare the exterior and interior dimensions on the pipe. Discuss the importance of understanding how to read dimensions on construction drawings.

7.0.0 ◆ DIMENSIONS

Dimensions are the parts of the blueprint that show the size and the placement of the objects that will be built or installed. Dimension lines usually have arrowheads at both ends, with the dimension itself written near the middle of the line. The dimension is a measurement written as a number, and it may be written in inches with fractions 6½", in feet with inches (1' - 2"), in inches with decimals (3.2"), or in millimeters (9mm) if the metric system is used. Feet are always in whole numbers on blueprints.

To do accurate work, you need to know how to read dimensions on construction drawings. This means you need to know whether the dimensions measure to the exterior or the interior of an object. To understand the difference, look at *Figure 39*, which shows a piece of pipe. There are two measurements you could take to get the pipe's dimensions.

Distribute sample blueprints with interior and exterior measurements. Divide the class into small groups according to the number of blueprints. Ask trainees to identify measurements that you specify and to indicate whether the measurements are interior or exterior.

The first measurement is from the pipe's exterior edge on one side directly across to its exterior edge on the other side. The second measurement is from the pipe's inside (interior) edge on one side directly across to its interior edge on the other side. Even though the difference between these two dimensions may be only a fraction of an inch (the thickness of the pipe), they are still two completely different dimensions. This is important to remember because any dimensioning inaccuracy or miscalculation in one place will affect the accuracy of calculations in other places.

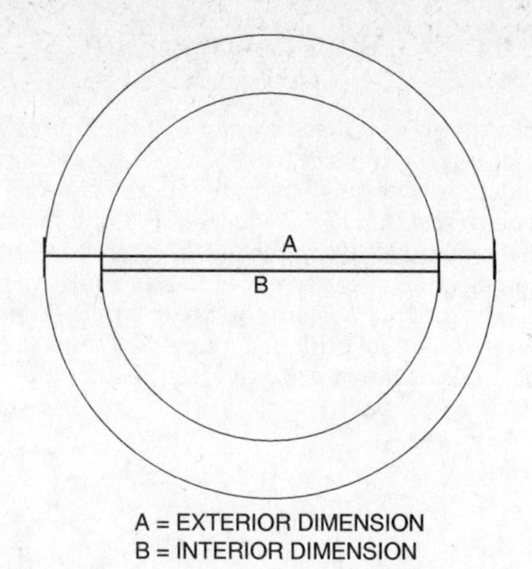

Figure 39 ◆ Exterior and interior dimensions on pipe.

Go over the Review Questions for Sections 5.0.0–7.0.0. Answer any questions trainees may have.

Review Questions

Sections 5.0.0–7.0.0

1. A(n) _____ is a number or letter (usually in a square or circle) with a leader and arrowhead that is used to identify a specific object.
   a. symbol
   b. keynote
   c. abbreviation
   d. notation

2. The abbreviation used to represent the face of a wall is _____.
   a. F.O. CONC.
   b. F.O.B.
   c. F.L.R.
   d. F.O.W.

3. Abbreviations _____.
   a. should be written in lower-case letters
   b. should be written in capital letters
   c. should be noted on the last page of the plans
   d. do not vary in different regions of the country

4. Gridlines are used to _____.
   a. indicate a cross-section of material
   b. indicate that an object has been broken off at that point to save space
   c. indicate land boundaries on the site plan
   d. refer to specific locations on the plan

5. Dimensions show _____.
   a. the scale of the parts
   b. the size and placement of objects to be built or installed
   c. how the walls fit together
   d. where an object is on a blueprint

Instructor's Notes:

5.48 CORE CURRICULUM ◆ INTRODUCTORY CRAFT SKILLS

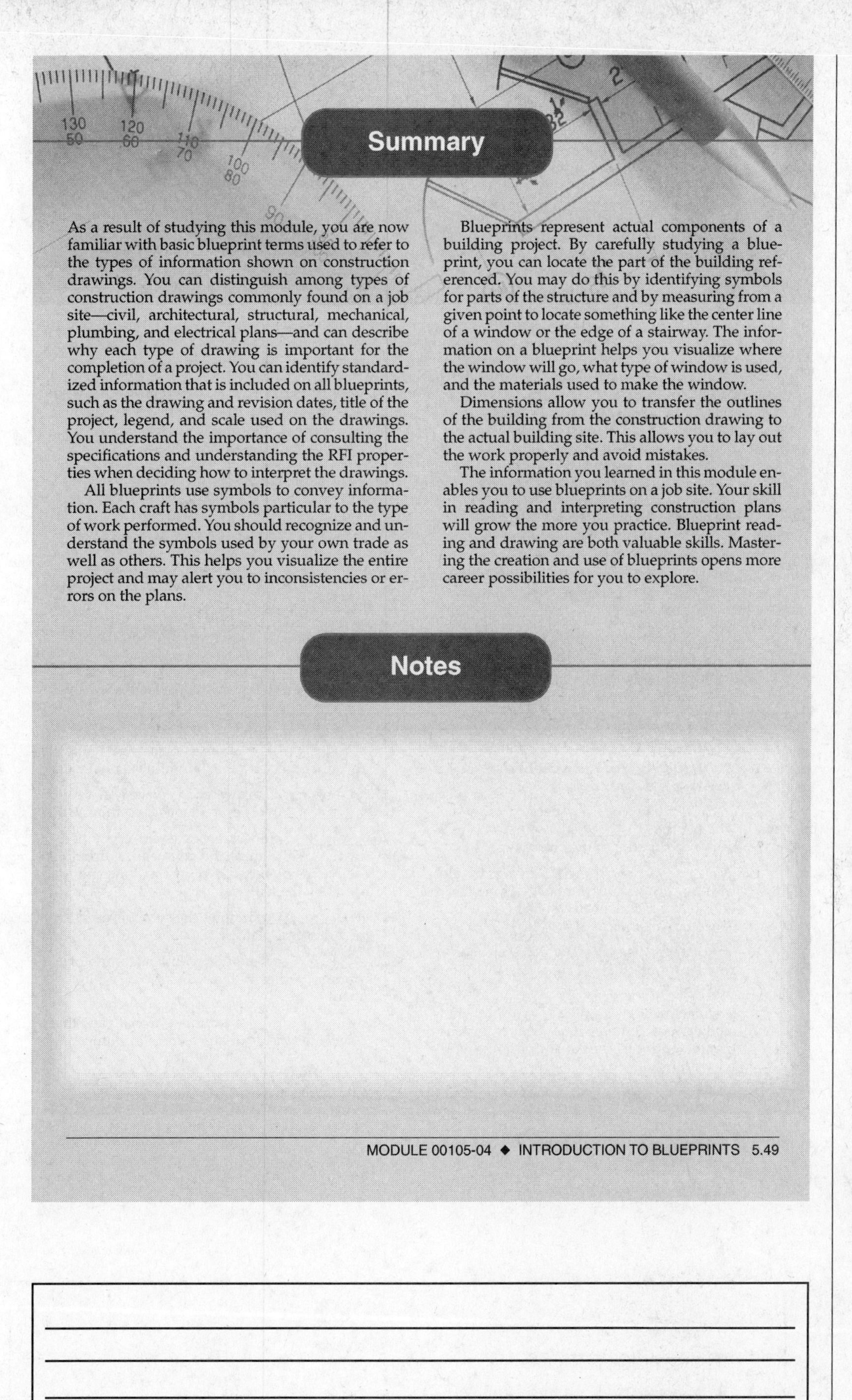

Summary

As a result of studying this module, you are now familiar with basic blueprint terms used to refer to the types of information shown on construction drawings. You can distinguish among types of construction drawings commonly found on a job site—civil, architectural, structural, mechanical, plumbing, and electrical plans—and can describe why each type of drawing is important for the completion of a project. You can identify standardized information that is included on all blueprints, such as the drawing and revision dates, title of the project, legend, and scale used on the drawings. You understand the importance of consulting the specifications and understanding the RFI properties when deciding how to interpret the drawings.

All blueprints use symbols to convey information. Each craft has symbols particular to the type of work performed. You should recognize and understand the symbols used by your own trade as well as others. This helps you visualize the entire project and may alert you to inconsistencies or errors on the plans.

Blueprints represent actual components of a building project. By carefully studying a blueprint, you can locate the part of the building referenced. You may do this by identifying symbols for parts of the structure and by measuring from a given point to locate something like the center line of a window or the edge of a stairway. The information on a blueprint helps you visualize where the window will go, what type of window is used, and the materials used to make the window.

Dimensions allow you to transfer the outlines of the building from the construction drawing to the actual building site. This allows you to lay out the work properly and avoid mistakes.

The information you learned in this module enables you to use blueprints on a job site. Your skill in reading and interpreting construction plans will grow the more you practice. Blueprint reading and drawing are both valuable skills. Mastering the creation and use of blueprints opens more career possibilities for you to explore.

Notes

See the Teaching Tip for the Summary at the end of this module.

Review the answers to the Key Terms Quiz. Go over any questions trainees may have.

Administer the Module Examination. Be sure to record the results of the Exam on Craft Training Report Form 200, and submit the results to the Training Program Sponsor.

Have each trainee perform Performance Tasks 1 through 3 to your satisfaction. Trainees will use the floor plans on the previous foldout page to complete the Performance Tasks. Fill out a Performance Profile Sheet for each trainee.

Ensure that all Performance Tests have been completed and Performance Profile Sheets for each trainee are filled out. Be sure to record the results of the testing on Craft Training Report Form 200, and submit the results to the Training Program Sponsor.

Key Terms Quiz

Fill in the blank with the correct key term that you learned from your study of this module.

1. A(n) _________ is a side view that shows height.
2. A(n) _________ usually has an arrowhead at both ends, with the measurement written near the middle of the line.
3. A(n) _________ is a qualified, licensed person who creates and designs drawings for a construction project.
4. _________, which show the design of the project, include many parts, such as the floor plan, roof plan, elevation drawings, and section drawings.
5. _________, together with the set of specifications, detail what is to be built and what materials are to be used.
6. Also called site plans or survey plans, _________ show the location of the building from an aerial view, as well as the natural contours of the earth.
7. Almost all blueprints today are made by _________.
8. _________ are solid or dashed lines showing the elevation of the earth on a civil drawing.
9. _________ are enlarged views of some special features of a building, such as floors and walls.
10. A(n) _________ applies scientific principles in design and construction.
11. To do accurate work, you need to know whether the _________ measure to the exterior or the interior of the object.
12. _________, or engineered drawings for electrical supply and distribution, include locations of the electric meter, switchgear, and convenience outlets.
13. Structural plans may show this large, horizontal support made of concrete, steel, stone, or wood. _________
14. Schematic drawings called _________ show all the equipment, pipelines, valves, instruments, and controls needed to operate a piping system.
15. An element of architectural drawings, _________ refers to the height above sea level or other defined surface.
16. Architectural, civil and structural engineering, mechanical, and plumbing _________ may be used on a drawing to tell what material is required for that part of the project.
17. Use a(n) _________ when an area is too large to be represented by the usual scale.
18. Also called a plan view, a(n) _________ is an aerial view of the layout of each room.
19. A(n) _________ is a one-line drawing showing the flow path for electrical circuitry.
20. Part of the structural plans, the _________ shows the lowest level of the building.
21. Piping system plans for gas, oil, or steam heat may be included in the _________ plan.
22. When a plan is marked _________, it means that the drawing gives approximate positions and sizes.
23. A(n) _________ is a dashed line on a plan showing an object obstructed from view by another object.
24. Known also as a pictorial illustration, a(n) _________ lets you see an object as it really is, rather than as a flat, two-dimensional view.
25. In drafting, an arrowhead is placed on a(n) _________ in order to identify a component.
26. _________ are written statements provided by the architectural and engineering firm to define the quality of work to be done and to describe the materials to be used.
27. The _________ defines the symbols used in architectural plans.
28. _________ are engineered plans for motors, pumps, piping systems, and piping equipment.
29. A(n) _________ is a cross-sectional view that shows the inside of an object or building.

Instructor's Notes:

5.50 CORE CURRICULUM ◆ INTRODUCTORY CRAFT SKILLS

30. A(n) _________, which is used to draw or measure lines on a blueprint, is divided into centimeters, which are each divided into 10 millimeters or 20 half-millimeters.

31. The term _________ refers to both water supply and all liquid waste disposal.

32. If you notice a discrepancy in the plans, notify the foreman, who will write up a(n) _________, explaining the problem and noting the date and time.

33. A(n) _________ is divided into feet and inches and is used on all plans other than site plans.

34. A(n) _________ shows the shape of the roof and the materials that will be used to finish it.

35. The _________ of a drawing tells the size of the object drawn compared with the actual size of the object represented.

36. _________ show the layout for the plumbing system that supplies hot and cold water, for the sewage disposal system, and for the location of plumbing fixtures.

37. The _________, which are used to support the architectural design, include the general notes, a foundation plan, a roof framing plan, and structural section drawings.

38. Part of the blueprint, the _________ gives information about the structure and is numbered for easy filing.

Key Terms

Architect
Architect's scale
Architectural plans
Beam
Blueprints
Civil plans
Computer-aided drafting (CAD)
Contour lines
Detail drawings
Dimension line
Dimensions
Electrical plans
Elevation (EL)
Elevation drawing
Engineer
Engineer's scale
Floor plan
Foundation plan
Heating, ventilating, and air conditioning (HVAC)
Hidden line
Isometric drawing
Leader
Legend
Mechanical plans
Metric scale
Not to scale (NTS)
Piping and instrumentation drawings (P&IDs)
Plumbing
Plumbing plans
Request for information (RFI)
Roof plan
Scale
Schematic
Section drawing
Specifications
Structural plans
Symbol
Title block

MODULE 00105-04 ◆ INTRODUCTION TO BLUEPRINTS 5.51

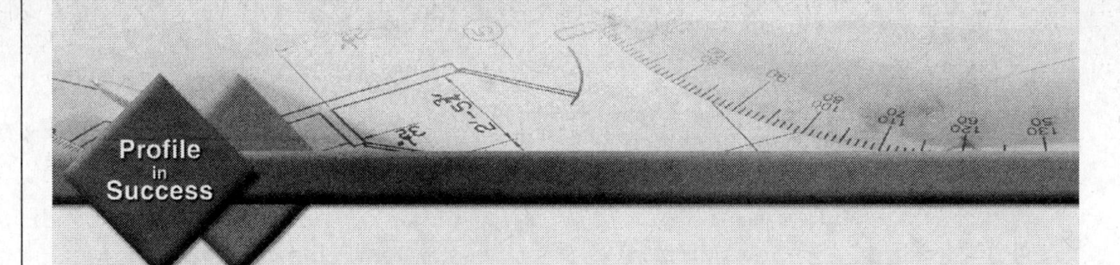

Charlie Haas

General Superintendent
Hensel Phelps Construction Company, Inc.
Austin, Texas

Charlie Haas was born in Fort Morgan, Colorado. His father was a contractor, and Charlie operated bulldozers and other heavy equipment for him at the age of 14. After earning his BS in Civil Engineering from the University of Colorado, Charlie went to work for the state of California. There he received his Professional Engineer's license. He joined Hensel Phelps in 1968 as a lead field engineer and worked his way up to carpenter superintendent, project superintendent, and project manager. After a 10-year hiatus running his own small construction business, Cornerstone Builders, Charlie returned to Hensel Phelps as General Superintendent.

How did you become interested in the construction industry?
I grew up in a construction family. Construction was what I always wanted to do. My mom was also from a construction family, but she wasn't too happy about all the moving around that goes with it, so she encouraged me to stay in school and get a degree before I did anything else. I honored her wishes and did exactly what she asked me to do. But the day I got my Professional Engineer's license, I applied for work as a contractor. That's because, as a civil engineer, I designed things but I didn't build them. I really wanted to build things with my own hands.

How did you decide on your current position?
All the time I was a civil engineer in California, I knew that I wanted to be a contractor and not a designer. Hensel Phelps was a contractor on a bridge project that I worked on as a design engineer. I noticed that at the bottom of every Hensel Phelps form or letterhead there was one word: *Performance.* They did their tasks right, and they did what they said they'd do. I think that's what drew me to them originally.

I went out on my own in 1978 because my children were in high school. My small contracting business worked on projects in a 150-mile radius around Greeley, Colorado, up to Wisconsin and also around Amarillo up to Oklahoma. But I missed working on the big contracts, so I returned to Hensel Phelps in 1989 as General Superintendent.

What are some of the things you do in your job?
I run the crews on big projects. I've worked on projects for the Army Corps of Engineers, and I've built airports and prisons. My current projects include the Austin City Hall and a building consisting of a three-level garage, a 150,000-square-foot store on the ground floor, and five stories of office space above it. For three years I worked with a team of 350 craftsmen to build the people mover at Dallas–Fort Worth International Airport (DFW). DFW was a cotton field when I started working there, back in 1971; I built the airport's first terminal. And that's what's fun about my job! I get to start from scratch and build really big things.

How important to your company is training?
Good training is very important. We have a program where we hire kids out of high school and put them through the NCCER curricula. We find that we get better superintendents out of the NCCER program than out of college programs. Many college students don't want to get dirty; they want to end up with a nice office job. The NCCER curricula reinforce the good work ethic that these young people get from their family and their coaches. I go to schools looking for kids who are involved in extracurricular activities like sports or band, kids who are self-starters. I also talk to a lot of pastors about students who are involved in church programs. They don't have to be "A" students. Maybe they're not the stars, but they are working as hard as they can. Those are the people who become good superintendents.

5.52 CORE CURRICULUM ◆ INTRODUCTORY CRAFT SKILLS

Instructor's Notes:

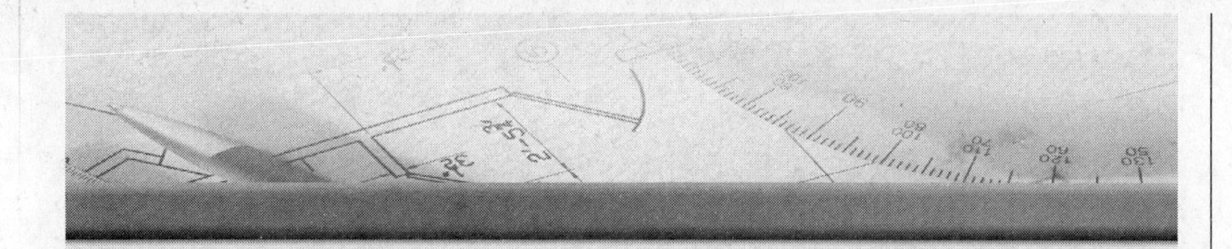

I think we're seeing a real need to bring young kids in from high school and train them through the NCCER curricula. Many kids coming out of college don't want to work in the field. A lot of contractors are becoming construction managers. We try to control project schedules by developing superintendents who want to be out there building things. We'll teach them how to build; they don't have to have the experience beforehand. We don't hire superintendents, we hire young people and train them to be superintendents. Most of our project managers started as carpenters or as field engineers. We promote from within.

We have found that this approach works very well in the industry. In fact, a lot of people call Hensel Phelps "The Academy" because we train people who then go to other companies as superintendents who are ready and willing to work in the field. And a lot of us are what we call "retreads"—people like me who went away and then come back because we wanted to work here again. It's that kind of a place.

What do you like most about your job?

What I love first and foremost is teaching kids, both the field engineers and the young people who begin as laborers and work their way up. That's how I started, after all. Probably second on my list is placing concrete. I just really like placing and forming concrete. I told my wife, "If I ever go five days in a row and I'm not having fun, then I'm going to quit." And in all the years I've been working, that hasn't happened yet. Sure, I've had two or three bad days in a row, but never more than that. I get up every day looking forward to going to work. And as long as that's the case, I'm going to keep doing what I love to do.

I'm a coach, basically. I teach people, and I get paid more than if I were coaching football or wrestling in school. That's all a superintendent is—a coach and actually a cheerleader, too.

I like working with all kinds of people. I love carpentry and large form work. I've worked in a lot of states and traveled all over this land. I've owned more houses than cars. I don't object to moving around, and my wife and four kids have enjoyed it, too. And now three of my kids are involved with construction. One has his own contracting business. And my grandson is currently going through the superintendent program, taking NCCER classes in Baltimore. He's going to be a field engineer, building a new terminal at Baltimore Washington Airport. It's a great business because it has made those things possible, and I'm just so glad to be part of it.

What do you think it takes to be a success in your trade?

It takes perseverance and the willingness to be on time every day. Be honest and be a self-starter. You don't have to be a straight "A" student to succeed. You have to try as hard as you can, give it your all, and people will see that and you will be rewarded for it.

What would you say to someone entering the trades today?

The best way for me to answer this is to say that my family and I are one of my company's biggest recruiting tools! I'm an old guy who's been in construction his whole life; my son owns his own contracting company; and now my grandson is in construction. If it wasn't the kind of field I'd be proud to have my kids in, to have my grandkids in, then I wouldn't let them do it.

There are two posters in my office that say what I'm trying to teach. One shows a young boy holding a blueprint and looking up at the empty sky. The caption reads, "I like to see lines in the sky." The other is a picture of a child's hand, with a man's hand holding a nail and guiding the boy's hand. The caption says, "I can do nothing unless the father helps." I look at those two posters a lot.

MODULE 00105-04 ◆ INTRODUCTION TO BLUEPRINTS 5.53

Civil Engineer Is Not Afraid to Get Her Hands Dirty

By Lia Steakley

Maureen Mathias first picked up a hammer at age five and has been designing and building structures and systems ever since. "My dad put an addition on our house when I was five and finished when I was about ten, and I loved helping him with simple things like pounding nails, painting, and plastering. I've developed an interest in a variety of activities like woodworking and stripping and refinishing furniture," the 30-year-old civil engineer says.

Mathias wanted a career that gave back to society and the environment. A love of math, science, and general tinkering led her to the field of engineering. She was drawn to civil engineering because of the "feeling of giving back to the community and the earth" associated with the concentration. "I saw this discipline as an opportunity to provide services in promoting environmental responsibility and assisting in public health technologies," says Mathias.

After graduating from the University of Wisconsin, Madison in 1996 with a B.S. degree in civil engineering, Mathias worked for the city of Madison where she spent her first year doing computer-aided design and drafting for streets and sewer systems. During the next two years, she worked for the city's environmental special projects division and took on senior engineer responsibilities.

Mathias later enrolled in Oregon State University in pursuit of a masters in civil engineering. Funded by the Strategic Environmental Research and Development Program, a combined effort of federal agencies including the Environmental Protection Agency and Department of Energy, Mathias studied bioremediation technologies.

In 2002, Mathias graduated and landed a job with Golder Associates Inc., which specializes in ground and environmental engineering, in the firm's Roseville, Calif. office. Although she has been with Golder less than a year, Mathias has worked on an array of projects including analyzing waste in landfills, designing and managing alternative covers for sewage solids disposal, and monitoring landfill gas generation and migration and developing landfill liner performance demonstrations.

This month, Mathias will finish up work on a Spill Control & Countermeasures Plan and Stormwater Pollution Prevention Plan developed for a power generation facility starting up near Las Vegas. She is also completing a landfill investigation and geotechnical design for an abandoned landfill in the San Francisco Bay Area.

Next fall, Mathias will wrap up work on a landfill gas monitoring plan for an elementary school in Sacramento, Calif. that is believed to be housed on a portion of a landfill. The work has involved installation and monitoring of landfill gas probes, data assessment, and reporting to the regulatory agencies, she says.

In the next two years, Mathias will continue her work on a California sanitation district project to design and implement innovative technology for closing dedicated land disposal units in central California that have accepted biosolids generated through wastewater treatment for several decades and have developed very high salinity levels.

Instructor's Notes:

5.54 CORE CURRICULUM ◆ INTRODUCTORY CRAFT SKILLS

Trade Terms Introduced in This Module

Architect: A qualified, licensed person who creates and designs drawings for a construction project.

Architect's scale: A measuring device used for laying out distances, with scales indicating feet, inches, and fractions of inches.

Architectural plans: Drawings that show the design of the project. Also called *architectural drawings.*

Beam: A large, horizontal structural member made of concrete, steel, stone, wood, or other structural material to provide support above a large opening.

Blueprints: Architectural or working drawings used to represent a structure or system.

Civil plans: Drawings that show the location of the building on the site from an aerial view, including contours, trees, construction features, and dimensions.

Computer-aided drafting (CAD): The making of a set of blueprints with the aid of a computer.

Contour lines: Solid or dashed lines showing the elevation of the earth on a civil drawing.

Detail drawings: Enlarged views of part of a drawing used to show an area more clearly.

Dimension line: A line on a drawing with a measurement indicating length.

Dimensions: Measurements such as length, width, and height shown on a drawing.

Electrical plans: Engineered drawings that show all electrical supply and distribution.

Elevation (EL): Height above sea level, or other defined surface, usually expressed in feet.

Elevation drawing: Side view of a building or object, showing height and width.

Engineer: A person who applies scientific principles in design and construction.

Engineer's scale: A straightedge measuring device divided uniformly into multiples of 10 divisions per inch so drawings can be made with decimal values.

Floor plan: A drawing that provides an aerial view of the layout of each room.

Foundation plan: A drawing that shows the layout and elevation of the building foundation.

Heating, ventilating, and air conditioning (HVAC): Heating, ventilating, and air conditioning.

Hidden line: A dashed line showing an object obstructed from view by another object.

Isometric drawing: A three-dimensional drawing of an object.

Leader: In drafting, the line on which an arrowhead is placed and used to identify a component.

Legend: A description of the symbols and abbreviations used in a set of drawings.

Mechanical plans: Engineered drawings that show the mechanical systems, such as motors and piping.

Metric scale: A straightedge measuring device divided into centimeters, with each centimeter divided into 10 millimeters.

Not to scale (NTS): Describes drawings that show relative positions and sizes.

Piping and instrumentation drawings (P&IDs): Schematic diagrams of a complete piping system.

Plumbing: A general term used for both water supply and all liquid waste disposal.

Plumbing plans: Engineered drawings that show the layout for the plumbing system.

Request for information (RFI): A means of clarifying a discrepancy in the blueprints.

Roof plan: A drawing of the view of the roof from above the building.

Scale: The ratio between the size of a drawing of an object and the size of the actual object.

Schematic: A one-line drawing showing the flow path for electrical circuitry.

Section drawing: A cross-sectional view of a specific location, showing the inside of an object or building.

Specifications: Precise written presentation of the details of a plan.

MODULE 00105-04 ◆ INTRODUCTION TO BLUEPRINTS 5.55

Structural plans: A set of engineered drawings used to support the architectural design.

Symbol: A drawing that represents a material or component on a plan.

Title block: A part of a drawing sheet that includes some general information about the project.

Instructor's Notes:

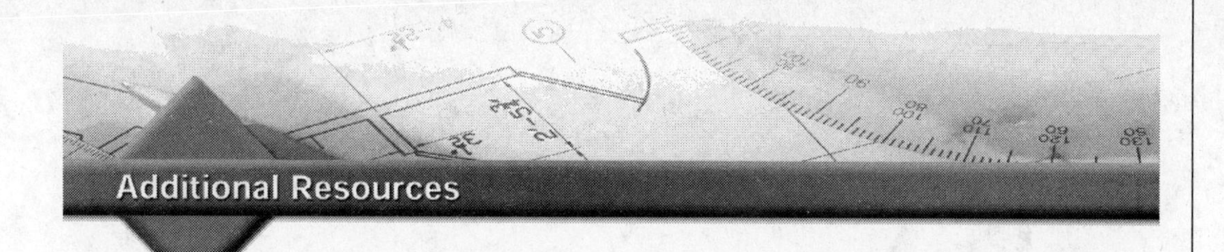

Additional Resources

This module is intended to present thorough resources for task training. The following reference works are suggested for further study. These are optional materials for continued education rather than for task training.

Blueprint Reading for the Building Trades. John Traister. Carlsbad, CA: Craftsman Book Co.

Blueprint Reading for Construction. James Fatzinger. Upper Saddle River, NJ: Prentice Hall.

Construction Blueprint Reading. Robert Putnam. Englewood Cliffs, NJ: Prentice Hall.

Print Reading for Construction: Residential and Commercial. Walter C. Brown. Tinley Park, IL: Goodheart-Willcox Co.

Reading Architectural Plans for Residential and Commercial Construction. Ernest R. Weidhaas. Englewood Cliffs, NJ: Prentice Hall Career & Technology.

MODULE 00105-04 ◆ INTRODUCTION TO BLUEPRINTS 5.57

MODULE 00105-04 — TEACHING TIPS

The following are suggested activities or instructional methods to help you teach the material in this AIG.

General

When you call on someone to answer a question, the rest of the class relaxes or even tunes out because they expect that the question and answer will take place only between you and the trainee you called on. Instead, use this technique to involve more trainees in answering questions and to keep them on their toes.

1. Ask trainees to define a term or explain a concept.
2. After one trainee has answered, ask a trainee seated nearby if the answer is right. Then ask whether a trainee in the back of the room agrees.
3. Ask trainees to explain why they think an answer is right or wrong.
4. Use the session to clear up incorrect ideas, and encourage trainees to learn from their mistakes.

Sections 1.2.0–1.2.6

Plans

You will need a variety of blueprints for this exercise, which will help trainees understand how a project's various plans relate to each other. Trainees will need their Trainee Guides and pencil and paper to take notes for discussion. Allow 30 minutes for this exercise.

1. Bring in and display a variety of blueprints for one project.
2. Ask trainees to correctly identify each type of plan. Have trainees write a brief statement explaining how the plan is used and how it relates to other blueprints in the set of plans.
3. Discuss the importance of double-checking all the plans for a project.
4. Ensure that trainees understand how to look for and locate missing dimensions.

Section 5.0.0

Abbreviations, Symbols, and Keynotes

This informal quiz will familiarize trainees with some of the commonly used symbols. For this exercise, trainees will need their Trainee Guides and pencil and paper to take notes for discussion. Allow 20 minutes for this exercise.

1. On the whiteboard/chalkboard, draw a variety of commonly used symbols.
2. Ask trainees to identify the symbols. Have them identify the set to which each belongs.
3. Review and discuss trainees' answers.
4. Emphasize that these symbols do not have to be memorized, but encourage trainees to refer to their code books, manufacturers' brochures, and specifications to verify meanings.

Summary

Introduction to Blueprints

This exercise will allow trainees to practice deciphering information on a variety of blueprints that you provide. Trainees will need their Trainee Guides and pencil and paper to take notes for discussion. Allow between 30 and 45 minutes for this exercise.

1. Divide the class into small groups according to the number of blueprints.
2. Have each group review its blueprint, noting the title block, abbreviations, symbols, and dimensions used on its blueprint.
3. Ask each group to classify the drawing, present a summary of the standardized information on its plan, and specify the scale used on the plan.
4. Ask each group to locate a specific part of the building using the gridline system.
5. Ensure that trainees properly care for their blueprints following the tips in "On-Site: Care of Blueprints."

MODULE 00105-04 — ANSWERS TO REVIEW QUESTIONS

Section 1.0.0

1. b
2. d
3. a
4. b
5. c

Section 2.0.0

1. a
2. c
3. c
4. c
5. b

Section 3.0.0

1. c
2. a
3. b
4. a
5. d

Section 4.0.0

1. c
2. a
3. b
4. d
5. c

Sections 5.0.0–7.0.0

1. b
2. d
3. b
4. d
5. b

MODULE 00105-04 — ANSWERS TO KEY TERMS QUIZ

1. elevation drawing
2. dimension line
3. architect
4. Architectural plans
5. Blueprints
6. civil plans
7. computer-aided drafting (CAD)
8. Contour lines
9. Detail drawings
10. engineer
11. dimensions
12. Electrical plans
13. beam
14. piping and instrumentation drawings (P&IDs)
15. elevation (EL)
16. symbols
17. engineer's scale
18. floor plan
19. schematic
20. foundation plan
21. heating, ventilating, and air conditioning (HVAC)
22. not to scale (NTS)
23. hidden line
24. isometric drawing
25. leader
26. Specifications
27. legend
28. Mechanical plans
29. section drawing
30. metric scale
31. plumbing
32. request for information (RFI)
33. architect's scale
34. roof plan
35. scale
36. Plumbing plans
37. structural plans
38. title block

CONTREN® LEARNING SERIES — USER FEEDBACK

The NCCER makes every effort to keep these textbooks up-to-date and free of technical errors. We appreciate your help in this process. If you have an idea for improving this textbook, or if you find an error, a typographical mistake, or an inaccuracy in NCCER's *Contren®* textbooks, please write us, using this form or a photocopy. Be sure to include the exact module number, page number, a detailed description, and the correction, if applicable. Your input will be brought to the attention of the Technical Review Committee. Thank you for your assistance.

Instructors – If you found that additional materials were necessary in order to teach this module effectively, please let us know so that we may include them in the Equipment/Materials list in the Annotated Instructor's Guide.

Write: Product Development
National Center for Construction Education and Research
P.O. Box 141104, Gainesville, FL 32614-1104

Fax: 352-334-0932

E-mail: curriculum@nccer.org

Craft ________ Module Name ________

Copyright Date ________ Module Number ________ Page Number(s) ________

Description ________

(Optional) Correction ________

(Optional) Your Name and Address ________

Basic Rigging

00106-04

NCCER STANDARDIZED CRAFT TRAINING PROGRAM

The National Center for Construction Education and Research (NCCER) provides a standardized national program of accredited craft training. Key features of the program include instructor certification, competency-based training, and performance testing. The program provides trainees, instructors, and companies with a standard form of recognition through a National Craft Training Registry. The program is described in full in the *Guidelines for Accreditation*, published by the NCCER. For more information on standardized craft training, contact the NCCER by writing us at P.O. Box 141104, Gainesville, FL 32614-1104; calling 352-334-0911; or e-mailing info@nccer.org. More information may be found at our Web site, www.nccer.org.

HOW TO USE THIS ANNOTATED INSTRUCTOR'S GUIDE

Each page presents two sections of information. The larger section displays each page exactly as it appears in the Trainee Module. The narrow column ties suggested trainee and instructor actions to each page and provides icons (detailed below) to call your attention to material, safety, audiovisual, or testing requirements. The bottom of each page includes space for your notes.

The **Audiovisual** icon indicates an appropriate time to show a transparency or other audiovisual aid.

The **Classroom** icon prompts you to define a term, stress a point, ask trainees to explain a concept, or give examples.

The **Demonstration** icon directs you to show trainees how to perform tasks.

The **Examination** icon tells you to administer the written module examination.

The **Homework** icon is placed where you may wish to assign reading for the next class, to assign a project, or to advise trainees to prepare for an examination.

The **Laboratory** icon is used when trainees are to practice performing tasks.

The **Materials** icon is a reminder for you to gather materials needed for classes, labs, and testing.

The **Performance Testing** icon tells you to administer a performance test or a portion thereof.

The **Safety** icon is used to emphasize safety issues. It is often keyed to *Caution* and *Warning* statements in the Trainee Module.

The **Teaching Tip** icon indicates additional guidance is available, such as how to conduct an exercise, get the most educational value from a field trip, or encourage class participation. Teaching Tips may expand on a feature (*Think About It*, *Did You Know?*) or provide Quick Quizzes or similar exercises. You will be referred to the Teaching Tips section at the back of the module if there is additional material.

The **Combination** icon indicates that the laboratory listed corresponds with a performance task. If desired, you can note the proficiency of the trainees during the laboratory and use it to satisfy performance testing requirements.

PREPARATION

Before teaching this module, you should review the Objectives, Performance Tasks, Materials and Equipment List, and the Module Outline. Be sure to allow ample time to prepare your own training or lesson plan and gather all required materials and equipment.

MODULE OVERVIEW

This module introduces the uses of slings and common rigging hardware. Trainees will learn basic inspection techniques, hitch configurations, and load-handling safety practices, as well as how to use American National Standards Institute hand signals.

PREREQUISITES

Prior to training with this module, it is recommended that the trainee shall have successfully completed the following: *Core Curriculum: Introductory Craft Skills,* Modules 00101-04 through 00105-04. This module is an elective. To receive a successful completion, you must take this module or Modules 00107-04 and 00108-04.

OBJECTIVES

Upon completion of this module, the trainee will be able to:

1. Identify and describe the use of slings and common rigging hardware.
2. Describe basic inspection techniques and rejection criteria used for slings and hardware.
3. Describe basic hitch configurations and their proper connections.
4. Describe basic load-handling safety practices.
5. Demonstrate proper use of American National Standards Institute (ANSI) hand signals.

PERFORMANCE TASKS

Under the supervision of the instructor, the trainee should be able to:

1. Select and inspect appropriate slings for a lift.
2. Given various loads, determine the proper hitch to be used.
3. Select and inspect appropriate hardware and/or lifting equipment.
4. Demonstrate and/or simulate the proper techniques for connecting hitches.
5. Demonstrate the proper use of all hand signals according to *ANSI B30.2* and *B30.5.*
6. Describe or demonstrate pre-lift safety checks.
7. Demonstrate and/or simulate how to lift the load level.
8. Describe and/or demonstrate loading and disconnecting safety precautions.

MATERIALS AND EQUIPMENT LIST

Transparencies
Markers/chalk
Blank acetate sheets
Transparency pens
Pencils and scratch paper
Overhead projector and screen
Whiteboard/chalkboard
Appropriate personal protective equipment
Copies of your local code
OSHA 29 CFR 1926
Identification tags for slings
Copies of Figure 15 with the callouts covered
Damaged slings or photos of damaged slings
Two glasses
One liter of water
Anchor shackles and chain shackles
Various types of pins, including:
  Screw pin shackle
  Round pin or straight pin shackle
  Safety shackle
Damaged shackles and pins
Damaged and undamaged eyebolts
Undamaged lifting clamps
Rusty or corroded lifting clamps
Wire brush to clean lifting clamps
Damaged and undamaged rigging hooks
Module Examinations*
Performance Profile Sheets*

*Located in the Test Booklet.

SAFETY CONSIDERATIONS

Ensure that the trainees are equipped with appropriate personal protective equipment. Always work in a clean, well-lit, appropriate work area.

NOTE

Due to liability issues, trainees under the age of 18 should not perform hoisting maneuvers; therefore, trainees under 18 should not perform the demonstration aspect of Performance Task numbers 4, 7, and 8. The instructor may choose to have trainees simulate the concepts underlying the tasks by using alternative methods.

If you do not have access to rigging hardware or equipment, there are many resources available to you including local contractors, rigging equipment manufacturers, or even your local Training Program.

ADDITIONAL RESOURCES

This module is intended to present thorough resources for task training. The following reference works are suggested for both instructors and motivated trainees interested in further study. These are optional materials for continued education rather than for task training.

Bob's Rigging and Crane Handbook, Latest Edition. Bob DeBenedictis. Leawood, KS: Pellow Engineering Services, Inc.

High Performance Slings and Fittings for the New Millennium, 1999 Edition. Dennis St. Germain. Aston, PA: I & I Sling, Inc.

Mobile Crane Manual, 1999. Donald E. Dickie, D. H. Campbell. Toronto, Ontario, Canada: Construction Safety Association of Ontario.

Rigging Manual, 1997. Toronto, Ontario, Canada: Construction Safety Association of Ontario.

TEACHING TIME FOR THIS MODULE

An outline for use in developing your lesson plan is presented below. Note that each Roman numeral in the outline equates to one session of instruction. Each session has a suggested time period of 2½ hours. This includes 10 minutes at the beginning of each session for administrative tasks and one 10-minute break during the session. Approximately 20 hours are suggested to cover *Basic Rigging.* You will need to adjust the time required for hands-on activity and testing based on your class size and resources. Because laboratories often correspond to Performance Tasks, the proficiency of the trainees may be noted during these exercises for Performance Testing purposes.

Topic	Planned Time
Session I. Slings, Part One	
A. Introduction	____________
B. Synthetic Slings	____________
Session II. Slings, Part Two	
A. Alloy Steel Chain Slings	____________
B. Wire Rope Slings	____________
C. Performance Testing (Task 1)	____________
Session III. Hitches	
A. Vertical Hitch	____________
B. Choker Hitch	____________
C. Basket Hitch	____________
D. Performance Testing (Task 2)	____________
Session IV. Rigging Hardware, Part One	
A. Shackles	____________
B. Eyebolts	____________

Session V. Rigging Hardware, Part Two

A. Lifting Clamps ______
B. Rigging Hooks ______
C. Performance Testing (Task 3) ______

Session VI. Sling Stress and Hoists

A. Sling Stress ______
B. Chain Hoists ______

Session VII. Rigging Operations and Practice, Part One

A. Rated Capacity ______
B. Sling Attachment ______
C. Hardware Attachment ______
D. Performance Testing (Task 4) ______

Session VIII. Rigging Operations and Practice, Part Two

A. Load Control ______
B. ANSI Hand Signals ______
C. Review ______
D. Module Examination ______
   1. Trainees must score 70% or higher to receive recognition from NCCER.
   2. Record the testing results on Craft Training Report Form 200, and submit the results to the Training Program Sponsor.
E. Performance Testing (Tasks 5–8) ______
   1. Trainees must perform each task to the satisfaction of the instructor to receive recognition from NCCER. If applicable, proficiency noted during laboratory exercises can be used to satisfy the Performance Testing requirements.
   2. Record the testing results on Craft Training Report Form 200, and submit the results to the Training Program Sponsor.

Excellence in Construction Winner— Institutional, $25–99 Million

Skanska USA Building, Inc. built the second U.S. courthouse in the country designed to meet the General Service Administration's blast-resistant criteria. The Jacksonville, Florida, $84-million, 14-story courthouse encompasses 457,416 square feet and has a four-story rotunda.

00106-04
Basic Rigging

Topics to be presented in this unit include:

Overview

Rigging is a common activity on a construction site. It involves transporting materials and equipment from one location to another. Rigging can be dangerous because of the weight, size, or awkwardness of the load. Elements such as training, planning, and using proper procedures will increase the success of a rigging operation.

As with all jobs on a site, rigging operations require specialized hardware. The hardware used in rigging operations works together as a system. Each piece is connected to another. For example, slings are used as links between the load and the lifting device, and hitches are used to arrange the sling to hold the load. The sling and the hitch, therefore, are a connected part of the rigging system.

The ability to control a load is a critical part of rigging operations. Items such as rated capacity, sling alignment, and hardware attachments should be checked before an operation begins. Also, the use of hand signals is critical to the success of the operation. Without proper direction, the operator could lose control of the load and cause property damage or injuries.

Rigging operations affect everyone on site, even if there is no direct involvement in the operation. Always be aware of the rigging operations that are taking place on your work site.

Instructor's Notes:

Objectives

When you have completed this module, you will be able to do the following:

1. Identify and describe the use of slings and common rigging hardware.
2. Describe basic inspection techniques and rejection criteria used for slings and hardware.
3. Describe basic hitch configurations and their proper connections.
4. Describe basic load-handling safety practices.
5. Demonstrate proper use of American National Standards Institute (ANSI) hand signals.

Prerequisites

Before you begin this module, it is recommended that you successfully complete the following:

Core Curriculum: Introductory Craft Skills, Modules 00101-04 through 00105-04. Modules 00106-04 through 00108-04 are electives. To receive a successful completion, you must take Module 00106-04 or Modules 00107-04 and 00108-04.

This course map shows all of the modules in *Core Curriculum: Introductory Craft Skills.* The suggested training order begins at the bottom and proceeds up. Skill levels increase as you advance on the course map. The local Training Program Sponsor may adjust the training order.

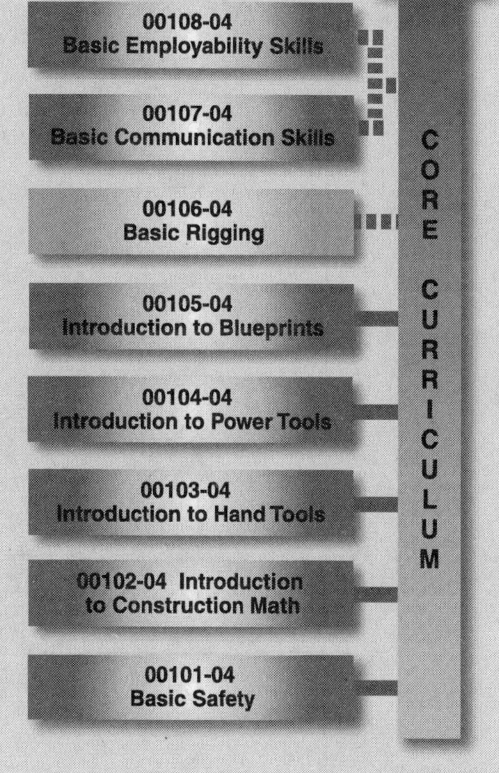

106CMAP.EPS

Key Trade Terms

ANSI hand signals	Rejection criteria
Block and tackle	Rigging hook
Bridle	Shackle
Bull ring	Sheave
Core	Side pull
Cribbing	Sling
Eyebolt	Sling angle
Grommet sling	Sling legs
Hitch	Sling reach
Hoist	Sling stress
Lifting clamp	Spliced
Load	Strand
Load control	Stress
Load stress	Tag line
Master link	Tattle-tail
One rope lay	Threaded shank
Pad eye	Warning-yarn
Plane	Wire rope
Rated capacity	

Required Trainee Materials

1. Appropriate personal protective equipment
2. Sharpened pencils and paper

Ensure that you have everything required to teach the course. Check the Materials and Equipment List at the front of this module.

See the general Teaching Tip at the end of this module.

Show Transparency 1, Course Objectives.

Show Transparency 2, Performance Tasks.

Explain that terms shown in bold (blue) are defined in the Glossary at the back of this module.

Refer to Figures 1 and 2. Compare the two basic types of cranes.

Emphasize that completing this module does not authorize trainees to make decisions about rigging.

Review the fundamentals of rigging. Emphasize that a qualified person must supervise rigging operations.

1.0.0 ◆ INTRODUCTION

Rigging is the planned movement of material and equipment from one location to another using **slings, hoists,** or other types of equipment. Some rigging operations use a loader to move materials around on a job site. Other operations require cranes to lift such **loads.** There are two basic types of cranes: overhead (*Figure 1*) and mobile (*Figure 2*).

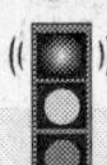

WARNING!

Although this module introduces you to the criteria that are used to make rigging decisions, you will not be making the decisions yourself at this stage in your training. If you ever have any questions about whether a synthetic sling is defective or unsafe, ask your instructor or your immediate supervisor. Always refer to the manufacturer's instructions for the sling rejection criteria for that particular sling.

ON-SITE

World's Largest Mobile Crane

The world's largest crane was built by Mammoet in the Netherlands. The crane is called the Platform Twinring. There are actually three of them in the fleet. The other two are newer and referred to as PTCs. PTC stands for Platform Twinring Containerised. These two can be disassembled into 20- and 40-ft. shipping containers, and every piece of the crane can be lifted with a normal container crane on any dock anywhere in the world. The fittings are already molded into the ends so the PTCs can go from any container crane onto any truck trailer that normally handles shipping containers. These giant cranes have a maximum load capacity of 1,600 tons.

106SA01.TIF

Instructor's Notes:

6.2 CORE CURRICULUM ◆ INTRODUCTORY CRAFT SKILLS

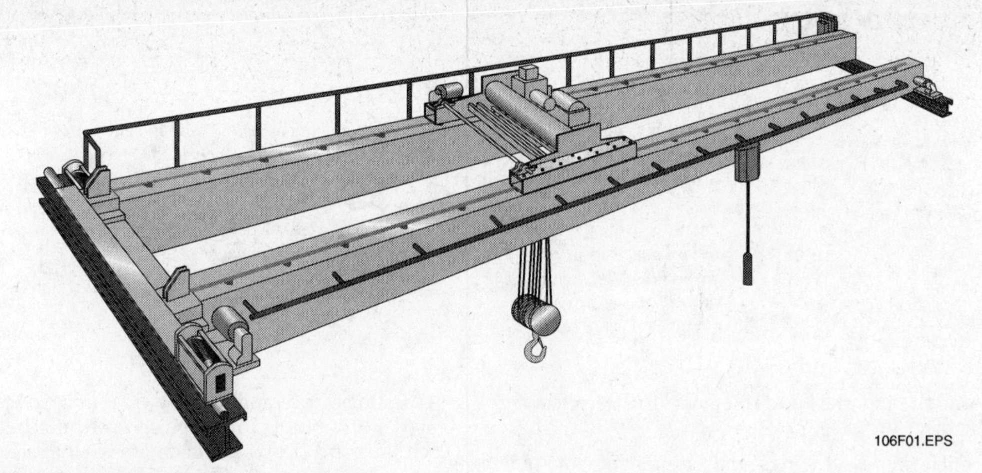

106F01.EPS

Figure 1 ◆ Overhead crane.

106F02A.EPS

106F02B.EPS

Figure 2 ◆ Mobile cranes.

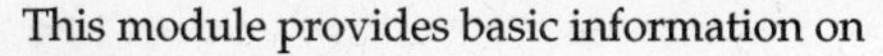

This module provides basic information on

- Slings
- **Hitches**
- Rigging hardware
- The principles and practices behind safe and efficient rigging operations

At this early stage in your career, you will not be asked to perform rigging tasks without supervision. You must not attempt any rigging operations on your own. All rigging operations must be done under the supervision of a competent person. Even so, it is important that you

MODULE 00106-04 ◆ BASIC RIGGING 6.3

Discuss the three types of slings, and explain the information that must appear on a sling identification tag.

Demonstrate how to complete an identification tag for the three types of slings covered in this module. Review the tagging requirements for the three types of slings.

Regulations and Site Procedures

The information in this module is intended as a general guide. The techniques shown here are not the only methods you can use to perform a lift. Many techniques can be safely used to rig and lift different loads (the total amount being lifted).

Some of the techniques for certain kinds of rigging and lifting are spelled out in requirements issued by federal government agencies. Some will be provided at the job site, where you will see written site procedures that address any special conditions that affect lifting procedures on that site. If you have questions about any of these procedures, ask the supervisor at the site.

understand the fundamentals of the following aspects of rigging:

- The types of slings and equipment you may eventually be using
- How to determine whether your equipment is fit for use
- The proper use of the most common types of slings and hardware
- How to select appropriate rigging equipment and how factors such as **load stress** affect that selection

Core-level rigging is designed to introduce you to the basic principles of rigging. These basic principles are true whether you are rigging a single piece of pipe, a bundle of lumber, or an 850-ton steam generator.

Rigging operations can be extremely complicated and dangerous. Do not experiment with rigging operations, and never attempt a lift on your own, without the supervision of an officially recognized qualified person. A lift may appear simple while it is in progress. That is because the people performing the lift know exactly what they are doing. There is no room for guesswork!

No matter whether rigging operations involve a simple vertical lift, a powered hoist, or a highly complicated apparatus, only qualified persons may perform them without supervision.

2.0.0 ◆ SLINGS

During a rigging operation, the load being moved must be connected to the device, such as a crane, that is doing the moving. The connector—the link between the load and the lifting device—is often a sling made of synthetic, chain, or **wire rope** materials. In this section, you will learn about three types of slings:

- Synthetic slings
- Alloy steel chain slings
- Wire rope slings

All slings are required to have identification tags. An identification tag must be securely attached to each sling and clearly marked with the information required for that type of sling. For all three types of slings, that information will include the manufacturer's name or trademark and the **rated capacity** of the type of hitch used with that sling. The rated capacity is the maximum load weight that the sling was designed to carry. An example of a rated capacities chart for synthetic web slings appears in the Appendix. Synthetic, alloy steel chain, and wire rope slings are covered in this module.

The following are the tagging requirements for synthetic slings:

- Manufacturer's name or trademark
- Manufacturer's code or stock number (unique for each sling)
- Rated capacities for the types of hitches used
- Type of synthetic material used in the manufacture of the sling

The following are the tagging requirements for alloy steel chain slings:

- Manufacturer's name or trademark
- Manufactured grade of steel
- Link size (diameter)
- Rated load and the angle on which the rating is based
- New *2003 ASME Standard* requires clarification as to the angle upon which the capacity is based
- **Sling reach**
- Number of **sling legs**

The following are the tagging recommendations for wire rope slings:

- Manufacturer's name or trademark
- Rated capacity in a vertical hitch (other hitches optional)
- Size of wire rope (diameter)
- Manufacturer's code or stock number (unique for each sling)

Instructor's Notes:

6.4 CORE CURRICULUM ◆ INTRODUCTORY CRAFT SKILLS

DID YOU KNOW?

Sling Identification Tags

The identification tags that are required on all slings must be securely attached to the slings and clearly marked with the information required for each type of sling.

WARNING!

All slings and hardware have a rated capacity, also called lifting capacity, working capacity, working load limit (WLL), or safe working load (SWL). Under no circumstances should you ever exceed the rated capacity! Overloading may result in catastrophic failure. Rated capacity is defined as the maximum load weight a sling or piece of hardware or equipment can hold or lift.

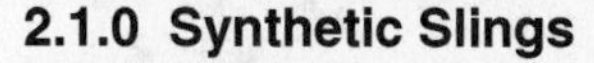

2.1.0 Synthetic Slings

Synthetic slings are widely used to lift loads, especially easily damaged ones. In this section, you will learn about two types of synthetic slings: synthetic web slings and round slings.

2.1.1 Synthetic Web Sling Design and Characteristics

Synthetic web slings provide several advantages over other types of slings:

- They are soft and wider than wire rope or chain slings. Therefore, they do not scratch or damage machined or delicate surfaces (*Figure 3*).
- They do not rust or corrode and therefore will not stain the loads they are lifting.
- They are lightweight, making them easier to handle than wire rope or chain slings. Most synthetic slings weigh less than half as much as a wire rope that has the same rated capacity. Some new synthetic fiber slings weigh one-tenth as much as wire rope.
- They are flexible. They mold themselves to the shape of the load (*Figure 4*).
- They are very elastic, and they stretch under a load much more than wire rope does. This stretching allows synthetic slings to absorb shocks and to cushion the load.
- Loads suspended in synthetic web slings are less likely to twist than those in wire rope or chain slings.

106F03.EPS

Figure 3 ◆ Surface protection.

106F04.EPS

Figure 4 ◆ Synthetic web sling shaping.

Synthetic web slings should not be exposed to temperatures above 180°F. They are also susceptible to cuts, abrasions, and other wear-and-tear damage. To prevent damage to synthetic web slings, riggers use protective pads (*Figure 5*).

Discuss the consequences of exceeding the rated capacity.

Discuss the advantages of using synthetic web slings.

Refer to Figure 4. Explain how web slings mold themselves to the shape of the load.

MODULE 00106-04 ◆ BASIC RIGGING 6.5

Discuss options for protecting the sling if protective pads are not included. Explain problems that can result if pads are not used.

Ask a trainee to explain how warning yarns indicate damage to the sling.

Refer to Figure 7. Review the uses of the most common types of synthetic web slings.

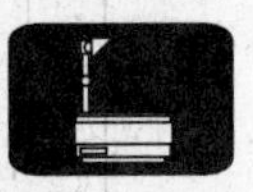

Show Transparency 3 (Figure 9). Compare male and female hardware end fittings.

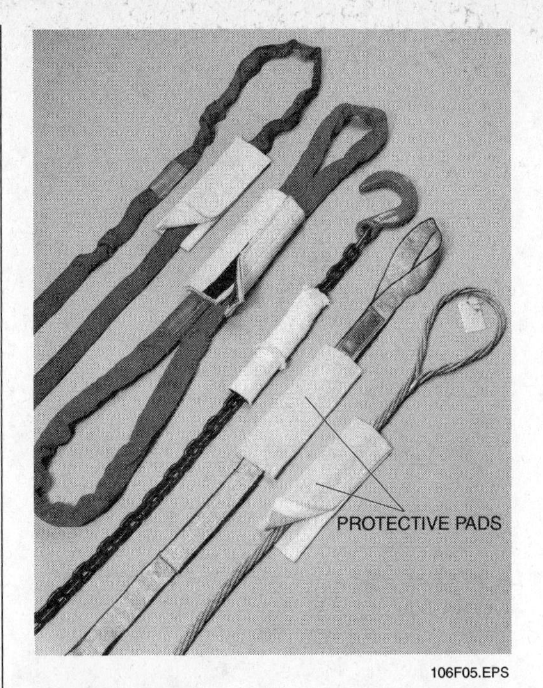

Figure 5 ◆ Protective pads.

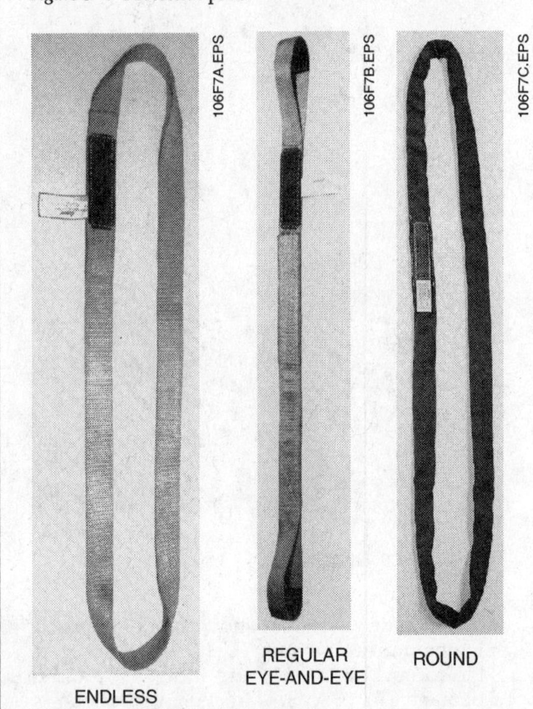

Figure 7 ◆ Synthetic web slings.

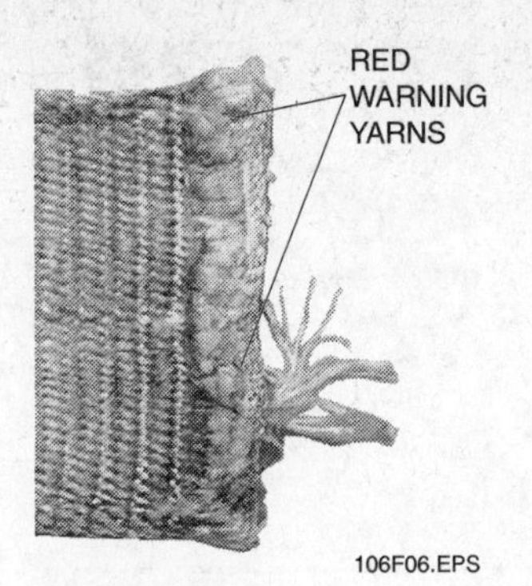

Figure 6 ◆ Synthetic web sling warning-yarns.

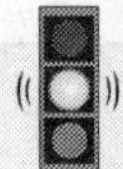

CAUTION

If the sling does not come with protective pads, use other kinds of softeners of sufficient strength or thickness to protect the sling. Pieces of old sling, fire hose, canvas, or rubber can be used.

Most synthetic web slings are manufactured with red core **warning-yarns.** These are used to let the rigger know whether the sling has suffered too much damage to be used. When the yarns are exposed, the synthetic web sling should not be used (*Figure 6*). Red core yarns should not be used exclusively.

2.1.2 Types of Synthetic Web Slings

Synthetic web slings are available in several designs. The most common are the following (*Figure 7*):

- Endless web slings, which are also called **grommet slings.**
- Round slings, which are endless and made in a continuous circle out of polyester filament yarn. The yarn is then covered by a woven sleeve.
- Synthetic web eye-and-eye slings, which are made by sewing an end of the sling directly to the sling body. Standard eye-and-eye slings have eyes on the same **plane;** twisted eye-and-eye slings have eyes at right angles to each other (*Figure 8*) and are primarily used for choker hitches (which you will learn about later).

Synthetic web eye-and-eye slings are also available with hardware end fittings instead of fabric eyes. The standard end fittings are made of either aluminum or steel. They come in male and female configurations (*Figure 9*).

Instructor's Notes:

6.6 CORE CURRICULUM ◆ INTRODUCTORY CRAFT SKILLS

Figure 8 ◆ Eye-and-eye synthetic web slings.

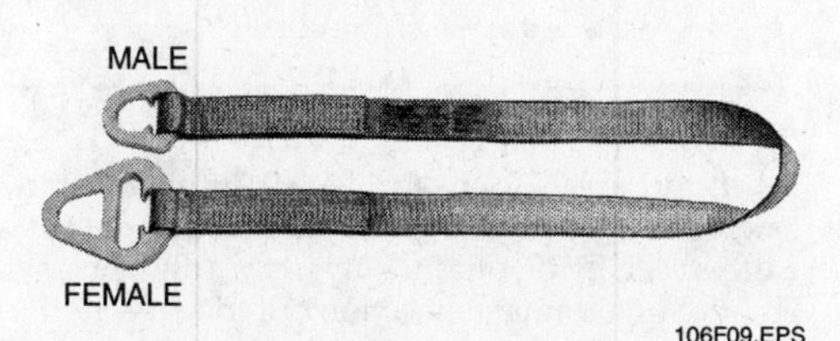

Figure 9 ◆ Synthetic web sling hardware end fittings.

2.1.3 Round Sling Design and Characteristics

Twin-Path® slings are made by wrapping a synthetic yarn around a set of spindles to form a loop. A protective jacket encases the core yarn (*Figure 10*).

Twin-Path® slings are made of a synthetic fiber, such as polyester. The material used to make up the **strands** and the number of wraps in a loop determine the rated capacity of the sling, or how much weight it can handle. Aramid round slings, for example, have a greater rated capacity for their size than the web-type slings do.

Twin-Path® slings are also available in a design with two separate wound loops of strand jacketed together side-by-side (*Figure 11*). This design greatly increases the lifting capacity of the sling.

The jackets of these slings are available in several materials for various purposes, including heat-resistant Nomek®, polyester, and bulked nylon (Covermax™) (*Figure 12*). Twin-Path® slings featuring K-Spec® yarn weigh at least 50 percent less than a polyester round sling of the same size and capacity.

Twin-Path® slings are equipped with **tattle-tail** yarns to help the rigger determine whether the

Figure 10 ◆ Synthetic endless-strand jacketed sling.

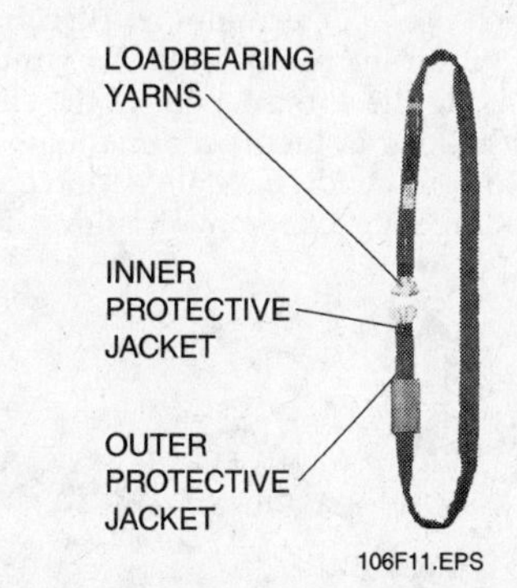

Figure 11 ◆ Twin-Path® sling.

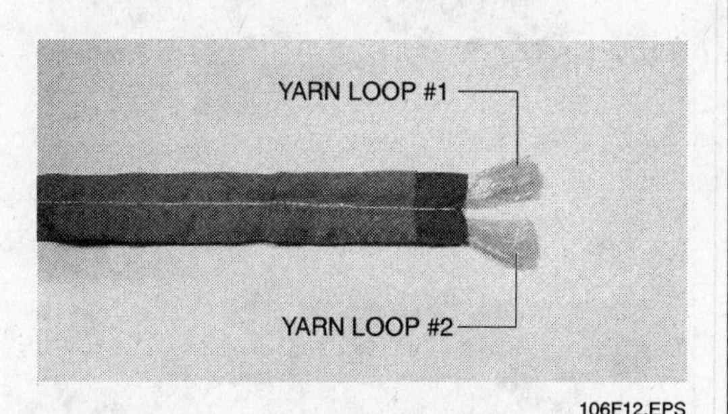

Figure 12 ◆ Twin-Path® sling makeup.

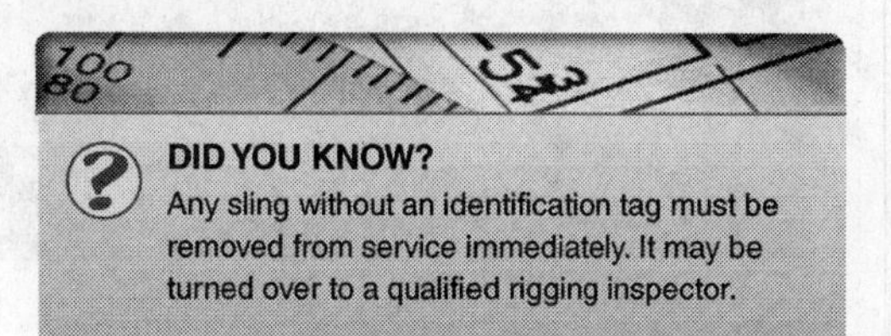

DID YOU KNOW?
Any sling without an identification tag must be removed from service immediately. It may be turned over to a qualified rigging inspector.

Refer to Figure 11. Explain how the rated capacity of a round sling is determined.

Review the types of materials used to make jackets for round slings. Go over any questions trainees may have.

Refer to Figure 13. Explain how a tattle-tail is used to determine whether a sling is overloaded.

Emphasize the importance of inspecting slings before every use.

Refer to Figure 15. Discuss the rejection criteria for synthetic slings.

Distribute copies of Figure 15 with the callouts covered. Ask trainees to identify each type of sling damage illustrated in the figure.

sling has become overloaded or stretched beyond a safe limit (*Figure 13*). These slings are also available with a fiber-optic inspection cable running through the strand (*Figure 14*). When you direct a light at one end of the fiber, the other end will light up to show that the strand has not been broken.

The way the sling will be used determines what material is used for the jackets of round slings. Polyester is normally used for light- to medium-duty sling jackets. Covermax™, a high-strength material with much greater resistance to cutting and abrasion, is used to make a sturdier jacket for heavy-duty uses.

2.1.4 Synthetic Sling Inspection

Like all slings, synthetic slings must be inspected before each use to determine whether they are in good condition and can be used. They must be inspected along the entire length of the sling, both visually (looking at them) and manually (feeling them with your hand). If any **rejection criteria** are met, the sling must be removed from service.

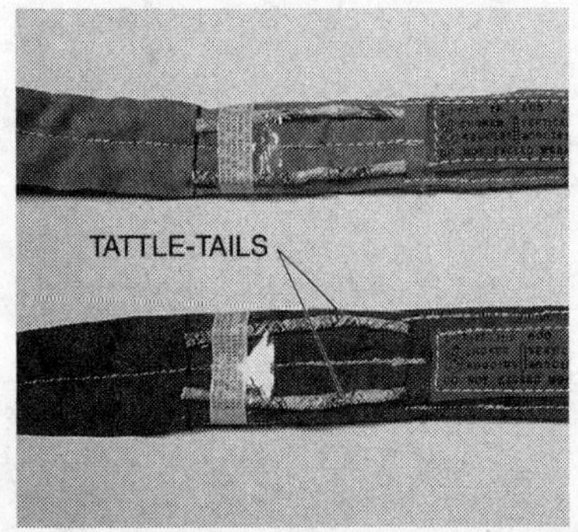

106F13.EPS

Figure 13 ◆ Tattle-tails.

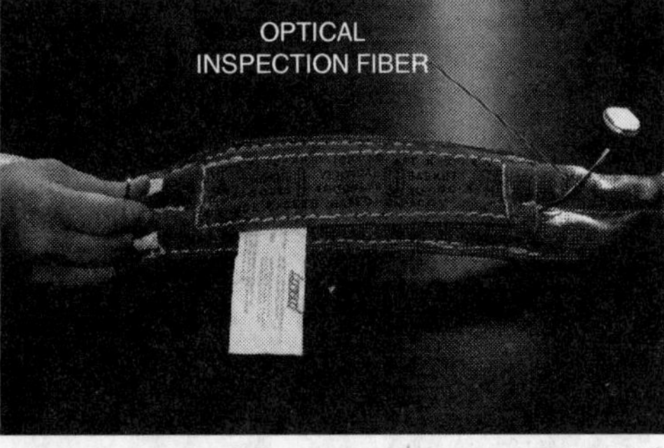

106F14.EPS

Figure 14 ◆ Fiber-optic inspection cable.

2.1.5 Synthetic Sling Rejection Criteria

If any synthetic sling meets any of the rejection criteria presented in this section, it must be removed from service immediately. In addition, the rigger has to exercise sound judgment. Along with looking for any single major problem, the rigger needs to watch for combinations of relatively minor defects in the synthetic sling. Combinations of minor damage may make the sling unsafe to use, even though such defects may not be listed in the rejection criteria.

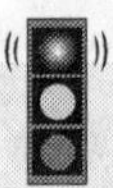

WARNING!

The rigger must always inspect synthetic slings before using them, every time. The rigger must inspect them by looking at them as well as by feeling them, because sometimes you can feel damage that you cannot see.

If you are helping to inspect synthetic slings, alert your supervisor if you suspect any defects at all, especially if you find any of the following synthetic sling damage rejection criteria (*Figure 15*):

- A missing identification tag or an identification tag that cannot be read. Any synthetic sling without an identification tag must be removed from service immediately.
- Abrasion that has worn through the outer jacket, or that has exposed the loadbearing yarn of the sling, or that has exposed the warning-yarn of a web sling.
- A cut that has severed the outer jacket or exposed the loadbearing yarn (single-layer jacket) of a round sling or that has exposed the warning-yarn of a web sling.
- A tear that has exposed the inner jacket or the loadbearing yarns (single-layer jacket) of a round sling or that has exposed the warning-yarn of a web sling.
- A puncture.
- Broken or worn stitching in the **splice** or stitching of a web sling.
- A knot that cannot be removed by hand in either a web or round sling.
- A snag in the sling that reveals the warning-yarn of a web sling or a snag that tears through the outer jacket or exposes the loadbearing yarns of a round sling.
- Crushing of either a web sling or a round sling. Crushing in a web sling feels like a hollow pocket or depression in the sling. Crushing in a round sling feels like a hard, flat spot underneath the jacket.

Instructor's Notes:

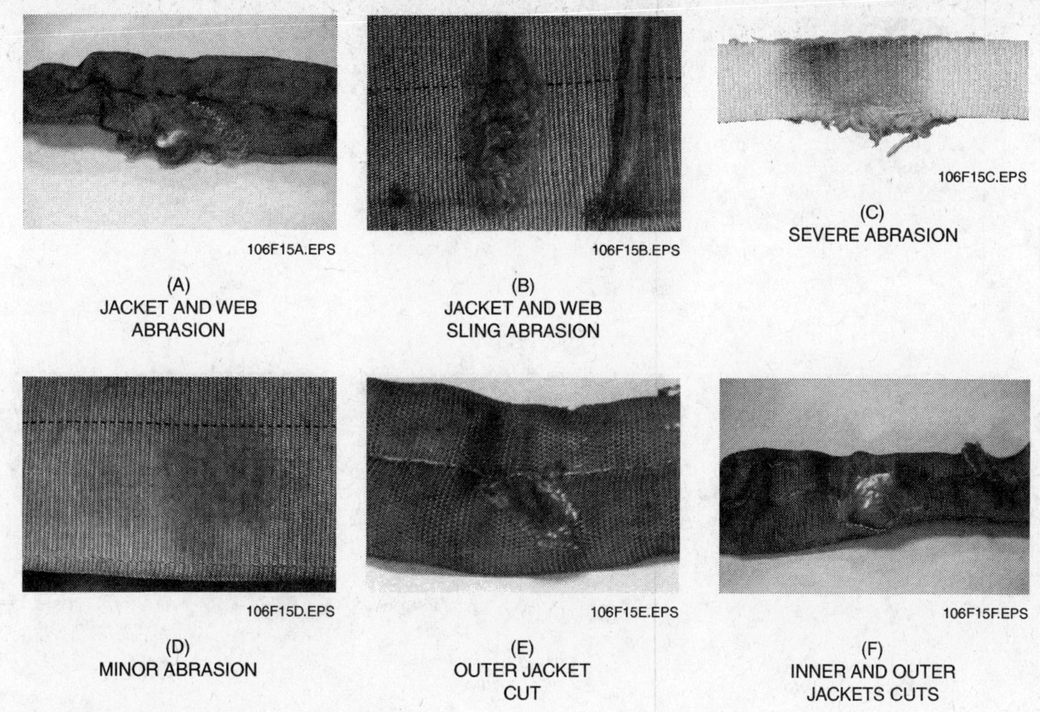

Figure 15 ◆ Sling damage rejection criteria. (1 of 2)

- Damage from overload (called tensile damage or overstretching) in a web or round sling. Tensile damage in a web sling is evident when the weave pattern of the fabric begins to pull apart. Twin-Path® slings have tattle-tails. When the tail has been pulled into the jacket, it indicates that the sling may have been overloaded.
- Chemical damage, including discoloration, burns, and melting of the fabric or jacket.
- Heat damage, ranging from friction burns to melting of the sling material, the loadbearing strands, or the jacket. Friction burns give the webbing material a crusty or slick texture. Heat damage to the jacket of a round sling looks like glazing or charring. In round slings with heat-resistant jackets, you may not be able to see heat damage to the outside of the sling; however, the internal yarns may have been damaged. You can detect that damage by carefully handling the sling and feeling for brittle or fused fibers inside the sling or by flexing and folding the sling and listening for the sound of fused yarn fibers cracking or breaking.
- Ultraviolet (UV) damage. The evidence of UV damage is a bleaching-out of the sling material, which breaks down the synthetic fibers. Web slings with UV damage will give off a powder-like substance when they are flexed and folded. Round slings, especially those made of Cordura nylon and other specially treated synthetic fibers, have a much greater resistance to UV damage. In round slings, UV damage shows up as a roughening of the fabric texture where no other sign of damage, such as abrasion, can be found.
- Loss of flexibility caused by the presence of dirt or other abrasives. You can tell that a sling has lost flexibility when the sling becomes stiff. Abrasive particles embedded in the sling material act like tiny blades that cut apart the internal fibers of the sling every time the sling is stretched, flexed, or wrapped around a load.

MODULE 00106-04 ◆ BASIC RIGGING 6.9

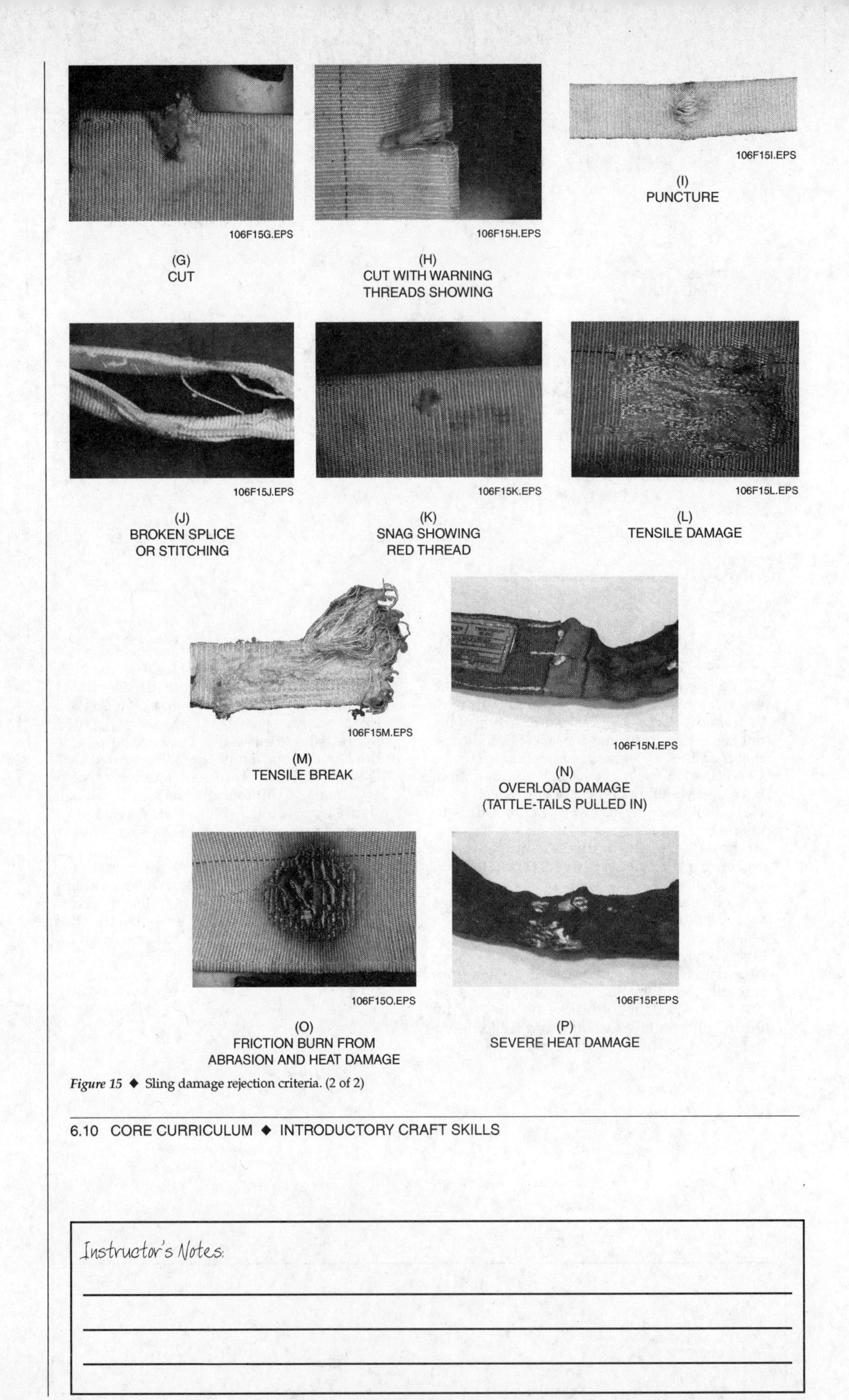

(G)
CUT

(H)
CUT WITH WARNING
THREADS SHOWING

(I)
PUNCTURE

(J)
BROKEN SPLICE
OR STITCHING

(K)
SNAG SHOWING
RED THREAD

(L)
TENSILE DAMAGE

(M)
TENSILE BREAK

(N)
OVERLOAD DAMAGE
(TATTLE-TAILS PULLED IN)

(O)
FRICTION BURN FROM
ABRASION AND HEAT DAMAGE

(P)
SEVERE HEAT DAMAGE

Figure 15 ◆ Sling damage rejection criteria. (2 of 2)

Instructor's Notes:

6.10 CORE CURRICULUM ◆ INTRODUCTORY CRAFT SKILLS

Sling Jackets

If only the outer jacket of a sling is damaged, under the rejection criteria, that sling can be sent back to the manufacturer for a new jacket. Before it is returned, it must be tested and certified before it is used again. It is much less costly to do this than to replace the sling. However, slings that are removed because of heat- or tension-related jacket damage cannot be returned for repair and must be disposed of by a qualified person.

Go over the Review Questions for Sections 2.0.0–2.1.5. Answer any questions trainees may have.

Have trainees review Sections 2.2.0–2.3.3 for the next session.

Review Questions

Section 2.0.0–2.1.5

1. Identification tags for slings must include the _____.

a. type of protective pads to use
b. type of damage sustained during use
c. color of the tattle-tail
d. manufacturer's name or trademark

2. A sling that exhibits _____ must be removed from service immediately.

a. a missing identification tag
b. any kind of abrasion
c. flexibility
d. paint on the sling

3. An advantage of synthetic slings is that they are _____.

a. narrower than wire rope or chain slings
b. lightweight and flexible
c. more rigid than other types of slings
d. less stretchable than wire rope

4. Synthetic slings must be inspected _____.

a. the first time they are ever used
b. before every use
c. every fifth time they are used
d. by sight only

5. When a tattle-tail has been pulled into the jacket, the sling has been _____.

a. chemically damaged
b. overheated
c. overloaded
d. crushed

MODULE 00106-04 ◆ BASIC RIGGING 6.11

Ensure that you have everything required for testing during this session.

Explain why the use of chain slings is discouraged.

Explain how the alloy grade affects the durability of the chain.

Show Transparencies 4 and 5 (Figures 17 and 18). Compare single- and double-basket slings with chain bridle slings.

Discuss the advantage of using alloy steel chain slings in high temperatures. Review problems that can occur when working with alloy steel chain slings.

Explain problems that can result if alloy steel chain slings are dragged across hard surfaces.

2.2.0 Alloy Steel Chain Slings

Alloy steel chain, like wire rope (which you will learn about later), can be used in many different rigging operations. Chain slings are often used for lifts in high heat or rugged conditions.

The use of chain slings is discouraged, however. A chain sling weighs much more than a wire rope sling of the same capacity. It is also harder to inspect and can fail without warning. Synthetic slings are preferred. But you will encounter chain slings in the field, so you need to know how to use them.

2.2.1 Alloy Steel Chain Sling Design and Characteristics

Steel chain slings used for overhead lifting must be made of alloy steel. The higher the alloy grade, the safer and more durable the chain is. Alloy steel chains commonly used for most overhead lifting are marked with the number 8, the number 80, the number 800, or the letter A (*Figure 16*).

Steel chain slings have two basic designs with many variations:

- Single- and double-basket slings (*Figure 17*) do not require end-fitting hardware. The chain is attached to the **master link** in a permanent basket hitch or hitches.
- Chain **bridle** slings are available with two to four legs (*Figure 18*). Chain bridle slings require some type of end-fitting hardware—eye hooks, grab hooks, plate hooks, sorting or pipe hooks (*Figure 19*), or links.

Alloy steel chain slings are used for lifts when temperatures are high or where the slings will be subjected to steady and severe abuse. Most alloy steel chain slings can be used in temperatures up to 500°F with little loss in rated capacity.

Even though alloy steel chain slings can withstand extreme temperature ranges and abusive working conditions, these slings can be damaged if loads are dropped on them, if they are wrapped around loads with sharp corners (unless you use softeners), or if they are exposed to intense temperatures.

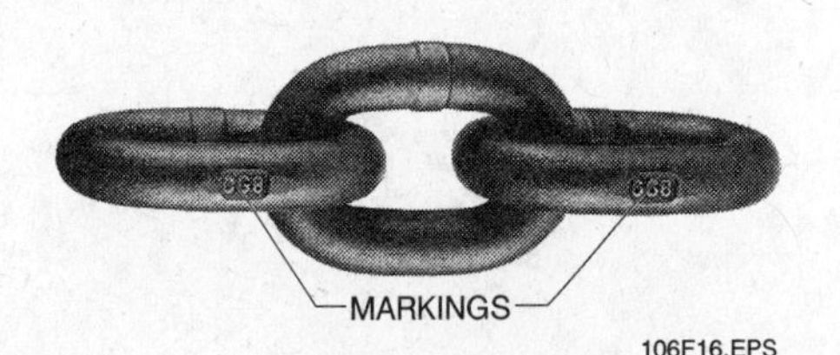

Figure 16 ◆ Markings on alloy steel chain slings.

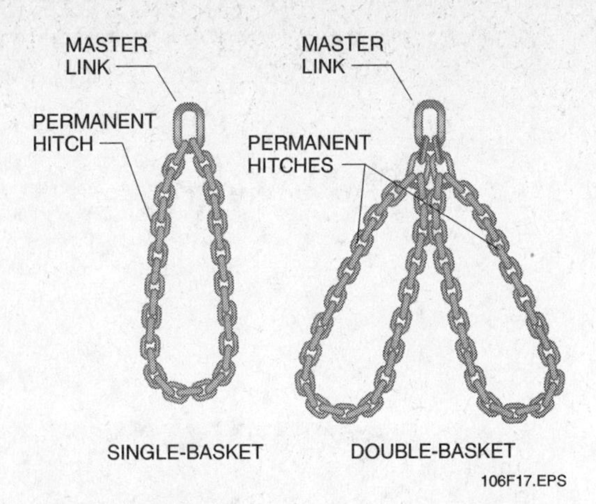

Figure 17 ◆ Chain slings.

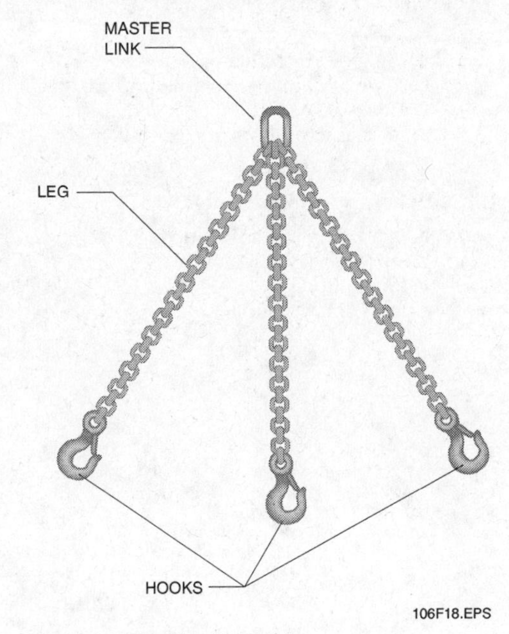

Figure 18 ◆ Three-leg chain bridle.

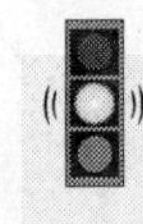

CAUTION

Never drag alloy steel chain slings across hard surfaces, especially concrete. Friction with abrasive surfaces wears out the chain and weakens the sling.

Instructor's Notes:

106F19A.EPS

EYE HOOK WITH GATE

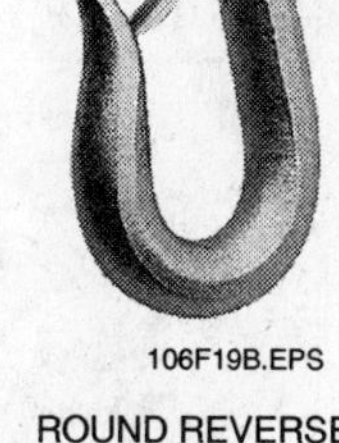

106F19B.EPS

ROUND REVERSE EYE HOOK

106F19C.EPS

SORTING HOOK

Figure 19 ◆ Eye and sorting hooks.

2.2.2 Alloy Steel Chain Inspection

Before each use, the rigger must inspect the alloy steel chain sling to determine whether it is safe to use. This module introduces you to the criteria involved in making this decision, but at this point in your training you will not make the decision yourself. If you have any questions about whether a sling is defective or unsafe and should not be used, ask your instructor or your immediate supervisor.

Steel chain slings must be visually inspected before each lift. A chain must be removed from service if a deficiency in the chain matches any of the rejection criteria. A chain may also need to be removed from service if something shows up that does not exactly match the rejection criteria.

2.2.3 Alloy Steel Chain Rejection Criteria

An alloy steel chain sling must be removed from service for any of the following defects (*Figure 20*):

- Missing or illegible identification tag
- Cracks
- Heat damage
- Stretched links. Damage is evident when the link grows long and when the barrels—the long sides of the links—start to close up
- Bent links
- Twisted links
- Excessive rust or corrosion, meaning rust or corrosion that cannot be easily removed with a wire brush
- Cuts, chips, or gouges resulting from impact on the chain
- Damaged end fittings, such as hooks, clamps, and other hardware
- Excessive wear at the link-bearing surfaces
- Scraping or abrasion

2.3.0 Wire Rope Slings

Wire rope slings (*Figure 21*) are made of high-strength steel wires formed into strands wrapped around a supporting **core.** They are lighter than chain, can withstand substantial abuse, and are easier to handle than chain slings. They can also withstand relatively high temperatures. However, because wire rope slings can slip, the use of synthetic slings is preferred. Wire rope is still being used, though, so you need to learn about the design and characteristics, applications, inspection, and maintenance of wire rope slings.

2.3.1 Wire Rope Sling Design and Characteristics

There are many types of wire ropes. They all consist of a core that supports the rope, and center strand wires, each with many high-grade steel wires wound around them to form the strands.

Wire rope is designed to operate like a self-adjusting machine. This means a wire rope has moving parts, as a machine does. A wire rope's moving parts are the wires that make up the strands and the core of the rope itself. These moving parts interact with one another by sliding and adjusting. This sliding and adjusting compensates for the ever-changing **stresses** placed upon a working rope. Because of this, a wire rope's rated capacity depends on it being in good condition. This means its wires must be able to move the way they were designed to.

Show Transparency 6 (Figure 20). Review the rejection criteria for alloy steel chain slings. Explain that, at this point in their training, trainees will not make the decision to reject a steel chain sling on their own.

Refer to Figure 21. Discuss the advantages and disadvantages of wire rope slings.

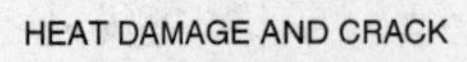

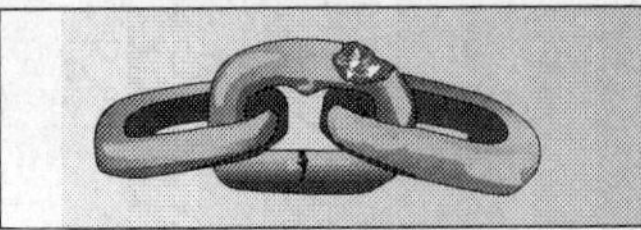

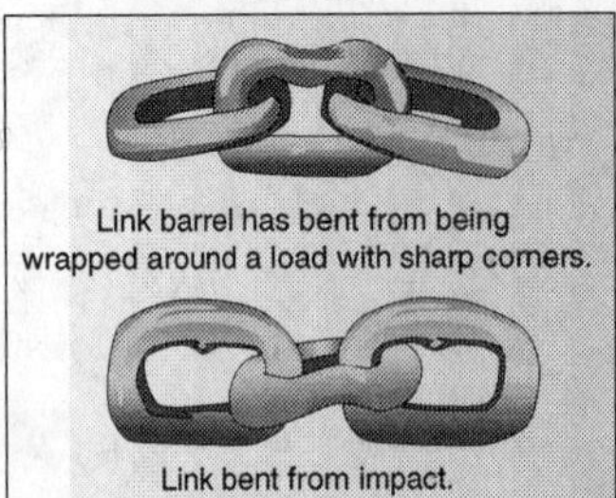

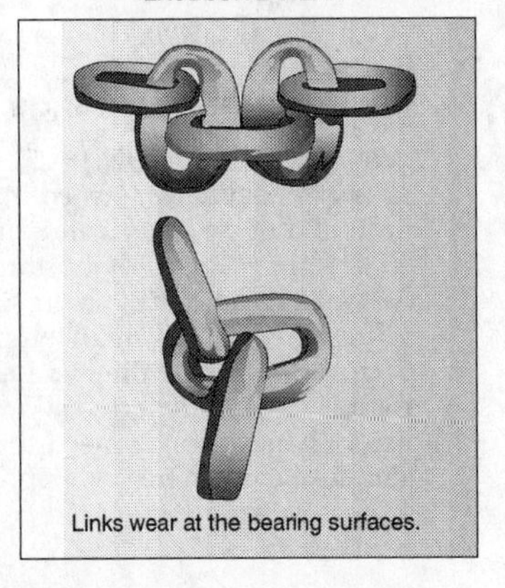

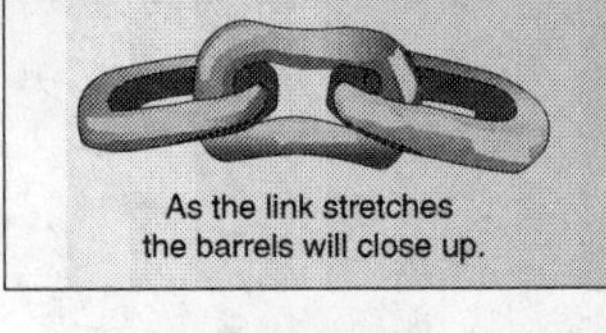

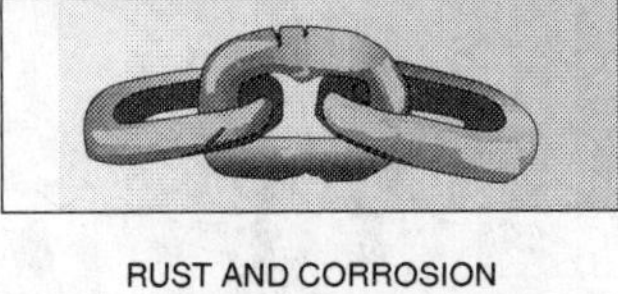

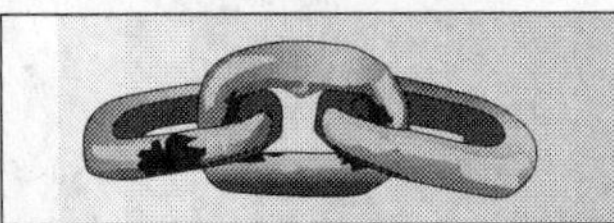

106F20.EPS

Figure 20 ◆ Damage to chains.

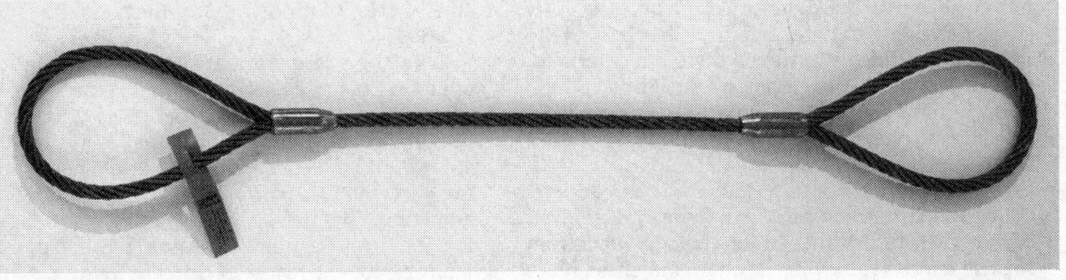

106F21.EPS

Figure 21 ◆ Wire rope sling.

Instructor's Notes:

6.14 CORE CURRICULUM ◆ INTRODUCTORY CRAFT SKILLS

The three basic components of a wire rope are as follows (*Figure 22*):

- A supporting core
- High-grade steel wires
- Multiple center wires

There are three basic types of supporting cores for wire rope (*Figure 23*): fiber cores, strand cores, and independent wire rope cores. Fiber cores are usually made of synthetic fibers, but they also can be made of natural vegetable fibers, such as sisal. Strand cores are made by using one strand of the same size and type as the rest of the strands of rope. Independent wire rope cores are made of a separate wire rope with its own core and strands. The core rope wires are much smaller and more delicate than the strand wires in the outer rope.

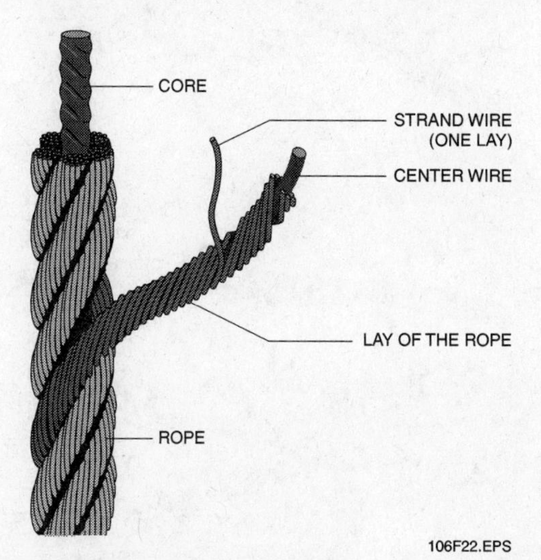

Figure 22 ◆ Wire rope components.

The various materials used to form the supporting core of a wire rope have both desirable characteristics and drawbacks, depending on how they are to be used. Fiber core ropes, for example, may be damaged by heat at relatively low temperatures (180°F to 200°F) as well as by exposure to caustic chemicals.

The function of the core in a wire rope is to support the strands so that the strands keep their original shapes. When the core supports the strands, their high-grade steel wires can slide and adjust against each other. The ability to adjust makes them less likely to be damaged by stress when the rope is bent around **sheaves** or loads, or when it is placed at an angle during rigging.

2.3.2 Wire Rope Sling Inspection

Like other slings, wire rope slings must be inspected before each use. If the wire rope is damaged, it must be removed from service. Always ask your instructor or immediate supervisor if you have any questions about whether a wire rope sling is suitable for use.

2.3.3 Wire Rope Sling Rejection Criteria

Following an inspection, a sling may be rejected based on several common types of damage (*Figure 24*), including broken wires, kinks, birdcaging, crushing, corrosion and rust, and heat damage. At this stage in your training, you need to be able to demonstrate that you understand the rejection criteria for wire ropes. Only a qualified person can actually make the decision to use a wire rope in a rigging operation or to discard it if it is damaged. If you think a wire rope may be damaged, bring it to the attention of your instructor or immediate supervisor. Additional information about the possible types of damage appears on page 6.16.

Refer to Figure 22. Review the basic components of a wire rope sling. Ask whether trainees have any questions.

Show Transparency 7 (Figure 23). Discuss the characteristics and drawbacks of the three types of rope supporting cores. Explain the function of the core in a wire rope chain.

Refer to Figure 24. Review the rejection criteria for wire rope slings.

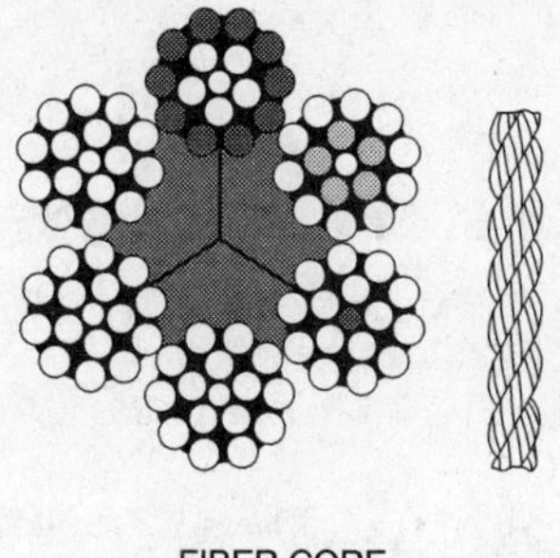
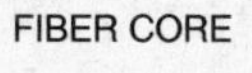

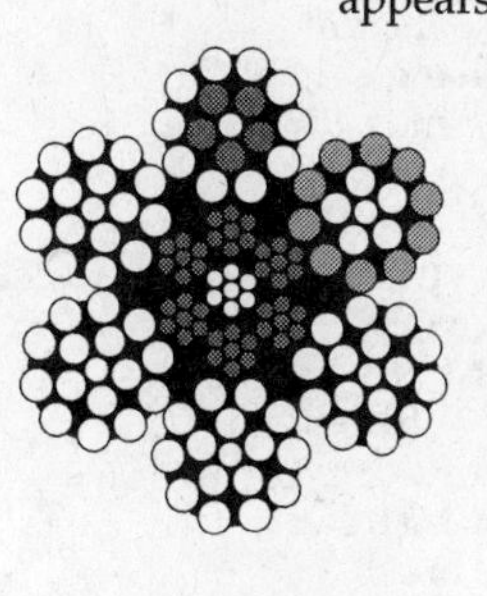

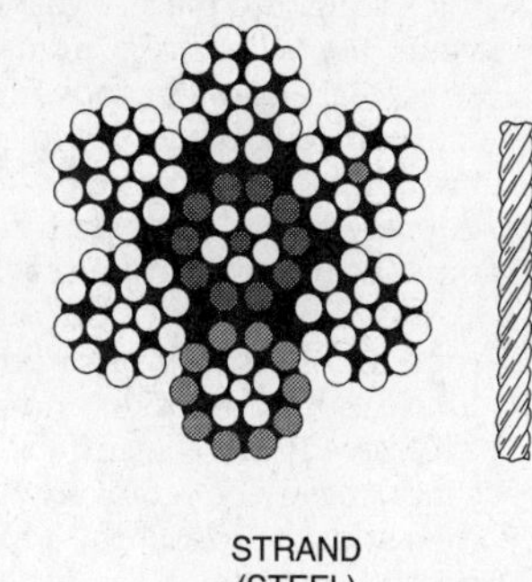

Figure 23 ◆ Wire rope supporting cores.

MODULE 00106-04 ◆ BASIC RIGGING 6.15

See the Teaching Tip for Sections 2.0.0–2.3.3 at the end of this module.

Go over the Review Questions for Sections 2.2.0–2.3.3. Answer any questions trainees may have.

Have each trainee perform Performance Task 1 to your satisfaction. Fill out a Performance Profile Sheet for each trainee.

- *Broken wires* – Broken wires in the strands of a wire rope lessen the material strength of the rope and interfere with the interaction among the rope's moving parts. External broken wires usually mean normal fatigue, but internal or severe external breaks should be investigated closely. Internal or severe breaks in a wire rope mean it has been used improperly.

  Rejection criteria for broken wires consider how many wires are broken in one lay length of rope, or **one rope lay.** One rope lay is a term that defines the lengthwise distance it takes for one strand of wire to make one complete turn around the core (*Figure 25*). Different wire ropes have different one rope lays, so it is important to inspect each wire rope closely when looking for broken wires.
- *Kinks* – Kinking, or distortion of the rope, is a very common type of damage. Kinking can result in serious accidents. Sharp kinks restrict or prevent the movement of wires in the strands at the area of the kink. This means the rope is damaged and must not be used. Ropes with kinks in the form of large, gradual loops in a corkscrew configuration must be removed from service.
- *Birdcaging* – This damage occurs when a load is released too quickly and the strands are pulled or bounced away from the supporting core. The wires in the strands cannot compensate for the change in stress level by adjusting inside the strands. The built-up stress then finds its own release out through the strands.

  Birdcaging usually occurs in an area where already-existing damage prevents the wires from moving to compensate for changes in stress, position, and bending of the rope. Any sign of birdcaging is cause to remove the rope from service immediately.
- *Crushing* – This results from setting a load down on a sling or from hammering or pounding a sling into place. Crushing of the sling prevents the wires from adjusting to changes in stress, changes in position, and bends. A crushed sling usually results in the crushing or breaking of the core wires directly beneath the damaged strands. If crushing occurs, the sling must be removed from service immediately.
- *Corrosion and rust* – Corrosion and rust of wire rope are the result of improper or insufficient lubrication. Corrosion and rust are considered excessive if there is surface scaling or rust that cannot easily be removed with a wire brush, or if they occur inside the rope. If corrosion and rust are excessive, the rope must be removed from service.
- *Heat damage* – Heat damage embrittles wires in the strands and core in the area directly affected by the heat's contact, but also in a surrounding area up to 12" in each direction. Heat damage appears as discoloration and sometimes the actual melting of the wire rope. A wire rope that has been damaged by heat must be removed from service.

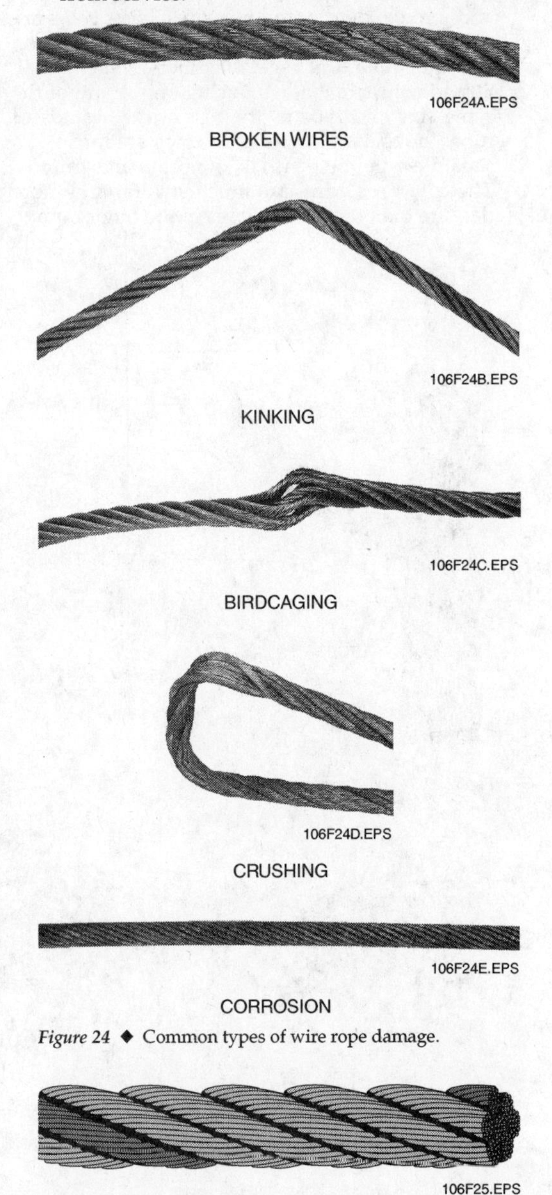

Figure 24 ◆ Common types of wire rope damage.

106F25.EPS

Figure 25 ◆ One rope lay.

6.16 CORE CURRICULUM ◆ INTRODUCTORY CRAFT SKILLS

Instructor's Notes:

Review Questions

Section 2.2.0–2.3.3

1. The best sling to use in lifts where high temperatures or abusive conditions are expected is _____.
   a. grommet steel chain
   b. round steel chain
   c. alloy steel chain
   d. carbon steel chain

2. A chain sling that _____ must not be used.
   a. exhibits links stretched from overloading
   b. exhibits paint on more than half of the links
   c. is more than one year old
   d. has an identification tag

3. In wire rope, the rope component that supports the strands is the _____.
   a. high-grade steel wire
   b. core
   c. center wire
   d. sling

4. The type of rope core that is susceptible to heat damage at relatively low temperatures is the _____.
   a. fiber core
   b. strand core
   c. independent wire rope core
   d. supporting core

5. The term used to describe the damage done when the strands bounce away from the core is _____.
   a. kinking
   b. corkscrewing
   c. shock-loading
   d. birdcaging

3.0.0 ◆ HITCHES

As you have learned, the link between the load and the lifting device is often a sling made of synthetic, alloy steel chain, or wire rope material. The way the sling is arranged to hold the load is called the rigging configuration or hitch. Hitches can be made using just the sling or by using connecting hardware as well. There are three basic types of hitches:

- Vertical
- Choker
- Basket

One of the most important parts of the rigger's job is making sure that the load is held securely. The type of hitch the rigger uses depends on the type of load to be lifted. Different hitches are used to secure, for example, a load of pipes, a load of concrete slabs, or a load of heavy machinery.

Controlling the movement of the load once the lift is in progress is another extremely important part of the rigger's job. Therefore, the rigger must also consider the intended movement of the load when choosing a hitch. For example, some loads are lifted straight up and then straight down. Other loads are lifted up, turned 180 degrees in midair, and then set down in a completely different place. In this section, you will learn about how each of the three basic types of hitches—vertical, choker, and basket—is used to both secure the load and control its movement.

3.1.0 Vertical Hitch

The single vertical hitch is used to lift a load straight up. It forms a 90-degree angle between the hitch and the load. With this hitch, some type of attachment hardware is needed to connect the sling to the load (*Figure 26*). The single vertical hitch allows the load to rotate freely. If you do not want the load to rotate freely, some method of **load control** must be used, such as a **tag line.** These methods are explained later in this module.

WARNING!
All rigging operations are dangerous, and extreme care must be used at all times. A straight-up-and-down vertical lift is every bit as dangerous as a lift that involves rotating a load in midair and moving it to a different place. Only a qualified person may select the hitch to be used in any rigging operation.

Have trainees review Sections 3.0.0–3.3.0 for the next session.

Ensure that you have everything required for the laboratory and testing during this session.

Review the three basic types of hitches. Explain how riggers determine which type of hitch to use.

Show Transparency 8 (Figure 26). Explain how a vertical hitch operates.

Emphasize that only a qualified person may select the hitch to be used.

MODULE 00106-04 ◆ BASIC RIGGING 6.17

Refer to Figures 27 and 28. Explain how a bridle hitch eliminates the need for a spreader beam.

Show Transparency 9 (Figure 29). Explain how a multiple-leg bridle hitch provides increased stability.

Show Transparency 10 (Figure 30). Explain how a choker hitch uses a shackle to form a loop around the load.

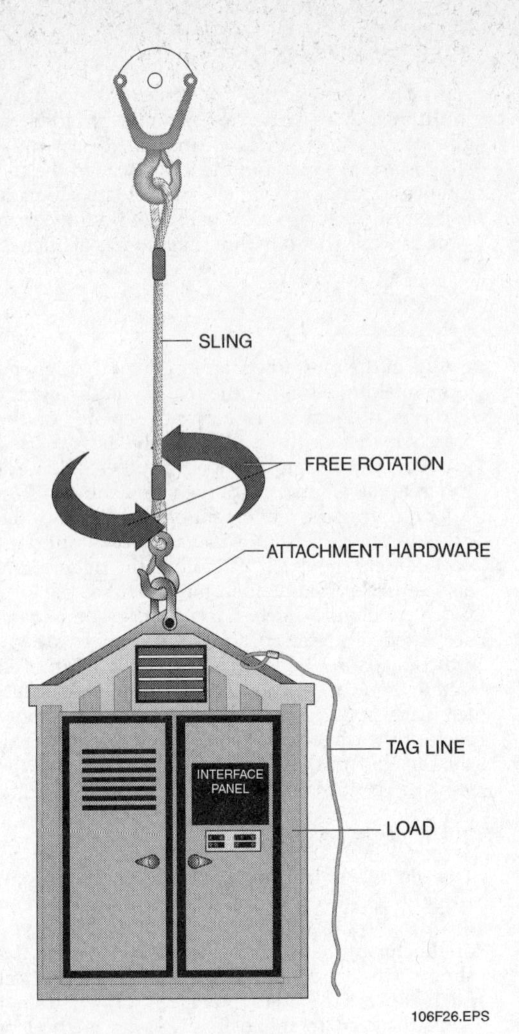

Figure 26 ◆ Single vertical hitch.

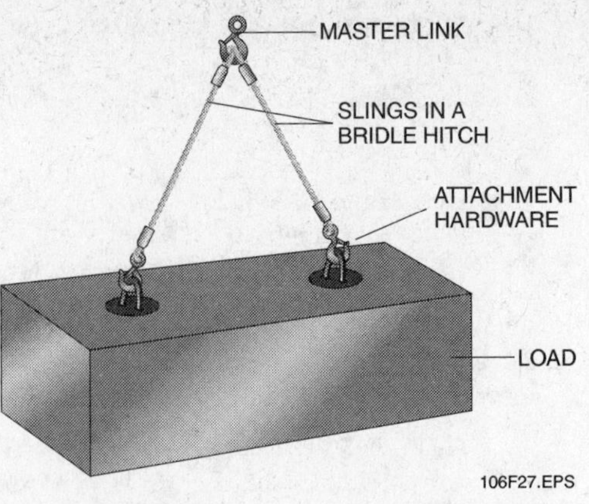

Figure 27 ◆ Bridle hitch.

Figure 28 ◆ Spreader beam.

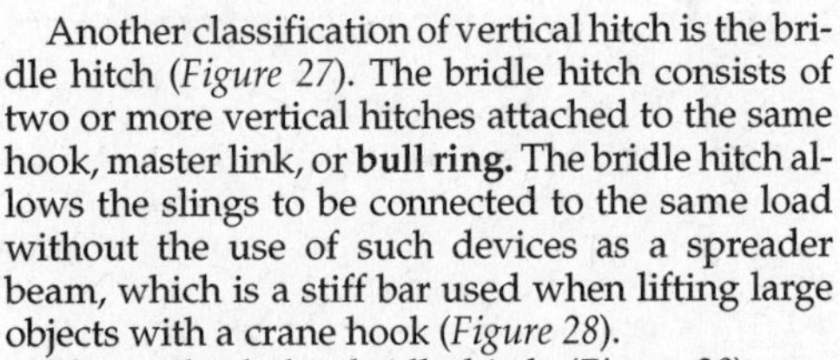

Another classification of vertical hitch is the bridle hitch (*Figure 27*). The bridle hitch consists of two or more vertical hitches attached to the same hook, master link, or **bull ring.** The bridle hitch allows the slings to be connected to the same load without the use of such devices as a spreader beam, which is a stiff bar used when lifting large objects with a crane hook (*Figure 28*).

The multiple-leg bridle hitch (*Figure 29*) consists of three or four single hitches attached to the same hook, master link, or bull ring. Multiple-leg bridle hitches provide increased stability for the load being lifted. A multiple-leg bridle hitch is always considered to have only two of the legs supporting the majority of the load and the rest of the legs balancing it.

3.2.0 Choker Hitch

The choker hitch is used when a load has no attachment points or when the attachment points are not practical for lifting (*Figure 30*). The choker hitch is made by wrapping the sling around the load and passing one eye of the sling through a **shackle** to form a constricting loop around the load. It is important that the shackle used in a choker hitch be oriented properly, as shown in

Instructor's Notes:

6.18 CORE CURRICULUM ◆ INTRODUCTORY CRAFT SKILLS

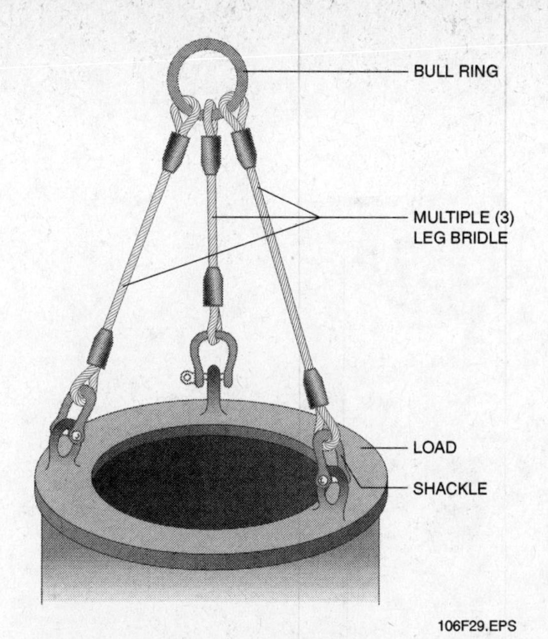

Figure 29 ◆ Multiple-leg bridle hitch.

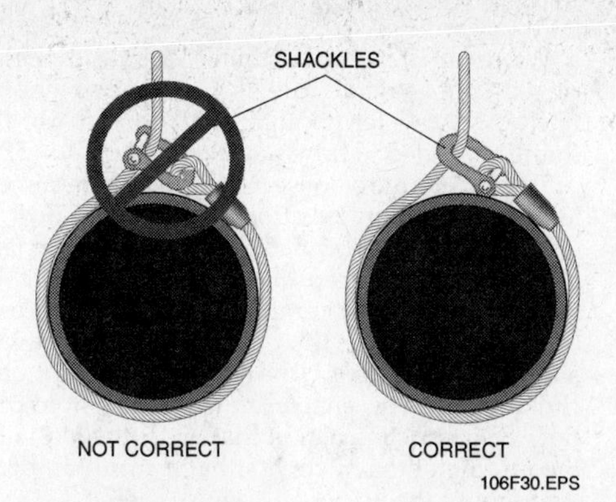

Figure 30 ◆ Choker hitch.

Figure 30. The choker hitch affects the capacity of the sling, reducing it by a minimum of 25 percent. This reduction must be considered when choosing the proper sling.

The choker hitch does not grip the load securely. It is not recommended for loose bundles of materials because it tends to push loose items up and out of the choker. Many riggers use the choker hitch for bundles, mistakenly believing that forcing the choke down provides a tight grip. In fact, it serves only to drastically increase the stress on the choke leg (*Figure 31*).

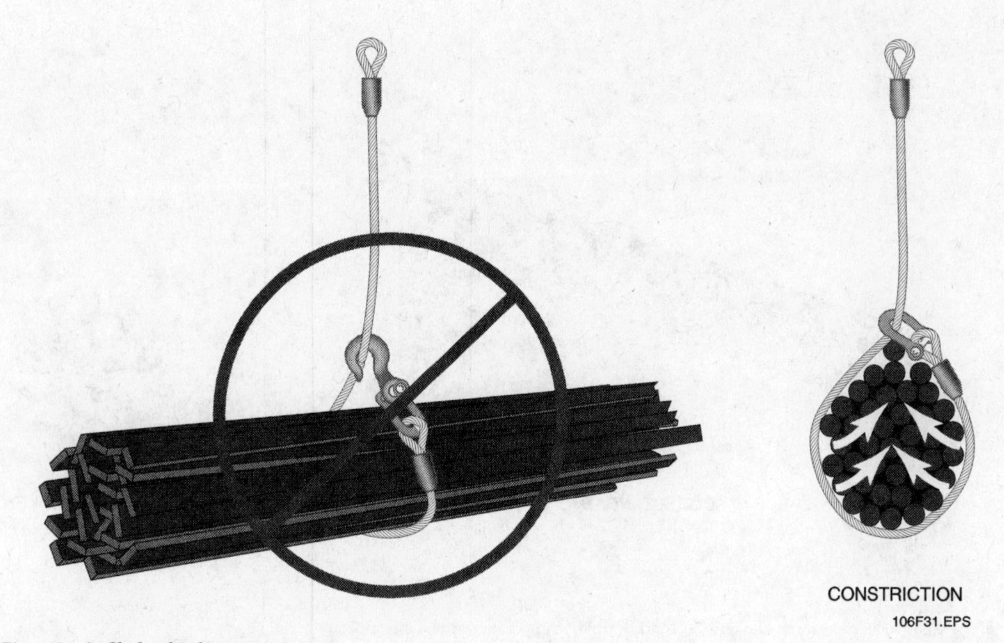

Figure 31 ◆ Choker hitch constriction.

MODULE 00106-04 ◆ BASIC RIGGING 6.19

Refer to Figure 32. Explain why double choker hitches are necessary to support loads longer than 12 feet.

Have a trainee explain why choker hitches are not recommended for loose bundles. Explain how to safely lift a bundle of loose items. Go over any questions trainees may have.

When an item more than 12′ long is being rigged, the general rule is to use two choker hitches spaced far enough apart to provide the stability needed to transport the load (*Figure 32*).

To lift a bundle of loose items, or to maintain the load in a certain position during transport, a double-wrap choker hitch (*Figure 33*) may be useful. The double-wrap choker hitch is made by wrapping the sling completely around the load, and then wrapping the choke end around again and passing it through the eye like a conventional choker hitch. This enables the load weight to produce a constricting action that binds the load into the middle of the hitch, holding it firmly in place throughout the lift.

Forcing the choke down will drastically increase the stress placed on the sling at the choke point. To gain gripping power, use a double-wrap choker hitch. The double-wrap choker uses the load weight to provide the constricting force, so there is no need to force the sling down into a tighter choke.

A double-wrap choker hitch is ideal for lifting bundles of items, such as pipes and structural steel. It will also keep the load in a certain position, which makes it ideal for equipment installation lifts. Lifting a load longer than 12′ requires two of these hitches.

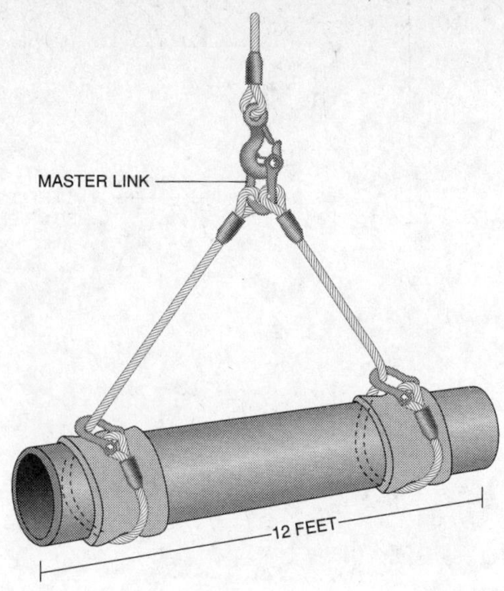

Figure 32 ◆ Double choker hitch.

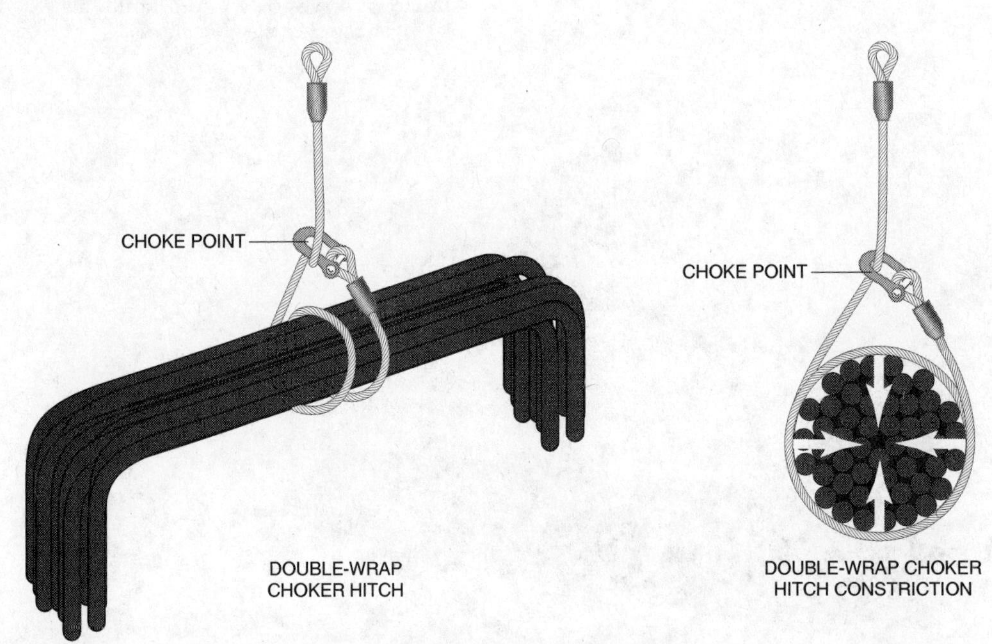

Figure 33 ◆ Double-wrap choker hitch and double-wrap choker hitch constriction.

Instructor's Notes:

6.20 CORE CURRICULUM ◆ INTRODUCTORY CRAFT SKILLS

3.3.0 Basket Hitch

Basket hitches are very versatile and can be used to lift a variety of loads. A basket hitch is formed by passing the sling around the load and placing both eyes in the hook (*Figure 34*). Placing a sling into a basket hitch effectively doubles the capacity of the sling. This is because the basket hitch creates two sling legs from one sling.

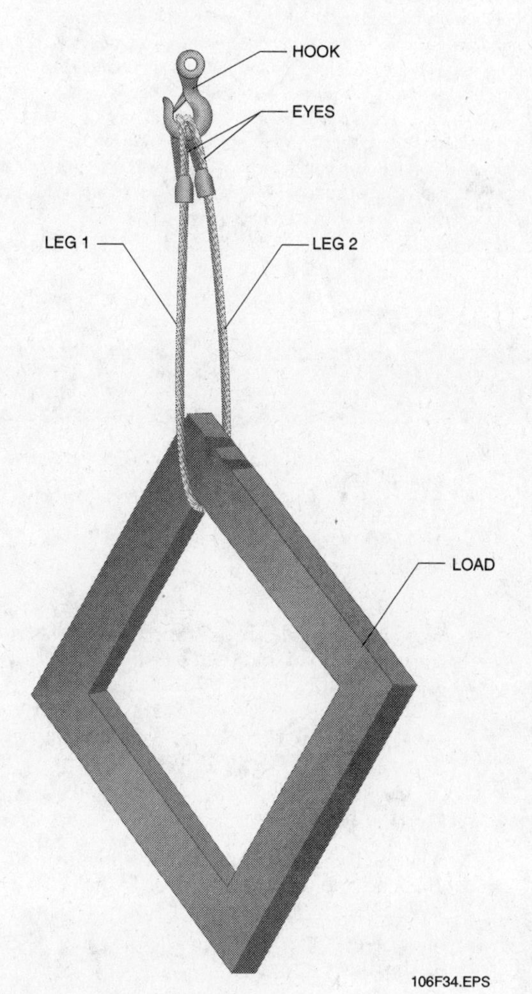

Figure 34 ◆ Basket hitch.

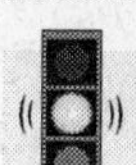

CAUTION

A basket hitch should not be used to lift loose materials. Loads placed in a basket hitch should be balanced.

The double-wrap basket hitch combines the constricting power of the double-wrap choker hitch with the capacity advantages of a basket hitch (*Figure 35*). This means it is able to hold a larger load more tightly.

The double-wrap basket hitch requires a considerably longer sling length than a double-wrap choker hitch. If it is necessary to join two or more slings together, the load must be in contact with the sling body only, not with the hardware used to join the slings. The double-wrap basket hitch provides support around the load. Just as with the double-wrap choker hitch, the load weight provides the constricting force for the hitch.

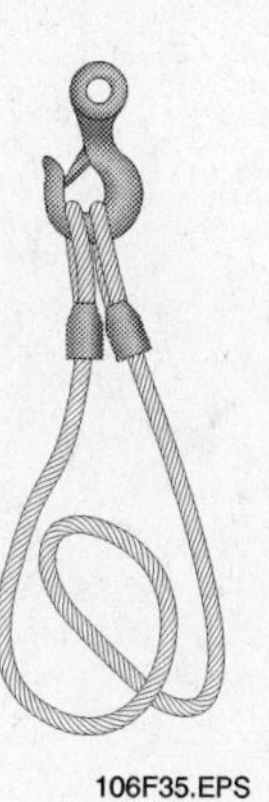

Figure 35 ◆ Double-wrap basket hitch.

Show Transparency 11 (Figure 34). Explain how using a sling in a basket hitch can double the capacity of a sling.

Ask a trainee to explain why a basket hitch should not be used to lift loose materials.

Discuss the advantages of a double-wrap basket hitch. Explain how to join two or more slings together.

See the Teaching Tip for Sections 3.0.0–3.3.0 at the end of this module.

MODULE 00106-04 ◆ BASIC RIGGING 6.21

Place a glass of water in the center of a table. Ask four trainees to each place only one of their fingers under their corner of the table and together try to walk the table across the room. Next ask the trainees to move the table, each person using both hands under their corner of the table. Have trainees observe the experiment and explain how it pertains to load oscillation.

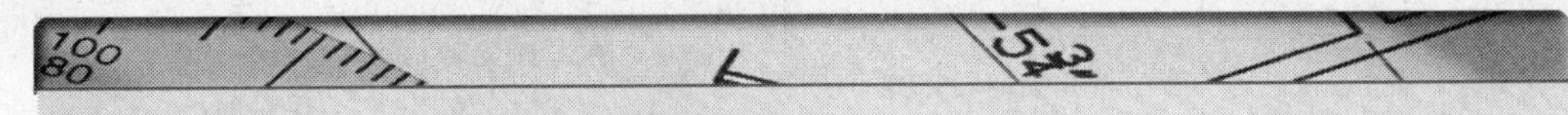

Load Oscillation

Because only three or four small points (the points where the hitches connect to the load) transfer the entire weight of the load, a load will oscillate (swing back and forth like a pendulum) as it is moved. Because of this load oscillation, at any given moment it is impossible to tell which slings are supporting the weight and which ones are providing stability.

To understand load oscillation, imagine the following example. A glass of water filled to the rim is placed on a table. Four people each place one finger under each corner of the table. They try to lift the table and walk across the room without spilling any of the water. The table under the glass oscillates as the people move the table, and a large amount of the water is probably spilled on the way across the room. This is because only a small amount of surface area of their fingers is in contact with the table, which causes the weight of the table to shift back and forth, or oscillate.

Now imagine that each person is allowed to use both hands on a corner of the table, thereby spreading the contact stress over a larger surface area. Now they could probably make it all the way across the room without spilling much, if any, of the water.

Go over the Review Questions for Section 3.0.0. Answer any questions trainees may have.

Have each trainee perform Performance Task 2 to your satisfaction. Fill out a Performance Profile Sheet for each trainee.

Have trainees review Sections 4.0.0–4.2.5 for the next session.

Review Questions

Section 3.0.0

1. The multiple-leg _____ hitch uses three or four single hitches to increase the stability of the load.
   a. vertical
   b. choker
   c. bridle
   d. basket

2. The _____ allows slings to be connected to the same load without using a spreader beam.
   a. pendulum
   b. choker hitch
   c. basket hitch
   d. bridle hitch

3. A _____ is ideal for lifting bundles of items such as pipes.
   a. choker hitch
   b. double-wrap choker hitch
   c. bull ring
   d. vertical hitch

4. The hitch made by wrapping the sling around the load and then passing one eye of the sling through a shackle is called the _____ hitch.
   a. basket
   b. bridle
   c. choker
   d. grommet

5. The hitch made by passing the sling around the load and placing both eyes in the hook is called the _____ hitch.
   a. basket
   b. bridle
   c. vertical
   d. choker

Instructor's Notes:

6.22 CORE CURRICULUM ◆ INTRODUCTORY CRAFT SKILLS

4.0.0 ◆ RIGGING HARDWARE

Rigging hardware is as crucial as the crane, the slings, or any specially designed lifting frame or hoisting device. If the hardware that connects the slings to either the load or the master link were to fail, the load would drop just as it would if the crane, hoist, or slings were to fail. Hardware failure related to improper attachment, selection, or inspection contributes to a great number of the deaths, serious injuries, and property damages in rigging accidents. The importance of hardware selection, maintenance, inspection, and proper use cannot be stressed enough. The regulations and requirements for rigging hardware are as stringent as those governing cranes and slings.

4.1.0 Shackles

A shackle is an item of rigging hardware used to attach an item to a load or to couple slings together. A shackle can be used to couple the end of a wire rope to eye fittings, hooks, or other connectors. It consists of a U-shaped body and a removable pin.

4.1.1 Shackle Design and Characteristics

Shackles used for overhead lifting should be made from forged steel, not cast steel. Quenched and tempered steel is the preferred material because of its increased toughness, but at a minimum, shackles must be made of drop- or hammer-forged steel.

All shackles must have a stamp that is clearly visible, showing the manufacturer's trademark, the size of the shackle (determined by the diameter of the shackle's body, not by the diameter of the pin), and the rated capacity of the shackle.

4.1.2 Types of Shackles

Shackles are available in two basic classes, identified by their shapes: anchor shackles and chain shackles (*Figure 36*). Both anchor and chain shackles have three basic types of pin designs, each with a unique function, as shown in *Figure 36*: the screw pin shackle, the round pin or straight pin shackle, and the safety shackle. The screw pin shackle design has become the most widely used type in general industry.

4.1.3 Specialty Shackles

Specialty shackles are available for specific applications where a standard shackle would not work well. For example, wide-body shackles (*Figure 37*) are for heavy lifting applications.

Synthetic web sling shackles (*Figure 38*) are designed with a wide throat opening and a wide bow that is contoured to provide a larger, nonslip surface area to accommodate the wider body of synthetic web slings.

4.1.4 Shackle Inspection and Rejection Criteria

Shackles, like any other type of hardware, must be inspected by the rigger before each use to make sure there are no defects that would make the shackle unsafe. Each lift may cause some degree of damage or may further reveal existing damage.

If any of the following conditions exists, a shackle must be removed from use:

- Bends, cracks, or other damage to the shackle body
- Incorrect shackle pin or improperly substituted pin
- Bent, broken, or loose shackle pin
- Damaged threads on threaded shackle pin
- Missing or illegible capacity and size markings

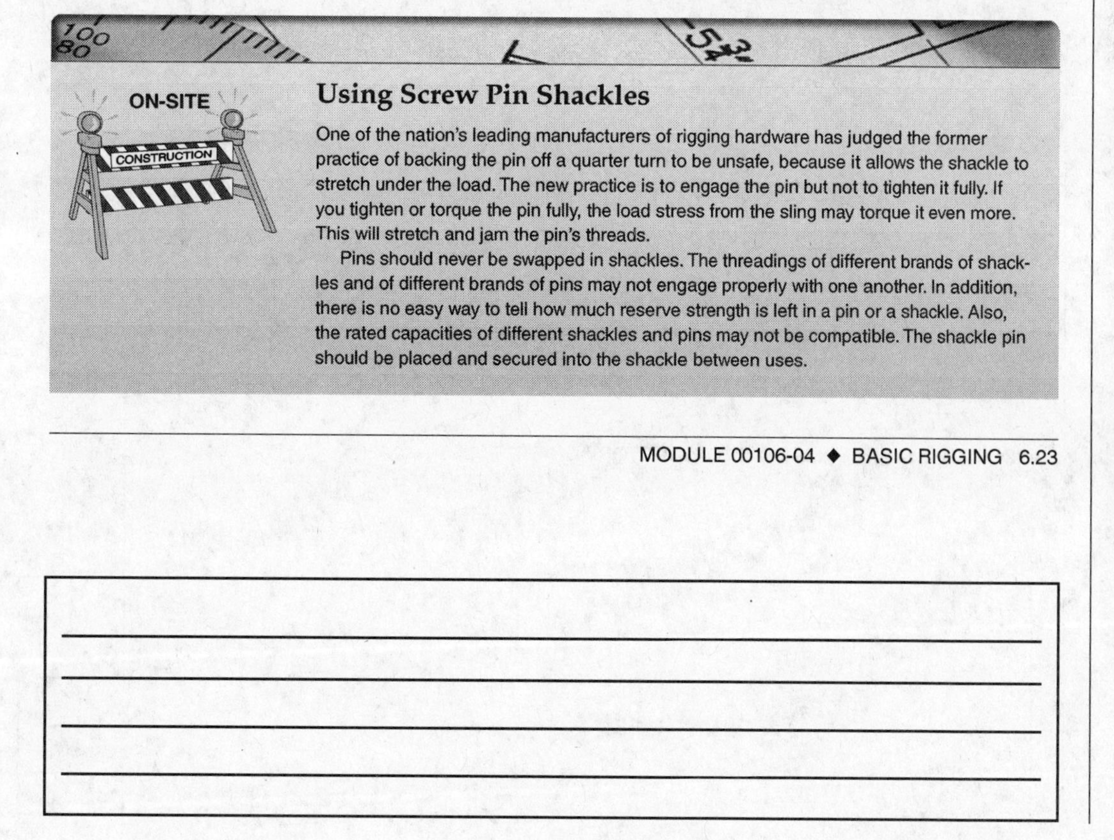

Using Screw Pin Shackles

One of the nation's leading manufacturers of rigging hardware has judged the former practice of backing the pin off a quarter turn to be unsafe, because it allows the shackle to stretch under the load. The new practice is to engage the pin but not to tighten it fully. If you tighten or torque the pin fully, the load stress from the sling may torque it even more. This will stretch and jam the pin's threads.

Pins should never be swapped in shackles. The threadings of different brands of shackles and of different brands of pins may not engage properly with one another. In addition, there is no easy way to tell how much reserve strength is left in a pin or a shackle. Also, the rated capacities of different shackles and pins may not be compatible. The shackle pin should be placed and secured into the shackle between uses.

Ensure that you have everything required for the demonstration, laboratories, and testing during this session.

Discuss the importance of hardware selection, maintenance, and inspection. Review the consequences of failing to use rigging hardware properly.

Explain how a shackle is used, and review the information that should be stamped on a shackle.

Refer to Figure 36. Review the different classes of shackles and the different types of pin designs.

Demonstrate how to properly engage a shackle pin. Explain why pins should no longer be backed off one-quarter of a turn.

Refer to Figures 37 and 38. Explain when it may be appropriate to use specialty shackles.

Bring in a variety of damaged shackles and pins. Have trainees inspect the shackles and determine the rejection criteria for each.

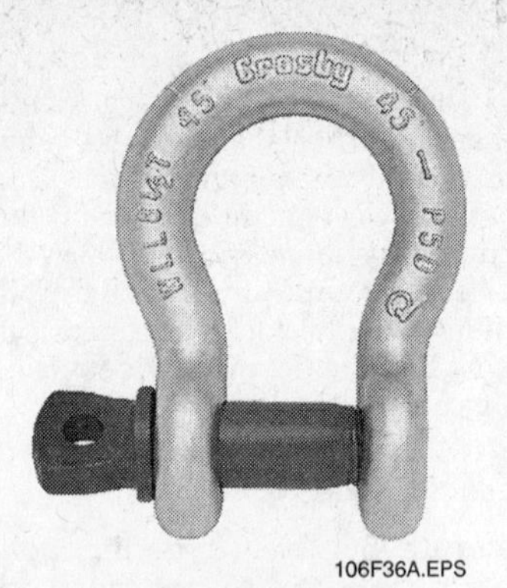

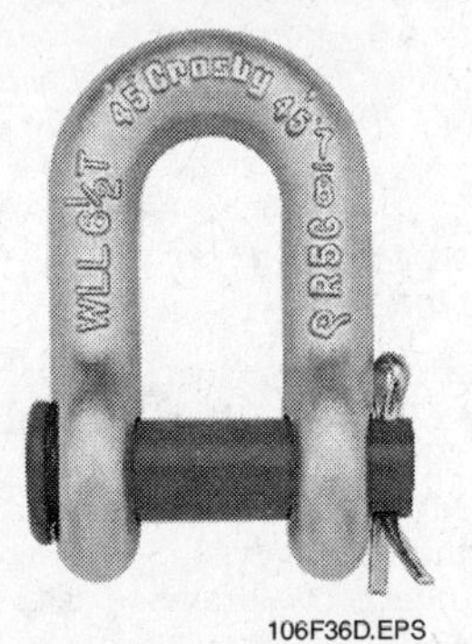

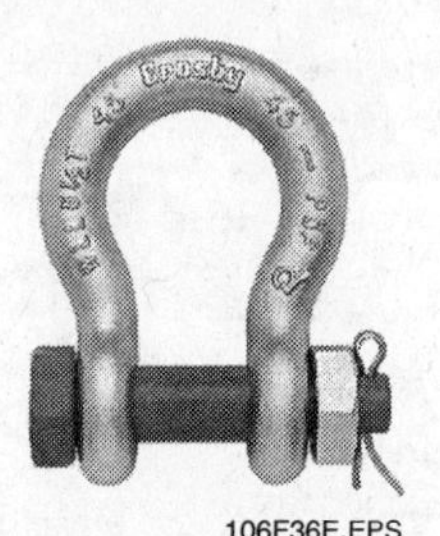

Figure 36 ◆ Shackles.

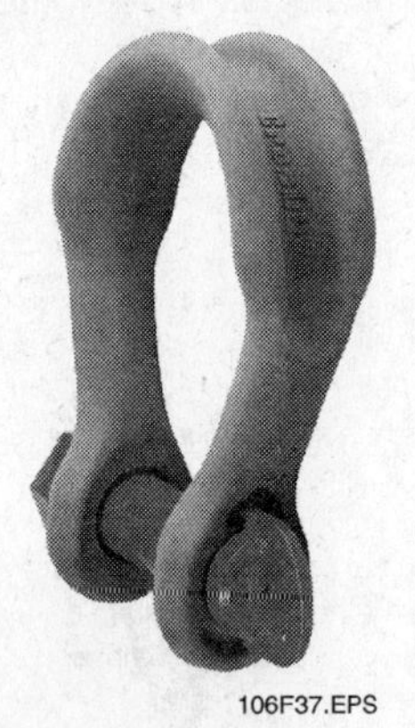

Figure 37 ◆ Wide-body shackle.

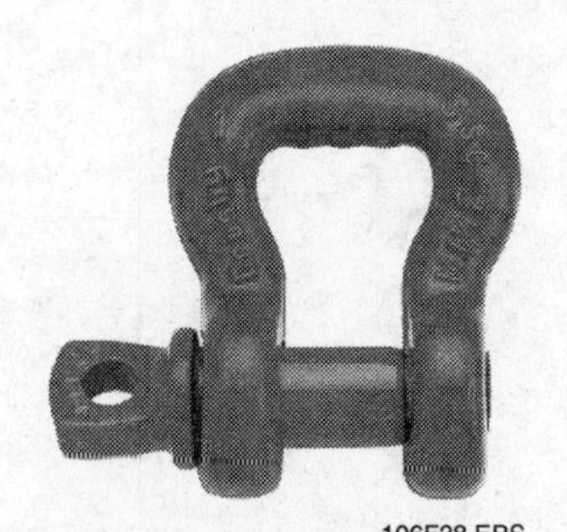

Figure 38 ◆ Synthetic web sling shackle.

Instructor's Notes:

6.24 CORE CURRICULUM ◆ INTRODUCTORY CRAFT SKILLS

4.2.0 Eyebolts

An **eyebolt** is an item of rigging hardware with a **threaded shank.** The eyebolt's shank end is attached directly to the load, and the eyebolt's eye end is used to attach a sling to the load.

4.2.1 Eyebolt Design and Characteristics

Eyebolts for overhead lifting should be made of drop- or hammer-forged steel. Eyebolts are available in three basic designs with several variations (*Figure 39*):

- Unshouldered eyebolts, designed for straight vertical pulls only
- Shouldered eyebolts, with a shoulder that is used to help support the eyebolt during pulls that are slightly angular
- Swivel eyebolts, also called hoist rings, designed for angular pulls from 0 to 90 degrees from the horizontal plane of the load

4.2.2 Unshouldered Eyebolts

Unshouldered eyebolts are designed for vertical pulls only, not angular pulls. In the installation of an unshouldered eyebolt, the sling must not pull perpendicular to the eyebolt (*Figure 40*). This will cause the eyebolt to bend and probably break off.

106F39A.EPS

UNSHOULDERED

SHOULDER

106F39B.EPS

106F39C.EPS

SHOULDERED

SWIVEL

Figure 39 ◆ Eyebolt variations.

4.2.3 Shouldered Eyebolts

Shouldered eyebolts can be used for some angular pulls. It is important to remember that the limit for angular pulling is 45 degrees from the horizontal plane of the load. Shouldered eyebolts lose capacity as the angle of the pull deviates from the vertical (*Figure 41*).

Refer to Figure 39. Explain how eyebolts are used, and compare their three basic designs.

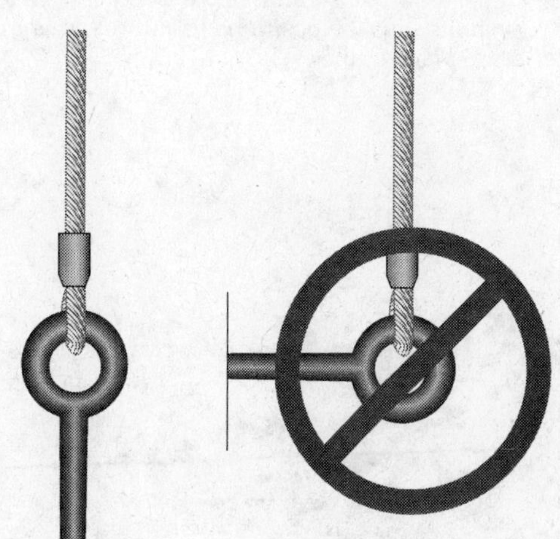

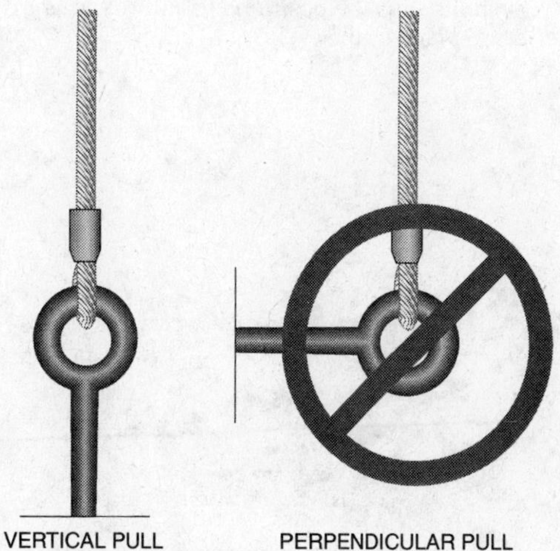

Figure 40 ◆ Installing an unshouldered eyebolt.

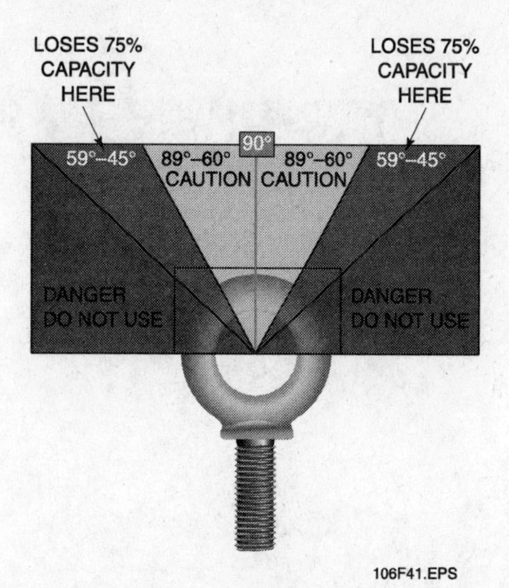

Figure 41 ◆ Effects of angular pull on shouldered eyebolts.

MODULE 00106-04 ◆ BASIC RIGGING 6.25

Discuss the installation requirements for unshouldered, shouldered, and swivel eyebolts. Ask whether trainees have any questions.

Most of the installation requirements for shouldered eyebolts are similar to the installation requirements for unshouldered eyebolts. However, the limit for angular pulling depends on proper installation. For example, in order for a shouldered eyebolt to support the load weight, it has to be installed so the eyebolt's shoulder is flush with the load surface (*Figure 42*). If the shoulder is not flush with the load surface, the eyebolt acts the same as an unshouldered eyebolt. Shouldered eyebolts must be positioned so the sling is in the same plane as the eyebolt.

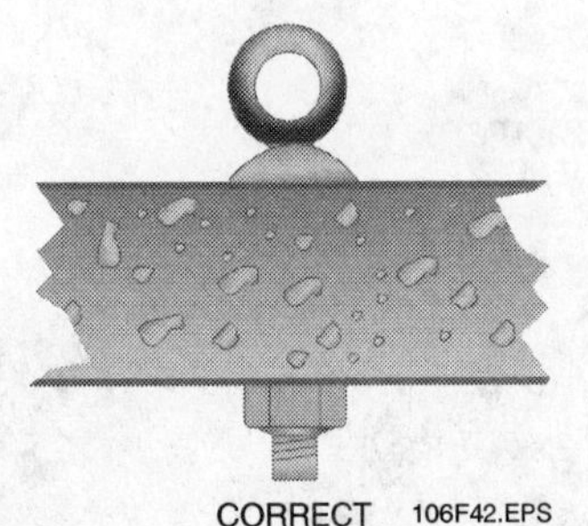

Figure 42 ◆ Proper installation of shouldered eyebolts.

4.2.4 Swivel Eyebolts

Swivel eyebolts are specially designed for angular pulls. They are available in through-bolt and machine-bolt styles. A through-bolt swivel eyebolt is designed so that the shank passes completely through the members it connects. A machine-bolt swivel eyebolt has a straight shank with a conventional head, such as a square or hexagonal head.

A swivel eyebolt's seating surface must be flush with the surface of the load or it will not function any better than an unshouldered eyebolt. Because the seating base of the swivel eyebolt is considerably larger than that of the unshouldered or shouldered eyebolt, it often will not install into the same area that a shouldered or unshouldered eyebolt would. Swivel eyebolts are self-aligning; their bases rotate 360 degrees, and the bail swivels 180 degrees. The rated capacity of a swivel eyebolt is based upon its 0- to 90-degree pull in any direction. There is no reduction factor for angular pulls (*Figure 43*).

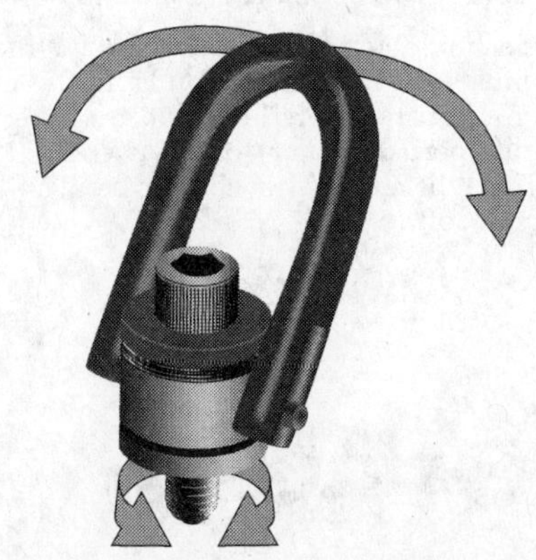

THREAD SIZE (inches)	SAMPLE RATED CAPACITY FROM 0°–180° (pounds)
½"	2,500
⅝"	4,000
¾"	5,000
⅞"	8,000
1"	10,000
1½"	24,000
2"	30,000

BOLT SPECIFICATION IS GRADE 8 ALLOY SOCKET HEAD CAP SCREW TO ASTM A 574.

106F43.EPS

Figure 43 ◆ Effects of angular pull on swivel eyebolts.

Instructor's Notes:

6.26 CORE CURRICULUM ◆ INTRODUCTORY CRAFT SKILLS

4.2.5 Eyebolt Inspection and Rejection Criteria

Eyebolts must be visually inspected before each use, and a qualified rigging inspector must inspect eyebolts once a year. They should be in like-new condition, although minor surface rust, superficial scraping, or minor nicks are permissible. Any degree of defect beyond these may be reason to remove eyebolts from service, even if the defect is not described in the rejection criteria below. Rejection criteria for eyebolts include the following (*Figure 44*):

- Scraping or abrasion that results in a noticeable loss of material
- A bent shank or distorted threads
- Stress cracks
- Rust or corrosion that cannot be easily removed with a wire brush
- Elongation from overload of the eye or bottle-necking of the shank
- Damaged or worn threads
- Deformation or twist from side loading
- Wear

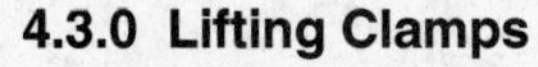

4.3.0 Lifting Clamps

Lifting clamps are used to move loads such as steel plates or concrete panels without the use of slings. Loads are moved one item at a time only.

4.3.1 Lifting Clamp Design and Characteristics

Lifting clamps are designed to bite down on a load and use the jaw tension to secure the load during transport (*Figure 45*). They are available in a wide variety of designs. Lifting clamps must be made of forged steel, and they must be stamped with their rated capacity.

Some clamps use the weight of the load to produce and sustain the clamping pressure. Others use an adjustable cam that is set and tightened to maintain a secure grip on the load.

Lifting clamps are designed to carry one item at a time, regardless of the capacity or jaw dimensions of the clamp or the thickness or weight of the item being lifted. Most clamps have a cam, which is adjustable, and a jaw, which is fixed. The item to be lifted is placed in the jaw of the clamp, which bites down onto the item as load stress is applied or as the adjustable cam is tightened.

In order for the clamp to hold an item securely, the cam and the jaw must bite or grip both sides of a single item. Placing more than one plate or sheet into the clamp prevents both the cam and the jaw from securing both sides of the plate or sheet.

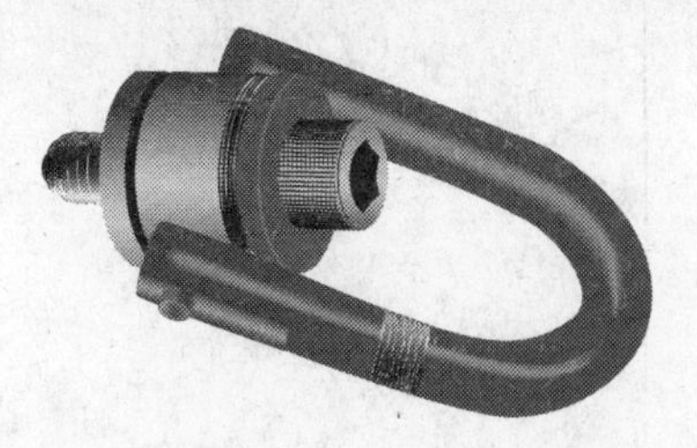

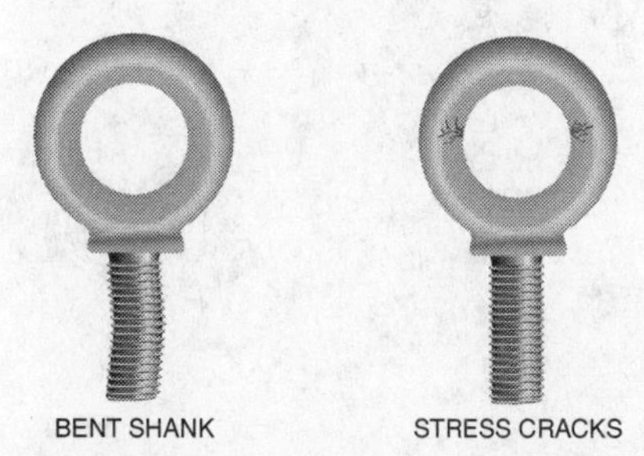

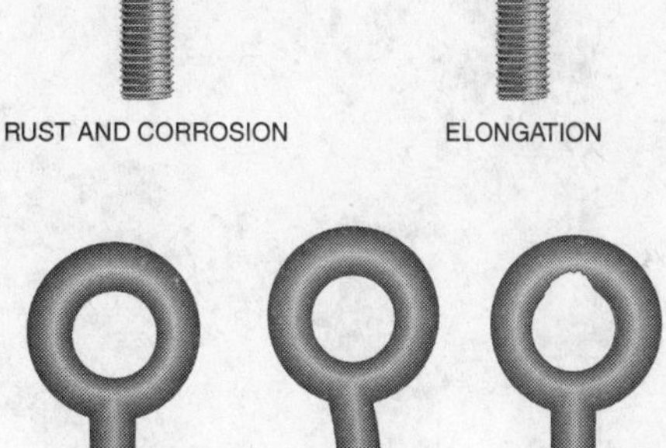

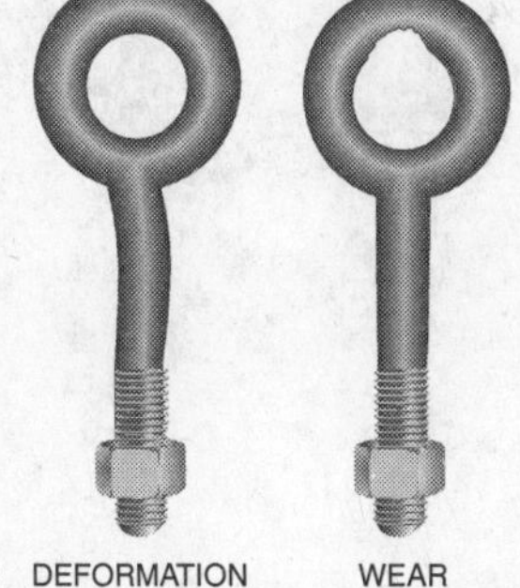

Figure 44 ◆ Rejection criteria for eyebolts.

Refer to Figure 44. Review the rejection criteria for eyebolts.

Set up stations with a variety of damaged and undamaged eyebolts. Ask trainees to identify the type of eyebolt and to note any rejection criteria.

Have trainees review Sections 4.3.0–4.4.2 for the next session.

Ensure that you have everything required for the laboratories and testing during this session.

Refer to Figure 45. Explain when and how lifting clamps are used to move loads.

Discuss the relationship between the jaw dimension of the clamp and the thickness or weight of the item being lifted. Explain how to ensure that the clamp is holding the item securely.

Refer to Figure 46. Review the types of lifting clamps and the lifting applications for which each is best suited.

Figure 45 ◆ Basic nonlocking clamp.

Clamp designs vary with the application, so it is important to match the design to the intended application. Lifting clamps must function smoothly and adjust with no mechanical difficulty. At least two lifting clamps should always be used when lifting an item. Lifting clamps must be placed to ensure that the load remains balanced. Loads lifted with both non-marring clamps and plate clamps must be lifted slowly and smoothly. These types of clamps are designed to grip the load gently yet securely. Their cams and jaws do not bite down too forcefully into the load, which might damage it. Therefore, you must be very careful to avoid any sudden movements that could jar the load out of these types of clamps.

4.3.2 Types of Lifting Clamps

Lifting clamps must be carefully selected for the specific lifting application, based on their design. There are four types of specialized lifting clamps and several variations in each type (*Figure 46*): linkage-type cam clamps, locking cam clamps, screw-adjusted cam clamps, and non-marring clamps.

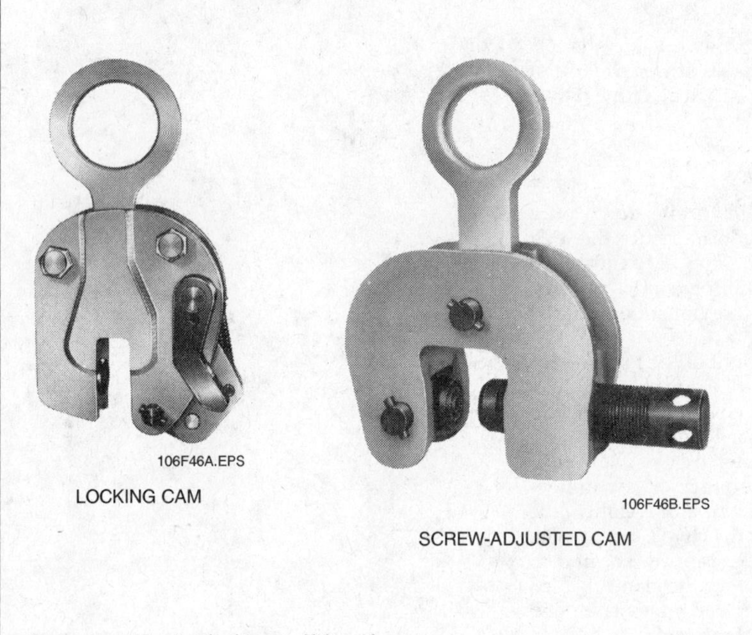

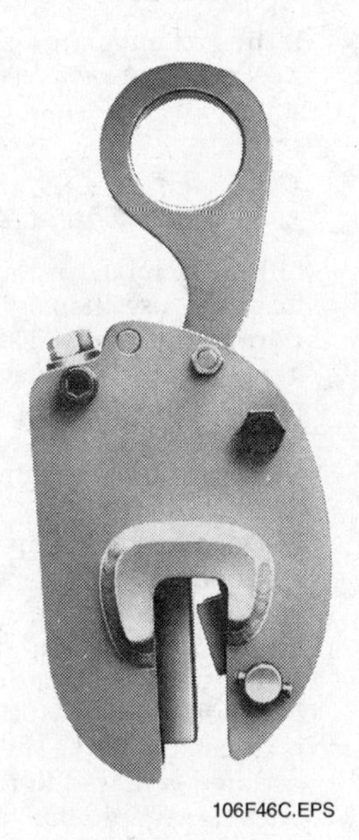

Figure 46 ◆ Nonstandard types of lifting clamps.

Instructor's Notes:

6.28 CORE CURRICULUM ◆ INTRODUCTORY CRAFT SKILLS

4.3.3 *Lifting Clamp Inspection and Rejection Criteria*

Lifting clamps, like any other type of rigging hardware, must be inspected by the rigger before every use to make sure there are no defects that would make the clamp unsafe. Each lift may cause some degree of damage or may further reveal existing damage.

If any of the following conditions exists, a clamp must be removed from use (*Figure 47*):

- Cracks
- Abrasion, wear, or scraping
- Any deformation or other impact damage to the shape that is detectable during a visual examination
- Excessive rust or corrosion, meaning rust or corrosion that cannot be removed easily with a wire brush
- Excessive wear of the teeth
- Heat damage
- Loose or damaged screws or rivets
- Worn springs

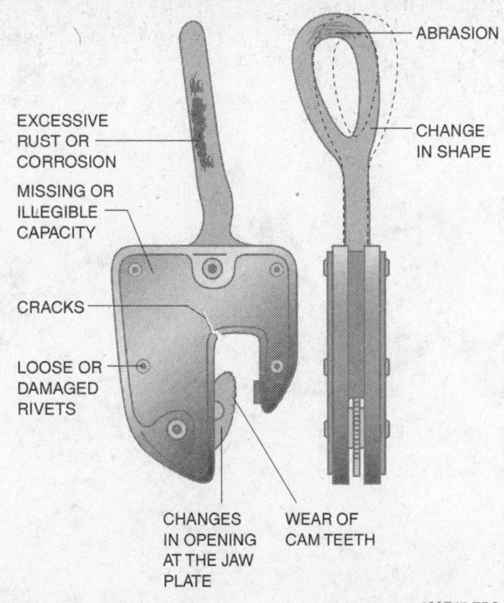

Figure 47 ◆ Rejection criteria for lifting clamps.

Show Transparency 12 (Figure 47). Review the rejection criteria for lifting clamps. Go over any questions trainees may have.

Have trainees practice using a wire brush to remove rust or corrosion from a lifting clamp. Ask a trainee to explain how to determine whether a rusty clamp should be rejected.

4.4.0 Rigging Hooks

A **rigging hook** is an item of rigging hardware used to attach a sling to a load. Although there are many classes of rigging hooks used for hoist hooks and rigging, there are only six basic types of rigging hooks (*Figure 48*):

- Eye hooks are the most common type of end fitting hook. These hooks may or may not have safety latches or gates.
- Sorting hooks, also called pipe hooks, are used to lift pipe sections or containers by inserting the hook into the load, thereby avoiding the need for shackles or additional hardware.
- Reverse eye hooks position the point of the hook perpendicular to the eye.
- Sliding choker hooks are installed onto the sling when it is made. The hooks, which can be positioned anywhere along the sling body, are used to secure the sling eye in a choker hitch. Sliding choker hooks are available for steel chain slings.
- Grab hooks are used on steel chain slings. These hooks fit securely in the chain link, so that choker hitches can be made and chains can be shortened.
- Shortening clutches, a more efficient version of the grab hooks, provide a secure grab of the shortened sling leg with no reduction in the capacity of the chain because the clutch fully supports the links.

Refer to Figure 48. Review the different types of rigging hooks, and explain how each is used.

MODULE 00106-04 ◆ BASIC RIGGING 6.29

Discuss the importance of ensuring that the safety latch is in good working condition.

Show Transparency 13 (Figure 49). Review the rejection criteria for rigging hooks.

106F48A.EPS

EYE HOOK

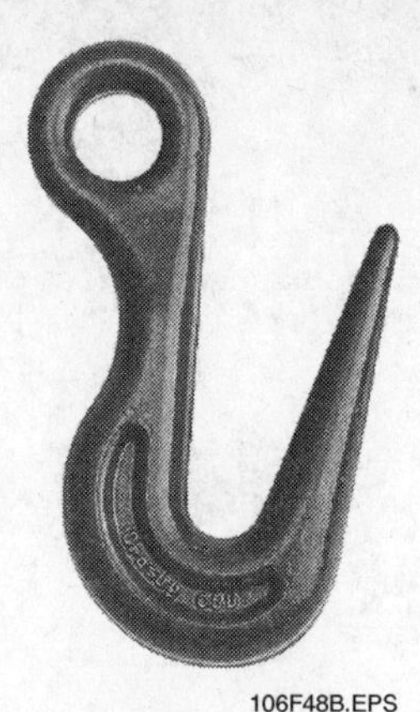

106F48B.EPS

SORTING HOOK

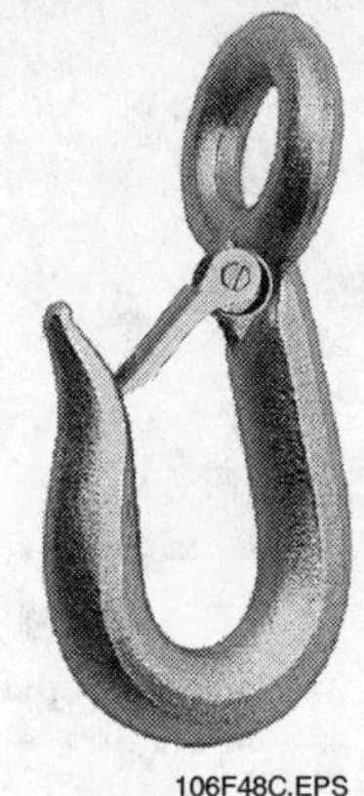

106F48C.EPS

ROUND REVERSE EYE HOOK

106F48D.EPS

SLIDING CHOKER HOOK

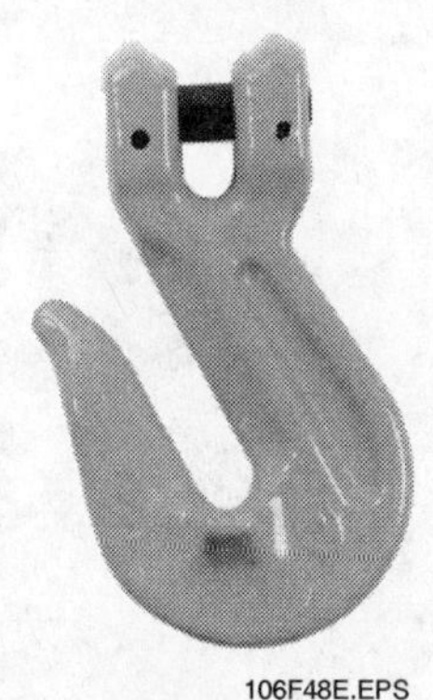

106F48E.EPS

GRAB HOOK

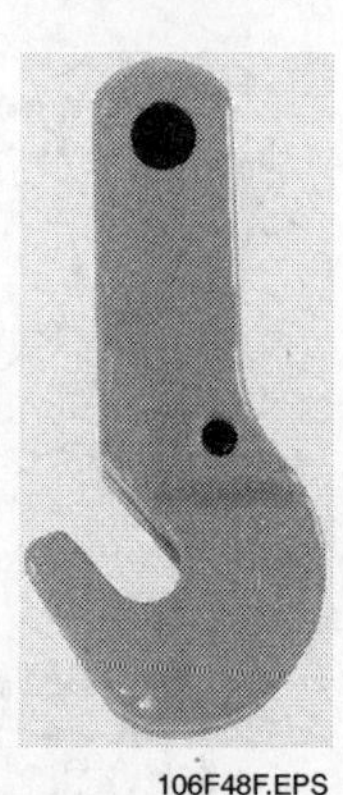

106F48F.EPS

SHORTENING CLUTCH

Figure 48 ◆ Rigging hooks.

4.4.1 Rigging Hook Design and Characteristics

Hooks used for rigging must be made of drop- or hammer-forged steel. All hooks used in rigging operations must meet the characteristics and performance criteria described in this module.

Safety latches or gates are installed in rigging hooks to prevent a sling from coming out of a hook or load when the sling is slackened. If a safety latch is installed in a rigging hook, the latch must be in good working condition. Damaged safety latches can be easily replaced. Report any damage you detect in a safety hook to your instructor or immediate supervisor.

4.4.2 Rigging Hook Inspection and Rejection Criteria

When hooks are installed as end fittings, they must be inspected along with the rest of the sling before each use. Slings with hook-type end fittings need to be removed from service for any of the following defects (*Figure 49*):

- Wear, scraping, or abrasion
- A broken or missing safety latch

Instructor's Notes:

6.30 CORE CURRICULUM ◆ INTRODUCTORY CRAFT SKILLS

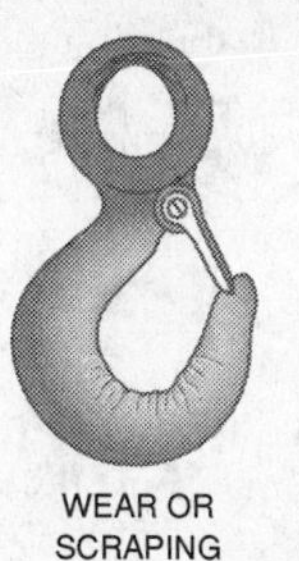
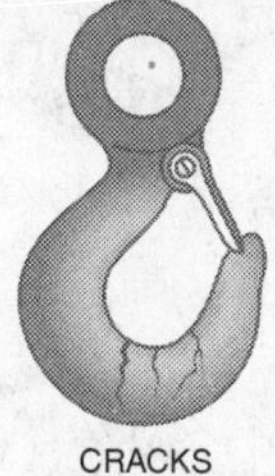
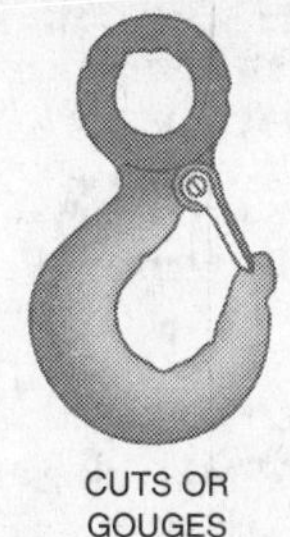

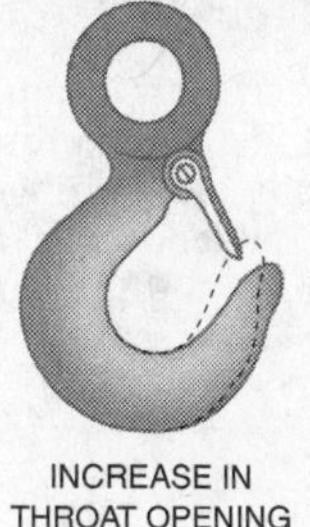

Figure 49 ◆ Rejection criteria for rigging hooks.

- Cracks
- Cuts, gouges, nicks, or chips
- Excessive rust or corrosion, meaning rust or corrosion that cannot be easily removed with a wire brush
- A twist of 10 degrees or more from the unbent plane of the hook
- An increase in the throat opening of the hook of 15 percent or more—easy to detect if the hook is equipped with a safety latch, because the latch will no longer bridge the throat opening
- An increase of 5 percent or more in the shank or overall elongation of the hook

Review Questions

Section 4.0.0

1. The piece of rigging hardware used to couple the end of a wire rope to eye fittings, hooks, or other connections is a(n) _____.
   a. eyebolt
   b. shackle
   c. clamp
   d. U-bolt

2. The type of bolt used only for straight vertical pulls is a(n) _____.
   a. U-bolt
   b. swivel eyebolt
   c. unshouldered eyebolt
   d. shouldered eyebolt

3. The piece of rigging hardware designed to move loads such as steel plates or concrete panels without the use of slings is called a(n) _____.
   a. shackle clamp
   b. C-clamp
   c. lifting clamp
   d. U-bolt clamp

4. Excessive rust or corrosion on an item of rigging hardware is rust or corrosion that cannot easily be removed with _____.
   a. a wet towel
   b. a chisel
   c. your hand
   d. a wire brush

5. _____ are installed in rigging hooks to prevent a sling from coming out of a hook when the sling is slackened.
   a. Safety latches
   b. Locking cam clamps
   c. Swivel eyebolts
   d. Unshouldered eyebolts

Set up stations with a variety of damaged and undamaged rigging hooks. Ask trainees to identify the type of rigging hooks and note any rejection criteria.

See the Teaching Tip for Sections 4.0.0–4.4.2 at the end of this module.

Go over the Review Questions for Section 4.0.0. Answer any questions trainees may have.

Have each trainee perform Performance Task 3 to your satisfaction. Fill out a Performance Profile Sheet for each trainee.

Have trainees review Sections 5.0.0–6.2.0 for the next session.

MODULE 00106-04 ◆ BASIC RIGGING 6.31

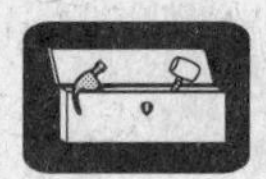

Ensure that you have everything required for the laboratory during this session.

Discuss the consequences of failing to accurately calculate the amount of stress placed on rigging components.

Show Transparency 14 (Figure 50). Explain how weight is distributed on a straight up and down sling.

Show Transparency 15 (Figure 51). Explain how side pull adds stress to the sling.

Show Transparencies 16 and 17 (Figures 52 and 53). Ask a trainee to explain what happens to side pull when a sling's angle is decreased. Review the relationship between sling angle and sling stress.

Refer to the Appendix. Review the steps for determining sling stress with the trainees. Ask trainees to determine the sling stress of hypothetical loads based on sling angles that you specify.

5.0.0 ◆ SLING STRESS

When **sling angle** decreases, **sling stress** increases. This is one of the most important facts you need to know to conduct rigging safely. It is essential that you understand this concept, because not understanding it—or misunderstanding it—could get you or someone else killed.

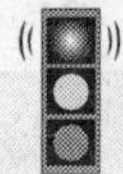

WARNING!

To rig and move a load safely, the rigger must understand the specific type and amount of stress placed on rigging components such as slings, hooks, and loads. If calculations are not made accurately, the rigging procedure could fail and damage the load and injure or kill the workers.

Here's a good way to understand sling stress. Imagine there are four loads, each weighing 2,000 pounds. If the first load (*Figure 50*) has two slings straight up and down (90 degrees), then each sling is itself holding a weight of 1,000 pounds.

The second load has slings at 60 degrees (*Figure 51*). Each sling still has the original 1,000 pounds of weight pulling down on it. However, because the slings are at 60-degree angles, the **side pull** now adds more stress to the sling. Side pull means the slings are being pulled in, or sideways, by the load's weight as well as down.

The third load (*Figure 52*) also weighs 2,000 pounds. Each sling still has the same 1,000 pounds of weight pulling down on it. But because the sling angle has decreased to 45 degrees, the amount of side pull has increased. This decrease in sling angle adds another 414 pounds of side pull, increasing the total weight the slings must hold.

The fourth load (*Figure 53*) also weighs 2,000 pounds. Each sling still has the same 1,000 pounds of weight pulling down on it. But here the sling angle has decreased to 30 degrees, which causes side pull to increase the sling stress to 1,000 extra pounds. This decrease in sling angle means each sling now has to hold up a total of 2,000 pounds.

As you can see from examples 1 through 4, as the sling angle decreases, the sling stress increases. By the time the sling angle gets to 30 degrees, the sling stress has doubled. Although the total weight of the load has stayed the same in each example, the stress on the sling has increased with each reduction in the sling's angle. Below 45 degrees, sling stress increases dramatically. The greater the sling stress, the greater the effect on the lift's safety.

See the Appendix in the back of this module for more information on calculating sling stress.

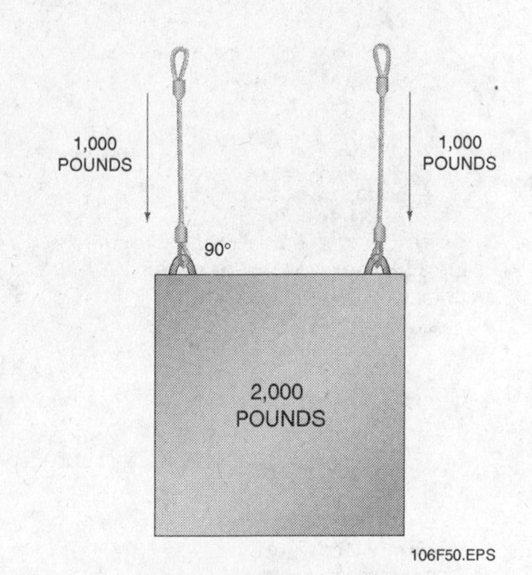

Figure 50 ◆ Sling stress example 1.

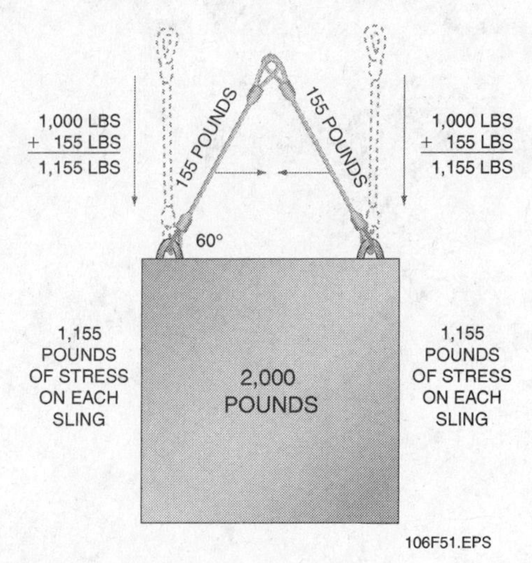

Figure 51 ◆ Sling stress example 2.

Instructor's Notes:

6.32 CORE CURRICULUM ◆ INTRODUCTORY CRAFT SKILLS

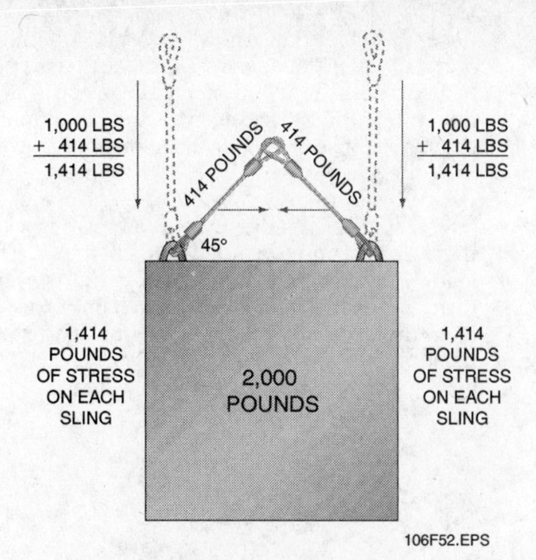

Figure 52 ◆ Sling stress example 3.

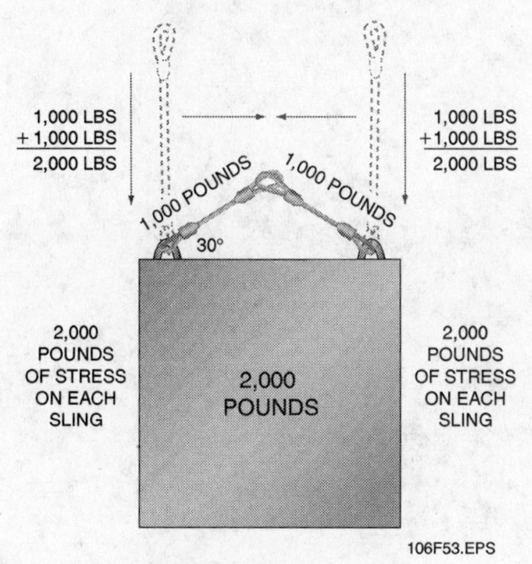

Figure 53 ◆ Sling stress example 4.

Review Questions

Section 5.0.0

1. When sling angle decreases, sling stress _____.
   a. stays the same
   b. decreases
   c. increases
   d. becomes zero

2. Side pull means the _____.
   a. load is being pulled from its side
   b. same as sling angle
   c. load stress has doubled
   d. slings are being pulled both in and down by the load's weight

3. A decrease in sling angle can add more side pull, increasing the total weight the slings must hold.
   a. True
   b. False

4. When a sling's angle goes from 90 degrees to 30 degrees, the _____.
   a. sling stress doubles
   b. sling stress becomes zero
   c. sling stress increases 30 pounds
   d. amount of side pull decreases

5. The greater the sling stress, _____.
   a. the less you have to worry about the lift's safety
   b. the greater the effect on the lift's safety
   c. the safer the lift
   d. the more slings you must use

Go over the Review Questions for Section 5.0.0. Answer any questions trainees may have.

Show Transparency 18 (Figure 54). Explain how a rigger can raise and lower a load using a block and tackle.

Refer to Figure 55. Compare the three types of chain hoists.

6.0.0 ◆ HOISTS

A hoist gives you a mechanical advantage for lifting a load, allowing you to move objects that you cannot lift manually. All hoists use a pulley system to transmit power and lift a load. Some hoists are mounted on trolleys and use electricity or compressed air for power. In this section, you will learn about a simple hoisting mechanism called a **block and tackle** and about a more complex hoisting mechanism called a chain hoist.

- *Block and tackle* – A block and tackle is a simple rope-and-pulley system used to lift loads (*Figure 54*). By using fixed pulleys and a wire rope attached to a load, the rigger can raise and lower the load by winding the wire rope around a drum.
- *Chain hoists* – Chain hoists may be operated manually or mechanically. There are three types of chain hoists (*Figure 55*): manual, electric, and pneumatic. Because electric and pneumatic chain hoists use mechanical, not manual, power, they are known as powered chain hoists. All chain hoists use a gear system to lift heavy loads (*Figure 56*). The gearing is coupled to a sprocket that has a chain with a hook attached to it. The load is hooked onto a chain and the gearing turns the sprocket, causing the chain to travel over the sprocket and moving the load. The hoist can be suspended by a hook connected to an appropriate anchorage point, or it can be suspended from a trolley system (*Figure 57*).

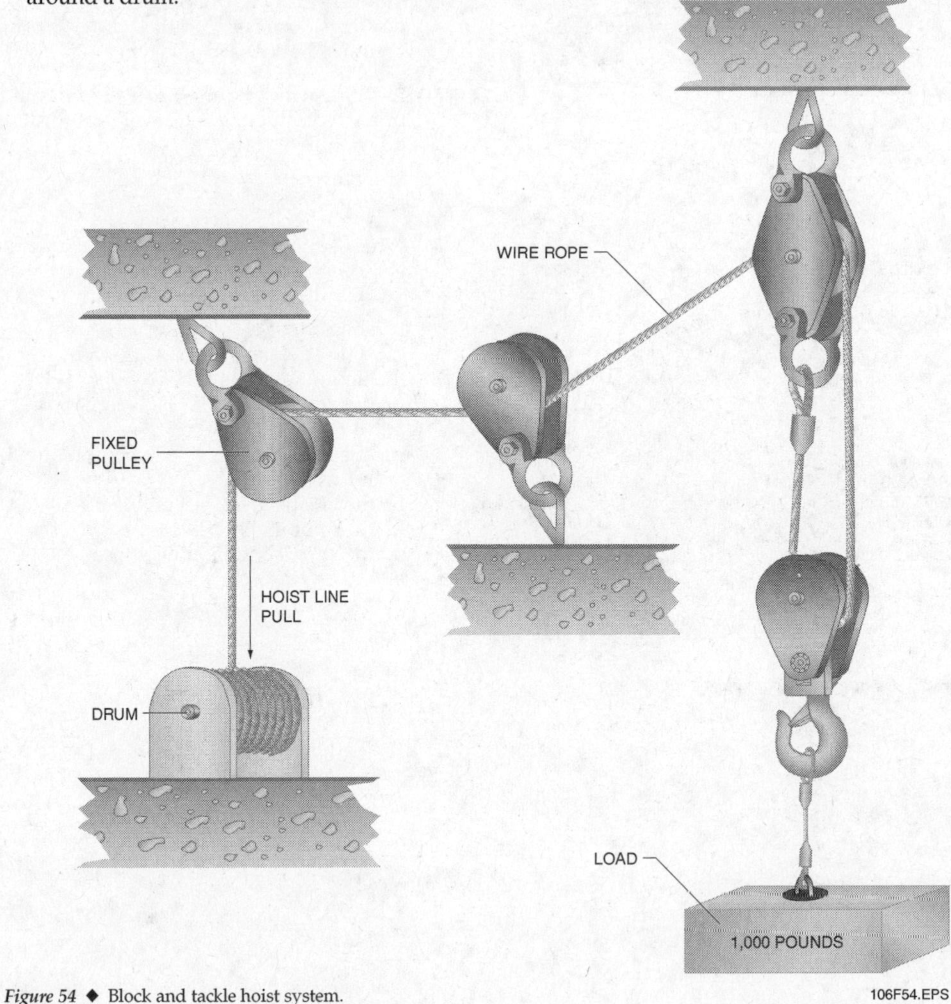

Figure 54 ◆ Block and tackle hoist system.

106F54.EPS

Instructor's Notes:

6.34 CORE CURRICULUM ◆ INTRODUCTORY CRAFT SKILLS

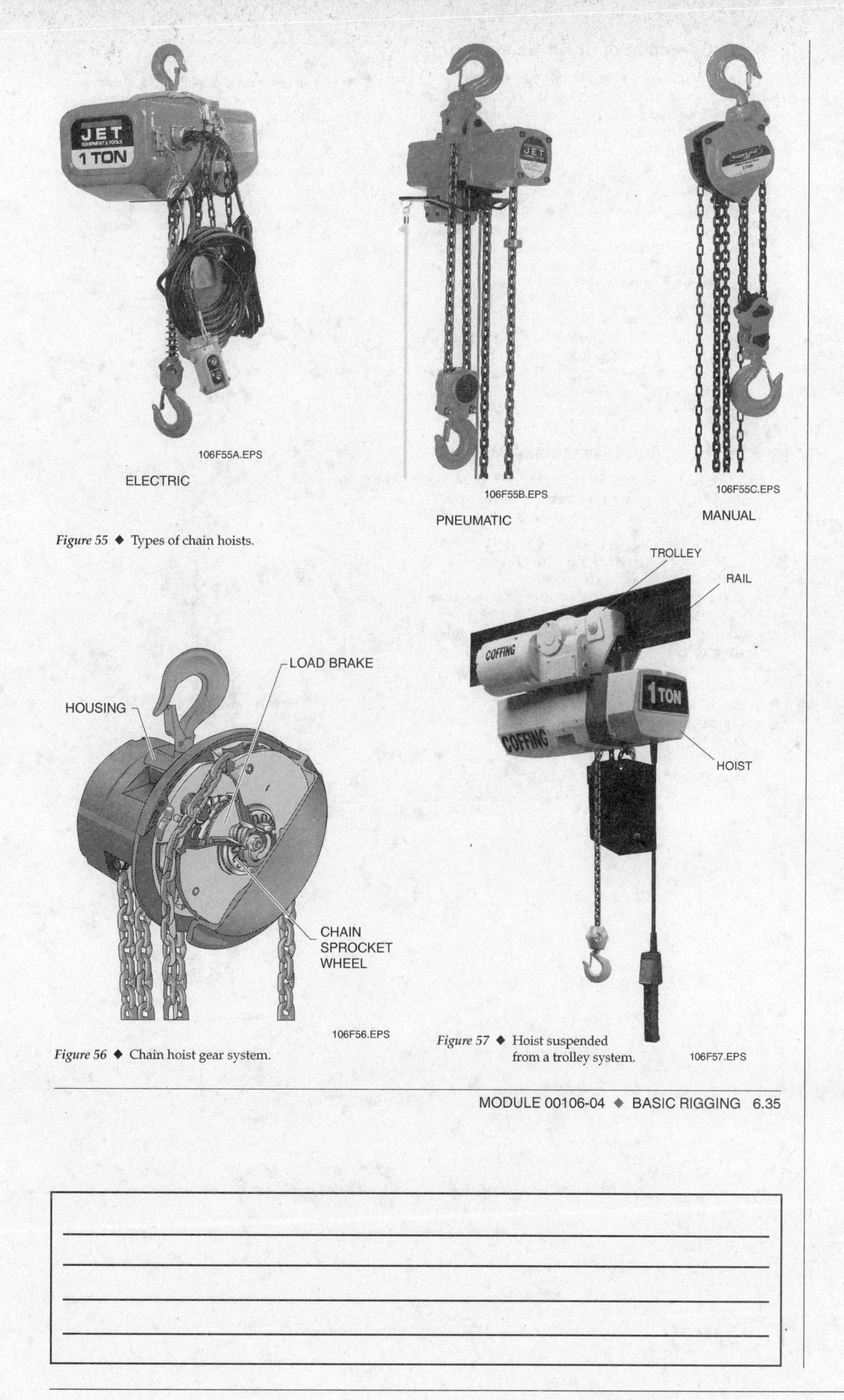

Figure 55 ◆ Types of chain hoists.

Figure 56 ◆ Chain hoist gear system.

Figure 57 ◆ Hoist suspended from a trolley system.

MODULE 00106-04 ◆ BASIC RIGGING 6.35

Review the basic operating procedures for hand and power chain hoists. Ask whether trainees have any questions.

Explain problems that can result from using a come-along for vertical overhead lifting.

See the Teaching Tip for Sections 6.0.0–6.2.0 at the end of this module.

6.1.0 Operation of Chain Hoists

Chain hoists are operated manually (by hand) or by electric or pneumatic power. In this section, you will learn some of the fundamental operating procedures for using both hand chain hoists and powered chain hoists.

- *Hand chain hoist* – To use a hand chain hoist, the rigger suspends the hoist above the load to be lifted, using either the suspension hook or the trolley mount. The rigger then attaches the hook to the load and pulls the hand chain drop to raise the load. The load will either rise or fall, depending on which side of the chain drop is pulled.
- *Powered chain hoist* – To use an electric or a pneumatic powered chain hoist, the rigger positions the chain hoist on the trolley above the load to be lifted, attaches the hook to the load, and uses the control pad to raise the load. Only qualified persons may use powered chain hoists.

6.2.0 Hoist Safety and Maintenance

In addition to the general safety rules you learned in the *Basic Safety* module, there are some specific safety rules for working with hoists.

- Always use the appropriate personal protective equipment when working with and around any lifting operations.
- Make sure that the load is properly balanced and attached correctly to the hoist before you attempt the lift. Unbalanced loads can slide or shift, causing the hoist to fail.
- Keep gears, chains, and ropes clean. Improper maintenance can shorten the working life of chains and ropes.
- Lubricate gears periodically to keep the wheels from freezing up.
- Never perform a lift of any size without proper supervision.

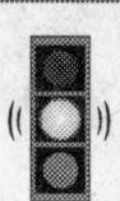

CAUTION

Never use a come-along for vertical overhead lifting. Use a come-along only to move loads horizontally over the ground. Be careful not to confuse a come-along with a ratchet lever hoist. Ratchet lever hoists have both a friction-type holding brake and a ratchet-and-pawl load control brake. Come-alongs have only a spring-load ratchet that holds the pawl in place. If the ratchet and pawl fail, the overhead load falls.

106SA02A.EPS

COME-ALONG

106SA02B.EPS

RATCHET LEVER HOIST

Instructor's Notes:

6.36 CORE CURRICULUM ◆ INTRODUCTORY CRAFT SKILLS

Review Questions

Section 6.0.0

1. All hoists use _____ to transmit power and lift a load.
   a. pneumatic gears
   b. a pulley system
   c. powered chains
   d. trolley mounts

2. The three types of chain hoists are _____.
   a. block, tackle, and pulley
   b. hook, gear, and sprocket
   c. anchor, suspension, and ratchet-and-pawl
   d. manual, electric, and pneumatic

3. To use a hand chain hoist, the rigger _____ to raise and lower the load.
   a. pulls the hand chain drop
   b. winds the steel rope
   c. uses the control panel
   d. ratchets the lever handle

4. Powered chain hoists must be used only by _____.
   a. those who can't lift loads manually
   b. mechanical engineers
   c. qualified persons
   d. rigging trainees

5. Make sure the load is _____ before you attempt the lift.
   a. heavy enough
   b. properly balanced
   c. detached from the hoist
   d. connected to an anchorage point

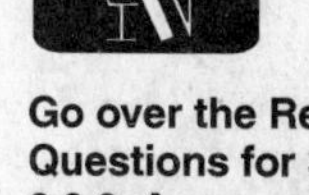

Go over the Review Questions for Section 6.0.0. Answer any questions trainees may have.

Have trainees review Sections 7.0.0–7.3.3 for the next session.

Ensure that you have everything required for the laboratory and testing during this session.

7.0.0 ◆ RIGGING OPERATIONS AND PRACTICES

In this module, you have learned some of the basics about tools used to perform rigging operations. You have also learned some of the basic rigging practices used to conduct lifts. Now you will learn how to apply this basic knowledge toward participating in a safe and efficient rigging operation.

7.1.0 Rated Capacity

All slings, hardware, and rigging devices are required to be clearly marked with their rated capacity. Paper tags cannot be used. Rated capacity means the same thing as the working load limit (WLL) or a safe working load (SWL). The rated capacity of slings and rigging hardware must be determined to make sure that loads are safely and effectively lifted and transported.

> **WARNING!**
> Under no circumstances should you ever exceed the rated capacity! Overloading may result in catastrophic failure.

7.2.0 Sling Attachment

Before attaching slings to a load, riggers must remember several important points. The rigger must select the most appropriate type of sling (synthetic, alloy steel, or wire rope) for the load to be lifted and inspect the sling for damage. Once the type of sling is chosen, the rigger determines the best rigging configuration or hitch (vertical [bridle], choker, or basket) to lift the load. Knowing the type of sling and the best hitch, the rigger then chooses the appropriate rigging hardware to connect the sling to the load. To make the safest selection, the rigger must know the total weight of the load, the angle at which the sling(s) will attach to the load, and the total sling stress applied during the lift. Only then can a sling be rigged safely to a load.

When using slings, follow these safety guidelines:

- Never try to shorten a sling by wrapping it around the hoist hook before attaching the eye to the hook.
- Never try to shorten the legs of a sling by knotting, twisting, or wrapping the slings around one another.
- Never try to shorten a chain sling by bolting or wiring the links together.

Explain how slings, hardware, and rigging devices are marked with their rated capacity.

Discuss the consequences of exceeding rated capacity.

Review the safety guidelines for using slings.

Explain problems that can result if the incorrect length sling is used.

Refer to Figure 58. Ask a trainee to explain why a pin should never be overtightened.

Show Transparency 19 (Figure 59). Review the guidelines for aligning and installing eyebolts. Go over any questions trainees may have.

> **WARNING!**
> Use the correct length sling for each lift. Trying to shorten a sling may cause an accident.

- Make sure all personnel are clear of the load before you take full load strain on the slings.
- Never try to adjust the slings while a strain is being taken on the load.
- Make sure all personnel keep their hands away from the slings and the load during hoisting.
- Use sling softeners whenever possible. Sling softeners protect the sling from abrasion, cuts, heat, and chemical damage. Whether the softeners or pads are manufactured or made in the field, using them regularly will considerably extend the life of the sling. If no standard sleeves are available, use rubber belting or sections of old slings. Softeners prevent kinking in wire rope. Corner buffers are specially made for this purpose, but if they are not available, the rigger can use whatever is handy to protect the rope. Wood, old web slings, and factory-made buffers will provide some protection. Softeners for chain slings protect the links contacting the corners of the load and keep the chain links from scraping or crushing the load itself.

7.3.0 Hardware Attachment

Most rigging hitches require some type of hardware to attach the slings to a piece of equipment or crane. You have learned about shackles, eyebolts, and hooks. This section summarizes the basics of how this equipment is attached to the load.

7.3.1 Shackles

Shackles have several installation requirements to prevent damage or failure during the lift. For instance, the shackle must remain in line with the sling so that the shackle does not become side-loaded and get pulled apart by the load stress. Also, when a screw pin type of shackle is installed, the pin should not be overtightened or it will stretch the threads. Overtightening makes the shackle difficult to remove and may damage the shackle so that it has to be removed from service (*Figure 58*).

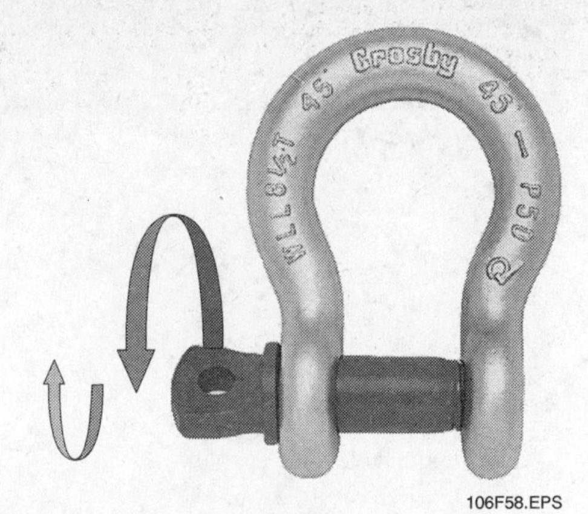

Figure 58 ◆ Don't overtighten the pin.

7.3.2 Eyebolts

Eyebolts must be safely and properly installed. Use the following guidelines:

Eyebolts must be installed so that the plane of the eye is in direct line with the plane of the sling when the sling is positioned at an angle other than vertical (*Figure 59*). Otherwise, the eyebolt may fold over under the load stress, which could severely deform the eyebolt or even lead to a fracture or a complete failure of the eye.

Eyebolts with shoulders or swivel bases must be installed so that the shoulder or base is flush with the surface of the load and makes positive contact around the entire circumference of the shoulder or base.

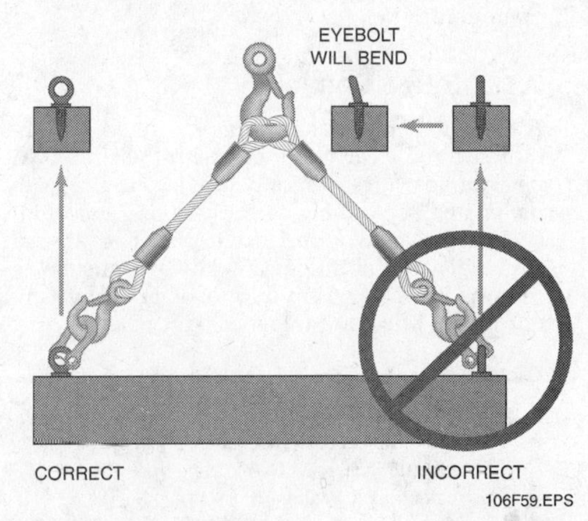

Figure 59 ◆ Eyebolt orientation.

Instructor's Notes:

6.38 CORE CURRICULUM ◆ INTRODUCTORY CRAFT SKILLS

CAUTION

Use a pair of shackles in the eyebolts instead of threading a sling through hardware around a load. Threading the sling through eyebolts as shown and applying stress will cause the eyebolts to pull toward each other and shear off the bolts.

CAUTION

The severe side stress of low sling angles may cause a load to buckle into the center because of side pull. Use extreme caution when attaching a sling to swivel eyebolts, even though the swivel eyebolt does not have restrictions on sling angles.

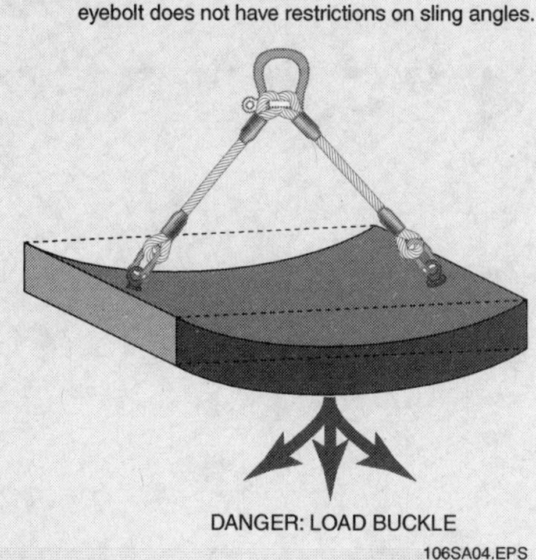

7.3.3 Hooks

Hooks must be carefully attached to the load to prevent binding of the hook. When a hook becomes bound in a **pad eye** or eyebolt, it can become point-loaded. A point-loaded hook will easily stretch open and slip out of the load. When a hook becomes point-loaded, the capacity of the hook decreases drastically the closer the load moves toward the point of the hook (*Figure 60*). Using a shackle instead of placing the hook directly into the pad eye or eyebolt will prevent this from happening.

7.4.0 Load Control

Safe and efficient load control (*Figure 61*) involves communication, the use of physical load control techniques, the safe handling of loads, taking landing zone precautions, and following sling disconnection and removal practices.

7.4.1 American National Standards Institute (ANSI) Hand Signals

Load navigation is accomplished by using common signals—either verbal signals given by radio or hand signals—to direct a load's movements.

Discuss the consequences of threading a sling through the eyebolts.

Emphasize that severe side stress can cause a load to buckle.

Show Transparency 20 (Figure 60). Explain how to prevent a hook from becoming point-loaded.

Bring in a variety of shackles, eyebolts, and hooks. Ask several trainees to simulate the techniques for connecting hitches to a load. Have the other trainees observe their techniques and comment on what is done correctly or incorrectly.

Have each trainee perform Performance Task 4 to your satisfaction. Fill out a Performance Profile Sheet for each trainee.

Have trainees review Sections 7.4.0–7.4.3 for the next session.

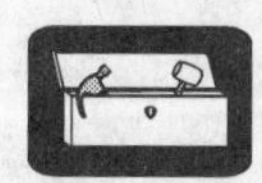

Ensure that you have everything required for the demonstration, laboratory, and testing during this session.

Discuss the elements of a pre-lift safety check for efficient load control.

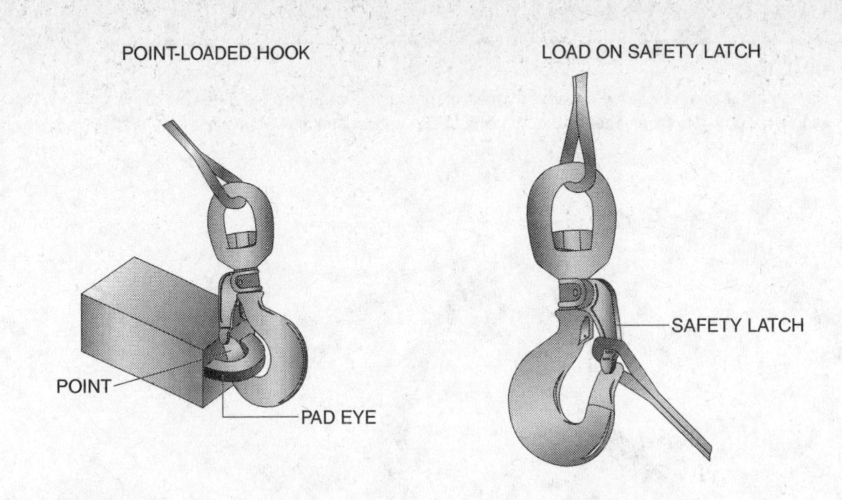

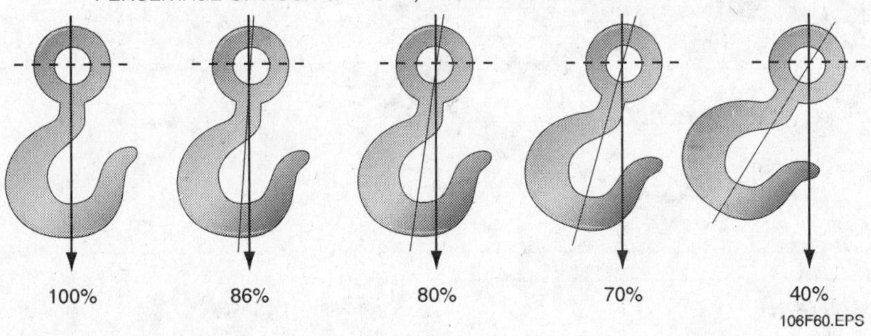

Figure 60 ◆ Point-loaded capacity reductions.

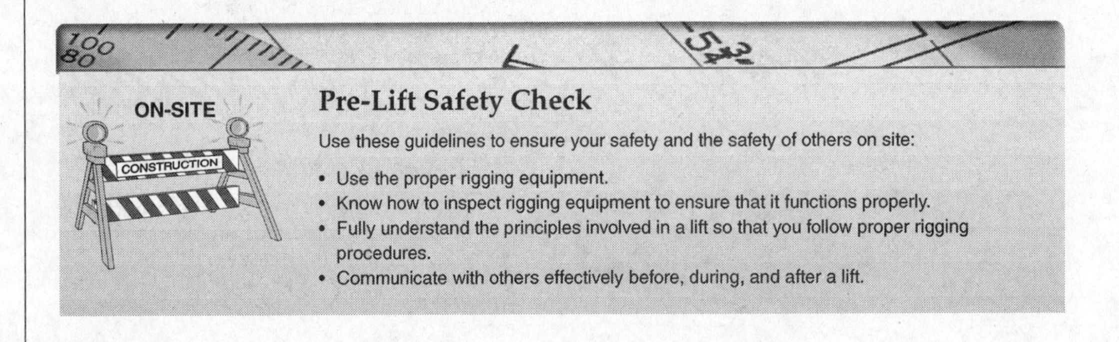

Pre-Lift Safety Check

Use these guidelines to ensure your safety and the safety of others on site:

- Use the proper rigging equipment.
- Know how to inspect rigging equipment to ensure that it functions properly.
- Fully understand the principles involved in a lift so that you follow proper rigging procedures.
- Communicate with others effectively before, during, and after a lift.

Instructor's Notes:

6.40 CORE CURRICULUM ◆ INTRODUCTORY CRAFT SKILLS

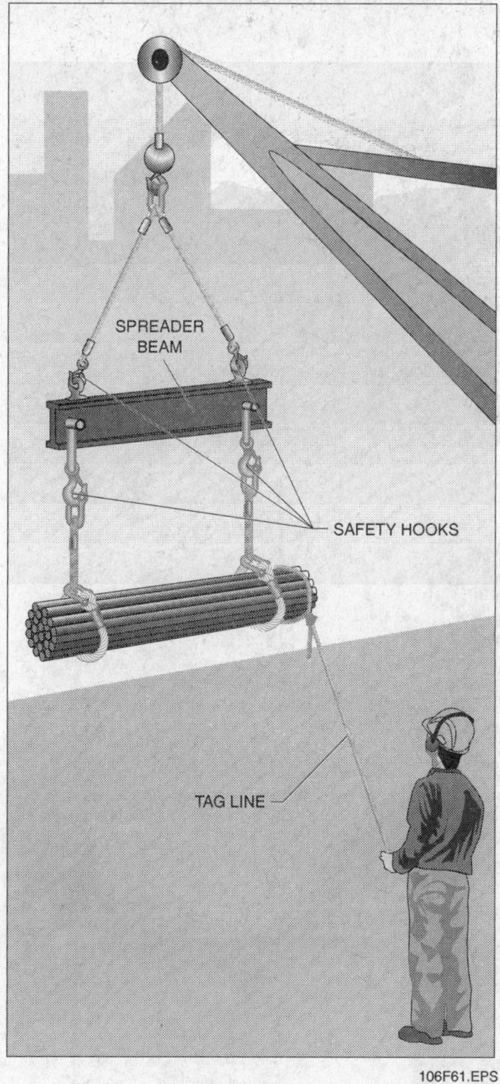

Figure 61 ◆ Safe and efficient load control.

Using hand signals to communicate load navigation directions is an integral element in the safe transport and control of all loads. Established hand signals are used for communicating load navigation directions from the signal person to the crane operator. The hand signals used in rigging operations have been developed and standardized by the American National Standards Institute, or ANSI. **ANSI hand signals** have been developed and standardized for use with mobile cranes (*Figure 62*) and for use with overhead cranes (*Figure 63*). Standard hand signals, if used correctly and clearly understood by both the person signaling and the crane operator, provide clear and unmistakable instructions for the various combinations of movements required for the crane to transport the load safely.

7.4.2 Load Path, Load Control, and Tag Lines

Physical control of the load beyond the ability of the crane may be required. Tag lines are used to limit the unwanted or inadvertent movement of the load as it reacts to the motion of the crane or to allow the controlled rotation of the load for positioning. Tag lines are attached after the rigger verifies that the load is balanced.

Show Transparency 21 (Figure 62, 1 of 2). Discuss the necessity for standardized hand signals. Review the signals used for mobile crane operation.

Show Transparency 22 (Figure 62, 2 of 2). Ask a trainee to explain how clear hand signals effectively communicate instructions.

Demonstrate a variety of potentially confusing hand signals. Discuss problems that can result if standardized hand signals are not used.

Ask a trainee to demonstrate mobile crane hand signals, and ask a second trainee to demonstrate overhead crane hand signals. Have trainees observe the hand signals and identify the instruction each communicates.

Explain how tag lines are used to limit unwanted movement of a load. Have a trainee explain when tag lines should be used. Ask whether trainees have any questions.

MODULE 00106-04 ◆ BASIC RIGGING 6.41

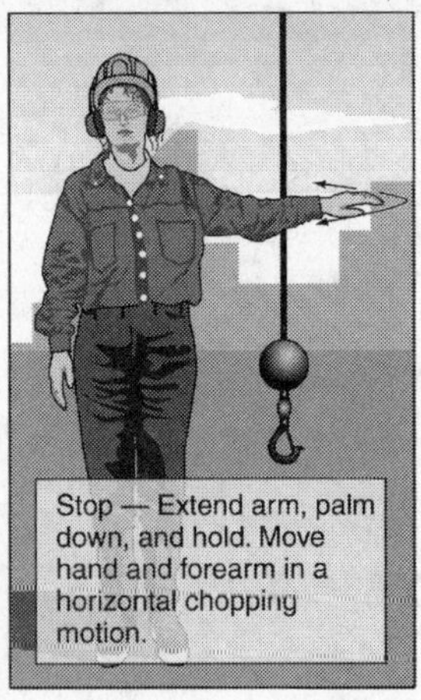

106F62A.EPS

Figure 62 ◆ Mobile crane hand signals. (1 of 2)

6.42 CORE CURRICULUM ◆ INTRODUCTORY CRAFT SKILLS

Instructor's Notes:

Figure 62 ◆ Mobile crane hand signals. (2 of 2)

MODULE 00106-04 ◆ BASIC RIGGING 6.43

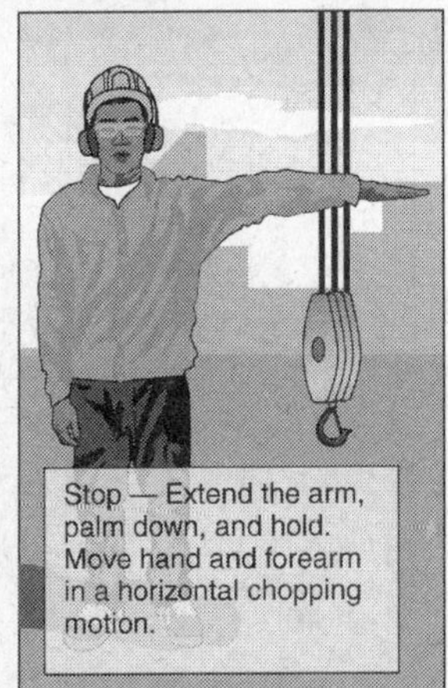

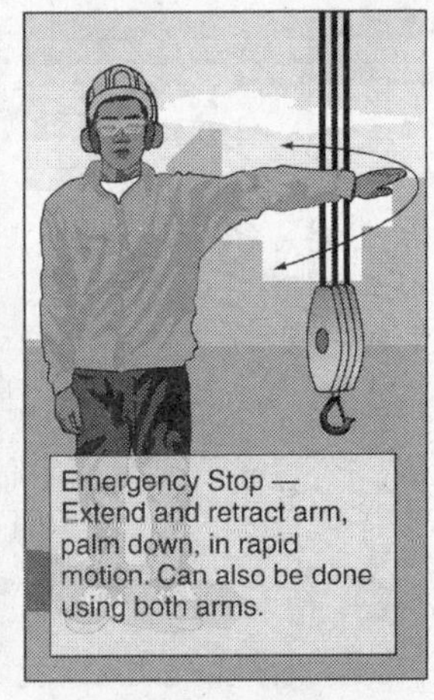

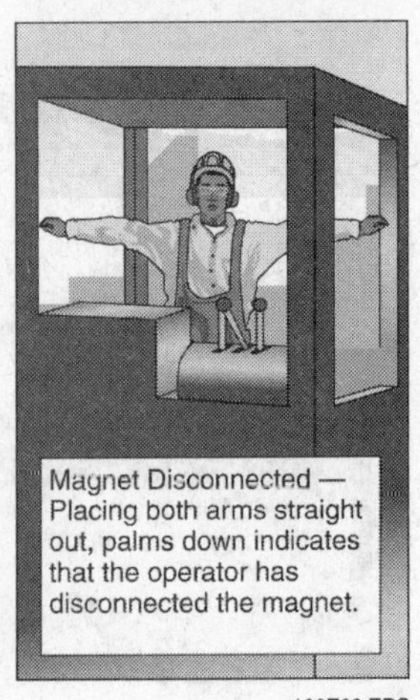

106F63.EPS

Figure 63 ◆ Overhead crane hand signals.

Instructor's Notes:

6.44 CORE CURRICULUM ◆ INTRODUCTORY CRAFT SKILLS

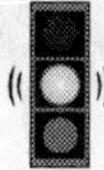

CAUTION

Keep the swing path or load path clear of personnel and obstructions. Crane or hoist operators should watch the load when it is in motion. Never work under a suspended load. For more information about swing path safety, please refer to OSHA's *CFR 1926*.

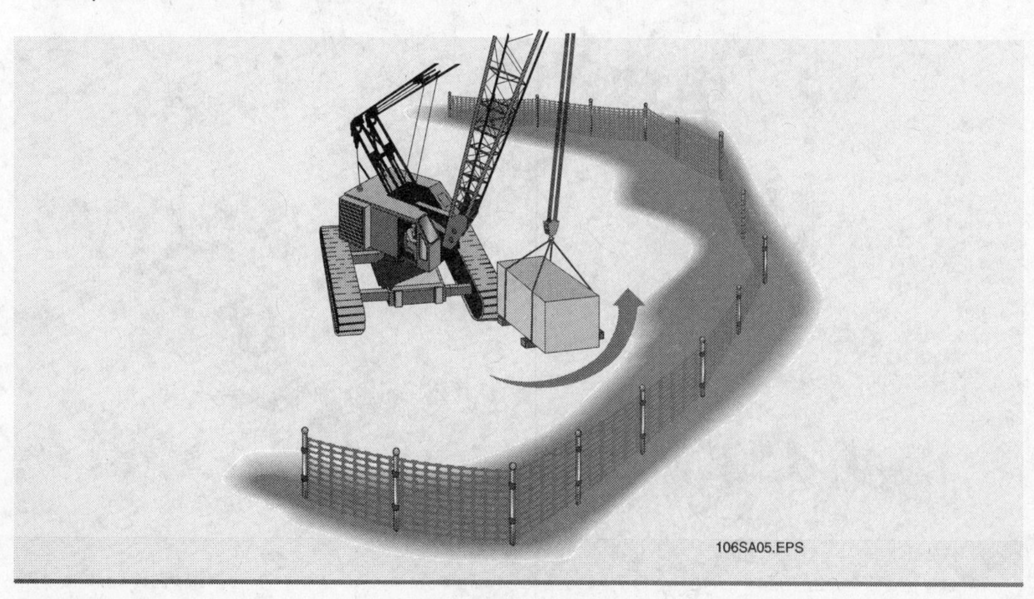

106SA05.EPS

7.4.3 Load-Handling Safety

The safe and effective control of the load involves the rigger's strict observance of load-handling safety requirements:

- Keep the front and rear swing paths of the crane clear throughout the lift (*Figure 64*). Most people watch the load when it is in motion, and this prevents them from seeing the back end of the crane coming around. Using safety flags and roping off the area will help prevent accident and injury.
- Keep the landing zone clear of personnel other than the tag line tenders.
- Be sure that the necessary blocking and **cribbing** for the load are in place and set before you position the load for landing. Lowering the load just above the landing zone and then placing the cribbing and blocking is dangerous because the riggers must pass under the load. The layout of the cribbing can be completed in the landing zone before you set the load. Blocking of the load may have to be done after the load is set. In this case, do not take the load stress off the sling until the blocking is set and secured.
- Do not attempt to position the load onto the cribbing by manhandling it.
- Account for the speed and direction of any wind. To see which way the wind is blowing, look at the wind sock.

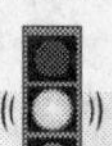

CAUTION

Landing the load and disconnecting the rigging are the final steps in any lift. Many rigging accidents and injuries occur at this stage because many riggers mistakenly consider the lift completed just before landing. They do not give their complete attention to the safe landing requirement.

CAUTION

Serious back injuries have resulted from personnel attempting to manually force a load into position or stop even a slight swing of the load. Injuries resulting in crushed hands and other limbs can easily occur should the load shift or swing during the final stages of landing. If the load must be landed with precise control, tag poles, additional tag lines, or hoisting devices should be used.

Refer to OSHA's *CFR 1926*. Stress the importance of keeping the swing and load paths clear.

Show Transparency 23 (Figure 64). Explain how to keep the rear swing path clear throughout a lift.

Explain when and how the necessary blocking and cribbing should be completed.

Explain problems that can result during the final steps of a lift.

Discuss the consequences of manually forcing a load into position.

See the Teaching Tip for Sections 7.0.0–7.4.3 at the end of this module.

Go over the Review Questions for Section 7.0.0. Answer any questions trainees may have.

Figure 64 ◆ Rear swing path.

106F64.EPS

Review Questions

Section 7.0.0

1. A safe guideline for using a sling is to _____.
   a. shorten a sling by wrapping it around the hoist hook
   b. select the correct length of sling for the lift
   c. adjust the sling legs by knotting the length
   d. wire the links of a chain together to shorten the leg

2. Softeners for chain slings can _____.
   a. protect the load
   b. make the landing gentler
   c. make the load easier to handle
   d. decrease sling stress

3. A point-loaded hook's capacity _____ the closer the load moves toward the point of the hook.
   a. decreases drastically
   b. increases drastically
   c. stays the same
   d. decreases slightly

4. Additional physical control of a load may be achieved by using _____.
   a. manual labor
   b. sling angles
   c. tag lines
   d. wire rope

5. A navigational signal of one or both arms extended, palms down, and retracted in rapid motion indicates _____.
   a. hoist up
   b. move slowly
   c. retract boom
   d. emergency stop

Instructor's Notes:

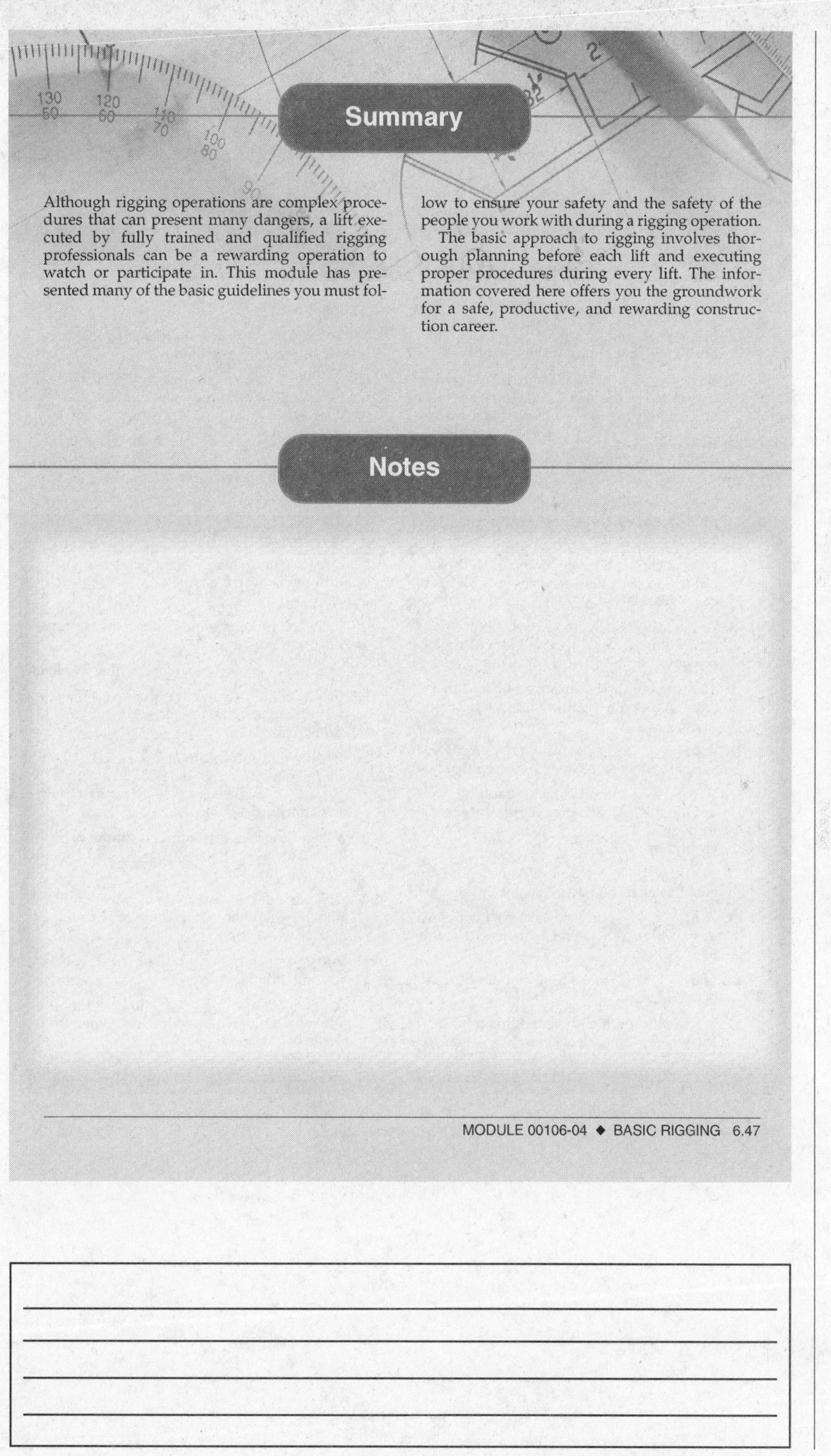

Summary

Although rigging operations are complex procedures that can present many dangers, a lift executed by fully trained and qualified rigging professionals can be a rewarding operation to watch or participate in. This module has presented many of the basic guidelines you must follow to ensure your safety and the safety of the people you work with during a rigging operation.

The basic approach to rigging involves thorough planning before each lift and executing proper procedures during every lift. The information covered here offers you the groundwork for a safe, productive, and rewarding construction career.

Notes

Ask trainees to complete the Key Terms Quiz. Review trainees' answers, and go over any questions trainees may have.

Administer the Module Examination. Be sure to record the results of the Exam on Craft Training Report Form 200, and submit the results to the Training Program Sponsor.

Have each trainee perform Performance Tasks 5–8 to your satisfaction. Fill out a Performance Profile Sheet for each trainee.

Ensure that all Performance Tests have been completed and Performance Profile Sheets for each trainee are filled out. If desired, trainee proficiency noted during laboratory sessions may be used to complete the Performance Tests. Be sure to record the results of the testing on Craft Training Report Form 200, and submit the results to the Training Program Sponsor.

Key Terms Quiz

Fill in the blank with the correct key term that you learned from your study of this module.

1. A simple rope-and-pulley system called a(n) _________ is used to lift loads.
2. _________, a type of rigging hardware, are available in three basic designs: unshouldered, shouldered, and swivel.
3. Use a single ring called a(n) _________ to attach multiple slings to a hoist hook.
4. The _________ is the distance between the master link of the sling to either the end fitting of the sling or the lowest point on the basket.
5. Also called blocking, _________ is material used to allow removal of slings after the load is landed.
6. The signal person uses _________ to communicate load navigation directions to the crane operator.
7. Using a choker hitch and forcing the choke down on loose bundles serves only to drastically increase the _________ on the choke leg.
8. An endless-loop synthetic web sling, also called a(n) _________, is made of a single-ply or multiple-ply sling formed into a loop.
9. The way the sling is arranged to hold the load is called the rigging configuration, or _________.
10. A(n) _________ hitch uses two or more slings to connect a load to a single hoist hook.
11. To form a choker hitch, wrap the sling around the load and pass one eye of the sling through a(n) _________ to form a loop around the load.
12. A(n) _________ uses a pulley system to give you a mechanical advantage for lifting a load.
13. A(n) _________ is used to move loads such as steel plates or concrete panels without the use of slings.
14. The _________ is the total amount of what is being lifted.
15. It is important to understand this concept: When _________ decreases, _____ increases.
16. The tension applied on the rigging by the weight of the suspended load is called the _________.
17. The _________ is the main connection fitting for chain slings.
18. To form an endless-loop web sling, the ends are _________ together.
19. _________ equals the lengthwise distance it takes for one strand of wire to make one complete turn around the core.
20. Standard eye-and-eye slings have eyes on the same _________, whereas twisted eye-and-eye slings have eyes at right angles to each other.
21. The _________ is the link between the load and the lifting device.
22. The maximum load weight that a sling is designed to carry is called _________.
23. Examples of _________ for synthetic slings include a missing identification tag, a puncture, and crushing.
24. Use a(n) _________ to attach a sling to a load.
25. Often found on a crane, a grooved pulley-wheel for changing the direction of a rope's pull is called a(n) _________.
26. Be sure to carefully attach hooks to the load to prevent binding of the hook in a(n) _________, which is a welded structural lifting attachment.
27. When slings are being pulled sideways by the load's weight, this _________ adds more stress to the sling.
28. The parts of the sling that reach from the attachment device around the object being lifted are called the _________.
29. A(n) _________ is a group of wires wound around a center core.
30. Riggers use a(n) _________ to limit the unwanted movement of the load when the crane begins moving.

Instructor's Notes:

6.48 CORE CURRICULUM ◆ INTRODUCTORY CRAFT SKILLS

31. If the __________ is showing, the sling is not safe for use.

32. A wire rope sling consists of high-strength steel wires formed into strands wrapped around a supporting __________.

33. An eyebolt is a piece of rigging hardware with a(n) __________, which means it has a series of spiral grooves cut into it.

34. To prevent the load from rotating freely, you must use some method of __________.

35. __________ slings are made of high-strength steel wires formed into strands wrapped around a core.

Key Terms

ANSI hand signals
Block and tackle
Bridle
Bull ring
Core
Cribbing
Eyebolt
Grommet sling
Hitch
Hoist
Lifting clamp
Load
Load control
Load stress
Master link
One rope lay
Pad eye
Plane
Rated capacity
Rejection criteria
Rigging hook
Shackle
Sheave
Side pull
Sling
Sling angle
Sling legs
Sling reach
Sling stress
Spliced
Strand
Stress
Tag line
Tattle-tail
Threaded shank
Warning-yarn
Wire rope

MODULE 00106-04 ◆ BASIC RIGGING 6.49

Profile in Success

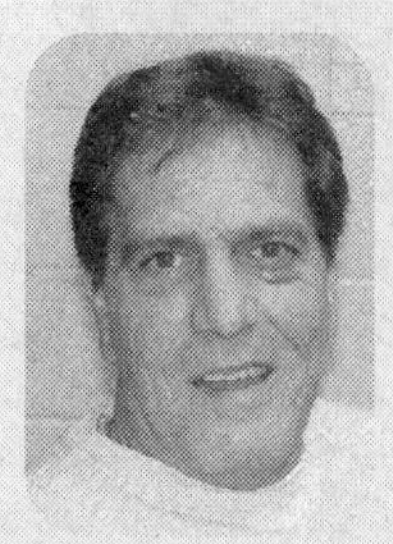

Richard W. Carr

Lead Technician
Constellation Energy Group, Power Source Generation Division
Baltimore, Maryland

Dick was born and raised in Baltimore. He majored in business administration at a local college preparatory school. After graduating, he took a job with Baltimore Gas and Electric (BG&E, now Constellation Energy Group) and has been with the company for 28 years. He has worked primarily as a machinery mechanic and plant technician, and recently became lead technician.

How did you become interested in the construction industry?
I started out as a business administration major, and I went on to become a technician in a power plant—you'll have to figure out that career path yourself!

I've always enjoyed fixing things at home, working with my hands, figuring out how radios and things like that worked. Growing up, though, I didn't have a career focus, a clear idea of what I wanted to do. Essentially, what drew me to BG&E was that it was a quality job with a quality company, in a field that I thought I could excel in. I liked the idea of a job for life, and the benefits were great. That sounded good to me, and the type of work was right up my alley.

When I started, I didn't really know what a power plant was or how electricity was made. I was only 21, and I didn't really know what I was getting myself into! But I'm still here. I've done well by it, and I think that they've done well by me.

I'm on the maintenance and mechanical side of the industry, doing fixes and repairs, as opposed to building things. It's much more integral with machinery and moving parts. The work involves taking care of things that need attention, as opposed to the other side of the construction industry where all that work is done ahead of time. I didn't have anything to do with the construction of the place where I work, but I keep the place working!

What are some of the things you do in your job?
I supervise roughly 20 men in a mechanical maintenance shop within a power plant. Our duties involve performing a schedule of maintenance work on the power-generating machinery. We have a shop where we dismantle smaller pieces for rebuild or inspection, but by and large most of the equipment—pumps, fans, motors, dampers, drives, and accessory equipment—is stationary. That means we have to go to it and work on it where it sits. This is a very hands-on type of job.

We have a very good preventive maintenance program that keeps us out of fighting fires all the time. Our job is to get to the problem before it gets to you, the customer. I'm sure you know the old saying, "If it ain't broke, don't fix it." Well, nowadays our motto is, "If it ain't broke, don't let it get broke!" Safety is our code, our number one concern.

What do you like most about your job?
I would say right off the top it is the sense of accomplishment, the pride of a job well done. My job is to make a machine work the way it was designed to. When I accomplish that, it's a good feeling. After all, it's kind of a strange industry we work in; we make a product you can't see. You can't wrap it up or put it in a box. I've been in this field for going on 30 years now, and I have always found it to be a very interesting business.

Another positive aspect of my job is the teamwork. We have scheduled outages, when a turbine or a boiler is taken offline for eight to twelve weeks for routine maintenance. We also have forced outages—for example, when a boiler develops a leak so large that it can't keep up operating pressure and it trips the generator. When an outage happens, whether it's scheduled or not, it takes a large group of people working toward that common goal to bring it back up online. I really enjoy working with a group of

Instructor's Notes:

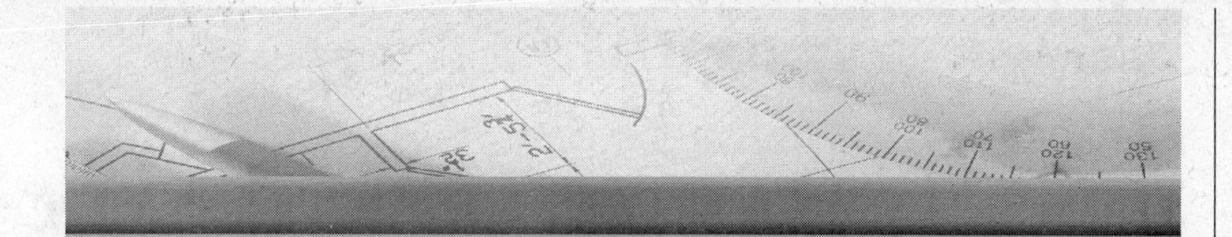

well-trained, interesting, and colorful people. When push comes to shove, everybody's there to do his or her part.

I really enjoy working with my hands. And you can use the knowledge and skills you pick up on the job in your household, on your car, or wherever you need to fix things. I get a lot of compliments around the house when I fix things or build a cabinet, for example. These skills can be applied to lots of things.

You mentioned the distinction between maintenance work and new construction. Can you elaborate a little about that?

Repair and maintenance is a relatively invisible activity. Most people don't stop to think about where electricity comes from. The only time they notice us is when our machines aren't working. But the lack of awareness goes beyond just the general public. We're finding that there's a lack of interest among students in this line of work as well. I can understand that, to some extent. You see people climbing utility poles or working down in a manhole in the street. When power lines go down in a storm, you see people out there putting them back up. But then when everything is fixed, they disappear and you forget about the generation end of the business. Power plants are tucked away out of sight.

Even if you end up in the construction end of the business, knowledge of the maintenance aspect can be really helpful. You should try to design and build things in such a way that they can be repaired. I don't know how many times we've been trying to repair a piece of machinery and someone has said in frustration, "Who on earth designed this thing?" Try to remember that people will be coming after you, and they're going to need to get access to the parts behind the wall.

I think that the maintenance part of our industry doesn't get as much exposure with younger people to get their interest. But I believe that if they knew that this kind of work was here, they'd probably be drawn to it. For a young person who likes working on cars or who is mechanically inclined, this type of work is right up their alley. And it's not like you only learn how to do one task or fix one type of machine. Years ago we had specialists, but now with cross-training we have people who can do many different tasks.

BG&E works with the school systems to promote the company's work-study program. Right now, we have two students who go to a tech school part time and get work experience here the rest of the time. The young people who have come through here as part of that program think it's really neat! To see students taking to it and enjoying it was good for me and for a lot of the other people here too. The average age of people in this field is going up, and hiring is down, so we're doing more with less. The need for more people is very real.

What do you think it takes to be a success in your trade?

I can relate my answer to the company where I've worked. Success is determined in large part by the basic job requirements because this company is very much focused on the basics: be prompt, be on time, have good work ethics, be willing to learn. I think those basics hold true in any field. You need to have a willingness to experience a new and interesting field of work.

What would you say to someone entering the trades today?

In my case, the best answer to that question is not so much what I would say, I think, as much as what people have said to me about this field. They tell me that they get a real sense of being part of something important when they belong to a group of talented, task-oriented people. For me, and for a lot of the people I work with, we don't do this job on a "need to do" basis, but on a "want to do" basis. The work is so multifaceted. You will face a new challenge every day, or develop different fixes for recurring problems. It challenges your skill.

The utility industry in general is a very good way to make a living. It's a challenging, thought-provoking job. It's very interesting, it's exciting, and it can make you tired, too. It can be really exhilarating to know that you're creating a product for hundreds of thousands of people who are depending on it. There's very seldom a dull day!

MODULE 00106-04 ◆ BASIC RIGGING 6.51

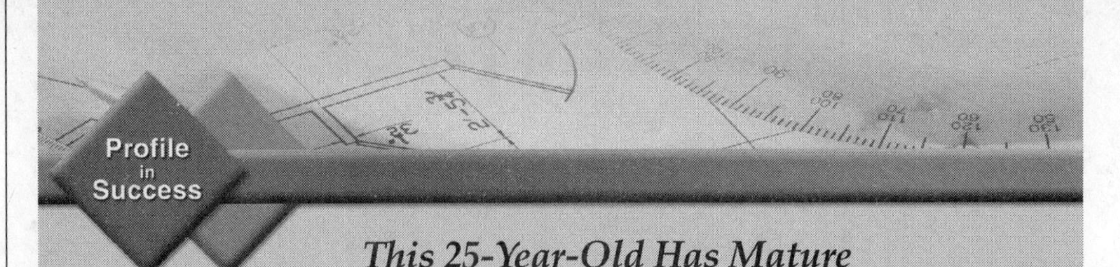

This 25-Year-Old Has Mature Responsibilities on Power Plant Project

By Leah Hitchings

From building concrete canoes to building a $350-million power plant, Jason Krueger has performed all kinds of construction and engineering work, and all by the age of 25. Hired by Boldt Construction, the largest construction company in Wisconsin, while still attending University of Wisconsin, Madison, Krueger values his few years in the field more than his 17 years of education.

"Experience is what everyone is looking for," Krueger says.

Krueger's construction education began in high school, when he took weekend classes at UW Madison and participated in an Explorers program on construction. In college, he earned a degree in civil engineering while emphasizing a construction engineering and management program that is currently being made into a major. While busy building concrete canoes for the American Society of Civil Engineers' competitions, and participating in UW's Construction Club, Krueger decided to do a co-op program at Boldt Construction. For eight months, Krueger worked full-time for Boldt, where at first he remained in the office, "looking at drawings and getting familiar with their projects," he says.

After two months in the office, Krueger was placed on site at Appleton Medical Center in Appleton, Wis., where Boldt was building an ambulatory surgery addition.

"I felt pretty lost at first," Krueger says, "but I had a lot of support around me. They didn't treat me like some little kid."

Work on the medical center included a main drive-up canopy and an all-glass entrance, lobby, and waiting room. One of the first problems Krueger faced was laying out a radius for the large glass entrance. Because the architect had not detailed the radius well, Krueger used the skills he had learned in classes to lay it out himself.

Krueger's project manager for the medical center project and for all of his other projects at Boldt, Patrick Loughrin, remembers that even Krueger's first days on site showed that he was a good fit for Boldt.

"Jason was very enthusiastic and aggressive from the beginning," Loughrin says. "He was willing to learn and was always looking for things to accomplish. He didn't wait for things to be handed to him."

Learning was a big part of Krueger's first months at Boldt because most of Krueger's knowledge was from books, not field experience. However, Krueger quickly moved to being a project manager on another project, building a dining hall for a Boy Scout camp. Loughrin says that Krueger's first job as project manager went well, although he made some mistakes.

"Jason was a little too optimistic on some stuff," Loughrin says. "Some things needed closer attention to detail."

A project that followed, remodeling a shelter for abused women, was where Krueger "grew up a lot," Loughrin says.

Krueger believes that much of his early success was a result of his attitude with his supervisors. "A lot has to do with how you ask questions, and I always tried to respect them and be polite," he says.

When Boldt offered Krueger a full-time position, he eagerly accepted, even though he still had a year of college to finish. Boldt is headquartered in Appleton, Krueger's hometown, and he also liked their large size, "because there's a lot of opportunity to grow in the company." Krueger also liked that everyone in the company knows each other, and even top executives make a point of talking to everyone.

Now Krueger is focused on his largest project to date, a $350-million gas-fired turbine and steam generation facility in Beloit, Wis. Krueger is the civil field engineer for the plant, which is being built for Calpine Corp. Although he works 60 hours a week or more now, "It's well worth it," he says. "I'd never think of changing what I do."

Krueger still has the time to visit high schools and recruit for the construction industry, as well as return to UW for career fairs. Eventually Krueger would like to return to UW for mechanical and electrical training, but he does not plan on earning a master's degree.

"A master's degree won't help you get a job, while job experience is what's more important," he says. His business minor has been invaluable, however, because he deals with contracts and labor relations every day.

The long hours and challenging work have been rewarding for Krueger, because, he says, "I love what I'm doing."

Instructor's Notes:

Trade Terms Introduced in This Module

ANSI hand signals: Communication signals established by the American National Standards Institute (ANSI) and used for load navigation for mobile and overhead cranes.

Block and tackle: A simple rope-and-pulley system used to lift loads.

Bridle: A configuration using two or more slings to connect a load to a single hoist hook.

Bull ring: A single ring used to attach multiple slings to a hoist hook.

Core: Center support member of a wire rope around which the strands are laid.

Cribbing: Material used to either support a load or allow removal of slings after the load is landed. Also called blocking.

Eyebolt: An item of rigging hardware used to attach a sling to a load.

Grommet sling: A sling fabricated in an endless loop.

Hitch: The rigging configuration by which a sling connects the load to the hoist hook. The three basic types of hitches are vertical, choker, and basket.

Hoist: A device that applies a mechanical force for lifting or lowering a load.

Lifting clamp: A device used to move loads such as steel plates or concrete panels without the use of slings.

Load: The total amount of what is being lifted, including all slings, hitches, and hardware.

Load control: The safe and efficient practice of load manipulation, using proper communication and handling techniques.

Load stress: The strain or tension applied on the rigging by the weight of the suspended load.

Master link: The main connection fitting for chain slings.

One rope lay: The lengthwise distance it takes for one strand of a wire rope to make one complete turn around the core.

Pad eye: A welded structural lifting attachment.

Plane: A surface in which a straight line joining two points lies wholly within that surface.

Rated capacity: The maximum load weight a sling or piece of hardware or equipment can hold or lift. Also called lifting capacity, working capacity, working load limit (WLL), and safe working load (SWL).

Rejection criteria: Standards, rules, or tests on which a decision can be based to remove an object or device from service because it is no longer safe.

Rigging hook: An item of rigging hardware used to attach a sling to a load.

Shackle: Coupling device used in an appropriate lifting apparatus to connect the rope to eye fittings, hooks, or other connectors.

Sheave: A grooved pulley-wheel for changing the direction of a rope's pull; often found on a crane.

Side pull: The portion of a pull acting horizontally when the slings are not vertical.

Sling: Wire rope, alloy steel chain, metal mesh fabric, synthetic rope, synthetic webbing, or jacketed synthetic continuous loop fibers made into forms, with or without end fittings, used to handle loads.

Sling angle: The angle of an attached sling when pulled in relation to the load.

Sling legs: The parts of the sling that reach from the attachment device around the object being lifted.

Sling reach: A measure taken from the master link of the sling, where it bears weight, to either the end fitting of the sling or the lowest point on the basket.

Sling stress: The total amount of force exerted on a sling. This includes forces added as a result of sling angle.

Spliced: Having been joined together.

Strand: A group of wires wound, or laid, around a center wire, or core. Strands are laid around a supporting core to form a rope.

Stress: Intensity of force exerted by one part of an object on another; the action of forces on an object or system that leads to changes in its shape, strain on it, or separation of its parts.

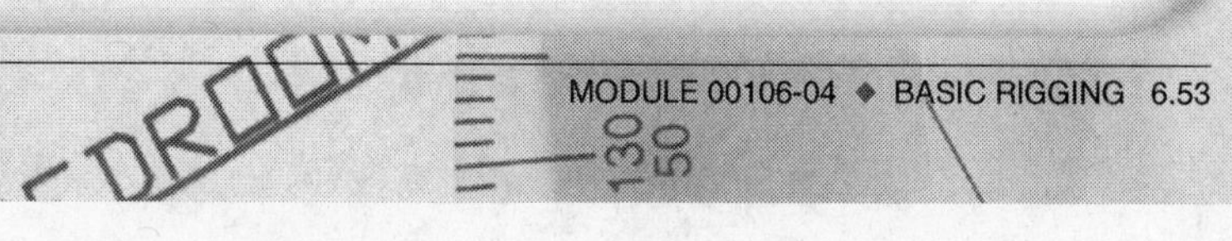

MODULE 00106-04 ◆ BASIC RIGGING 6.53

Tag line: Rope that runs from the load to the ground. Riggers hold on to tag lines to keep a load from swinging or spinning during the lift.

Tattle-tail: Cord attached to the strands of an endless loop sling. It protrudes from the jacket. A tattle-tail is used to determine if an endless sling has been stretched or overloaded.

Threaded shank: A connecting end of a fastener, such as a bolt, with a series of spiral grooves cut into it. The grooves are designed to mate with grooves cut into another object in order to join them together.

Warning-yarn: A component of the sling that shows the rigger whether the sling has suffered too much damage to be used.

Wire rope: A rope made from steel wires that are formed into strands and then laid around a supporting core to form a complete rope; sometimes called cable.

Instructor's Notes:

Appendix

Rated Capacities

You should never exceed a sling's rated capacity. The rated capacity varies depending upon the type of sling, the size of the sling, and the type of hitch. Rigging operators must know the capacity of the sling they are using. Sling manufacturers usually have charts or tables that contain this information. A sample chart of rated capacities for web slings is shown in *Table A-1.*

Table A-1 Rated Capacities of Some Web Slings

FLAT EYE AND TWISTED EYE WEB SLINGS

TYPE III - FLAT EYES EACH END

TYPE IV - TWISTED EYE EACH END

			RATED CAPACITIES IN POUNDS		
WIDTH	EYE WIDTH	EYE LENGTH	CHOKER	VERTICAL	BASKET
1"	1"	9"	1,280	1,600	3,200
1"	1"	9"	2,480	3,100	6,200
1"	1"	12"	4,960	6,200	12,400
2"	2"	9"	2,480	3,100	6,200
2"	2"	9"	4,960	6,200	12,400
2"	2"	12"	9,920	12,400	24,800
3"	1½"	9"	3,720	4,650	9,300
3"	1½"	12"	7,440	9,300	18,600
3"	1½"	18"	14,880	18,680	37,200
4"	1½"	12"	4,960	6,200	12,400
4"	1½"	12"	8,800	11,000	22,000
4"	2"	18"	17,600	22,000	44,000
6"	2"	14"	7,440	9,300	18,600
6"	2"	14"	13,200	16,500	33,000
6"	3"	20"	26,400	33,000	66,000
8"	3"	18"	9,920	12,400	24,800
8"	3"	18"	17,600	22,000	44,000
8"	4"	24"	35,200	44,000	88,000
10"	4"	22"	12,400	15,500	31,000
10"	4"	22"	22,000	26,500	55,000
10"	5"	28"	44,000	55,000	110,000
12"	5"	26"	14,880	18,600	37,200
12"	5"	26"	26,400	33,000	66,000
12"	6"	32"	52,800	66,000	132,000

106A01.EPS

MODULE 00106-04 ◆ BASIC RIGGING 6.55

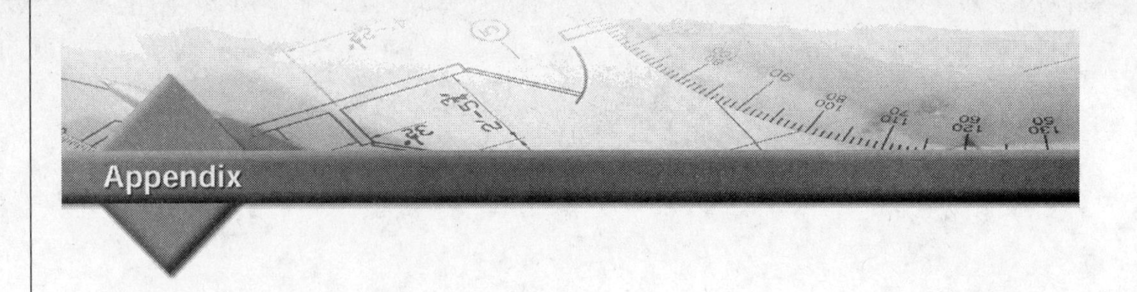

SLING ANGLES

The amount of tension on the sling is directly affected by the angle of the sling. *Figure A-1* shows the effect of sling angles on sling loading.

Figure A-1 ◆ Sling angles.

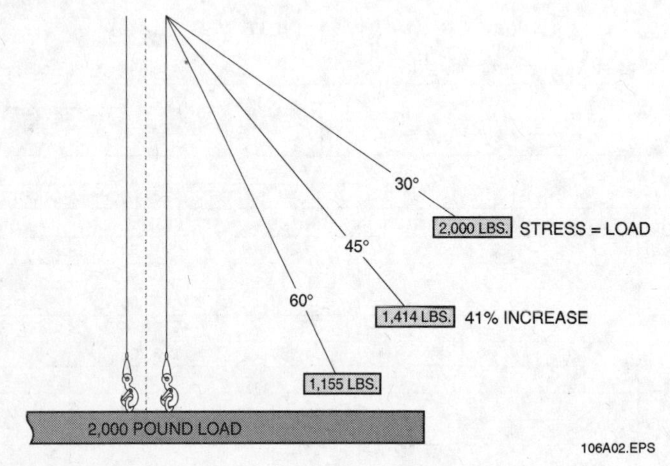

To actually determine the load on a sling, a factor table is used (See *Table A-2*).

Table A-2 Factor Table

SLING ANGLE	LOAD ANGLE FACTOR
30°	2.000
35°	1.742
40°	1.555
45°	1.414
50°	1.305
55°	1.221
60°	1.155
65°	1.104
70°	1.064
75°	1.035
80°	1.015
85°	1.004
90°	1.000

106A03.EPS

Instructor's Notes:

6.56 CORE CURRICULUM ◆ INTRODUCTORY CRAFT SKILLS

In the example shown in *Figure A-1,* two slings are being used to lift 2,000 pounds. When the slings are at a 45-degree angle, there is 1,414 pounds of tension on each sling. This can be determined mathematically by using the following equation:

$$\text{Sling Tension} = \frac{\text{Load (lbs)} \times \text{Load Angle Factor}}{\text{Number of Legs}}$$

Therefore:

$$\text{Sling Tension} = \frac{2{,}000 \times 1.414}{2}$$

$$\text{Sling Tension} = \frac{2{,}828}{2}$$

$$\text{Sling Tension} = 1{,}414$$

Use the weight of the load (2,000 pounds) and multiply it by the corresponding angle factor in *Table A-2.* Then, divide it by the number of slings (2). The result is 1,414 pounds of tension on each sling.

MODULE 00106-04 ◆ BASIC RIGGING 6.57

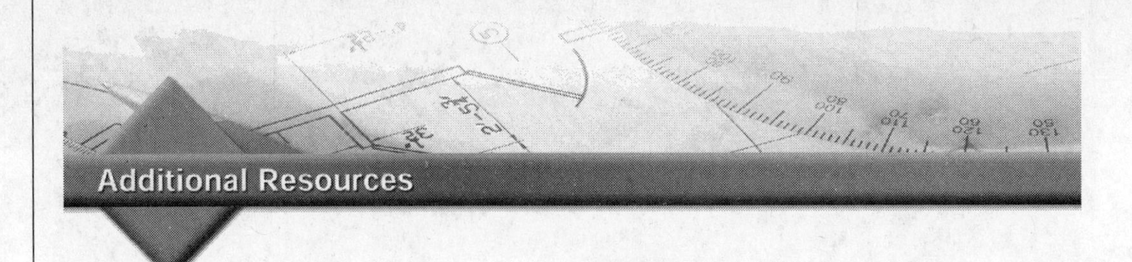

Additional Resources

This module is intended to present thorough resources for task training. The following reference works are suggested for further study. These are optional materials for continued education rather than for task training.

Bob's Rigging and Crane Handbook, Latest Edition. Bob DeBenedictis. Leawood, KS: Pellow Engineering Services, Inc.

High Performance Slings and Fittings for the New Millennium, 1999 Edition. Dennis St. Germain. Aston, PA: I & I Sling, Inc.

Mobile Crane Manual, 1999. Donald E. Dickie, D. H. Campbell. Toronto, Ontario, Canada: Construction Safety Association of Ontario.

Rigging Manual, 1997. Toronto, Ontario, Canada: Construction Safety Association of Ontario.

Instructor's Notes:

MODULE 00106-04 — TEACHING TIPS

The following are suggested activities or instructional methods to help you teach the material in this AIG.

General

When you call on someone to answer a question, the rest of the class relaxes or even tunes out because they expect that the question and answer will take place only between you and the trainee you called on. Instead, use this technique to involve more trainees in answering questions and to keep them on their toes.

1. Ask trainees to define a term or explain a concept.
2. After one trainee has answered, ask a trainee seated nearby if the answer is right. Then ask whether a trainee in the back of the room agrees.
3. Ask trainees to explain why they think an answer is right or wrong.
4. Use the session to clear up incorrect ideas, and encourage trainees to learn from their mistakes.

Sections 2.0.0–2.3.3

Slings

For this exercise, which will familiarize trainees with damaged slings, you will need either a variety of damaged slings or photos of damaged slings. Trainees will need their Trainee Guides and pencil and paper to take notes for discussion. Allow 30 minutes for this exercise.

1. Bring in damaged slings or photos of damaged slings.
2. Display the slings/photos at several stations.
3. Ask trainees to visit each station and inspect each sling/photo.
4. Have trainees identify the type of sling and determine the rejection criteria for each.
5. Discuss the consequences of using damaged slings.

Sections 3.0.0–3.3.0

Hitches

In this exercise, trainees will practice selecting the appropriate type of hitch for a variety of lifting procedures. Trainees will need their Trainee Guides and pencil and paper to take notes for discussion. Allow between 20 and 30 minutes for this exercise.

1. Describe a variety of scenarios for using hitches. Be sure to include the type of sling, the item being lifted, the weight of the load, and any other information that could be helpful in determining which type of hitch to use.
2. Ask trainees to consider the information and determine which type of hitch they would select for each scenario. Make sure that trainees justify their choices.
3. Review and discuss trainees' answers. Go over any questions trainees may have about hitches.

Sections 4.0.0–4.4.2

Rigging Hardware

This informal quiz will familiarize trainees with a variety of shackles, eyebolts, clamps, and hooks, which you will provide. Trainees will need their Trainee Guides and pencil and paper to take notes for discussion. Allow 30 minutes for this exercise.

1. Bring in a variety of shackles, eyebolts, clamps, and hooks.
2. Number and display each item.
3. On a sheet of paper, have trainees identify each item and write the following information for each piece of hardware:
   - Appropriate applications
   - Factors for proper use
   - Safety considerations
   - Maintenance and inspection
   - Design characteristics
   - Installation considerations
4. Go over trainees' answers, and ask whether trainees have any questions about rigging hardware.

Sections 6.0.0–6.2.0

Hoists

This exercise will enable trainees to observe a lift scheduled at a nearby construction site where you have arranged a visit. Trainees will need their Trainee Guides and pencil and paper to take notes for discussion. Allow 60 minutes for this exercise.

1. Ask a competent or qualified person to explain the scheduled lift procedure to the trainees. If possible, have the person explain what is being lifted, how much it weighs, and any details that may not be readily obvious.
2. Have trainees observe the lift and consider the following:
   - Is a hand chain hoist or powered chain hoist used?
   - Is the load properly balanced and attached?
   - Is the load properly positioned?
   - Is proper personal protective equipment used?
   - Is the equipment well maintained?
   - Is the lift properly supervised?
3. Following the procedure, ask trainees to discuss their observations with the competent or qualified person.
4. Answer any questions trainees may have about using hoists.

Sections 7.0.0–7.4.3

Rigging Operations and Practices

In this exercise, trainees will plan and describe a variety of rigging operations and practices. For this exercise, trainees will need their Trainee Guides and pencil and paper to take notes for discussion. Allow 30 minutes for this exercise.

1. Divide the class into small groups and provide each group with a hypothetical rigging operation.
2. After each group discusses the operation, ask group members to describe the necessary rigging operations and practices, including the following:
   - Techniques for connecting hitches
   - Use of all hand signals (demonstration)
   - Pre-lift safety checks
   - Techniques for lifting the load level
   - Safety precautions for loading and disconnecting
3. Have trainees observe the practices and comment on any procedures that may be demonstrated or described incorrectly.
4. Provide recommendations for safe rigging operations, and answer any questions trainees may have.

MODULE 00106-04 — ANSWERS TO REVIEW QUESTIONS

Sections 2.0.0–2.1.5

1. d
2. a
3. b
4. b
5. c

Sections 2.2.0–2.3.3

1. c
2. a
3. b
4. a
5. d

Section 3.0.0

1. c
2. d
3. b
4. c
5. a

Section 4.0.0

1. b
2. c
3. c
4. d
5. a

Section 5.0.0

1. c
2. d
3. a
4. a
5. b

Section 6.0.0

1. b
2. d
3. a
4. c
5. b

Section 7.0.0

1. b
2. a
3. a
4. c
5. d

MODULE 00106-04 — ANSWERS TO KEY TERMS QUIZ

1. block and tackle
2. Eyebolts
3. bull ring
4. sling reach
5. cribbing
6. ANSI hand signals
7. stress
8. grommet sling
9. hitch
10. bridle
11. shackle
12. hoist
13. lifting clamp
14. load
15. sling angle, sling stress
16. load stress
17. master link
18. spliced
19. One rope lay
20. plane
21. sling
22. rated capacity
23. rejection criteria
24. rigging hook
25. sheave
26. pad eye
27. side pull
28. sling legs
29. strand
30. tag line
31. warning-yarn
32. core
33. threaded shank
34. load control
35. Wire rope

CONTREN® LEARNING SERIES — USER FEEDBACK

The NCCER makes every effort to keep these textbooks up-to-date and free of technical errors. We appreciate your help in this process. If you have an idea for improving this textbook, or if you find an error, a typographical mistake, or an inaccuracy in NCCER's *Contren®* textbooks, please write us, using this form or a photocopy. Be sure to include the exact module number, page number, a detailed description, and the correction, if applicable. Your input will be brought to the attention of the Technical Review Committee. Thank you for your assistance.

Instructors – If you found that additional materials were necessary in order to teach this module effectively, please let us know so that we may include them in the Equipment/Materials list in the Annotated Instructor's Guide.

Write: Product Development
National Center for Construction Education and Research
P.O. Box 141104, Gainesville, FL 32614-1104

Fax: 352-334-0932

E-mail: curriculum@nccer.org

Craft ______ Module Name ______

Copyright Date ______ Module Number ______ Page Number(s) ______

Description ______

______

______

______

(Optional) Correction ______

______

______

(Optional) Your Name and Address ______

______

______

Basic Communication Skills

00107-04

NCCER STANDARDIZED CRAFT TRAINING PROGRAM

The National Center for Construction Education and Research (NCCER) provides a standardized national program of accredited craft training. Key features of the program include instructor certification, competency-based training, and performance testing. The program provides trainees, instructors, and companies with a standard form of recognition through a National Craft Training Registry. The program is described in full in the *Guidelines for Accreditation*, published by the NCCER. For more information on standardized craft training, contact the NCCER by writing us at P.O. Box 141104, Gainesville, FL 32614-1104; calling 352-334-0911; or e-mailing info@nccer.org. More information may be found at our Web site, www.nccer.org.

HOW TO USE THIS ANNOTATED INSTRUCTOR'S GUIDE

Each page presents two sections of information. The larger section displays each page exactly as it appears in the Trainee Module. The narrow column ties suggested trainee and instructor actions to each page and provides icons (detailed below) to call your attention to material, safety, audiovisual, or testing requirements. The bottom of each page includes space for your notes.

The **Audiovisual** icon indicates an appropriate time to show a transparency or other audiovisual aid.

The **Classroom** icon prompts you to define a term, stress a point, ask trainees to explain a concept, or give examples.

The **Demonstration** icon directs you to show trainees how to perform tasks.

The **Examination** icon tells you to administer the written module examination.

The **Homework** icon is placed where you may wish to assign reading for the next class, to assign a project, or to advise trainees to prepare for an examination.

The **Laboratory** icon is used when trainees are to practice performing tasks.

The **Materials** icon is a reminder for you to gather materials needed for classes, labs, and testing.

The **Performance Testing** icon tells you to administer a performance test or a portion thereof.

The **Safety** icon is used to emphasize safety issues. It is often keyed to *Caution* and *Warning* statements in the Trainee Module.

The **Teaching Tip** icon indicates additional guidance is available, such as how to conduct an exercise, get the most educational value from a field trip, or encourage class participation. Teaching Tips may expand on a feature (*Think About It, Did You Know?*) or provide Quick Quizzes or similar exercises. You will be referred to the Teaching Tips section at the back of the module if there is additional material.

The **Combination** icon indicates that the laboratory listed corresponds with a performance task. If desired, you can note the proficiency of the trainees during the laboratory and use it to satisfy performance testing requirements.

PREPARATION

Before teaching this module, you should review the Objectives, Performance Tasks, Materials and Equipment List, and the Module Outline. Be sure to allow ample time to prepare your own training or lesson plan and gather all required materials and equipment.

MODULE OVERVIEW

This module reviews basic communication skills. Trainees will learn how to interpret information in written and verbal form and how to communicate effectively using written and verbal skills.

PREREQUISITES

Prior to training with this module, it is recommended that the trainee shall have successfully completed the following: *Core Curriculum: Introductory Craft Skills,* Modules 00101-04 through 00105-04. This module is an elective. To receive a successful completion, you must take this module and Module 00108-04 or Module 00106-04.

OBJECTIVES

Upon completion of this module, the trainee will be able to:

1. Demonstrate the ability to interpret information and instructions presented in both written and verbal form.
2. Demonstrate the ability to communicate effectively in on-the-job situations using written and verbal skills.

PERFORMANCE TASKS

Under the supervision of the instructor, the trainee should be able to:

1. Write a daily report, ensuring it is as follows:
   - Clear (understandable)
   - Concise (to the point)
   - Correct (documents daily activities, manpower, equipment and materials, as well as safety issues; documents owner/architect instructions; and includes correct spelling and punctuation)
2. Fill out work-related forms (for example, accident reports, time and materials, training reports, time sheets, punch lists, change orders, and RFIs), ensuring that they are as follows:
   - Complete
   - Accurate
   - On time
3. Read instructions for how to properly don a safety harness, and orally instruct another person to don the apparatus.
4. Perform a given task after listening to oral instructions.

MATERIALS AND EQUIPMENT LIST

Transparencies
Markers/chalk
Blank acetate sheets
Transparency pens
Pencils and scratch paper
Overhead projector and screen
Whiteboard/chalkboard
Appropriate personal protective equipment
Copies of your local code
Examples of written materials commonly used on the job, including:
- Safety instructions
- Blueprints
- Manufacturer's installation instructions
- Materials lists
- Signs and labels
- Work orders and schedules
- Specifications
- Change orders
- Industry magazines and company newsletters

Trade manuals
Sets of instructions
Work-related forms, including:
- Accident reports
- Time and materials
- Training reports

Time sheets
Punch lists
Change orders
RFIs
Examples of well-written and poorly written emails
Module Examinations*
Performance Profile Sheets*

*Located in the Test Booklet.

SAFETY CONSIDERATIONS

Ensure that the trainees are equipped with appropriate personal protective equipment. Always work in a clean, well-lit, appropriate work area.

ADDITIONAL RESOURCES

This module is intended to present thorough resources for task training. The following reference works are suggested for both instructors and motivated trainees interested in further study. These are optional materials for continued education rather than for task training.

Communicating at Work, 1993. Tony Alessandra and Phil Hunsaker. New York, NY: Simon and Schuster.

Communicating in the Real World: Developing Communication Skills for Business and the Professions, 1987. Terrence G. Wiley and Heide Spruck Wrigley. Englewood Cliffs, NJ: Prentice Hall.

Communication Skills for Business and Professions, 1996. Paul R. Timm and James A. Stead. Upper Saddle River, NJ: Prentice Hall.

Elements of Business Writing, 1992. Gary Blake and Robert W. Bly. New York, NY: Collier.

Improving Business Communication Skills, 1998. Deborah Britt Roebuck. Upper Saddle River, NJ: Prentice Hall.

TEACHING TIME FOR THIS MODULE

An outline for use in developing your lesson plan is presented below. Note that each Roman numeral in the outline equates to one session of instruction. Each session has a suggested time period of 2½ hours. This includes 10 minutes at the beginning of each session for administrative tasks and one 10-minute break during the session. Approximately 5 hours are suggested to cover *Basic Communication Skills.* You will need to adjust the time required for hands-on activity and testing based on your class size and resources. Because laboratories often correspond to Performance Tasks, the proficiency of the trainees may be noted during these exercises for Performance Testing purposes.

Topic — **Planned Time**

Session I. Reading and Writing Skills

A. Reading on the Job ____________
B. Writing on the Job ____________
C. Performance Testing (Tasks 1–3) ____________

Session II. Listening and Speaking Skills

A. Active Listening on the Job ____________
B. Speaking on the Job ____________
C. Review ____________
D. Module Examination ____________
   1. Trainees must score 70% or higher to receive recognition from NCCER.
   2. Record the testing results on Craft Training Report Form 200, and submit the results to the Training Program Sponsor.
E. Performance Testing (Task 4) ____________
   1. Trainees must perform each task to the satisfaction of the instructor to receive recognition from NCCER. If applicable, proficiency noted during laboratory exercises can be used to satisfy the Performance Testing requirements.
   2. Record the testing results on Craft Training Report Form 200, and submit the results to the Training Program Sponsor.

**Build America Winner—
Building Renovation**

Centex Construction Company renovated the Children's Medical Hospital in Dallas, Texas, and finished the large-scale building five months early. Finishing the project under budget, Centex was able to give money back to their client. In addition to completing this amazing facility, Centex raised $280,000 for the hospital, installed an MRI and Cath lab, organized toy runs for the children in the hospital, replaced roofs, added a skywalk, and added a new level to the top of an existing parking garage.

00107-04
Basic Communication Skills

Topics to be presented in this unit include:

Overview

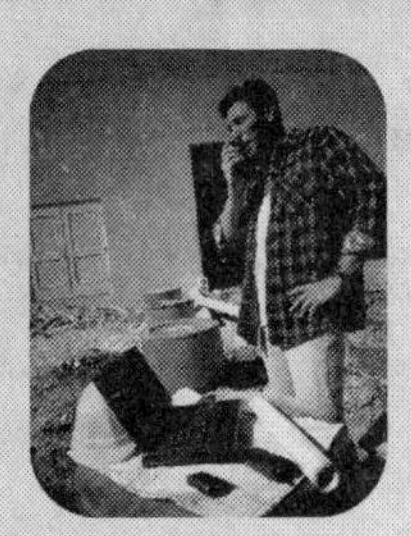

During your career in the construction trades, you will learn many skills. Effective communication skills, however, will always be your greatest asset. The ability to read, write, speak, and listen effectively will remain critical throughout your career as you interact with workers at all levels.

The types of reading materials that are typically read on a construction site include safety instructions, blueprints, signs, labels, change orders, and material lists. The ability to read these documents is important. For example, if a worker who is unable to read is given written safety instructions, it is likely that an accident or injury will occur. As you progress through your career, you may gain a leadership role. In this role, you will likely be responsible for writing documents related to the job. Make writing a strong skill in the beginning of your career. It is a skill that can only improve with time and experience.

The ability to speak and listen is as important as the ability to read and write. Because workers are often given instructions verbally, it is crucial that the person giving the instructions speaks clearly and concisely and that he or she recognizes any language or intellectual barrier that may exist. The ability to listen is also important in this process. There can be serious consequences if the instructions that are being given are unclear or misunderstood. Always make sure that you understand what is being said to you and that those that you speak to understand what you are saying.

Instructor's Notes:

Objectives

When you have completed this module, you will be able to do the following:

1. Demonstrate the ability to interpret information and instructions presented in both written and verbal form.
2. Demonstrate the ability to communicate effectively in on-the-job situations using written and verbal skills.

Prerequisites

Before you begin this module, it is recommended that you successfully complete the following: *Core Curriculum: Introductory Craft Skills,* Modules 00101-04 through 00105-04. Modules 00106-04 through 00108-04 are electives. To receive a successful completion, you must take Module 00106-04 or Modules 00107-04 and 00108-04.

This course map shows all of the modules in the first level of *Core Curriculum: Introductory Craft Skills.* The suggested training order begins at the bottom and proceeds up. Skill levels increase as you advance on the course map. The local Training Program Sponsor may adjust the training order.

00108-04 Basic Employability Skills
00107-04 Basic Communication Skills
00106-04 Basic Rigging
00105-04 Introduction to Blueprints
00104-04 Introduction to Power Tools
00103-04 Introduction to Hand Tools
00102-04 Introduction to Construction Math
00101-04 Basic Safety
CORE CURRICULUM

107CMAP.EPS

Key Trade Terms

Active listening	Index
Appendix	Italics
Body language	Jargon
Bullets	Memo
Cell phone	Memorandum
Electronic signature	Permit
Font	Punch list
Glossary	Table
Graph	Table of contents

Ensure that you have everything required to teach the course. Check the Materials and Equipment List at the front of this module.

See the general Teaching Tip at the end of this module.

Show Transparency 1, Course Objectives.

Show Transparency 2, Performance Tasks.

Explain that terms shown in bold (blue) are defined in the Glossary at the back of this module.

MODULE 00107-04 ◆ BASIC COMMUNICATION SKILLS 7.1

Discuss the importance of being able to communicate clearly on the job site.

Review the different types of written documents. Provide examples of documents that are frequently used.

Refer to Figure 2. Explain the importance of reading and understanding instruction manuals in the construction trade.

Discuss the consequences of not being able to read on the job. Go over any questions trainees may have.

Bring in a variety of examples of written materials commonly used on the job. Demonstrate tips and techniques that will help trainees read more efficiently on the job.

Distribute a variety of trade manuals. Have trainees review the special features that help locate information. Ask trainees to locate chapters, tables, graphs, and appendixes using the table of contents.

1.0.0 ◆ INTRODUCTION

Every construction professional learns how to use tools. Depending on your trade, the tools you use could include welding machines and cutting torches, press brakes and plasma cutters, or surveyor's levels and pipe threaders. However, some of the most important tools you will use on the job are not tools you can hold in your hand or put in a toolbox. These tools are your abilities to read, write, listen, and speak.

At first, you might say that these are not really construction tools. They are things you already learned how to do in school, so why do you have to learn them all over again? The types of communication that take place in the construction workplace are very specialized and technical, just like the communications between pilots and air traffic control. Good communications result in a job done safely—a pilot hears and understands the message to change course to avoid a storm, and a construction worker hears and understands the message to install a water heater according to the local code requirements.

In a way, you are learning another language—a special language that only trained professionals know how to use. A good communications toolbox is a badge of honor; it lets everyone know that you have important skills and knowledge. And like a physical toolbox, the ability to communicate well verbally and in writing is something that you can take with you to any job. You will find that good communications skills can help you advance your career. This module introduces you to the techniques you will need to read, write, listen, and speak effectively on the job.

2.0.0 ◆ READING AND WRITING SKILLS

The construction industry depends on written materials of all kinds to carry on business—from routine office paperwork to blueprints and building codes (*Figure 1*). Written documents allow workers to follow instructions accurately, help project managers ensure that work is on schedule, and enable the company to meet its legal obligations. As a construction professional, you need to be able to read and understand the written documents that apply to your work, whether they are printed in a book or letter, or on a computer. And as you gain professional experience, you will end up taking on responsibilities for writing documents.

Another form of written construction document that you are very familiar with is the textbook, like the curriculum developed by NCCER (*Figure 2*). A textbook is an instructional document that contains information that a reader needs to know to carry out a task or a series of tasks. Instruction manuals for tools and installation manuals from manufacturers are other common examples of this type of document. This section reviews some techniques that you may find helpful in reading and, later, writing construction documents.

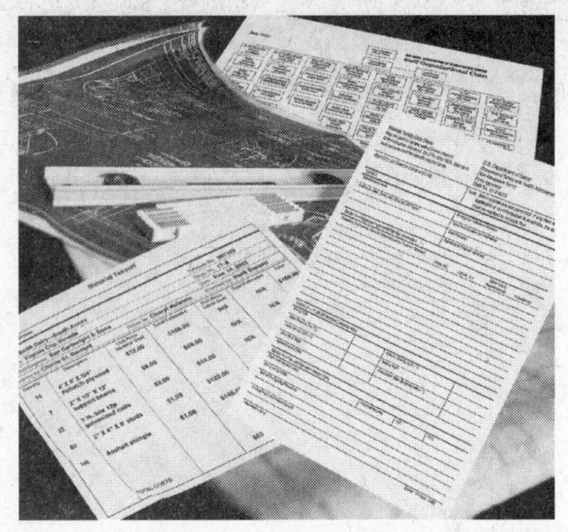

107F01.EPS

Figure 1 ◆ The backbone of the construction industry is the written document.

2.1.0 Reading on the Job

Say your company has just issued new safety guidelines, or the latest version of the local building code contains new standards that affect how you do your job. You need to be able to understand these documents to perform your job safely and effectively. You should not rely on someone else to tell you how to do your job. Reading is central to what construction professionals do.

You may think that you don't have to read on the job, but in fact the written word is at the center of the construction trade. The following are some typical examples of things construction workers read on the job every day:

- Safety instructions
- Blueprints
- Manufacturer's installation instructions
- Materials lists
- Signs and labels
- Work orders and schedules
- **Permits**
- Specifications
- Change orders
- Industry magazines and company newsletters

Instructor's Notes:

7.2 CORE CURRICULUM ◆ INTRODUCTORY CRAFT SKILLS

Explain how appendixes are used to provide detailed or specific information.

107F02.EPS

Figure 2 ◆ Textbooks teach you to become a construction professional.

This section reviews some simple tips and techniques that will help you read faster and more efficiently on the job. And, because someone wrote everything that you read, some of these tips and techniques will help you become a better writer as well. Because you have been reading for many years now, it is probably something you do without even thinking about it. With practice, these guidelines will become second nature too.

First of all, you should always have a purpose in mind when you read. This will help you find the information faster. For example, say you are looking for a specific installation procedure in a manual. Do not waste time reading other parts of the manual; go straight to the section that deals with that procedure. Read slowly enough to be able to concentrate on what you are reading. Often, technical publications are packed with detailed information, and you do not want to misunderstand it because you rushed.

Most books have special features that can help you locate information, including the following:

- **Table of contents**
- **Index**
- **Glossary**
- **Appendixes**
- **Tables** and **graphs**

Tables of contents are lists of chapters or sections in a book (*Figure 3*). They are usually at the front of a book. In books, indexes are alphabetical listings of topics with page numbers to show where those topics appear. Indexes are also used to list information in documents other than books—for example, in blueprints (*Figure 4*). Glossaries are alphabetical lists of terms used in a book, along with definitions for each term.

Appendixes are sources of additional information placed at the end of a section, chapter, or book. Appendixes are separate because the information in them is more detailed or specific than the information in the rest of the book. If the information were placed in the main part of the book, it could distract readers or slow them down too much. Tables and graphs summarize important facts and figures in a way that lets you understand them at a glance. Tables usually contain text or numbers, and graphs use images or symbols to convey their meaning.

The noise and activity on a construction site or in a shop can easily distract you while you are

MODULE 00107-04 ◆ BASIC COMMUNICATION SKILLS 7.3

Review tips for accurately reading instructions and scanning unfamiliar books.

107F03.TIF

Figure 3 ◆ A table of contents lists the chapters or sections in a book.

reading. Try to avoid distractions while you're reading. Radios, televisions, conversations, and nearby machinery and power tools can all affect your concentration. Take notes or use a highlighter to help you remember and find important text.

When reading an unfamiliar book, skim or scan the chapter titles and section headings before you start to read. This will help you organize the material in your mind. Look for visual clues that indicate important material. For example, a bold **font** or **italics** indicate important words or information. Bold fonts are letters and numbers that are heavier and darker than the surrounding text. Italics are letters and numbers that lean to the right, rather than stand straight up.

Instructor's Notes:

7.4 CORE CURRICULUM ◆ INTRODUCTORY CRAFT SKILLS

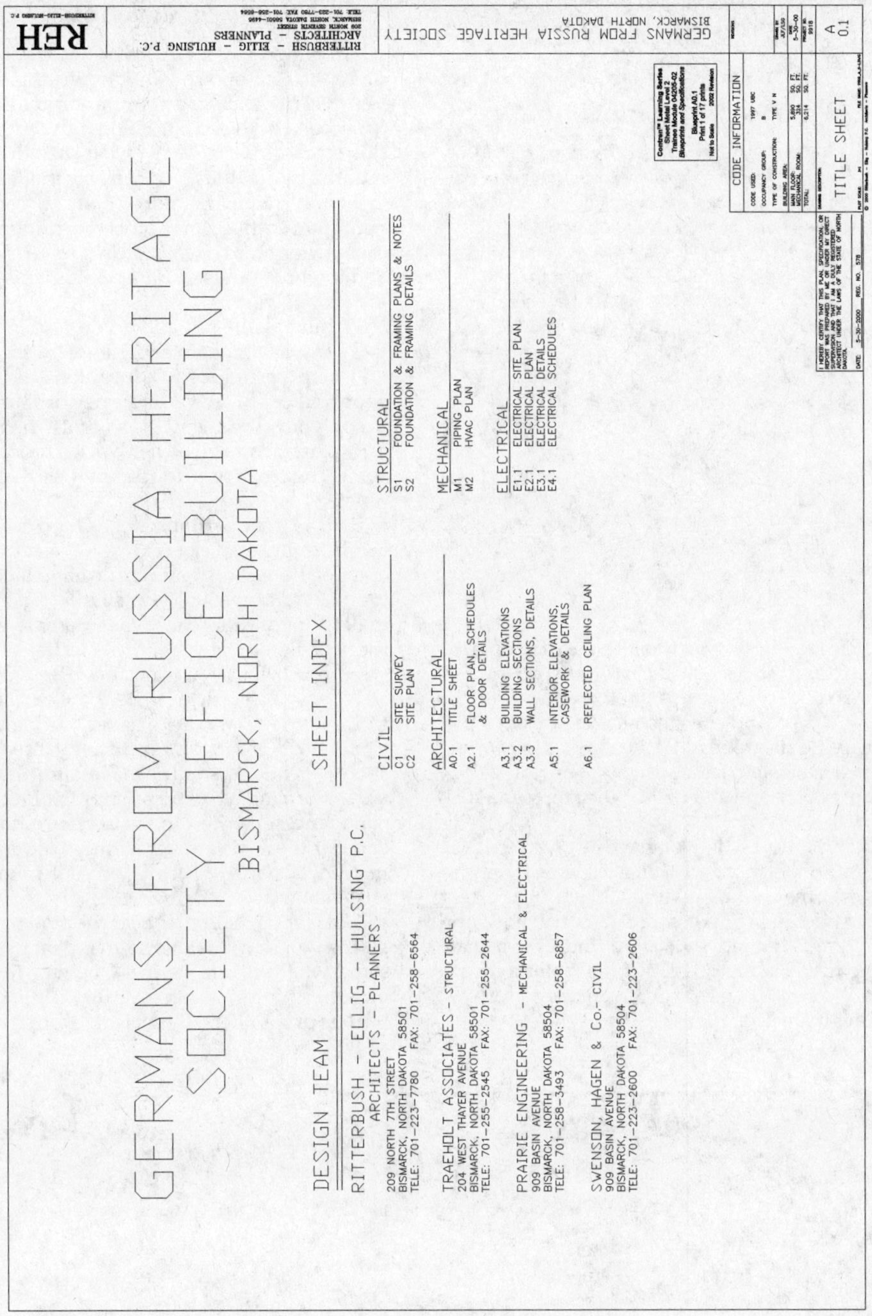

GERMANS FROM RUSSIA HERITAGE
SOCIETY OFFICE BUILDING
BISMARCK, NORTH DAKOTA

DESIGN TEAM

RITTERBUSH - ELLIG - HULSING P.C.
ARCHITECTS - PLANNERS
209 NORTH 7TH STREET
BISMARCK, NORTH DAKOTA 58501
TELE: 701-223-7780 FAX: 701-258-6564

TRAEHOLT ASSOCIATES - STRUCTURAL
204 WEST THAYER AVENUE
BISMARCK, NORTH DAKOTA 58501
TELE: 701-255-2545 FAX: 701-255-2644

PRAIRIE ENGINEERING - MECHANICAL & ELECTRICAL
909 BASIN AVENUE
BISMARCK, NORTH DAKOTA 58504
TELE: 701-258-3493 FAX: 701-258-6857

SWENSON, HAGEN & Co.- CIVIL
909 BASIN AVENUE
BISMARCK, NORTH DAKOTA 58504
TELE: 701-223-2600 FAX: 701-223-2606

SHEET INDEX

CIVIL
C1 SITE SURVEY
C2 SITE PLAN

ARCHITECTURAL
A0.1 TITLE SHEET
A2.1 FLOOR PLAN, SCHEDULES & DOOR DETAILS
A3.1 BUILDING ELEVATIONS
A3.2 BUILDING SECTIONS
A3.3 WALL SECTIONS, DETAILS
A5.1 INTERIOR ELEVATIONS, CASEWORK & DETAILS
A6.1 REFLECTED CEILING PLAN

STRUCTURAL
S1 FOUNDATION & FRAMING PLANS & NOTES
S2 FOUNDATION & FRAMING DETAILS

MECHANICAL
M1 PIPING PLAN
M2 HVAC PLAN

ELECTRICAL
E1.1 ELECTRICAL SITE PLAN
E2.1 ELECTRICAL PLAN
E3.1 ELECTRICAL DETAILS
E4.1 ELECTRICAL SCHEDULES

Contren™ Learning Series
Sheet Metal Level 2
Trainee Module 04205-02
Blueprints and Specifications
Blueprint A0.1
Print 1 of 17 prints
Not to Scale 2002 Revision

CODE INFORMATION

CODE USED:	1997 UBC
OCCUPANCY GROUP:	B
TYPE OF CONSTRUCTION:	TYPE V N
BUILDING AREA:	
MAIN FLOOR:	5,890 SQ. FT.
MECHANICAL ROOM:	324 SQ. FT.
TOTAL:	6,214 SQ. FT.

I HEREBY CERTIFY THAT THIS PLAN, SPECIFICATION, OR REPORT WAS PREPARED BY ME OR UNDER MY DIRECT SUPERVISION AND THAT I AM A DULY REGISTERED ARCHITECT UNDER THE LAWS OF THE STATE OF NORTH DAKOTA.
DATE: 5-30-2000 REG. NO. 578

TITLE SHEET

A
0.1

GERMANS FROM RUSSIA HERITAGE SOCIETY
BISMARCK, NORTH DAKOTA

RITTERBUSH - ELLIG - HULSING P.C.
ARCHITECTS - PLANNERS

REH

107F04.EPS

Figure 4 ◆ Indexes are often used in construction drawings to identify key information.

MODULE 00107-04 ◆ BASIC COMMUNICATION SKILLS 7.5

Distribute sets of instructions. Have trainees review and follow the directions. Ask trainees to identify any steps that are confusing or misleading.

Show Transparencies 3 and 4 (Figure 5). Explain that writing skills are essential to success in the construction industry. Review the Five C's of good writing.

Discuss techniques for better writing. Emphasize that longer is not necessarily better. Ask whether trainees have any questions.

Distribute a variety of work-related forms. Ask trainees to complete the forms accurately and completely. Discuss the importance of completing and submitting forms on time.

Discuss visual means that make main points easy for the reader to find.

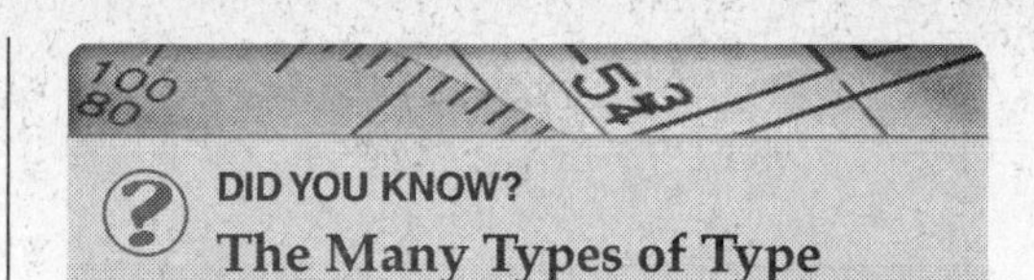

DID YOU KNOW?

The Many Types of Type

The printed word can be shown in many different ways. Depending on the message that the writer is trying to get across, the font (the shape of the letters), size, and color can all be changed. The shape and style of the letters are often used to show the structure of the text. For example, look at the text used to label the titles, sections, and subsections of this module. You will see that each one has its own style that is used consistently throughout.

Some common styles used in text include:

ALL CAPITAL LETTERS
SMALL CAPITAL LETTERS
Bold
Italics
<u>Underline</u>

When reading instructions or a series of steps for performing a task, such as turning on a welding machine, imagine yourself performing the task. You may find the steps easier to remember that way. Be sure to read all the directions through before you begin to follow them. Then you will be able to understand how all of the steps work together.

Always reread what you have just read to make sure you understand it. This is especially important when the reading material is complex or very long. And finally, take a break every now and then. If you read slowly or you find that you are getting tired or frustrated, put the material down and do something else to give your mind a break. Reading to understand is hard work!

2.2.0 Writing on the Job

At this stage in your career, you will probably do more reading on the job than writing. Nevertheless, writing skills are very important if you want to succeed in the construction industry. Construction workers write work orders, health and safety reports, **punch lists** (written lists that identify deficiencies requiring correction at completion), **memoranda** (informal office correspondence, or **memos** for short), work orders or change orders, and a whole range of other documents as part of their job (*Figure 5*).

Writing skills will allow you to perform these tasks too, as you gain experience on the job. They will help you move into positions of greater responsibility. It is never too early to start brushing up on your writing skills, so that when opportunity comes, you will be ready. This section reviews some common writing techniques that you can practice.

Being a good writer is not as hard as you might think. You do not need to be a master of English composition and grammar or have a college degree to write well. A few simple guidelines can make a lot of difference when you are writing on the job. Before you begin to write, organize in your mind what you want to write.

When you start to write, be clear and direct. More words do not mean better writing. Instead of writing "In order to prevent unwanted electrical shocks, workers should make sure they disconnect the main power supply before work commences," write "To avoid being shocked, disconnect the main power supply before starting." Instead of writing "at the present time," simply write "now."

Make your main point easy to find and include all the necessary details so the reader can understand what you are saying. One way to highlight important information is to list it. Lists can use **bullets** or numbers. Bullets are large dots that line

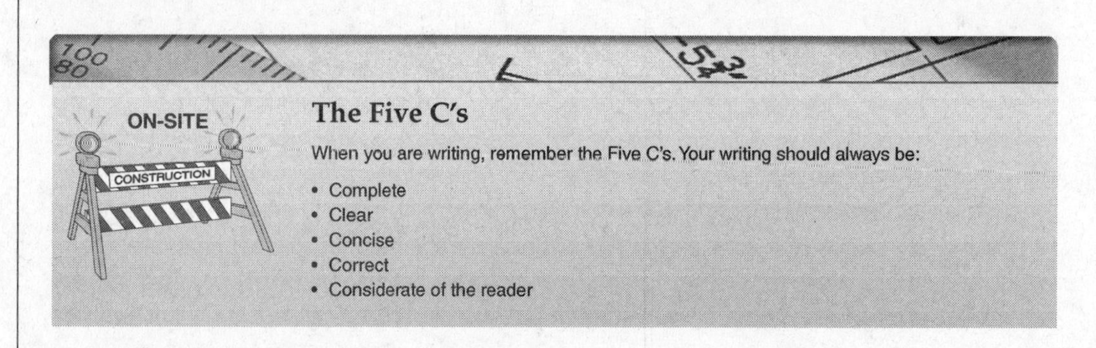

The Five C's

When you are writing, remember the Five C's. Your writing should always be:

- Complete
- Clear
- Concise
- Correct
- Considerate of the reader

Instructor's Notes:

7.6 CORE CURRICULUM ◆ INTRODUCTORY CRAFT SKILLS

Burning - Welding - Hot Work Permit

Valid from _____ to _____, _________ Master Card No.________________
(am/pm) (am/pm) DATE

1. Work Description
Equipment Location or Area ________________________________
Work to be done:

2. Gas Test

☐ None Required				
☐ Instrument Check	Test Results	Other Tests	Test Results
☐ Oxygen 20.8% Min			
☐ Combustible % LFL			

Gas Tester Signature Date Time

3. Special Instructions ☐ None ☐ Check with issuer before beginning work

4. Hazardous Materials ☐ None What did the line / equipment last contain?

5. Personal Protection ☐ Standard Equipment: welders hood with long sleeves; cutting goggles

☐ Goggles or Face Shield ☐ Respirator ☐ Forced Air Ventilation

☐ Standby Man ☐ Other, specify: ________________

6. Fire Protection ☐ None Required ☐ Portable Fire Extinguisher

☐ Fire Watch ☐ Fire Blanket ☐ Other, specify: ________________

7. Condition of Area and Equipment

Required Yes	No		THESE KEY POINTS MUST BE CHECKED
		a.	Lines disconnected & blanked or if disconnecting is not possible, blinds installed?
		b.	Lines steamed, purged, or otherwise properly cleared of combustibles?
		c.	Area and equipment satisfactorily clean of oil or combustibles?
		d.	Trenches, catch basins & sewer connections properly covered or sealed?
		e.	Immediate area and/or area under the work barricaded or roped off?
		f.	Adjoining equip. & operations checked to have any effect on the job?
		g.	Area fire suppression (fire water and sprinkler system) in service?

Comments

107F05A.EPS

Figure 5 ◆ A hot work permit is a typical written product in a construction project. (1 of 2)

MODULE 00107-04 ◆ BASIC COMMUNICATION SKILLS 7.7

Burning - Welding - Hot Work Permit

8. Approval	Permit Authorization			Permit Acceptance		
	Area Supv.	Date	Time	Maint. Supv./Engineer Contractor Supv.	Date	Time
Issued by						
Endorsed by						
Endorsed by						

9. Individual Review

I have been instructed in the proper Hot Work Procedures

Signed　　Signed

Persons Authorized to Perform Hot Work ______________ ______________

______________ ______________

______________ ______________

______________ ______________

______________ ______________

Fire Watch ______________ ______________

10. Job Completion

☐ Yes ☐ No Is the work on the equipment completed?

☐ Yes ☐ No Has the worksite been cleaned and made safe?

Workman answering above questions ______________________

Issuer's Acceptance ______________________

Forward to Production Superintendent within 7 days of job completion

107F05B.EPS

Figure 5 ◆ A hot work permit is a typical written product in a construction project. (2 of 2)

Instructor's Notes:

____

____

____

7.8 CORE CURRICULUM ◆ INTRODUCTORY CRAFT SKILLS

up vertically. When you finish writing, reread what you have written to make sure it is clear and accurate. Check for mistakes and take out words you don't need.

Accuracy is very important in the construction industry. If you are filling out a form, for example, to order supplies from a manufacturer's catalog, make sure you use the correct terms and quantities. When writing material takeoffs or other documents that will be used to purchase equipment and assign workers, you need to get it right. Errors cost money and time.

Bring in a variety of well-written and poorly written emails. Read aloud and compare both types of emails. Demonstrate how to improve the information in the poorly written emails.

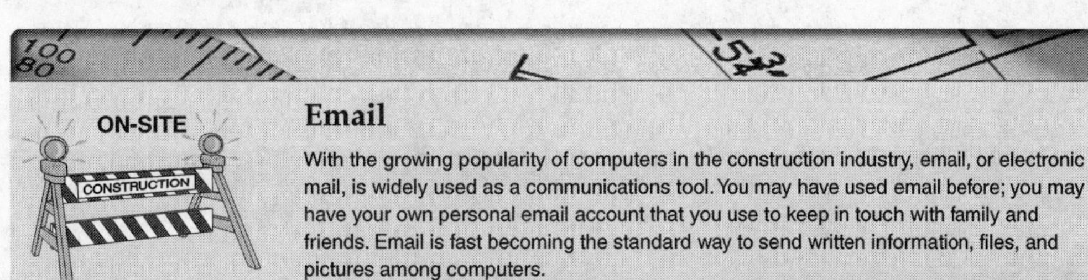

Email

With the growing popularity of computers in the construction industry, email, or electronic mail, is widely used as a communications tool. You may have used email before; you may have your own personal email account that you use to keep in touch with family and friends. Email is fast becoming the standard way to send written information, files, and pictures among computers.

Like printed letters and memos, email is used to ask and answer questions and to provide information. However, email differs from paper documents in several important ways. Most important of all, email is not private. You should not include private, sensitive, or confidential information in business emails. They can easily be sent by accident to people who are not authorized to read them. Also, in some cases email is not considered legally binding. Written documents with handwritten signatures remain the standard for contracts and other formal agreements. However, this situation is changing with the advent of **electronic signatures.**

```
From:
Date: Thu, Aug 24, 2000, 10:10 AM
To: <email_address> (E-mail 2)<email_address>
Cc: <email_address> (E-mail)<email_address> (E-mail)<email_address>
Subject: Task Status Report: Outstanding Issues

Updated August 24 with some M & E info.

> ------Original Task------
> Subject:      Outstanding Issues
> Priority:     High
>
> Status:       In Progress
> % Complete:   20  %
>
> Total work:   0 hours
> Actual work:  0 hours
>
> Requested by:
>
> ARCHITECTURAL:
>
> Snow Guards:
> *     Aug 23/00- AAA company has men in Talita, I spoke to Alex and he
> will have his men complete the snow guards this week.
>
> Cleanup of Non-Slip Treads in Stair #1:
> *     Aug 23/00-I left a voice mail for Mike asking him to complete the
> treads with the linseed oil as agreed to between Mike and Cleve.
> Mike called back, will proceed.
>
> Metal Shelving:
> *     Aug 23/00-Rick confirmed that the shelving was stolen from
> site. ABCD to advise.
>
> Water Leak in Atrium:
> *     Aug 18/00-Called Demetrium, John is in Talita, he will
> review leaks and repair the damage to the wall.
> *     Aug 23/00-Spoke to Alex, they will send a person on the roof
> to recheck caulking and caulk everything in the area of the leak and
```

107SA01.EPS

See the Teaching Tip for Section 2.0.0 at the end of this module.

Have each trainee perform Performance Tasks 1–3 to your satisfaction. Fill out a Performance Profile Sheet for each trainee.

Have trainees review Section 3.0.0 for the next session.

MODULE 00107-04 ◆ BASIC COMMUNICATION SKILLS 7.9

Go over the Review Questions for Section 2.0.0. Answer any questions trainees may have.

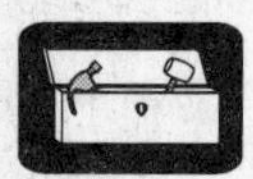

Ensure that you have everything required for the demonstration, laboratories, and testing during this session.

Refer to Figure 6. Provide examples of ways that verbal teaching occurs on the job.

Emphasize that people learn by listening, not by speaking.

Review Questions

Section 2.0.0

1. Textbooks, instruction manuals, and installation manuals are all examples of instructional documents.
   a. True
   b. False

2. Each of the following is an important reason for reading on the job *except* _____.
   a. it allows you to improve your employability skills
   b. it allows you to understand how other people's work affects yours
   c. it allows you to perform a task however you want
   d. it allows you to take on additional responsibilities as you gain experience

3. When you are looking for a specific installation procedure in a manual, the best way to find the information is to _____.
   a. identify the correct section and go straight there
   b. start reading the book from the beginning
   c. study the terms in the glossary
   d. skim the book and take notes on all important information

4. Skimming or scanning chapter titles and section headings before you start to read helps you _____.
   a. explain the book's contents to your instructor
   b. find specific information right away
   c. highlight important passages
   d. organize the material in your mind

5. Say you read the following sentence in an instruction manual: "To prevent the accidental occurrence of a crown venting situation, it should be noted that continuous vents should be installed at a sufficient distance from fixture traps so as to prevent the occurrence of such an event." You could rewrite the sentence more clearly as _____.
   a. "Crown venting situations can be avoided by the installation of continuous vents."
   b. "To prevent crown venting, install fixture traps close to continuous vents."
   c. "Install continuous vents far enough away from fixture traps to avoid crown venting."
   d. "Install crown vents far from fixture traps to avoid the continuous vent."

3.0.0 ◆ LISTENING AND SPEAKING SKILLS

Every day on the job can be a learning experience. The more you learn, the more you will be able to help others learn too (*Figure 6*). One of the most effective methods of learning and teaching is through verbal communication—that is, through speaking and listening. As a construction professional, you need to be able to state your ideas clearly. You also need to be able to listen to and understand ideas that other people express. The following are some of the ways that verbal teaching and learning takes place on the job:

- Giving and taking instructions
- Offering and listening to presentations
- Participating in team discussions
- Talking with your teammates and your supervisor
- Talking with clients

At this stage in your career, you will probably do more listening than speaking. You may be wondering why it is so important to be a good listener. The answer is simple: experience. People learn by listening, not by speaking. You are only beginning to learn how the construction industry

PHOTO COURTESY OF OAK RIDGE NATIONAL LABORATORY 107F06.TIF

Figure 6 ◆ Teaching and learning are often accomplished by speaking and listening.

Instructor's Notes:

7.10 CORE CURRICULUM ◆ INTRODUCTORY CRAFT SKILLS

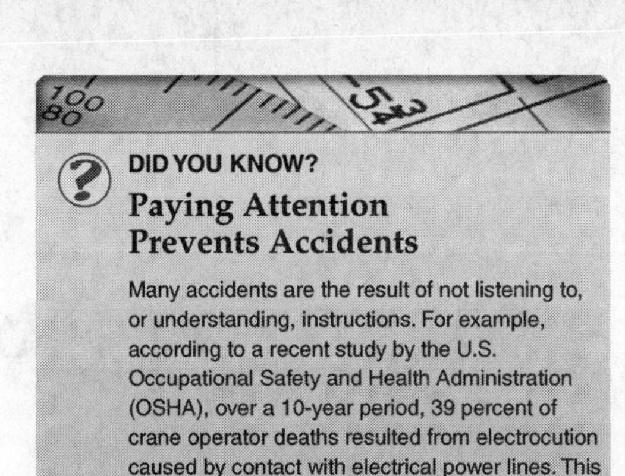

DID YOU KNOW?

Paying Attention Prevents Accidents

Many accidents are the result of not listening to, or understanding, instructions. For example, according to a recent study by the U.S. Occupational Safety and Health Administration (OSHA), over a 10-year period, 39 percent of crane operator deaths resulted from electrocution caused by contact with electrical power lines. This was the single largest cause of death in the study. How many of those accidents could have been prevented if the operator had paid attention to safety instructions before climbing into the cab? Do not become a statistic like that.

works, and there is a lot to learn! Teachers, supervisors, and experienced workers can guide you to make sure you are learning what you need to know (*Figure 7*).

3.1.0 Active Listening on the Job

You might think that listening just happens automatically, that someone says something and someone else hears it. However, real listening, the process not only of hearing, but of understanding what is said, is an active process. You have to be involved and paying attention to really listen. Understanding comes from **active listening** (*Figure 8*). You must develop good listening skills to be able to listen actively. This section presents some tips and suggestions that you can use to develop good listening skills.

107F07.EPS

Figure 7 ◆ Your supervisor can help you learn what you need to know.

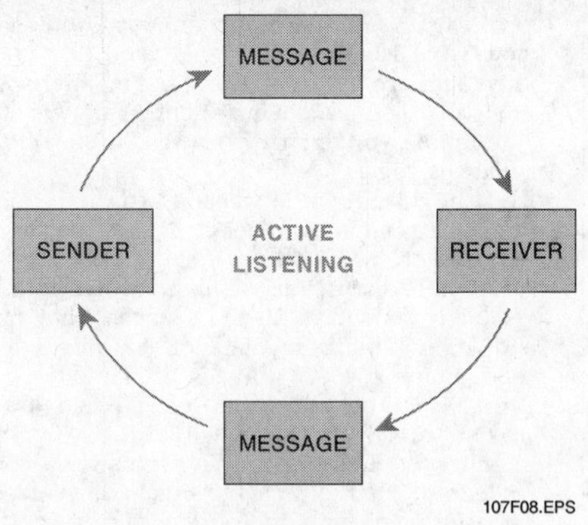

107F08.EPS

Figure 8 ◆ Active listening process.

First of all, you should understand the possible consequences of not listening. Poor listening skills can cause mistakes that waste time and money (*Figure 9*). Stay focused and do not let your mind wander. One way to do this is to make eye contact with the person who is speaking. Try to keep an

107F09.EPS

Figure 9 ◆ Safety violations can result from not understanding consequences.

Show Transparency 5 (Figure 8). Discuss active listening skills, and explain what it means to be listening actively.

Refer to Figure 9. Review accidents that can result from poor listening skills.

Discuss 10 techniques for active listening skills. Review common signs of barriers to listening. Go over any questions trainees may have.

Ask trainees to take turns telling you details about a specific task. Demonstrate skills for active listening.

Refer to Figure 12. Discuss the importance of speaking clearly.

Review techniques for speaking effectively. Ask whether trainees have any questions.

Stress the importance of communicating in a professional manner.

open mind; never tune out because you think you know what is being said. Make sure that your **body language,** or your posture and mannerisms, shows that you are paying attention (*Figure 10*). For example, you can nod your head to show that you are listening.

If you do not understand a word or trade term, or if something is not clear, ask questions. Take notes to help you remember better. At the end of the discussion, summarize everything you have heard back to the speaker. That will help you find out immediately if you misunderstood anything.

Just as there are things that you can do to listen effectively, there are things that act as barriers to listening. Just like the warning signs on a construction site that indicate hazards to be avoided, there are signs that indicate problems in the process of listening and understanding (*Figure 11*). Here are some common warning signs:

- Talking instead of listening
- Tuning out when someone is speaking
- Getting distracted
- Interrupting the speaker
- Finishing the speaker's sentences
- Letting ego get in the way

3.2.0 Speaking on the Job

Effective listening depends on effective speaking. After all, you cannot be expected to understand what has not been made clear to you. Look at the following three examples of sentences spoken by one worker to another. Which one is the clearest and most effective? Which one would you like to hear if you were the listener?

- "Hand me that tool there."
- "Hand me the grinder on that bench."
- "Hand me the 4-inch angle grinder that's on the bench behind you."

107F10.EPS

Figure 10 ◆ Body language shows whether you are paying attention.

107F11.EPS

Figure 11 ◆ Learn to read the warning signs of listening problems.

The third example has enough information for you to identify the correct tool and its location. You do not have to ask the speaker, "Which tool? Where is it?" You will not accidentally give the other worker the wrong tool. As a result, time is not wasted trying to clear up confusion (*Figure 12*). The time it takes for someone to stop what he or she is doing and explain something again because it was not clear the first time is time that the job is not getting done. Time lost this way can add up very quickly.

One of the best ways to learn to speak effectively is to listen to someone who speaks well. Think about what makes that person such an effective speaker. Is it the person's choice of words? Or perhaps the body language? Or their ability to

107F12.EPS

Figure 12 ◆ Use clear and effective statements when speaking on the job.

Instructor's Notes:

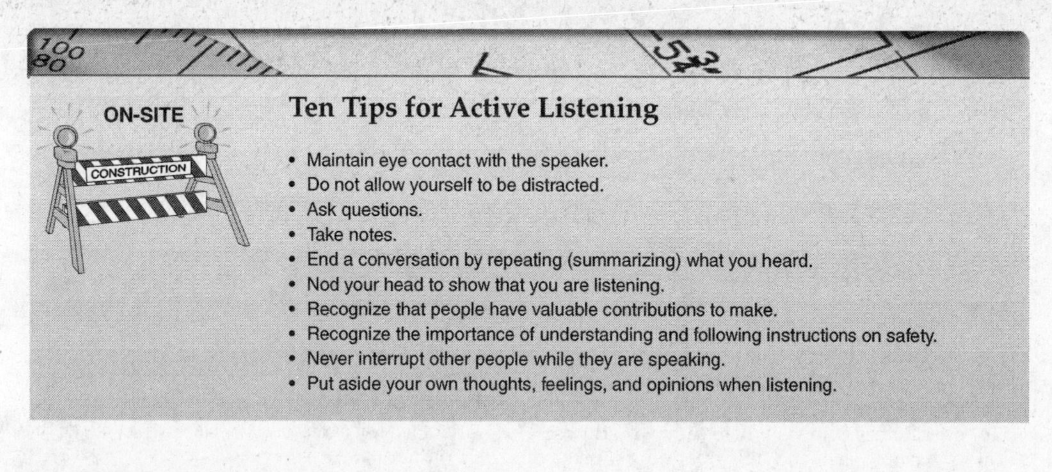

ON-SITE

Ten Tips for Active Listening

- Maintain eye contact with the speaker.
- Do not allow yourself to be distracted.
- Ask questions.
- Take notes.
- End a conversation by repeating (summarizing) what you heard.
- Nod your head to show that you are listening.
- Recognize that people have valuable contributions to make.
- Recognize the importance of understanding and following instructions on safety.
- Never interrupt other people while they are speaking.
- Put aside your own thoughts, feelings, and opinions when listening.

Demonstrate a variety of body language postures. Ask trainees to identify what the postures and mannerisms say about how you are (or are not) listening.

Have two trainees demonstrate a work-related telephone call. Ask trainees to take notes of the call. Have trainees share the details from the call.

make something complex sound simple? Keep those things in mind when you speak, and they will make a difference for you.

Think about what you are going to say before you say it. As with writing, take time to organize your topic logically. Choose an appropriate place and time. For example, if you need to give detailed assembly instructions to your team, pick a quiet place, and do not hold the meeting just before lunch.

Encourage your listeners to take notes if necessary. Do not overexplain if people are already familiar with the topic. Always speak clearly, and maintain eye contact with the person or people you are speaking to. When using **jargon,** or terms that people outside the industry may not understand, be sure that everyone knows what the terms mean. Give your listeners enough time to ask questions, and take the time to answer questions thoroughly. Finally, when you are finished, make sure that everyone understands what you were saying.

3.2.1 Placing Telephone Calls

You may remember when telephones were anchored to walls and desks. To make or receive a phone call, you had to stop what you were doing and go to the telephone. Today, **cell phones** allow you to make and receive phone calls from just about anywhere on a construction site. Cell phones can distract you from your job, so never make or

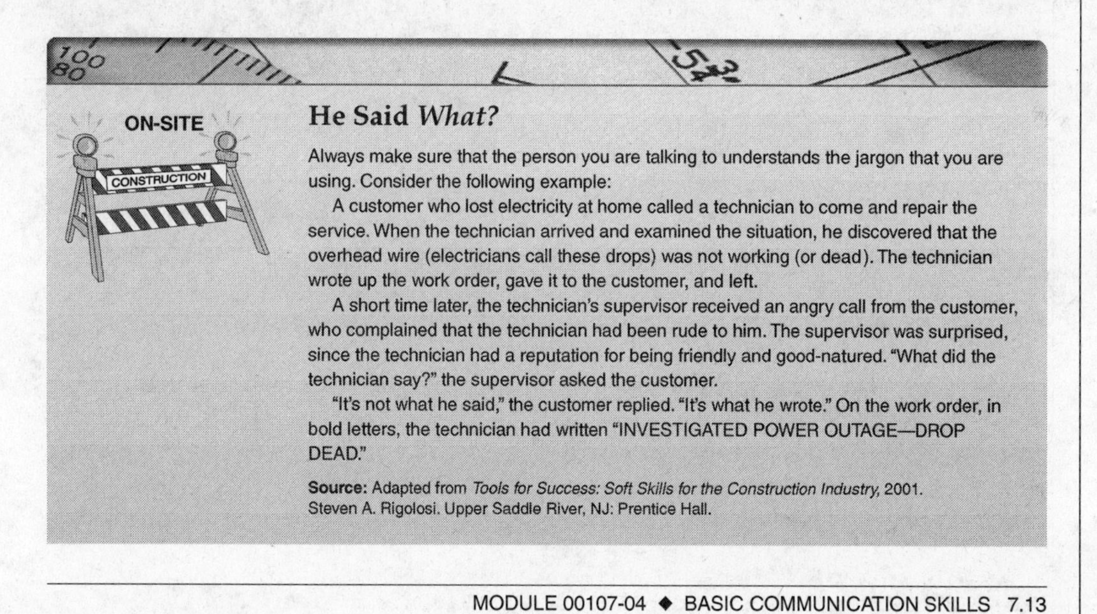

ON-SITE

He Said *What?*

Always make sure that the person you are talking to understands the jargon that you are using. Consider the following example:

A customer who lost electricity at home called a technician to come and repair the service. When the technician arrived and examined the situation, he discovered that the overhead wire (electricians call these drops) was not working (or dead). The technician wrote up the work order, gave it to the customer, and left.

A short time later, the technician's supervisor received an angry call from the customer, who complained that the technician had been rude to him. The supervisor was surprised, since the technician had a reputation for being friendly and good-natured. "What did the technician say?" the supervisor asked the customer.

"It's not what he said," the customer replied. "It's what he wrote." On the work order, in bold letters, the technician had written "INVESTIGATED POWER OUTAGE—DROP DEAD."

Source: Adapted from *Tools for Success: Soft Skills for the Construction Industry,* 2001. Steven A. Rigolosi. Upper Saddle River, NJ: Prentice Hall.

MODULE 00107-04 ◆ BASIC COMMUNICATION SKILLS 7.13

Emphasize that workers should never use a phone while driving or operating heavy equipment.

Refer to Figure 14. Review proper phone etiquette for receiving a telephone call.

See the Teaching Tip for Section 3.0.0 at the end of this module.

A Sense of Humor

A special tip: Maintain a sense of humor! A good sense of humor will get you through many situations. As a construction professional, you should always take yourself seriously. This means speaking well and conveying the proper professional attitude. But your listeners will always appreciate you more if you show them that you have a sense of humor. Remember, though, never to tell off-color or offensive jokes or play practical jokes.

receive personal calls while working. Wait until a designated break time to make or receive calls. Do not operate cell phones where they would pose a safety hazard, such as while operating a piece of machinery or a power tool or driving a vehicle.

When you speak to people face-to-face, you can see them and judge how they react to what you say. When you are on the telephone, you don't have these clues. Effective speaking is all the more important in such cases. When making a call, start by identifying yourself and ask whom you are speaking to. Speak clearly and explain the purpose of your call. Take notes to help you remember the conversation later (*Figure 13*).

If you leave a message for someone, keep it brief. Prepare your message ahead of time so you will know what to say. Be sure to leave a number where you can be reached and the best time to reach you.

WARNING!

Never make or receive phone calls while driving or operating heavy equipment. Not only is it extremely dangerous, it is also illegal in many states.

3.2.2 Receiving Telephone Calls

When you receive a call, identify yourself immediately. Don't keep people on hold, either on the phone or in person. People resent it. Transfer calls courteously and introduce the caller to the recipient. Keep your calls brief. Finally, never talk on the phone in front of teammates, supervisors, or customers (*Figure 14*). This is rude and unprofessional behavior.

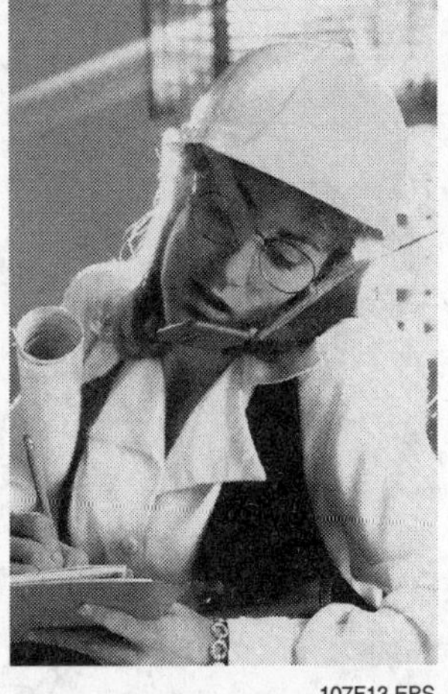

107F13.EPS

Figure 13 ◆ Take notes to help you remember important details.

107F14.EPS

Figure 14 ◆ Never talk on the phone in front of teammates, supervisors, or customers.

Instructor's Notes:

7.14 CORE CURRICULUM ◆ INTRODUCTORY CRAFT SKILLS

Review Questions

Section 3.0.0

1. Each of the following is an example of positive verbal communication on the job *except* _____.
   a. giving and taking instructions
   b. tuning out when someone is speaking
   c. offering and listening to presentations
   d. participating in team discussions

2. Real listening is a passive process.
   a. True
   b. False

3. A supervisor wants an apprentice to insert a plug into a water supply line for a pressure test. The clearest way to instruct the apprentice would be for the supervisor to say _____.
   a. "Put the plug in."
   b. "Insert that one plug into the pipe."
   c. "Get the water supply line ready for the pressure test."
   d. "Insert the plug into the end of the water supply line."

4. A teammate asks you to hand her a tool, but does not specify which tool she needs. The most appropriate response would be _____.
   a. "Get it yourself."
   b. "There are a lot of tools over here; how would I know which one you want?"
   c. "I'm sorry, which tool do you need?"
   d. "I don't understand what you mean."

5. It is rude to talk on the telephone in front of other people.
   a. True
   b. False

Go over the Review Questions for Section 3.0.0 and the Key Terms Quiz. Answer any questions trainees may have.

Administer the Module Examination. Be sure to record the results of the Exam on Craft Training Report Form 200, and submit the results to the Training Program Sponsor.

Have each trainee perform Performance Task 4 to your satisfaction. Fill out a Performance Profile Sheet for each trainee.

Ensure that all Performance Tests have been completed and Performance Profile Sheets for each trainee are filled out. If desired, trainee proficiency noted during laboratory sessions may be used to complete the performance test. Be sure to record the results of the testing on Craft Training Report Form 200, and submit the results to the Training Program Sponsor.

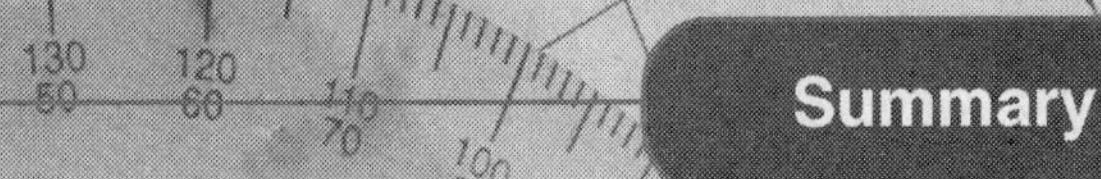

Summary

Communications skills—the ability to read, write, listen, and speak effectively—are essential for success in the construction workplace. Effective communication ensures that work is done correctly, safely, and on time. The construction industry relies on a wide variety of written documents. You must be able to read documents that apply to your job. As your responsibilities increase, you will be called on to write some of these documents yourself. This module introduced you to simple tips and techniques that you can use every day to read and write efficiently.

Construction professionals state their ideas clearly, and they also listen to the ideas of others. This means not only hearing, but understanding what is said. Real listening is an active process. Listeners are involved with the conversation, and they pay attention to the person who is speaking. Not listening causes mistakes, which waste time and money. Of course, effective listening depends on effective speaking. You should know how to speak effectively, even though at this stage in your career you will probably be listening more than speaking. No one can understand what is not expressed clearly. When using jargon, be sure that your listeners understand the terms that you are using. Once you master these basic communication skills and add them to your mental toolbox, you will find that you can succeed at more things than you ever thought possible, both inside and outside of the construction industry.

Notes

Instructor's Notes:

____

____

____

7.16 CORE CURRICULUM ◆ INTRODUCTORY CRAFT SKILLS

Key Terms Quiz

Fill in the blank with the correct key term that you learned from your study of this module.

1. Often, books will include additional detailed information in a(n) _________.
2. In a list, a large dot is called a(n) _________.
3. Words and numbers can be printed using many different _________, or styles.
4. A(n) _________ includes the definition of terms used in a book.
5. Information can be presented in picture form using a(n) _________.
6. To find an alphabetical listing of topics in a book, consult the _________.
7. _________ are slanted print used to emphasize text.
8. Construction workers write _________ to list deficiencies requiring correction at completion.
9. You can use a(n) _________ to send an informal written message to someone in your company.
10. You can find a book's chapters and headings listed in a(n) _________.
11. Numerical or written information can be presented for quick visual scanning in a(n) _________.
12. A(n) _________ is another way to sign documents.
13. Good listening is an essential part of the _________ process.
14. Your _________ silently communicates whether or not you are paying attention to a speaker.
15. Never make or receive calls on a(n) _________ while operating heavy machinery.
16. As you gain experience, you will learn more of the special _________ that you can use to communicate with other construction professionals.

Key Terms

Active listening
Appendix
Body language
Bullets
Cell phone
Electronic signature
Font
Glossary
Graph
Index
Italics
Jargon
Memo
Memorandum
Permit
Punch list
Table
Table of contents

MODULE 00107-04 ◆ BASIC COMMUNICATION SKILLS 7.17

Profile in Success

Steven Miller

Construction Trades Consultant
North Carolina Department of Public Instruction
Raleigh, North Carolina

Steven grew up on a Pennsylvania farm. From an early age Steven knew that he wanted to follow in the footsteps of his father, who taught metalworking, woodworking, and carpentry. After high school Steven served in the army as a combat construction specialist, where he built obstacle courses, minefields, and bridges. The army provided him with both money for college and valuable construction experience. After graduating from Millersville University, Steven moved to North Carolina, where he did residential renovation and restoration work and also worked as a machinist in a tool and die shop.

Steven taught architectural drafting and construction courses in North Carolina public schools before getting his masters in school administration. He is currently completing his doctorate in education leadership. He is an NCCER Master Trainer.

How did you become interested in the construction industry?

My first introduction to the construction field was as a go-fer for my dad! From a very early age, I have had a sense of crafts and trades work. My life and my careers have followed the two tracks of education and construction work. They are very closely linked for me. For example, when I was at Millersville University in Pennsylvania, I studied industrial arts, technology education, and industrial technology. I chose to stay an extra semester to do student teaching, and I also took some additional classes in those areas. When I moved to North Carolina, there was nothing available in teaching, so I worked in industry for two years before I got a teaching job in a public school.

What are some of the things you do in your job?

Being a construction trades consultant means that I travel all over the state visiting schools and program sites. I work with all the teachers in North Carolina who teach carpentry, commercial and residential electrical wiring, masonry, and furniture and cabinetmaking. I also work with student organizations and help facilitate student contests. I do professional development, training, testing, and accountability for technical educators. I do curriculum development as well. Most of our construction trades curriculum is developed from the different NCCER curricula.

What do you like most about the construction work that you've done?

I have always liked the feeling of being worn out at the end of the day. When I build something there is a sense of accomplishment, the knowledge that I've created something with my hands, something that would not have been there if it had not been for my work. And I have the same feeling for my job in the Department of Public Instruction. I get to keep busy. I don't ever seem to stay still!

What do you think it takes to be a success in your trade?

I've always found that, for a person coming into the trades, if they are capable of thinking through the process and keeping one step ahead of the people they are working with, and if they are prepared for hard work, they will succeed. You know the old saying, "an honest day's wage for an honest day's work?" Well, it's a very true saying.

Instructor's Notes:

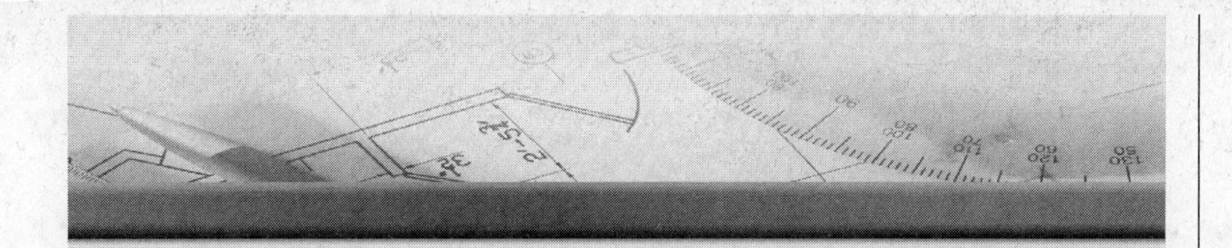

Successful workers tend to rise to the top because they are good at thinking things through. If they encounter new things that they don't know, they will tend to grasp those things more intuitively. The change is just something they deal with.

What would you say to someone entering the trades today?
You always have to push yourself to do new and better things. Along with that comes education and a willingness to try new things and conquer them. If you realize that things are going to change, and you educate yourself to be able to work through those changes, then that kind of positive outlook will serve you well. Also, remember to work safely. Accidents happen so swiftly that you just don't have time to react. But if you are working safely, then the accident won't happen in the first place.

MODULE 00107-04 ◆ BASIC COMMUNICATION SKILLS 7.19

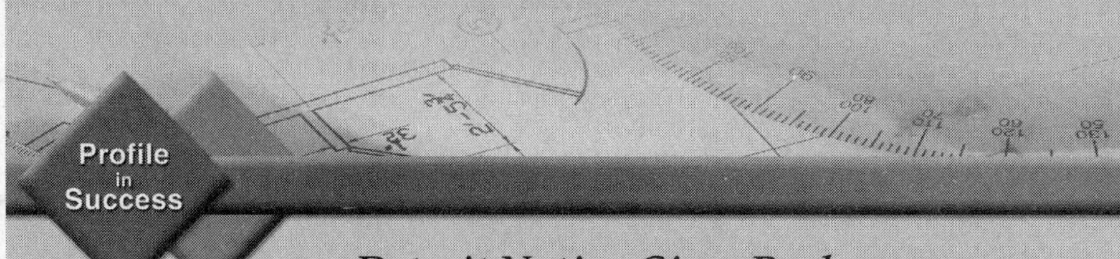

Detroit Native Gives Back to the Community

By Lia Steakley

Michigan ranked dead last in interstate bridge upkeep and near the bottom in pavement repair on interstates, despite record spending on road improvements, according to a report by the Road Information Program. However, Edzra Gibson is working to change that.

A graduate of the University of Michigan, Gibson went to work for the Detroit office of Howard, Needles, Tammen, & Bergendoff, where he has a hand in several highway and bridge improvement projects. "Being from Detroit, I'm studying things I've been driving all my life. It's one of the oldest freeway systems in the country and a lot of stuff is falling down and now they are rebuilding a lot of it," says the 28-year-old construction engineer/inspector. Starting as an intern with Howard, Needles, Tammen, & Bergendoff, Gibson assisted with a study of the Interstate 94 corridor and maintenance and traffic issues involving the detour for the M39 project, which wrapped up construction in 2001.

Currently, he is working on the reconstruction of Interstate 75 in the metropolitan Detroit area, which will be completed by September. He is also part of the design team on a reconstruction project involving North America's number one international border crossing, the Ambassador Bridge. Opened in 1929, it spans the Detroit River to connect Detroit with Windsor, Canada and carries about 8 million cars, 2.7 million trucks, and $60 billion in trade a year. The reconstruction will allow better security and truck passing by adding a second span to increase the capacity of bridge traffic, which is expected to double as early as 2012.

Gibson graduated from the University of Michigan in 1998 with a B.S. degree in civil/environmental engineering, but he did more than build models as a student. The Detroit native was a member of the university's varsity track team and competed in the 100-, 200-, and 400-meter events. In 1995, he and a handful of students created the Black Volunteer Network, a nonprofit organization that helps the less fortunate in surrounding neighborhoods. It assists Habitat for Humanity, mentors high school students, lends a hand at homeless shelters, and organizes an annual basketball tournament for the community. "Some friends that were on the track team and I felt we were privileged because we made it to college," says Gibson. "My parents sacrificed so my two sisters and I always had what we needed. We decided we needed to give back to the community." Gibson also found time to be involved with the National Black Engineers Association.

The son of a Ford Motor Corp. engineer, Gibson says he has always been interested in the technical side of things, which led him to a career in engineering. During his time with Howard, Needles, Tammen, & Bergendoff, he has worked as both a designer and on the construction site.

"With construction, the gratification is right there. You can see it in a day, whereas design is a longer process," says Gibson. "Working at the job site is more exciting and I think at the end of the day the fulfillment is a little more."

Over the next couple of years, Gibson plans to continue on the path he's started and expand his industry portfolio to include work on buildings, highways, and power plants.

"I'm trying to diversify my design package," says Gibson.

Source: Reprinted from *Engineering News-Record*, June 17, 2003.

Instructor's Notes:

Trade Terms Introduced in This Module

Active listening: A process that involves respecting others, listening to what is being said, and understanding what is being said.

Appendix: A source of detailed or specific information placed at the end of a section, a chapter, or a book.

Body language: A person's physical posture and gestures.

Bullets: Large, vertically aligned dots that highlight items in a list.

Cell phone: A mobile radiotelephone that uses a network of short-range transmitters located in overlapping cells throughout a region, with a central station making connections to regular telephone lines.

Electronic signature: A signature that is used to sign electronic documents by capturing handwritten signatures through computer technology and attaching them to the document or file.

Font: The type style used for printed letters and numbers.

Glossary: An alphabetical list of terms and definitions.

Index: An alphabetical list of topics, along with the page numbers where each topic appears.

Italics: Letters and numbers that lean to the right rather than stand straight up.

Jargon: Specialized terms used in a specific industry.

Memo: Another term for memorandum.

Memorandum: Informal written correspondence. The plural of memorandum is memoranda.

Permit: A legal document that allows a task to be undertaken.

Punch list: A written list that identifies deficiencies requiring correction at completion.

Table: A way to present important text and numbers so they can be read and understood at a glance.

Table of contents: A list of book chapters or sections, usually located at the front of the book.

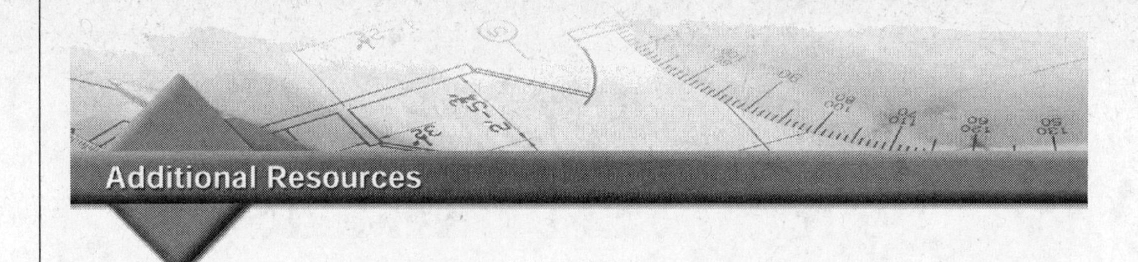

Additional Resources

This module is intended to present thorough resources for task training. The following reference works are suggested for further study. These are optional materials for continued education rather than for task training.

Communicating at Work. Tony Alessandra and Phil Hunsaker. New York, NY: Simon and Schuster.

Communicating in the Real World: Developing Communication Skills for Business and the Professions. Terrence G. Wiley and Heide Spruck Wrigley. Englewood Cliffs, NJ: Prentice Hall.

Communication Skills for Business and Professions. Paul R. Timm and James A. Stead. Upper Saddle River, NJ: Prentice Hall.

Elements of Business Writing. Gary Blake and Robert W. Bly. New York, NY: Collier.

Improving Business Communication Skills. Deborah Britt Roebuck. Upper Saddle River, NJ: Prentice Hall.

Instructor's Notes:

7.22 CORE CURRICULUM ◆ INTRODUCTORY CRAFT SKILLS

MODULE 00107-04 — TEACHING TIPS

The following are suggested activities or instructional methods to help you teach the material in this AIG.

General

When you call on someone to answer a question, the rest of the class relaxes or even tunes out because they expect that the question and answer will take place only between you and the trainee you called on. Instead, use this technique to involve more trainees in answering questions and to keep them on their toes.

1. Ask trainees to define a term or explain a concept.
2. After one trainee has answered, ask a trainee seated nearby if the answer is right. Then ask whether a trainee in the back of the room agrees.
3. Ask trainees to explain why they think an answer is right or wrong.
4. Use the session to clear up incorrect ideas, and encourage trainees to learn from their mistakes.

Section 2.0.0

Reading and Writing Skills

You will need a written set of instructions for this exercise, which will allow trainees to practice their reading and writing skills. Trainees will need their Trainee Guides and pencils and paper. Allow 30 minutes for this exercise.

1. Distribute a set of instructions for a commonly performed construction task.
2. Ask trainees to read the set of instructions. Then divide the class into small groups.
3. Have each group review the instructions and write either a memo or an email to a co-worker explaining the task that needs to be accomplished and what items will be needed to perform the task.
4. Compare the groups' memos and emails and see which contain the most details. Evaluate the accuracy of the written communication.

Section 3.0.0

Listening and Speaking Skills

This exercise will allow trainees to practice their listening and speaking skills. Trainees will need their Trainee Guides and pencils and paper. Allow 30 minutes for this exercise.

1. Divide the class into four groups: speakers, listeners, writers, and readers.
2. Assign the speakers a list of points to communicate to the trainees.
3. Have the speakers practice communicating the points.
4. Have the listeners take notes on the points communicated and summarize the information for the writers.
5. Have the writers draft a memo outlining the points.
6. Have the readers review what was communicated and assess the success of the process.

MODULE 00107-04 — ANSWERS TO REVIEW QUESTIONS

Section 2.0.0

1. a
2. c
3. a
4. d
5. c

Section 3.0.0

1. b
2. b
3. d
4. c
5. a

MODULE 00107-04 — ANSWERS TO KEY TERMS QUIZ

1. appendix
2. bullet
3. fonts
4. glossary
5. graph
6. index
7. Italics
8. punch lists
9. memo or memorandum
10. table of contents
11. table
12. electronic signature
13. active listening
14. body language
15. cell phone
16. jargon

The NCCER makes every effort to keep these textbooks up-to-date and free of technical errors. We appreciate your help in this process. If you have an idea for improving this textbook, or if you find an error, a typographical mistake, or an inaccuracy in NCCER's *Contren®* textbooks, please write us, using this form or a photocopy. Be sure to include the exact module number, page number, a detailed description, and the correction, if applicable. Your input will be brought to the attention of the Technical Review Committee. Thank you for your assistance.

Instructors – If you found that additional materials were necessary in order to teach this module effectively, please let us know so that we may include them in the Equipment/Materials list in the Annotated Instructor's Guide.

Write: Product Development
National Center for Construction Education and Research
P.O. Box 141104, Gainesville, FL 32614-1104

Fax: 352-334-0932

E-mail: curriculum@nccer.org

Craft ____________ Module Name ____________

Copyright Date ____________ Module Number ____________ Page Number(s) ____________

Description ____________

____

____

____

(Optional) Correction ____________

____

____

(Optional) Your Name and Address ____________

____

____

Basic Employability Skills

00108-04

NCCER STANDARDIZED CRAFT TRAINING PROGRAM

The National Center for Construction Education and Research (NCCER) provides a standardized national program of accredited craft training. Key features of the program include instructor certification, competency-based training, and performance testing. The program provides trainees, instructors, and companies with a standard form of recognition through a National Craft Training Registry. The program is described in full in the *Guidelines for Accreditation*, published by the NCCER. For more information on standardized craft training, contact the NCCER by writing us at P.O. Box 141104, Gainesville, FL 32614-1104; calling 352-334-0911; or e-mailing info@nccer.org. More information may be found at our Web site, www.nccer.org.

HOW TO USE THIS ANNOTATED INSTRUCTOR'S GUIDE

Each page presents two sections of information. The larger section displays each page exactly as it appears in the Trainee Module. The narrow column ties suggested trainee and instructor actions to each page and provides icons (detailed below) to call your attention to material, safety, audiovisual, or testing requirements. The bottom of each page includes space for your notes.

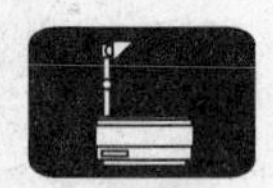

The **Audiovisual** icon indicates an appropriate time to show a transparency or other audiovisual aid.

The **Classroom** icon prompts you to define a term, stress a point, ask trainees to explain a concept, or give examples.

The **Demonstration** icon directs you to show trainees how to perform tasks.

The **Examination** icon tells you to administer the written module examination.

The **Homework** icon is placed where you may wish to assign reading for the next class, to assign a project, or to advise trainees to prepare for an examination.

The **Laboratory** icon is used when trainees are to practice performing tasks.

The **Materials** icon is a reminder for you to gather materials needed for classes, labs, and testing.

The **Performance Testing** icon tells you to administer a performance test or a portion thereof.

The **Safety** icon is used to emphasize safety issues. It is often keyed to *Caution* and *Warning* statements in the Trainee Module.

The **Teaching Tip** icon indicates additional guidance is available, such as how to conduct an exercise, get the most educational value from a field trip, or encourage class participation. Teaching Tips may expand on a feature (*Think About It*, *Did You Know?*) or provide Quick Quizzes or similar exercises. You will be referred to the Teaching Tips section at the back of the module if there is additional material.

The **Combination** icon indicates that the laboratory listed corresponds with a performance task. If desired, you can note the proficiency of the trainees during the laboratory and use it to satisfy performance testing requirements.

PREPARATION

Before teaching this module, you should review the Objectives, Performance Tasks, Materials and Equipment List, and the Module Outline. Be sure to allow ample time to prepare your own training or lesson plan and gather all required materials and equipment.

MODULE OVERVIEW

This module discusses basic employability skills. Trainees will learn how to effectively use critical thinking, computer, and relationship skills in the construction industry. This module will also increase trainee awareness of such workplace issues as sexual harassment, stress, and substance abuse.

PREREQUISITES

Prior to training with this module, it is recommended that the trainee shall have successfully completed the following: *Core Curriculum: Introductory Craft Skills,* Modules 00101-04 through 00105-04. This module is an elective. To receive a successful completion, you must take this module and Module 00107-04 or Module 00106-04.

OBJECTIVES

Upon completion of this module, the trainee will be able to:

1. Explain the construction industry, the role of the companies that make up the industry, and the role of individual professionals in the industry.
2. Demonstrate critical thinking skills and the ability to solve problems using those skills.
3. Demonstrate knowledge of computer systems, and explain common uses for computers in the construction industry.
4. Demonstrate effective relationship skills with teammates and supervisors, the ability to work on a team, and appropriate leadership skills.
5. Be aware of workplace issues such as sexual harassment, stress, and substance abuse.

PERFORMANCE TASKS

Under the supervision of the instructor, the trainee should be able to:

1. Prepare and submit a complete employment application.
2. Demonstrate the ability to access, retrieve, and print from the following basic software programs:
   - Email
   - Databases
   - Internet
3. Divide into teams of three or more students, elect a team leader/presenter and a recorder, and develop a list detailing how employees can affect their company's profitability and its ability to reward employees. The recorder should scribe the punch list. The leader should present the list to the class and to the instructor.

MATERIALS AND EQUIPMENT LIST

Transparencies
Markers/chalk
Blank acetate sheets
Transparency pens
Pencils and scratch paper
Overhead projector and screen
Whiteboard/chalkboard
Copies of your local code
Company mission statement
Job listings from newspapers, trade magazines, and the internet
Copies of blank job applications from area companies
Figure 10 with the callouts covered
Computer reference books
Laptop computer (if available)
Excerpts from federal laws prohibiting job discrimination
News articles highlighting workplace incidents, including:
  Harassment
  Stress
  Drug and alcohol abuse
Module Examinations*
Performance Profile Sheets*

*Located in the Test Booklet.

SAFETY CONSIDERATIONS

Ensure that the trainees are equipped with appropriate personal protective equipment. Always work in a clean, well-lit, appropriate work area.

ADDITIONAL RESOURCES

This module is intended to present thorough resources for task training. The following reference works are suggested for both instructors and motivated trainees interested in further study. These are optional materials for continued education rather than for task training.

Art and Science of Leadership, 2000. Afsaneh Nahavandi. Upper Saddle River, NJ: Prentice-Hall.

Computer Numerical Control, 1997. Jay Stenerson. Upper Saddle River, NJ: Prentice-Hall.

Introduction to Computer Numerical Control, 2002. James Valentino. Upper Saddle River, NJ: Prentice-Hall.

Tools for Teams: Building Effective Teams in the Workplace, 2003. Craig Swenson, ed. Leigh Thompson, Eileen Aranda, Stephen P. Robbins. Boston, MA: Pearson Custom Publishing.

Your Attitude Is Showing, 1999. Elwood M. Chapman. Upper Saddle River, NJ: Prentice-Hall.

TEACHING TIME FOR THIS MODULE

An outline for use in developing your lesson plan is presented below. Note that each Roman numeral in the outline equates to one session of instruction. Each session has a suggested time period of 2½ hours. This includes 10 minutes at the beginning of each session for administrative tasks and one 10-minute break during the session. Approximately 15 hours are suggested to cover *Basic Employability Skills.* You will need to adjust the time required for hands-on activity and testing based on your class size and resources. Because laboratories often correspond to Performance Tasks, the proficiency of the trainees may be noted during these exercises for Performance Testing purposes.

Topic	Planned Time
Session I. The Construction Business	
A. Entering the Construction Workforce	________
B. Entrepreneurship	________
C. Performance Testing (Task 1)	________
Session II. Critical Thinking Skills	
A. Barriers to Problem Solving	________
B. Solving Problems Using Critical Thinking Skills	________
C. Problems with Planning and Scheduling	________
Session III. Computer Skills	
A. Computer Terms	________
B. Basic Software Packages	________
C. Electronic Mail (Email)	________
D. Computers in the Construction Industry	________
E. Performance Testing (Task 2)	________
Session IV. Relationship Skills, Part One	
A. Self-Presentation Skills	________
B. Conflict Resolution	________
C. Giving and Receiving Criticism	________
Session V. Relationship Skills, Part Two	
A. Teamwork Skills	________
B. Leadership Skills	________
C. Performance Testing (Task 3)	________

Session VI. Workplace Issues

A. Harassment ____________

B. Stress and Drug and Alcohol Abuse ____________

C. Review ____________

D. Module Examination ____________

1. Trainees must score 70% or higher to receive recognition from NCCER.
2. Record the testing results on Craft Training Report Form 200, and submit the results to the Training Program Sponsor.

E. Performance Testing ____________

1. Trainees must perform each task to the satisfaction of the instructor to receive recognition from NCCER. If applicable, proficiency noted during laboratory exercises can be used to satisfy the Performance Testing requirements.
2. Record the testing results on Craft Training Report Form 200, and submit the results to the Training Program Sponsor.

Basic Employability Skills 00108-04

Excellence in Construction Winner— Historical Restoration, $10–99 Million

Skanska USA Building, Inc. completed the restoration on the Ca'D'Zan Museum in Sarasota, Florida. The work included restoration of terra cotta, stucco, plaster, painted surfaces, marble, and stained glass; a new barrel tile roof; a museum-quality HVAC system; and a state-of-the-art fire protection system.

00108-04
Basic Employability Skills

Topics to be presented in this module include:

Overview

Everyone wants a job. It's the workers who present themselves in a professional manner, however, who usually get the job. Being professional means getting along with co-workers, following the rules, behaving appropriately, being honest, and completing your work on time. All of these skills are considered basic employability skills.

Basic employability skills require you to think critically and make appropriate decisions. They also require you to be prepared and knowledgeable about the job for which you are applying. You are most prepared for finding a job when you have a good resume and references. It's also important to ask appropriate questions of the potential employer to find out if they are a good match for your needs.

Computer skills can also contribute to your employability. More and more companies are using computer software for administrative, design, and fabrication functions. Developing your computer skills will provide you with an advantage over the competition.

Instructor's Notes:

Objectives

When you have completed this module, you will be able to do the following:

1. Explain the construction industry, the role of the companies that make up the industry, and the role of individual professionals in the industry.
2. Demonstrate critical thinking skills and the ability to solve problems using those skills.
3. Demonstrate knowledge of computer systems and explain common uses for computers in the construction industry.
4. Demonstrate effective relationship skills with teammates and supervisors, the ability to work on a team, and appropriate leadership skills.
5. Be aware of workplace issues such as sexual harassment, stress, and substance abuse.

Prerequisites

Before you begin this module, it is recommended that you successfully complete the following: *Core Curriculum: Introductory Craft Skills,* Modules 00101-04 through 00105-04. Modules 00106-04 through 00108-04 are electives. To receive a successful completion, you must take Module 00106-04 or Modules 00107-04 and 00108-04.

This course map shows all of the modules in the first level of Core Curriculum: Introductory Craft Skills. The suggested training order begins at the bottom and proceeds up. Skill levels increase as you advance on the course map. The local Training Program Sponsor may adjust the training order.

Key Trade Terms

Absenteeism	Initiative
Amphetamine	Keyboard
Barbiturate	Laptop computer
Browser	Lateness
CAD	Leadership
CD-ROM	Mission statement
CNC	Monitor
Compromise	Mouse
Computer literacy	Operating system
Confidentiality	PDA
Constructive criticism	Plotter
Critical thinking skills	Printer
Database	Processor
Desktop computer	Professionalism
Desktop publishing	Reference
Disk drive	Scanner
Documentation	Self-presentation
EDM	Sexual harassment
Email	Software
Entrepreneur	Spreadsheet
Goal-oriented	Tactful
Hallucinogen	Teamwork
Handheld computer	User
Harassment	Wireless
Hard drive	Word processor
Hardware	Work ethic

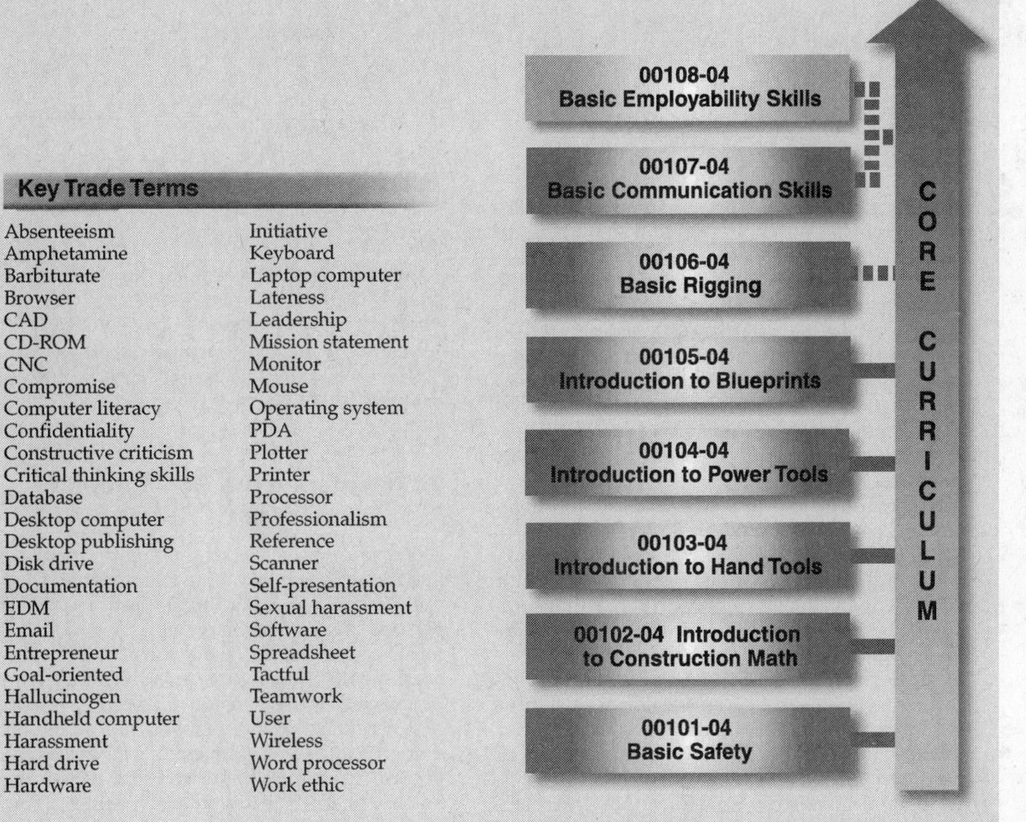

108CMAP.EPS

Ensure that you have everything required to teach the course. Check the Materials and Equipment List at the front of this module.

See the general Teaching Tip at the end of this module.

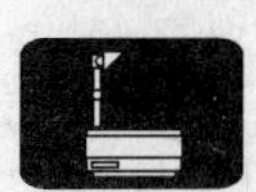

Show Transparency 1, Course Objectives.

Show Transparency 2, Performance Tasks.

Explain that terms shown in bold (blue) are defined in the Glossary at the back of this module.

Review the types of skills that make employees more employable in the construction industry.

Discuss the role of the construction industry in the national economy. Review the types of work completed by construction companies.

Explain the cost of one employee to a company. Review ways that employees can help a company be profitable.

Bring in and read a copy of a company's mission statement. Explain how to incorporate the company's business philosophy into the employees' work ethic.

1.0.0 ◆ INTRODUCTION

To succeed in the construction industry, you need to know how to do your job well. This means more than just using tools, machines, and equipment with skill. You must also know how to do many things that at first might not seem to have anything to do with being a construction professional. For example, understanding how a construction business works, thinking critically, and being able to solve problems are vital skills. So are working safely, presenting yourself well, getting along with your teammates and supervisors, and being flexible. Computer skills are also important.

At first, these skills might seem unrelated to construction work, but if you think about it, every good role model exhibits some or all of them. These skills are just as real as your ability to operate a power tool: People can see whether you know how to use it, and they can see the results. The same goes for your behavior on the job. This module introduces the nontechnical skills that you must master to succeed in the construction industry.

DID YOU KNOW?

How much does it cost your boss to employ you? If your base pay is $10.00 per hour, you are costing your employer $13.50 per hour. The following factors add to the cost of employing you:

1. Workers' compensation
2. Insurance
3. Social Security
4. Equipment
5. Vacation days
6. Sick leave

These factors make hourly costs to your employer 35 percent higher than your salary. The additional 35 percent must come from company profits. Therefore, if your employer does not make a profit, it's unlikely that you will get a raise. You may even lose your job. You can help your company earn a profit by doing the following:

- Avoiding accidents
- Following company rules and policies
- Working quickly and efficiently
- Using company property appropriately
- Meeting deadlines
- Showing up for work on time

2.0.0 ◆ THE CONSTRUCTION BUSINESS

The construction industry creates the environment in which we live and work. It is made up of a wide variety of specialized skills that all share a single goal: to make our lives more comfortable (see *Figure 1*). The construction industry is the second largest industry in the United States, larger than the steel and automotive industries combined. The construction industry employs more than 5 million people. Recently, the total value of new construction in this country was estimated to exceed $500 billion, and this figure grows every day.

The construction industry consists of independent companies of all sizes that specialize in one or more types of work. For example, a company might install heating, ventilation, and air conditioning systems in residences. Another company might oversee the construction of an entire office complex. Sheet metal shops, civil engineering firms, well and septic system installation specialists, welding and cutting specialists—all these and more form the construction industry.

Your company might have a **mission statement,** which explains how it does business. The company may describe its philosophy in an employee handbook or other materials you read when you joined the company. However your company describes its role in the construction industry, you should make sure that you understand not only its mission but also your role in the company. In addition, you need to be familiar with your company's policies and procedures. Knowing these things will help you to be a better employee.

2.1.0 Entering the Construction Workforce

When you complete your training, you will need to find a job where you can put your new skills to good use. Perhaps your training is part of your job, or perhaps you have a job waiting for you when you finish your training. Even if you already have a job, brushing up on basic job search skills is a good idea. You never know when you will want to use them.

When you look for a job, try to find one that is a good match for your skills and experience. If you are a carpenter with two years' experience, do not apply for a journey-level position. You can find jobs listed in newspapers and trade magazines. You can also search for them on the internet.

Instructor's Notes:

108F01.EPS

Figure 1 ◆ The construction industry offers many career options.

Show Transparency 3 (Figure 2). Discuss the importance of maintaining an up-to-date resume.

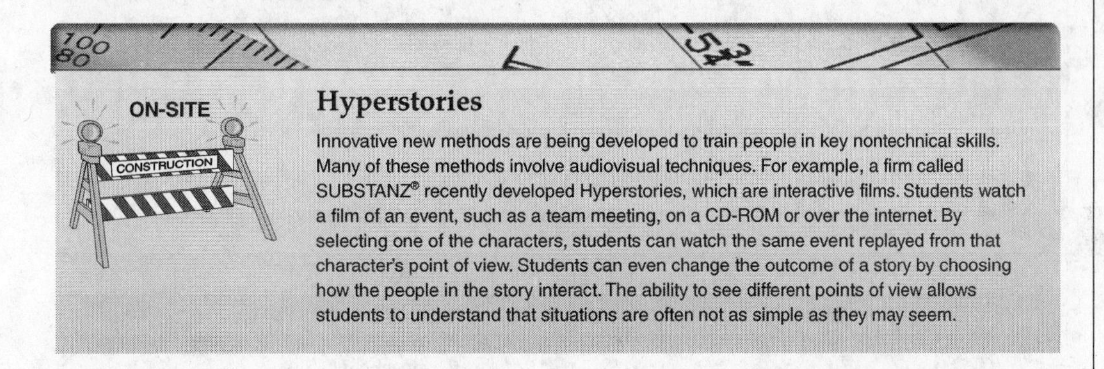

Hyperstories

Innovative new methods are being developed to train people in key nontechnical skills. Many of these methods involve audiovisual techniques. For example, a firm called SUBSTANZ® recently developed Hyperstories, which are interactive films. Students watch a film of an event, such as a team meeting, on a CD-ROM or over the internet. By selecting one of the characters, students can watch the same event replayed from that character's point of view. Students can even change the outcome of a story by choosing how the people in the story interact. The ability to see different points of view allows students to understand that situations are often not as simple as they may seem.

Many companies that announce jobs on the internet allow you to apply online.

Ask your current supervisor and your teammates if they would act as **references** for you. A reference is a person who can vouch for your skills, experience, and work habits. Make sure that your resume is up to date, well organized, and easy to read (see *Figure 2*). Apply the effective writing techniques that you learned elsewhere in this curriculum when writing your resume. You can find excellent sources in print and online to help you write a resume.

By carefully selecting jobs that you are qualified for and by submitting an accurate, well-written

MODULE 00108-04 ◆ BASIC EMPLOYABILITY SKILLS 8.3

Craftworker

5676 Golden Road, Somewhere, Alberta TSX 5T4
(780) 555-4537 • mdsf@internet.com

Maintenance Project Manager

• Willing to travel and relocate •

Self-motivated, peak-performing professional with 20+ years of progressive supervisory experience in all phases of the maintenance industry. Effective problem solver who enjoys the challenge of achieving goals and accomplishing objectives; career history of consistent advancement based on achievements and motivational tactics. Reputation for taking technical, complex projects from inception to completion. Ability to streamline procedures that improve safety levels, productivity, and control costs through expertise in

- Team Building & Leadership
- Troubleshooting
- Production Planning & Scheduling
- Blueprint Analysis
- Organization & Time Management
- Budgeting & Finance
- Safety & Compliance Management
- Strategic Planning/Implementation
- Vendor Management
- Preventative Management

Professional Experience

XYZ Corporation, Somewhere, AB 1980 to Present
Maintenance Manager

Fast-track promotion through increasingly responsible positions directing maintenance and project planning for a Fortune 500 food manufacturer totalling $40 billion annual sales and employing 50,000 employees nationally. Lead, coordinate, and supervise 120 maintenance personnel to ensure preventative maintenance (mechanical and electrical), breakdown repair, safety awareness, new project implementation, and equipment maintenance and complete within budget.

- Leadership, dedication, and teamwork earned promotion to maintenance manager within one year.
- Planned, measured, purchased, and coordinated installation of critical high-tech warehouse system, the largest, most expensive project in the company's history—$5 million.
- Authored and implemented comprehensive maintenance and safety program—acknowledged by company president and implemented companywide.
- Set and maintained high standards. Provided daily motivation and training to ensure that employees realized that their work was meaningful. Influenced and encouraged personnel to participate in self-improvement processes.
- Instituted a strong focus on safety through creative incentives, education, and employee empowerment that decreased Workers' Compensation premiums by 30%.

108F02.EPS

Figure 2 ◆ Your resume can be your ticket to an exciting career in the construction industry.

Instructor's Notes:

8.4 CORE CURRICULUM ◆ INTRODUCTORY CRAFT SKILLS

resume, you improve your chances of being called for interviews (see *Figure 3*). The effective communication skills you learned elsewhere in this curriculum will help you present yourself well. In this module, you will learn more skills that will make a good impression at your interviews.

Good resumes and interviews lead to job offers. You will need to evaluate the offers to select the one that works best for you. The following are some of the questions to consider when selecting a job:

- Is the salary enough to meet my needs?
- Does the company offer a benefits package that covers what I need?
- Will the work be interesting and challenging but not more than I can handle?
- Does the company have a good reputation in the industry?
- Do the people appear to be nice to work with?
- Does the company offer training?
- What is the company's safety record?

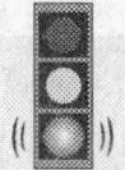

NOTE

Some employers require applicants to apply online. If you don't have a computer or access to the internet, check local businesses or the library, or ask if a friend or relative has internet access.

108F03.EPS

Figure 3 ◆ Job interviews are an important part of the hiring process.

2.2.0 Entrepreneurship

Many of the companies in the construction industry are considered small businesses. You may work for a small business right now, or you may even be thinking about starting your own someday. A small business is defined as follows:

- It is independently owned and operated.
- It is not the main or largest company in its field.
- It has fewer than 500 employees.
- It makes less than a certain amount of money per year, depending on the type of work it does (less than $27.5 million for general and heavy construction companies; less than $11.5 million for special trade contractors).

People who start and run their own businesses are called **entrepreneurs** (see *Figure 4*). Before starting a business, an entrepreneur must evaluate the need for such a business. Do other companies already fill that need? If so, will a new

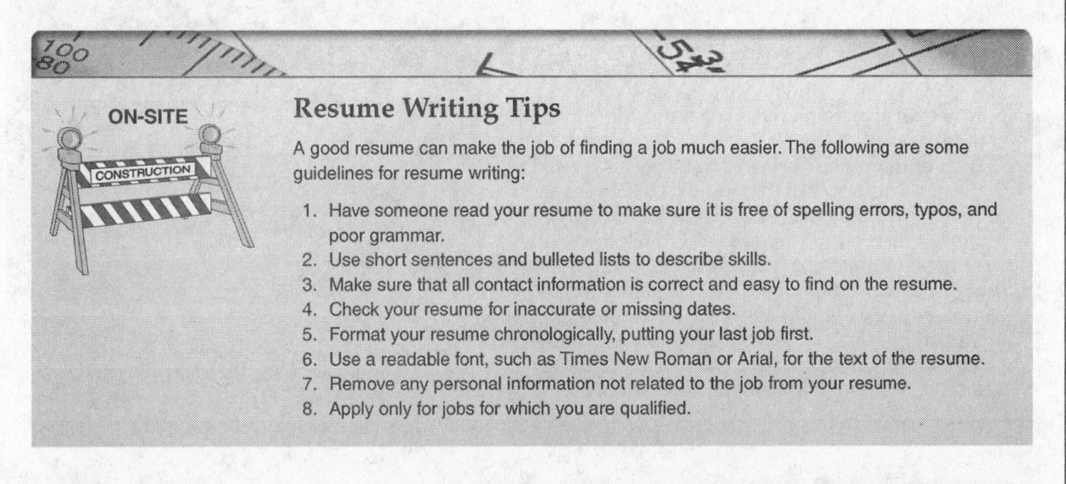

Resume Writing Tips

A good resume can make the job of finding a job much easier. The following are some guidelines for resume writing:

1. Have someone read your resume to make sure it is free of spelling errors, typos, and poor grammar.
2. Use short sentences and bulleted lists to describe skills.
3. Make sure that all contact information is correct and easy to find on the resume.
4. Check your resume for inaccurate or missing dates.
5. Format your resume chronologically, putting your last job first.
6. Use a readable font, such as Times New Roman or Arial, for the text of the resume.
7. Remove any personal information not related to the job from your resume.
8. Apply only for jobs for which you are qualified.

Bring in a variety of job listings from newspapers and trade magazines. Working in pairs, have trainees practice interviewing for the positions. Have trainees assess their partner's interview skills.

See the Teaching Tip for Section 2.1.0 at the end of this module.

Ask trainees to explain when a job might not be suitable and when it might be appropriate to turn down a job offer.

Explain that entrepreneurship can be a rewarding yet challenging endeavor. Review criteria for determining whether a new company is viable.

Bring in sample job applications from area companies. Have trainees practice completing the applications. Emphasize the importance of completing applications accurately and legibly.

Refer to Figure 5. Explain how creating an organizational chart can help owners understand the structure of a company.

Have each trainee perform Performance Task 1 to your satisfaction. Fill out a Performance Profile Sheet for each trainee.

Have trainees review Sections 3.0.0–3.3.0 for the next session.

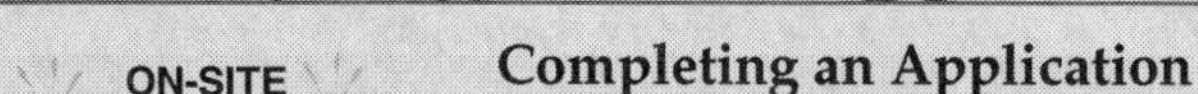

Completing an Application

Completing an application is one of the first steps toward getting a job. Applications require specific information about your work and training history. Be sure to have the following information available when applying for a job:

1. Personal identification, such as your driver's license, a passport, or state or military ID
2. Your contact information, including your current phone number and address
3. Your Social Security number
4. The names and locations of the school(s) or training classes you attended
5. The dates you attended the school(s) or classes and the subject(s) you studied
6. The names, addresses, and phone numbers of previous employers
7. The dates of employment for each employer listed
8. The name and contact information for a nonfamily member who can vouch for your personal and professional abilities

108F04.TIF

Figure 4 ◆ Entrepreneurs are an important part of the construction industry.

company be able to offer something different or better than existing businesses? A new company may not survive long unless it stands out from the competition.

If an entrepreneur believes that there is a need that a new business can fill, the next challenges are to select a location for the business and set up plans for marketing, finances, and management. An organizational chart can help the owner and the employees understand the structure of the company (see *Figure 5*). Only when all these elements are in place can the business open. A successful entrepreneur needs to have a lot of energy and be able to adapt and respond to changes in the business situation.

108F05.TIF

Figure 5 ◆ An organizational chart for a construction business.

Instructor's Notes:

8.6 CORE CURRICULUM ◆ INTRODUCTORY CRAFT SKILLS

3.0.0 ◆ CRITICAL THINKING SKILLS

Suppose you are team leader on a construction crew that is building a shopping mall. The job will last approximately 18 months. The parking area for site workers is half a mile from the job site, and your team members will have to carry their heavy toolboxes from their cars to the work area every day. Delays caused by late arrivals and team members arriving to work tired from the walk threaten to derail the tight job schedule. As the team leader, you need to solve the problem. What are your options?

Throughout your construction career, you will encounter problems like this one that you will have to solve. **Critical thinking skills** allow you to solve such problems effectively. Critical thinking means evaluating information and then using it to reach a conclusion or to make a decision. Critical thinking allows you to draw sound conclusions and make correct decisions using the following steps:

- Evaluating new information
- Identifying the options and alternatives
- Weighing the merits of each option and alternative
- Accepting or rejecting options and alternatives based on their merits

Consider the credibility, or trustworthiness, of a source of information when deciding how to treat it. For example, information provided by your supervisor will be more reliable than gossip you overhear at lunch. Also think about the expertise or experience of people who give you information. For example, if you need to know about the electrical installation on a project, would you take the carpenter's advice or the electrician's? You'd be better off listening to the electrician. Feel free to ask experts and people you trust for their advice.

Compare information with what you already know. If it does not fit in with what you know, question it. For example, suppose you read in a manual that workers do not need to wear hard hats in the construction area. During your training, you have learned that hard hats should always be worn in a construction area. Knowing that the information in the manual is incorrect, you can safely decide not to follow that instruction.

Sometimes, however, your personal feelings can get in the way of evaluating information. For example, if you don't like working with computers, you may not want to draft blueprints with them. But if your shop uses computers to draw blueprints, then you may disrupt the workflow if you hold on to your bias. Put your personal feelings aside and remain open-minded when evaluating information.

3.1.0 Barriers to Problem Solving

When you are searching for the solution to a problem, you may fall into a trap that prevents you from making the best possible decision. The following are the most common barriers to effective problem solving:

- Closed-mindedness
- Personality conflicts
- Fear of change

To be closed-minded is to distrust any new ideas. Effective problem solving, however, requires you to be open to new ideas (see *Figure 6*). Sometimes the best solution is one that you would

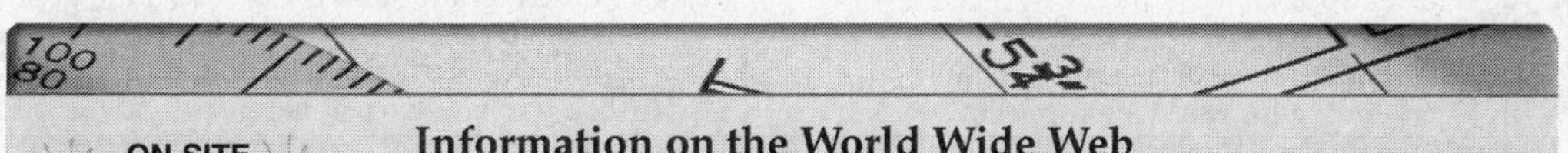

ON-SITE

Information on the World Wide Web

Everybody knows that the World Wide Web has information about almost everything. The trick is to know whether the answers you find there are the right ones for you. How do you evaluate sources that you find on the World Wide Web?

One of the easiest and most effective ways is to look at the two- or three-letter extension at the end of the website address. Addresses ending in .gov or .us are official local, state, or federal government web pages. Use those pages to find accurate information about codes and regulations that apply to your work. Addresses ending in .org are nonprofit organizations. If the organization is affiliated with your industry or sets the standards for it, you can trust its information.

Manufacturers put a wide variety of information, such as product specifications and catalogs, on their websites. When looking for this type of information, go straight to the manufacturer's website. The websites of companies that sell products from many different manufacturers may not be as complete or up to date as the manufacturer's own site.

Ensure that you have everything required for the laboratories and demonstration during this session.

Explain how critical thinking skills can be used to solve problems effectively.

Explain how critical thinking can help you draw sound conclusions.

Review techniques for evaluating sources and weighing information.

If possible, schedule time at your local library's computer lab. Explain how to use the internet to find information that is otherwise difficult to locate. Have trainees practice locating information on the Web.

Review barriers to effective problem solving. Discuss the importance of being open to new ideas.

Explain that, although change can be difficult, it is a necessity. Have trainees discuss a time when they faced a difficult professional change and how they dealt with it.

MODULE 00108-04 ◆ BASIC EMPLOYABILITY SKILLS 8.7

Show Transparency 4 (Figure 7). Review the five-step problem-solving procedure, and explain how to apply each procedure.

108F06.TIF

Figure 6 ◆ Be open to new ideas.

have never considered on your own. Remember that other people have good ideas too. You should be willing to listen to them.

Sometimes you may fail to appreciate the value of information or advice simply because you do not get along with the person offering it. Maybe someone has an abusive or insensitive way of talking that offends you. Or maybe someone acts superior or bossy. One of the most important skills you can master is the ability to separate the message from the messenger. Weigh the value of the information separately from your feelings about the individual. This ability will show people that you are a real professional.

People often fear change when they believe it will threaten them, but as the old saying goes, "the only constant is change." Very few changes turn out to be as threatening as people fear. Often, the lack of change is the problem. In the construction industry, new tools, machines, techniques, and materials appear every day. As a construction professional, you need to stay informed about technical advances in your field. If you are open to change, you will never stop learning better ways to solve problems.

3.2.0 Solving Problems Using Critical Thinking Skills

Problems arise when there is a difference between the way something is and the way you would like it to be. Sometimes, you might feel frustrated or intimidated by a problem, or you might feel that you do not have enough time to solve it. Your reaction might be to simply ignore the problem. Instead, try to look at it as an opportunity to demonstrate your skills. By actively seeking solutions to problems, you will demonstrate to your colleagues and supervisors that you are responsible and capable.

To solve problems that arise on the job, use the following five-step procedure (*Figure 7*):

Step 1 Define the problem.

Step 2 Analyze and explore the alternatives.

Step 3 Choose a solution and plan its implementation.

Step 4 Put the solution into effect and monitor the results.

Step 5 Evaluate the final result.

Reviewing each of these steps in detail will help you understand how to apply them. Before you can solve a problem, you need to know exactly what it is. This step might seem obvious, but people often forget it. For example, suppose you are working at a job site when the generator goes out and your team members cannot use their power tools. Is the real problem the fact that the generator quit, or the fact that your team is not able to work? Most likely it's the second issue. You can easily bring in auxiliary generators that will allow your team to keep working. However, your team's inability to work could throw off the entire project's schedule. The generator failure was not the problem but rather the cause of the problem.

Once you've defined the problem, consider different ways to solve it. Collect information from a wide range of sources, and ask people for their opinions. Teamwork is an important part of this step. The more suggestions you get from experts and co-workers, the better your chances are of finding the right answer. Identify and compare the alternatives. Look for solutions that will be the most cost-effective, take the least time, and ensure the highest quality. Try to identify the short- and long-term consequences, both good and bad, of each possible solution.

Instructor's Notes:

8.8 CORE CURRICULUM ◆ INTRODUCTORY CRAFT SKILLS

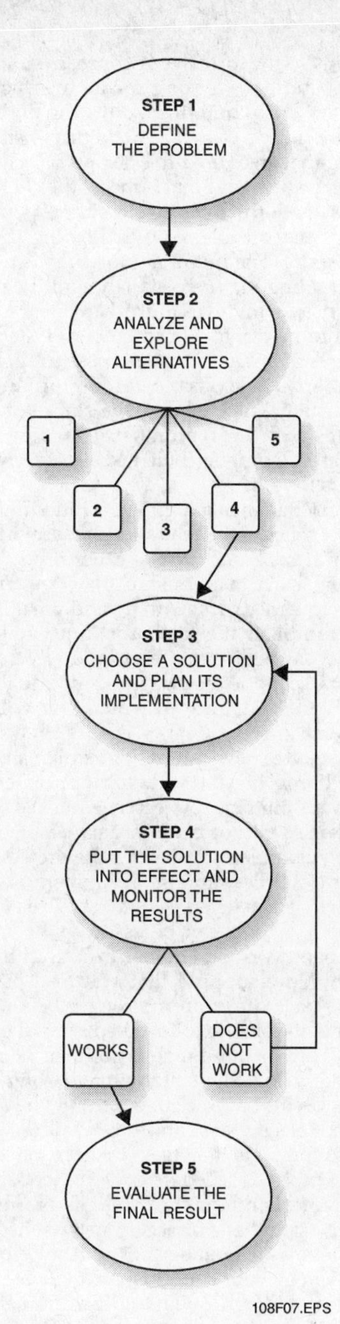

Figure 7 ◆ The five-step problem-solving procedure.

Then, choose the solution that you think will work the best. Develop a plan for carrying it out. The plan should include all necessary tools and materials. It should also specify all the tasks involved and who is responsible for them, and estimate how long each task will take.

When you are ready to put the plan into effect, follow it closely to ensure that it is bringing about the desired results. If it is not, go back to Step 2 and choose another possible solution. Develop and implement a plan suited to the new solution.

When the problem is finally solved, look back over the steps you took and see what lessons you can learn. Perhaps there was something you could have done better. Or perhaps you could have saved time or money by changing part of the plan. When you are satisfied with the solution, remember what worked and what did not. When you face a similar problem in the future, your experience will help you find a solution.

Go back to the beginning of this section and review the hypothetical case discussed there. You are responsible for figuring out a way to help your teammates carry their toolboxes to and from the work area from a parking lot half a mile away. How would you apply the five-step process to solve that problem? You might approach it the following way:

Step 1 *Define the problem* – Because the parking area is far from the job site, your team members are forced to waste time walking to and from their cars. They are also forced to waste energy carrying their heavy toolboxes.

Step 2 *Analyze and explore the alternatives* – You could arrange a shuttle service from the parking area to the job site. You could provide lock-up space for tools and toolboxes near the work area. Or you could designate a drop-off point for the tools at the job site. What are some of the benefits and drawbacks of each alternative?

Step 3 *Choose a solution and plan its implementation* – After weighing each option, you decide that the shuttle service is the best one. It allows workers to take their tools home each night. This eliminates the possibility of theft, which was a drawback of the other options. You present your plan to the site superintendent, who agrees with you and announces that every morning at 8:00, the company van will pick up the construction crew in the parking area. Every afternoon at 5:30, the van will make the return trip.

MODULE 00108-04 ◆ BASIC EMPLOYABILITY SKILLS 8.9

Using the example from the beginning of this section, ask trainees to follow the five-step process to arrive at potential solutions. Compare trainees' solutions, and discuss the merits of each.

Explain how each person's role on a project team affects other aspects of the project.

Using a hypothetical task, demonstrate how to divide it into steps placed in the sequence in which you will perform the work.

Review common problems that can disrupt construction jobs, and explain how each can affect work progress.

Step 4 *Put the solution into effect and monitor the results* – At first, the new solution seems to work. However, you realize that the van cannot hold everyone. The van will have to make more than one trip. You discuss this with the site superintendent and get permission for three trips in the morning, at 7:45, 8:00, and 8:15, and three trips in the afternoon, at 5:15, 5:30, and 5:45. The site superintendent agrees to schedule the workers' starting times to correspond with the different trips.

Step 5 *Evaluate the final result* – The new plan works well. Everyone gets to the job site on time and ready to work. The schedule is back on track. This cost- and time-effective solution makes everybody happy.

Remember, the word critical means important or indispensable. Critical thinking skills are important and indispensable tools in your personal toolkit (see *Figure 8*). And as you would with any other tool, you must learn how to use them correctly and safely. When you do, you will find that one of the most rewarding experiences you can have as a trade professional is to face a problem and solve it yourself.

3.3.0 Problems with Planning and Scheduling

As a member of a project team, you are a very important link in a chain that stretches from you all the way to the top of your company. Suppose you are a plumber working for a construction company on a new building. At the top of the chain is the company's owner, who is responsible for ensuring that an entire project stays on schedule and within budget and that it meets the client's needs. Underneath the owner is the job superintendent, who is responsible for coordinating the work at the job site. The plumbing foreman is responsible for coordinating the installation of the plumbing, if the project involves more than one trade, and for assigning work to each member of the plumbing crew. And finally, but no less important, the plumber is responsible for performing specific tasks, such as locating waste stacks or installing fixtures. Successful plumbers learn how to undertake and complete their tasks in the right order and on time.

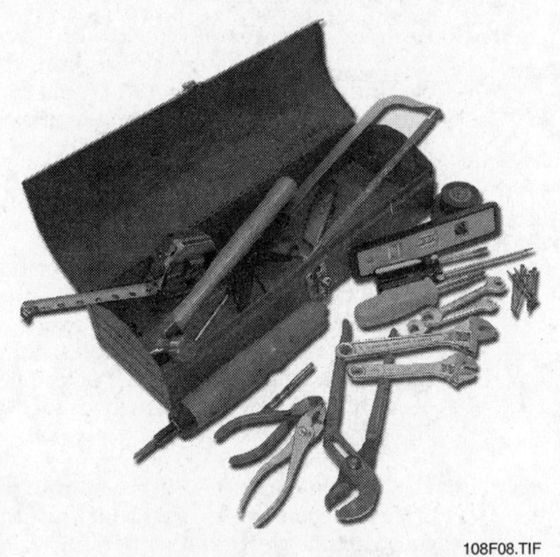

108F08.TIF

Figure 8 ◆ Critical thinking skills are just as important as any of these tools.

Every task has a logical position in the overall project schedule. Each task must occur before, after, or at the same time as other tasks. Before beginning each task, ensure that you understand what you are expected to do and what is required to accomplish the task. You must know where your work begins and ends and which materials to install. Divide the task into individual steps and decide the sequence in which you will perform the work.

No matter how carefully you plan ahead, unexpected problems can suddenly appear. Usually, extra time is built into project schedules so that simple problems will not cause a delay. However, more complex problems can throw the project off track.

For example, suppose a shipment of bathroom fixtures was delayed by one day. This is a simple problem, and it can be resolved by having the plumbers work on another task until the fixtures arrive. Now suppose that you find out that the bathroom fixtures are no longer being manufactured in the size or color called for in the specifications. This is a more serious problem, because if the plumbers do not get alternative fixtures, they cannot finish the bathroom. As a result, the drywall installers, electricians, and other trades will not be able to complete their tasks in that room either. The whole project's schedule could be jeopardized.

Foremen and superintendents are responsible for ensuring that a project stays on schedule. Nevertheless, as a member of the team, you may be called on by your supervisor to evaluate and solve some problems. You should be familiar with the types of problems that can disrupt a job. Generally, problems will fall under one or more of the following categories:

- Materials
- Equipment
- Tools
- Labor

Instructor's Notes:

8.10 CORE CURRICULUM ◆ INTRODUCTORY CRAFT SKILLS

The materials required for a job are identified during the planning stages of a project. They are ordered from suppliers and delivered according to a prearranged schedule. Problems with materials can include errors of quantity or type, delays in delivery, and unavailability due to backorders or to business closure. Shortage of materials due to waste, loss of materials due to theft and vandalism, and inability to locate materials in storage are other sources of project delays.

Construction equipment is usually selected and scheduled before the project begins. The site superintendent is responsible for ensuring that equipment is on site at the right time. Often, two pieces of equipment must be scheduled at the same time; for example, backhoes to excavate a foundation and dump trucks to haul away the excavated dirt. Problems with equipment can include unavailability on the scheduled day or days, lack of qualified operators, mechanical breakdown, and extended maintenance.

Supervisors are responsible for ensuring that their workers have the appropriate tools. Workers may need specialized tools that they do not have in their personal toolkits. Or they may be learning how to use such tools on the job, under proper supervision. Common problems related to tools include breakage or damage, loss or theft, or lack of skilled users.

Labor—the men and women who perform the work on the job site—is the most important component of a project. Project planners estimate the number of people needed every day and identify the range of skills required. Supervisors ensure that the right people are available and working on the right tasks. Common problems related to labor include **lateness** and **absenteeism,** lack of experience or qualifications to perform a given task, and not enough available workers. Idleness due to lack of instructions or because of laziness can also contribute to delays.

Always keep your eyes open for possible delays as you carry out your assigned tasks. If you see a potential source of delay, notify your supervisor immediately. Always know clearly what your responsibilities are in such situations; be prepared to solve the problem yourself if that is what you are called on to do. Always keep your supervisor up to date on your progress. Be sure to let your supervisor know when you have finished a task.

Review Questions

Sections 1.0.0–3.0.0

1. The construction industry is the fourth largest single industry in the United States.

a. True
b. False

2. A(n) _____ can vouch for your skills, experience, and work habits.

a. mission statement
b. entrepreneur
c. interviewer
d. reference

3. A small business generally has no more than _____ people working for it.

a. 250
b. 500
c. 750
d. 1,000

4. Critical thinking is the process of first _____ and then _____.

a. evaluating information; reaching a conclusion or making a decision
b. searching for jobs; applying and interviewing
c. identifying problems; solving them
d. evaluating the need; developing business and marketing plans

Put the following problem-solving steps in the correct order.

5. *Step 1:* ____________________
6. *Step 2:* ____________________
7. *Step 3:* ____________________
8. *Step 4:* ____________________
9. *Step 5:* ____________________

a. Analyze and explore the alternatives.
b. Put the solution into effect and monitor the results.
c. Define the problem.
d. Evaluate the final result.
e. Choose a solution and plan its implementation.

Ask trainees to suggest ways they can help prevent work delays.

Go over the Review Questions for Sections 1.0.0–3.0.0. Answer any questions trainees may have.

Have trainees review Sections 4.0.0–4.4.0 for the next session.

MODULE 00108-04 ◆ BASIC EMPLOYABILITY SKILLS 8.11

Ensure that you have everything required for the laboratories, demonstration, and testing during this session.

Refer to Figure 9. Discuss the role of computers in the modern construction industry. Emphasize the importance of computer literacy.

Review computer terms. Distribute copies of Figure 10 with the callouts covered. Ask trainees to identify the components and briefly explain how each enables the computer to perform its functions.

Refer to Figure 12. Provide examples of software packages commonly used in the construction industry.

Bring in a variety of computer reference books that help users understand software and hardware, and allow trainees to browse through them.

4.0.0 ◆ COMPUTER SKILLS

Computers are everywhere. The most familiar types of computers are probably the **desktop computers** that you see on office desks and the **laptop computers** that people use on the road. These are not the only type of computers, however. Computers small enough to hold in your hand, called **handheld computers** or **PDAs** (for personal digital assistants), are now on the market. Small computers are also used in cars, televisions, microwave ovens, and even many alarm clocks and coffee makers.

Computers play an important part in the modern construction industry. Just a few years ago, fast, powerful computers were too expensive for the average business to afford. Now, computers are far more affordable. Industries rely on computers for everything from payroll and billing to design and fabrication (see *Figure 9*). As part of your workplace skills, you need at least a basic understanding of computers and the work they can do. This understanding is called **computer literacy.**

This section will introduce some basic computer terms, components, and applications. Because computer technology changes very fast, the topics will be covered generally. Some of the terms and technologies described here may be out of date by the time you read about them. Your instructor will be able to suggest sources that contain more specific information.

4.1.0 Computer Terms

Computer systems consist of **hardware, software,** and **operating systems.** Together, these three elements allow a computer operator or **user** to tell the computer to do something, observe the results, make changes, and give new instructions. Hardware is the set of physical components that make up a computer (see *Figure 10*). Basic hardware components include the following:

- The **processor,** sometimes called a central processing unit or CPU, which contains the chips and circuits that enable the computer to perform its functions
- The **monitor,** which is a television-like screen that shows information, text, and pictures to the user
- The **keyboard,** which allows computer users to type in text and instructions to the processor
- A **mouse** or other device that allows users to enter data without having to type on the keyboard
- A **hard drive** that stores software and electronic files, plus a **disk drive** or **CD-ROM** drive for transferring software and files to and from the computer
- **Printers** and **plotters** that print out text, labels, graphics, and drawings on paper, plastic, and other materials
- **Scanners** that allow printed text or pictures to be copied into an electronic format so they can be manipulated on the computer

Software tells a computer how to perform one task or a whole series of tasks (see *Figure 11*). Software is usually installed on a computer from one or more CD-ROMs. CD-ROMs look like music CDs, but they contain electronic files instead of songs. Operating systems allow the hardware and software to communicate. Popular operating systems include Microsoft® Windows, Apple® Mac OS®, and Linux.

108F09.TIF

Figure 9 ◆ Computers are an essential part of the modern construction industry.

4.2.0 Basic Software Packages

The following are some of the software packages that are commonly used in office computers in the construction industry (see *Figure 12*):

- **Word processors,** used to write text documents, such as letters, reports, and forms
- **Spreadsheets,** used to perform math calculations, such as for project budgets and company payrolls
- **Databases,** used to store large amounts of information, such as employee lists and material inventories, in a way that can be easily retrieved
- **Desktop publishing** software, used to lay out text and graphics for publications, such as brochures and newsletters
- **Computer-aided design (CAD)** systems, used to draw and print civil, architectural,

Instructor's Notes:

8.12 CORE CURRICULUM ◆ INTRODUCTORY CRAFT SKILLS

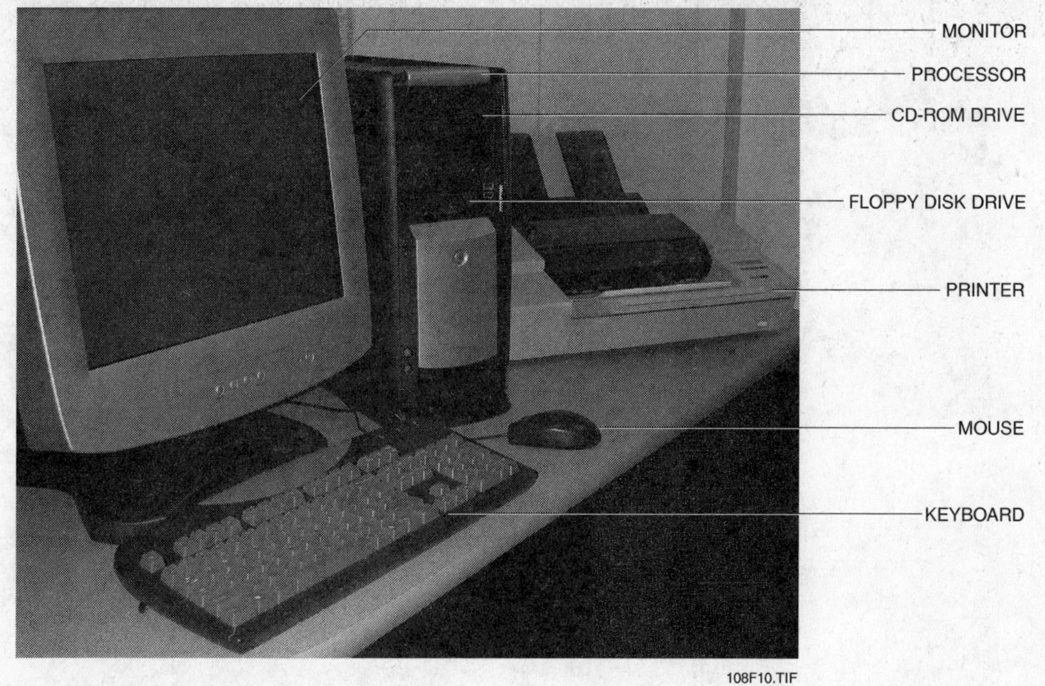

Figure 10 ◆ Computer hardware.

Figure 11 ◆ Software is loaded on a computer to perform tasks.

mechanical, structural, plumbing, and electrical drawings
- **Browsers,** used to search for information on the World Wide Web
- **Electronic mail** programs, used to send and receive electronic mail messages (called **email** for short)

Many different companies produce and sell software. Some software is sold separately from hardware so that it can be installed on hardware of the user's choice. Other software is designed to operate on only one model of hardware. Some software is designed for use in many industries; other software is developed specifically to perform a single task in one industry. Sometimes, software is even designed to meet the needs of one company.

Software usually comes with instructional manuals and other information, called **documentation.** In addition, you can buy how-to books for installing software at your local bookstore or from vendors on the World Wide Web.

4.3.0 Electronic Mail (Email)

Email is used to send and receive messages electronically over a computer network. Today, it is a common way to communicate, both professionally and personally. Email is typically generated in an email software program on an individual

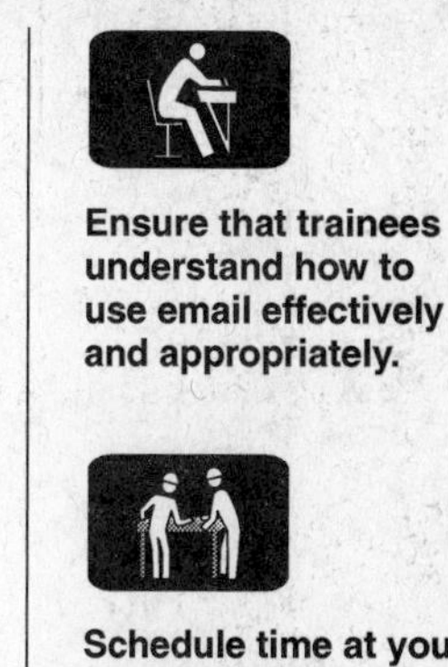

Ensure that trainees understand how to use email effectively and appropriately.

Schedule time at your local library's computer lab. Have trainees practice sending and receiving emails. If any of the trainees do not have an email account, provide assistance in setting up a free account such as those provided by Hotmail® or Yahoo!®.

MODULE 00108-04 ◆ BASIC EMPLOYABILITY SKILLS 8.13

AutoCAD PROPERTY PALETTE

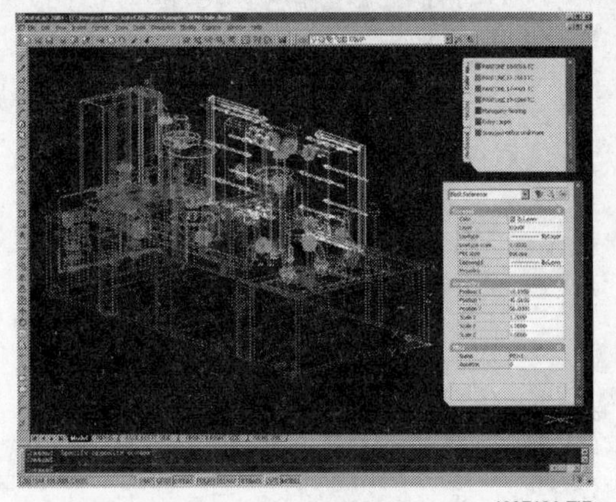

108F12A.TIF

AutoCAD STANDARDS CHECKING

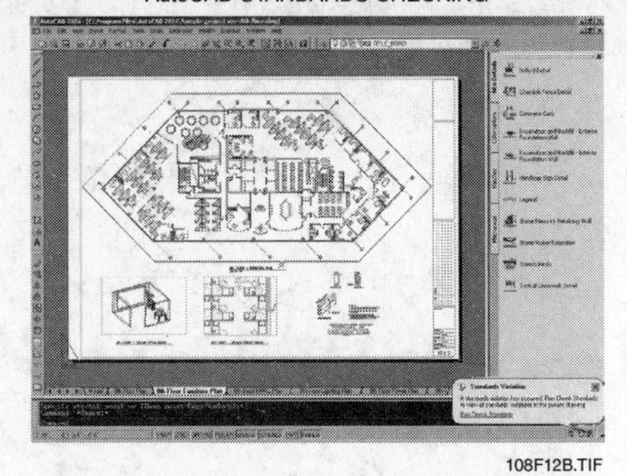

108F12B.TIF

AutoCAD TOOL PALETTE

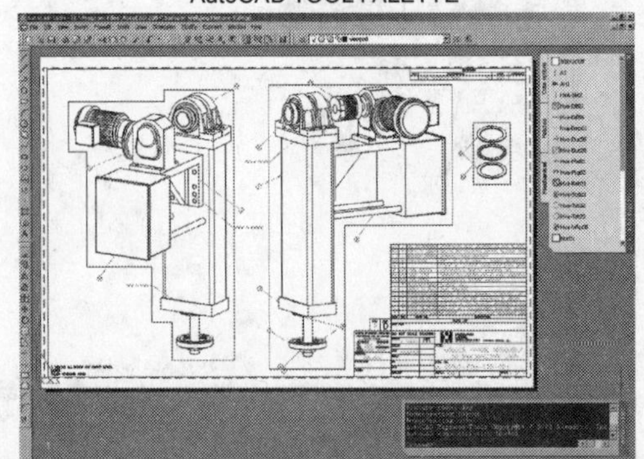

108F12C.TIF

PRIMAVERA EXPEDITION 9.0

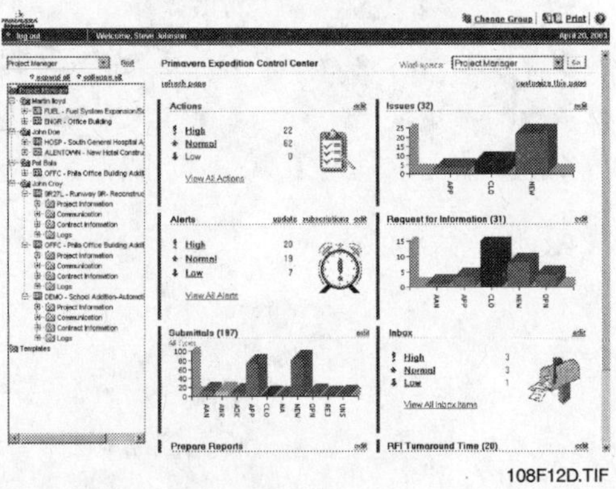

108F12D.TIF

PRIMAVERA P3e/c FOR CONSTRUCTION

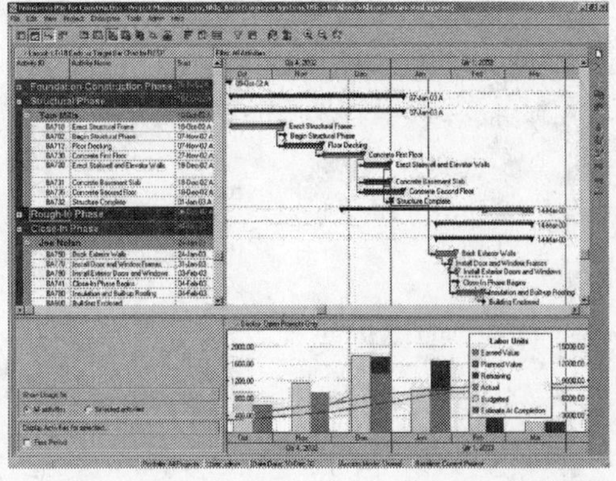

108F12E.TIF

Figure 12 ◆ Many different types of software packages are available for use on office computers in the construction industry, including AutoCAD and Primavera.

Instructor's Notes:

8.14 CORE CURRICULUM ◆ INTRODUCTORY CRAFT SKILLS

computer and then transmitted to the recipient via a network connection, such as modem, cable modem, local area network, or the internet.

4.4.0 Computers in the Construction Industry

As you have learned, companies use office computers to track time and labor costs, control inventory, and plan and control projects. However, most of the work being done in the construction industry is not done in an office. Work is usually done either on a job site or in a shop. Nowadays, you will find plenty of computers in both of these places.

Field workers can use laptop and handheld computers that are specifically made for the construction industry. Many of them are designed to be operated in the field; they use long-lasting batteries and **wireless** communications technology. Wireless allows you to communicate with other computers without a telephone line or other physical connection. Supervisors can send daily reports, forms, notes, and even photographs to the home office. The project manager can track a job's progress without having to travel back and forth to the job site. Many companies that use wireless technology are reporting increased productivity and reduced costs.

If you were to walk into a typical machine shop, you would probably see many different types of machines—lathes, milling machines, and drill presses, for example—that are operated by computers (see *Figure 13*). Machine tools that are controlled by computers are often referred to as **CNC**, or computer numerical control, machines. Machine operators enter electronic design drawings into a CNC machine, and the machine creates a finished product of that exact design. CNC machines can be used to mill materials such as wood, metal, and plastic and to fabricate tools, dies, and fittings.

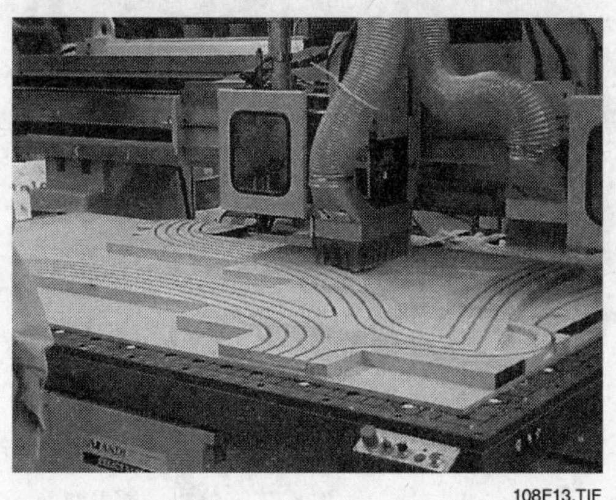

108F13.TIF

Figure 13 ◆ Many modern tools and machines are computer controlled.

Computer-controlled electrical discharge machines, or **EDMs,** are used to cut and form parts that cannot be easily handled by other types of machines. EDMs are used on hard metals and on parts that have very complex shapes. Among other things, EDMs are used to make slots and holes in metal, repair damaged dies, fabricate jigs to hold parts, and thread pipe.

5.0.0 ◆ RELATIONSHIP SKILLS

A relationship is the process of interacting with another person or with a group of people. Relationships are affected by both actions and perceptions. Every day, you interact with teammates, supervisors, and members of the public who see you working. No one wants to work with, hire, or hang out with an unprofessional person.

You need to be aware of the appropriate professional conduct for work situations, and you must follow that conduct at all times. Your actions reflect on your own professional status, that of your colleagues, the company you work for, and the image of the profession in the public eye.

5.1.0 Self-Presentation Skills

Relationships are like equations. They happen between you and another person or between you and a group of people. Because you are half of that equation, proper **self-presentation**—the way you dress, speak, act, and interact—is a vital part of any successful work relationship. Effective self-presentation involves developing good personal and work habits. Personal habits apply to your appearance and general behavior. Work habits, called your **work ethic,** apply to how you do your job. Your personal habits and work ethic make powerful impressions on your colleagues, supervisors, and potential employers. Once formed, impressions are hard to change—so you want to make sure that you make good impressions.

In this section, you will learn which personal habits and work ethics make good impressions and build strong work relationships. Take the time to read these carefully and think about how they apply to you. Proper self-presentation is not only about respecting others but also about respecting yourself.

Explain how wireless technology enables field workers to use computers on the job site.

Refer to Figure 13. Review the types of machines operated by computers in a typical machine shop. Explain how computer-controlled electrical discharge machines (EDMs) are used.

See the Teaching Tip for Section 4.0.0 at the end of this module.

Have each trainee perform Performance Task 2 to your satisfaction. Fill out a Performance Profile Sheet for each trainee.

Have trainees review Sections 5.0.0–5.2.2 for the next session.

Ensure that you have everything required for the laboratories during this session.

Emphasize the importance of maintaining positive working relationships with coworkers.

Discuss elements of self-presentation. Ask trainees to explain how appearance and behavior can make positive or negative impressions on colleagues or supervisors.

MODULE 00108-04 ◆ BASIC EMPLOYABILITY SKILLS 8.15

Discuss problems that could result if work areas and tools are not kept clean and organized.

Discuss the importance of professionalism, and review professional work habits.

Emphasize the importance of respecting company confidentiality.

Encourage trainees to understand and follow their company dress codes.

Robots on the Job

High-tech computer-controlled machines and equipment are becoming more popular on construction sites. Robots can safely perform dangerous jobs, without putting the health and safety of workers at risk.

For example, the Brokk company of Sweden has developed a line of remote-controlled demolition robots used for breaking roads, walls and floors, and other pavement or concrete structures. Traditionally, workers use jackhammers or water jets to break up old pavement, putting them at risk from noise, dust, stress injury, and flying debris. In contrast, robot controls are located on a tethered console, allowing workers to operate equipment at a safe distance from the point of impact (up to 20 feet, depending on the robot). Robot concrete breakers are also used in confined spaces or where access is difficult. Computers make possible safer and more efficient tools, and they can reduce the risk of injury or death on the job.

108SA01.TIF

5.1.1 Personal Habits

Teammates and supervisors like people who are dependable. Teammates know that they can trust dependable people to pull their own weight, take their responsibilities seriously, and look after one another's safety. Supervisors know that a dependable person will do a job correctly and on time. Being dependable means showing up for work on time, every day (see *Figure 14*). It also means not stretching out lunch hours and breaks. When dependable workers say they will do a task, they follow through on their promise.

Instructor's Notes:

8.16 CORE CURRICULUM ◆ INTRODUCTORY CRAFT SKILLS

108F14.TIF

Figure 14 ◆ Always show up on time and put in a full day's work.

108F15.TIF

Figure 15 ◆ Respect your company's confidentiality.

Refer to Figure 16. Explain what it means to have a strong work ethic.

Explain the difference between taking initiative and making inappropriate decisions without approval.

Organizational skills are important as well. Keep your tools and your work areas clean and organized. Remember, craftworkers are judged by their tools, so know which tools you are responsible for, and keep them in good working order. Always know what you need to do each day before you begin. Approach your work in an organized fashion. Follow the schedule that you have been assigned.

Ensure that you are technically qualified to perform your tasks. This means that you know how to use tools, equipment, and machines safely. Take advantage of opportunities to expand your technical knowledge through classes, books and trade periodicals, and mentoring by experienced colleagues.

Offer to pitch in and help whenever you can. The best workers are willing to take on new tasks and learn new ideas. Supervisors notice employees who are willing. Be careful not to take on more tasks than you can safely handle. You will not impress anyone if you cannot deliver!

Honesty is one of the most important personal habits you can have. Do not abuse the system, such as by calling in sick just to take a day off or by leaving early and asking someone else to punch you off the clock. Never steal from the company or from your co-workers. This includes everything from tools and equipment to the simplest of office supplies. If you are struggling with a problem, don't hide it. Speak with your supervisor about it truthfully. Be willing to look actively for a solution to problems you are facing.

Professionalism means that you approach your work with integrity and a professional manner. As you learned in *Basic Safety,* there is no place for horseplay or irresponsible behavior on the construction site. Employers want workers who respect the rules and understand that they exist to keep people safe and to keep a project on schedule. A professional employee always respects company **confidentiality.** This means not sharing with other people any information that belongs to the company (see *Figure 15*).

Most companies have a dress code, and often they have additional special requirements for specific jobs. Follow these requirements. Finally, do not forget to attend to the basics of good grooming—comb your hair, brush your teeth, and wash your clothes. People who pay attention to their appearance and who develop positive personal habits will almost always be considered for a job over people who lack these habits. In the field, your personal habits reflect not just your own professionalism, but also the professionalism of your trade.

5.1.2 Work Ethic

Along with good personal habits, a strong work ethic is essential for getting hired and being promoted. Employers look for people who they believe will give them a fair day's work for a fair day's pay (see *Figure 16*). Having a strong work ethic means that you enjoy working and that you always try to do your best on each task. When work is important to you, you believe that you can make a positive contribution to any project you are working on.

Construction work requires people who can work without constant supervision. When there is a problem that you can solve, solve it without waiting for someone to tell you to do it. If you finish your task ahead of time, look for another task that you can do in the time remaining. This type of positive action is called **initiative.** Colleagues and supervisors will respect you more when you demonstrate initiative on the job.

MODULE 00108-04 ◆ BASIC EMPLOYABILITY SKILLS 8.17

Discuss the adverse effects lateness and absenteeism can have on construction schedules.

Ask trainees to review the list of suggestions for improving and maintaining their records of punctuality and attendance. Have trainees evaluate their own habits and consider potential areas of improvement.

Review techniques for appropriate conflict resolution on the job site. Discuss problems that could result if an employee reacted negatively to another employee's behavior.

Discuss options for preventing conflicts. Ask trainees to share examples from their experiences of preventing or resolving conflicts.

108F16.TIF

Figure 16 ◆ Show your work ethic when you are on the job.

An important part of taking initiative, however, is to know when and how to take it. Suppose your supervisor tells you to perform a task. He shows you the five steps required to complete the task. Later, as you perform the work, you realize that one of the steps is unnecessary. Leaving out the step could save time. Should you?

No. That would not be the right initiative. Instead, tell your supervisor about your discovery, and ask what you should do. Bringing the options to your supervisor is showing initiative. Your supervisor may realize that you are right and allow you to leave out the extra step. After that, every time anyone performs that task, it will take less time and save the company money. Or, your supervisor may explain why the step is important. Then you will have learned something more about the work you are doing. Either way, the result of taking the initiative is a positive one.

5.1.3 Lateness and Absenteeism

The two most common problems supervisors face on the job are lateness and absenteeism. Lateness is when a worker habitually shows up late to work. Absenteeism is when a worker consistently fails to show up for work at all, with or without excuses. People with a strong work ethic are not often late or absent.

To make a profit and stay in business, construction companies operate under tight schedules (see *Figure 17*). These schedules are built around workers. If you are often late or do not show up regularly, immediate and expensive adjustments become necessary. Your employer will likely decide that one of the best ways to save money is to stop wasting any more of it on you.

Consider the following suggestions for improving or maintaining your record of punctuality and attendance:

- Think about what would happen if everyone on the job were late or absent frequently.
- Think about how being late or absent will affect teammates.
- Know and follow your company's policy for reporting legitimate absences or lateness.
- Keep your supervisor informed if you need to be out for more than a day.
- Allow yourself enough time to get to work.
- Explain a late arrival to your supervisor as soon as you get to work.
- Do not abuse lunch and break-time privileges.

5.2.0 Conflict Resolution

Suppose you show up to work one day with a new haircut, and one of your teammates starts to tease you about it. Or you suspect that a co-worker has been stealing tools, but you cannot prove it. These situations are examples of conflicts that you might face on the job. How do you resolve, or handle, a conflict between you and someone else? You may be the type of person who can resolve a disagreement quickly. Or you may prefer to avoid addressing conflicts directly. Or you may get angry and resent the other person.

Conflict resolution is an important relationship skill, because conflict can happen anywhere, anytime. Conflict can happen when people disagree, and people disagree all the time. Conflicts between you and your teammates can arise because of disagreements over work habits; different attitudes about the job or the company; differences in personality, appearance, or age; or distractions caused by problems at home. Conflicts between you and your supervisor can happen because of a disagreement over workload, lateness or absenteeism, or criticism of mistakes or inefficiencies.

Most of the time, people are not trying to turn disagreements into conflicts. People are often unaware of the effects of their behavior. Before reacting negatively to a teammate's behavior, remember to be **tactful.** This means considering how the other person will feel about what you say or do. Do not accuse, embarrass, or threaten the person. Try to have a sense of humor about your

Instructor's Notes:

8.18 CORE CURRICULUM ◆ INTRODUCTORY CRAFT SKILLS

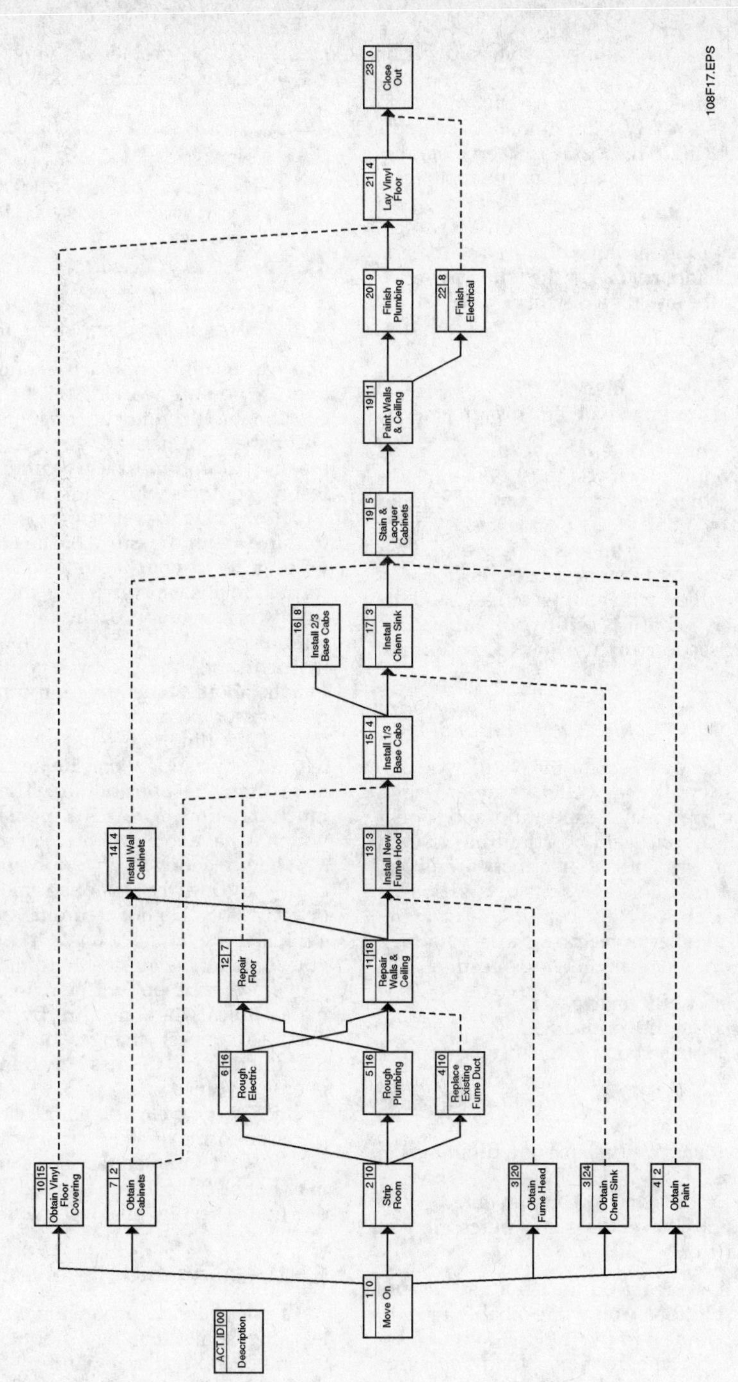

Figure 17 ◆ Construction companies operate under tight schedules.

MODULE 00108-04 ◆ BASIC EMPLOYABILITY SKILLS 8.19

Review the factors that trainees should consider as they analyze conflict problems.

Explain the importance of being able to compromise.

Review techniques for resolving conflicts with supervisors.

Divide the class into small groups, and present each group with a hypothetical problem. Ask trainees to brainstorm potential resolutions to the conflict using techniques discussed in this section.

Have trainees review Sections 5.3.0–5.5.0 for the next session.

Ensure that you have everything required for the laboratories and testing during this session.

Explain why it is important to be able to give and receive criticism.

situation. Try to avoid calling in your supervisor except as a last resort.

Never let a disagreement or conflict affect job performance, site safety, team morale, or an individual's well-being. The goal is to keep disagreements from turning into conflicts in the first place. If that is not possible, then the next best thing is to address the conflict quickly and resolve it professionally. If a disagreement with a co-worker is getting out of hand, try one of the following techniques to cool the situation down:

- Think before you react.
- Walk away.
- Do not take it personally.
- Avoid being drawn into others' disagreements.

If, despite your best efforts, a disagreement turns into a conflict, then you must take immediate action to resolve it. Do not let a conflict simmer and then boil over before you decide to act. That is not professional. Keep in mind, however, that there are important differences between the way you resolve conflicts with your teammates and the way you resolve conflicts with your supervisor. The following sections discuss how to handle each type of conflict.

5.2.1 Resolving Conflicts with Teammates

Remember to have respect for the people you disagree with. After all, they believe they are right, too. Be clear, rational, respectful, and open-minded at all times. Begin by admitting to each other that there is a conflict. Then analyze and discuss the problem. Allow everyone to describe his or her own perception of the conflict. You may realize that the whole problem was simply a miscommunication. Ask the following questions:

- How did the conflict start?
- What is keeping the conflict going?
- Is the conflict based on personality issues or a specific event?
- Has this problem been building up for a while, or did it start suddenly?
- Did the conflict start because of a difference in expectations?
- Could the problem have been prevented?
- Do both sides have the same perception of what's going on?

Once you have analyzed the situation, discuss the possible solutions. You will probably have to **compromise,** or meet in the middle, to find a solution that everyone agrees on. When you agree on the solution, act on it, and see if it works. If it does not, then consult your supervisor for help in resolving the conflict. Notice that this process is similar to the problem-solving techniques discussed elsewhere in this module.

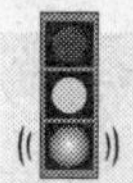

NOTE

Resolve your conflicts quickly. Keep your focus on getting your work done. Avoid prolonged arguments.

5.2.2 Resolving Conflicts with Supervisors

You can usually approach teammates as equals, because you are working on the same job. However, on the job, your supervisor is not your equal. He or she is in charge of you and your work. This means that you must use a different approach to resolve conflicts with your supervisor.

Before going to your supervisor, take some time to think about the cause of the conflict. Consider writing down your thoughts. Organizing the information this way often puts things into perspective. When you approach your supervisor, do so with respect. Wait until your supervisor has a free moment, and then ask if you could arrange a time to talk about something important. Or leave a message or note for your supervisor asking to meet. Be willing to meet at a time that is convenient for your supervisor. Remember, supervisors have many responsibilities. They do not have much free time during the regular workday, and you may have to meet before or after work.

When meeting with your supervisor, speak calmly and clearly. Do not be emotional, sarcastic, or accusatory. Do not confront your supervisor in a threatening or angry way. State only the facts as you see them; never say anything that you cannot prove. Do not mention the names of other teammates unless they are directly involved. If you want to suggest changes or solutions, explain them clearly and discuss how they will benefit the people involved.

Once you have made your case, allow your supervisor to make a decision. You should accept and respect your supervisor's final decision. It may not be the one you wanted, but it will be the one that is best for all concerned.

5.3.0 Giving and Receiving Criticism

As a construction professional, you are always learning something new about your job. New technologies, materials, and methods appear all the time. Talking to an experienced construction worker can help you learn a new way to perform a task. As your skills improve with practice, you

Instructor's Notes:

8.20 CORE CURRICULUM ◆ INTRODUCTORY CRAFT SKILLS

will be able to use tools that you were not able to use before. All of these are common ways of learning on the job (see *Figure 18*).

Another way to learn on the job is through **constructive criticism.** Constructive criticism is advice designed to help you correct a mistake or improve an action. Constructive criticism can improve your job performance and relations with teammates. You have probably heard the word criticism used in a negative way to indicate fault or blame. That is not the type of criticism discussed here. On-the-job criticism does not mean that colleagues and supervisors think little of you. In fact, it means exactly the opposite. Colleagues and supervisors who offer constructive criticism do so because they believe it is worth their time to help you improve your skills.

As you gain experience on the job, you will be able to give constructive criticism as well as receive it. To make sure that someone does not mistake your constructive criticism for blame, you need to know how to both offer and receive it in the spirit intended. The following sections offer some general advice on offering and receiving constructive criticism.

5.3.1 Offering Constructive Criticism

When you are training a less-experienced person or working with teammates, you might find yourself offering some constructive criticism. How should you offer it? Before you say anything, think about the rules of effective speaking that you learned in *Communication Tools for Success.* Use positive, supportive words, and offer facts, not opinions (see *Figure 19*).

Constructive criticism works best when offered occasionally. Do not constantly comment on other people's work or methods. They may block out your criticism or even become angry. Never criticize people in front of other teammates or supervisors. They may feel embarrassed and will probably resent you. Criticism should include suggested alternatives. Do not criticize how a teammate does something unless you can suggest another way. Above all, remember to limit your comments to people's work or methods, not the people themselves.

Remember to compliment the person you are criticizing. Compliments have a genuinely positive effect on the person receiving them, and they are easy to offer. You do not have to wait to compliment someone until you offer constructive criticism, either. Try to offer compliments on a regular

108F18.TIF

Figure 18 ◆ A good employee never stops learning on the job.

108F19.TIF

Figure 19 ◆ Offer constructive criticism using supportive words and facts.

Compare destructive criticism with constructive criticism.

Discuss the communication techniques necessary to criticize constructively.

Discuss the benefits of offering genuine compliments on a regular basis.

MODULE 00108-04 ◆ BASIC EMPLOYABILITY SKILLS 8.21

Emphasize that how well an employee receives criticism can affect how that person is perceived by co-workers.

Ask a trainee to explain how to respond to constructive criticism without appearing to be defensive.

Discuss the concept of teamwork, and explain how teamwork is integral to construction work.

basis. Your appreciative and respectful approach will help keep team spirit high. Your supervisors will appreciate your professional attitude.

NOTE

Point out improper behavior or incorrect work techniques the first time they occur. If not addressed, the behavior or technique could become a bad habit. Bad habits, in turn, can lead to accidents. Gently and firmly criticize improper behavior as soon as you see it.

5.3.2 Receiving Constructive Criticism

Not only can criticism make the difference between success and failure, but in cases where workplace safety is involved, it can also make the difference between life and death. Think of criticism as a chance to learn and to improve your skills. Remember to take criticism from your supervisor seriously. Treat criticism from experienced co-workers with the same respect; even though they may not be your supervisors, their experience gives them a different type of authority.

Always take responsibility for your actions. You may want to explain why you did what you did to the person who is offering the criticism. Try not to do this. The person you are talking to may think that you are making excuses or trying to shift responsibility onto someone else. Never be defensive or dispute criticism. If you respond negatively, you will offend the other person. As a result, that person will either lose respect for you or will no longer trust you—or both. If that happens, the quality of your work, and that of everybody who works with you, will suffer.

When someone offers you constructive criticism, demonstrate your positive personal habits and work ethic (see *Figure 20*). If the criticism is vague, ask the person to suggest specific changes that you should try to make. If you do not understand the criticism, ask for more details. Always be respectful to the people you receive criticism from, and take their criticism seriously. The deepest respect you can show them is to improve your performance based on their advice.

You have a right to disagree with criticism that is unacceptable or incorrect. Someone might honestly misunderstand your situation and offer incorrect advice. Or a co-worker might criticize you simply because he thinks he performs a task better than you do. In such cases, clearly and respectfully give your reasons for disagreeing, and explain why you believe that you are correct.

108F20.TIF

Figure 20 ◆ Receive constructive criticism with professional behavior.

5.4.0 Teamwork Skills

No matter what your job is, chances are that you work with other people. Every day, you interact with members of your work crew and with your supervisor. As you have learned elsewhere in this module, the ability to get along with people is an important skill in the construction industry. But simply "working and playing well with others" is not enough. The success of any team depends on all of its members doing their parts. Everyone must contribute to the team to ensure that it achieves its goals. This cooperation is called **teamwork.**

From the moment you began your construction career, you've been on different teams for different projects. Teams can be as small as two people and as large as your entire company. Teams are often made up of people from different trades who work together to complete a task. For example, a team that is assigned the task of burying

Instructor's Notes:

8.22 CORE CURRICULUM ◆ INTRODUCTORY CRAFT SKILLS

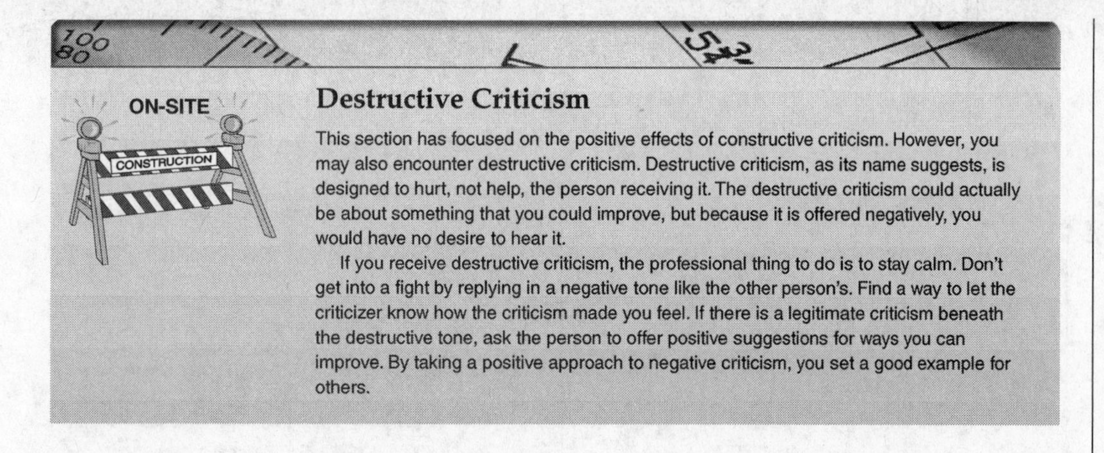

Destructive Criticism

This section has focused on the positive effects of constructive criticism. However, you may also encounter destructive criticism. Destructive criticism, as its name suggests, is designed to hurt, not help, the person receiving it. The destructive criticism could actually be about something that you could improve, but because it is offered negatively, you would have no desire to hear it.

If you receive destructive criticism, the professional thing to do is to stay calm. Don't get into a fight by replying in a negative tone like the other person's. Find a way to let the criticizer know how the criticism made you feel. If there is a legitimate criticism beneath the destructive tone, ask the person to offer positive suggestions for ways you can improve. By taking a positive approach to negative criticism, you set a good example for others.

Explain what it means to be goal-oriented. Emphasize the importance of being respectful to co-workers.

Explain how to be a good team member. Provide examples from your own experience when you have had to work with others who were not team players.

Explain problems that can result from allowing personal problems to affect you at work. Emphasize the importance of appropriately communicating serious personal problems with a supervisor.

water and electrical lines might include a backhoe operator to dig the trench, plumbers to join and lay the pipe, and electricians to run the electrical cable.

Always show respect for the other members of your team. You can do this by allowing them to share their ideas with you and with the rest of the team. Though you have a specific job to do, always be willing to help a teammate. Never say that something is "not my job." Someday, on another project, it might be! Support your teammates when they need it, and they will support you when you need it.

Good team members are **goal-oriented.** This may sound like management jargon, but it is actually a well-known and accepted concept in the construction industry. Being goal-oriented means making sure that all activities focus on the team's final objective. Whatever the end result is—fabricating fittings, laying pipe, or framing a house, for example—everyone on the team knows that result in advance and concentrates on achieving it.

As a team member, you should always try to use your skills and strengths while at the same time accepting your limitations and those of others. To be a good team member, try to do the following:

- Follow your team leader's and/or supervisor's directions.
- Accept that others might be better at some tasks than you.
- Keep a positive attitude when you work with other people.
- Recognize that the work you do is for the benefit of the entire company, not for you personally.
- Learn to work with people who work at different speeds.
- Accept goals that are set by someone else, not by you.
- Trust other members of the team to perform their tasks, just as you perform yours.
- Appreciate the work of others as much as you appreciate your own.

Keep in mind that you are not the only person who is working hard. Everyone on your team is focusing on the goal too. Offer praise and encouragement to your teammates, and they will do the same for you. This mutual respect will help you feel confident that your team will reach its goal. Share the credit for good work, and be willing to take blame for your mistakes and errors. These actions will help you earn the respect of your teammates.

If you practice goal-oriented teamwork, you should be able to meet your deadlines and keep projects on schedule. This translates into time and money saved by your company. Sometimes, your teammates or people in other trades cannot start their tasks until you have completed yours. Out of respect for them, and for your company's reputation, always finish your work in a timely manner. You do not like to be kept waiting; do not make someone else wait.

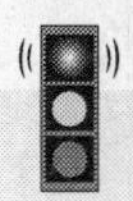

WARNING!

Distractions caused by personal problems can lead to injury and death. If you are concentrating on a problem instead of paying attention to your job, you can easily make a careless mistake or overlook a vital safety precaution. Stay focused on your job, and take whatever steps are necessary to eliminate distractions.

MODULE 00108-04 ◆ BASIC EMPLOYABILITY SKILLS 8.23

Have trainees review the chart of unpopular behaviors and complete an action plan for improvement. Stress that this is a self-assessment tool that can be used any time trainees feel they may need guidance.

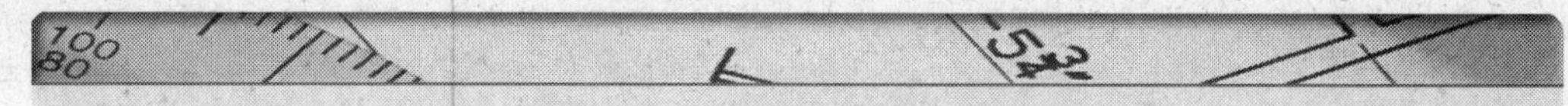

How Do Your Teammates See You?

Do you exhibit any of the following unpopular behaviors on the job? If you do, they may be keeping you from being an effective member of the team.

If you do exhibit any of these behaviors, develop an action plan to deal with them. Your action plan should be a two- or three-step process that will allow you to correct the behavior. Be honest!

Being a loner
Taking yourself too seriously
Being uptight
Always needing to be the best at everything
Holding a grudge
Arriving late to work
Being inconsiderate
Taking breaks that are too long
Gossiping about others
Sticking your nose into other people's business
Acting as a "spy" and reporting on the behavior of others to your supervisor
Saying or doing things to create tension or unhappiness
Complaining constantly
Bad-mouthing the company or your boss
Being unable to take a joke
Taking credit for others' work
Bragging
Being sarcastic
Refusing to listen to other people's ideas
Looking down on other people
Being unwilling to pitch in and help
Horsing around when others are trying to work
Thinking you work harder than everyone else
Being stingy with a compliment
Having a chip on your shoulder
Manipulating people

Action Plan for Improvement

Example:

Problem: Gossiping about others

Action Plan: 1. Walk away when people start gossiping or bad-mouthing others.

2. Don't repeat what I hear.

3. Focus on my own work, not on that of others.

Problem: ______________________

Action Plan: ______________________

____

Problem: ______________________

Action Plan: ______________________

____

Problem: ______________________

Action Plan: ______________________

____

108SA02.EPS

Instructor's Notes:

____

____

____

Keep Your Problems at Home

Few things are more challenging to deal with than people who take their personal problems to work with them. Being a professional means keeping your personal life and your work life separate. This does not mean that you have to keep your personal life secret from your teammates. It means that you should not take out your personal frustrations, fears, or hostilities on your colleagues. As you gain more experience on the job, you will discover an appropriate comfort level for discussing personal issues with your teammates.

Some events—for example, the death of a loved one—can affect you so deeply that it becomes impossible for you to separate your personal life from work. Let your teammates know that you are having difficulties. Discuss the situation with your supervisor. Your friends and colleagues can help when you are having a problem.

Review techniques for teaching someone in a productive manner.

See the Teaching Tip for Section 5.4.0 at the end of this module.

Review characteristics of successful leaders. Provide examples from your own leadership experience.

An important part of teamwork is training. As an apprentice, you received guidance and advice from more experienced colleagues. As you gain your own experience, you will be able to help less-experienced colleagues. Training offers an excellent opportunity to build good relationships. You probably remember at least one teacher who was patient and understanding and who made a difference in your life. On a team, you can be that person for someone else!

Being asked to teach someone is an honor. It means that someone believes you do your job well enough to teach it to other people. Such confidence should inspire you to be the best teacher you can be. Remember to be patient with the person you are teaching, and teach by example. Offer encouragement and give constructive advice. You learned how to do that elsewhere in this module. Teach people the same way that you would like to be taught.

5.5.0 Leadership Skills

As you gain experience and earn credentials, you will assume positions of greater responsibility in the construction industry. You will earn these positions through your hard work and dedication. Starting as an apprentice, you can work your way

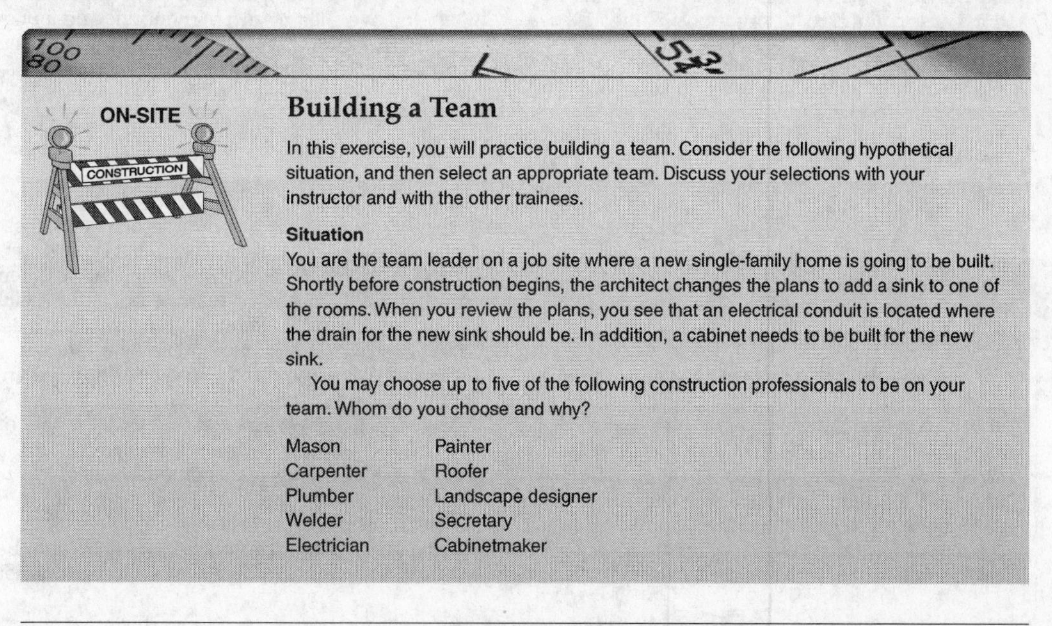

Building a Team

In this exercise, you will practice building a team. Consider the following hypothetical situation, and then select an appropriate team. Discuss your selections with your instructor and with the other trainees.

Situation

You are the team leader on a job site where a new single-family home is going to be built. Shortly before construction begins, the architect changes the plans to add a sink to one of the rooms. When you review the plans, you see that an electrical conduit is located where the drain for the new sink should be. In addition, a cabinet needs to be built for the new sink.

You may choose up to five of the following construction professionals to be on your team. Whom do you choose and why?

Mason	Painter
Carpenter	Roofer
Plumber	Landscape designer
Welder	Secretary
Electrician	Cabinetmaker

MODULE 00108-04 ◆ BASIC EMPLOYABILITY SKILLS 8.25

Discuss functions common to all situations requiring leadership.

Review the various leadership styles, and present a variety of hypothetical situations. Ask trainees to determine which leadership style they feel would be most successful in each situation.

Emphasize the importance of ethical conduct. Ask trainees to provide examples of unethical conduct they may have witnessed on the job site.

Review techniques for motivating people on the job.

Have each trainee perform Performance Task 3 to your satisfaction. Fill out a Performance Profile Sheet for each trainee.

Have trainees review Section 6.0.0 for the next session.

Ensure that you have everything required for the laboratory and testing during this session.

up to team leader, foreman, supervisor, and project manager. Someday, with enough hard work and dedication, you might even be able to run your own company.

To progress steadily through your career, you will need to develop **leadership** skills and learn how to use them. Leaders set an example for others to follow. Because of their skills, leaders are trusted not only with the authority to make decisions, but also with the responsibility to carry them out.

The typical image of a leader is that of a boss who has many people working for him or her. But you can be a leader at any stage in your career. As an apprentice, you have the authority to perform the task given to you by your supervisor, and your supervisor expects you to be a responsible worker. By carrying out your task quickly, correctly, and independently, you are setting an example for others to follow. In doing so, you are demonstrating leadership skills.

People with the ability to become leaders often exhibit the following characteristics:

- They lead by example.
- They have a high level of drive, determination, and persistence.
- They are effective communicators.
- They can motivate their team to do its best work.
- They are organized planners.
- They have self-confidence.

The functions of a leader will vary with the environment, the group of workers being led, and the tasks to be performed. However, certain functions are common to all situations. Some of these functions include the following:

- Organizing, planning, staffing, directing, and controlling
- Empowering team members with authority and responsibility
- Resolving disagreements before they become problems
- Enforcing company policies and procedures
- Accepting responsibility for failures as well as for successes
- Representing the team to different trades, clients, and others

Leadership styles vary widely, and they can be classified in many ways. If you classify leadership styles according to the way a leader makes decisions, for example, you end up with three broad categories of leader: autocratic, democratic, and hands-off. An autocratic leader makes all decisions independently, without seeking recommendations or suggestions from the team. A democratic leader involves the team in the decision-making process. Such a leader takes team members' recommendations and suggestions into account before making a decision. A hands-off leader leaves all decision-making to the team members themselves.

Select a leadership style that is appropriate to the situation. You will need to consider your authority, experience, expertise, and personality. Leaders need to have the respect of people on the team; otherwise, they will be unable to set an example worth following.

Leaders have to make sure that the decisions they make are ethical. The construction industry demands the highest standards of ethical conduct. The three types of ethics that you will encounter on the job are business or legal ethics, professional ethics, and situational ethics. Business or legal ethics involve adhering to all relevant laws and regulations. Professional ethics involve being fair to everybody. Situational ethics involve appropriate responses to a particular event or situation.

Effective leaders motivate, or inspire, people to do their best. People are motivated by different things at different times. The following are some common ways to motivate people on the job:

- Recognize and praise a job well done.
- Allow people to feel a sense of accomplishment.
- Provide opportunities for advancement.
- Encourage people to feel that their job is important.
- Provide opportunities for change to prevent boredom.
- Reward people for their efforts.

Leaders who can motivate people are more likely to have a team with high morale and a positive work attitude. Morale and attitude are key components of a successful company with satisfied workers. And such a company is more likely to be successful.

6.0.0 ◆ WORKPLACE ISSUES

The modern construction workplace is a cross-section of our society (see *Figure 21*). A typical construction project will involve men and women from all walks of life, of many ethnic and racial backgrounds, and from many different countries. Many workers speak more than one language. Workers have grown up in, and currently live in, many different income brackets.

Construction workers face a wide range of mental and physical demands every day on the job. These demands can sometimes feel overwhelming, and people may seek escape through illegal drugs or alcohol abuse.

As a construction professional, you need to be aware of these issues. You also need to know what

Instructor's Notes:

8.26 CORE CURRICULUM ◆ INTRODUCTORY CRAFT SKILLS

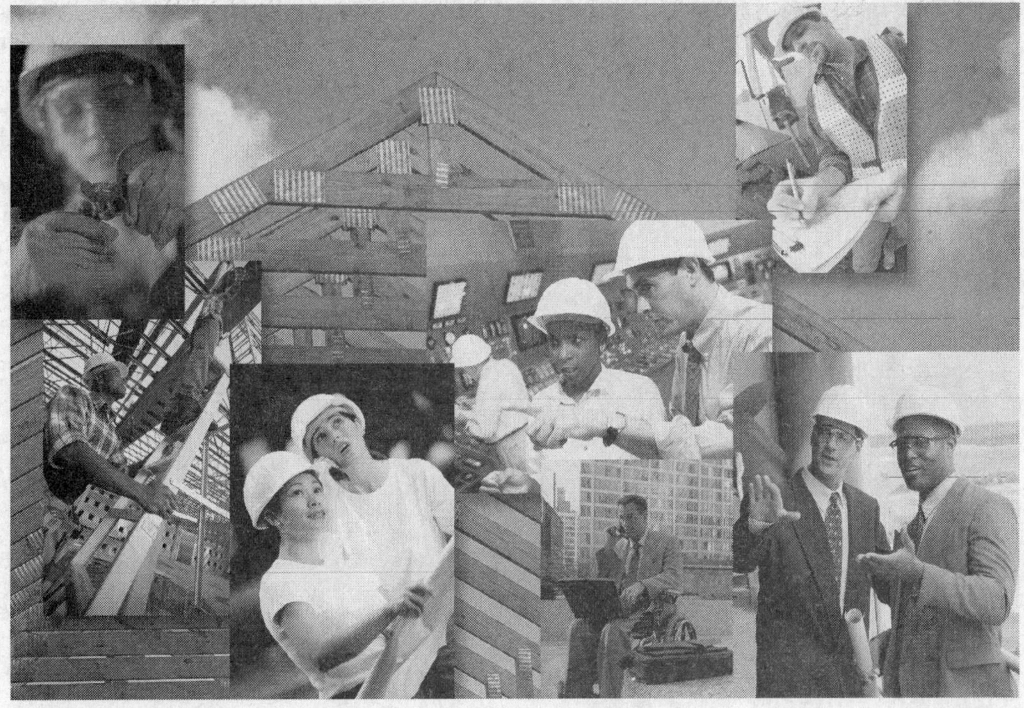

108F21.TIF

Figure 21 ◆ Today's construction workplace is a cross-section of our society.

to do if you see someone behaving inappropriately. This section briefly reviews some of the most common workplace issues you are likely to encounter:

- **Harassment** (sexual, age, height, weight, religious, cultural, disability)
- Stress
- Drug and alcohol abuse

Harassment is a type of discrimination that can be based on race, age, disabilities, sex, religion, cultural issues, health, or language barriers. Harassment often takes the form of ethnic slurs, racial jokes, offensive or derogatory comments, and other verbal or physical conduct. It can create an intimidating, hostile, or offensive working environment. It can also interfere with an individual's work performance. Harassment can take on many forms. The type most commonly reported and talked about is **sexual harassment.** But just because the other forms of harassment are not talked about as much or as openly doesn't mean they don't happen.

When someone makes unwelcome sexual advances or requests or exhibits other verbal or physical behavior with sexual overtones, he or she is guilty of sexual harassment. Sexual harassment can happen between members of the opposite or the same gender. The harasser can threaten to fire the victim to keep him or her quiet. Sexual harassment is illegal, and if you experience it, you should report it to your supervisor immediately. Usually, sexual harassment is a pattern of behavior repeated over a period of time, so if you do not report it, you run the risk of experiencing it again.

Stress is the tension, anxiety, or strain that you feel when you face unexpected events or things that are outside your control. Stress can cause headaches, irritability, mental and physical exhaustion, and even health problems (see *Figure 22*). Stresses at home can make themselves felt at work, and work stresses can affect home life.

You can take steps to prevent stress from happening and reduce the stresses that you cannot prevent. To prevent stress, keep your tools and

Emphasize that trainees will work with a variety of people from diverse ethnic and racial backgrounds and that it is always necessary to respect those differences.

Discuss the types of harassment that may be present on the work site. Emphasize that workplace harassment of any type has serious consequences.

Review techniques for preventing and reducing stress in a healthy manner.

Bring in excerpts from federal laws prohibiting job discrimination. Ask trainees to review the excerpts, and discuss the relevance of each. Discuss the consequences of discrimination, and ask trainees to consider whether they may have unknowingly discriminated.

Discuss the various types of alcohol and drug abuse. Emphasize that all abuse has adverse effects.

DID YOU KNOW?

Several federal laws prohibit job discrimination:

- *Title VII of the Civil Rights Act of 1964,* which prohibits employment discrimination based on race, color, religion, sex, or national origin.
- *Equal Pay Act of 1963,* which protects men and women who perform substantially equal work in the same establishment from sex-based wage discrimination.
- *Age Discrimination in Employment Act of 1967,* which protects individuals who are 40 years of age or older.
- *Title I and Title V of the Americans with Disabilities Act of 1990,* which prohibit employment discrimination against qualified individuals with disabilities in the private sector and in state and local governments.
- *Sections 501 and 505 of the Rehabilitation Act of 1973,* which prohibit discrimination against qualified individuals with disabilities who work in the federal government.
- *Civil Rights Act of 1991,* which, among other things, provides monetary damages in cases of intentional employment discrimination.

Source: U.S. Equal Employment Opportunity Commission.

equipment in good working order, and be neat. Plan your work, and try to finish the most difficult tasks first, while you have the energy and attention. Always eat properly and get plenty of exercise. Manage your time and money wisely so that you are not surprised by sudden shortages in either.

For stresses that you cannot prevent, make a plan to deal with your problems. Control your anger and frustration by relaxing in a way that you enjoy—whether it is taking a hike or getting a massage. Talk with someone who will help you get a better perspective on your situation.

Drinking is a common way to avoid or forget about stress. Alcohol is an accepted part of modern social life. Moderate use of alcohol is socially acceptable and is generally believed to cause little or no harm to most people. Alcohol abuse, which means habitually drinking to excess, not only is socially unacceptable, but also poses a serious health and safety risk for yourself, your colleagues, and your family.

Unlike alcohol, illegal drug use is never socially acceptable. Misusing legal prescription drugs is also not acceptable. There are three general classes of drugs:

- **Amphetamines**
- **Barbiturates**
- **Hallucinogens**

Amphetamines are stimulants. Also called uppers, they are used to prolong wakefulness and endurance, and they produce feelings of euphoria, or excitement. They can disturb your vision, cause dizziness and an irregular heartbeat, cause you to lose your coordination, and can even cause you to collapse. Caffeine and tobacco are common and legal amphetamines; cocaine and crack are illegal amphetamines.

Barbiturates are sedatives, which means they cause you to relax. They are also called downers. They can create slurred speech, slow your reactions, cause mood swings, and cause a loss of inhibition, which means that they make you feel less shy or self-conscious. Alcohol and the prescription drug Valium® are barbiturates.

Hallucinogens distort your perception of reality to the point where you see things that are not there (hallucinations). The effects of hallucinogens can range from ecstasy to terror. When you are hallucinating, you put yourself and others at great risk because you cannot react to situations the right way. Hallucinogens can cause chills, nausea, trembling, and weakness. Mescaline and LSD are two illegal hallucinogens.

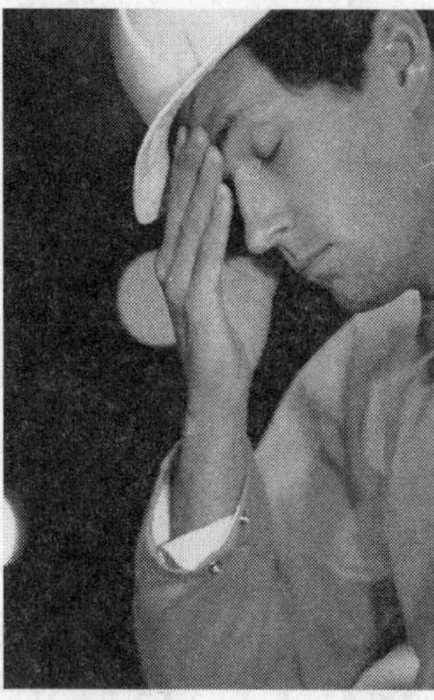

108F22.TIF

Figure 22 ◆ Deal with stress before it wears you down.

Instructor's Notes:

8.28 CORE CURRICULUM ◆ INTRODUCTORY CRAFT SKILLS

If you abuse drugs or alcohol, you will probably not be able to get or keep the job you want. If you are discovered using illegal drugs, you can be fired on the spot because of the safety risks that drug use poses. Addiction affects your well-being and state of mind, and you cannot do your best on the job or at home. It can also jeopardize your relationships with friends, family, and co-workers. If people you know are addicted to alcohol or drugs, seek help for them or encourage them to seek it for themselves. The appendix contains a list of organizations that can help people cope with addictions.

Review Questions

Sections 4.0.0–6.0.0

1. PDA stands for _____.
   a. personal drafting aid
   b. personality disorder analysis
   c. portable document assistance
   d. personal digital assistant

In questions 2–5, match the computer hardware term with its definition.

2. A _______ shows information, text, and pictures to the user.

3. A _______ allows users to enter data without having to type.

4. A _______ contains the chips and circuits that enable the computer to perform its functions.

5. A _______ allows users to type in text and instructions.
   a. processor
   b. monitor
   c. keyboard
   d. mouse

6. Spreadsheets are used to perform math calculations, such as for project budgets and company payrolls.
   a. True
   b. False

7. Each of the following is a way to prevent disagreements from getting out of hand *except* _____.
   a. thinking before reacting
   b. not taking it personally
   c. walking away
   d. being resentful and holding a grudge

8. Illegal drug use is considered socially acceptable in moderation but not when it becomes an addiction.
   a. True
   b. False

Refer trainees to the Appendix for a list of organizations that can help people cope with addictions. Offer to assist trainees in addressing any addiction problems they may have.

See the Teaching Tip for Section 6.0.0 at the end of this module.

Go over the Review Questions for Sections 4.0.0–6.0.0. Answer any questions trainees may have.

Have trainees complete the Key Terms Quiz. Ask whether trainees have any questions.

MODULE 00108-04 ◆ BASIC EMPLOYABILITY SKILLS 8.29

Administer the Module Examination. Be sure to record the results of the Exam on Craft Training Report Form 200, and submit the results to the Training Program Sponsor.

Ensure that all Performance Tests have been completed and Performance Profile Sheets for each trainee are filled out. If desired, trainee proficiency noted during laboratory sessions may be used to complete the Performance Tests. Be sure to record the results of the testing on Craft Training Report Form 200, and submit the results to the Training Program Sponsor.

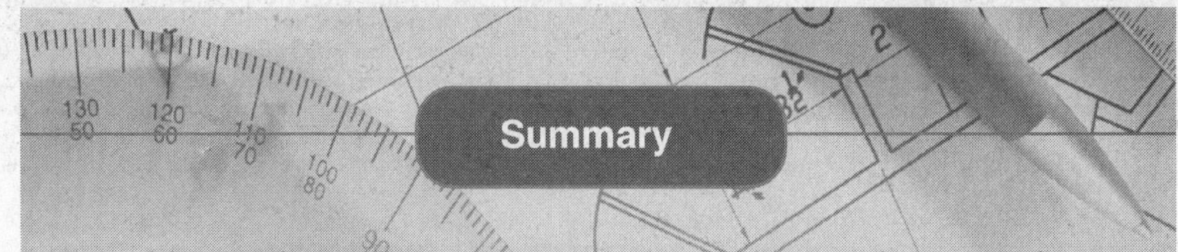

Summary

This module reviewed some of the important nontechnical skills that you must learn to be successful as a construction professional. The construction industry, the second largest industry in the country, consists of independent companies of all sizes that specialize in one or more types of construction work. Whatever positions you may achieve throughout your career, these skills are vital.

Good construction workers understand the industry and their place in it. You need to understand not only your company's roles in the construction industry, but also your own role in the company. You also need to know company policies and procedures.

Whether just starting out or moving on to new opportunities and challenges, you should select a job for which you are qualified, write a top-notch resume, and prepare for interviews. Develop skills at evaluating different job offers to select the ones that work best for you. Eventually, many construction professionals become entrepreneurs and start their own small businesses. Entrepreneurs are energetic people who can adapt to a changing business environment.

Everyone has to work together, from senior-level employees to apprentices. Problems arise for everyone, no matter who they are. The ability to solve problems using critical thinking skills is important for any employer and employee. Critical thinking involves evaluating and using information to reach conclusions or make decisions. Solving problems using critical thinking skills is a five-step process: Begin by defining the problem, then explore alternatives, choose a solution, implement and monitor that solution, and evaluate the result.

Computer literacy is also important for construction professionals. Computers play many important roles in the construction industry, where most work is done on a job site or in a shop. Nowadays, you will find plenty of computers in both these places. As part of your workplace skills, you need to have at least a basic understanding of computers and the work they perform. This module introduced you to the different types of hardware, software, and operating systems.

You should develop strong relationships with your teammates and supervisors. Self-presentation skills, including personal habits, work ethic, and timeliness, along with conflict resolution skills—the ability to give and receive constructive criticism, the willingness to work as part of a team, and the courage to demonstrate leadership traits—are all types of relationship skills. Leaders set an example for others and are trusted with authority and responsibility. You can demonstrate leadership at any stage in your career.

You also need to know what to do if you see a person behaving inappropriately. Sexual harassment and drug and alcohol abuse should not be tolerated by anyone in the construction industry. Those who abuse drugs or alcohol jeopardize their teammates and can be fired.

The modern construction industry demands the best from its people. With such a wide range of skills and talents, the men and women who work in the industry can live up to such a demand with pride. You are choosing a career that will bring out the best in you, but only if you are willing to give it your best to begin with.

Instructor's Notes:

8.30 CORE CURRICULUM ◆ INTRODUCTORY CRAFT SKILLS

Key Terms Quiz

Fill in the blank with the correct key term that you learned from your study of this module.

1. A company uses a(n) _________ to state how it does business.
2. Someone who can vouch for your skills, experience, and work habits is a(n) _________.
3. A(n) _________ is someone who starts and runs a business.
4. Problems can be effectively resolved using _________.
5. Computers that are primarily used on desks are _________.
6. The portability of a(n) _________ makes it easy to take a computer on the road.
7. _________ are small enough to hold in your hand.
8. The basic understanding of how computers work is called _________.
9. _________, _________, and _________ are the three elements that make it possible for a computer to operate.
10. The person using the computer is called the _________.
11. A(n) _________ is sometimes called a CPU.
12. Text and pictures are shown on the screen of a(n) _________.
13. Instructions for the processor are typed on a(n) _________.
14. Another method for getting information to the processor is to use a(n) _________.
15. The _________ on a computer stores software and electronic files.
16. A(n) _________ is a device that allows the processor to save and transfer software and electronic files that are not on the hard drive.
17. A portable storage device that allows software and electronic files to be transferred from one computer to another is a(n) _________.
18. Both a(n) _________ and a(n) ____ are used to transfer text, labels, graphics, and drawings onto paper and other materials.
19. Printed text or pictures can be changed into electronic files by using a(n) _________.
20. Letters, reports, and forms can be written using a(n) _________.
21. Math problems can be calculated using a software program that utilizes _________.
22. Employee lists and material inventories are often organized and stored in a(n) _________.
23. _________ is the process that is used to lay out text and graphics.
24. Structural, plumbing, and electrical plans can be designed and printed using a(n) _________ system.
25. The World Wide Web is typically searched using a(n) _________.
26. Electronic messages are sent and received using _________.
27. The instructional manuals that come with software are called _________.
28. _________ communication allows computers to communicate without being physically connected.
29. Machine tools controlled by computers are called _________ machines.
30. Slots and holes in metal can be made using a(n) _________.
31. _________ is the way you act, speak, and dress.
32. Another term for work habit is _________.
33. _________ is closely related to self-presentation, but it relies more on your integrity and manner.
34. When you break _________, you share with other people information that belongs to the company.

MODULE 00108-04 ◆ BASIC EMPLOYABILITY SKILLS 8.31

35. You will gain the respect of your supervisor if you regularly take the _________ to complete additional tasks when your work is done.

36. _________ and _____ are the two most common problems supervisors face with workers.

37. A(n) _________ approach should be used when approaching a co-worker about negative behavior.

38. _________ is often necessary to resolve some situations.

39. Advice that is given to point out or correct a mistake is called _________.

40. _________ is often called cooperation.

41. Workers who focus on reaching final outcomes are considered _________.

42. A person who sets an example for other people to follow is generally demonstrating good _________ skills.

43. Racial jokes are considered a form of _________.

44. Unwelcome sexual advances or requests are considered _________.

45. Uppers are another word for _________.

46. _________ cause you to feel sedated.

47. The drug LSD is considered a(n) _________.

Key Terms

Absenteeism
Amphetamine
Barbiturate
Browser
CAD
CD-ROM
CNC
Compromise
Computer literacy
Confidentiality
Constructive criticism
Critical thinking skills
Database
Desktop computer
Desktop publishing
Disk drive
Documentation
EDM
Email
Entrepreneur
Goal-oriented
Hallucinogen
Handheld computer
Harassment
Hard drive
Hardware
Initiative
Keyboard
Laptop computer
Lateness
Leadership
Mission statement
Monitor
Mouse
Operating system
PDA
Plotter
Printer
Processor
Professionalism
Reference
Scanner
Self-presentation
Sexual harassment
Software
Spreadsheet
Tactful
Teamwork
User
Wireless
Word processor
Work ethic

Instructor's Notes:

8.32 CORE CURRICULUM ◆ INTRODUCTORY CRAFT SKILLS

Profile in Success

Connell Linson

President and CEO
TABACON Systems, Inc.
Houston, Texas

Connell was born in Nashville, Tennessee, but is a long-time Texas resident. His connection to the construction industry dates from high school, when he worked as a plumber's assistant for an independent contractor in Houston. He taught electronics in the Air Force and later worked on the Apollo space program as an engineer for Singer General Precision. There, Connell helped design and build a vehicle that was used by astronauts to simulate moon landings. Connell retired from Southwestern Bell as an area manager engineer in 1990 and formed TABACON Systems, Inc., to help young people find careers in the construction industry.

What led you to establish your company?

I came back to the construction industry because of my experience as a literacy volunteer. I got to know students, and I listened to them talk about their experiences. They said to me, "We can't see a future. We don't know what to do." I saw that simply teaching them to read was not enough. Once they are literate, what's the next step to helping them become employed and put that literacy to use? I wanted them to experience the dignity of earning their living.

I saw that the construction career path was being ignored. So I did a lot of research. I looked at Bureau of Labor Statistics data, local area employment statistics, and other sources. The more I found out about career opportunities in construction, the more I wondered why nobody in the industry was shouting at the top of their voice that if you go to school and take advantage of the educational opportunities open to you, you could be a journeyman in five years and be successful. So I formed TABACON to preach the opportunities of construction employment to young folk.

What is TABACON's mission?

TABACON is a construction industry advocate. We actively recruit students for large construction firms in the area. We are an affiliate of the Houston chapter of the Associated Builders and Contractors, which teaches some of our classes as well. I have a core staff of five, including my son and my wife. We offer consulting; recruiting; training, using the NCCER curriculum; placement services; and safety training. Depending on the size of the project, we outsource many of those activities. We also offer training packages to inner-city youth programs.

My goal is to create a group of self-sufficient young folk, and construction is a way to do that. The critical thinking skills that you get as a journeyman are absolutely essential for success in life, not just in your construction career. People at the top of their game in this field are absolutely artistic.

We are a 100 percent drug-free company. Every employee, every client, every student who comes through our program must sign an agreement. I think that's important because it's a serious issue with employers. For example, in Texas, if an employee is in an on-the-job accident and they test positive for drugs, they can't draw unemployment compensation. We want to make sure our kids understand that. So we are doubly cautious and conscious. We work with an organization called Drug Free Business Houston, and this past year we were awarded a plaque for our work with investing in young workers. We are really proud of the fact that we've been recognized by an organization of that scope.

As well as advocating on behalf of the industry, TABACON also does a lot of advocating to the industry itself. The industry doesn't seem to recognize that it has so much to offer to young folk who are looking for a career. Salaries in the construction industry are comparable to those in information technology, and at the base levels it's

MODULE 00108-04 ◆ BASIC EMPLOYABILITY SKILLS 8.33

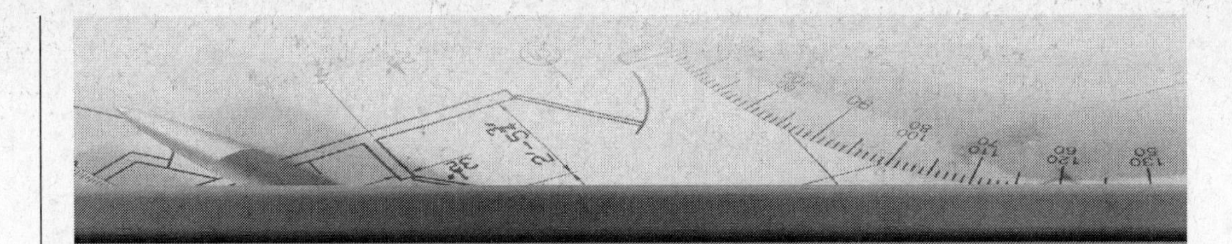

actually more than competitive. I did the research, and I've seen the numbers. I am always asking industry people, "Why are you hiding your light under a bushel?"

Our motto on our business cards is, "Since you work to live, why not earn a living?" I think that says it all.

What do you like most about your job?
I like going out to churches, social organizations, community groups, and other organizations to talk about the advantages of construction careers. Many of them partner with TABACON. For example, a project called Beam Up uses modules from the NCCER Core Curriculum and Tools for Success with inner-city kids. When they finish, employers are really excited about grabbing these kids up. I really like that. We have kids who are great success stories. I'm committed to turning out a group of successful adults.

I love coordinating with companies, but I love being in the classroom most of all. When I am teaching a class and all of a sudden I see the light go on in a young person's eye, and I know that they got it, there's no greater joy than that. Sometimes they'll even say to me, "Wait! Stop! I think I get it!" and then they explain it back to me. That's what it is all about.

What would you say to someone entering the trades today?
I think the most important thing to remember is that you must be persistent and not quit. One former student of mine had a job that involved a lot of shoveling. He told me that the shoveling was killing him, but instead of quitting he just kept reminding himself what I told him about persistence, and he stayed with it. He started out as a laborer and has moved up to being helper. His hourly wage has gone up three dollars an hour. He's now taking the NCCER Core Curriculum, and when he's done he will be able to go anywhere in the industry.

Stay open-minded. Listen to what your supervisor says, because your supervisor wants you to succeed. I'm a living example of that!

Instructor's Notes:

______

______

______

ASU Students Up at Dawn— Even on Saturdays

By Leah Hitchings

For students of Arizona State University's Del E. Webb School of Construction, Saturdays aren't for sleeping late. Students in faculty associate Christopher Aulerich's surveying courses start Saturday labs at 5 A.M. However, Aulerich's evaluations are consistently among the highest at the School of Construction, according to Director Bill Badger. That's because as president and owner of Brady Aulerich & Associates, Aulerich brings students into the field to use the cutting-edge equipment his company uses for projects.

"When Badger called me in to look at ASU's equipment in 1989, their equipment was from the 1960s, not what was actually being used in the industry," Aulerich says. "I offered our equipment, and he asked me to teach."

Teaching about 100 students a year means that Aulerich has taught 1,300 students in surveying, and half of those students now own construction companies.

"As vice presidents of companies, these kids think of me when they need surveying done. My company hasn't had to do marketing since I started teaching the classes," Aulerich says.

Aulerich teaches skills that are necessary in the general contracting industry, from reading blueprints to laying out projects. Because the prints and laying out are for real projects at Brady Aulerich & Associates, students are immediately able to do work in the field upon graduation.

"We lay out parking lots, buildings, and utilities," Aulerich says. "I try to incorporate the real world."

For Peter Trowbridge, a former student of Aulerich's, the surveying course was the most important class he took at ASU. Now owner of general contracting company TrowTierce Precision Construction, in Phoenix, Ariz., Trowbridge uses Aulerich's company for much of his surveying needs.

Source: Reprinted from *Engineering News-Record,* January 14, 2003.

Although he makes students come to class at 5 A.M. on Saturdays, Aulerich isn't always working, according to Trowbridge.

"I remember Chris would ask the students for a pack of cigarettes, and would hold it horizontally for ten seconds," Trowbridge says. "If a drop of rain fell on the pack, then class was cancelled for the day. On a Notre Dame football game day, no rain fell, but he cancelled class."

Another former student of Aulerich's, Cal Detwiler, says that the course was one of the toughest at the School of Construction.

"Even though he's a tougher teacher, it taught me a lot about surveying, where you can't afford to make mistakes," Detwiler says. As a project manager at Kitchell Contractors in Phoenix, Detwiler uses Aulerich almost exclusively for surveying.

Badger says that Aulerich makes a great professor, because "he's young, dynamic, and loves being in front of a class. The students like getting their hands on survey equipment, and they like the high tech aspect of the work," he says. This semester, "high tech" includes a new $65,000 GPS system and an HP 48 GX data collector and tripod data system. Aulerich even donates part of his salary back to the School of Construction, according to Badger.

Aulerich has always been fascinated with land surveying and being in the field. After interning with an engineering firm at the age of 18, Aulerich spent ten years in the field and has been in management ever since. He says, "I have something where 90% of Monday mornings, I love going to work."

ASU's School of Construction includes classes taught by 15 other faculty associates, industry professionals who also take the time to teach classes. Beyond college, Aulerich offers his students help.

He says, "I tell them I'll help them with problems they come across, they just have to call me."

MODULE 00108-04 ◆ BASIC EMPLOYABILITY SKILLS 8.35

Trade Terms Introduced in This Module

Absenteeism: Consistent failure to show up for work.

Amphetamine: A class of drugs that causes mental stimulation and feelings of euphoria.

Barbiturate: A class of drugs that induces relaxation.

Browser: Software that allows users to search the World Wide Web.

CAD: Software that is used to draw and print design drawings.

CD-ROM: Abbreviation for compact disk read-only memory. Hardware that stores software and allows it to be transferred between computers.

CNC: Abbreviation for computer numerical control. A general term used to describe computer-controlled machine tools.

Compromise: When people involved in a disagreement make concessions or meet in the middle to reach a solution that everyone agrees on.

Computer literacy: An understanding of how computers work and what they are used for.

Confidentiality: Privacy of information.

Constructive criticism: A positive offer of advice and help intended to help someone correct mistakes or improve actions.

Critical thinking skills: The skills required to evaluate and use information to make decisions or reach conclusions.

Database: Software that stores, organizes, and retrieves information.

Desktop computer: A computer that is placed on a large flat surface, such as a desk or tabletop.

Desktop publishing: Software used to lay out text and graphics for publication.

Disk drive: Hardware that allows software and files to be transferred between computers.

Documentation: Instruction manuals and other information for software.

EDM: Abbreviation for electrical discharge machines. Computer-controlled machine tools that cut and form parts that cannot be easily fabricated otherwise.

Email: Electronic mail messages.

Entrepreneur: A person who starts and runs his or her own business.

Goal-oriented: To be focused on an objective.

Hallucinogen: A class of drugs that distort the perception of reality and cause hallucinations.

Handheld computer: A computer that is designed to be small and light enough to be carried.

Harassment: A type of discrimination that can be based on race, age, disabilities, sex, religion, cultural issues, health, or language barriers.

Hard drive: Hardware that stores software and electronic files.

Hardware: The physical components that make up a computer.

Initiative: The ability to work without constant supervision and solve problems independently.

Keyboard: Hardware that allows users to enter text and instructions and send them to the processor.

Laptop computer: A computer that is larger than a handheld computer but smaller than a desktop computer.

Lateness: Habitually showing up late for work.

Leadership: The ability to set an example for others to follow by exercising authority and responsibility.

Mission statement: A statement of how a company does business.

Monitor: Hardware that shows information, text, and pictures on a television-style screen.

Mouse: Hardware that allows a user to enter data without a keyboard.

Operating system: A complex set of commands that enables hardware and software to communicate.

PDA: Abbreviation for personal digital assistant. Another name for a handheld computer.

Plotter: Hardware that prints large architectural and construction drawings and images.

8.36 CORE CURRICULUM ◆ INTRODUCTORY CRAFT SKILLS

Instructor's Notes:

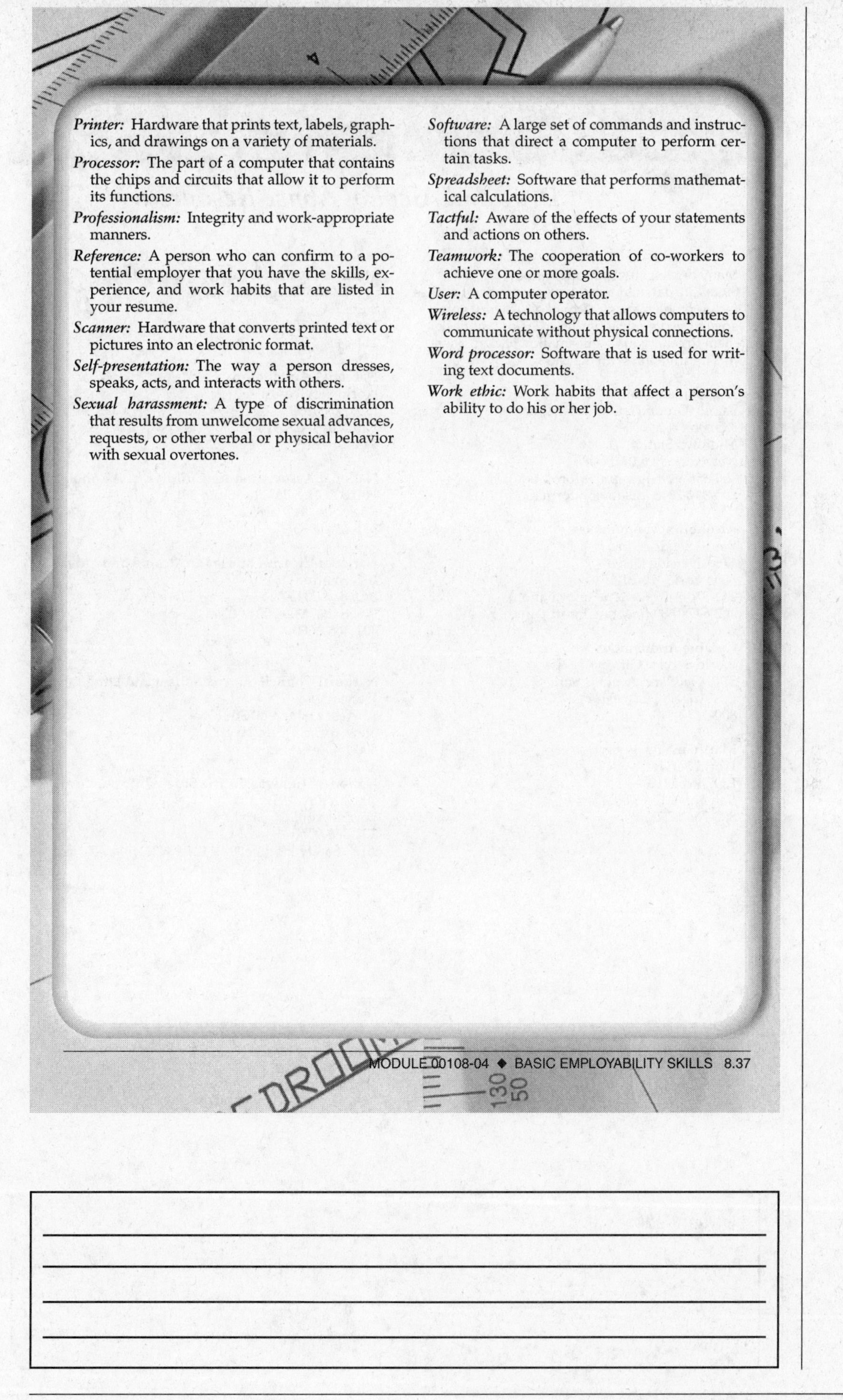

Printer: Hardware that prints text, labels, graphics, and drawings on a variety of materials.

Processor: The part of a computer that contains the chips and circuits that allow it to perform its functions.

Professionalism: Integrity and work-appropriate manners.

Reference: A person who can confirm to a potential employer that you have the skills, experience, and work habits that are listed in your resume.

Scanner: Hardware that converts printed text or pictures into an electronic format.

Self-presentation: The way a person dresses, speaks, acts, and interacts with others.

Sexual harassment: A type of discrimination that results from unwelcome sexual advances, requests, or other verbal or physical behavior with sexual overtones.

Software: A large set of commands and instructions that direct a computer to perform certain tasks.

Spreadsheet: Software that performs mathematical calculations.

Tactful: Aware of the effects of your statements and actions on others.

Teamwork: The cooperation of co-workers to achieve one or more goals.

User: A computer operator.

Wireless: A technology that allows computers to communicate without physical connections.

Word processor: Software that is used for writing text documents.

Work ethic: Work habits that affect a person's ability to do his or her job.

MODULE 00108-04 ◆ BASIC EMPLOYABILITY SKILLS 8.37

Drug and Alcohol Abuse Resources

Many organizations throughout the United States are devoted to helping people and families facing problems with alcohol or drugs. Contact any of the organizations listed below, and ask them to send you an information packet. Share this information with the class.

Alanon/Alateen
Family Group Headquarters, Inc.
P.O. Box 862
Midtown Station
New York, NY 10018-0862
800-356-9996 (general information)
800-344-2666 (meeting information)

Alcoholics Anonymous
World Services, Inc.
475 Riverside Drive
New York, NY 10115
212-870-3400 (general information)
212-647-1680 (meeting information)

Cocaine Anonymous
World Service Office
3740 Overland Avenue, Suite C
Los Angeles, CA 90034
800-347-8998

Marijuana Anonymous
World Services
P.O. Box 2912
Van Nuys, CA 91404
800-766-6779

Mothers Against Drunk Driving (MADD)
511 E. John Carpenter Freeway
Suite 700
Irving, TX 75062
214-744-6233

Narcotics Anonymous
World Service Office
P.O. Box 9999
Van Nuys, CA 91409
818-773-9999

National Association for Children of Alcoholics
11426 Rockville Pike, Suite 301
Rockville, MD 20852
301-468-0985

National Clearinghouse for Alcohol and Drug Information
P.O. Box 2345
Rockville, MD 20847-2345
301-468-2600
800-729-6686

National Council on Alcoholism and Drug Dependence
12 West 21st Street, 7th Floor
New York, NY 10011
800-NCA-CALL

National Highway Traffic Safety Information
400 7th Street, SW
Washington, DC 20590
202-366-9550
Auto Safety Hotline: 800-424-9393

Instructor's Notes:

8.38 CORE CURRICULUM ◆ INTRODUCTORY CRAFT SKILLS

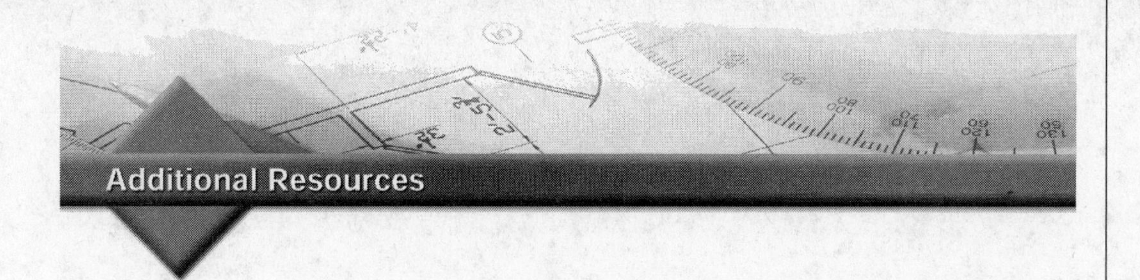

Additional Resources

This module is intended to present thorough resources for task training. The following reference works are suggested for further study. These are optional materials for continued education rather than for task training.

Art and Science of Leadership. Afsaneh Nahavandi. Upper Saddle River, NJ: Prentice-Hall.

Computer Numerical Control. Jay Stenerson. Upper Saddle River, NJ: Prentice-Hall.

Introduction to Computer Numerical Control. James Valentino. Upper Saddle River, NJ: Prentice-Hall.

Tools for Teams: Building Effective Teams in the Workplace. Craig Swenson, ed. Leigh Thompson, Eileen Aranda, Stephen P. Robbins. Boston, MA: Pearson Custom Publishing.

Your Attitude Is Showing. Elwood M. Chapman. Upper Saddle River, NJ: Prentice-Hall.

MODULE 00108-04 ◆ BASIC EMPLOYABILITY SKILLS 8.39

MODULE 00108-04 — TEACHING TIPS

The following are suggested activities or instructional methods to help you teach the material in this AIG.

General

When you call on someone to answer a question, the rest of the class relaxes or even tunes out because they expect that the question and answer will take place only between you and the trainee you called on. Instead, use this technique to involve more trainees in answering questions and to keep them on their toes.

1. Ask trainees to define a term or explain a concept.
2. After one trainee has answered, ask a trainee seated nearby if the answer is right. Then ask whether a trainee in the back of the room agrees.
3. Ask trainees to explain why they think an answer is right or wrong.
4. Use the session to clear up incorrect ideas, and encourage trainees to learn from their mistakes.

Section 2.1.0 *Entering the Construction Workforce*

In this exercise, trainees will create or update a professional resume. They will need their most current resume or an outline of their work history, their Trainee Guides, and pencils and paper. Allow between 45 and 60 minutes for this exercise.

1. Ask trainees to bring in a copy of their current resume or an outline of their work history.
2. Have trainees take a few moments to review the resume/work history and make necessary updates.
3. Ask trainees to review the guidelines for resume writing in On-Site: Resume Writing Tips and to edit their resumes based on the recommendations.
4. Have trainees exchange resumes with another trainee and make suggestions for improvement.

Section 4.0.0 *Computer Skills*

This exercise will allow trainees to practice their computer skills. You will need to schedule time at your local library's computer lab. Trainees will need their Trainee Guides and pencils and paper. Allow 60 minutes for this exercise.

1. Schedule a visit to your local library's computer lab.
2. Ask trainees to demonstrate their ability to complete a variety of basic computer skills. Suggested tasks include the following:
   - Access a word processing application.
   - Create and save a document.
   - Locate and retrieve a document.
   - Print a document.
   - Access the internet.
   - Navigate to a website that you specify.
   - Send and receive an email with an attachment.
3. Review and assess trainees' abilities. Offer suggestions for more efficient computer use.

Section 5.4.0 *Teamwork Skills*

This exercise will give trainees the opportunity to practice building a team. Trainees will need their Trainee Guides and pencils and paper. Allow 45 minutes for this exercise.

1. Ask trainees to read the hypothetical situation in "On-Site: Building a Team."
2. After reviewing the situation, have trainees choose up to five construction professionals to be on a team. Ask trainees to explain whom they chose and why.
3. Have trainees discuss their selections with the other trainees.
4. Compare trainees' selections and justifications for selecting their team members.

Section 6.0.0 *Workplace Issues*

This exercise will emphasize the severity of harassment, stress, and substance abuse. You will need a variety of news articles highlighting these subjects as they apply to the workplace. Trainees will need their Trainee Guides and pencils and paper. Allow 60 minutes for this exercise.

1. Bring in and distribute a variety of news articles covering harassment, stress, and substance abuse in the workplace.
2. Divide the class into small groups according to the number of articles.
3. Ask trainees to review the articles and consider the following:
   - In the sexual harassment articles, were charges filed against the harassers? Did the harassers lose their jobs? Was the company held responsible for the harassers' actions?
   - In regard to stress, what other health problems are associated with it? What are the consequences of workplace stress? What problems may have caused the stress? How could the stress have been dealt with?
   - In regard to substance abuse, what other health problems are associated with it? What are the consequences of substance abuse? Was a company program available to the substance abuser? What other problems did the substance abuse cause?
4. Discuss the severity of each issue. Review solutions to dealing with harassment, stress, and substance abuse.

MODULE 00108-04 — ANSWERS TO REVIEW QUESTIONS

Sections 2.0.0 and 3.0.0

1. b
2. d
3. b
4. a
5. c
6. a
7. e
8. b
9. d

Sections 4.0.0–6.0.0

1. d
2. a
3. b
4. d
5. a
6. c
7. d
8. b

MODULE 00108-04 — ANSWERS TO KEY TERMS QUIZ

1. mission statement
2. reference
3. entrepreneur
4. critical thinking skills
5. desktop computer
6. laptop computer
7. Handheld computers; PDAs
8. computer literacy
9. Hardware; software; operating systems
10. user
11. processor
12. monitor
13. keyboard
14. mouse
15. hard drive
16. disk drive
17. CD-ROM
18. printer; plotter
19. scanner
20. word processor
21. spreadsheets
22. database
23. Desktop publishing
24. CAD
25. browser
26. email
27. documentation
28. Wireless
29. CNC
30. EDM
31. Self-presentation
32. work ethic
33. Professionalism
34. confidentiality
35. initiative
36. Lateness; absenteeism
37. tactful
38. Compromise
39. constructive criticism
40. Teamwork
41. goal-oriented
42. leadership
43. harassment
44. sexual harassment
45. amphetamines
46. Barbiturates
47. hallucinogen

CONTREN® LEARNING SERIES — USER FEEDBACK

The NCCER makes every effort to keep these textbooks up-to-date and free of technical errors. We appreciate your help in this process. If you have an idea for improving this textbook, or if you find an error, a typographical mistake, or an inaccuracy in NCCER's *Contren®* textbooks, please write us, using this form or a photocopy. Be sure to include the exact module number, page number, a detailed description, and the correction, if applicable. Your input will be brought to the attention of the Technical Review Committee. Thank you for your assistance.

Instructors – If you found that additional materials were necessary in order to teach this module effectively, please let us know so that we may include them in the Equipment/Materials list in the Annotated Instructor's Guide.

Write: Product Development
National Center for Construction Education and Research
P.O. Box 141104, Gainesville, FL 32614-1104

Fax: 352-334-0932

E-mail: curriculum@nccer.org

Craft ______ Module Name ______

Copyright Date ______ Module Number ______ Page Number(s) ______

Description

______

______

______

______

(Optional) Correction

______

______

______

(Optional) Your Name and Address

______

______

______

Figure Credits

Module 00101-04

Accuform Signs, 101F01, 101F14
Associated General Contractors of America, Module Opener
Badger Fire Protection, 101F42
Charles Rogers, 101F05 (figure, face protection), 101F17 (face shields)
DBI/SALA and Protecta, 101F18
DEWALT Power Tools, 101F23
Courtesy Hornell Inc., 101F05 (respirator) *www.hornell.com. All rights reserved. Used with permission.*
Ed Gloninger, 101F44
Makita U.S.A., Inc., 101F43 (router)
Anne Meade, 101F21
Mid-State Manufacturing Corporation, 101F45
NAFED, 101F41
North Safety Products U.S.A., 101F16 (photos), 101F17 (glasses, goggles), 101F19, 101F20, 101F22, 101F24 (A, B, C)
Occupational Safety and Health Administration, 101F39
Ridge Tool Company/Louisville Ladder, 101F32, 101F33, 101F34
Gerald Shannon, 101F06
Becki Swinehart, 101F24 (D)
Werner Co., 101F35
Veronica Westfall, 101F46

Module 00102-04

Associated Builders and Contractors, Inc., Module Opener
Calculated Industries, Inc., 102F04
Courtesy Cooper Hand Tools, 102F06 (standard ruler)
Courtesy of Oak Ridge National Laboratory, 102F31
WIWA Wilhelm Wagner LP, 102F17

Module 00103-04

American Clamping Corp./Bessey, 103F27 (hand-screw)
Associated General Contractors of America, Module Opener
Channellock, Inc., 103F09 (tongue-and-groove)
Chicago Brand Industrial, 103SA02
Courtesy Cooper Hand Tools, 103F08, 103F09 (slip-joint, long-nose, Vise-Grip®), 103F12, 103F27 (locking), 103F29 (backsaw, compass, coping, handsaw), 103F31 (top six images), 103F40 (bottom), 103F48
Estwing Manufacturing Co., 103SA01
Futek Advanced Sensor Technology, 103F44 (digital)
Holloway Engineering, 103F44 (tensiometer)
IRWIN Industrial Tool Company, 103F27 (bar). *IRWIN is a registered trademark of IRWIN Industrial Tool Company*
JET Brand of WMH Tool Group, 103F27 (C-clamp), 103F42, 103F49, 103F50
Klein Tools, Inc., 103F38 (open-end, Allen), 103F44 (manual)
M-D Building Products, Inc., 103SA03
Porter Cable RoboToolz, 103SA04
RIGID/Ridge Tool Company, 103F04, 103F09 (lineman), 103F14, 103F15 (top), 103F26, 103F29 (hacksaw), 103F34 (top), 103F40 (top two images), 103F52
S•K Hand Tool Corporation, 103F38 (box-end)
The Stanley Works, 103F01, 103F06, 103F13, 103F15 (bottom), 103F17, 103F18, 103F22, 103F24, 103F27 (spring, web), 103F29 (dovetail), 103F34 (bottom), 103F36, 103F38 (combination), 103F43, 103F48
Ted Pella Inc./51906, 103F51

Module 00104-04

Associated Builders and Contractors, Inc., Module Opener
Bosch Power Tools and Accessories, 104F05 (right), 104F36 (rock drill)
DEWALT Power Tools, 104F01, 104F02 (twist, auger), 104F03, 104F04, 104F05 (left), 104F06, 104F07, 104F08, 104F09, 104F10 (A, B), 104F13, 104F14, 104F15, 104F16, 104F17, 104F18, 104F19, 104F20, 104F21, 104F22, 104F23, 104F24 (detail), 104F25, 104F26, 104F27, 104F28, 104F29, 104F30, 104F31, 104F33
JET™ Equipment & Tools, 104F11, 104F35
Milwaukee Electric Tool Corporation, 104F02 (forstner, paddle), 104F24 (shank-mounted points), 104F36 (pavement breaker, clay spade, attachments)
Benjamin D. Robb, 104F02 (masonry)
SENCO Products, Inc., 104SA02
Shinn Fu Company of America, 104F37
Stanley Fastening Systems, L.P./Stanley Bostitch, 104F32, 104SA01

Module 00105-04

Associated General Contractors of America, Module Opener
D&E Steel Services, Inc., 105F01 (B), 105F03, 105F04, 105F05, 105F06, 105F07, 105F08, 105F09, 105F10, 105F11, 105SA01, 105SA03
Reprinted by permission of Pearson Education, Inc., 105F30, 105F32, 105F33, 105F34, 105F35, 105F36

Module 00106-04

Associated Builders and Contractors, Inc., Module Opener
Coffing Hoists, 106F57
Columbus McKinnon Corporation, 106F56
The Crosby Group, Inc., 106F16, 106F19, 106F36, 106F37, 106F38, 106F39, 106F48 (A, B, C, D), 106F54, 106F58, 106F60 (top), 106SA04
Ed Gloninger, 106F15 (A, B, D, E, F, G, H, J, K, L, N, O, P)
Gunnebo Johnson Corporation, 106F48 (E, F)
ITNAC Corporation, 106F28
J.C. Renfroe & Sons, Inc., 106F45, 106F46
JET brand of WMH Tool Group, 106F55, 106SA02
Lift-All Company, Inc., 106F04, 106F05, 106F06, 106F07, 106F08, 106F09, 106F10, 106F15 (C, I, M), 106F21, 106F24
Lift-It Manufacturing Co., Inc., 106F03, 106F11, 106F12, 106F13, 106F14. *www.lift-it.com, 323-582-6076*
Mammoet USA, Inc., 106F02, 106SA01
North American Industries, Inc., 106F01

Module 00107-04

Arlington Central School District, Poughkeepsie, NY, 107F07
Associated General Contractors of America, Module Opener
© 2003-2004 ClipArt, www.clipart.com, 107F11
© 1997 Digital Vision, 107F12
Ed Gloninger, 107F01, 107F02, 107F10
Courtesy Oak Ridge National Laboratory, 107F06
Ritterbush-Ellig-Hulsing P.C., 107F03, 107F04

Module 00108-04

Associated Builders and Contractors, Inc., Module Opener
Associated General Contractors of America, 108F11 (software)
Autodesk, Inc., 108F11 (software), 108F12 (A, B, C)
Brokk, Inc., 108SA01
Ed Gloninger, 108F01 (montage), 108F03, 108F08, 108F09, 108F10, 108F21
League Manufacturing, 108F13
Microsoft Corporation, 108F11 (software)
National Association of Convenience Stores, 108F14 (left)
Courtesy Oak Ridge National Laboratory, 108F01 (background photo)
Primavera Systems, 108F12 (D, E)

Other Credits

Art Explosion; Corbis; Estwing Manufacturing Co.; EyeWire; PhotoDisc; Porter Cable RoboToolz; SENCO Products, Inc.; Stanley Fastening Systems, L.P./Stanley Bostitch

Glossary of Key Trade Terms

Abrasive: A substance—such as sandpaper—that is used to wear away material.

Absenteeism: Consistent failure to show up for work.

AC (alternating current): An electrical current that reverses its direction at regularly recurring intervals; the current delivered through wall plugs.

Active listening: A process that involves respecting others, listening to what is being said, and understanding what is being said.

Acute angle: Any angle between 0 degrees and 90 degrees.

Adjacent angles: Angles that have the same vertex and one side in common.

Allen wrench: A hexagonal steel bar that is bent to form a right angle. Also called a hex key wrench.

Angle: The shape made by two straight lines coming together at a point. The space between those two lines is measured in degrees.

ANSI hand signals: Communication signals established by the American National Standards Institute (ANSI) and used for load navigation for mobile and overhead cranes.

Amphetamine: A class of drugs that causes mental stimulation and feelings of euphoria.

Apparatus: An assembly of machines used together to do a particular job.

Appendix: A source of detailed or specific information placed at the end of a section, a chapter, or a book.

Arc: The flow of electrical current through a gas such as air from one pole to another pole.

Architect: A qualified, licensed person who creates and designs drawings for a construction project.

Architect's scale: A measuring device used for laying out distances, with scales indicating feet, inches, and fractions of inches.

Architectural plans: Drawings that show the design of the project. Also called *architectural drawings.*

Arc welding: The joining of metal parts by fusion, in which the necessary heat is produced by means of an electric arc.

Area: The surface or amount of space occupied by a two-dimensional object such as a rectangle, circle, or square. To calculate the area for rectangles and squares, multiply the length and width. To calculate the area for circles, multiply the radius squared and pi.

Auger: A tool with a spiral cutting edge for boring holes in wood and other materials.

Ball peen hammer: A hammer with a flat face that is used to strike cold chisels and punches. The rounded end—the peen—is used to bend and shape soft metal.

Barbiturate: A class of drugs that induces relaxation.

Beam: A large, horizontal structural member made of concrete, steel, stone, wood, or other structural material to provide support above a large opening.

Bell-faced hammer: A claw hammer with a slightly rounded, or convex, face.

Bevel: To cut on a slant at an angle that is not a right angle (90 degrees). The angle or inclination of a line or surface that meets another at any angle but 90 degrees.

Bisect: To divide into equal parts.

Blueprints: Architectural or working drawings used to represent a structure or system.

Block and tackle: A simple rope-and-pulley system used to lift loads.

Body language: A person's physical posture and gestures.

Booster: Gunpowder cartridge used to power powder-actuated fastening tools.

Borrow: To move numbers from one value column (such as the tens column) to another value column (such as units) to perform subtraction problems.

Box-end wrench: A wrench, usually double-ended, that has a closed socket that fits over the head of a bolt.

Bridle: A configuration using two or more slings to connect a load to a single hoist hook.

Browser: Software that allows users to search the World Wide Web.

Bullets: Large, vertically aligned dots that highlight items in a list.

Bull ring: A single ring used to attach multiple slings to a hoist hook.

CAD: Software that is used to draw and print design drawings.

Carbide: A very hard material made of carbon and one or more heavy metals. Commonly used in one type of saw blade.

Carpenter's square: A flat, steel square commonly used in carpentry.

Carry: To transfer an amount from one column to another column.

Cat's paw: A straight steel rod with a curved claw at one end that is used to pull nails that have been driven flush with the surface of the wood or slightly below it.

CD-ROM: Abbreviation for compact disk read-only memory. Hardware that stores software and allows it to be transferred between computers.

Cell phone: A mobile radiotelephone that uses a network of short-range transmitters located in overlapping cells throughout a region, with a central station making connections to regular telephone lines.

Chisel: A metal tool with a sharpened, beveled edge used to cut and shape wood, stone, or metal.

Chisel bar: A tool with a claw at each end, commonly used to pull nails.

Circle: A closed curved line around a central point. A circle measures 360 degrees.

Chuck: A clamping device that holds an attachment; for example, the chuck of the drill holds the drill bit.

Chuck key: A small, T-shaped steel piece used to open and close the chuck on power drills.

Circumference: The distance around the curved line that forms a circle.

Civil plans: Drawings that show the location of the building on the site from an aerial view, including contours, trees, construction features, and dimensions.

Claw hammer: A hammer with a flat striking face. The other end of the head is curved and divided into two claws to remove nails.

CNC: Abbreviation for computer numerical control. A general term used to describe computer-controlled machine tools.

Combination square: An adjustable carpenter's tool consisting of a steel rule that slides through an adjustable head.

Combination wrench: A wrench with an open end and a closed end.

Combustible: Capable of easily igniting and rapidly burning; used to describe a fuel with a flash point at or above 100°F.

Competent person: A person who can identify working conditions or surroundings that are unsanitary, hazardous, or dangerous to employees and who has authorization to correct or eliminate these conditions promptly.

Compromise: When people involved in a disagreement make concessions or meet in the middle to reach a solution that everyone agrees on.

Computer-aided drafting (CAD): The making of a set of blueprints with the aid of a computer.

Computer literacy: An understanding of how computers work and what they are used for.

Concealed receptacle: The electrical outlet that is placed inside the structural elements of a building, such as inside the walls. The face of the receptacle is flush with the finished wall surface and covered with a plate.

Confidentiality: Privacy of information.

Confined space: A work area large enough for a person to work, but arranged in such a way that an employee must physically enter the space to perform work. A confined space has a limited or restricted means of entry and exit. It is not designed for continuous work. Tanks, vessels, silos, pits, vaults, and hoppers are examples of confined spaces. See also *permit-required confined space.*

Constructive criticism: A positive offer of advice and help intended to help someone correct mistakes or improve actions.

Contour lines: Solid or dashed lines showing the elevation of the earth on a civil drawing.

Convert: To change from one unit of expression to another. For example, convert a decimal to a percentage: 0.25 to 25%; or convert a fraction to an equivalent: ¾ to ⁶⁄₈.

Core: Center support member of a wire rope around which the strands are laid.

Countersink: A bit or drill used to set the head of a screw at or below the surface of the material.

Crescent wrench: A smooth-jawed adjustable wrench used for turning nuts, bolts, and pipe fittings.

Cribbing: Material used to either support a load or allow removal of slings after the load is landed. Also called blocking.

Critical thinking skills: The skills required to evaluate and use information to make decisions or reach conclusions.

Cross-bracing: Braces (metal or wood) placed diagonally from the bottom of one rail to the top of another rail that add support to a structure.

Cubic: Measurement found by multiplying a number by itself three times; it describes volume measurement.

Database: Software that stores, organizes, and retrieves information.

DC (direct current): Electrical current that flows in one direction, from the negative (−) to the positive (+) terminal of the source, such as a battery.

Decimal: Part of a number represented by digits to the right of a point, called a decimal point. For example, in the number 1.25, .25 is the decimal part of the number.

Degree: A unit of measurement for angles. For example, a right angle is 90 degrees, an acute angle is between 0 and 90 degrees, and an obtuse angle is between 90 and 180 degrees.

Denominator: The part of a fraction below the dividing line. For example, the 2 in ½ is the denominator.

Desktop computer: A computer that is placed on a large flat surface, such as a desk or tabletop.

Desktop publishing: Software used to lay out text and graphics for publication.

Detail drawings: Enlarged views of part of a drawing used to show an area more clearly.

Diagonal: Line drawn from one corner of a rectangle or square to the farthest opposite corner.

Diameter: The length of a straight line that crosses from one side of a circle, through the center point, to a point on the opposite side. The diameter is the longest straight line you can draw inside a circle.

Difference: The result you get when you subtract one number from another. For example, in the problem $8 - 3 = 5$, the number 5 is the difference.

Digit: Any of the numerical symbols 0 to 9.

Dimension line: A line on a drawing with a measurement indicating length.

Dimensions: Measurements such as length, width, and height shown on a drawing.

Disk drive: Hardware that allows software and files to be transferred between computers.

Documentation: Instruction manuals and other information for software.

Dowel: A pin, usually round, that fits into a corresponding hole to fasten or align two pieces.

Dross: Waste material resulting from cutting using a thermal process.

EDM: Abbreviation for electrical discharge machines. Computer-controlled machine tools that cut and form parts that cannot be easily fabricated otherwise.

Electrical distribution panel: Part of the electrical distribution system that brings electricity from the street source (power poles and transformers) through the service lines to the electrical meter mounted on the outside of the building and to the panel inside the building. The panel houses the circuits that distribute electricity throughout the structure.

Electrical plans: Engineered drawings that show all electrical supply and distribution.

Electric tools: Tools powered by electricity. The electricity is supplied by either an AC source (wall plug) or a DC source (battery).

Electronic signature: A signature that is used to sign electronic documents by capturing handwritten signatures through computer technology and attaching them to the document or file.

Elevation (EL): Height above sea level, or other defined surface, usually expressed in feet.

Elevation drawing: Side view of a building or object, showing height and width.

Email: Electronic mail messages.

Engineer: A person who applies scientific principles in design and construction.

Engineer's scale: A straightedge measuring device divided uniformly into multiples of 10 divisions per inch so drawings can be made with decimal values.

English ruler: Instrument that measures English measurements; also called the standard ruler. Units of English measure include inches, feet, and yards.

Entrepreneur: A person who starts and runs his or her own business.

Equilateral triangle: A triangle that has three equal sides and three equal angles.

Equivalent fractions: Fractions having different numerators and denominators, but equal values, such as ½ and 2/4.

Excavation: Any man-made cut, cavity, trench, or depression in an earth surface, formed by removing earth. It can be made for anything from basements to highways. See also *trench.*

Extension ladder: A ladder made of two straight ladders that are connected so that the overall length can be adjusted.

Eyebolt: An item of rigging hardware used to attach a sling to a load.

Fastener: A device such as a bolt, clasp, hook, or lock used to attach or secure one material to another.

Ferromagnetic: Having magnetic properties. Substances such as iron, nickel, cobalt, and various alloys are ferromagnetic.

Flammable: Capable of easily igniting and rapidly burning; used to describe a fuel with a flash point below 100°F.

Flash: A sudden bright light associated with starting up a welding torch.

Flashback: A welding flame that flares up and chars the hose at or near the torch connection. It is caused by improperly mixed fuel.

Flash burn: The damage that can be done to eyes after even brief exposure to ultraviolet light from arc welding. A flash burn requires medical attention.

Flash goggles: Eye protective equipment worn during welding operations.

Flash point: The temperature at which fuel gives off enough gases (vapors) to burn.

Flat bar: A prying tool with a nail slot at the end to pull nails out in tightly enclosed areas. It can also be used as a small pry bar.

Flats: The straight sides or jaws of a wrench opening. Also, the sides on a nut or bolt head.

Floor plan: A drawing that provides an aerial view of the layout of each room.

Font: The type style used for printed letters and numbers.

Foot-pounds: Unit of measure used to describe the amount of pressure exerted (torque) to tighten a large object.

Formula: A mathematical process used to solve a problem. For example, the formula for finding the area of a rectangle is side A times side B = Area, or $A \times B$ = Area.

Fraction: A number represented by a numerator and a denominator, such as ½.

Foundation plan: A drawing that shows the layout and elevation of the building foundation.

Glossary: An alphabetical list of terms and definitions.

Goal-oriented: To be focused on an objective.

Grit: A granular abrasive used to make sandpaper or applied to the surface of a grinding wheel to give it a nonslip finish. Grit is graded according to its texture. The grit number indicates the number of abrasive granules in a standard size (per inch or per cm). The higher the grit number, the finer the abrasive material.

Grommet sling: A sling fabricated in an endless loop.

Ground: The conducting connection between electrical equipment or an electrical circuit and the earth.

Ground fault circuit interrupter (GFCI): A circuit breaker designed to protect people from electric shock and to protect equipment from damage by interrupting the flow of electricity if a circuit fault occurs.

Ground fault protection: Protection against short circuits; a safety device cuts power off as soon as it senses any imbalance between incoming and outgoing current.

Guarded: Enclosed, fenced, covered, or otherwise protected by barriers, rails, covers, or platforms to prevent dangerous contact.

Hallucinogen: A class of drugs that distort the perception of reality and cause hallucinations.

Handheld computer: A computer that is designed to be small and light enough to be carried.

Hand line: A line attached to a tool or object so a worker can pull it up after climbing a ladder or scaffold.

Harassment: A type of discrimination that can be based on race, age, disabilities, sex, religion, cultural issues, health, or language barriers.

Hard drive: Hardware that stores software and electronic files.

Hardware: The physical components that make up a computer.

Hazard Communication Standard (HazCom): The Occupational Safety and Health Administration standard that requires contractors to educate employees about hazardous chemicals on the job site and how to work with them safely.

Hazardous materials: Materials (such as chemicals) that must be transported, stored, applied, handled, and identified according to federal, state, or local regulations. Hazardous materials must be accompanied by material safety data sheets (MSDSs).

Heating, ventilating, and air conditioning (HVAC): Heating, ventilating, and air conditioning.

Hidden line: A dashed line showing an object obstructed from view by another object.

Hitch: The rigging configuration by which a sling connects the load to the hoist hook. The three basic types of hitches are vertical, choker, and basket.

Hoist: A device that applies a mechanical force for lifting or lowering a load.

Hydraulic tools: Tools powered by fluid pressure. The pressure is produced by hand pumps or electric pumps.

Improper fraction: A fraction whose numerator is larger than its denominator. For example, $\frac{8}{4}$ and $\frac{6}{3}$ are improper fractions.

Inch-pounds: Unit of measure used to describe the amount of pressure exerted (torque) to tighten a small object.

Index: An alphabetical list of topics, along with the page numbers where each topic appears.

Initiative: The ability to work without constant supervision and solve problems independently.

Invert: To reverse the order or position of numbers. In fractions, to turn upside down, such as $\frac{3}{4}$ to $\frac{4}{3}$. When you are dividing by fractions, one fraction is inverted.

Isometric drawing: A three-dimensional drawing of an object.

Isosceles triangle: A triangle that has two equal sides and two equal angles.

Italics: Letters and numbers that lean to the right rather than stand straight up.

Jargon: Specialized terms used in a specific industry.

Joint: The point where members or the edges of members are joined. The types of welding joints are butt joint, corner joint, and T-joint.

Kerf: A cut or channel made by a saw.

Keyboard: Hardware that allows users to enter text and instructions and send them to the processor.

Lanyard: A short section of rope or strap, one end of which is attached to a worker's safety harness and the other to a strong anchor point above the work area.

Laptop computer: A computer that is larger than a handheld computer but smaller than a desktop computer.

Lateness: Habitually showing up late for work.

Leader: In drafting, the line on which an arrowhead is placed and used to identify a component.

Leadership: The ability to set an example for others to follow by exercising authority and responsibility.

Level: Perfectly horizontal; completely flat; a tool used to determine if an object is level.

Legend: A description of the symbols and abbreviations used in a set of drawings.

Lifting clamp: A device used to move loads such as steel plates or concrete panels without the use of slings.

Load: The total amount of what is being lifted, including all slings, hitches, and hardware.

Load control: The safe and efficient practice of load manipulation, using proper communication and handling techniques.

Load stress: The strain or tension applied on the rigging by the weight of the suspended load.

Lockout/tagout: A formal procedure for taking equipment out of service and ensuring that it cannot be operated until a qualified person has removed the lockout or tagout device (such as a lock or warning tag).

Long division: Process of writing out each step of a division problem until you reach the answer and identify any remainder that can no longer be divided by the divisor.

Machinist's ruler: A ruler that is marked so that the inches are divided into 10 equal parts, or tenths.

Management system: The organization of a company's management, including reporting procedures, supervisory responsibility, and administration.

Masonry: Building material, including stone, brick, or concrete block.

Master link: The main connection fitting for chain slings.

Material safety data sheet (MSDS): A document that must accompany any hazardous substance. The MSDS identifies the substance and gives the exposure limits, the physical and chemical characteristics, the kind of hazard it presents, precautions for safe handling and use, and specific control measures.

Maximum intended load: The total weight of all people, equipment, tools, materials, and loads that a ladder can hold at one time.

Mechanical plans: Engineered drawings that show the mechanical systems, such as motors and piping.

Memo: Another term for memorandum.

Memorandum: Informal written correspondence. The plural of memorandum is memoranda.

Meter: The base unit of length in the metric system; approximately 39.37 inches.

Metric ruler: Instrument that measures metric lengths. Units of measure include millimeters, centimeters, and meters.

Metric scale: A straightedge measuring device divided into centimeters, with each centimeter divided into 10 millimeters.

Mid-rail: Mid-level, horizontal board required on all open sides of scaffolding and platforms that are more than 14 inches from the face of the structure and more than 10 feet above the ground. It is placed halfway between the toeboard and the top rail.

Mission statement: A statement of how a company does business.

Miter joint: A joint made by fastening together usually perpendicular parts with the ends cut at an angle.

Mixed number: A combination of a whole number with a fraction or decimal. For example, mixed numbers are 3⁷⁄₁₆, 5.75, and 1¼.

Monitor: Hardware that shows information, text, and pictures on a television-style screen.

Mouse: Hardware that allows a user to enter data without a keyboard.

Nail puller: A tool used to remove nails.

Negative numbers: Numbers less than zero. For example, −1, −2, and −3 are negative numbers.

Not to scale (NTS): Describes drawings that show relative positions and sizes.

Numerator: The part of a fraction above the dividing line. For example, the 1 in ½ is the numerator.

Obtuse angle: Any angle between 90 degrees and 180 degrees.

Occupational Safety and Health Administration (OSHA): An agency of the U.S. Department of Labor. Also refers to the Occupational Safety and Health Act of 1970, a law that applies to more than 111 million workers and 7 million job sites in the country.

One rope lay: The lengthwise distance it takes for one strand of a wire rope to make one complete turn around the core.

Open-end wrench: A nonadjustable wrench with an opening at each end that determines the size of the wrench.

Operating system: A complex set of commands that enables hardware and software to communicate.

Opposite angles: Two angles that are formed by two straight lines crossing. They are always equal.

Pad eye: A welded structural lifting attachment.

PDA: Abbreviation for personal digital assistant. Another name for a handheld computer.

Peening: The process of bending, shaping, or cutting material by striking it with a tool.

Percent: Of or out of one hundred. For example, 8 is 8 percent (%) of 100.

Perimeter: The distance around the outside of any closed shape, such as a rectangle, circle, or square.

Permit: A legal document that allows a task to be undertaken.

Permit-required confined space: A confined space that has been evaluated and found to have actual or potential hazards, such as a toxic atmosphere or other serious safety or health hazard. Workers need written authorization to enter a permit-required confined space. See also *confined space.*

Personal protective equipment (PPE): Equipment or clothing designed to prevent or reduce injuries.

Pi: A mathematical value of approximately 3.14 (or 22/7) used to determine the area and circumference of circles. It is sometimes symbolized by π.

Pipe wrench: A wrench for gripping and turning a pipe or pipe-shaped object; it tightens when turned in one direction.

Piping and instrumentation drawings (P&IDs): Schematic diagrams of a complete piping system.

Place value: The exact quantity of a digit, determined by its place within the whole number or by its relationship to the decimal point.

Plane: A surface in which a straight line joining two points lies wholly within that surface.

Planed: Describing a surface made smooth by using a tool called a plane.

Planked: Having pieces of material 2 or more inches thick and 6 or more inches wide used as flooring, decking, or scaffolding.

Pliers: A scissors-shaped type of adjustable wrench equipped with jaws and teeth to grip objects.

Plotter: Hardware that prints large architectural and construction drawings and images.

Plumb: Perfectly vertical; the surface is at a right angle (90 degrees) to the horizon or floor and does not bow out at the top or bottom.

Plumbing: A general term used for both water supply and all liquid waste disposal.

Plumbing plans: Engineered drawings that show the layout for the plumbing system.

Pneumatic tools: Air-powered tools. The power is produced by electric or fuel-powered compressors.

Points: Teeth on the gripping part of a wrench. Also refers to the number of teeth per inch on a handsaw.

Positive numbers: Numbers greater than zero. For example, 1, 2, and 3 are positive numbers.

Printer: Hardware that prints text, labels, graphics, and drawings on a variety of materials.

Processor: The part of a computer that contains the chips and circuits that allow it to perform its functions.

Professionalism: Integrity and work-appropriate manners.

Proximity work: Work done near a hazard but not actually in contact with it.

Punch: A steel tool used to indent metal.

Punch list: A written list that identifies deficiencies requiring correction at completion.

Qualified person: A person who, by possession of a recognized degree, certificate, or professional standing, or by extensive knowledge, training, and experience, has demonstrated the ability to solve or prevent problems relating to a certain subject, work, or project.

Radius: The distance from a center point of a circle to any point on the curved line, or half the width (diameter) of a circle.

Rafter angle square: A type of carpenter's square made of cast aluminum that combines a protractor, try square, and framing square.

Rated capacity: The maximum load weight a sling or piece of hardware or equipment can hold or lift. Also called lifting capacity, working capacity, working load limit (WLL), and safe working load (SWL).

Reciprocating: Moving back and forth.

Rectangle: A four-sided shape with four 90-degree angles. Opposite sides of a rectangle are always parallel and the same length. Adjacent sides are perpendicular and are not equal in length.

Reference: A person who can confirm to a potential employer that you have the skills, experience, and work habits that are listed in your resume.

Rejection criteria: Standards, rules, or tests on which a decision can be based to remove an object or device from service because it is no longer safe.

Remainder: The leftover amount in a division problem. For example, in the problem 34 ÷ 8, 8 goes into 34 four times (8 × 4 = 32) and 2 is left over, or, in other words, it is the remainder.

Request for information (RFI): A means of clarifying a discrepancy in the blueprints.

Respirator: A device that provides clean, filtered air for breathing, no matter what is in the surrounding air.

Revolutions per minute (rpm): The number of times (or rate) a motor component or accessory (drill bit) completes one full rotation every minute.

Rigging hook: An item of rigging hardware used to attach a sling to a load.

Right angle: An angle that measures 90 degrees. The two lines that form a right angle are perpendicular to each other. This is the angle used most in the trades.

Right triangle: A triangle that includes one 90-degree angle.

Ring test: A method of testing the condition of a grinding wheel. The wheel is mounted on a rod and tapped. A clear ring means the wheel is in good condition; a dull thud means the wheel is in poor condition and should be disposed of.

Ripping bar: A tool used for heavy-duty dismantling of woodwork, such as tearing apart building frames or concrete forms.

Roof plan: A drawing of the view of the roof from above the building.

Round off: To smooth out threads or edges on a screw or nut.

Scaffold: An elevated platform for workers and materials.

Scale: The ratio between the size of a drawing of an object and the size of the actual object.

Scalene triangle: A triangle with sides of unequal lengths.

Scanner: Hardware that converts printed text or pictures into an electronic format.

Schematic: A one-line drawing showing the flow path for electrical circuitry.

Section drawing: A cross-sectional view of a specific location, showing the inside of an object or building.

Self-presentation: The way a person dresses, speaks, acts, and interacts with others.

Sexual harassment: A type of discrimination that results from unwelcome sexual advances, requests, or other verbal or physical behavior with sexual overtones.

Shackle: Coupling device used in an appropriate lifting apparatus to connect the rope to eye fittings, hooks, or other connectors.

Shank: The smooth part of a drill bit that fits into the chuck.

Sheave: A grooved pulley-wheel for changing the direction of a rope's pull; often found on a crane.

Shoring: Using pieces of timber, usually in a diagonal position, to hold a wall in place temporarily.

Side pull: The portion of a pull acting horizontally when the slings are not vertical.

Signaler: A person who is responsible for directing a vehicle when the driver's vision is blocked in any way.

Slag: Waste material from welding operations.

Sling: Wire rope, alloy steel chain, metal mesh fabric, synthetic rope, synthetic webbing, or jacketed synthetic continuous loop fibers made into forms, with or without end fittings, used to handle loads.

Sling angle: The angle of an attached sling when pulled in relation to the load.

Sling legs: The parts of the sling that reach from the attachment device around the object being lifted.

Sling reach: A measure taken from the master link of the sling, where it bears weight, to either the end fitting of the sling or the lowest point on the basket.

Sling stress: The total amount of force exerted on a sling. This includes forces added as a result of sling angle.

Software: A large set of commands and instructions that direct a computer to perform certain tasks.

Specifications: Precise written presentation of the details of a plan.

Spliced: Having been joined together.

Spreadsheet: Software that performs mathematical calculations.

Spud wrench: An adjustable wrench used for fittings on drain traps, sink strainers, toilet connections, and odd-shaped nuts.

Square: (1) A special type of rectangle with four equal sides and four 90-degree angles. (2) Exactly adjusted; any piece of material sawed or cut to be rectangular with equal dimensions on all sides; a tool used to check angles. (3) The product of a number multiplied by itself. For example, 25 is the square of 5; 16 is the square of 4.

Standard ruler: An instrument that measures English lengths (inches, feet, and yards). See *English ruler.*

Stepladder: A self-supporting ladder consisting of two elements hinged at the top.

Straight angle: A 180-degree angle or flat line.

Straight ladder: A nonadjustable ladder.

Strand: A group of wires wound, or laid, around a center wire, or core. Strands are laid around a supporting core to form a rope.

Stress: Intensity of force exerted by one part of an object on another; the action of forces on an object or system that leads to changes in its shape, strain on it, or separation of its parts.

Striking (or slugging) wrench: A nonadjustable wrench with an enclosed, circular opening designed to lock onto the fastener when the wrench is struck.

Strip: To damage the threads on a nut or bolt.

Structural plans: A set of engineered drawings used to support the architectural design.

Sum: The total in an addition problem. For example, in the problem $7 + 8 = 15$, 15 is the sum.

Switch enclosure: A box that houses electrical switches used to regulate and distribute electricity in a building.

Symbol: A drawing that represents a material or component on a plan.

Table: A way to present important text and numbers so they can be read and understood at a glance.

Table of contents: A list of book chapters or sections, usually located at the front of the book.

Tactful: Aware of the effects of your statements and actions on others.

Tag line: Rope that runs from the load to the ground. Riggers hold onto tag lines to keep a load from swinging or spinning during the lift.

Tang: Metal handle-end of a file. The tang fits into a wooden or plastic file handle.

Tattle-tail: Cord attached to the strands of an endless loop sling. It protrudes from the jacket. A tattle-tail is used to determine if an endless sling has been stretched or overloaded.

Teamwork: The cooperation of co-workers to achieve one or more goals.

Tempered: Treated with heat to create or restore hardness in steel.

Tenon: A piece that projects out of wood or another material for the purpose of being placed into a hole or groove to form a joint.

Threaded shank: A connecting end of a fastener, such as a bolt, with a series of spiral grooves cut into it. The grooves are designed to mate with grooves cut into another object in order to join them together.

Title block: A part of a drawing sheet that includes some general information about the project.

Toeboard: A vertical barrier at floor level attached along exposed edges of a platform, runway, or ramp to prevent materials and people from falling.

Top rail: A top-level, horizontal board required on all open sides of scaffolding and platforms that are more than 14 inches from the face of the structure and more than 10 feet above the ground.

Torque: The turning or twisting force applied to an object, such as a nut, bolt, or screw, using a socket wrench or screwdriver to tighten it. Torque is measured in inch-pounds or foot-pounds.

Trench: A narrow excavation made below the surface of the ground that is generally deeper than it is wide, with a maximum width of 15 feet. See also *excavation.*

Trencher: An excavating machine used to dig trenches, especially for pipeline and cables.

Triangle: A closed shape that has three sides and three angles.

Trigger lock: A small lever, switch, or part that you push or pull to activate a locking catch or spring. Activating the trigger lock causes the trigger to stay in the operating mode even without your finger on the trigger.

Try square: A square whose legs are fixed at a right angle.

User: A computer operator.

Vertex: A point at which two or more lines or curves come together.

Vise: A holding or gripping tool, fixed or portable, used to secure an object while work is performed on it.

Volume: The amount of space occupied in three dimensions (length, width, and height/depth/thickness).

Warning-yarn: A component of the sling that shows the rigger whether the sling has suffered too much damage to be used.

Weld: To heat or fuse two or more pieces of metal so that the finished piece is as strong as the original; a welded joint.

Welding shield: (1) A protective screen set up around a welding operation designed to safeguard workers not directly involved in that operation. (2) A shield that provides eye and face protection for welders by either connecting to helmet-like headgear or attaching directly to a hard hat; also called a welding helmet.

Whole numbers: Complete units without fractions or decimals.

Wind sock: A cloth cone open at both ends mounted in a high place to show which direction the wind is blowing.

Wireless: A technology that allows computers to communicate without physical connections.

Wire rope: A rope made from steel wires that are formed into strands and then laid around a supporting core to form a complete rope; sometimes called cable.

Word processor: Software that is used for writing text documents.

Work ethic: Work habits that affect a person's ability to do his or her job.

Index

The designation "*f*" refers to figures.

C

D

N

O

P

Q

R

Y